Mindfulness in Behavioral Health

Series editor

Nirbhay N. Singh
Medical College of Georgia
Georgia Regents University
Augusta, Georgia, USA

More information about this series at http://www.springer.com/series/8678

Ronald E. Purser · David Forbes
Adam Burke
Editors

Handbook of Mindfulness

Culture, Context, and Social Engagement

 Springer

Editors
Ronald E. Purser
Department of Management
San Francisco State University
San Francisco, CA
USA

Adam Burke
Institute for Holistic Health Studies
San Francisco State University
San Francisco, CA
USA

David Forbes
Brooklyn College
CUNY Graduate Center
Brooklyn, NY
USA

ISSN 2195-9579   ISSN 2195-9587   (electronic)
Mindfulness in Behavioral Health
ISBN 978-3-319-44017-0   ISBN 978-3-319-44019-4   (eBook)
DOI 10.1007/978-3-319-44019-4

Library of Congress Control Number: 2016947204

Printed on acid-free paper

This Springer imprint is published by Springer Nature
The registered company is Springer International Publishing AG Switzerland

Preface

This volume is a critical inquiry into the meaning of mindfulness today. It explores the extent to which classic and modern concepts and practices of mindfulness clash, converge, and influence each other, and what that exchange holds for the future. The problematic, as the Venerable Bhikkhu Bodhi has said, is that mindfulness as a concept has become "so vague and elastic that it serves almost as a cipher into which one can read virtually anything we want" (Bodhi 2011). Indeed, the increasing popularity of mindfulness in the West has led to it being called a "movement." *Time* magazine's cover article went so far to declare a "Mindful Revolution" was sweeping the country (Pickert 2014). The launch of the glossy new magazine, *Mindful*, is a signal for a growing market demand for what was once considered a strange and foreign "Eastern religious" practice. Indeed, secular mindfulness has situated itself as a new brand within a self-help industry, promising to offer a panacea for the existential angst of mainly the white middle and upper classes. In fact, in 2007, the National Institute of Health (NIH) estimated that consumers spent $4 billion on meditation (Barnes et al. 2008).

The mindfulness movement received a great deal of media attention that has, until recently, been uncritically celebratory and positive. Even among prominent clinicians, researchers, and scientists, the way scientific investigations have been reported, both in print and in public, has often overstated the benefits and efficacy of mindfulness interventions while downplaying a range of methodological weaknesses. The emerging field of contemplative studies and the burgeoning "science of mindfulness" has sought refuge in the fields of psychology and neuroscience, capitalizing on the West's cultural fascination with brain imagery. Neuroscientific studies using functional magnetic resonance imaging (fMRI) of meditators' brain states are frequently touted in the media as incontrovertible evidence that science has verified the efficacy of mindfulness. Whether it is increasing the size of gray matter, shrinking the amygdala, or quieting the default mode network, reports of functional and structural changes in the brain (even if the neuroscientists themselves are more circumspect about the actual significance of their findings) have come to symbolize an official stamp of scientific legitimacy.

Yet, the meteoric rise of the "mindfulness revolution" has led to growing chorus of criticism. Those who initially raised critical questions regarding the mindfulness movement were few and far between, and they were often rebuked or dismissed as either Buddhist fundamentalists, naysayers, or downright cranks. In 2013, Ron Purser and David Loy's article "Beyond McMindfulness" in the *Huffington Post* called into question the efficacy, ethics, and narrow interests of corporate mindfulness programs (Purser and Loy 2013). This scathing critique seemed to open the floodgates as a stream of critical commentaries appeared in a scattered corpus of writings found on Internet blogs, social media outlets, as well as in a number of academic journals and books. Such was the beginning of what the media termed the "mindfulness backlash" (North 2014; Roca 2014).

Buddhist scholars and teachers began comparing and contrasting Jon Kabat-Zinn's definition of mindfulness as "paying attention in a particular way: on purpose, in the present moment, and non-judgmentally" (the gold standard for secular mindfulness-based interventions) to various Buddhist conceptualizations of mindfulness. Numerous scholars took issue with Kabat-Zinn's bold claims and rhetoric, calling into question the reductionistic and mystifying assertion that "meditation as being the heart of Buddhism," and mindfulness-based stress reduction (MBSR) is "Buddhist meditation without the Buddhism." Kabat-Zinn even went so far to claim that MBSR is the "universal dharma that is co-extensive, if not identical, with the teachings of the Buddha, the Buddhadharma" (Kabat-Zinn 2011, p. 290).

This backlash also included a number of contemplative scientists who began raising questions regarding the media hype and exaggerated scientific claims about the validity and reliability of mindfulness research studies (Heuman 2014a; Purser and Cooper 2014). Scientific claims of mindfulness research studies are also being examined now with greater scrutiny. A meta-analytic study on the efficacy of mindfulness meditation was recently published in the *Journal of the American Medical Association (JAMA), Internal Medicine*. Dr. Madhav Goyal and his colleagues from Johns Hopkins University searched databases using a set of key meditation terms. They obtained 18,753 citations of which 47 matched their inclusion criteria, such as being randomized controlled trials. They found that mindfulness was moderately effective in treating a variety of conditions, but was not found to be more effective than other active treatments, such as drugs or exercise (Goyal et al. 2014).

"Public enthusiasm is outpacing scientific evidence," says Brown University researcher Willoughby Britton (Heuman 2014b). And "experimenter allegiance," she goes on to say, which is a factor when the researcher also happens to be a creator of the therapy, "can count for a larger effect than the treatment itself. People are finding support for what they believe rather than what the data is actually saying." Moreover, there is convincing evidence that mindfulness studies suffer from positive reporting bias (Coronado-Montoya et al. 2016). A team of researchers at McGill University recently found that authors of mindfulness studies tend to spin their positive results, downplaying negative results. Given the small sample size and weak statistical power of the pool of studies examined, McGill researchers were concerned by the skewed results.

A number of Buddhist scholars, teachers, and practitioners have become increasingly concerned about the long-term implications of the mindfulness movement, and whether the rush toward secularization may lead to a gradual denaturing and banalization of the Buddhist path of awakening. Some Buddhist teachers believe that the West is moving too quickly to appropriate Buddhist mindfulness practices, diluting and adapting the teachings to fit our consumerist society. Other teachers and practitioners, usually those who also have a professional investment in promoting mindfulness, have advocated that such rapid secularization of mindfulness is necessary if it is to be made more widely available and relevant to a modern society.

Clearly, extracting a spiritual and meditative discipline from its social and historical contexts in which it originates has radically changed the meaning, function, and fruition of mindfulness practices in the West. On the one hand, Buddhism must change as it takes root in the West. Traditional concerns for preserving the authenticity, integrity, and canonical authority with regard to Buddhist conceptions of mindfulness, while admirable, have failed to take into account the pluralistic nature of Western society. In addition, such a defensive and reactionary posture also fails to address the inevitable migration and transformation of Buddhism in its encounter with modernity. As David Loy has argued, the East and the West need each other, and this meeting has already begun to come about. However, we must ask what the relationship is between the two, what is problematic about that relationship, and how can they be of mutual benefit. Buddhism will change and is changing, as it mixes with the dominant values of modern Western cultures. A significant question addressed in this volume is what actually happens to Buddhist mindfulness teachings and practices as they are decontextualized, adapted, and applied in secular contexts? What is gained and what is lost?

Another equally important question and central concern of this volume is what is mindfulness for? Are mindfulness-based interventions limited to a palliative for individual stress relief and mental hygiene, or can mindfulness programs develop in ways that call into question deeply rooted cultural assumptions which have been the source of so much misery, injustice, and unnecessary suffering in the modern Western world? Or will mindfulness be used to accommodate to those cultural assumptions? What is the relation between the efficacy of mindfulness practice and the contexts that inform its pedagogical goals and applications? Is mindfulness practice (or any meditative discipline) the main reductive ingredient that can function as a neutral tool or technique independent of its context?

Numerous contributors to this volume show how mindfulness in the West, under the claim that it is derived from Buddhism, has become severed from not only Buddhist ethical contexts, but also its roots in Buddhist philosophy and soteriology. Advocates of secular mindfulness have for the most part downplayed questions of ethics and what constitutes the good life by insisting that ethical development is simply intrinsic or "built-into" the practice. Such a claim is also an appeal to a universal view of human beings that transcends culture and context. A perennialist view underlies the discourse that mindfulness is a "free agent"—a universal human capacity–unbeholden to any historical contingency or cultural context. This laissez-faire "innatist" philosophy puts mindfulness programs at risk of being employed as a technology to accommodate people to individualistic, consumerist, and corporate values. Rather than developing a critical pedagogical framework for mindfulness programs which could potentially challenge, interrogate, and transform our deeply rooted Western cultural values and assumptions, the majority of clinical, school-based, and corporate mindfulness training programs are informed by biomedical models of stress and well-being. The medicalization of mindfulness has limited program curricula to essentialist constructs that explain stress as an individual pathology, deflecting attention away from culture and context. Indeed, the cultural dominance of the biomedical paradigm has reinforced the notion that disease (including psychosomatic symptoms such as chronic stress, depression, and anxiety), along with interventions for enhancing health and well-being is a matter for autonomous individuals. Because mindfulness practice has succumbed to an individualistic worldview, it has "overstated internal pathology while understating environmental stressors" (Goddard 2014, p. 212). Individualistic, laissez-faire oriented mindfulness programs, perhaps unwittingly, are preserving the status quo and maintaining institutional structures that contribute to social suffering. Moreover, considering mindfulness as simply a form of "mental fitness" analogous to autonomous forms of physical exercise such as weight-lifting or running reinforces reductionist conceptions of psychological distress.

In broad terms, the Buddhist practice of mindfulness is concerned with the interior, or first-person perspective. It values higher states of consciousness that are historically intended to lead to deep and irreversible insights into the nature of reality, including a dissolution of a separate sense of self as a real and permanent identity. However, the Buddhist practice of mindfulness is also a socially engaged endeavor and insists on a commitment to the fulfillment of ethical awareness and practices such as right speech, intention, action, and livelihood. Buddhism offers a soteriological solution to human suffering based on a deep and embodied insight into the nature of reality. The fruition of full realization is the outcome of an integrated path of ethical and moral development, conjoined with the meditative training and the cultivation of insight that leads to seeing the truth of impermanence, the illusoriness of a permanent and separate sense of self, and that all conditioned phenomena has the nature of suffering (the "three marks" in Buddhist teachings). "Seeing things as they truly are" is simultaneously seeing there is no ultimate split between one's experience and all others. This is liberation from suffering, a non-dual wisdom that manifests as spontaneous and uncontrived universal compassion for all sentient beings.

Buddhism, however, as a religion must find its way in a secular society that relies on scientific evidence and the study of cultural and historical contexts as manifestations of the forms of everyday life. Toward this end, it is arguable that Buddhism and mindfulness can adapt to and gain from the West's social scientific (e.g., developmental and clinical psychology, sociology), historical, and neuroscientific knowledge and practices and make it more widely available without diluting its foundational premises and approach. In this regard, Wilber (2014) suggests Buddhism is ripe for a "Fourth Turning" that includes the best wisdom of the West.

The West tends to emphasize exterior, objective, or third-person perspectives that promote the historical progress of society and social institutions through science, technology, and economic growth (materialism, consumerism). This tendency minimizes the development of interior and moral wisdom which Buddhism provides and which can benefit the West. There is no disputing the fact that mindfulness-based interventions have been shown to have salutary health benefits and have alleviated psychological suffering, helping thousands of people reduce and manage chronic pain. While this has occurred to an extent, this volume is critically concerned with the numerous ways the West employs the Buddhist-derived practice of mindfulness out of context and in ways that reinforce its problematic tendencies. While there have been attempts to have dialogues between Buddhism and cognitive/neuroscience, as well as between Buddhism and Western psychiatry and psychology, these dialogues have often privileged Western metaphysical assumptions based on scientific materialism and a narrow focus on biophysical explanations of mental health and illness (Kirmayer 2015, p. 451). As Kirmayer and Crafa (2014) have pointed out, the dialogue between Buddhism and neuroscience has not only been limited by the narrow focus on neural correlates of meditation, but brain-based explanations have occluded giving equal attention to "social, contextual, and value-based aspects" of such practices (Kirmayer 2015, p. 451).

The contributions in this volume situate the mindfulness movement within broader philosophical, historical, and cultural contexts. The theory and practice of mindfulness and its various manifestations in health care, education, contemplative neuroscience, and corporations are examined in terms of how mindfulness is being influenced and shaped by cultural assumptions, institutional structures, economic systems, and political forces. Given that the mindfulness movement has spread to practically all domains of society, as editors, we have solicited and selected a wide range of contributions from authors in order to offer a more transdisciplinary perspective. Indeed, this handbook includes contributions from prominent Buddhist scholars and teachers, clinicians and contemplative scientists, as well as scholars in such fields as philosophy, educational counseling, sociology, anthropology, social psychology, media and cultural studies, and management. What these differing perspectives share is a core concern with the ways in which the nexus between the mindfulness revolution in the West and Buddhism is shaping and being shaped by each other. Further, each of the contributors of this volume *deeply care* about the dissemination and practice of mindfulness in society; their varied breadth and depth of professional and personal experience provides a multitude of voices that provoke, question, and challenge the status quo.

We hope that this handbook volume will help establish the foundations for an emerging field of critical mindfulness studies. It is intended for academics, clinicians, scientists, and Buddhist teachers and scholars, social activists, and university students, as well as mindfulness practitioners who are sympathetic to the need for more critical inquiry and cultural analyses of the mindfulness movement. Readers will find this handbook to offer a comprehensive compendium of social criticism that is aimed at excavating and exposing hidden assumptions, misconceptions, and ideologies that have remained below the surface of modern mindfulness discourse. The purpose of such critiques is grounded in the faith that secular mindfulness practices can be reformed and reoriented to enhance the common good. This passion for critique among the contributors of this handbook is matched by their passion for truth-telling, often going against the mainstream narrative with its self-help rhetoric and psychological-neurospeak explanations that have characterized the benefits of mindfulness. Because mindfulness practices are intended for the relief of human suffering in society, the questions our contributors raise are significantly ethical and political ones. A medicalization of mindfulness limits the practices and programs to the symptomatic relief of individuals' distress, essentially a highly privatized and individualistic approach that has favored neurological and psychological reductive explanations of meditation. The effects of social, political, and economic factors, as well as the situational stressors caused by our major institutions themselves, are left out of such mainstream accounts. The emancipatory potential of mindfulness for addressing social suffering will remain neutered and limited so long as "critique is turned inward," as Davies (2015) so eloquently stated in his book *The Happiness Industry*. It is in this spirit that criticism plays a role in fostering civic or social mindfulness—where those teaching and practicing mindfulness turn critical attention outward to include institutions, histories, socioeconomic, and cultural influences that contribute to, and are often causes of, social suffering.

The handbook consists of thirty-three chapters organized into four parts: (1) "Between Tradition and Modernity," (2) "Neoliberal Versus Critical Mindfulness," (3) "Genealogies of Mindfulness-Based Interventions," and (4) "Mindfulness as Critical Pedagogy." Now, we move on to a preview of the chapters.

Part I

Part I, "Between Tradition and Modernity," sets out to define key issues of concern and contested meanings of mindfulness as those teachings and practices have migrated from traditional Buddhist settings into a modern and Western context. A number of scholars have questioned whether the dominant meaning of modern mindfulness of "paying attention to the present moment" by cultivating nonjudgmental "bare attention" (Bazzano 2013; Bodhi 2011; Brazier 2013; Dreyfus 2011; Purser 2015; Sharf 2015; Wallace 2007) forecloses the wider ethical aspects of the practice, along with omitting the cultivation of compassion commitment to social welfare. Mindfulness training represents only a sliver of the plethora of Buddhist meditation methods (Lopez 2012). In addition, even within Buddhism, there are varied conceptions of mindfulness across various schools and traditions (Dunne 2011; Sharf 2015).

Within a Buddhist context, the term "mindfulness" first appeared in 1881 in Max Müller's book, *Buddhist Suttas*, translated by Thomas W. Rhys

Davids. Sir Monier Monier-Williams, a Sanskrit scholar at Oxford University, also deliberately used the term in his 1889 book, *Buddhism in its Connexion with Brahmanism and Hinduism*. According to Buddhist scholars, the modern translation of mindfulness from *sati* (*smṛti* in Sanskrit) is derived from the verb, "to remember," or the act of "calling to mind" (Anālayo 2010; Davids 1881; Gethin 2001; Nanamoli and Bodhi 2005; Ṭhānissaro 2012). The establishment of mindfulness in meditation, however, is not merely a function of memory, nor merely a passive and nonjudgmental attentiveness to the present moment exclusively, but an actively engaged and discerning awareness that is capable of recollecting various teachings, ethical commitments, and the eradication of greed, ill will, and delusion.

It is also worth noting here that the Buddhist tradition is not monolithic. The affinities between modern therapeutic mindfulness-based interventions and "Buddhist conceptions of mindfulness" have often been over-generalized, linking more often than not to recent modernized versions of Theravada Buddhist vipassana insight meditation practices that have their origins in the Theravada revival movement of the nineteenth and twentieth centuries (Braun 2013). In other Buddhist schools and traditions, such as Tibetan Buddhism, mindfulness has never been foregrounded or relegated such central status as a core practice. Many of the Tibetan Buddhist schools first required students to engage in intensive analytical and textual studies, philosophical meditations, combined with devotional and purification practices, along with a progressively being introduced to preliminary reflective practices before a student is exposed to formal meditative methods and somatic and energetic yogic trainings. This progressive and graduated approach is considered foundational to providing the educational and values-based framework for contemplative practice.

Modern cultural translations of mindfulness practices have also excluded and downplayed the vast array of contextual and cultural mediated forms of understanding, considering such practices as "culturally laden forms of baggage." However, it is precisely this comprehensive and cultural framing of contemplative experience that provides the interpretative frameworks for guiding, making sense of, and enacting meditative insights on progress of the path of liberation.

As Germano (2016) has asked, "If the preliminary practices create a context for meditation in Tibet, what creates the context for meditation in the West?" The current contemporary fascination with mindfulness as a therapeutic intervention, what Richard King, a contributor to his handbook refers to as the "mindfulness-only" school, is a relatively recent phenomenon. This is understandable given that the goals of therapeutic mindfulness diverge from traditional Buddhist soteriological aims for total and complete liberation from suffering. Indeed, the mainstreaming and medicalization of mindfulness has often been conjoined with enhancing sensual pleasures, intensifying appreciation for present-moment aesthetic experience, and seeking happiness in various mundane worldly concerns (career success, relationships, better sex, weight control, and so on) (Wilson 2014). The recontextualization and cultural transmission of modern mindfulness has often failed to illuminate or take into account how such practices and interventions are themselves

Westernized "forms of life that are social, embodied, and enacted in social contexts" (Kirmayer 2015). In this respect, the "mindfulness-only" school with its universalizing rhetoric has situated itself within the individualistic norms of Western consumer capitalism as its de facto educational context.

The Venerable Bhikkhu Bodhi's chapter "The Transformation of Mindfulness" leads off Part I by offering his very personal account of how mindfulness took the route it did in America over the course of the past forty years. Having been an American Theravada Buddhist monk since his ordination in 1972 in Sri Lanka, as well the foremost scholar and translator of Buddhist texts from the Pāli Canon, Ven. Bodhi is able to describe how early Western Buddhist teachers severed the explicit connections between insight meditation and Buddhist spirituality. These transformations significantly altered the practice of mindfulness by reframing it in psychological terms, eventually undergoing a major overhaul with regard to its objectives and goals.

Next, in Chap. 2, David Loy addresses how we need both individual and social transformation, and how the best ideals from the Western tradition with its concern for social justice and human rights can join forces with the most important goal for traditional Buddhism—to put an end to one's *dukkha* ("suffering" in the broadest sense), especially that associated with the delusion of a separate self. Loy calls on the mindfulness movement to go beyond its current individualistic, consumerist orientation in order to mitigate the causes of collective and organizational *dukkha*.

In Chap. 3, Richard King examines the role of intellectual analysis and ethical judgment in ancient Indian Buddhist accounts of *sati* and contemporary discourses about "mindfulness." King draws on sources from the Abhidharma and early Mahāyāna philosophical discussions in India, which informed the cultivation of *sati*, comparing and contrasting these ancient understandings with modern discourses of mindfulness. He offers a cogent analysis of how the rise of modern mindfulness is linked to the processes of detraditionalization, the global spread of capitalism, and widespread adoption of new information technologies. In addition, King explores the modern history of attention, tracing how these trajectories have produced divergent contemporary accounts of mindfulness. The history of attention, King argues, cannot be separated from the history of mindfulness given how both streams are implicated in the rise of digital technologies and neoliberalism as cultural phenomenon.

Geoffrey Samuel undertakes the task in Chap. 4 of first providing an overview of how the early stages of the mindfulness movement were defined mainly by the meditation practices from the nineteenth-century Theravada reform movement, what is now often referred to a strand of Buddhist modernism. Samuel describes the early research on mindfulness meditation as it was focused mainly on therapeutic efficacy and how this was key to situating modern mindfulness within contemporary scientific thought and biomedical practice. He goes on to explore for consideration a much wider range of meditative forms that exist within Asian Buddhist traditions which could themselves stimulate and expand our Western modes of scientific thought and aid us to develop a more varied and productive range of therapeutic applications.

Next, in Chap. 5, David Brazier distinguishes modern, utilitarian mindfulness from traditional, Buddhist mindfulness. He examines and critiques a number of the cultural factors that have shaped utilitarian forms of modern mindfulness, including what he describes as "here-and-now-ism" and the overvaluation of consciousness. Brazier questions whether the modern version will prove to be simply a weak variant, or a step on the way to a more wide-ranging transformation of our cultural values.

In Chap. 6, Candy Gunther Brown disputes a major claim that "secular" mindfulness programs teach a purely secular, universal technique. She argues that so-called secular mindfulness programs instill culturally and religiously specific and contested worldviews, epistemologies, and values. Her chapter critically examines Jon Kabat-Zinn's mindfulness-based stress reduction (MBSR) in terms of three common patterns: (1) code-switching (skillful means, stealth Buddhism, Trojan horse, and scripting), (2) unintentional indoctrination, and (3) religious and spiritual effects. She goes on to argue that in particular cultural contexts, mindfulness programs could explicitly or implicitly convey religious meanings or facilitate religious and spiritual experiences. Despite the use of secularizing rhetoric, she contends the separation of mindfulness from its religious worldview and values may not be entirely possible.

Part I concludes with a contribution by Jack Petranker as he introduces a novel "field-centered" mindfulness, a practice that focuses on the fullness of space rather than the present-centered immediacy of time. According to Petranker, field-centered mindfulness builds on present-centered mindfulness, but introduces a fundamentally different orientation to the stream of experiences and appearances we encounter. He points out that the currently popular practice of present-centered mindfulness does little to challenge the standard subject/object framework. His proposal for "field-centered mindfulness" is consonant with the sensibilities of modern secular practitioners who need not study and accept Buddhist doctrines or a Buddhist worldview, yet it still offers a way of seeing that is congruent with key Buddhist insights.

Part II

In Part II, "Neoliberal Mindfulness Versus Critical Mindfulness," the chapters address a range of issues and concerns with regard to how neoliberal discourse and capitalist imperatives have influenced and exploited the way mindfulness is utilized as modern behavioral technology of the self (Foucault 1998). Stress, disengagement, and discontent are pathologized as an individual-level phenomenon within the majority of mindfulness programs. This is particularly true in corporations where mindfulness programs aim at the formation of an entrepreneurial self that is willfully productive and responsible for their own self-care. The contributions in this section help to expose how contemporary mindfulness programs are both compatible and complicit with neoliberal values which frame mindfulness primarily as an instrumental and privatized practice. This framing essentially depoliticizes mindfulness training curricula by foreclosing alternative pedagogical encounters that could foster critical engagement with the causes and conditions of social suffering that are implicated in power structures and economic systems of capitalist society. A number of contributions draw on the work of Michel Foucault, particularly his 1979 lectures where he explained how neoliberal modes of governing amount to a form of "biopolitics" and "biopower" which infuse self-disciplinary regimes into the embodied and social domains of modern society. Mindfulness then can be envisioned as a form of embodied mental cultivation that is employed productively in the workings of power. In attempting to account for the processes of subjectification in capitalist societies, Foucault (2008) introduces the concept of "governmentality" which he often referred to as the "conduct of conduct."

Conforming to the logic of governmentality, the project of contemporary mindfulness is a conservative one: The mindful subject is constituted as being free to choose happiness or misery, stress or well-being. It is important to point out that this mode of control is not repressive or coercive, nor is it a sinister form of mind control or brainwashing as some mindfulness proponents have misrepresented recent critiques of contemporary mindfulness. Rather, the recontextualization of mindfulness in late capitalist society is a cultural and political translation that relays neoliberal values in the formation of a new subject that is freely choosing to control his or her own freedom. It is in this sense that form of disciplinary power is productive; mindfulness practice can then be viewed as a technology for reflexive self-formation, shaping and producing the behavior of a conservative "mindful subject."

The popular interest and widespread acceptance of contemporary mindfulness programs might partially be explained by the fact such programs are conducive to an instrumental reformulation of all spheres of life, those which were previously impervious to the market and institutions. In this respect, mindfulness also represents a new form of *biopower* where both the mind and body become sites for self-disciplinary control, self-surveillance, and self-optimization. As a disciplinary apparatus, mindfulness can also serve to ensure that subjects are constituted as private and atomistic individuals that not only voluntarily participate in their own governance, but also come to forget and forfeit bonds of solidarity and collectivity. This ideology of individual autonomy strongly resonates with neoliberal values of freedom, choice, authenticity, entrepreneurialism, and competitiveness. When viewed through the lens of biopower, mindfulness is also constituted as a lifestyle choice, fully symmetrical with market imperatives for consumption, efficiency, productivity, and social order.

Jeff Wilson begins this section with his chapter that describes how mindfulness meditation has been shaped and influenced by capitalist values and marketed as a commodity to Western consumers. Wilson provides a detailed analysis of the popular magazine *Mindful*, paying particular attention to its advertising policies and featured advertisers. His chapter provides insight into the forces at work in the commodification and diversification of the mindfulness movement.

Next, in Chap. 9, Richard Payne examines how American self-improvement culture has shaped the propagation and ethos of mindfulness training. Payne argues the driving ethic of that culture is the moral imperative to improve oneself, rooted in Puritan theology. Tracing these historical influences, Payne shows how the ethic of self-improvement has infected the ideology of American popular religious culture and how this moral imperative is linked to neoliberalism and foundational to the marketing and promotion of mindfulness.

In Chap. 10, Edwin Ng explores a style of thought that he aptly calls "critical mindfulness." He describes how the adaptation of mindfulness across multiple domains has to negotiate the dominant logics of the present neoliberal capitalist order of things. Ng argues that neoliberalism is not a sinister ploy that hides the truth, but is a regime of truth that functions as a political ontology. It is within this everyday, uncritical acceptance of neoliberalism that conditions how we come to make reasonable judgments and conduct our own lives and behavior. Drawing on the work of Michel Foucault, Ng explores how mindfulness might function as a disruptive technology of the self within and against these dominant logics. Ng makes use of Foucault's analytic of governmentality as a means for developing this style of thought and explains Foucault's work is not restricted to Engaged Buddhist concerns.

In Chap. 11, Zack Walsh presents a discourse analysis of mindfulness critiques circulating in online media, identifying the key contested issues that have framed the public debate on mindfulness. Walsh not only provides a coherent summary of critics' concerns, but he also outlines the conditions for renegotiating how mindfulness can be reframed. Arguing that neoliberalism has transformed mindfulness into a variety of depoliticized and commodified self-help techniques, Walsh explains why universal, asocial, and ahistorical views of mindfulness should be replaced by critical, socially aware, and engaged forms of mindfulness. Walsh's chapter must be considered in conjunction with the chapter that follows. Here, Per Drougge identifies many of the same issues as Walsh, drawing even further attention to the upsurge in critical engagement with mindfulness and the mindfulness industry. Drougge offers a penetrating critique on the marketing and presentation of mindfulness, its relation to the Buddhist tradition and cultural appropriation, its conceptual fuzziness and exaggerated claims, methodological insufficiencies in studies of meditation and mindfulness, and the ideological function of mindfulness practices. His chapter summarizes and discusses a number of critical articles that have appeared on Web sites and in popular media during the past few years and the responses they have elicited.

Longtime Buddhist meditation teacher Christopher Titmuss explores the recent development of mindfulness in the West since the late 1970s, focusing particularly on the growth of mindfulness programs in large corporations. Titmuss raises a number of concerns and questions pertaining to whether corporate mindfulness programs are offering a comprehensive application of mindfulness and/or whether such programs are quietly subservient to the productivity and efficiency goals of corporations. In addition, Titmuss calls for the application of a modern variant of the Four Noble Truths to business.

Continuing with a critique of corporate mindfulness programs, in Chap. 14, Alex Caring-Lobel provides an in-depth historical account of the ideological drivers of corporate mindfulness initiatives, viewing such management-driven programs as part of an evolutionary response to the specific needs of capital. Caring-Lobel explains that corporate mindfulness programs have been enthusiastically embraced because they offer a way of mitigating the psychological collapse of postindustrial knowledge workers without confronting the social and economic causes of their discontent. In particular, noteworthy is how his chapter connects the corporate mindfulness movement to the work of past management science gurus going back to Frederick Taylor and Elton Mayo. He calls for a repoliticization of the forms of worker stress and discontent that workplace mindfulness rhetoric and praxis obfuscate by framing them in purely psychological terms.

In the concluding chapter in this section, Massimo Tomassini begins by reviewing and critiquing the dominant conceptions and applications of mindfulness within corporations. Going beyond these corporate-driven approaches, Tomassini considers a different approach to mindfulness at work, one that is not simply a form of stress reduction or attention enhancement technique, but a liberating communal practice that can occur outside of the normal performance-driven work culture, incorporating more reflective types of practices that are self-determined by the participants themselves.

Part III

Part III of the handbook, "Genealogies of Mindfulness-Based Interventions," turns to critical examinations of mindfulness-based interventions (MBIs), along with the scientific and public discourse that has served to establish the legitimacy of MBIs as a psychotherapeutic technique. Collectively, these chapters constitute a genealogy of the mainstreaming of MBIs, and each attempts to historicize and contextualize the emergence of mindfulness within the helping professions and healthcare institutions. It is in this section that authors examine the medicalization of mindfulness and how the behavioral medicine paradigm has been used as an explanatory narrative for making individuals responsible for their own stress and healing. One of the basic assumptions of MBSR and MBIs is that our failure to pay attention to the present moment, that is, our mindlessness and mind-wandering, is the main reason underlying of dissatisfaction and disease. This etiological explanation for stress as being a deficit of an individual's attention is a common trope. But Kabat-Zinn takes it even a step further by claiming that our cultural malaise is also the result of an attention disorder *en masse*; capitalist societies are themselves suffering from attention deficit disorders (ADDs). As Kabat-Zinn (2005, p. 143) states our "…entire society is suffering from attention disorder-big time." Apparently, widespread societal stress and social suffering are not the result of massive inequalities, material conditions, nefarious corporate business practices, or political corruption, but an individual-level psychic dysfunction—a "thinking disease" (Barker 2014; Goto-Jones 2013).

The unspoken assumption here is that there is nothing inherently dysfunctional with capitalism itself; rather, we simply are not mindful or resilient enough as individuals to be fully functioning, authentic, and happy human beings. The mindfulness revolution promises to bring relief and resolution to individuals debilitated by the demands of late capitalism, but without any political agenda, or any substantial challenge to the institutional structures which enable capitalism to inject its toxicity system-wide. And, as Goto-Jones (2013) points out, the mindfulness revolution also functions as a type of secular, quasi-religion within capitalism, especially in such regions as Silicon Valley where corporate mindfulness programs have become the rage.

The solution for addressing the ills of society and for social change will come about not through any form of political struggle or grassroots political revolution, but through a conservative mindful revolution—training individuals in mindfulness (Goto-Jones 2013). This is also known as the "Trojan horse" hypothesis that individuals who are more mindful, compassionate, and authentic themselves will slowly and peacefully ensure the emergence of a humane and compassionate capitalist society. The mindfulness revolution then is essentially a therapeutic not a political project. As we saw in Part II, neoliberal mindfulness emphasizes the sovereignty of autonomous individuals who can navigate the vicissitudes of late capitalist society by becoming self-regulating and self-compassionate, governing themselves, and by freely choosing their own welfare, well-being, and security.

In this narrative, moderns are disenchanted, suffering from an obsession with "doing" rather "being." Kabat-Zinn's famous initiation rite for MBSR programs—to slowly savor and mindfully eat a raisin—is symbolic of the mindfulness cure, to appreciate the present moment in all its fullness. Such appreciative apprehending, for Kabat-Zinn, "coming to our senses" by dwelling in the "being versus doing" mode, draws its phenomenological inspiration directly from the American transcendentalists (McMahan 2008). It is supposedly through non-striving and non-doing that a magical reenchantment occurs, countering the iron cage of rationalization and frantic pace of our 24/7 digital economy. Barker (2014) points out that Kabat-Zinn's social admonition to rest in the mindful being mode as a cure for our thinking disease is contradicted by his own opposing disciplinary injunction of the need to be mindful as one goes about all of one's daily activities. Indeed, one of Kabat-Zinn's most favorite public quips is "mindfulness is the hardest thing to do."

The popular portrayal of the mindful subject as one who must be constantly in a mode of self-surveillance is reflective of what Nikolas Rose characterizes as the "genealogy of subjectification" (Rose 1998). Rose (1998, p. 23) elaborates:

> A geneaology of subjectification takes [this] individualized, interiorized, totalized, and psychologized understanding of what it is to be human as the site of a historical problem, not as the basis of a historical narrative.

In this respect, MBSR and MBIs can be understood as rationalized schemes that are what Foucault referred to as "self-steering mechanisms" that shape our behavior (the conduct of conduct). Mindfulness as a regulatory mode of thought is one of the most recent additions to what Rose has called the "psy-sciences." Rose situates the psy-disciplines as a historical project, problematizing their emergence in relation to the crises of capitalism, political economies, and institutional structures. Offering a critical history of the psychological sciences, Rose is able to describe and articulate how psychology is a form of technology which has provided answers to contemporary society by legitimizing expert claims to authoritative knowledge production (Doran 2011, p. 23).

> The modern self is impelled to make life meaningful through the search for happiness and self-realization in his or her individual biography: the ethics of subjectivity are inextricably locked into the procedures of power (Rose 1998, p. 79).

Kabat-Zinn's proclamations that the problems of society can be traced to mindless individuals suffering from a disease of thinking is a continuation of the psy-sciences predilection for producing expert knowledge that construes our lives in psychological terms, and reduces the problems of economic and social life to the calculability of individuals. It is important to point out that the regulatory and disciplinary functions of mindfulness that Kabat-Zinn professes are not necessarily conscious aims. As part of the psy-sciences, mindfulness as a liberation technology of the self is a system of expert thought for governing certain forms of thinking, or mental ruminations, as governable by individuals themselves. The contemporary regime of the free individual in capitalist society is now the mindful individual.

Brooke Lavelle begins this section with a chapter that examines three modern secularized mindfulness and compassion-based contemplative programs, namely mindfulness-based stress reduction (MBSR), cognitively-based compassion training (CBCT), and sustainable compassion training (SCT). Lavelle challenges the rhetoric that such programs have universal applicability, along with pointing out how the underlying assumption of universality has created a cultural blind spot and bias that has had the result of privileging theory over context. Her chapter provides a useful framework for understanding how certain Buddhist contemplative frames (i.e., innatism and constructivism) and modern cultural frames (i.e., individualism, scientific reductionism, and secularization) both limit and permit different possibilities for health and healing.

Next, David Lewis and Deborah Rozelle closely examine mindfulness-based interventions (MBIs) and in particular mindfulness-based stress reduction (MBSR), comparing these psychological treatments to the fundamental tenets and ultimate goals of the Buddhist path of liberation, which they refer to as the Buddhadharma. Their critique takes aim at the claim that MBIs (and MBSR) embody the essence of Buddhadharma. Their analysis employs a unique analogical methodology to compare key aspects of both MBIs and the Buddhadharma teachings and practices, focusing on such commonly used terms as suffering (dukkha), impermanence, and no-self. Lewis and Rozelle are able to demonstrate that many of the claims put forth by Jon Kabat-Zinn—that MBIs embody the essence of the Dharma—actually have the result of reducing the Buddhadharma to the psychological level, while inflating MBIs to a transcendent level. By providing a cogent analogical framework, they are able to show that MBIs are actually a psychological analog of the transcendental realm, with a similar structure but at a very different ontological level.

In Chap. 18, Paul Moloney takes the critique a step further by exposing the limits, methodological weaknesses, and unsubstantiated claims of mindfulness-based interventions. Moloney critiques the popular mindfulness movement by situating its discourse within a much wider historical context originating in the psychotherapy industry. Examining the exuberant claims of mindfulness-based cognitive therapy (MBCT) and the Mindful Nation report, Moloney scrutinizes the scientific methodologies of psychotherapy research. His chapter illustrates how mindfulness is the latest phase in the privatization of the self that has been underway from the middle of the twentieth century, and in which the applied psychology professions have been instrumental.

Manu Bazzano begins Chap. 19 by noting that our age, in terms of Buddhism's "three treasures" (the Buddha, Dharma and the Sangha), is that of the Sangha, or spiritual community. Bazzano instructs us that creative engagement with these three treasures requires a form of active adaptation, rather than simply defending tradition or passively adapting to it. Active adaptation requires going beyond the reductionism that has characterized the mainstreaming of neoconservative mindfulness practices as they have been propagated through the proliferation of the contemporary neuroscience literature. His chapter goes on to explore the desirability of a *fourth treasure*, psychotherapy and its relation to the Dharma—a potential pathway away from the current mindfulness brand with its communal deficits. Drawing on humanistic psychology and Zen, Bazzano affirms the value of inquiry, social solidarity, and the ability to perceive the elusive dimension of affect.

Next, Steven Stanley and Charlotte Longden report on their research on mindfulness courses using a combination of discourse and conversation analysis of language used within these courses. Their chapter begins by situating mindfulness historically within therapeutic culture, discussing how both the medicalization and psychologization of mindfulness practices as forms of self-help have strong affinity to the "psy-complex" and psychological styles of "governmentality" (Rose 1998). Their findings describe how affective–discursive and inquiry practices in mindfulness courses, particular the interactions between teachers and participants, function to practically produce mindful subjects who can monitor, govern, and take care of themselves. Mindful subjectivity is produced through the application of liberal power and negotiation of ideological dilemma within inquiry sequences, functioning as technologies of the self.

In Chap. 21, Jenny Eklöf examines the ways in which the scientific meaning of mindfulness is communicated in public and to the public. Her chapter shows how experts in the field of mindfulness neuroscience seek to communicate to the public at large the imperative of brain fitness for the promotion of health, well-being, and happiness. Through her analysis of the claims being made in popular outlets such as self-help books, Web sites, and online videos, Eklöf identifies what she describes as *personalized* science communication, demonstrating that the boundary between science and pop-ularized science is the outcome of human negotiations. Her analysis also shows how prominent contemplative neuroscientists have used personalized communications as a way to infuse their scientific findings with subjective meaning, turning their communication with the public into a moral vocation.

Part III concludes with a chapter by Lisa Dale Miller that examines the mental and emotional suffering involved in what Buddhist psychology identifies as "self-cherishing." Her chapter compares Western and Buddhist psychological models of self, Buddhist theories of not-self, and conventional and ultimate self-cherishing. In addition, she outlines a clinical approach that can help individuals to recognize self-cherishing mentation, illustrating through examples of therapist–client dialogue how such individuals strug-gling with depressive, anxious, trauma-related symptoms and addictions can lessen its deleterious effects.

Part IV

Part IV, "Mindfulness as Critical Pedagogy," discusses how mindfulness programs are employed in K-12 and postsecondary education. As with cor-porations, mindfulness programs in schools arise within and are influenced by broader neoliberal structures and ideologies. Although the aim of public education is not intended to be about profitability, productivity, and con-sumption per se, it is nevertheless a contested site that is subject to market forces and demands. Within an undertheorized neoliberal climate, mindful-ness programs in schools become a form of governmentality that helps shape individuals to adjust to the needs of a society that must compete in a global economy. Mindfulness practices in many school programs encourage both students and educators to self-regulate and become the kind of self-sufficient, emotionally adjusted entities that can function and thrive in a market-based and consumer society. What is often omitted from such programs is the critical cultivation of awareness, appreciation, and employment of the cul-tural context and cultural capital of both students and educators; this omis-sion contributes to reinforcing racist systems within education that in turn reproduces racism in the larger social structure. A number of articles in this section point to ways in which mindfulness can be embedded within edu-cation programs that are informed by critical pedagogy, interconnectedness, awareness of structural inequities, and engaged practices that promote inclusive and universal social justice. Others also build bridges between classical Buddhist wisdom and contemporary scientific and practical knowledge, suggesting new directions for mindfulness education programs.

In Chap. 23, David Forbes provides an overview of the problematic of mindfulness education programs that do not address contested social, developmental, and cultural contexts within which such programs are practiced in schools. He defends the merits of social critique and those critics who have called out McMindfulness, the use of mindfulness for self-aggrandizement and adjustment to social institutions that promote greed, delusion, and ill will. Forbes critically employs concepts from integral metatheory with an emphasis on cultural meanings, optimal human development, and universal social justice within schools. He offers directions toward a critical integral contemplative education that promotes full individual, interpersonal, and social development.

Next, in Chap. 24, Funie Hsu looks further at secular mindfulness programs in schools within the contexts of neoliberalism and race. In particular, she focuses on the ideology of white conquest that makes invisible the enduring efforts of Asian and Asian American Buddhists in maintaining the legacy of mindfulness practices. She shows how mindfulness curricula discipline students through neoliberal self-regulation and the racial conditioning of white superiority. Hsu calls for secular mindfulness to be part of a broader paradigm shift in education that enhances the value of education as a public good.

Terry Hyland examines in Chap. 25 mindfulness-based applications within education against the background of the ethical and educational shortcomings of the McMindfulness models of practice. He argues for the need to foreground educational and moral components of mindfulness programs related to personal and social transformation in order to avoid the limitations of McMindfulness. Hyland recommends that mindfulness-based interventions be firmly grounded in Buddhist ethical foundations in order to achieve the full objectives of the transformative project of the dharma.

Jennifer Cannon continues in Chap.26 viewing the mindfulness education movement through a social justice and antiracist lens and develops a constructive critique that calls for a socially engaged mindfulness. She analyzes a film that promotes mindfulness in schools that unwittingly demonstrates the white savior trope. Cannon offers a social justice framework that shifts the deficit discourse of school failure and troubled communities to a collaborative practice that critically considers the social conditions that create suffering, and that promotes mindfulness as a practice of freedom rather than a technology of compliance.

In Chap. 27, Joy L. Mitra and Mark T. Greenberg seek to create a secular ethical framework for interpersonal forms of compassion that reflect the relational nature of the self and mental processes. The relational nature is supported by both classical teachings and contemporary evidence-based research in many disciplines. They are critical of mindfulness approaches that do not account for the illusory boundaries of the separate self. Mitra and Greenberg argue there is an urgent need to instead create new modes of secular education, such as a curriculum of Right Mindfulness that is based on softened boundaries between self and other. These would support nonviolent and sustainable communities and can be applied to educational settings.

Rhonda Magee makes the case in Chap. 28 that social justice concerns are inherent to mindfulness and secular Buddhist practices and describes an approach she terms community-engaged mindfulness. She first discusses how mindfulness practices cultivate a felt sense of solidarity among people with a common purpose such as working together for a more just world. Magee provides an exploratory case study involving a community that was dealing with evidence of racial bias within the local police department. She describes two workshops she co-facilitated that included mindfulness-and-compassion-based-practices that enabled participants to feel heard, build on their own community resources, and begin to heal and initiate policy changes.

Next, in Chap. 29, Natalie Flores brings a critical approach to mindfulness to bear on early childhood education settings. She investigates how mindfulness is used with respect to school readiness and schoolification. Flores examines three popular mindfulness programs that have been used with young children and argues that these aim to provide educators with tools to more effectively implement school readiness and schoolification. She also makes recommendations that would assist educators to implement a more holistic approach to mindfulness in early childhood education settings.

Next, in Chap. 30, Joshua Moses and Suparna Choudhury investigate some mindfulness meditation programs in schools, including ones that emphasize neuroscience, and remain ambivalent about their benefits. They note that good contemplative programs touch on interconnectedness and social relationships, as Mitra and Greenberg also point out. They argue that all programs, even ones that focus on neuroscience, have implicit moral assumptions and that they could enable children to become more socially engaged and critically examine their circumstances. Moses and Choudhury suggest an ideological underpinning for the popularity of mindfulness programs that combine neuroscience with a secularized spiritual–moral discourse: They expand the hopeful scientific narrative about human nature that people are social, benevolent, and evolving toward better futures.

In Chap. 31, Adam Burke describes a course he developed that employs mindfulness practices in helping underrepresented college students improve their rates of retention, graduation, and academic success. Unlike many programs that apply mindfulness in education, Burke is aware of the structural and systemic forces both at the societal level and university level that impede many students of color, women, and those from working-class backgrounds. He notes that although it does not impact structural change, the focus on classroom instruction, including mindfulness practices, does provide students with awareness tools they can use at an individual level to navigate diverse institutional settings.

Commentary

The final segment of this volume then turns to invited commentaries by Rick Repetti and Glenn Wallis. Repetti's chapter aims to defend secular mindfulness programs against the "McMindfulness" critique. He first argues that mindfulness is a form of metacognitive awareness that is intrinsic to, and a universal property of, human consciousness—independent of any religious or secular context, or ethical commitments. Thus, mindfulness can be put to use—it is a tool, but a tool for enhancing metacognitive awareness, not changing the world. And because mindfulness can be viewed simply as context-free form of mental cultivation, analogous to weight-lifting or physical exercise, expecting anything more than individual mental enhancement from mindfulness training is both unrealistic and misplaced. Thus, the objections raised by the McMindfulness critique are nothing more than hand-waving hyperbole.

In the concluding chapter, Glenn Wallis offers a cogent rebuttal by first noting how Repetti's chapter actually is reflective of how secular mindfulness advocates have failed to respond to criticisms by resorting to what he refers to the rhetorical strategies of "conceptual shape-shifting and covert idealism." Wallis points out that Repetti sidesteps the fact that an ideological edifice has been erected around Jon Kabat-Zinn's operational definition of mindfulness, turning it into a system of thought and practice that is embedded within a social–economic–political context, and which produces a very particular form of subjectivity and world. Indeed, Wallis argues that Repetti's reactionary stance to the McMindfulness critique amounts to a faithful valorization of the diminished neoliberal subject who utilizes mindfulness practice as essentially a self-help technique for enhancing our (natural) capacities for adaptation, acceptance, and resilience.

Taken in its totality, this handbook provides a wide-ranging overview and introduction to the emerging field of critical mindfulness studies. As Edwin Ng, one of our contributors points out, "When we speak of 'critical mindfulness,' we are following Foucault in performing critique not simply to decry that things are not right as they are. Rather, it is 'to show that things are not as self-evident as one believed, to see that what is accepted as self-evident will no longer be accepted as such.'" Each of our contributors has engaged in critical inquiries, examining and interrogating the ideologies, cultural context, and institutional interests that have shaped and framed our contemporary understanding of mindfulness. At a time when the hype, commercialization, and popularity of mindfulness are at its peak, critical mindfulness has much to offer by challenging the dominant frames that have informed contemplative programs and concomitant scientific research. For students and professionals wishing to go beyond universalist, ahistorical, and decontextualized treatments of mindfulness, and for scholars seeking new frames that take into account historical, cultural, social, political, economic, racial, and ethical dimensions of contemplative practice, this handbook will provide both insight, inspiration, and direction.

San Francisco, CA, USA   Ronald E. Purser
Brooklyn, NY, USA   David Forbes
San Francisco, CA, USA   Adam Burke

References

Anālayo. (2010). *Satipaṭṭhāna: The direct path to realization*. Birmingham: Windhorse Publications.

Barker, K. (2014). Mindfulness meditation: Do-it-yourself medicalization of every moment. *Social Science & Medicine, 106*, 168–176.

Barnes, P. M., Bloom, B., & Nahin, R. L. (2008). Complementary and alternative medicine use among adults and children: United States, 2007.

Bazzano, M. (2013). In praise of stress induction: Mindfulness revisited. *European Journal of Psychotherapy and Counseling, 15*(2), 174–185.

Bodhi, B. (2011). What does mindfulness really mean? A canonical perspective. *Contemporary Buddhism, 12*(1), 19–39.

Braun, E. (2013). *The Birth of Insight*. Chicago, IL: University of Chicago Press.

Brazier, D. (2013). Mindfulness reconsidered. *European Journal of Psychotherapy and Counseling, 15*(2), 116–126.

Coronado-Montoya, S., Levis, A. W., Kwakkenbos, L., Steele, R. J., Turner, E. H., & Thombs, B. D. (2016). Reporting positive results in randomized controlled trials of mindfulness-based mental health interventions. *PLOS One, April*. http://dx.doi.org/10.1371/journal.pone.0153220

Davids, T. W. R. (1881). *Buddhist Suttas*. Oxford, UK: Clarendon Press.

Davies, W. (2015). *The happiness industry: How the Government and big business sold us well-being*. London: Verso Books.

Doran, P. (2011). Is there a role for contemporary practices of askNsis in supporting a transition to sustainable consumption? *International Journal of Green Economics, 5*(1), 15–40.

Dreyfus, G. (2011). Is mindfulness present-centered and non-judgmental? A discussion of the cognitive dimensions of mindfulness.*Contemporary Buddhism, 12*(1), 41–54.

Dunne, J. (2011). Toward an understanding of non-dual mindfulness. *Contemporary Buddhism, 12*(1), 71–88.

Foucault, M. (1998). Technologies of the self. In L. H. Martin, H. Gutman & P. H. Hutton (Eds.), *Technologies of the self: A seminar with Michel Foucault*. Amherst, MA: University of Massachusetts Press.

Foucault, M. (2008). *The birth of biopolitics: Lectures at the Collège de France 1978-1979* (M. Senellart, Ed., G. Burchell, Trans.). Houndmills, Basingstoke, Hampshire; New York: Palgrave Macmillan.

Germano, D. (2016). Tibetan meditation and the Modern World: Lesser Vehicle. *Coursera*. Retrieved from https://www.coursera.org/learn/buddhist-meditation/home/welcome

Gethin, R. (2001). *The Buddhist path to awakening: A study of the Bodhi-Pakkhiya Dhamma* (previously published in 1992). Oxford: Oneworld Publications.

Goddard, M. (2014). Critical psychiatry, critical psychology, and the behaviorism of B.F. Skinner. *Review of General Psychology, 18*(3), 208–215.

Goto-Jones, C. (2013). Zombie apocalypse as mindfulness manifestor (after Žižek). *Postmodern Culture, 24*(1), 1–14.

Goyal, M., Singh, S., Sibinga, E. M. S., Gould, N. F., Rowland-Seymore, A., Sharma, R., et al. (2014). Meditation programs for psychological stress and well-being: A systematic review and meta-analysis. *JAMA Internal Medicine, 174*(3), 357–368.

Heuman, L. (2014a). Don't belive the hype. *Tricycle*. Retrieved October 1, 2014, from http://tricycle.org/trikedaily/dont-believe-hype/

Heuman, L. (2014b). Meditation nation. *Tricycle*. Retrieved April 25, 2015, from http://www.tricycle.com/blog/meditation-nation

Kabat-Zinn, J. (2005). *Coming to our senses: Healing ourselves and our world through mindfulness*. London: Piatkus.

Kabat-Zinn, J. (2011). Some reflections on the origins of MBSR, skillful means, and the trouble with maps. *Contemporary Buddhism, 12*(01), 281–306.

Kirmayer, L. J. (2015). Mindfulness in cultural context. *Transcultural Psychiatry, 52*(4), 447–469.

Kirmayer, L. J., & Crafa, D. (2014). What kind of science is psychiatry? *Frontiers in Human Neuroscience, 8*, 435. doi:10.3389/fnhum.2014.00435

Lopez, D. (2012). *The scientific Buddha*. New Haven, CT: Yale University Press.

McMahan, D. (2008). *The making of Buddhist modernism*. New York: Oxford University Press.

Nanamoli, B., & Bodhi, B. (2005). *The middle length discourses of the Buddha: A translation of the Majjhima Nikaya* (3rd ed.). Boston, MA: Wisdom Publications.

North, A. (2014). The mindfulness backlash. *The New York Times*. Retrieved June 30, 2014, from http://op-talk.blogs.nytimes.com/2014/06/30/the-mindfulness-backlash/?_r=0

Pickert, K. (2014). The mindful revolution. *Time*, pp. 34–38, February 3, 2014.

Purser, R. (2014). The myth of the present moment. *Mindfulness*. Retrieved from http://download.springer.com/static/pdf/548/art%253A10.1007%252Fs12671-014-0333-z.pdf?auth66=1416427673_5de7e2d80e58dc6ce81fabdcb4c0d3cf&ext=.pdf

Purser, R., & Loy, D. (2013). Beyond McMindfulness. *Huffington Post*. Retrieved from http://www.huffigtonpost.com/ron-purser/beyond-mcmindfulness_b_35919289.html

Purser, R., & Cooper, A. (2014). Mindfulness' "truthiness" problem: Sam Harris, science and the truth about the Buddhist tradition. *Salon.com*. Retrieved December 6, 2014, from http://www.salon.com/2014/12/06/mindfulness_truthiness_problem_sam_harris_science_and_the_truth_about_buddhist_tradition/

Roca, T. (2014). The dark night of the soul. *The Atlantic*. Retrieved June 25, 2014, from http://www.theatlantic.com/health/archive/2014/06/the-dark-knight-of-the-souls/372766/

Rose, N. (1998). *Inventing our selves: Psychology, power, and personhood*. Cambridge, U. K.: Cambridge University Press.

Sharf, R. H. (2015). Is mindfulness Buddhist? (and why it matters). *Transcultural Psychiatry, 52*(4), 470–484.

Ṭhānissaro, B. (2012). *Right mindfulness: Memory & ardency on the Buddhist path*. Retrieved from http://www.accesstoinsight.org/lib/authors/thanissaro/rightmindfulness.pdf

Wallace, B. A. (2007). A mindful balance: What did the Buddha really mean by 'mindfulness'? http://buddhanet.net/budas/ebud/ebdha344.htm

Wilber, K. (2014). *The fourth turning: Imagining the evolution of an integral Buddhism*. Boulder, CO: Shambala.

Contents

Editors and Contributors

About the Editors

Ronald E. Purser, Ph.D. is a professor of management at San Francisco State University where he has taught the in the MBA and undergraduate business programs, as well in the doctoral program in the College of Education. Prior to his current appointment, he was an associate professor of organization development at Loyola University of Chicago. His recent research has been exploring the challenges and issues of introducing mindfulness into secular contexts, particularly critical perspectives of mindfulness in corporate settings. Author and co-editor of five books, including *24/7: Time and Temporality in the Network Society* (Stanford University Press, 2007). He sits on the editorial board of the *Mindfulness* journal, as well as the executive board of the Consciousness, Mindfulness and Compassion (CMC) International Association. A student and Buddhist practitioner since 1981, he was recently ordained as a Dharma instructor in the Korean Zen Buddhist Taego order. His article "Beyond McMindfulness" (with David Loy) in the *Huffington Post* went viral in 2013.

David Forbes, Ph.D. (U.C. Berkeley), LMHC, is an associate professor in the School Counseling program in the School of Education at Brooklyn College/CUNY and affiliate faculty in the Urban Education doctoral program at the CUNY Graduate Center. He was a co-recipient of a program grant from the Center for Contemplative Mind in Society and wrote *Boyz 2 Buddhas: Counseling Urban High School Male Athletes in the Zone* (Peter Lang, 2004) about counseling and teaching mindfulness meditation to a Brooklyn high school football team. Forbes teaches critical and integral approaches to mindfulness. He writes on the social and cultural context of mindfulness in education and wrote "Occupy Mindfulness" and "Search Outside Yourself: Google Misses a Lesson in Wisdom 101" with Ron Purser. He consults with schools in New York on developing integral mindfulness programs and practices meditation with a group from the New York Insight Meditation Center.

Adam Burke, Ph.D., M.P.H. is a professor in health education and director of the Institute for Holistic Health Studies at San Francisco State University. He holds advanced degrees in the social/behavioral sciences from UCLA and UC Santa Cruz, and is a licensed acupuncturist, trained in San Francisco and

Sichuan, China. Meditation training and practice commenced in the 1970's and continues across diverse traditions. Research and publication interests focus on student achievement and education equity, meditation and imagery, and cross-cultural studies of traditional health practices. Recent published works include *Learning Life* (Rainor Media, 2016). He has served on the American Public Health Association's Governing Council, as chair of the California Acupuncture Board, editor-in-chief of the *American Acupuncturist*, and as an Advisory Council member of the NIH's National Center for Complementary and Integrative Health (NCCIH).

Contributors

Manu Bazzano is a writer, psychotherapist, and supervisor in private practice, a primary tutor at Metanoia Institute, and a visiting lecturer at Roehampton University and various other schools and colleges. He facilitates workshops and seminars internationally. Among his books: *Buddha is Dead: Nietzsche and the Dawn of European Zen* (2006); *Spectre of the Stranger: Towards a Phenomenology of Hospitality* (2012); *After Mindfulness: New Perspective on Psychology and Meditation* (2014); *Therapy and the Counter-tradition: The Edge of Philosophy*, co-edited with Julie Webb (Routledge) and the forthcoming *Zen & Therapy: A Contemporary Perspective* (Routledge), and *Nietzsche and Psychotherapy* (Karnac) www.manubazzano.com.

Ven. Bhikkhu Bodhi is an American Buddhist monk, scholar, and translator of Buddhist texts from the Pāli Canon. After completing a Ph.D. in philosophy at Claremont Graduate School (1972), he traveled to Sri Lanka, where he received novice ordination in 1972 and full ordination in 1973, both under the renowned scholar-monk Ven. Balangoda Ānanda Maitreya. Ven. Bodhi lived in Asia for 24 years, primarily in Sri Lanka. He returned to the USA in 2002. In 2008, he founded Buddhist Global Relief, a nonprofit dedicated to helping communities afflicted by chronic hunger and malnutrition. He now lives and teaches at Chuang Yen Monastery near Carmel, New York.

David Brazier, Ph.D. is president of the Instituto Terapia Zen Internacional, Head of the Order of Amida Buddha, a Buddhist priest, an author of nine previous books, a psychotherapist, and a social worker. He was fortunate to encounter leading Buddhist teachers at the beginning of his adult life, and their teachings spoke to his condition. He travels widely and has been the creator of aid, education, and social work projects in Europe, India, and elsewhere and of training programs in Buddhist psychology, Zen Therapy, and Buddhist ministry.

Candy Gunther Brown (Ph.D. Harvard University) is a professor of religious studies at Indiana University. Brown is author of *The Word in the World: Evangelical Writing, Publishing, and Reading in America, 1789–1880* (University of North Carolina Press, 2004); *Testing Prayer: Science and Healing* (Harvard University Press, 2012); and *The Healing Gods: Complementary and Alternative Medicine in Christian America* (Oxford University Press, 2013). She is editor of *Global Pentecostal and Charismatic*

Healing (Oxford University Press, 2011) and co-editor (with Mark Silk) of *The Future of Evangelicalism in America* (Columbia University Press, 2016). She is writing a book about mindfulness meditation and yoga in public schools.

Jennifer Cannon is a Ph.D. candidate in the Department of Teacher Education and Curriculum Studies at the University of Massachusetts Amherst and holds a graduate certificate in Social Justice Education from the same institution. Her areas of scholarship include critical pedagogy, women of color feminism, decolonial theory, and contemplative pedagogy. She has been teaching in the field of social justice education for over 20 years and currently works as an educational consultant. Jennifer is a UCLA-certified mindfulness facilitator and is committed to integrating antiracism education and training with mindfulness education.

Alex Caring-Lobel is a writer and editor from New York City, currently based in Mexico City. He holds a Bachelor's degree in comparative literature from Emory University and is a graduate of its Tibetan Studies Program, where he wrote his thesis on Nagarjuna's two truths based on research in Dharamsala, India. He served as associate editor of the quarterly *Tricycle: The Buddhist Review* until 2015.

Suparna Choudhury is an assistant professor at the Division of Social & Transcultural Psychiatry, McGill University, where she works on the adolescent brain at the intersection of anthropology and cognitive neuroscience. Her doctoral research in cognitive neuroscience at UCL investigated the development of the social brain during adolescence. During her postdoctoral work at McGill, she founded the Critical Neuroscience program which focuses on perspectives of science studies and medical anthropology, examining how neuroscientists construct their objects of inquiry, and how findings are transformed into popular knowledge and public policy. Her current work investigates how the dissemination of cognitive neuroscience may shape the ways in which researchers, clinicians, patients, and laypeople understand themselves, their mental health, and their illness experiences.

Per Drougge is a Ph.D. candidate at Stockholm University's Department of Social Anthropology. His research has focused on monasticism as ideal and (occasionally) lived experience within contemporary, laicized, and "modernist" Buddhism. He has also written about the relation between soteriological and therapeutic goals in mindfulness practice and *naikan* introspection, two very different kinds of "Buddhist psychotherapy." Drougge's other research interests include medical anthropology, ritual studies, globalization and religion, and Western Buddhism.

Jenny Eklöf is an associate professor and senior lecturer in the history of science and ideas at Umeå University, Sweden. Her research lies at the intersection of science and technology studies and history of science. Dr. Eklöf has focused on topics such as the political history of gene technology and biofuel promotion, as well as the emergence of mindfulness as a scientific and medical field in the 1990s and onward. She has long-term experience of teaching on the science journalism bachelor program at Umeå University, and many of her publications address questions of science communication and the ongoing medicalization of science. Science of journalism? Or teaching in the science journalism bachelor program?

Natalie Flores is a doctoral student in the department of Curriculum and Teaching at Teachers College, Columbia University. She has been a bilingual educator of young children aged 0–8 for several years in both Los Angeles and New York City. Her experiences in diverse classroom settings have lead her to not only research, but also advocate for more equitable learning experiences for children living in disenfranchised communities. Natalie's interest and experience in critical approaches to mindfulness began over a decade ago, when she developed her own mindfulness practice and shared it with her students.

Mark T. Greenberg, Ph.D. holds The Bennett Endowed Chair in Prevention Research in Penn State's College of Health and Human Development, and he is the founding director of the Prevention Research Center for the Promotion of Human Development. He is a board member of the Collaborative for Academic, Social, and Emotional Learning (CASEL). One of his current interests is how to help nurture awareness and compassion in our society.

Funie Hsu is an assistant professor of American studies at San José State University. Her research interests include race, gender, education, US empire, language, mindfulness, and Engaged Buddhism. A past U.C. President's Postdoctoral Fellow, she received her Ph.D. in Education from U.C. Berkeley with a Designated Emphasis in Women, Gender and Sexuality, and an Ed.M. from the Harvard Graduate School of Education. She is a former elementary school teacher and serves on the board of the social justice nonprofit, Buddhist Peace Fellowship. Her work on mindfulness, neoliberalism, and education integrates her research with her community and spiritual work on Engaged Buddhism.

Terry Hyland, Ph.D., M.A. (distinction), BEd (Hons), Cert Ed is a fellow of the Higher Education Academy and has over forty years experience of teaching in schools, further, adult and higher education. He retired as a professor of education and training at the University of Bolton, UK, in 2009 and now spends his time on academic research and writing, research student supervision, teaching mindfulness and philosophy courses at the Free University of Ireland, and Ph.D. external examining at UK universities.

Richard King is a professor of Buddhist and Asian studies at the University of Kent, UK. He works on the history of Hindu and Buddhist philosophy and postcolonial critiques of the study of religion. He is the author of a number of books including: *Early Advaita Vedanta and Buddhism* (SUNY Press, 1995); *Orientalism and Religion: India, Postcolonial Theory and "The Mystic East"* (1999, Routledge), and *Indian Philosophy. An Introduction to Hindu and Buddhist Thought* (Edinburgh and Georgetown University Presses), and *Selling Spirituality* (co-authored with Jeremy Carrette, Routledge, 2005). His current research explores ethical issues in both classical apophatic literature and the impact of global capitalism on representations of mindfulness.

Brooke D. Lavelle holds a Ph.D. in religious studies and cognitive science from Emory University. Her academic work focuses on the confluence of Buddhist contemplative theory and cognitive science, as well as the cultural contexts that shape the transmission, reception, and secularization of Buddhist contemplative practices in America. She is the co-founder of the Courage of Care Coalition, a nonprofit dedicated to providing sustainable compassion training to educators, healthcare professionals, and others in social service. Brooke also serves as a Senior Education Consultant to the Mind and Life Institute and the Greater Good Science Center at the University of California, Berkeley.

David J. Lewis is a retired computer scientist, mathematician, faculty at Brown University, Cornell University, and Ithaca College, and software engineer and manager. He is a longtime Buddhist practitioner under the tutelage of Gelek Rimpoche and a student and independent researcher of Western and Eastern philosophical and psychological traditions at The Center for Trauma and Contemplative Practice.

Charlotte Longden is an undergraduate student in the School of Social Sciences at Cardiff University. In recent research, she has used discourse analysis to explore postfeminist subjectivities, focusing on issues of gender and power in the construction of femininity in Gothic subculture. With a budding interest in the broader discipline of critical social psychology, Charlotte is currently working on research concerning the interactional construction of subjectivity in mindfulness-based interventions.

David R. Loy is a professor of Buddhist and comparative philosopher and a teacher in the Sanbo Zen tradition of Japanese Buddhism. He has written or edited eleven books, as well as many academic papers and more recently articles in popular journals such as *Tricycle* and *Buddhadharma*. He lectures nationally and internationally on various topics, focusing primarily on the encounter between Buddhism and modernity: what each can learn from the other. He is especially concerned about social and ecological issues. See: www.davidloy.org.

Rhonda V. Magee [J.D./M.A. Sociology (University of Virginia)] is a professor of law at the University of San Francisco, School of Law who teaches mindfulness. She has been a full professor since 2004, teaching torts, race, law and policy, and courses in contemplative/mindful lawyering. She trained with Jon Kabat-Zinn and the University of Massachusetts's School of Medicine's Center for Mindfulness and now serves on the CFM Board of Advisors. A fellow of the Mind and Life Institute, she recently wrote *The Way of ColorInsight: Understanding Race and Law Effectively Using Mindfulness-Based ColorInsight Practices*, Georgetown J. of Mod. Crit. Race Perspectives (forthcoming, 2016).

Lisa Dale Miller, LMFT, LPCC, SEP is a private practice psychotherapist in Los Gatos, CA, specializing in mindfulness psychotherapy and somatic experiencing therapy for depression, anxiety, trauma, addiction, and chronic pain. Lisa is the author of a highly regarded textbook on Buddhist psychology for mental health professionals, *Effortless Mindfulness: Genuine Mental Health Through Awakened Presence*. Training clinicians in the practical application of Buddhist psychology is her greatest joy. Lisa has been a yogic and Buddhist meditation practitioner for four decades. For more information visit: www.lisadalemiller.com.

Joy L. Mitra, M.A. has served as researcher, facilitator, and consultant in the areas of mindfulness and human development, including studies on mindfulness with youth and families at the Edna Bennett Pierce Prevention Research Center of Penn State University. Previously, she served as a consultant and lecturer in the areas of organizational development and effectiveness to international educational and healthcare organizations and as a research scholar at the Indian Institute of Technology. Her current areas of interest include ethical implications of the Eightfold Path, the capacity for compassion and applications of mindfulness in aging populations.

Paul Moloney is a counseling psychologist based in an NHS Adult Learning Disability service in Shropshire, UK. He has a background in community and social work, alcohol and drugs counseling, and university teaching. His clinical interests include the social and material roots of distress and well-being, and how psychological practice can elucidate or conceal or these influences. He is the author of *The Therapy Industry* (2013) and a member of the Midlands Psychology Group: Founded by the late David Smail, this is a collection of academic and therapeutic psychologists dedicated to questioning the dominant assumptions of therapeutic psychology.

Joshua Moses engages in multidisciplinary research focuses on disaster and complex socio-ecological change. He has worked on religious response to the attacks of September 11 and Hurricane Katrina, studying the formation of disaster expertise in what he calls the current "New Age of Anxiety." Joshua has worked with Nunatsiavut Inuit communities in northern Labrador on inequality, dispossession, community well-being, migration, and identity in the context of recent land claim settlements and large-scale resource extraction. He has also conducted research in the Northwest Territories on migration, housing, and homelessness and is also interested in the response of higher education to climate change.

Edwin Ng, Ph.D. is a cultural theorist and teaches media and communication studies at Deakin University, Australia. His interest in the cultural translation of Buddhism and mindfulness is motivated by the lived tensions of straddling multiple cultural and intellectual heritages and of attempting to bring mindfulness to support scholarship, pedagogy, and activism within an increasing corporatized academic regime. His first book *Buddhism and Cultural Studies: A Profession of Faith* (Palgrave 2016) reconceptualizes and explores the (unacknowledged) role of faith as well as intellectual hospitality in dialogical scholarship. He is working on a second book on Buddhist critical theory and mindfulness (Bloomsbury).

Richard K. Payne (Ph.D., History and Phenomenology of Religion, Graduate Theological Union, 1985) is currently the Yehan Numata Professor of Japanese Buddhist Studies at the Institute of Buddhist Studies, Berkeley. His engagement with Buddhist practices began in the late 1960s and has spanned a variety of traditions, including Soto Zen, Vipassana, Nyingma, and Shingon (Japanese tantric Buddhism). During his doctoral research on the fire ritual in Shingon, he completed training on Mt. Koya, Japan, and was ordained as a Shingon priest in 1983. A special interest has been popular American religious culture and its influence on the development of Buddhism in the West.

Jack Petranker, M.A., J.D. is director of the Mangalam Research Center for Buddhist Languages and the Center for Creative Inquiry. He is the faculty of the Tibetan Nyingma Institute and Dharma College and a senior editor for Dharma Publishing. His scholarly interests included consciousness studies, organizational change, the encounter of Buddhism with modernity, and philosophy as a way of life. In addition to teaching on Buddhism, he offers programs that explore the power of inquiry grounded in first-person experience. His most recent book is *Inside Knowledge*. He lives and works in Berkeley, California.

Rick Repetti, Ph.D. (CUNY 2005) is a professor of philosophy at Kingsborough Community College. He is a member of the Association for Contemplative Mind in Higher Education and a member of the editorial boards of the journal *Science, Religion & Culture* and the *APA Newsletter on Two-Year Colleges*. Rick founded and facilitates a weekly meditation group on campus, open to faculty, staff, and students, and co-founded the CUNY Contemplatives Network. Rick has published two books and several articles on Buddhism, meditation, free will, contemplative pedagogy, and philosophy of religion. Rick is a multiple-decades practitioner and teacher of meditation and yoga.

Deborah Rozelle is a clinical psychologist who trains widely on psychological trauma and its relation to contemplative practice. She is co-director of the Jewel Heart Buddhist Chaplaincy Program and co-editor of *Mindfulness-Oriented Interventions for Trauma: Integrating Contemplative Practices*. She has been senior fellow for the Initiative for Transforming Trauma at Garrison Institute and faculty at The Trauma Center at JRI and the Institute for Meditation and Psychotherapy. Dr. Rozelle is a longtime Buddhist practitioner under the tutelage of Gelek Rimpoche. She has a clinical psychology practice in Lexington, Massachusetts.

Geoffrey Samuel is Emeritus Professor in the School of History, Archaeology and Religion at Cardiff University and director of the Body, Health and Religion (BAHAR) Research Group and an honorary associate of the Department of Indian Sub-Continental Studies at the University of Sydney,

Australia. He is a president of the International Association for the Study of Traditional Asian Medicine (IASTAM) and was until recently co-editor of IASTAM's journal *Asian Medicine: Tradition and Modernity*.

Steven Stanley, Ph.D. is a critical psychologist in the School of Social Sciences at Cardiff University, Wales, UK. He is particularly interested in the potentials of early Pāli Buddhist ideas and practices, as well as modern retreat practice, for potentially reorienting our relationship to life in capitalism. In his research, he has investigated historical changes in meanings of mindfulness and meditation, ethics and politics of the mindfulness movement, and mindfulness meditation as a psychosocial research methodology. His current research concerns the production of subjectivity within mindfulness courses.

Christopher Titmuss a senior Dharma teacher in the West offers, retreats, and facilitates pilgrimages and leads Dharma events worldwide. His teachings focus on insight meditation (vipassana), the expansive heart and enquiry into emptiness and liberation. Poet, photographer, and social critic, he is the author of numerous books including *Light on Enlightenment, The Mindfulness Manual*, and *Poems from the Edge on Time*. More than 100 of his talks are freely available on www.archive.org. A former Buddhist monk in Thailand and India, he is the founder of the online mindfulness training course. He teaches in Australia, India, Israel, France, and Germany every year. Christopher has been teaching annual retreats in India since 1975. He has lived in Totnes, Devon, England since 1982.

Massimo Tomassini teaches organizational learning as a contract professor at the Department of Education Science (University of Roma 3). He is in charge of the module "mindfulness in organizations" within the "mindfulness: practice, clinic and neuroscience" advanced course, at the Department of Psychology (University of Roma La Sapienza). He is research fellow—engaged in international projects—at the Institute for Employment Research (University of Warwick, UK). A private counseling practitioner, he holds a diploma of mindfulness counselor awarded by Mindfulness Project (Istituto Lama Tzong Khapa, Pisa).

Glenn Wallis holds a Ph.D. in Buddhist Studies from Harvard University. He is the author of several books and articles exploring various facets of ancient and medieval Indian Buddhism. His most recent book, *Cruel Theory/Sublime Practice: Toward a Revaluation of Buddhism* (with Tom Pepper and Matthias Steingass), offers a critical theory of contemporary Buddhism. Wallis is currently completing a book for Bloomsbury Press titled *A Critique of Western Buddhism*. Wallis is chair of the Critical Contemplative Studies program at the Won Institute of Graduate Studies near Philadelphia.

Zack Walsh is a teaching assistant and Ph.D. student of religion specializing in process studies at Claremont School of Theology. He is also a steering committee member of Toward Ecological Civilization and a research fellow at the Institute for Advanced Sustainability Studies and the Institute for the Postmodern Development of China. His research currently focuses on both deconstructive and constructive approaches to Engaged Buddhism, eco-theology, and contemplative studies with a practical orientation toward social–ecological transformation. In future work, Zack plans to develop theoretical and practical resources for advancing climate justice and transitioning us to a postcapitalist, ecological civilization.

Jeff Wilson is an associate professor of religious studies and East Asian studies at Renison University College. He is the author of *Mindful America: The Mutual Transformation of Buddhist Meditation and American Culture* (Oxford University Press, 2014), *Dixie Dharma: Inside of Buddhist Temple in the American South* (University of North Carolina Press, 2012), and *Mourning the Unborn Dead: A Buddhist Ritual Comes to America* (Oxford University Press, 2009). He founded the Buddhism in the West Group at the American Academy of Religion in 2007 and serves as an editorial board member for the *Journal of Global Buddhism*.

Part I
Between Tradition and Modernity

Ven. Bhikkhu Bodhi

A Parting of the Ways

I first learned to practice Buddhist meditation in 1967, during my first year at Claremont Graduate School, where I was enrolled in a doctoral program in philosophy. At the beginning of my second term, a Buddhist monk from Vietnam came to study at Claremont and was assigned to the same residence hall where I was living. I had become interested in Buddhism a year or two earlier, while I was still in college, and had even tried to meditate on my own, without success. But now that there was a monk living on the floor just below, I called on him to learn more about Buddhism and was soon practicing meditation under his guidance. He initially instructed me in meditation on the breath, and from there he led me on to the observation of thoughts and feelings. During this early stage of my practice, I did not know of a precise word to describe the process I was learning. I could see that an interesting psychological phenomenon was at play, a kind of "bending back" of awareness upon its own contents. But lacking the word, I thought of it simply as "meditation."

Several months after I began to meditate, I came across a book titled *The Heart of Buddhist Meditation*, published by Rider in London. The author's name was given as Nyanaponika Thera, but the book did not provide a biographical note about the author. Since the introduction was signed "Nyanaponika Thera, Forest Hermitage, Kandy, Ceylon," I assumed the author was a monk from Ceylon, as the country was then known before 1972, when it changed its name to Sri Lanka. Only years later did I learn the author was originally a Jew from Germany who had left his native land in the early years of the Nazi regime, intent on entering the Buddhist order in Ceylon. Through the strange workings of fate, some fifteen years later, I came to live with him at the same Forest Hermitage in Kandy, where I attended on him until his death in 1994.

It was this book that put a name on the method I had learned from my Vietnamese teacher. The word was "mindfulness," which Nyanaponika singled out as the key to the practice of Buddhist meditation. He described "the way of mindfulness"—not mindfulness itself but "the way of mindfulness," a broader concept—as "the heart of Buddhist meditation," and he explained in some detail the fourfold application of mindfulness to the contemplation of body, feelings, states of mind, and mental phenomena. Having learned the name for the endeavor that I had been engaged with, and having seen that the process of mental cultivation was minutely analyzed by the texts of Early Buddhism, I felt a stronger sense of confidence in the path I had entered.

Ven. Bhikkhu Bodhi (✉)
Chuang Yen Monastery, Carmel, NY, USA
e-mail: venbodhi@gmail.com

© Springer International Publishing Switzerland 2016
R.E. Purser et al. (eds.), *Handbook of Mindfulness*,
Mindfulness in Behavioral Health, DOI 10.1007/978-3-319-44019-4_1

My Vietnamese teacher at Claremont belonged to the Mahāyāna branch of Buddhism, which is often contrasted with the more conservative Theravāda branch that predominates in southern Asia. However, in the decades before he came to the USA, the Buddhist revival in southern Asia had opened cross-cultural contacts between Buddhist traditions, and as a result of this he had come to realize the importance of the Chinese Āgamas and Pāli Nikāyas, the seminal texts of Early Buddhism, for understanding the Buddha's original teachings. While he remained firmly committed to the spirit of the Mahāyāna, he took these ancient texts as the foundation for his own understanding and practice and urged me to learn them as well. He also gave me the Three Refuges—refuge in the Buddha, the Dhamma, and the Sangha—and stressed the need to bring faith, understanding, and meditation practice together into harmonious balance. In his view, these three strands of Buddhist spirituality were inextricably interwoven, such that none could be separated from the others without becoming itself enfeebled while weakening the whole to which it belonged.

When I arrived in Sri Lanka and entered the Theravāda monastic order in 1972, I found that my ordination teacher, Ven. Balangoda Ānanda Maitreya, had a very similar attitude. Though as a scholar, he emphasized doctrinal and linguistic study above strict meditation practice, he was himself a meditator who had practiced both concentration and insight meditation with some degree of facility. He also had deep personal devotion to the Buddha and had written a biography of the Buddha, in the Sinhala language, which was used as the classic textbook on the subject in the Sri Lankan monastic institutes. The same ideas and attitude were shared by the other teachers in Sri Lanka under whom I studied. Some put more emphasis on doctrinal understanding, others on meditation practice, but what they had in common was the conviction that knowledge and practice go together like the left foot and the right foot. And just as both feet rest on the ground, my teachers insisted that both learning and practice should be solidly planted on the ground of reverence for the Three Jewels,

upright moral conduct, and an aspiration to achieve the supreme goal set by the Dhamma.[1]

After I returned to the USA for a five-year stay (1977–82), I began to hear about other Westerners—both Americans and Europeans—who had trained in Asia around the same time I was living in Sri Lanka. Some had been bhikkhus but had since disrobed, while others had trained as lay meditators. Now, back in the West, they were conducting intensive meditation retreats of ten days, a month, and even three months for people who had virtually no prior acquaintance with the Buddha's teachings. Initially what I heard perplexed me, since this approach differed quite markedly from the guidance I had received from my own teachers, who held that intensive meditation was appropriate for those who have already gone for refuge, established a firm foundation in virtuous conduct, and possessed a clear understanding of the Buddha's teachings. But, I pondered, perhaps I was taking too conservative a stance. After all, I thought, Buddhism itself has evolved differently in different cultures and eras, and skilled teachers must make use of the *upāya*, expedient means, appropriate for the time in which they live, applying them as they see fit. Perhaps, I thought, in our own era—this *kali yuga* or degenerate age—when we were living in the shadows of the Vietnam War, the Watergate scandal, the Iran hostage crisis, and the Reagan presidency, a pressure cooker approach to meditation was the most effective way to rescue those whose minds were being buffeted by a consumerist culture driven by nothing higher than the pursuit of money and power.

Occasionally, to escape my duties as a resident monk at the Washington Buddhist Vihara, the Sri Lankan temple in D.C. where I lived from 1979 to 1982, I would attend retreats at the Insight Meditation Society in Barre, Massachusetts. This gave me the chance to see firsthand the adaptations that Buddhism was

---

[1]I will generally use the Pāli forms of technical Buddhist terms except when I am citing or referring to those who use the Sanskrit forms. Thus, I generally use "Dhamma," but "Dharma" when referring to those who have adopted this form of the word.

making as it sent down roots in American soil. While I found the actual meditation instructions to be quite similar to those I had received from my teachers in Sri Lanka, the evening "Dharma talks"—and the other garnishings with which the teachings were embellished—sometimes left me disoriented. Among a myriad of impressions of those times, three stand out in my memory.

One was that the evening talks seldom related the practice of mindfulness meditation we were engaged in during the day to the actual teachings of the Buddha. Exception made of some excellent talks in a more traditional style by Joseph Goldstein, the teachers said virtually nothing about the backdrop to the practice of mindfulness meditation as we find it described in the Pāli Canon, which I had studied in Sri Lanka. There was no talk about our bondage to the beginningless cycle of rebirths; nothing about the role of kamma, understood as the impact of our volitional actions from one life to the next; nothing about the goal of the practice as release from the round of rebirths. All these topics, central to the Dhamma, were simply passed over in silence, or at most treated as metaphors. The Buddha's discourses were seldom taken up as themes for the evening talks, and if on occasion the Buddha was quoted, it was only by selecting snippets from the suttas, individual lines that would be cited out of context and freely interpreted by the speakers, somewhat in the manner a jazz musician might improvise on a tune by Cole Porter.

This leads into my second recollection, that the talks were extremely eclectic. Not only would the Buddha be quoted infrequently and with little context, but on any evening we might be treated to an assortment of readings from Ramana Maharshi, Krishnamurti, Ram Dass, Lao Tzu, Japanese Zen masters, and Sufi sages. It seemed to me that the teachers did not fully realize the implications of the passages they were citing or the way they differed from the Buddha's teachings. Numerous times I heard things that even jarred my sensibility, such as: "The Buddha didn't teach Buddhism; he taught the Dharma." The implication of this, it seemed, was that all the other sages and saints being quoted were teaching the same thing as the Buddha, and

despite the vast diversity in their expressions, what they were all teaching could be reduced to present-moment awareness.

The statement that the Buddha did not teach Buddhism, by the way, is only half true, which means that it is also half false. The Buddha certainly did not teach the historical–cultural–institutional religion that we now know as "Buddhism." However, in numerous passages, he refers to his teaching as "the Dhamma and discipline proclaimed by the Tathāgata," thereby indicating that what he teaches is a unique doctrine without a counterpart elsewhere. It is not merely that he expresses the one truth differently, but that he teaches things that are, in principle and not merely in words, incompatible with many of the pivotal ideas of other spiritual systems.

The third thing that I recall from those talks—reconfirmed for me over the years, as I read the books and magazines emerging from the mindfulness movement—is that the practice itself was undergoing a major overhaul with regard to its objectives and goal.[2] While it may have preserved the same formal elements as had been transmitted in Asia through the centuries—the specific practical instructions on how to set up attention on an object, how to deal with distractions, how to intensify one's practice, and so forth—I found that the framing of the practice was undergoing some subtle shifts. The new context had led to changes in primary function. In its classical role, as an integral component of the Buddhist path, the purpose of mindfulness meditation is to eradicate the mind's deep defilements and uproot the belief in a substantial self. This objective is in turn determined by the ultimate goal of the Buddha's teaching, the attainment of nibbāna, liberation from the cycle of rebirths.

---

[2]I use the expression "mindfulness movement" guardedly and only as a matter of convenience. I do not intend to suggest by this term that there was any concerted effort to propagate mindfulness meditation around the country. Rather, at roughly the same time, different people with different backgrounds were teaching Buddhist meditation as they had learned it in Asia, and the teaching styles were too diverse to constitute anything resembling a movement.

During the years I lived in Washington, I seldom came across references to this goal in the talks I heard at the lay-oriented centers or in the books and magazines emerging from the centers, their teachers, and affiliated groups.[3] Rather, it seemed that the purpose in maintaining present-moment awareness, in so far as it was directed toward any goal beyond itself, was to enhance appreciation of the present moment. The actual purpose for which the Buddha taught the way of mindfulness, I learned, was to help us to live in the present, to savor each moment in its immediacy, to ride the ever-changing flow of events with uncluttered minds, letting whatever arises take its course without clinging to anything.

The new wave of meditation teachers recognized, of course, that the practice they were teaching had auxiliary benefits, and these were highlighted in the talks I heard and the conversations that came afterward. The practice could aid self-understanding and self-acceptance. It could disentangle the oppressive coils of memories, worries, fears, desires, plans, and pursuits. It could counteract greed and hatred and nurture such qualities as generosity, patience, kindness, and compassion. The practice promoted inner peace, and if only enough people would learn mindfulness, it even had the potential to bring world peace.

It is possible, of course, that these particular points of emphasis were not entirely innovations of the Western pioneers of insight meditation. The pioneers may have picked up just such an approach from their own Asian teachers, who in the 1960s and early 1970s were already

---

[3]In time, two major lines of transmission would emerge, which were quite distinct. One, which stemmed from teachers trained in the Mahasi Sayadaw style of practice, was based at the Insight Meditation Society in Barre. It later spread to California with the establishment of Spirit Rock in the Bay Area. These became the East Coast and West Coast focal points of "Mahasi style" insight meditation. The other, which stemmed from Goenka, had its own centers in the USA. While Western teachers in the Mahasi system tended to be syncretic, the Goenka lineage did not caucus with followers of other forms of Buddhist meditation, much less with other spiritual traditions, but strictly adhered to Goenka's teaching, as it still does today.

emphasizing the immediate experiential benefits of the practice of mindfulness. And they may have had their own reasons for doing so. They may have taken such a tack as a defensive maneuver, to demonstrate that the Buddha Dhamma, unlike the Christianity being foisted on their countries by the Christian missionaries, was tough, experiential, and realistic. Or they may have thought such a style of exposition was better suited to the minds of skeptical inquirers from the West, who were not yet ready to take on board the whole package of Buddhist doctrine. Or they may have even assumed that their Western students, in taking up Buddhist meditation, had already adopted the classical Buddhist worldview.

As I heard the Dhamma being expounded as a teaching fully applicable to our present life here and now, I found much that I agreed with and thought worthy of respect. What I felt to be missing, however, was the larger framework of the Buddha's teachings as I had encountered them in the suttas. In the classical teaching, the cultivation of wholesome qualities is harnessed to the task of realizing the ultimate goal, "the taintless liberation of mind, liberation by wisdom" that is won with the utter eradication of all defilements. It is harnessed to breaking the bonds that tie us to the cycle of birth and death. It is harnessed to a transcendent goal that is birthless and deathless. But in the modernist adaptation, it seemed that the practice was no longer integrally tied to the system of Buddhist faith and doctrine that had contained it for some 2500 years. Lifted from its source, the Dhamma had been reduced to the practice of a particular style of meditation, and the meditation itself had been reduced to the technique of present-moment mindfulness just to win purely "immanent" goals such as peace of mind and a more stable grounding in immediate experience.

The Division Widens

After five years back in America, I returned to Sri Lanka in 1982, and in the years that followed I continued to read about the development of Buddhism in the USA and elsewhere in the West. The trajectory that I could detect from my base in Sri Lanka was one I might well have predicted from my experiences in the early 1980s. I saw mindfulness meditation becoming increasingly psychologized. This was hardly surprising, given that many of the practitioners at the insight meditation retreats were psychologists who took up the practice, not to attain release from saṃsāra, but to gain a deeper, first-person perspective on the human mind, a perspective that would make them more effective in their professional careers. Gradually, some would begin to incorporate the techniques of mindfulness meditation into their therapies, until the boundaries between psychotherapy and contemplative practice would become permeable and new hybrids would appear.

From the realization that mindfulness could be used to help people suffering from chronic illness and unbearable pain, a new system of palliative care took shape, designed by the redoubtable Jon Kabat-Zinn. This system, called "Mindfulness Based Stress Reduction," soon spread to hospitals and treatment centers throughout the USA and all over the world. Before long, MBSR expanded beyond the walls of the medical establishment and metamorphosed into an autonomous practice advocated for people in normal bodily and mental health. It was even championed as a universal Dharma, as *the essential message* of the Buddha and all great spiritual masters, now freed from the baggage of religion including the Buddhist religion itself.

But it did not take long for the next wave of practitioners to realize that this ancient method of mind training had still more potentials waiting to be tapped. Their efforts, spread out across a wide spectrum of disciplines, utterly changed the face of mindfulness. Where the Western pioneers of insight meditation had openly acknowledged the Buddhist roots of mindfulness training, occasionally referred to the Buddha, and even displayed Buddha statues at the meditation centers, and where the next generation had called it a universal Dharma, the new wave of innovators boldly stripped away the remaining tendrils that connected mindfulness to Buddhism and everything else that might have been redolent with the

smell of cumin and turmeric. They saw in mindfulness a free-floating variable that could be attached to virtually any human endeavor, somewhat as salt can be added to any dish to enhance its flavor. Mindfulness was even hailed as the flowering of American democracy, the natural culmination of the Declaration of Independence, offering every citizen life, liberty, and the realization (not merely the pursuit) of happiness.

In the early stage of this process, mindfulness was used for purposes that would generally be considered commendable. There was mindfulness for school children to help them concentrate better, mindfulness for pregnant women to keep them calm through their delivery, mindfulness for moms to help them better raise their kids, mindfulness for couples to help resolve the strains in their relationships, mindfulness for addicts to help them break free from addiction, mindful exercise to improve health, and mindful eating to curb harmful food habits. But the process of divestment did not stop there. The trendsetters of culture and commerce soon saw in mindfulness an effective marketing tool that could be used to turn a quick profit. Thus, like rain on a summer afternoon, new applications of mindfulness soon began pouring down on us in an incessant patter. We began to hear about such things as mindful business strategies, mindful shopping, mindful dating, mindful sex, mindful investing, mindful sports, mindful politics, and mindful military training.

One would never have imagined that mindfulness would travel so far from the ancient monasteries where it was first proclaimed as "the direct way for the purification of beings and the realization of nibbāna." While for some 2500 years, it had remained a staid and steady pillar of Buddhist mind training, in record time—in a mere two decades—it had taken on more forms than an Amazonian shapeshifter. And outside the meditation centers, hardly a trace remained of any connection between mindfulness and Buddhism. It seemed that mindfulness had first been born *ex nihilo* in the mind of some twenty-first century genius.

Why Did Mindfulness Take This Route?

At this point, I want to raise the question why the practice of Buddhist meditation, and in particular the practice of mindfulness, followed the particular trajectory it did in the West. Multiple factors, woven together into a complex tapestry, contributed to this development, including the American spirit of pragmatism, the declining influence of theistic religion, the triumph of the therapeutic, the human potential movement, the quest for authenticity, the reaction against technological impersonality, and crass American commercialism. However, I want to go back to an early stage in the process of transmission and single out one shift that took place as meditation practice moved from East to West. This was a transfer in the "custodianship" of the Dhamma— that is, in teaching authority—from the monastic Sangha to Western lay teachers, from ordained monks to young men and women who had received their training in Asian monasteries and meditation centers without taking monastic vows, or who may have been ordained but returned to lay life after setting out to teach in the West.

I believe that this shift in teaching authority played a monumental role in the revamping of mindfulness and thus in extending it into new domains never found in the Buddhist traditions of Asia. In the Buddhist countries of the Theravāda tradition, religious life revolves around the monks and the monasteries, and it is the monks who are regarded as the custodians of the Dhamma. The grounding of spiritual authority in the monastic establishment is in some respects stultifying, binding the Dhamma to a conservative institution stubbornly bound to upholding conventional observances against the pressures of modernity. However, despite its faults, this tradition has ensured that all modes of Buddhist practice—whether scholarly, ritualistic, or contemplative—are imbued with veneration for the Three Jewels and rooted in a worldview based on the Buddha's discourses.

While throughout Buddhist history, laypeople have often engaged in meditation, until recently most lay meditation had focused on the devotional practices such as recollection of the Buddha, Dhamma, and Sangha, and the "immeasurable" meditations on loving-kindness and compassion. Often practice was undertaken in short sessions on the lunar observance days, performed in a semi-ritualistic manner according to simplified versions of the instructions provided by such manuals as the *Visuddhimagga*.[4] However, starting early in the twentieth century, several Burmese meditation masters—most prominent among them Ledi Sayadaw and Mahasi Sayadaw—opened up the gates of meditation practice to laypeople, and it was through these gates that itinerant young Westerners, curious about "the wisdom of the East," stepped when they arrived in Asia in the late 1960s and early 1970s. It was only natural that in their encounter with the Dhamma they would bring along the questions and problems that reflected their cultural backgrounds and personal needs. Inevitably, they took away from Buddhism answers that corresponded to these needs, and when they began to teach, their own understanding of the Dhamma shaped the way they would communicate the teachings to others. This became the legacy they would transmit to their own students and down the line to future generations.

While much water has flowed beneath the bridge since the practice of mindfulness was first introduced to the West, the basic shape the teaching received at the hands of this early generation of Western teachers is still discernible. I already discussed several distinctive features earlier when writing about the impressions I gathered from the Dhamma talks I heard at the insight meditation centers I attended in the early 1980s. Now I want to discuss these in greater detail.[5] To get my points across, I will have to oversimplify. I do so in the recognition that such oversimplification risks obscuring significant differences among Western lay teachers and

---

[4] Translated by Bhikkhu Ñāṇamoli as *The Path of Purification* (Kandy, Sri Lanka: Buddhist Publication Society, 1991).

[5] Though what I say relates particularly to the early generation of Western pioneers, who went to Asia in the late 1960s and early 1970s, I will frame my discourse in the present tense, as relating to present-day practitioners.

nuances in their teaching styles. The teachers fall at different points along a wide spectrum, ranging from those who are quite traditional and well versed in the canonical texts to those who are more adaptive, eclectic, and experimental. Nevertheless, despite the potential pitfalls, such generalizations can still bring dominant tendencies to light.

As I see it, what motivates most Westerners to seek out the Dhamma is an acute sense of what we might call "existential unease." By this expression, I am not referring to clinical depression, a morbid disposition, or any other type of psychopathology, but to a gnawing sense of lack, a feeling of incompleteness that cannot be appeased by easy answers or the pursuit of worldly distractions. This sense of existential unease can coexist with a personality that is, by all other criteria, quite sound and healthy. Those troubled by existential unease come to the Dhamma to resolve the anguish, to plug this hole that has opened up at the bottom of their being. Most are not seeking a new religion, a new system of worship and beliefs, or a new conceptual model for understanding the world. What they are seeking above all is a *practice*, a set of clear and pragmatic instructions that they can take up to transform and enrich the felt quality of their lives. Since they approach the Dhamma seeking a way to induce concrete changes in their life experience, this is exactly what they get from it. And if they teach others, this is what they will teach. They will present the Dhamma as a practice, a way, a path, that can ameliorate the disturbing sense of existential lack and infuse our lives with joy, zest, and meaning. They will present it as a radical, pragmatic, existential therapy that does not require any belief commitments, as a "Buddhism without beliefs" that does not ask for any more faith than a readiness to apply the method and see what one can get from it.

Now what I have called existential unease, the sense of lack, the feeling of alienation, is not unique to contemporary Western civilization. The sense of lack or insufficiency seems to be a universal feature of human experience, which different peoples will seek to redress in ways that

are close at hand. Thus, the Christian will turn to God, the Hindu to Shiva or Krishna or meditation on the supreme self, the Jew to the Torah, and so forth. This sense of lack also underlies the quest for liberation in Buddhism. Despite certain similarities, however, there is a difference in how this sense of lack operates in classical Buddhism and in the modern mindfulness movement.[6] This difference, I believe, takes us to the crux of the matter. In classical Buddhism, this sense of lack or voidness is seen as *emblematic*, that is, as *pointing beyond itself* to the intrinsic unsatisfactoriness of existence itself, to the pervasive and ever-present fact of *dukkha*. The solution, therefore, lies beyond the innately flawed, deficient, and perilous world in which we are immersed, in a state, dimension, or condition that is secure, peaceful, and free from all deficiencies—that is, in nibbāna, the deathless. To win this state requires that one turn away from the world and step out in the direction of renunciation and transcendence. We see this pattern articulated in the legend of the Buddha's renunciation, where his encounter with an old man, an ill man, and a corpse shattered his complacency and the figure of the ascetic showed him what he must do to resolve his inner crisis.

For the felt sense of existential suffering to trigger a clear acknowledgment of what I call "the intrinsic and ever-present unsatisfactoriness of existence," two additional factors are needed. One is faith (*saddhā*) and the other right view (*sammā diṭṭhi*). These two factors not only turn the felt sense of existential unease into a recognition of the inherently flawed nature of conditioned existence, but in the classical Buddhist model they motivate and sustain the practice of meditation from start to finish. Thus, for the

---

[6]While the expression "classical Buddhism" is problematic, I prefer it to "traditional Buddhism" and "religious Buddhism," which both suggest the Buddhism of rituals, ceremonies, and devotional observances. By "classical Buddhism," I have in mind the doctrines and practices of Theravāda Buddhism as derived from the Pāli Canon. Other schools of contemporary Buddhism have their own classical forms, which could be compared with modern adaptations. Here, however, I am concerned with the school from which the prevailing systems of mindfulness meditation directly stem.

practitioner of classical Buddhism, the ultimate purpose for which mindfulness meditation is taken up is not to quell the feeling of existential angst and gain peace, joy, and equanimity in this present life—though these will naturally come as by-products of the practice—but to win the state of unshakable liberation that lies beyond the pale of repeated birth and death.

In classical Buddhism, faith or *saddhā* is specifically tied to the Three Jewels. *Saddhā* means faith in *the Buddha* as the fully enlightened teacher, the one who has arrived at complete enlightenment; it means faith in *the Dhamma* as expressing the Buddha's realization, the full Dhamma and not merely selected quotations; and it means faith in *the Sangha*, that is, faith in *the ariyan Sangha*, the invisible spiritual community made up of those who have realized the Dhamma, and reverence for *the monastic Sangha* as the visible, embodied, communal representation of the ariyan Sangha. For classical Buddhism, faith in the Three Jewels is specific to its objects. It is not an open variable that can attach itself to anything worthy of respect. As the traditional Pāli chant puts it: "For me there is no other refuge; the Buddha, Dhamma, and Sangha are my only refuge."

The other factor in classical Buddhism that guides and motivates the practice of mindfulness meditation is "right view" (*sammā ditthi*). Right view has multiple facets, but following the Pāli texts, we can speak of two kinds of right view. The *foundational type* is acceptance of the principle of kamma, the lawful relationship that holds between volitional deeds and their consequences, such that unwholesome deeds bring suffering and wholesome deeds bring happiness. For followers of classical Buddhism, the operation of this law, repeatedly emphasized by the Buddha, is taken as axiomatic, beyond doubt and dispute; it is understood literally and not treated as a metaphor or symbol. Moreover, since kamma can only be truly effective if it operates through a sequence of many lives, the corollary to the right view of kamma and its fruit is acceptance of rebirth, the

recognition that any single life is but a link in a series of lives that has been going on without discernible beginning and, unless sufficient effort is made, will continue without end. This too is understood literally and not treated as a metaphor.

The second level of right view is *the wisdom that understands the four noble truths*. This higher right view begins as a conceptual understanding of the four noble truths, which are grasped through study and reflection, and as the practice unfolds, it matures into direct insight into the truths and finally into penetration of them as an inseparable whole, with each truth interwoven with and reflective of the others. While in modernist adaptations of Buddhism, the four noble truths are often taught as a diagnosis of the psychology of suffering—of sorrow, discontent, worry, and fear—in classical Buddhism the four truths build upon the right view of kamma and rebirth and offer not merely a psychological diagnosis of suffering but a comprehensive *existential diagnosis* of our saṃsāric predicament. *Dukkha*, the first noble truth, is epitomized by the factors of mental and bodily experience that are "acquired" again at each new birth and then discarded at each death, the "five aggregates subject to clinging." The second noble truth, the cause of *dukkha*, is craving (*taṇhā*), described as *ponobhavikā*, "productive of renewed existence," that is, as capable of generating a new birth consisting of the five aggregates. The elimination of craving culminates not only in the extinction of sorrow, anguish, and distress, but in the unconditioned freedom of nibbāna, which is won with the ending of repeated birth.

It is these two factors—faith in the Three Jewels and right view—that I see as marking the dividing line between, on the one hand, classical Buddhism and, on the other, the various forms of Buddhist modernism, secular Buddhism, and the teaching of Buddhist meditation practices separated from their Buddhist roots. These different relationships to faith and right view are not inconsequential. They determine the vision that sustains and inspires the practice of mindfulness meditation, the expectations about the benefits to be derived, and the way the practice unfolds in actual personal experience.

What seems to be happening today, in many circles of Western Buddhism, is that the Dhamma is being taught primarily on the basis of the equation: "Dhamma equals mindfulness meditation equals bare attention." Mindfulness meditation has thus been lifted out of its original context, the context of faith in the Three Jewels and the full noble eightfold path headed by right view, including the "mundane right view" of karmic causality, and taught in a way that fits seamlessly with the secular outlook of contemporary Western society. It is thus taught not for the purpose of winning liberation from the ever-repeated cycle of birth and death, which is perilous and fraught with misery, but for the purpose of allaying existential distress simply by being attentive to what is occurring in the present moment. This is how the riddle of existence is being solved; this is how the alienation from direct experience is being overcome, namely by using mindfulness meditation as a bridge to take us back to the living experience of the present moment. It does not aim at transcendent liberation, but at healing inner divisions and at enhancing the appreciation of life through sustained attention to immediate experience.

As an adherent of "classical Buddhism," I have pondered whether this mode of practice is intrinsically capable of leading to the full enlightenment and ultimate liberation that the Buddha's way of mindfulness is intended to bring. And the answer that I have come to, based on my reflections and reading of the texts, is that on its own it cannot. The Buddha made right mindfulness a factor of the noble eightfold path, and thus to unfold its full potential and culminate in the ultimate goal, it would seem that it must be guided by right view and accompanied by the other path factors such as right intention, right speech, right action, right livelihood, and right effort. It must lead on to the following steps, to right concentration and right cognition, culminating in liberation.

A Case Study: The Contemplation of Impermanence

I want to exemplify my point by considering how classical Buddha Dhamma and the modern mindfulness movement diverge in their perspectives on the contemplation of impermanence. Both share the understanding that the fact of impermanence entails the injunction: "Don't cling, for if you cling to what is impermanent, you will eventually suffer." However, the two approaches to the Dhamma draw different conclusions from this maxim—indeed, almost contrary conclusions. For classical Buddhism, insight into impermanence is the passageway to a radical understanding of the second characteristic, the *dukkha-lakkhaṇa*, the mark of suffering: "Whatever is impermanent is *dukkha*." This does not mean, of course, that whatever is impermanent is a mass of misery, but rather that whatever is impermanent is inherently flawed, inadequate, and defective, unable to provide lasting happiness and security. The first two characteristics—impermanence and *dukkha*—jointly entail the third, the selfless nature of phenomena, the absence of genuine selfhood in all the bases of self-identification, summed up in the five aggregates: bodily form, feeling, perception, volitional activities, and consciousness.

The suttas then marshal impermanence and *dukkha* together to expose the third characteristic, the non-self nature of all the constituents of individual being. Again and again they hammer home the message: "Whatever is impermanent is *dukkha*. Whatever is impermanent, *dukkha*, and subject to change, should be seen as it really is with correct wisdom thus: 'This is not mine, this I am not, this is not my self.'" This module of contemplation is applied to all five aggregates, thereby breaking the identification with them.

Contemporary teachers of insight meditation also dwell on the teaching of non-self, often hailing it as the core of the Dhamma. This focus has opened up avenues of dialogue between proponents of insight meditation and neuroscientists, cognitive psychologists, and psychotherapists. Debates have even been waged between those who see Buddhist meditation and psychotherapy as pointing in the same direction —toward healthy ego function marked by the reduction in narcissistic self-obsession—and those who see them as pointing in opposite directions. Some take the contrast between them as antithetical, so that one must be jettisoned in favor of the other; others take them to be different but complementary.

Nevertheless, a significant difference can still be discerned between the perspectives on impermanence advocated by teachers of modern mindfulness meditation and by classical Buddhism. Proponents of modern mindfulness meditation often see impermanence as imbued with positive significance. They admit that clinging to what is impermanent brings suffering, but take this connection to mean, not that one should renounce the impermanent in favor of the imperishable nibbāna, but that one should learn to live in the world with an open mind and loving heart, capable of experiencing everything with awe and wonder. The practice of mindfulness thus leads through the door of impermanence and selflessness to a new affirmation and appreciation of the world, so that one can joyfully savor each fleeting event, each relationship, each undertaking in its wistful evanescence, unperturbed when it passes.

This attitude, though it has some resonances with Zen Buddhism particularly as expressed by Thich Nhat Hanh, is quite at odds with the Buddhism of the Pāli Canon, the tradition from which mindfulness meditation originates. In classical Buddhism, the fact of impermanence is viewed as a sign of deficiency, a warning signal that the things we turn to for happiness are unworthy of our ultimate concern. As the Buddha says: "Conditioned things, monks, are impermanent, unstable, unreliable. It is enough to be disenchanted with all conditioned things, enough to be dispassionate toward them, enough to be liberated from them" (SN 15:20, II 193).

The process of contemplation that leads from impermanence to non-self does not come to a stop with insight into the non-self nature of things but serves a purpose beyond itself. It is designed to put an end to identification and appropriation, to eliminate the ingrained tendencies to take things to be "I" and "mine." This insight leads to disenchantment, dispassion, and liberation: "Seeing the five aggregates thus [as not mine, not I, not my self], one becomes disenchanted with them. Being disenchanted, one becomes dispassionate. Through dispassion, the mind is liberated. When the mind is liberated, one directly knows: 'It is liberated.'"[7] And liberation (*vimutti*) here means the release of the mind from the taints (*āsavas*) and fetters (*samyojanas*), the primordial forces that drive the cycle of rebirths. When they are eliminated, the cycle itself comes to an end and one knows the task is done: "One understands: 'Birth is finished; the spiritual life has been lived; what had to be one has been done; there is no further coming back to this state of being.'"[8]

The Trajectory in Retrospect

The trajectory that mindfulness has followed over the past forty or fifty years demonstrates that context determines function. Looked at in the abstract, mindfulness appears to be completely transparent as to its function. It is simply the bending back of the beam of awareness upon oneself, to clearly illuminate one's bodily and mental experience. As such it can be used for diverse purposes, spiritual and worldly, lofty and mundane, humble and profound. In classical Buddhism, the purposes to which it is applied are determined by the parameters of the Buddha's teaching. It is cultivated to enhance the ability to sustain attention on an object, which leads to *samādhi* or concentration and *paññā* or wisdom. However, though utilized in the Buddhist path, the bare act of mindfulness is context-neutral. As modern exponents of mindfulness meditation are

[7]The formula appears numberless times. See for example SN 22:49, V 49–50; SN 22:59, V 67–68; and SN 22:82, V 104. SN = Saṃyutta Nikāya, translated by me under the title *Connected Discourses of the Buddha* (Boston: Wisdom Publications, 2000). The references give chapter and sutta numbers, followed by the volume and page number of the Pali Text Society's Roman-script edition.

[8]This formula, too, appears numberless times. See for example SN 22:12–20, V 21–24; SN 22:63–72, V 74–81, and so forth.

fond of saying, mindfulness does not carry around a banner stating that it is inherently Buddhist.

In the late twentieth century, Theravāda Buddhist mindfulness meditation was fissured along several lines, leading to new and unexpected bends and twists in the destiny of this ancient practice. In the initial phase, the Western seekers who returned home after training under Asian Buddhist masters taught the way of mindfulness as a non-religious discipline that could be as relevant and beneficial to non-Buddhist practitioners as to those who placed faith in the Buddha and his Dhamma. To justify this approach, they appealed to an adage that became very popular: "The Buddha didn't teach Buddhism; he taught the Dhamma." Though at first glance such a claim appeared innocuous, in time it amounted to a virtual "declaration of independence" severing insight meditation from its anchorage in Buddhist religious faith. Thereby it propelled the practice along a new trajectory.

To understand, from a traditional perspective, what has happened to mindfulness in the course of its transition, I find it helpful to set it in relation to a scheme I employed in my anthology of the Buddha's discourses, *In the Buddha's Words* (Boston: Wisdom Publications, 2005). In preparing this anthology, I organized my selection of suttas by way of the three benefits to which the practice of the Dhamma is said to lead: (1) welfare and happiness visible in this present life; (2) welfare and happiness pertaining to future lives; and (3) the supreme good, which is nibbāna. The means to "the welfare and happiness visible in this present life" is generosity, ethical conduct, and other acts that lead to interpersonal and communal harmony. The "welfare and happiness pertaining to future lives" is the attainment of a fortunate rebirth. The practices that lead to this kind of well-being are essentially the same as those that lead to welfare and happiness in this present life, but they are viewed from a higher standpoint rooted in the acceptance of kamma as the determinant of human destiny and rebirth into various planes as the natural result of kamma. The third type of benefit, the supreme good, is nibbāna, liberation from the entire cycle of rebirths. This cannot be won simply by virtuous conduct and meritorious deeds but

requires the development of the noble eightfold path, with particular emphasis on the cultivation of concentration and wisdom.

It is in relation to this third type of benefit that the way of mindfulness plays a central role. The four establishments of mindfulness are said to be "the direct path for the purification of beings, for the overcoming of sorrow and lamentation, for the extinction of pain and dejection, for the achievement of the true way, and for the realization of nibbāna" (SN 47:1, V 141). They are "noble and emancipating, and lead the one who practices them outward to the complete destruction of suffering" (SN 47:17, V 166). When developed and cultivated, they "lead to utter disenchantment, to dispassion, to cessation, to peace, to superior knowledge, to enlightenment, to nibbāna" (SN 47:32, V 179). In other words, in its original context, the cultivation of mindfulness is an integral part of a contemplative path to world-transcending liberation. The practice builds upon the second level of teaching, on kamma, rebirth, and the round of birth and death. It presupposes a critical insight into the intrinsic flaws of the human condition and a transcendent vision of the ultimately worthy goal of human endeavor. To lift the practice out from this context and transfer it to another context governed by a secular worldview and mundane ends is to alter its function in crucial ways. It transforms the function of mindfulness from the spiritually liberative to the therapeutic, from the sacred to the ordinary, from the life-transcending to the life-affirming.

It seems to me that in the West this is just what has happened with the practice of mindfulness, and the process already started with the pioneering teachers of insight meditation as they sought to disseminate the practice in the new cultural setting. Since these teachers did not emphasize the Buddhist worldview of rebirth or inculcate faith in the Three Jewels, they marginalized the second level of teaching that aimed at the good in future lives, which disappeared from view. To underscore the therapeutic capacity of the practice, they merged the first and third levels of teachings, so that practices prescribed for attaining the supreme good, liberation from the round of birth and death, were presented as a means for attaining well-being and happiness here and now. Mindfulness, concentration, and wisdom became not the means for breaking the fetters that bind us to saṃsāra, but qualities that "free the heart" so that we can live meaningfully, happily, peacefully in the present, acting on the basis of our perception of the interconnectedness of all life. The aim of the practice was still said to be freedom, but it was an immanent freedom, really more a kind of inner healing than liberation (*vimutti*) in the classical sense of the word. This reconceptualization of the training may have made the practice of mindfulness much more palatable than would have been the case if it were taught in its original context. But the omission may have set in motion a process that, for all its advantages, is actually eviscerating mindfulness from within.

In making these observations, I do not wish to demean in any way the efforts of the early pioneers who brought insight meditation to the West. They discovered the practice of mindfulness at a time when it was sorely needed and skillfully molded it to the situation at hand. The West was floundering in a morass of spiritual and moral emptiness. Materialism and commercialism were rampant, and sensitive people suffered from unbearable stress, confusion, and inner conflict. Under such conditions, the mode of mindfulness training the new teachers designed may have had inestimable value. It fostered such precious qualities as contentment, joy, gentleness, kindness, patience, equanimity, and compassion. It helped people learn to live at peace with themselves, to cherish the natural world, and to live more amicably with others.

What is worrisome, however, is the subsequent trajectory that mindfulness took once they had inserted a wedge between meditation practice on the one side and its supportive envelope of Buddhist faith, ethics, and understanding on the other. The first act of separation was followed by still sharper divisions between classical Buddhism and the mindfulness movement. In Act Two, mindfulness meditation came to be taught, deliberately and emphatically, as a non-religious discipline. Initially, this took place with the

emergence of "Mindfulness Based Stress Reduction," which was offered in a gesture of compassion to help people crushed by stress and chronic pain regain their hope and inner dignity. Before long, MBSR was given a still broader mandate, reformatted to teach ordinary people, weighed down by the dull routines of their daily lives, how to find sources of meaning and joy through sustained attention to the present.

In Act Three, mindfulness was aligned with other "caring professions," which led to multiple mergers and marriages that gave birth to children of their own. As it flowed downstream still further it became a secular form of inner hygiene, similar to yoga but with a more distinctive psychological flavor. At the mouth of the river, where we stand now, mindfulness has become a handy buzzword that can be attached at random to virtually any product or skill in order to invest it with a spiritual aura or increase its market appeal: mindful romance, mindful birthing, mindful athletics, mindful exercise, mindful business strategies, mindful warfare. Perhaps, just over the horizon, we will find some entrepreneurs pushing mindful mindfulness.

The severing of mindfulness practice from the Buddha Dhamma may well bring unforeseen benefits. It has already proved effective in helping people deal with chronic pain and illness and has opened doors to personal growth for people in all walks of life, folks who would never have passed through the gates of a Buddhist monastery. In scientific circles, the study of mindfulness is opening fresh avenues in the understanding of the relationship between the mind and the brain. In the caring professions, it has revealed its potentials for fostering healthier personal attitudes and better human relationships. Perhaps mindfulness will even lead to a new era of peace and international cooperation. But it is also possible that in its new secular role mindfulness will turn out to be just one more fad to be pushed and promoted in the global marketplace, to flourish for a while, like so many others, and then fade into oblivion. We are at a stage in the history of mindfulness when it is still far too early to make any sure predictions.

The Challenge of Mindful Engagement

David R. Loy

> The mercy of the West has been social revolution. The mercy of the East has been individual insight into the basic self/void. We need both.
>
> —*Gary Snyder, "Buddhist Anarchism"*

Another way to say it: the highest ideal of the Western tradition has been the concern to restructure our societies so that they are more socially just. The most important goal for traditional Buddhism has been to awaken and put an end to one's *dukkha* ("suffering" in the broadest sense), especially that associated with the delusion of a separate self. Today it has become obvious that we need both: Not just because these ideals complement each other, but because each project needs the other.

Snyder's essay on "Buddhist Anarchism" was published over 50 years ago. Now there is a new kid on the block: the mindfulness movement, which straddles West (it is a modern development...) and East (... based on early Buddhist teachings). Yet if "individual insight into the basic self/void" refers to enlightenment, that is not what mindfulness practice is about—it leaves all that religious mumbo jumbo behind, right? And it is certainly not concerned about social revolution, either. So where does it fit into Gary Snyder's contrast—if at all?

The answer, I think, is that the mindfulness revolution is a psychological movement still in its infancy and evolving very quickly. One of the important dimensions that remain to be developed is its relationship to the social justice issues that Snyder alludes to. Mindfulness practices address the way my mind works. By becoming more attentive, more aware of persistent patterns of thinking and feeling, I can free myself from the discomfort that those patterns often cause. But what about the "discomfort" caused by inequitable economic and social relations?

Mindfulness meditation is often marketed as a method for personal self-fulfillment, a reprieve from the ordeals of corporate life. Although such an individualistic, consumerist orientation to the practice may be effective for self-preservation and self-advancement, it is essentially impotent for mitigating the causes of collective and organizational *dukkha*. After a mindfulness program, individual employees in a company may feel that their stress, unhappiness, and doubts are self-made. Such training promotes a tacit acceptance of the status quo and can become an instrumental tool for keeping attention focused on institutional goals. When mindfulness practice is compartmentalized in this way, however, there is a disconnection between one's own personal transformation and the kind of organizational restructuring that might address the causes and conditions of suffering in the broader environment. Such a colonization of mindfulness reorients the practice to the needs of the company, rather than

D.R. Loy (✉)
Niwot, Colorado, USA
e-mail: davidrobertloy@gmail.com

mindfulness encouraging a critical reflection on the causes of our collective suffering, or *social dukkha*. Bhikkhu Bodhi, one of the foremost American Buddhist monastics, has warned: "absent a sharp social critique, Buddhist practices could easily be used to justify and stabilize the status quo, becoming a reinforcement of consumer capitalism."

But the mindfulness revolution is still very new, and its future possibilities bring us back to the relationship between East and West that Gary Snyder highlights: "We need both." Both individual transformation and social transformation. And a closer look at both of those processes will reveal why each needs the other.

The Western conception of justice largely originates with the Abrahamic traditions, particularly the Hebrew prophets, who fulminated against oppressive rulers for afflicting the poor and powerless. Describing Old Testament prophecy, Walter Kaufmann writes that "no other sacred scripture contains books that speak out against social injustice as eloquently, unequivocally, and sensitively as the books of Moses and some of the prophets."

Is there a Buddhist equivalent? The doctrine of *karma* understands something like justice as an impersonal moral law built into the fabric of the cosmos, but historically karma has functioned differently. Combined with the doctrine of rebirth (a corollary, since evil people sometimes prosper this life) and the belief that each of us is now experiencing the consequences of actions in previous lifetimes, the implication seems to be that we do not need to be concerned about pursuing justice, because sooner or later everyone gets what they deserve. In practice, this has often encouraged passivity and acceptance of one's situation, rather than a commitment to promote social justice.

Does the Buddhist emphasis on *dukkha* provide a better parallel with the Western conception of justice? *Dukkha* is unquestionably the most important Buddhist concept: according to the Pali Canon, Gautama Buddha said that what he taught was *dukkha* and how to end it. The best

known summary of the Buddha's teachings, the four noble (or ennobling) truths, is all about *dukkha*, its cause, its extinction, and how to end it. Historically, Asian Buddhism has focused on individual *dukkha* and personal karma, a limitation that may have been necessary in autocratic societies whose rulers could and sometimes did repress Buddhist institutions. Today, however, the globalization of democracy, human rights, and freedom of speech opens the door to new ways of responding to social and structural causes of *dukkha*. In response, a more socially engaged Buddhism has been developing, which also raises an important question for the mindfulness movement: What are the social implications of mindfulness practice?

The Abrahamic emphasis on justice, in combination with the classical Greek realization that society is a collective construct that can be restructured, has resulted in our modern concern to reform political and economic institutions. This has involved, most obviously, a variety of human rights movements (the abolition of slavery, the civil rights movement, women's rights, LGBT liberation, etc.), none of which has been an important concern of traditional Asian Buddhism, and none of which is an important concern of the burgeoning mindfulness movement today.

As valuable as these social reforms have been, however, the limitations of such an institutional approach, by itself, are becoming evident. Even the best possible economic and political system cannot be expected to function well if the people within that system remain motivated by greed, aggression, and delusion—the "three fires" or "three poisons" that Buddhism encourages us to transform into their more positive counterparts: generosity, loving-kindness, and the wisdom that recognizes our interdependence.

Today, in our globalizing world, the traditional Western concern for social transformation encounters not only the traditional Buddhist focus on individual awakening but also the psychological focus of the mindfulness movement. In what ways do these movements need each other in order to actualize their own ideals?

Good Versus Evil

The Abrahamic religions are the primary examples of what is often called "ethical monotheism" because they emphasize most of all ethical behavior. God's main way of relating to us, his creatures, is instructing us how to live. To be a good Jew, Christian, or Muslim is to follow his moral commandments. The fundamental issue is *good versus evil*: Going what God wants us to do (in which case we will be rewarded) and not doing what he does not want us to do (to avoid punishment).

Even the supposed origin of human history, in the *Genesis* story of Adam and Eve, is understood as an act of disobedience against God the Father. Later, because of the wickedness and corruption of the human race—in other words, because people were not living the way God wanted them to—God sends a great flood that destroys all humans (and most animals) except those in Noah's ark. Eventually God formalizes his moral covenant with humanity by giving the Decalogue to Moses: "Thou shalt not…" Jesus's additional emphasis on love does not abrogate the importance of living according to God's commands: of our will submitting to his will.

Although many people in the modern world no longer believe in an Abrahamic God, the struggle between good and evil remains our favorite story. It is the main theme in most popular novels, films, and television shows (think of James Bond, Star Wars, Harry Potter, the Lord of the Rings, not to mention every detective novel and TV crime series). From a Buddhist perspective, however, our preoccupation with that theme is … well, both good and evil.

The duality between good and evil is a good example of the problem that often occurs with dualistic concepts, when we think in terms of bipolar opposites such as high and low, big and small. In many cases, we want one pole and not the other, but because the meaning of each is the opposite of the other (you do not really know what "high" means unless you know what "low" means), we cannot have one without the other. This is true not only logically but also psychologically. If it is really important for you to live a *pure* life (however you understand purity), you will inevitably be preoccupied with (avoiding) *impurity*.

The relationship between good and evil is arguably the most problematical example of bipolar thinking, because their interdependence means that we do not know what good is until we determine what evil is (being good means avoiding evil), and we feel good when we are struggling against that evil—an evil *outside* ourselves, of course. Hence, the inquisitions, witchcraft and heresy trials that plagued Christian Europe.

The tragic paradox is that, historically, one of the main causes of evil has been the attempt to destroy evil, or what we have understood as evil. What was Hitler trying to do? Eliminate the evil elements that pollute the world: Jews, homosexuals, Roma gypsies, and so forth. Stalin attempted to do the same with landowning peasants (*kulaks*), and Mao Zedong with Chinese landlords.

That is the problematic aspect of the duality between good and evil, yet there is also a beneficial side, which brings us back to the Hebrew prophets. Isaiah is a good example when he complains about those "who write oppressive laws, to turn aside the needy from justice and to rob the poor of my people of their right, that widows may be your spoil, and that you may make the orphans your prey" (*Isaiah* 10:2). He speaks on behalf of God, to rulers who abuse their authority. Speaking truth to power, such prophets called for social justice for the oppressed, who suffered from what might be called *social dukkha*.

The other source of Western civilization was classical Greece, which discovered the momentous distinction between *physis* (the natural world) and *nomos* (social convention). In effect, this was the realization that *whatever is social constructed can be changed:* we can reorganize our own societies and in that way (attempt to) determine our own collective destiny. This discovery challenged the archaic religious worldview that embedded the traditional social order within the natural order. Now humans could consciously determine for themselves how to live, which led to Athen's experiment with direct democracy, although a very limited one by today's standards (women and slaves did not participate). The various revolutions that for better and worse have reconstructed our modern world—English, American, French, Russian, Chinese, etc.—all took for granted such an understanding: if a political regime is unjust and oppressive, it should be challenged, because social structures are collective human creations that can be recreated.

Bringing together the Hebrew concern for social justice with the Greek realization that society can be restructured has resulted in the highest ideal of the West, actualized in revolutions, reform movements, the development and spread of democracy, human rights, etc.—in sum, social progress.

So, with such lofty ideals, everything is fine now, right? Well, not exactly…. Even with the best goals (what might be called our "collective intentions"), our societies have not become as socially just as most of us would like, and in some ways, they are becoming more unjust. The obvious economic example is the growing gap between rich and poor in the United States and in much of the rest of the world as well. How shall we understand this discrepancy between ideal and reality? One obvious reply is that our economic system, as it presently operates, is still unjust because wealthy people and powerful corporations manipulate our political systems, for their own self-centered benefit.

I would challenge that explanation, but by itself is it sufficient? Is the basic difficulty that our economic and political institutions are not structured well enough to avoid such manipulations, or might it be the case that they *cannot* be structured well enough—in other words, that we cannot rely only on an institutional solution to structural injustice? Is it possible to create a social order so perfect that it will function well regardless of the personal motivations of the people socially ordered, or do we also need to find ways to address those motivations?

The Greek experiment with democracy failed for the same reasons that our modern experiment with democracy is in danger of failing: unless social reconstruction is accompanied by personal reconstruction, democracy merely liberates the ego-self. So long as the illusion of a discrete self, separate from others, prevails, democracy simply provides different types of opportunities for individuals to take advantage of other individuals.

If we can never have a social structure so good that it obviates the need for people to be good (in Buddhist terms, to make efforts not be motivated by greed, aggression, and delusion), then our modern emphasis on social transformation—restructuring institutions to make them more just—is necessary but not sufficient. That brings us to the issue of personal transformation.

Ignorance Versus Awakening

Of course, ethical behavior is also important in Buddhism. Lay Buddhists are expected to follow the five precepts (to avoid harming living beings, stealing, sexual misconduct, improper speech, and intoxicants) and hundreds of additional rules and regulations are prescribed for monastics. But if we view them in an Abrahamic fashion, we are liable to miss the main point. Since there is no Buddhist God telling us that we must live this way, the precepts are important because living in accordance with them means that the circumstances and quality of our own lives will improve. They can be understood as exercises in mindfulness, to train ourselves in a certain way, like the training wheels on the bicycle of a young child.

In the *Brahmajala Sutta*—one of the earliest and most important Buddhist texts—the Buddha distinguishes between what he calls "elementary, inferior matters of moral practice" and "other matters, profound, hard to see, hard to understand … experienced by the wise" that he has realized. He makes that distinction because for Buddhism, the fundamental axis is not between good and evil, but between ignorance/delusion and awakening/wisdom. The primary challenge is cognitive in the broad sense: becoming more aware of the way things really are. In principle, someone who has awakened to the true nature of the world (including the true nature of oneself) no longer needs to follow an external moral code because he or she *naturally* wants to behave in a way that does not violate the spirit of the precepts.

The Buddha emphasized that he taught *dukkha* suffering and how to end it. Did he have in mind only individual *dukkha* and personal karma—that resulting from our own thoughts and actions—or did he have a wider social vision that encompassed structural *dukkha*: the suffering caused by oppressive rulers and unjust institutions? A few scholars such as Trevor Ling (1985) and Nalin Swaris (2011) have argued for the latter, that the Buddha may have intended to start a movement that would transform society, rather than merely establish a monastic order with alternative values to the mainstream. Certainly his attitudes toward women and caste were extraordinarily progressive for his day—more progressive than many if not most of his followers, even today.

Regardless of what Gautama Buddha may or may not have intended, what apparently happened is that early Buddhism as it institutionalized came to an accommodation with the state, relying on not only the tolerance of rulers but also their material support, to some extent. And if you want to be supported by the powers-that-be, you'd better support the powers-that-be. Because no Asian Buddhist society was democratic, that placed limits on what types of *dukkha* Buddhist teachers could emphasize.

The result was that the tradition as it developed could not address structural *dukkha*—for example, the exploitative policies of many rulers —that ultimately could only be resolved by some institutional transformation. On the contrary, the karma and rebirth teaching could easily be used, and was, to legitimate the power of kings and princes, who must be reaping the fruits of their benevolent actions in past lifetimes, and to rationalize the disempowerment of those born poor or disabled, who must also be experiencing the consequences of (unskillful) actions in previous lifetimes.

The coming of Buddhism to the West—more precisely, the globalization of Buddhism—challenges such mystifications. Secularism and democracy are liberating Buddhism from any need to cozy up to autocratic rulers. In most locales, Buddhists and Buddhist institutions are no longer subject to oppressive governments, and we also have a much better understanding of the structural causes of *dukkha*. This opens the door to expanded possibilities for the tradition, which can now develop more freely the social implications of its basic perspective. As Buddhist emphasis on impermanence and insubstantiality suggests, history need not be destiny.

Another way to express the relationship between the Western ideal of social transformation (social justice that addresses social *dukkha*) and the Buddhist goal of personal transformation (an awakening that addresses individual *dukkha*) is in terms of different types of *freedom*. The emphasis of the modern West has been on individual freedom from oppressive institutions, a prime example being the Bill of Rights appended to the US Constitution. The emphasis of Buddhism (and now the mindfulness movement) has been on what might be called psycho-spiritual freedom. Freedom for the self, or freedom from the (ego)self? What have I gained if I am free from external control but still at the mercy of my own greed, aggression, and delusions? And awakening from the delusion of a separate self will not by itself free me, or all those with whom I remain interdependent in so many ways, from the *dukkha* perpetuated by an exploitative economic system and an oppressive government. Again, we need to actualize both ideals to be truly free.

One might conclude from this that contemporary Buddhism and the mindfulness movement simply need to incorporate a Western concern for social justice. Yet that would overlook the distinctive implications of the Buddhist understanding of suffering, craving, and delusion. To draw out some of those implications, the next section offers a Buddhist-type perspective on our economic situation today.

The Economic Challenge

Despite many optimistic reports about economic recovery—for banks and investors, at least—in the United States, the disparity between rich and poor continues to widen. For example, at the time of writing this, the 20 wealthiest billionaires in America have more total wealth than the poorest half of Americans—about 152 million people.

"It's not *fair!*" Increasingly, citizen movements are calling for social justice—in this case, for distributive justice. Why should the wealthy have so much, and the rest of us so little? It is not difficult to imagine what the Hebrew prophets might say about this situation. But does the Buddhist emphasis on delusion-versus- awakening provide an alternative perspective to supplement such a concern for social justice?

Two implications of Buddhist teachings stand out here. One of them focuses on our individual predicament, and the other considers the structural or institutional dimensions of that system.

Arguably, the single most important teaching of the Buddha is about the relationship between *dukkha* "suffering" and *anatta* "not-self" or "nonself." In more contemporary language, our sense of self is a psychological and social construction that does not have any *svabhava* "self-existence" of its own. Being composed of mostly habitual ways of thinking, feeling, acting, intending, remembering, and so forth—processes that are impermanent and insubstantial—such a construct is inevitably haunted by *dukkha*: inherently insecure, because not only ungrounded but ungroundable.

In other words, the sense of separate self is normally haunted by a sense of *lack*: The feeling that something is wrong with me, that I'm not good enough, or that something is not quite right about my life. Usually, however, we misunderstand the source of our discomfort, and believe that what we are lacking is something outside ourselves. And this brings us back to our individual economic predicament, because in the "overdeveloped" world, we often grow up conditioned to understand ourselves as consumers and to understand the basic problematic of our lives as getting more money in order to acquire more things, because they are what will eventually make us happy.

There is an almost perfect fit between this fundamental sense of *lack* that unenlightened beings have, according to Buddhism, and our present economic system, which uses advertising and other devices to persuade us that the next thing we buy will make us happy—which it never does, at least not for long. In other words, a consumerist economy exploits our sense of lack, instead of helping us understand and address the root problem. The system generates profits by perpetuating our discontent in a way that aggravates it and leaves us wanting more.

What does this imply about our economic institutions, the structural aspect? The Buddha had little to say about evil per se, but he had a lot to say about the three "roots of evil": greed, aggression, and delusion. When what I do is motivated by any of these three (and they tend to overlap), I create problems for myself (and often for others too, of course). Yet we not only have individual senses of self, we also have collective selves: I am a man not a woman, an American not a Chinese, and so forth. Do the problems with the three poisons apply to collective selves as well?

To further complicate the issue, we also have much more powerful institutions than in the Buddha's time. These constitute another type of collective self that often assumes a life of its own, in the sense that such institutions can have their own motivations built into them. Elsewhere I have argued that in the United States, our present economic system can be understood as institutionalized greed; that our militarism institutionalizes aggression; and that our (corporate) media institutionalize delusion, because their primary focus is profiting from advertising and consumerism, rather than informing us about the crucial issues of our day.

Here, let us consider only the first poison: How our economic system promotes structural *dukkha* by institutionalizing greed.

One definition of greed is "never enough," which functions institutionally as well as personally: Corporations are never large enough or profitable enough, share values are never high enough, our national GDP is never big enough…. In fact, we cannot imagine what "big enough" might be. It is built into these systems that they must keep growing, or else they tend to collapse.

Consider the stock market, high temple of the economic process. On the one side are many millions of investors, most anonymous and mostly unconcerned about the details of the corporations they invest in, except for their profitability and its effects on share prices. Such an attitude is not considered disreputable, of course: On the contrary, investment is a highly respectable endeavor, and the most successful investors are idolized. So Warren Buffet is "the sage of Omaha."

On the other side of the stock market, however, the desires and expectations of those millions of investors become transformed into an impersonal and unremitting pressure for growth and increased profitability that every CEO must respond to, and preferably in the short run. Consider, as an unlikely example, the CEO of a large fossil fuel corporation, who one morning wakes up to the imminent dangers of climate change and wants to do everything he (it is usually a he) can to address this challenge. If what he wants to do threatens corporate profits, however, he is likely to lose his job. And if that is true for the CEO, how much more true it is for everyone else further down the corporate hierarchy. Corporations are legally chartered so that their first responsibility is not to their employees or customers, nor to other members of the societies they operate within, nor to the ecosystems of the earth, but to the individuals who own them, who with very few exceptions are concerned primarily about return on investment.

Who is responsible for this collective fixation on growth? The important point is that the system has attained not only a life of its own but its own in-built motivations, quite apart from the motivations of the individuals who work for it and who will be replaced if they do not serve those institutional motivations. And all of us participate in this process in one way or another, as workers, consumers, investors, pensioners, and so forth, usually with little if any sense of personal responsibility for the collective result. Everyone is just doing their job, playing their role.

From this Buddhist perspective, any genuine solution to the economic crisis will require something more than some redistribution of wealth, necessary as that is. The issue of structural *dukkha* implies an alternative evaluation of our economic situation, which focuses on the consequences of individual and institutionalized delusion: The *dukkha* of a sense of a self that feels separate from others, whose sense of *lack* consumerism exploits and institutionalizes into economic structures that assume a life of their own. It has become evident that what is beneficial for those institutions (in the short run) is very different from what is beneficial for the rest of us and for the earth's ecosystems.

To sum up, we cannot expect social transformation to succeed without personal transformation as well, and the history of Buddhism shows that the opposite is also true: Teachings that promote individual awakening cannot avoid being affected by social structures that promote collective delusion and docility. As the sociological paradox puts it, people create society, yet society also creates people.

Western attempts at collective social reconstruction have had limited success because they have been compromised by ego-driven individual motivations. The Asian Buddhist traditions, and of course the mindfulness movement today, have also had limited success at eliminating *dukkha* and delusion, because up until now they have not been able to challenge successfully the *dukkha* and delusion built into unjust social hierarchies that mystify themselves as necessary and beneficial. The convergence of those two projects in our times opens up fresh possibilities. They need each other. Or, more precisely, we need both.

The New Bodhisattva

The Western (modern) ideal of a collective transformation that institutionalizes social justice has achieved much, but not as much as we need. Climate breakdown … mass extinction of species … a dysfunctional economic system with a growing gap between rich and poor … corporate domination of government … overpopulation … It's a critical time in human history, and the collective decisions to be made during the next few years may set the course of events for generations to come. The problems are so enormous and intimidating—where to start?

For those inspired by Buddhist teachings, or the mindfulness movement today, an important issue is how much they can help us respond to these crises. Of course, we cannot expect to find precise answers to contemporary difficulties in ancient Buddhist texts. The Buddha lived in Iron Age India, and his society faced a different set of problems, most notably aggressive monarchies competing to swallow up smaller states. Pre-modern teachings cannot help us decide whether to rein in growth-obsessed capitalism or to replace it with some alternative economic system. We cannot depend on the Buddha to advise us whether a revitalized representative democracy can work well enough or whether we should push for more local, decentralized governance.

Nevertheless, Buddhism—and by extension, the mindfulness movement—opens up the possibility of a new model of activism that connects inner and outer practice: a fresh version of the *bodhisattva ideal*.

Within Buddhism, the bodhisattva concept became a sectarian and divisive issue. According to one account, there was a conspicuous difference between the Buddha and his followers: The Buddha devoted himself to helping everyone awaken. This perception led to the development of a more altruistic model of practice, in which one vows to awaken in order to help everyone else. Today we can understand the bodhisattva path as a nonsectarian archetype that offers a new vision of the relationship between spiritual practice and social engagement—an alternative to rampant self-centered individualism, including versions of what might be called "spiritual materialism" preoccupied solely with one's own personal development.

According to the traditional understanding, bodhisattvas are self-sacrificing because they could choose to escape this world by entering into *nirvana*, but instead they take a vow to hang around here in order to help the rest of us. Yet there is a better way to understand what motivates the bodhisattva, if awakening includes the realization that I am not separate from others. Then, the bodhisattva's preoccupation with helping "others" is not a personal sacrifice but a further stage of personal development. Because one's realization does not automatically eliminate habitual self-centered ways of thinking and acting, following a bodhisattva path becomes important for re-orienting my relationship with the world. Instead of asking "what can I get out of this situation?" one asks: "What can I contribute to this situation, to make it better?"

One of the most important attributes of a bodhisattva is equanimity, due to nonattachment to the fruits of one's action. That is not the same as detachment from the state of the world or the fate of the earth. Nonattachment does not mean that one is unconcerned about the results of one's activism, yet it is essential in the face of the inevitable setbacks and disappointments that activism involves, which otherwise lead to simmering anger, despair, and burnout. Given the urgency of the crises that confront us, we work as hard as we can. When our efforts do not bear fruit in the ways that we hoped, we naturally feel frustrated—but one does not remain stuck there, because Buddhist meditators and mindfulness practitioners have an inner practice that helps them not to hold on to such feelings.

In other words, the path of the bodhisattva is to do the best one can, without knowing what the consequences will be. Have we already passed ecological tipping points and human civilization is doomed? Frankly we do not know—yet rather than being overawed by the unknown the bodhisattva embraces "do not know mind," because meditation practice opens us up to the awesome mystery of an impermanent world where everything is changing whether or not we notice. If we do not really know what is happening, do we really know what is possible, until we try?

The bodhisattva archetype is a way of emphasizing the important distinction between two basic ways of understanding the path: Do I practice in order to end my own suffering, or to help end the suffering of everyone?

This question is as important for mindfulness meditators as it is for contemporary Buddhists. It speaks directly to an important tension today between "self-help" practice and a socially engaged path. Meditation can provide much-needed relief from the pressures of daily life. Nevertheless, and without denigrating the importance of such practice, we need to ask: Does any approach that focuses solely on our own individual development help to develop an awakened society that is socially just and eco-logically sustainable, or does it tend to maintain the present social order? Are Western Buddhism and the mindfulness movement being commod-ified and co-opted into stress-reduction programs that adapt to institutionalized *dukkha*, leaving practitioners atomized and powerless? Or are they opening up new perspectives and possibili-ties that challenge us to transform our societies as well as ourselves?

Appendix

The letter that follows is self-explanatory. I want to emphasize that the issue is not personal. The basic problem, it seems to me, is that one can be well-intentioned and yet play an objectionable role in an economic system that has become unjust and unsustainable—in fact, a challenge to the well-being of all life on this planet. Mr. George is an important figure in the "mind-fulness in business" movement: as well as being a professor in Harvard's MBA program, he has written some influential books that emphasize the importance of ethics and mindfulness in the marketplace. His position therefore highlights some concerns expressed in my article about the role of the "mindfulness movement," and also has broad implications for socially engaged Buddhism generally.

16 October 2012

William George
George Family Office
1818 Oliver Ave.
S. Minneapolis, Minnesota 55405

Dear Mr. George,

We have not met, but I'm taking the liberty of contacting you because you are in a position to contribute in a valuable way to an important debate that is developing within the Buddhist community in North America. (I'm a professor of Buddhist and comparative philosophy, and also a Zen student/teacher.)

The UK *Financial Times* magazine of August 25–26 included an article on "The Mind Busi-ness" that begins: "Yoga, meditation, 'mindful-ness'… Some of the west's biggest companies are embracing eastern spirituality—as a path which can lead to bigger profits." You are men-tioned on p. 14:

> William George, a current Goldman Sachs board member and a former chief executive of the healthcare giant Medtronic, started meditating in 1974 and never stopped. Today, he is one of the main advocates for bringing meditation into cor-porate life, writing articles on the subject for the Harvard Business Review. "The main business case for meditation is that if you're fully present on the job, you will be more effective as a leader, you will make better decisions and you will work better with other people," he tells me [the author, David Gelles]. "I tend to live a very busy life. This keeps me focused on what's important."

I was initially struck by your position (since 2002) as a board member of Goldman Sachs, one of the largest and most controversial investment banks. Researching online, I learned that you have also been on the corporate board of Exxon Mobil since 2005 and Novartis since 1999. I also read that you participated in a "Mind & Life" conference with the Dalai Lama and Yongey Mingyur Rinpoche, on "Compassion and Altru-ism in Economic Systems." These discoveries led to my decision to contact you, in order to get your perspective on what is becoming a crucial issue for Western Buddhists.

The debate within American Buddhism focu-ses on how much is lost if mindfulness as a technique is separated from other important aspects of the Buddhist path, such as precepts, community practice, awakening, and living compassionately. Traditional Buddhism understands all these as essential parts of a spiritual path that leads to personal transformation. More recently, there is also concern about the social implications of Buddhist teachings, especially given our collective ecological and economic situation. The Buddha referred to the "three poisons" of greed, ill will, and delusion as unwholesome motivations that cause suffering, and some of my own writing argues that today those three poisons have become institutionalized, taking on a life of their own.

I do not know how your meditation practice has affected your personal life, nor, for that matter, what type of meditation or mindfulness you practice. Given your unique position, my questions are the following: How has your practice influenced your understanding of the social responsibility of large corporations such as Goldman Sachs and Exxon Mobil? And what effects has your practice had personally on your advisory role within those corporations?

Those questions are motivated by the controversial—I would say problematical—role of those two corporations recently in light of the various ecological, economic, and social crises facing us today. As you know, the pharmaceutical giant Novartis has also received much criticism. (In 2006, Novartis tried to stop India developing affordable generic drugs for poor people; in 2008, the FDA warned it about deceptive advertising of focalin, an ADHD drug; in 2009, Novartis declined to follow the example of GlaxoSmithKline and offer free flu vaccines to poor people in response to a flu epidemic; in May 2010, a jury awarded over $253 million in compensatory and punitive damages for widespread sexual discrimination, a tentative settlement that may increase to almost $1 billion; in September 2010, Novartis paid $422.5 million in criminal and civil claims for illegal kickbacks.) However, my main interest is with your role on the corporate board of Goldman Sachs and Exxon Mobil, and how your meditation practice may or may not have influenced that.

Since you have been on the Goldman Sachs board for a decade, you are no doubt very aware of the controversies that have dogged it for many years, and especially since the financial meltdown of 2008. There are so many examples that one hardly knows where to begin. In July 2010, Goldman paid a record $550 million to settle an SEC civil lawsuit, but that is only the tip of the iceberg. In April 2011, a Senate Subcommittee released an extensive report on the financial crisis alleging that Goldman Sachs appeared to have misled investors and profited from the mortgage market meltdown. The chairman of that subcommittee, Carl Levin, referred this report to the Justice Department for possible prosecution; later he expressed disappointment when the Justice Department declined to do so, and said that Goldman's "actions were deceptive and immoral." Perhaps this relates to an ongoing issue: A "revolving door" relationship with the federal government, in which many senior employees move in and out of high-level positions, which has led to numerous charges of conflict of interest. It may be no coincidence that Goldman Sachs was the single largest contributor to Obama's campaign in 2008.

In July 2011, a suit to fire all the members of Goldman's board—including you—for improper behavior during the financial crisis was thrown out of court, for lack of evidence.

Controversy ignited again this year when a senior Goldman employee, Greg Smith, published an OpEd piece in the *New York Times* on "Why I Am Leaving Goldman Sachs" (March 14, 2012), writing that "the environment [at Goldman Sachs] now is as toxic and destructive as I have ever seen it." He blames poor leadership for a drastic decline in its moral culture—which is especially interesting, given your own teaching emphasis on the importance of leadership. In just the few months since that OpEd, however, Goldman has been fined in the UK for manipulating oil prices, and in separate US cases has paid $22 million for favoring select clients, $16 million for a pay-to-play scheme, $12 million for improper campaign donations, and $6.75 million to settle claims about how it handled option claims. Such fines seem to be acceptable as simply another cost of business, rather than a spur to change how the company conducts business.

Please understand that I'm not criticizing you for these illegal activities. Being on the board, you are not usually involved in day-to-day management. However, I would like to know how you view the "toxic environment" at Goldman Sachs, and the larger social responsibilities of such a powerful firm, in light of your own meditation practice. And since you have been on the Goldman board since 2002, how do you understand the responsibility of a board member in such a situation, and what role have you been able to play in affecting its problematical culture?

I am also curious about your position as a board member of ExxonMobil since 2005. It is reportedly the world's largest corporation ever, both by revenue and by profits. According to a 2012 article in *The Daily Telegraph*, it has also "grown into one of the planet's most hated corporations, able to determine American foreign policy and the fate of entire nations." It is regularly criticized for risky drilling practices in endangered areas, poor response to oil spills (such as the Exxon Valdez in 1989), illegal foreign business practices, and especially its leading role in funding climate change denial.

ExxonMobil was instrumental in founding the first skeptic groups, such as the Global Climate Coalition. In 2007, a Union of Concerned Scientists report claimed that between 1998 and 2005 ExxonMobil spent $16 million supporting 43 organizations that challenged the scientific evidence for global warming and that it used disinformation tactics similar to those used by the tobacco industry to deny any link between smoking and lung problems, charges consistent with a leaked 1998 internal ExxonMobil memo.

In January 2007, the company seemed to change its position and announced that it would stop funding some climate-denial groups, but a July 2009 *Guardian* newspaper article revealed that it still supports lobbying groups that deny climate change, and a 2011 Carbon Brief study concluded that 9 out of 10 climate scientists who deny climate change have ties to ExxonMobil.

Even more important, the corporation's belated and begrudging acknowledgment that global change is happening has not been accompanied by any determination to change company policies to address the problem. Although there has been some recent funding for research into biofuels from algae, ExxonMobil has not moved significantly in the direction of renewable sources of energy such as solar and wind power. According to its *2012 Outlook for Energy: A View to 2040*, petroleum and natural gas will remain its main products: "By 2040, oil, gas and coal will continue to account for about 80 % of the world's energy demand" (p. 46). This is despite the fact that many of the world's most reputable climate scientists are claiming that there is already much too much carbon in the atmosphere and that we are perilously close to "tipping points" that would be disastrous for human civilization as we know it.

In response to this policy, I would like to learn how, in light of your meditation practice, you understand the relationship between one's own personal transformation and the kind of economic and social transformation that appears to be necessary today, if we are to survive and thrive during the next few critical centuries. How does your concern for future generations express itself in your activities as a board member of these corporations (among others)? Are you yourself skeptical about global warming? If not, how do you square that with your role at ExxonMobil?

Let me conclude by emphasizing again that this letter is not in any way meant to be a personal criticism. From what I have read and heard, you are generous with your time and money, helping many nonprofits in various ways. What I'm concerned about is the "compartmentalization" of one's meditation practice, so that mindfulness enables us to be more effective and productive in our work and provides some peace of mind in our hectic lives, but does not encourage us to address the larger social problems that both companies (for example) are contributing to. Today the economic and political power of such corporations is so great that, unless they became more socially responsible, it is difficult to be hopeful about what the future holds for our grandchildren and their grandchildren.

What is the role of a corporate board member in critical times such as ours? I would much appreciate your reflections and your experience on this issue.

Sincerely yours,
David Loy
www.davidloy.org

(Mr. George never replied to this letter.)

'Paying Attention' in a Digital Economy: Reflections on the Role of Analysis and Judgement Within Contemporary Discourses of Mindfulness and Comparisons with Classical Buddhist Accounts of *Sati*

Richard King

Introduction

By the beginning of the twenty-first century, building upon the development of reformist-oriented Buddhist modernisms in the previous century (McMahan 2008), Asian philosophies and meditative practices have increasingly been adopted as means of reducing stress and adjusting to life in a fast-paced world of a globalizing and capitalist economy. This can be seen in the extraordinary popularity and spread of Jon Kabat-Zinn's *Mindfulness-Based Stress Reduction* (MBSR) techniques, itself drawing directly upon the revivalist *vipassana*-only movement of Burma's Mahasi Sayadaw (1904–1982), within Western health-care systems, corporate 'stress-relief' management classes and even within the USA and Korean military. That there are considerable disparities between the techniques and aims of these practices (and their emphasis upon immediate stress-*relief*) and traditional Buddhist meditational teachings and practices, which seek to intensify one's awareness of *duḥkha*, is a subject requiring rigorous and critical attention by scholars of Buddhism.

What is *new* about modern discourses of mindfulness and how might they relate or not to the ancient Buddhist discourses about mental training/development (*bhāvanā*) to which they often appeal? How does an ancient set of practices designed to cultivate a spiritual awareness of radical impermanence (*anitya*) and existential strife (*duḥkha*) become a globally accepted secular technique for stress reduction and well-being? What issues are involved when a set of ancient meditative practices, designed to achieve a state of liberation (*nirvāṇa*) from rebirth and embedded in Buddhist monastic rituals, institutional practices and an ethic of non-violence, are transformed into a modern, secularized therapeutic intervention widely adopted in Western health-care systems, corporate boardrooms and military training regimes?

Mindfulness and Attention

A history of mindfulness is simultaneously a history of attention. According to the late nineteenth-century French psychologist Théodule Ribot, attention can be characterized as 'progress towards unity of consciousness'. In this regard, Ribot argues attention 'is an exceptional, abnormal state, which cannot last a long time, for

R. King (✉)
University of Kent, Kent, UK
e-mail: R.E.King@kent.ac.uk

© Springer International Publishing Switzerland 2016
R.E. Purser et al. (eds.), *Handbook of Mindfulness*,
Mindfulness in Behavioral Health, DOI 10.1007/978-3-319-44019-4_3

the reason that it is in contradiction to the basic condition of psychic life; namely, change'.[1]

Using Ribot's designation we can go someway to understand what classical Buddhist literature means by *sati* (Sanskrit: *smṛti*), the Pali word now almost universally translated into English as 'mindfulness'. Attention involves the adverting of consciousness towards an object of experience but to 'hold one's attention' upon that object also requires a certain 'unity of consciousness'. In classical Buddhist accounts of mental training (*bhāvanā*), overcoming the oscillating nature of consciousness and achieving mental equipoise are associated with techniques designed to facilitate concentration (*samādhi*) and calm (*samatha*). The standard account that emerged within the Buddhist literature tended to emphasize the conjoining of techniques designed to facilitate awareness and attention (*vipassanā*) and those which facilitated an ever greater unity of consciousness (*samādhi*), although it is likely that the precise balance between these two varied in different circumstances, traditions and individual practices (Cousins 1973).

Although classical Buddhist literature might agree with some of Ribot's characterization of attention, it would not necessarily agree with his description of it as an 'abnormal' state of mind. Arguably, the Buddhist—and generally yogic—diagnosis of our mental condition is that the so-called everyday, distracted (*vikṣepa*) states of mind are themselves the aberration or problem to be overcome. However, most of our everyday experience is indeed a history of repeated distraction (what the Buddhists describe as our 'monkey mind'). Similarly, Ribot's account implies that attention is a fleeting matter undermined by the fluctuating nature of experience. For Buddhists, focused attention leads to a much greater awareness of the fact of change, but in advanced practitioners, this is not seen as preventing the cultivation of attention as a stabilizing mode of continued awareness. Indeed, prolonged attention is seen, in many Buddhist

accounts as a much greater awareness of that flux.

Nevertheless, it is clear that as the Buddhist tradition developed two different characterizations of consciousness emerged: one focused on the reality of impermanence, and the Buddhist emphasis on no-abiding-self (*anātman*) emphasized the processual nature of consciousness. The path of mental training involves disciplining the mind to avoid distraction and to remain present to one's experience of the radical impermanence of reality. However, another strand of thought is also present in the early Buddhist literature which resonated more strongly with the prevailing 'yogic' philosophical opinion in India. This second strand postulated an innate unity and purity of consciousness and saw the achievement of mental equipoise and calmness as a return of consciousness to its natural state—like a pond once the ripples of a pebble have dispersed or the ocean below the waves. On this view, our prevailing everyday experience of dispersed and distracted states of mind constituted the stirring up or 'whirring' of consciousness (*citta-vṛtti*) from its natural state and was indicative of life in the saṃsāric realm for those not yet awakened and liberated from the cycle of rebirths. This notion of an underlying unity of consciousness behind our changing states of mind was the model that predominated in the Brahmanical yogic traditions associated with Sāṃkhya, Yoga and the *Upaniṣads* (Vedānta) where it was associated with a non-agential and pure 'witness consciousness' (*sākṣin*) standing 'behind' the changing flow of experiences. Although the dominant conception of consciousness in Buddhist philosophical thought in India however remained the processual model, as outlined in the Abhidharma literature, the 'innate purity' model continued to find vehicles for expression, most overtly in the 'Buddha nature' (*tathāgatagarbha*) strand of the Mahāyana (emerging in the fourth/fifth century CE) and in subsequent debates about the sudden or gradual nature of enlightenment.[2]

---

[1]Ribot (1898: 2).

[2]For further discussion of this see Faure 1991; Sharf 2014a, b.

'Meditation' and the Role of Intellectual Analysis

The Buddhist tradition has long had a specific association with what we have come to call in the West 'meditation'. Use of this English word carries an ambiguity within it since it is often used to denote a set of specific practices linked to pacifying the analytic processes of the mind and achieving a state of concentrated calmness, practices that, in the Buddhist tradition, are associated with the *jhānas* (Sanskrit; *dhyāna*) and the cultivation of concentration and calm (*samādhi/samatha*). However, the English word meditate is also used as a synonym of the exercise of sustained mental reflection upon something as in 'I shall meditate on that question and get back to you'. In a Buddhist context, the exercise of reflective cognition is associated with the cultivation of insight (Pali: *vipassanā*; Sanskrit: *vipaśyanā*) and wisdom or 'analytical insight' (*paññā/prajñā*). The potential elision between this second aspect of 'mental training' (*bhāvanā*, what we now routinely translate into English as 'meditation') and the general application of analytic reasoning/mental reflection produced a similar ambiguity within Buddhist circles, akin to the two senses of 'meditation' in an Anglophone context. Although, as we shall see, the mainstream Abhidharmic account of Buddhist mental training presupposes a significant role for mental ratiocination and cognition, alternative views which characterize awakening (*bodhi*) as the quiescence of all mental activity continue to be expressed, especially in those strands of Buddhist thought which came to adopt a non-dualistic worldview (such as some forms of Ch'an/Zen (Sharf 2014a, b) and Tibetan *dzogchen* practice).[3]

The thorny question of the relationship of an intellectual analysis of the nature of reality and the systematic practice of disciplining and calming the mind is encapsulated by the combination of *sammā-sati* and *sammā-samādhi* as twin components of standard Buddhist accounts of the nature of mental development and training. As La Vallée Poussin first noted, a concrete instance of the tension between 'understanding the Dhamma' and disciplining the mind can be found in the example of two of the Buddha's disciples Musīla and Nārada (La Vallée Poussin 1937). Musīla is said to have acquired a detailed understanding of the teachings of the Buddha based upon mental comprehension and analysis but has not 'touched nirvāṇa with the body', that is not achieved a direct *experiential* realization of it.

> Friend, though I have clearly seen as it really is with correct wisdom 'Nibbāna is the cessation of existence,' I am not an arahant, one whose taints are destroyed. Suppose, friend, there was a well along a desert road, but it has neither a rope nor a bucket. Then a man would come along, oppressed and afflicted by the heat, tired parched, and thirsty. He would look down into the well and the knowledge would occur to him, 'There is water,' but he would not be able to make bodily contact with it (*na ca kāyena phusitvā vihareyya*). So too, friend, though I have clearly seen as it really is with correct wisdom, 'Nibbāna is the cessation of existence,' I am not an arahant, one whose taints are destroyed.[4]

Similarly, *Anguttara Nikāya* VI, 46 records discord within the community of the Buddha's disciples in the form of a distinction between the *jhāyin* (one who practices the *jhānas*) and the *dhammayogins* who are said to have an intellectual grasp of the teachings based upon the application of analytical insight (*prajñā*).

> Friends, there are monks who are keen on Dhamma (*dhammayogin*) and they disparage those monks who are meditators (*jhāyin*), saying: 'Look at those monks! They think, "We are meditating, we are meditating!" And so they meditate to and meditate fro, meditate up and meditate down. What, then, do they meditate about and why do they meditate?" Thereby neither these monks keen on Dhamma nor the meditators will be pleased, and they will not be practising for the welfare and happiness of the multitude, for the good of the multitude, for the welfare and happiness of *devas* and humans.[5]

---

[3]See Sharf (2014a, b) and Dunne (2013) for further discussion of this.

[4]*Kosambī Sutta, Saṃyutta Nikāya* II.68, translation in Bodhi (2000: 611).

[5]Anguttara Nikāya VI, 46, translation in Bodhi and Thera (1999: 163–164).

It is not immediately clear from this account if we are to take the *jhāyin* to denote a practitioner of techniques leading to the quiescence of 'mental whirring' (*citta-vṛtti*)[6] associated with *samādhi* training or if this also includes the systematic cultivation of insight (*vipassanā*) and 'mindfulness' (*sati*). Thus, we cannot be absolutely certain whether the term *dhammayogin* denotes a 'purely intellectual' and scholarly appreciation of the *Dhamma* or it relates to a conception of meditative practice that emphasizes the continued application (and even *enhancement*) of mental cognition, analytic reasoning through the cultivation of insight (*vipassanā*).

The discord recorded between these two groups perhaps reflects early ambiguities and tensions about the role of and relationship between 'insight-based' and 'concentration-based' techniques in the Pali Buddhist literature but may also reflect a difference of opinion over the role and importance of mental ratiocination in the achievement of liberation. As the traditional story of the Buddha's life coalesced, probably over many centuries, the standard resolution of this tension was to assign the practice of advanced stages of concentration, such as the achievement of the sphere of nothingness (*ākiñcaññāyatana*) and the sphere of neither perception nor non-perception (*nevasaññānāsaññāyatana*), to the training undertaken by Gotama under the guidance of Aḷara Kalama and Uddaka Rāmaputta prior to his full awakening (see Wynne 2007). The problem with following these methods alone, it came to be argued, is that while they pacify the thirst-driven motivational impulses to a significant extent and also train the aspirant in achieving a one-pointed (*ekāgatta*) state of mind, without the cultivation of insight and the development of a full existential appreciation of the four noble truths

(and three marks of existence), they do not lead to final awakening (*bodhi*).

A similar tension, I wish to argue, plays out in a new form and context in contemporary discourses about 'mindfulness' in the late twentieth and early twenty-first centuries. As 'mindfulness-based' practices become adapted and applied in non-Buddhist and 'secular' contexts, the dominant discourse has tended to characterize 'mindfulness' as a present-centred and non-judgemental awareness, seeking to curtail to a significant degree our usual processes of mental ratiocination and cultivating an attitude of calm acceptance and 'bare attention' free from analysis and judgement. Thus, as Jon Kabat-Zinn describes it, mindfulness is about 'paying attention in a particular way: on purpose, in the present moment, and non-judgementally'.[7] However, while this is perhaps the dominant characterization of mindfulness, it is by no means the only model of mindfulness in operation.

Many contemporary Buddhist accounts of mindfulness, drawing upon the Abhidharmic model, assert quite forcefully the role of cognition and ethical judgement in the context of mindfulness practice. This is most strikingly clear in accounts offered by proponents of what has come to be known as Engaged Buddhism. As we shall see, the traditional Abhidharmic emphasis upon analysing the causal conditions which produce suffering (*duḥkha*) and the clear role of ethical reflections and judgements upon one's experience in seeking to cultivate harmonious states of mind (*kuśala*) are emphasized and in fact quite radically extended in some engaged Buddhist accounts transforming mindfulness into a form of direct political 'consciousness-raising' in relation to the embedded structures of social and economic injustice that inform our everyday experience of the world. The distinction between these two characterizations of mindfulness, I shall argue, constitutes a still-emerging theoretical fault line

---

[6] I use this phrase because it resonates more generally with the trend in yogic philosophical circles to focus on techniques for pacifying mental vacillation in advanced states of concentration (*samādhi*). Note for instance how in the *Ur*-text of the Hindu Brahmanical yoga school, Patañjali defines yoga precisely as the 'cessation of mental whirring' (*cittavṛttinirodhāḥ*, YS1.2).

[7] Kabat-Zinn (1994), 4 For some insightful discussion of the modern emphasis on 'being in the moment non-judgementally': see Bodhi (2013: 27f) and also Dreyfus and Olendski.

within contemporary discourses of mindfulness and is thrown into relief by the rapidly changing context of early twenty-first-century life.

I will briefly discuss three factors of contemporary life that have precipitated this fault line in the late twentieth and early twenty-first centuries. They are as follows: the global spread of neoliberal forms of capitalism, growing concerns about climate change and social and economic disparities of wealth, and the impact of new digital technologies on human consciousness. First, however, it is important to be clear about some of the philosophical assumptions underlying traditional Buddhist accounts of *sati*.

Mind and Mindfulness in Ancient Indian Buddhist Thought

> We take the rendering 'mindfulness' so much for granted that we rarely inquire into the precise nuances of the English term, let alone the meaning of the original Pali word it represents and the adequacy of the former as a rendering for the latter. (Bodhi 2013: 22)

It is important to take a moment to look afresh at ancient Buddhist debates about techniques of mental development/training (*bhāvanā*) and resist their easy assimilation into a set of modern, Western assumptions and representations of what we now call 'Buddhist meditation'. This is especially important since Buddhist traditions have come to be associated in the West with a particular understanding of 'meditation', often conceived in terms of the 'pacification of the mind' because of the way that 'Buddhism' came to be associated with prevailing Orientalist stereotypes about 'the mystic East'. If "mysticism" is seen as the pre-eminently non-rational, then Buddhism, when viewed as a mystical tradition, comes to be framed in terms that reflect such cultural assumptions. As already noted, however, even in English the word 'meditation' carries an ambiguity—denoting either a pacification of the mind or a process of mental reflection. The association of 'Buddhism' with the former in the popular imagination has occluded the important role assigned to mental reflection and analysis in many traditional Buddhist accounts of the cultivation of *sati*.

Another way to illustrate this point is to consider the English phrase 'being philosophical'. There are two primary ways in which this phrase is used. Firstly, and probably more commonly, it denotes a form of relaxed detachment in the face of adversity, e.g. 'Her beloved piano fell down the stairs but she was philosophical about it'. There is a second use of the term however denoting a form of critical, intellectual reflection upon language and/or experience associated more specifically with the disciplined activity of philosophical analysis. Consider for instance the example of the sixth century BCE pre-Socratic philosopher Anaxamines. It is said that he once thought to blow on his hand in two ways: first with his mouth open and then with his lips pursed. When blowing with an open mouth, he experienced warmth, but with his lips pursed, his breath felt cold to his hand. Anaxamines then asked why this was so and in doing so sought to analyse his experience to understand the underlying cause of the change in sensations. Such examples as this have often been used to locate the origins of philosophy and even science as a whole in the thought experimentations of the pre-Socratics of ancient Greece.[8] However, it strikes me that on some classical Buddhist readings of *sati,* there is a similar emphasis upon a stepping back and observation of experience *combined* with an analytical reflection upon its antecedent causes. From this perspective, *sati* is much more about cultivating a 'philosophical approach' to the world—in both senses of the modern use of that term—on the one hand as a form of suspended emotional detachment ('being philosophical') but also in the sense of offering a meta-analytic perspective upon experience—a mental cogitation on what is presented in perceptions, the exercise, if you like, of critical thinking or a philosophical analysis of experience.

Modern accounts of mindfulness of the Kabat-Zinn variety tend to ignore this second dimension of *sati*. Mindfulness becomes primarily about witnessing without reacting, 'being philosophical' in the first sense but certainly not in the second. As we will see, in classical

---

[8]See for instance, Vernon (2015).

Abhidharma and early Mahāyāna accounts, *sati* is usually represented as exemplifying both dimensions—fostering a degree of emotional detachment—a 'standing back' from reactive habitual forms (emphasized in Nyanaponika's focus upon *sati* as a form of 'bare attention') but also by the disciplined exercise of analytical insight (*prajñā*) to that experience through an examination of its antecedent causes and conditions and an intention to direct consciousness towards ethically wholesome rather than unwholesome thoughts.

We must appreciate therefore that the political and cultural transformation involved in the translation of key terms and practices from their ancient Buddhist context and into a modern English conceptual frame, replete with its own cultural associations. As Talal Asad has noted:

> To put it crudely, because the languages of third world societies ... are seen as weaker in relation to Western languages (and today, especially to English), they are more likely to submit to forcible transformation in the translation process than the other way around.[9]

In this sense, one needs to revisit the standard translation of these terms in order to resist their easy assimilation to modern Anglophone assumptions about 'mindfulness', allowing them to retain a 'discomforting—even scandalous—presence within the received language' (Asad 1993: 199). To do this, we need to appreciate that there is an enormous complexity to ancient Buddhist philosophical discussions of consciousness and a rich vocabulary of technical terms encompassing what in an English language context would be called 'mind' or 'consciousness'. In the Indian traditions of Buddhist thought include Sanskrit terms such as *citta, manas and vijñāna* and cognate terms (such as *jñāna, prajñā, saṃjñā and dhyāna*) referring to different functions and modalities of awareness, representing affective, cognitive and conative dimensions of consciousness. Understanding these terms is crucial for an appreciation of the emergence and eventual consolidation of early

Buddhist accounts of the mental training (*bhāvanā*) required to achieve awakening (*bodhi*).

In the West, the material and the mental worlds have often been treated as two distinctive domains; however, in the ancient Indian context in which Buddhist notions of mental training first developed it is important to recognize the inadequacy of such dualisms. Although Buddhists texts frequently refer to '*nāma-rūpa*' (name and form, often glossed in English as 'mind' and 'body'), these are usually taken in unison as a compound form, reflecting a recognition of the 'psychosomatic' nature of human experience. It is also stated many times throughout the early Buddhist literature that mind or consciousness cannot arise without a material base and similarly that our experience of material objects is dependent upon the arising of a consciousness of them. Moreover, Indian Buddhist thought developed a complex array of terms to denote the different affective, cognitive and conative operations of consciousness.

Sensory awareness (*vijñāna*) arises as a result of contact between the sense organs and their specific sense objects. There are six sensory realms in classical Buddhist thought, what have traditionally been known as the five senses (sight, sound, touch, smell and taste), plus *mano-vijñāna*—mental consciousness, which apprehends internal states of mind, ideas, etc. The mental function of apperception (*mano-vijñāna*) came to be distinguished over time from *manas*—the mind as a centralizing and agential faculty that organizes the different arrays of sense data, thereby constructing a coherent mental picture out of these disparate sensory sources. Thus, it is quite common in an Indian Buddhist context to see *mano-vijñāna* described as a 'sixth sense'—an apprehender of 'mental' sensory data and for this to be clearly distinguished from the more analytical functions of consciousness (carried out by the *manas*). Thus, apart from a basic conscious awareness (*vijñāna*) of a sensation (*vedanā*), Buddhist thought also acknowledges the role of mental cognition in the classification of sensory impressions (*saṃjñā*), as well as the affective response that arises in relation to those impressions (the various *saṃskāras*). These factors then induce the arousal of intention

---

[9]Asad (1993:190).

(*cetanā*, the conative aspect) in the individual, reflecting a goal-directed response to one's environment.

The Pali word for 'mindfulness', *Sati*, and its Sanskrit equivalent, *smṛti*, have a primary meaning of memory or recollection. In a Hindu Brahmanical context, *smṛti* denotes the 'remembered traditions' (such as the *Mahābhārata* and the Rāmāyana), to be distinguished from *śruti*—'that which is heard,' namely the direct revelation of the Vedas. In the context of training of the mind (*bhāvanā*), the early Buddhist usage retains some of this sense, but, rather than focusing upon 'historical memory', relates more to the idea of a mental state of sustained attention—an awareness that *remains* present to the complex, evanescent and causally produced operations of consciousness and its objects, or to use John Peacock's preferred translation: 'present moment recollection' (Peacock 2014: 6).[10] Buddhaghosa (1950) characterizes *sati* as a form of 'remembering' (*saraṇa*) and says it is characterized by 'not wobbling' (*apilāpana*): 'Its function is not to forget. It is manifested as guarding, or it is manifested as the state of confronting an objective field' (*Visuddhimagga* XIV, 141).[11] As Gethin (2013: 264) notes, early English renditions of the term in its specifically Buddhist context include 'correct meditation' (for *sammā-sati*, Gogerley 1845); 'the faculty that reasons on moral subjects, the conscience' (Hardy 1850); and the 'ascertainment of truth by mental application' (Hardy 1853). It seems, however, that the first person to translate *sati* (Sanskrit: *smṛti*) as mindfulness was T. W. Rhys-Davids in 1910. He remarks:

Etymologically, Sati is memory. But as happened at the rise of Buddhism to so many other expressions in common use, a new connotation was then attached to the word, a connotation that have a new meaning to it, and renders 'memory' a most inadequate and misleading translation. It became the memory,

recollection, calling-to-mind, being aware of, certain specified facts. Of these the most important was the impermanence (the coming to be as the result of a cause, and the passing away again) of all phenomena, bodily and mental. And it included the repeated application of this awareness, to each experience of life, from the ethical point of view.[12]

It is clear that in classical Buddhist literature, *sati* involves an analytic awareness of the truth of the four noble truths leading to a deep appreciation of the impermanent, suffering and no-self marks of existence. This involves a clear comprehension (*sampajañña*) of causal relations (how things arise and cease), and part of the point in using a term like *sati* is to emphasize how this requires a 'memory of the present', a sustained attention to the present moment, including its causal history— that is, a recollection of past behavioural patterns and experiences that inform the present moment. In the Nikāya and Abhidharma discussions of *sati* then, such practice *requires* rather than suspends analytical reflection upon experience.[13] Moreover, the practice of *sati* is taken to be a practice integrated within the wider aspects of the eightfold path and includes ethical reflection upon the wholesome and unwholesome *dhammas* that arise within the mind and an explicit aim of cultivating the former and uprooting the latter. It seems quite clear then that from the Abhidharmic point of view, *sati* involves sustained ethical reflection and analysis of the processes of causation that lead to the rise of *dhammas*. Thus, drawing upon traditional Abhidharmic accounts of *sati*, Dreyfus (2013: 47) argues that

Mindfulness then is not the present-centred non-judgemental awareness of an object but the paying close attention to an object, leading to the retention of the data so as to make sense of the information delivered by our cognitive apparatus. Thus, far from being limited to the present and to a mere refraining from passing judgement, mindfulness is a cognitive activity closely connected to memory, particularly to working memory, the ability to keep relevant information active so that it can be integrated within meaningful patterns and used for goal-directed activities.[14]

---

[10]Peacock (2014). Referring in particular to *Dhammasaṅghaṇi* 16, Gethin (2013: 270) notes the following early Abhidhamma terms associated with *sati*: recollection (*annusati*), recall (*paṭissati*) remembrance (*saraṇatā*), keeping in mind (*dhāraṇatā*), absence of floating (*apilāpanatā*) and an absence of forgetfulness (*asammussanatā*).

[11]Translation in Ñāṇamoli (1975: 467).

[12]Rhys-Davids and Rhys-Davids (1910: 322).

[13]For a useful discussion of the role of mental cognition in Pali canonical Buddhist accounts of *sati* see Bodhi (2013)

[14]Dreyfus (2013)

The Centrality of Prajñā in Abhidharma and Early Mahāyāna Accounts

As a number of scholars have suggested (see for instance Gethin 2011; Cousins 1996), the singling out of 'insight meditation' as the distinctive element within Buddhist meditational practice does not seem to reflect a traditional Theravāda perspective which generally involves a conjunction of insight and concentration practices as symbiotic constituents of the eightfold path. Indeed, it is questionable whether one can speak accurately of 'insight meditation' in this way before the modern period. As Bhikkhu Anālayo notes:

> [I]n the thought-world of the early discourses the term *vipassanā* stands predominantly for insight as a quality to be developed. This thus differs from the modern day usage, where *vipassanā* often stands representative for a particular form of meditation, usually a specific technique whose practice marks off one insight meditation tradition from another.[15]

Nevertheless, in the stress placed upon the cultivation of mindfulness (*sati*) and wisdom (*paññā*) as a necessary component of the path to awakening, we see an important ideological marker of the distinctive contribution of the Buddha as a teacher when compared to the other yogically oriented movements of the India of his day. Indeed, in characteristically Indic fashion, concentration-inducing practices—and the prevailing hierarchical cosmologies associated with them—were incorporated into the Buddhist eightfold path (as *sammā samādhi*, 'right concentration') but characterized as singularly deficient unless symbiotically linked to the practice of *sammā sati* ('right mindfulness') and the cultivation of insight (*vipassanā*).

Within Indian Buddhist literature, therefore, the cultivation of wisdom or 'analytical insight' (*paññā/prajñā*) came to be seen as a crucial marker of a distinctively Buddhist path of mental development (*bhāvanā*) when compared to prevailing yogic systems in India. The cultivation or exercise of *prajñā* thus came to be used in

Buddhist circles as an indicator of the superiority of Buddhist mental training (*bhāvanā*) when compared to other systems of yogic discipline which also utilized the language of concentration (*samādhi*) and the goal of the unification of consciousness through meditative equipoise. The claim that *prajñā* and the cultivation of insight were specific features of the Buddhist approach to mental training is of course not one that was accepted by these rival schools. Patañjali's *Yoga-Sūtra* for instance sees the goal of yogic practice as the 'cessation of mental fluctuations' (*cittavṛttinirodhāh*, YS 1.2) but makes it abundantly clear that advanced forms of *samādhi* rather than being mere states of internalized concentration remain truth-bearing states that involve *prajñā* (YS I.48).[16] In contrast, many Buddhist accounts speak of *samādhi* as a state of inward concentration leading to calm, but not necessarily to insight. One of the thorny issues here is recognizing how different yogic literary traditions deploy the same technical terms (such as *samādhi* and *prajñā*) but with quite different implications.

It is worth dwelling briefly then upon the role and place of *prajñā* in the practice of 'mindfulness' (*smṛti/sati*). One of the challenges here is that because *prajñā* came to be seen as an indispensable component of an awakened mind, the term took on a level of significance within the Buddhist tradition which meant that while it could never be repudiated as central to the cultivation of mindfulness and the achievement of the Buddhist goal of awakening, its precise meaning often varied according to the context. This led Padmanabh Jaini to remark:

> It must be admitted … the precise meaning of *prajñā* itself remains obscure. One sometimes feels that nothing definite can be said beyond the statement that *prajñā* is something which was attained by the Buddha and is attainable by bodhisattvas.[17]

---

[15]Anālayo (2012: 214)

[16]What Patañjali means by '*prajñā*' here is of course up for discussion. Is it to be viewed as a general term for wisdom/insight or does it denote something like the Abhidharma technical usage of the term as analytical insight into the nature of things, that is, as a form of analytic cognition?

[17]Jaini (1977: 403).

Nevertheless, it is clear that the general understanding of the term within the Nikāya literature is that it is through *prajñā* that one sees things as they are (*yathābhūta*). Although the term is often translated generically as 'wisdom' in English (a vague rendition that works well in obscuring underlying philosophical technicalities and tensions sometimes operating across traditions), within the Abhidharma literature it is clear that *prajñā* is used in a more technically precise sense to denote the faculty of 'analytical insight', that is the mental power (*bāla*) of analysing entities and breaking them down into their more basic elemental components—the *dhammas* that constitutes the underlying, impermanent flow of evanescent moments (*kṣaṇa*) which constitute our experiences. In the Southern/Theravāda tradition, Buddhaghosa explains that *prajñā* (*paññā*) is that which penetrates the own nature of things (*dhamma-sabhava-pativedha*, *Visuddhimagga* XIV, 7). *Paññā* then is explicitly linked to the cultivation of *vipassanā*, usually translated as insight. This is seen as a profound realization of the impermanent and dependently originated nature of entities. As Nanayakkara (1993: 580) notes 'Insight is not knowledge in the general sense, but penetrative knowledge acquired as a result of not looking *at* but looking *through* things'.[18]

However, it is important to note that *prajñā* is considered an *occasional* mental factor according to the Pāli Abhidhamma tradition, whereas in the Northern Abhidharma literature of the Sarvāstivāda/Vaibhāṣika (and much of the subsequent Mahāyāna literature which inherited and responded to the Northern traditions), it is seen as a *universal* factor present in all experience (if developed to varying degrees).

With the emergence of Mahāyāna forms of Buddhism in India from the first century BCE, we see a reaction to the Abhidharma approach and its scholastic analysis of experience into momentary events (*dharmas*). However, in the *Prajñāpāramitā* literature this involves not a repudiation of the Abhidharma emphasis upon *prajñā*, but rather its intensification. *Prajñā* involves the analytic reduction of the conventionally real entities of phenomenal experience into their underlying (and for the Abhidharma, ultimately real), dharmic components. The exercise of the faculty of *prajñā* is crucial in an Abhidharma context for establishing the distinction between ultimate (*paramārtha*) and conventional (*saṃvṛti*) entities made by Vasubandhu (1967) in *Abhidharmakośa VI.4*:

> If the awareness of something does not operate after that thing is physically broken up or separated by the mind into other things, it exists conventionally like a pot or water; others exist ultimately.[19]

Thus, the *Prajñāpāramitā* literature accepted the Northern/Sarvāstivāda inclusion of *prajñā* as a universal factor in experience and indeed presupposed it as the basis for the universalization of the ideal of the *bodhisattva* and the goal of achieving full awakening for all sentient beings. However, it criticized the Abhidharmic enterprise for failing to take its own reductive analysis of experience to its final conclusion, that is a recognition of the emptiness of *dharmas* themselves. *Prajñā*, or analytical insight, required further intensification (to be achieved by 'practising the perfection of *prajñā*'). Within this context, wisdom (*jñāna*) in its most advanced forms came increasingly to be characterized as non-conceptual (*nirvikalpa*) in nature.

Mahāyāna and the Emergence of a Non-dualistic Understanding of Mindfulness

Within those strands of what became Mahāyāna Buddhism, we see the emergence of a more avowedly non-dualistic conception of reality. The dominant intellectual approaches in Indian Mahāyāna, building upon the *Prajñāpāramitā* worldview, emphasized the emptiness (*śūnyatā*) of all *dharmas*. Although the precise nature of this emptiness was conceived of slightly differently between early Mahāyāna schools such as the Madhyamaka and Yogācāra, they both

---

[18]Nayanakkara (1993). It is linked to a growing awareness of the three marks of existence.

[19]*Abhidharmakośa* VI.4, translation by Buescher (1982).

continued the radicalization of the no-abiding-self teaching (*anātman*) and accorded a central role to *prajñā* in Buddhist yogic practice. The non-dualistic spirit of these movements however opened up the possibility of a greater emphasis upon what Dunne (2013) calls the 'innateist' strand of Buddhist thought, that is an approach to awakening which sees it as the unveiling of a pure consciousness that *already exists* in a veiled form within each sentient being. Awakening (*bodhi*), on this model of consciousness, involves the realization of that which one already possesses, but which is hidden from view by the karmic defilements of consciousness. Buddhist mental training on this model became characterized as cleaning the mirror of consciousness so that it could directly reflect things as they are (*yathābhūta*). Indeed, as Olenzski suggests (2013: 67), the Northern Abhidharma tradition's inclusion of *prajñā* as a universal mental factor provided a theoretical rationale for the innateist view (that the mind already contains the factors pertaining to an already awakened consciousness) to emerge. As suggested earlier, this understanding of the Buddhist path is asserted most strongly in the *tathāgatagarbha* ('Buddha nature') literature that emerges from around the third/fourth centuries CE and is further consolidated by later Mahāyāna developments such as Tibetan notions of 'other emptiness' (*gzhan stong*, propounded especially but not exclusively by the Jo nan pas)[20] and in meditative practices such as *dzogchen* which seek to uncover the pristine nature of consciousness.

Dunne (2013: 75) has argued that the accounts given of mindfulness practice in MBSR and MBCT programmes seem more intellectually akin to the non-dualistic innateist position than to the constructivist position that generally prevails in mainstream Abhidharma literature. Thus, he suggests:

non-dual traditions, striking a stance deliberately contrary to *Abhidharma* scholasticism, remain highly sceptical about the utility of evaluative thought in practice. Instead, one must become released from the very structures of such thoughts, since they are a manifestation of ignorance itself.[21]

Although the historical roots of the modern 'mindfulness-only' movement spring from late colonial Burma and Theravāda reformism, as Dunne suggests, the theoretical framework for modern mindfulness discourse often bears a closer resemblance to some forms of non-dualistic Mahāyāna and Vajrayāna conceptions of meditative practice. Jon Kabat-Zinn, for instance, suggests that his own formulation of MBSR reflects influences not only from the Theravāda *vipassanā* movement but also from Korean Zen. In general terms, however, influence may have less to do with direct Mahāyāna influence than with the diffusion of a broadly non-dualistic conception of 'eastern spirituality' that emerged first with figures like Swāmi Vivekānanda (1863–1902) and then circulated more generally in Western popular culture throughout the twentieth century.

However, the curtailment of judgement and ethical reflection are by no means absent in many non-dualistic accounts because, as we shall see, even within Buddhist trends with a strongly non-dualistic philosophical orientation (such as in the Zen-inspired Engaged Buddhism of Thich Nhat Hanh and David Loy), the role of discernment and a deep cognition of the underlying causes of suffering remain central features of their conception of engaged mindfulness practice. In these accounts, the traditional emphasis upon the importance of *prajñā* in the cultivation of mindfulness is not only endorsed but also extended.

Buddhist Meditation: 'Capitalist Spirituality' or Anti-consumerist Resistance?

In a number of his writings, Slavoj Zizek, a *doyen* and *enfant terrible* of contemporary 'critical theory' circles but hardly any kind of expert

---

[20]For discussions of *gzhan stong* see Ruegg (1989); Hookham (1991); Kapstein (2000); Smith (2001). Nhat Hanh (1991), 'tation and activity.ultural associations of 'ization of the ideal of the bodhisatvva—the ka and Yoshe.

[21]Dunne (2013: 79).

in the history of Buddhism, has argued that 'New Age Asiatic thought' is 'establishing itself as the hegemonic ideology of global capitalism'. (Zizek 2001: 12). According to Zizek (2001: 13):

> the "Western Buddhist" meditative stance is arguably the most efficient way, for us, to fully participate in the capitalist dynamic while retaining the appearance of mental sanity. If Max Weber were alive today, he would definitely write a second, supplementary volume to his *Protestant Ethic*, entitled *The Taoist Ethic and the Spirit of Global Capitalism*.

Zizek's account however reflects a poor understanding of the rigour and diversity of the Buddhist traditions and practices that he so readily dismisses and is part of a wider agenda in his work in seeking to promulgate a 'non-religious Christianity' as the underlying cultural identity of the West and defend it from foreign importations and influences. Putting aside the considerable flaws in Zizek's polemical arguments for the moment,[22] the question of distinguishing between the rich diversity of Buddhist traditions in their historical context and the ways in which they are being deployed and represented in a modern 'late capitalist' context is an important issue to be addressed in any attempt to understand modern discourses of "mindfulness', their roots and their relationship to historical forms of Buddhism. What Zizek rather casually refers to as 'Western Buddhism' or 'New Age Asiatic thought' (and which he often conflates with 'Buddhism' and 'Taoism' as a whole) is really an aspect of what I have called elsewhere 'capitalist spirituality' (Carrette and King 2005). Indeed, it is the latest manifestation in a long history of Western Orientalist fantasies about 'the mystic East' (King 1999), generated and perpetuated by a continuous flow of corporate advertising, marketing and popular cultural images of 'eastern spirituality'. It is vital that we do not confuse these trends with the rich and diverse Buddhist traditions that they so actively misrepresent, not based upon some traditional Orientalist appeal to the authority of original forms, but rather to be able to understand

from the perspective of an informed history of ideas, the sense in which modern discourses of mindfulness carry forward and translate long-established debates and tensions about the nature of mental training (*bhāvanā*) in the Buddhist tradition, and also ways in which they represent significantly innovative developments in response to the demands and context of twenty-first-century life.

Just as the early Buddhist movement in India developed its conception of mind training in response to prevailing attitudes and practices of its day (what I am calling the 'yogic philosophical milieu' of classical Indian thought), contemporary discussions of 'mindfulness' are articulated in relation to their own cultural/intellectual influences. In seeking to identify some of the key cultural, social and political markers that are reframing the discourse of mindfulness in the early twenty-first century, I wish to draw attention to three factors: detraditionalization, capitalist globalization and the impact of new digital technologies on human consciousness.

'Eastern Spirituality' and the DeTraditionalization of Buddhism

Firstly, with regard to the process of the detraditionalization of Buddhist ideas and practices, the transformation of Asian religions into 'eastern spiritualities' in the late twentieth and early twenty-first centuries has of course also rendered such established cultural traditions as more readily exportable to the West, leading to the development of what Heelas (1996) has called the 'self-spiritualities' associated with the New Age and to the commodification and marketing of yoga (for instance) as a physicalized therapy and aid to 'lifestyle enhancement' in a late twentieth-century context alongside the popularity of MBSR practices. Zizek then is partly correct in that 'Buddhism' has indeed seen the greatest market potential for 'New Age Capitalists' in the West.

As many scholars have noted, the spread of modern 'mindfulness-only' practices is linked to

---

[22]For a critique of Zizek's arguments in this regard see Bowman (2007).

the twentieth-century revival of Theravāda meditation in Southeast Asia and to the impact of figures such as Burmese monk Mahāsī Sayādaw (1904–1982) and his student and translator Nyanaponika Thera (German-born Siegmund Feniger 1901–1994) in simplifying and codifying a form of 'insight-only' meditation accessible to the laity (see for instance Braun 2013). The roots of the modern mindfulness movement lie in the late colonial and twentieth-century period, where Western fascination with 'the mystic East' (King 1999) was consolidated and combined with claims about the scientific and/or humanistic nature of the Buddha and his teaching (Lopez 2009; McMahan 2008) to produce the conditions for the emergence of the *Mindfulness-Based Stress Reduction* (MBSR) program of Jon Kabat-Zinn (1990) that has become so popular today. This would have been impossible without the earlier contribution of figures such as Swāmi Vivekānanda (1863–1902) and D.T. Suzuki (1870–1966)) who sought to distil the 'universal' message of 'eastern spirituality' from it's specifically *Asian* cultural and religious underpinnings, thereby facilitating the migration and translation of classical Buddhist discussions of mental training into a modern psychologized discourse of 'experience' (Sharf 1995; King 1999; Carrette and King 2005). This is not a value-neutral decontextualization of Buddhist ideas, as is often claimed, but rather their *recontextualization* in terms of a new cultural, political and symbolic order (Sharf 1995; King 1999).

Building upon the rise of Buddhist modernisms in the last century, concepts, ideas and practices associated with Western conceptions of 'Buddhism' have become easily segregated from their cultural, cosmological and institutional origins through homogenizing discourses about 'eastern spirituality' (Carrette and King 2005) and MBSR practices that gain traction and popularity based upon the ancient and exotic cultural capital of 'Buddhism', but have a low level of engagement with Buddhist theories and practices. Moreover, since the dawn of European romanticism and then again since the 1960s, 'eastern philosophies' have been associated in the West with a kind of 'countercultural'

exoticism that makes them hip, fashionable and fresh for those seeking an alternative to mass consumerism but also as an 'alternative' and exotic 'spirituality' that offers an edge in the competitive world of marketing and business management. Thus, Kabat-Zinn is able to make a double move whereby the cultural authority provided by the ancient Buddhist origins of 'mindfulness' can be deployed to give social capital and credibility to his techniques at the same time as a rapid disavowal of the particularity of those Buddhist roots are asserted through a decontextualized universalization of 'mindfulness' as simply the practice of attention.

> Mindfulness is actually a practice. It is a way of being, rather than merely a good idea or a clever technique or a passing fad. Indeed, it is thousands of years old and is often spoken of as 'the heart of Buddhist meditation', although its essence, being about attention and awareness, is universal.[23]

However, to understand the explosion of interest in mindfulness-related practices and techniques in the contemporary period it is inadequate to focus exclusively upon changing modes of 'religiosity'. One must also consider what social, economic and political conditions have encouraged this popularity. What changes have precipitated the incredible demand for mindfulness-related practices in the early twenty-first century that have captured the attention of defenders and critics alike?

Digital Technologies, Distracted Attention and the Problem of 'Information Overload'

A 2015 study (*'Attention Spans'*), commissioned by Microsoft Corp., recently suggested that

---

[23]Jon Kabat-Zinn, Foreword to Williams and Penman (2011: 10). Indeed in an interview with the *Los Angeles Times* in 2010, Kabat-Zinn goes even further, remarking that 'Mindfulness, the heart of Buddhist meditation, is at the core of being able to live life as if it really matters. It has nothing to do with Buddhism. It has to do with freedom'. Cited by Morris (2010) http://articles.latimes.com/2010/oct/02/local/la-me-1002-beliefs-meditation-20101002

widespread use of digital media technologies is having a deleterious effect on sustained and selective attention and contributing to a reorientation of human consciousness where 'alternating attention' (as in multitasking and switching between devices) was becoming enhanced.

[What information consumes is] the attention of its recipients. Hence a wealth of information creates a poverty of attention. (Herbert Simon 1978 Nobel Prizewinner for Economics)

The fast-paced nature of contemporary digital communications, the 'information overload' that this creates, when combined with a neoliberal conception of the individual as a high-functioning 'entrepreneur of oneself' (Rose 1996, 1999) has arguably contributed to unprecedented levels of stress and depression. This phenomenon—what Jock Young (2007) has called the 'vertigo of late modernity'—has created a demand for techniques to master and control attention. For this reason, a critical analysis of the modern mindfulness movement, from the point of view of the history of ideas, must also examine the modern history of distraction (Löffler 2014), its mediatized intensification in an age of fast-paced digital technologies, the levels of stress and anxiety produced by continually dispersed attention in an age of perceived economic and social precarity and the requisite demand this has created for a variety of relaxation techniques such as yoga and mindfulness-related practices that seek to intensify self-awareness and promote a non-distracted sense of emotional integration, calmness and well-being.

We are moving from a world where computing power was scarce to a place where it now is almost limitless, and where the true scarce commodity is increasingly human attention.
(Satya Nadella, CEO of Microsoft)

In an era of digital 'information overload' delivered through multiple devices (multichannel 24-hour television, smart phones, computers, tablets), the emphasis has shifted away from *advertising* products to *adverting* the attention of human beings towards those products. Thus, in a data-saturated marketplace, capturing the attention of the potential consumer has now become the emergent issue for corporate marketing strategies looking to gain a competitive edge over their opponents in the marketplace:

In post-industrial societies, attention has become a more valuable currency than the kind you store in bank accounts. The vast majority of products have become cheaper and more abundant as the sum total of human wealth increases. Venture capital dollars have multiplied like breeding hamsters. The problems for businesspeople lie on both sides of the attention equation: how to get and hold the attention of consumers, stockholders, potential employees and the like, and how to parcel out their own attention in the face of overwhelming options. People and companies that do this succeed. The rest fail. *Understanding and managing attention is now the single most important determinant of business success. Welcome to the attention economy'* (my italics for emphasis).[24]

This new frontline in the global economy of proliferated advertising has precipitated a corporate-driven demand for techniques that seek to capture, master and control attention. Similarly, longer lifespan, population growth and the spread of a neoliberal conception of the state as increasingly withdrawn from providing public services and social welfare have led to a widespread privatization of health and social welfare provision. This has generated a demand in health-care systems worldwide for effective, non-invasive and above all 'cost-efficient' techniques for enhancing patient health and well-being. Thus, a critical understanding of the emergence of the modern mindfulness movement must consider *not only the impact of consumer capitalism and new digital technologies, but also the modern history of mediatised distraction* (Löffler 2014) and the levels of stress and anxiety engendered by changing lifestyles, occupational patterns and new technologies (such as email) that demand a state of continually dispersed rather than sustained attention. This cognitive 'switching' demanded by these aspects of modern life has led to a growing demand for relaxation techniques such as yoga and 'mindfulness' that soothe a purposely displaced mind and seek to intensify self-awareness and promote a

---

[24]Davenport and Beck (2001: 3).

non-distracted sense of emotional integration, calmness and well-being.

I wish to argue that this context is producing a discursive split between two significantly new developments within what has been called 'Buddhist modernism' (see McMahan 2008) and related secular proponents of 'mindfulness' practice. At the same time, as some see 'Buddhism' as the perfect customizable 'spirituality' for the contemporary 'entrepreneur of the self' in a neoliberal social context,[25] Buddhist teachings and traditions of practice also continue to resonate with those interested in developing countercultural resistance to 'Western materialism' and consumerism, especially within what has become known as 'Engaged Buddhism'.

The Contemporary Reworking of an Ancient Debate: Does Mindfulness Involve Mental Analysis and Ethical Judgment?

The capitalist-oriented trend is exemplified in the business world by the proliferation of 'spiritual management' courses exploring 'Eastern' philosophical themes and meditative practices with the aim of promoting workplace productivity, short-term stress-relief for employees and profit generation, and also by various forms of 'prosperity Buddhism' such as the *Dhammakaya* movement in contemporary Thailand. The counter-consumerist trend manifests itself in contemporary Thai movements such as the Santi Asoke and in transnational trends such as the various forms of 'Engaged Buddhism' which seek to highlight social injustice and challenge what is usually seen as corporate-driven consumerism and materialism within contemporary society. The distinction between these two Buddhist strands is not always as clear cut as it might seem, but much of their cultural authority in the contemporary world resides in what they both share in common, namely a reliance upon a

history of Orientalist assumptions and stereotypes about Asian spirituality and philosophy that have circulated the globe in the last couple of centuries (King 1999; van der Veer 2013) and the development of transnational forms of 'Buddhist modernism' in the last century (Lopez 2009; McMahan 2008).

As a number of scholars have noted, this dominant popular trend, influenced by Mahāsi Sayadaw and Nyanaponika Thera, generally characterizes 'mindfulness' as a form of 'bare attention'—a witnessing of mental, emotional and physical changes without any judgement or disturbance by an inquiring or analytic mindset. In the contemporary context, this has been reinforced by widespread popular cultural associations of 'Zen' in the West with 'chilling out' and pacifying mental agitation and activity. The second trend linked to the rise of an overtly political wing of what has become known as 'Engaged Buddhism' sees mindfulness practice as a form of *consciousness-raising* with regard to social, political and economic injustice, driven by a conceptualization of *duḥkha* as having sociopolitical as well as individual dimensions. As Nhat Hanh himself notes:

> When I was in Vietnam, so many of our villages were being bombed. Along with my monastic brothers and sisters, I had to decide what to do. Should we continue to practice in our monasteries, or should we leave the meditation halls in order to help the people who were suffering under the bombs? After careful reflection, we decided to do both – to go out and help people and to do so in mindfulness. We called it engaged Buddhism. Mindfulness must be engaged. One there is seeing, there must be acting .... We must be aware of the real problems of the world. Then, with mindfulness, we will know what to do and what not to do to be of help.[26]

Nhat Hanh is quite explicit in noting that attention to the causal conditions out of which our everyday experiences emerge involves a mindful awareness of their interdependent origination (*pratītyasamutpāda*). This is pretty standard fare from a traditional Abhidharmic point of view. Of course, Nhat Hanh approaches mindfulness

---

[25]For a useful discussion of the rise of the 'entrepreneur of the self' in neoliberal contexts see the works of Rose (1996, 1999).

[26]Nhat Hanh (1991).

practice from the point of view of Mahāyāna--based Zen notions of emptiness (*śūnyatā*) and a non-dualistic worldview. He extends this philosophy through his notion of 'interbeing'. Thus,

> If you wish to have the insight of *Interbeing* you only need to look at a basket of fresh green vegetables which you have just picked. Looking deeply, you will see the sunshine, clouds, compost, gardener and hundreds of thousands of elements more. Vegetables cannot arise on their own, they can only arise when there is sun, clouds, earth etc. If you take the sun out of the basket of vegetables the vegetables will no longer be there. If you take the clouds away it is the same.[27]

Most of the time Nhat Hanh describes these kinds of mindful moments in a way that reflects a spirituality of ecological interdependence and perhaps a recognition of the impact of our individual patterns of consumption.[28] Other advocates of Engaged Buddhism such as David Loy, Steven Batchelor[29] and Phra Payutto[30] are also explicit about the crucial role that ethics and ethical judgements play in mindfulness practice. However, the recognition by engaged Buddhists that *duḥkha* in fact is not merely an *individual* experience of existential dissatisfaction, but are also formed by instances of social suffering and structural injustice, opens up the possibility that to be *truly* mindful of the causal conditions that produce, say, your experience of eating chocolate, would necessitate an awareness of the history of slavery and ongoing economic exploitation of populations in relation to the cocoa plantations out of which the chocolate was produced and transported. This intellectual move, it strikes me, takes mindfulness practice into a new dimension that of facilitating a geopolitical or global awareness of 'interdependence' and the ways in which the lives of others impact upon our most basic everyday experiences—especially in facilitating a remembrance of history (*smṛti*, traditionally translated) and a structural awareness of the economic, political and ecological dimensions of consumption.[31]

> Meditation is to be aware of what is going on—in our bodies, our feelings, our minds and the world. Each day 40,000 children die of hunger. The former superpowers still have more than 50,000 nuclear warheads, enough to destroy the Earth many times. Yes, the sunrise is beautiful, and the rose that bloomed this morning along the wall is a miracle. Life is both dreadful and wonderful. To practice meditation is to be in touch with both aspects.[32]

Note in the above quote how Nhat Hanh begins with the standard four objects of meditation as outlined in the *Mahā-saṭṭipathāna Sutta*, viz. the body, sensations, the mind and mental objects (*dhammas*, here glossed as 'the world') and then juxtaposes this to instances of mass-suffering and military capacities for state-induced violence. This is a clear extension of the range of 'awareness' from individual experience to a sociopolitical level and reflects an attempt to link individual spiritual practice with a geopolitical consciousness, a development that Raphäel Liogier has

---

[27]Nhat Hanh (2004), (see webpage: http://www. purifymind.com/ManNotEnemy.htm).

[28]See for instance, Nhat Hanh (2009).

[29]Steven Batchelor asserts that 'Ethics as practice beings by including ethical dilemmas in the sphere of meditative awareness- to be mindful of the conflicting impulses that invade consciousness during meditation. Instead of dismissing these as distractions *(which would be quite legitimate when cultivating concentration),* one recognizes them as potentials for actions that may result in one's own or others' suffering.' (my italics for emphasis). See Batchelor (1993).

[30]Payutto, for instance asserts that '*Buddhadhamma* emphasizes the importance of *sati* at every level of ethical conduct. Mindfully conducting your life and your practice of the *Dhamma* is called *appamāda*, or conscientiousness [and is] of central importance to progress in the Buddhist system of ethics'. Reciprocally, ' proper ethics have value because they because they nurture and improve the quality of the mind'. Payutto (1995).

[31]The best example I have found of this in Nhat Hanh's writings are his reflections on his poem '*Please Call Me By My True Names'* where Nhat Hanh makes explicit the link between the individual and the political: 'Do our daily lives have nothing to do with our government? Please meditate on this ... When we pick up a Sunday newspaper, we should know that in order to print that edition, which sometimes weights 10 or 12 lb, they had to cut down a whole forest. We are destroying our Earth without knowing it. Drinking a cup of tea, picking up a newspaper, using toilet paper, all of these thing to do with peace. Nonviolence can be called 'awareness' We must be aware of what we are, of who we are, and of what we are doing.' See Nhat Hanh (1988: 31–39).

[32]Nhat Hanh (1987).

described as the 'individuo-globalist ideology' of such engaged forms of Buddhism.[33] From this kind of vantage point, mindfulness practice explicitly involves not only the exercise of ethical judgements and analysis of underlying causal processes but also the fostering of a 'deep' cognition of the geopolitical dimensions of individual experiences. Thus, Sulak Sivaraksa makes the claim that:

> On a political level, mindfulness can help in our work against consumerism, sexism, militarism, and the many other isms that undermine the integrity of life. It can be a tool to help us criticize positively and creatively our societies, nations and even cultural and religious traditions. Rather than hate our oppressors, we can dismantle oppressive systems. Is the international economic system that demands unlimited growth inherently defective? From a Buddhist perspective, the answer is yes.[34]

By contrast, as we have seen, building upon Nyanaponika Thera's focus upon 'bare attention', contemporary secular accounts of mindfulness practice tend to focus upon an attitude of passive acceptance and a suspension of critical reflection when practising mindfulness.[35] Thus, Mark Williams, Emeritus Professor of Clinical Psychiatry and former Director of the Oxford Mindfulness Centre at Oxford University and Danny Penman, a meditation teacher and journalist, in outlining the significance of Mindfulness-Based Cognitive Therapy (MBCT), make the claim that 'Mindfulness is about observation without criticism; being compassionate with yourself.'[36]

Conclusion

Both the MBSR/MBCT and Engaged Buddhist developments resonate with ancient strands within earlier Buddhist discussions of mental training (*bhāvanā*). The first, in the emphasis placed upon a

suspension of ratiocination, is arguably more closely associated with the path of concentration (*śamatha-yāna*) and the quiescence of cognition, but has a long history in Buddhist literature, reinforced by the emergence of non-dualistic interpretations of the Buddha's message which in some instances see the goal of mental training as the cultivation of a form of non-conceptual awareness (*nirvikalpa jñāna*) grounded in the cultivation of equanimity (*upekṣā*). It is perhaps ironic that the modern practice of 'mindfulness-only' is generally characterized by an abandonment of the long-standing emphasis upon the cultivation of 'concentration' techniques designed to stabilize and quieten the mind, when the characterization often provided of what such mindfulness practice entails bears more of a resemblance to the establishment of mental quiescence rather than achieving greater cognitive acuity. One explanation for this is that what is being discussed in many accounts of 'suspending judgement' during mindfulness practice corresponds to what would have been seen in a traditional Buddhist context as a fairly preliminary act of mental cleansing required for beginners (what Nyanaponika calls 'tidying up the mental household')[37] rather than the cultivation of a highly rarefied and concept-free state of awareness as in the advanced *samādhis*. As Dreyfus (2013: 52) notes:

> By over-emphasizing the non-judgemental nature of mindfulness and arguing that our problems stem from conceptuality, contemporary authors are in danger of leading to a one-sided understanding of mindfulness as a form of therapeutically helpful spacious quietness.

The second trend in modern accounts of mindfulness builds upon the emphasis in many Buddhist texts on the role of *paññā/prajñā*—analytical insight—as a deconstructive analysis of entities into the evanescent *dharmas* that are said to constitute the underlying complexity that makes up our experiences. This second approach places great emphasis on the role of judgement and discernment in 'witnessing' one's experiences, mental reflection upon the underlying causes of their emergence and an ethical consciousness to

---

[33]Liogier (2004).

[34]Sivaraksa (2011: 83).

[35]For an insightful discussion of Nyanaponika's focus on 'bare attention' as a characterisation of *sati* see the discussion in Bodhi (2013: 27f).

[36]Williams and Penman (2011: 5).

[37]Thera (1968: 1).

direct the mind gently towards ever more wholesome mental states (Sanskrit: *kuśalā dharmā*). In this second formulation of mindfulness, therefore, discernment, analysis and ethical judgement are part and parcel of the awakening experience. What is innovative however about the way this is being developed within some Engaged Buddhist literature and movements is the consideration of the geopolitical and economic dimensions of the causal nexus of the individual human experience. What we have then is an Engaged Buddhist reformulation of traditional discussions about *sati* in a way that reframes mindfulness as a geopolitical or planetary awareness of one's 'interbeing' (Thich Nhat Hanh) and the social, political and economic injustices that operate in the causal nexus of even our most everyday, subjective experiences. In this way, what we see emerging here is a Buddhist project for an ethical decolonization of consciousness in response to a perceived sense of growing global inequalities in an age characterized by neoliberal ideologies and capital-driven globalization. This, despite the claims of many Engaged Buddhists, is demonstrably new and an innovation in Buddhist discourses about mindfulness, as is the emphasis upon 'mindfulness-only' practices in general.

Our discussion has focused on two divergent trends in contemporary discourses of mindfulness. One trend, following Mahāsī Sayādaw and Nyanaponika Thera, represents 'mindfulness' as a form of 'bare attention'—a largely pacified 'witness consciousness' devoid of judgement or disturbance by an inquiring or analytic mindset (see Sharf 2014a; Dreyfus 2013) and is the dominant, popular characterization of mindfulness in the secular, scientific, military and business worlds. In contrast, the second trend, linked to what has become known as 'Engaged Buddhism', emphasizes an extensive role for ethical reflection and mental cognition, arguing that mindfulness denotes an awareness of our radical *interbeing* (as in Thich Nhat Hanh's (re-) formulation of the Buddhist teaching of *pratītyasamutpada*) and even a recognition of the *geopolitical* dimensions of individual experiences (such as awareness of the history of colonial exploitation and economic inequality of cocoa plantations as causal factors in

one's experience of eating chocolate). Both interpretations build upon ancient strands: the first in the emphasis placed upon an abandonment of ratiocination and the quiescence of cognition (Griffiths 1986; Sharf 2014b) and the second by resonating with the emphasis in many Buddhist texts on the role of *paññā/prajñā*—analytical insight (i.e. a deconstructive analysis of entities into the evanescent *dharmas* that constitute our experiences) and an ethical concern to direct the mind towards wholesome mental states (Sanskrit: *kuśalā dharmā*). Between these two characterizations, there are of course a multitude of practices and emphases and it is not my intention to suggest that all practices seeking to promote mindfulness meditation fall easily into either of these camps. The different characterizations of mindfulness practices over the question of mental reflection and ethical judgement have ancient roots but are today reflective of the struggle to represent the implications and importance of modern mindfulness practices in an age of economic and social anxiety about the impact of consumerism and rapid neoliberal globalization. Together, these two ends of the spectrum embody two sides of an emerging fault line about the meaning and significance of mindfulness practice in the twenty-first century.

References

Asad, T. (1993). *Genealogies of religion. Discipline and reasons of power in Christianity and Islam*, Baltimore and London: Johns Hopkins University Press.

Batchelor, S. (1993). The future is in our hands. In T. Nhat Hanh (Ed.) *For a future to be possible* (pp. 136–142). Berkeley CA: Parallax Press.

Bhikkhu, A. (2012). *Excursions in the thought-world of the Pāli discourses*. Onalaska, WA: Pariyatti Press.

Bodhi, B. (2000). *The connected discourses of the Buddha*, Boston: Wisdom Publications.

Bodhi, B., & Thera, N., (1999). *The numerical discourses of the Buddha*, Walnut Creek, California: Altamira Press, Rowman and Littlefield Publishers Inc.

Bodhi, B. (2013). What does mindfulness really mean? A canonical perspective. In M. G. Williams, & J. Kabat-Zinn (Eds.), *Mindfulness. Diverse perspectives on its meaning, origins and applications* (pp. 19–41), London and New York: Routledge, Originally published in *Contemporary Buddhism 12*(1), 19–39.

Bowman, P. (2007). The Tao of Zizek. In P. Bowman & R. Stamp (Eds.), *The truth of Zizek* (pp. 27–44). London and New York: Continuum.

Braun, E. (2013). *The birth of insight: Meditation. Modern Buddhism and the Burmese monk Ledi Sayadaw*. Chicago: University of Chicago Press.

Buddhaghosa. (1950). *Visuddhimagga*. In H. Warren (Ed.) revised D. Kosambi, Harvard.

Buescher, J. (1982). *The Buddhist doctrine of the two truths in the Vaibhāṣika and Theravāda schools*, Unpublished PhD thesis, University of Virginia.

Carrette, J., & King, R. (2005). *Selling spirituality. The silent takeover of religion*. London and New York: Routledge.

Cousins, L. (1973). Buddhist Jhāna: Its nature and attainment according to the Pali Sources. In *Religion III*, Part 2, pp. 115–131.

Cousins, L. (1996). The origins of insight meditation. In T. Skorupski (Ed.), *The Buddhist forum IV: Seminar papers*, 1994–1996 (pp. 35–58). London: School of Oriental and African Studies.

Davenport, T. H., & Beck, J. C. (2001). *The attention economy. Understanding the new currency of business*. Boston: Harvard Business School Press (Mass).

Dreyfus, G. (2013). Is mindfulness present-centred and non-judgemental? A discussion of the cognitive dimensions of mindfulness. In M. G. Williams & J. Kabat-Zinn (Eds.), *Mindfulness. Diverse perspectives on its meaning, origins and applications*, London and New York: Routledge, Originally published in *Contemporary Buddhism 12*(1), 41–54.

Dunne, J. (2013). Toward an understanding of non-dual mindfulness. In M. G. Williams & J. Kabat-Zinn (Eds.), *Mindfulness. Diverse perspectives on its meaning, origins and applications*, London and New York: Routledge, Originally published in *Contemporary Buddhism 12*(1), 71–88.

Faure, B. (1991). *The rhetoric of immediacy: A cultural critique of Chan/Zen Buddhism*. Princeton: Princeton University Press.

Gethin, R. (2013). On some definitions of mindfulness. In M. G. Williams & J. Kabat-Zinn (Eds.), *Mindfulness. Diverse perspectives on its meaning, origins and applications* (pp. 263–79), London and New York: Routledge, Originally published in *Contemporary Buddhism 12*(1), 263–279.

Gethin, R. (2011). On some definitions of mindfulness. *Contemporary Buddhism, 12*(1), 263–279.

Gogerley, D. (1845). On Buddhism. *Journal of the Ceylon Branch of the Royal Asiatic Society 1*, 7–28.

Hardy, R. S. (1850). *Eastern monachism*. London: Partridge and Oakey.

Hardy, R. S. (1853). *A manual of Budhism: in its modern development*. London: Patridge and Oakey.

Heelas, P. (1996), *The new age movement. The celebration of self and the sacralization of modernity*, Oxford and Cambridge (Mass): Blackwells.

Hookham, S. (1991). *The Buddha within: Tathagatagarbha doctrine according to the Shentong interpretation of the Ratnagotravibhaga*. Albany: SUNY Press.

Jaini, P. S. (1977). Prajñā and Dṛṣṭi in the Vaibhāsik Abhidharma. In L. Lancaster & L. Gomez (Eds.), *Prajñāpāramitā and related systems: Studies in honor of Edward Conze*, University of California Press.

King, R. (1999). *Orientalism and religion. Postcolonial theory, India and "the Mystic East"*, London and New York: Routledge.

Kabat-Zinn, J. (1990). *Full catastrophe living. Using the wisdom of your body and mind to face stress, pain, and illness*, Delta Trade Paperbacks.

Kabat-Zinn, J. (1994). *Wherever you go, there you are: Mindfulness meditation in everyday life*. New York: Hyperion.

Kapstein, M. (2000). We are all Gzhan stong pas: Reflections on the reflexive nature of awareness: A Tibetan madhyamaka defence, by Paul Williams, *Journal of Buddhist Ethics 7*, 105–125.

La Vallée Poussin, L. D. (1937). Le chemin de Nirvana. *Melanges chinois et bouddhiques, 5*, 189–190; partially reproduced in Gombrich, R. F. (2005). *How Buddhism began: The conditioned genesis of the early teachings*, (2nd ed.), appendix. London: Routledge Critical Studies in Buddhism, (pp. 133-134).

Liogier, Raphaël. (2004). *Le Bouddhisme Mondialisé. Une Perspective Sociologique sur la Globalisation du Religieux*. Ellipses: Paris.

Lopez Jr, D. (2009). *Buddhism and science: A guide for the perplexed*. Chicago: University of Chicago Press.

Löffler, P. (2014). *Verteilte Aufmerksamkeit. Eine Mediengeschichte der Zerstreuung* (English translation of title: *Distributed Attention: A Media History of Distraction*,) Zürich, Berlin: Diaphanes.

Microsoft Corporation Report. (2015) Attention Spans, Consumer Insights, Canada, Spring 2015.

Morris, N. (2010, October 2) Fully experiencing the present: A practice for everyone, religious or not, *Los Angeles times*. http://articles.latimes.com/2010/oct/02/local/la-me-1002-beliefs-meditation-20101002. Accessed January 4, 2016.

McMahan, D. L. (2008). *The making of Buddhist modernism*, Oxford University Press.

Ñāṇamoli, B. (1975). *Visuddhimagga, the path of purification*, Seattle WA: BPS Pariyatti Edition, 1991.

Nayanakkara, S. K., (1993). Insight. In W. G. Weeratne (Ed.), *Encyclopaedia of Buddhism*, Sri Lanka: Department of Buddhist Affairs, 5.40: 580–584.

Nhat Hanh, T. (1987). *'Suffering is not enough' from being peace* (p. 14). Berkeley CA: Parallax Press.

Nhat Hanh, T. (1988). Please call me by my true names. In F. Eppsteiner (Ed.), *The path of compassion*, Berkeley CA: Parallax Press.

Nhat Hanh, T. (1991). *Peace is every step*. New York: Bantam Books.

Nhat Hanh, T. (2004). Man is not our enemy, Dharma talk 2004, see webpage: http://www.purifymind.com/ManNotEnemy.htm

Nhat Hanh, T (2009). Diet for a mindful society. In *For a future to be possible* (pp. 62–79).

Olendzski, A. (2013). The construction of mindfulness. In M. G. Williams & J. Kabat-Zinn (Eds.),

Mindfulness. Diverse perspectives on its meaning, origins and applications (pp. 41–54), London and New York: Routledge, Originally published in *Contemporary Buddhism 12*(1), 55–70.

Payutto, P. P. (1995). *Buddhadharma: Natural laws and values for life, translation grant A.* Olson, Albany: State University of New York Press.

Peacock, J. (2014). Sati or mindfulness: Bridging the divide. In M. Bazzano (Ed.), *After mindfulness: New perspectives on psychology and meditation* (pp. 3–22). New York: Palgrave Macmillan.

Rhys-Davids, C. A. F. & Rhys-Davids, T. W. (1910). Introduction to the Mahā-satipatthana Suttanta. *Dialogues of the Buddha 3.*

Ribot, T. (1898). *The psychology of attention*, Open Court Publishing Company.

Rose, N. (1996), *Inventing our selves, psychology, power and personhood*, Cambridge University Press.

Rose, N. (1999). *Governing the soul: The shaping of the private self* (1st Edn. 1989, 2nd Edn. 1999). Free Association Books.

Ruegg, D. S. (1989). *Buddha nature, mind and the problem of gradualism in a comparative perspective: On the transmission and reception of Buddhism in India and Tibet.* London: School of Oriental and African Studies, University of London.

Sharf, R. (1995). Buddhist modernism and the rhetoric of meditative experience. *Numen 42*(3), 228–283.

Sharf, R. (2014a). Mindfulness and mindlessness in early Chan. *Philosophy East and West, 64*(4), 933–964.

Sharf, R. (2014b). Is Nirvāṇa the same as insentience? Chinese struggles with an Indian Buddhist ideal. In J. Kieschnick & M. Shara (Eds.), *Indian in the Chinese imagination: Myth, religion, and thought* (pp. 141–170). Philadelphia: University of Pennsylvania Press.

Sivaraksa, S. (2011). *The wisdom of sustainability. Buddhist economics for the 21st century.* In A. Kotler & N. Bennett (Ed.), Souvenir Press.

Thera, N. (1968). *The power of mindfulness.* Kandy, Sri Lanka: Buddhist Publication Society.

Smith, E. G. (2001). *Among Tibetan texts: History and literature of the Himalayan Plateau.* In K. R. Schaeffer (Ed.), Boston: Wisdom Publications.

Van der Veer, P. (2013). *The modern spirit of Asia: the spiritual and the secular in China and India.* Princeton: Princeton University Press.

Vasubandhu. (1967). *Abhidharmakośabhāṣya.* In P. Pradhan (Ed.), Patna.

Vernon, Mark. (2015). *The Idler guide to ancient philosophy.* London: Idler Books.

Williams, M. & Penman, D. (2011). *Mindfulness. A practical guide for finding peace in a frantic world*, Piatkus.

Wynne, A. (2007). *The origins of Buddhist meditation*, Routledge.

Young, J. (2007). *The Vertigo of late modernity.* London: Sage.

Zizek, S. (2001). *On belief.* London and New York: Routledge.

Mindfulness Within the Full Range of Buddhist and Asian Meditative Practices

4

Geoffrey Samuel

Introduction

The interest in mindfulness as a therapeutic modality within Western medicine got going in 1979, with Jon Kabat-Zinn's introduction of the Mindfulness-Based Stress Reduction program (MBSR) at the University of Massachusetts Medical Center (Kabat-Zinn 2003). There is by now a very substantial literature both on Mindfulness-Based Stress Reduction and on the family of techniques and therapies that derive from it, and also a substantial critical literature, including a couple of my own contributions (Samuel 2014, 2015a). Here I hope to take the argument further, but begin with a brief summary of points I made in the two earlier pieces, in particular to emphasize the extent to which MBSR and related techniques are already quite a long way distant from anything that might be labeled as 'mindfulness' in pre-modern Buddhist contexts.

Firstly, the core techniques of MBSR were derived from recent developments in Buddhist practice. They were drawn for the most part from

traditions of lay Buddhist meditation developed in the Theravāda Buddhist countries of Burma and Thailand in the early to mid-twentieth century. The specific techniques on which MBSR drew, particularly the Vipassanā practices associated with U Ba Khin, S.N. Goenka, Ajahn Cha, and others, had already been taught widely in Western Buddhist contexts, in North America and elsewhere, from the 1960s and 1970s onward. These Southeast Asian and North American approaches already represented what can be called a Buddhist modernism, or modernist Buddhism, particularly in their presentation of Buddhism as a philosophy or scientific teaching rather than a religion. Kabat-Zinn and his associates were aware of and influenced by other forms of Buddhism, particularly varieties of Ch'an (Zen) Buddhist meditation, but these too had undergone substantial rethinking in modern terms, both in their native context of Japan and in the West.

This 'modernist Buddhist' background provided the basis for a further modernization and secularization of Buddhism in the form of MBSR itself and the various associated and derivative techniques, such as Mindfulness-Based Cognitive Therapy (MBCT) developed in the UK by Mark Williams and others. Here I am using the term 'secularization' to refer to the removal of explicitly religious, spiritual and ethical content and to the presentation of the practices as 'scientific' and as explainable in materialist terms.

Admittedly ambivalence persists regarding how far MBSR and MBCT should still in some

This chapter is a revised version of my presentation at the conference, 'Mindfulness and Compassion: The Art and Science of Contemplative Practice,' San Francisco State University, June 3–7, 2015.

G. Samuel (✉)
School of Languages and Cultures, University of Sydney, Sydney 2006, NSW, Australia
e-mail: SamuelG@cardiff.ac.uk

© Springer International Publishing Switzerland 2016
R.E. Purser et al. (eds.), *Handbook of Mindfulness*,
Mindfulness in Behavioral Health, DOI 10.1007/978-3-319-44019-4_4

sense be seen as 'Buddhist.' Jon Kabat-Zinn himself is a committed Buddhist practitioner, as were many of those involved in developing and teaching the various Mindfulness techniques. Kabat-Zinn's writings tend to play it both ways: Mindfulness is not Buddhism, but it also in some sense represents the essence of Buddhism, the one significant core message of the tradition. However, it is clear that MBSR, MBCT, and most of the derivative techniques have been intended by their originators to be seen by both therapists and patients as non-religious, and indeed to be accessible because of this non-religious character to those who might reject an explicitly 'Buddhist' approach. In addition, while 'mindfulness' is an attractive label, its meaning in these contemporary contexts is significantly different from the various meanings of the Buddhist terms—*sati* in Pali, *smṛti* in Sanskrit, *dran pa* in Tibetan, and so on—which it purports to translate.[1]

The rejection of religion has led to some commentators raising the question of whether the new mindfulness-based approaches represent an authentic version of the Buddhist teachings. This question is worth taking seriously, if only because it alerts us to the distance between the Buddhist origins of these practices and their present form. However, in many ways the business of origins can be a distraction from the perhaps more significant questions of what our present culture and society is doing with these practices, what we are using them for, and what else we might be doing with them.

I think that we can accept that the introduction of the mindfulness-based therapies has, by and large, been a good thing. There are certainly arguments about whether they are as effective as is sometimes claimed, both in terms of the standard and quite restrictive procedures of evidence-based medicine, and in more general terms. In terms of official recognition, perhaps the high point came in 2009 when the UK's National Institute for Clinical Excellence brought

out new guidelines on depression which 'recommended MBCT for people who are currently well but have experienced three or more episodes of depression' (Williams and Kuyken 2012). However, while the mindfulness-based techniques are certainly being taught, and presumably practiced, on a very large scale, there have been few other major breakthroughs of this kind, and it is less than clear how good mindfulness actually is as a therapeutic technique.

For what it is worth, a series of systematic reviews and meta-analyses have so far been unable to demonstrate that mindfulness-based therapies are any better than other available therapies. Some of the problems here are methodological rather than substantive; despite the large number of studies, there are not very many high-quality studies of the right kind on the mindfulness-based therapies to evaluate their usefulness. Willoughby Britton, herself a significant researcher in this field, comments on the latest major systematic review, the Association for Health and Research Quality (AHRQ)'s 2014 report (Goyal et al. 2014), 'This review—and pretty much every one before it—has found that meditation is not any better than any other kind of therapy.' She goes on to say that 'The important thing to understand about the report is that they were looking for active control groups, and they found that only 47 out of over 18,000 studies had them, which is pretty telling: it suggests that there are fewer than 50 high-quality studies on meditation.'[2]

Evidence-based medicine, with its clinical trials, randomized double-blind protocols and meta-analyses, is a major minefield in its own right, and part of the problem is that meditation does not fit well into this frame. Neither, it might be said, do a lot of other things that are also probably quite good for us. Evidence-based medicine is primarily a way of making decisions about resource allocation. It cannot tell what is the best treatment for any individual person. But it should also be said that if the mindfulness-based techniques are no better than the alternatives in

---

[1]For this reason, I generally speak below of 'mindfulness-based' techniques or practices, rather than of 'mindfulness.'

[2]Britton's comments are in a 2014 interview with *Tricycle* (Britton 2014). For an earlier report, see Chiesa and Serretti (2010).

terms of their ability to meet specific therapeutic goals, they may at least be less harmful. The mindfulness-based techniques have helped to legitimate a substantial reduction in the massive default prescribing of psycho-active drugs for psychiatric illness. They have probably enabled significant numbers of people who might have become dependent on such drugs to gain some autonomy, agency, and control over their situation. If so, we should surely welcome these developments. The mindfulness-based techniques are also, I would suggest, of value in that they have allowed psychiatry, clinical psychology, and related disciplines to start taking consciousness and its role in human function seriously once more, and in a new way. I shall say some more about this later on.

The discussion which follows in this chapter takes the above points more or less for granted. To sketch the argument I am going to present: I will begin by emphasizing that, while there remains a tendency for the MBSR model to be taken as the default approach, and even to be regarded as a kind of panacea, it is far from appropriate for all situations and all people. In the following sections of my argument I shall note that we have in the Buddhist tradition, and in the various other Asian and alternative traditions of knowledge and healing, a vast range of potential therapeutic techniques. Importantly, these do not all operate in the same way, nor are they expected to have the same effects. We need to recognize—and I think that we are in fact beginning to recognize—the variety of resources that are available to us, and we urgently need to start building a fuller awareness of that range of resources.

It is not just a question of what techniques are available, however, but of how we use them. This brings up the question of the social context of mindfulness practice, and I shall discuss this briefly in the following section of the chapter. In the final section, I move to look more directly at the spiritual, religious, or transpersonal aspects of the traditions from which the contemporary practice of mindfulness derives. What did this mean for the practice of those traditions in their non-Western and pre-modern contexts? How might the presence or absence of these spiritual

or transpersonal aspects affect the practice of mindfulness-based techniques in the contemporary world? What sense can we make of these aspects within contemporary scientific contexts? More generally, if the growth of these techniques in their new global context offers us an unprecedented opportunity, and I believe that it does, then how do we grasp that opportunity in order to respond to the equally unprecedented problems that we, the human population living on this finite planet, will have to face over the years to come?

Mindfulness as Panacea and the Range of Approaches

I start, as I said, with the question of 'mindfulness as a panacea.' If MBSR encodes the essential message of the Buddhist teachings, as Kabat-Zinn and others have told us, then it should be good for everybody, since that message is of universal applicability. At least, this seems to have been the general orientation of many of the early proponents of mindfulness, particularly of MBSR, which in any case has a pretty generic remit. There are few of us who could not use a bit of stress reduction.

The problem here is not that the techniques do not have effects on a very wide range of people. It is rather that they do, and these effects are not necessarily desirable or easy to deal with. As Willoughby Britton among others has pointed out, a serious engagement with meditative practice is fully capable of leading to a major psychological and existential crisis. In fact, such episodes are a frequent and usual part of the Buddhist path.[3] While, for some of the reasons already noted, MBSR and MBCT are not the same things as a long-term engagement with Buddhist meditational practice, they can still have very real and major effects. Arguably the lack of a spiritual framework to give meaning

---

[3]See her 'Dark Night Project,' discussed in a 2014 *Atlantic* article, 'Dark Knight of the Soul' (Rocha 2014). See also http://www.buddhistgeeks.com/2011/09/bg-231-the-dark-side-of-dharma/ and its sequel, two podcasts in which she discusses the finding of the project further.

and structure to such experiences, or of a teacher or therapist capable of handling a major existential crisis, may leave the patient in a very vulnerable and dangerous situation. In fact, as Britton has noted, the vast majority of clinical studies of the mindfulness-based therapies do not even measure negative effects: after all, showing that MBCT has serious side effects 'is not really going to get you funding.'[4]

My own doubts about the universal applicability of the MBSR-derived therapies initially came from a research project at Cardiff University with people on the autism spectrum. The autism spectrum, and even the sub-category of Aspergers, which has dropped out of the latest iterations of the Bible of psychiatry, the American Psychiatric Association's *Diagnostic and Statistical Manual of Mental Disorders* (American Psychiatric Association 2013), includes quite a range of individuals with different kinds of personal issues. However, a key issue for many of these people is the overwhelming nature of immediate everyday experience, and the need to deal with the consequent sensory overload both by periodic withdrawal and by the building of complex ordered structures that make it possible to see reality as regular and predictable, and so to inhabit it without experiencing unbearable levels of anxiety and threat. People at the high-functioning end of the autism spectrum have generally established reasonably effective personal routines for dealing with sensory experience. Many of these are aimed at creating structure, order, and predictability in their lives; these are the kind of people who, stereotypically, know the bus or railway timetables by heart. In some cases, they have achieved very high levels of ability in mathematics, information technology, theoretical physics, and related areas. Thus, there are clearly positive abilities associated with the high-functioning autism group, many of whom used to be classed as having Aspergers, and the language preferred in this area nowadays

speaks of 'neurodiversity' rather than of deficits from some supposed normal condition.

A key process in MBSR is the encouraging of direct and unmediated awareness of the present, and one can see why this might be particularly threatening and difficult for people on the autism spectrum. In fact, word among the 'Aspie community' is quite divided on the subject of the mindfulness-based techniques. The Aspie community consists of people who would now generally be classed at the high-functioning end of the autism spectrum, and who are well represented on the Internet—in fact, for many of them the Internet has been a major boon, since it allows a much more controlled and unthreatening form of interaction than face-to-face speech. To quote one comment on the Aspie forum www. wrongplanet.net,

> Mindfulness DOES NOT help me. It makes things worse; actually: I already have problems with not being able to shut sensory stimuli out, and I'm supposed to pay MORE attention?[5]

Some of the problems here may be a question of the need to modify the standard mindfulness protocols so as to be more appropriate to the specific situation of people on the autism spectrum, and there have been moves in this direction, for example, by Annelies Spek in the Netherlands.[6]

> Specifically, the MBCT protocol of Segal et al. (2002) was used, but because of the information processing deficits that characterize autism, the cognitive elements were omitted. For example, exercises examining the content of ones thoughts were omitted. Also, the information processing deficits that characterize autism were taken into account. For example, because individuals with autism have the tendency to take language literally, the use of metaphors was avoided. Further, words or sentences that are ambiguous or that require imagination skills were avoided. In addition, the eight-week protocol was extended by one week, due to the relatively slow information processing in adults with ASD... For the same reason, the

---

[4]Britton made this comment at the conference, 'Mindfulness and Compassion: The Art and Science of Contemplative Practice,' San Francisco State University, June 3–7, 2015.

[5]http://www.wrongplanet.net/forums/viewtopic.php?f= 3&t=251820&start=0. She continues, 'The therapist who suggested that and CBT can go f**k herself.' Similar comments have been made to me by other people who have been diagnosed as on the autism spectrum.

[6]See also Mitchell (2009, 2013).

three-minute breathing exercise was changed into a five-minute breathing exercise (Spek et al. 2013: 249).

Spek's intentions here are surely positive, but a lot of people diagnosed as on the autism spectrum, or who have been involved with people on the spectrum, might be unhappy about the phraseology of deficit. What we seem to be getting is a watered-down version of MBCT with the bits that Aspies might find difficult taken out. However, it may also be that MBCT is really not the best technique for much of this community. We might be better off looking for approaches that work positively with the abilities of people on the spectrum.

I am not myself involved in providing therapy, so I cannot speak with authority here, but I suggested some years ago that Buddhist practices such as those of Tibetan deity yoga, with their highly ordered mandala structures, might provide an easier and more natural way for some of these people to move toward higher levels of awareness and self-control (Samuel 2008). To put it in rather basic terms, the array of deities of which the Tibetan mandala consists provides a very clearly structured, highly intelligible community of positive forces, with a complex philosophical underpinning and an elaborate liturgical structure, through which experience can be filtered and sorted and by which one can gradually come to terms with the complexity of reality. Rather than forcing a direct confrontation with sensory overload, they provide a structure through which that overload might be gradually tamed and integrated. However, we are here a long way away from the secularized, simplified, and scientific world of MBSR and MBCT.

Elsewhere I have heard suggestions that meditation practices based on loving-kindness (Paili *metta*) were more suitable for the autism community than the standard mindfulness approaches. That would also make sense, and I shall have a little more to say about these practices later. I suspect however that much depends on individuals, on the particular form of their 'neurodiversity' and the specific techniques they

have acquired in order to deal with it. But that is really the point I would like to make more generally. An important corollary to moving away from seeing MBSR as a panacea is to start looking at a much wider range of approaches, learning what they can do, and building training programs for therapists that equip them with a range of different techniques, and alert them to the positive and negative aspects of each approach.

The Range of New Consciousness Techniques

In one respect, this is quite straightforward. If MBSR derives from one specific modernist Buddhist tradition, then why not look at the wider range of forms of practice within Theravāda Buddhism, East Asian Buddhism, Tibetan Buddhism, and so on, or for that matter within other Asian religious contexts, such as Hinduism or Daoism? Many of these practices are already well-established within Western societies, in some cases going back more than a century. In many cases, there is also a substantial body of research concerning the physiological and/or psychological effects on practitioners. In some ways, this is precisely what I am suggesting.

However, there are some catches here. One is that, while MBSR went through a self-conscious and deliberate process of 'secularization' and was in any case based on a form of Buddhist practice that had already been to a significant degree stripped of its religious context, many of these other practices are much more explicitly religious. Do we try to secularize them? Can we? Is it even appropriate to do so?

To go back to my suggestion of Tibetan deity yoga and mandala visualization for people at the high-functioning end of the autism spectrum, would this practice retain any meaning if one left out the deities, the progressive approach to the deities through liturgy and mantra, and the question of the Buddhist awakening or Enlightenment which the deities encode and represent? But is the Buddhist awakening what we are really aiming at here?[7] Are we trying to achieve enlightenment, or healing? This is not a difficulty in Tibetan culture, where there are indeed many deity yoga practices which aim at healing of various kinds, but then for Tibetans it is taken for granted that the Buddha and the Buddhist awakening underlies and empowers the entire structure of therapeutic applications.[8] Within Western therapeutic contexts, this assumption is likely to be quite alien and even unwelcome to a significant number of therapists and patients.

So far, as far as I know, there have been no attempts at the therapeutic use of secularized versions of the deity and mandala process in the Western medical context, though the idea is not implausible. There are certainly analogies to other kinds of therapeutic process on which one might build here, as with the well-known Jungian use of mandala structures (Jung 1968; Slegelis 1987). We might though start by looking at some of the practices that have been developed for therapeutic purposes in recent years but involve a less drastic departure from the assumptions of secular scientific thought.

Loving-Kindness, Compassion, and Mindful Self-Compassion Meditations

Loving-kindness meditation (LKM) and Compassion Meditation (CM) are terms which have been devised for forms of meditation derived from the well-known canonical set of Buddhist meditations known as the *brahmavihāras*. In fact, these are not exclusively Buddhist practices, but seem to have been common to the various early ascetic traditions. Loving-kindness (*metta* in Pali) is the wish that all sentient beings be happy; compassion (*karuṇā*) is the wish for all sentient beings to be free from suffering. Here 'sentient beings' is Buddhist terminology for all beings possessing consciousness, a class that includes animals and several classes of spirit-beings as well as humans. Both loving-kindness and compassion are practices that can be seen as fostering and encouraging social connection.[9] One advantage of these practices is that while they have rather more analytical content than MBSR in its original form, what the patient or practitioner is asked to do is fairly easy for many people to take on without feeling that they are being drawn into a complex web of religious assumptions.

There have been a number of attempts to develop clinical interventions based on these Buddhist practices (see Hofmann et al. 2011; Shonin et al. 2015 for two recent reviews). Most attention has been paid so far to LKM, which is derived from *metta* (loving-kindness is one of the standard translations of *metta*). LKM has been used both as a stand-alone procedure and as an adjunct to MBSR and related techniques.

> This practice, in which one directs compassion and wishes for well-being toward real or imagined others, is designed to create changes in emotion, motivation, and behavior in order to promote positive feelings and kindness toward the self and others (Hutcherson et al. 2008: 720).

An initial stage of *metta* and *karuṇā* practice as described in Buddhist texts and in modern practice is developing kindness or compassion toward the self, and the Mindful Self-Compassion (MSC) program developed by Kristin Neff and Christopher Germer builds these stages into a systematic training:

> In this program…, participants meet for 2.5 h once a week for 8 weeks, and also attend a half-day silent meditation retreat. The MSC program teaches a variety of meditations (e.g., loving-kindness, affectionate breathing) and informal practices for use in daily life (e.g., soothing touch, self-compassionate

---

[7]It might also be worth looking at Navajo sand-painting, where mandala-type structures are being explicitly deployed for healing. Here too though the religious identity of the forces involved would appear to be essential (cf. Samuels 1995).

[8]See, e.g., Samuel (2013, 2016 in press), on the longevity practices.

[9]Whether this was their intention in their original context is perhaps another question. In the Visuddhimagga, *metta* and *karuṇā* appear to be introduced primarily as techniques for entering the trance-states (Pali *jhāna*) or meditative absorptions (Buddhaghosa 2010: 291–302; for the *jhāna* see, e.g., Cousins 1973). However, there is no doubt that the socially positive effects of these practices have been recognized and appreciated by Buddhist practitioners over the centuries.

letter writing). Self-compassion is evoked during the classes using experiential exercises, and home practices are taught to help participants develop the habit of self-compassion. Participants are encouraged to practice these techniques for a total of 40 min per day, either in formal sitting meditation or informally throughout the day. (Germer and Neff 2013: 859)

Practices such as LKM and MSC can be presented in a relatively secularized form without major difficulties. Other compassion-oriented practices, such as the Cognitively Based Compassion Training (CBCT) developed at Emory University in 2004 with the participation of the Tibetan teacher Geshe Lobsang Tenzin Negi, and the Compassion Cultivation Training subsequently developed at Stanford University in conjunction with another Tibetan teacher, Geshe Thubten Jinpa, remain closer to their religious origins. These practices are both based on a well-known set of Tibetan practices generically referred to as *lojong*, mind-training, or mental purification. More specifically, they include elements from the two classic *lojong* texts, by the twelfth-century Tibetan teachers Langri Tangpa, author of the *Eight Stanza Mental Purification* text, and Chekawa, author of the *Seven Topic Mental Purification* text (Sweet 1996).

CBCT is currently taught through six 'curriculum modules' (Aspen Centre for Living Peace 2016). The first two are aimed at 'stabilizing attention and developing present-moment awareness'; the remaining four modules 'use analytical practices to increase well-being and unbiased compassion toward others.' The individual modules are described as follows:

Module I: Developing Attentional Stability and Clarity
This initial practice trains attentional stability in order to improve mental stability and clarity; typically this is done by placing and retaining focus on the unfolding sensations of the breath and by learning to notice and release distractions as they arise.

Module II: Cultivating Insight into the Nature of Mental Experience
Still rooted in the present moment, the focus shifts to how mental experiences unfold from moment to moment, neither pushing away such experiences or becoming overly involved in them. This practice improves calmness of mind and provides insight into habitual mental patterns.

Module III: Self-compassion
Using insights from Module II, this self-care practice examines the basic nature of distress and dissatisfaction and cultivates more realistic and positive approaches to difficult life circumstances. When done with kindness, these practices strengthen the determination to replace unhelpful attitudes with constructive ones, leading to realistic optimism and greater self-determination.

Module IV: Cultivating Impartiality
As humans are innately social creatures, relationships are central to well-being. This practice examines habitual ways of thinking about others. Seeing that all people, despite apparent differences, shares a fundamental desire to seek well-being and to avoid distress and dissatisfaction, this practice leads to a greater capacity to see others as similar to one's self on the most basic level, opening the door to a more inclusive compassion.

Module V: Appreciation and Affection for Others
By examining how all things that are beneficial depend upon others, this module cultivates an appreciation for this basic kindness, intended or unintended. Considering the drawbacks of an unrealistic attitude of independence and isolation, the practitioner reflects on the daily and long-term gifts of the broader society, and a deepening affection is cultivated for others.

Module VI: Empathy and Engaged Compassion
With the perspectives of seeing each person as equally deserving of happiness and as having great value in their own right, practitioners focus on the difficulties and distress experienced by so many, which naturally invokes an empathic response. When supported by the inner strength developed in earlier modules, this empathy leads to the strong wish to see others free of difficulties and distress and to orient one's core motivation toward the alleviation of the suffering of others. (Aspen Centre for Living Peace 2016)

As can be seen, there is much more emphasis on analytic thought in this set of practices than in practices such as LKM or MSC: 'The goal of CBCT is to challenge unexamined assumptions regarding feelings and actions toward others, with a focus on generating spontaneous empathy and compassion for the self as well as others.' (Pace et al. 2013: 294).

What is also evident to anyone familiar with the Tibetan *lojong* practices is that the arguments employed are closely modeled on those in *lojong*. An article on CBCT from 2012 explains that it incorporates elements from two different 'styles of practice' used in Tibetan *lojong*, the 'seven-limb cause and effect' method, 'which involves principally generating a strong sense of gratitude toward one's mother or another loved one by reflecting upon their kindness (…), cultivating that into love and compassion, and then gradually extending that love and compassion toward others' and the method of 'equalizing and exchanging oneself and others,' by reflecting on how we are all fundamentally the same in wishing for happiness and wishing to be free from suffering; and how oneself and others are equal in deserving happiness and to be free from suffering. One further reflects on the disadvantages of an excessive self-centered view and the benefits of a view that recognizes interdependence and our need for others and then 'exchanges' one's self-cherishing for other-cherishing (…) (Ozawa-de Silva et al. 2012: 155).

When asked how CBCT differed from *lojong*, Geshe Lobsang Tenzin responded,

> It's not different. It's based on *lojong*, except that it excludes Buddhist beliefs like reincarnation and things like that. It is secular so you can practice it without having to learn in the context of the belief of rebirth and previous and future lives. Wanting to be happy and free are universal aspirations. (Hawthorn 2013)

While at one level the Geshe is right in noting that the specifically Buddhist concept of rebirth has been omitted, much of the analytic content reflects Tibetan approaches closely, including the strong emphasis on the kindness of one's mother, and the importance of rejecting involvement in the so-called negative emotions, which in fact include all forms of attachment. As Brendan Ozawa-de Silva, one of the main teachers and researchers in this system, has pointed out (Ozawa-de Silva 2015c),[10] this rejection, which

is styled 'self-compassion' in CBCT (Module 3 in the program structure presented above), corresponds to the Tibetan *ngejung*, the renouncement of *samsaric* life, which is a radical step for anyone engaged in everyday secular life in a modern Western society. In Tibetan terms, *ngejung* marks a serious commitment to something that could be called a spiritual life, and normally coincides with entry to a monastery or other full-time spiritual center. *Ngejung* is also closely linked to the whole idea of the cycle of rebirth. Normatively, in Tibet, it corresponds to the point at which someone has realized the pointlessness of continual rebirth and has decided to take on the Buddhist path, which will lead, normally in some future rebirth, to the attainment of Buddhahood and so transcendence of the cycle of rebirth. It is precisely the cycle of rebirth, associated with everyday secular life, that is being renounced.

Ozawa-de Silva noted that the parallel Stanford program, Compassion Cultivation Training or CCT, developed by Geshe Thubten Jinpa, tones down this step.[11] While it also includes a step called 'self-compassion,' this is understood more in the sense of developing loving-kindness and compassion toward oneself, rather as is done in LKM and MSC, and not as a full-on renunciation of emotional involvement with others. The two programs cooperate with each other, but it seems that the less demanding Stanford program, which makes more concessions to Western modes of thought and feeling, has so far been more successful in training teachers and establishing itself. CBCT may omit explicit reference to rebirth or the Tantric deities, but it would seem that it is still pushing the limits for an intervention that might be taken up on a large scale within the Western secular context. In part, this is because it relies less on imagery or the direct training of emotion, and more on analytical reasoning, so that the areas of conflict with standard Western modes of thought become more explicit.

---

[10]In discussion following his paper (Ozawa-de Silva 2015c) at the 5th Annual Tung Lin Kok Yuen Canada Foundation Conference ('Buddhism and Wellbeing:

(Footnote 10 continued)
Therapeutic Approaches to Human Flourishing'), University of British Columbia, May 28–30, 2015.
[11]http://ccare.stanford.edu/

Another relatively 'analytical' approach focusing on the self-other relation is the Japanese Naikan therapy, which has been studied by Brendan and Chikako Ozawa-de Silva at Emory, and also by Clark Chilson at the University of Pittsburgh (Ozawa-de Silva 2007, 2015a, b; Ozawa-de Silva and Ozawa-de Silva 2010; Chilson 2015). Naikan (the name literally means 'looking inside,' self-examination) was developed in Japan in the 1940s, again as a somewhat toned-down and secularized version of a self-examination practice (*mishirabe*) within Shin Buddhism. *Mishirabe* involves fasting and sleep deprivation, but this element was removed in Naikan. Naikan was originally introduced in prisons, but has been developed further and is now seen as applicable for alcoholism, addictions of other kinds, and a variety of other disorders, including eating disorders, mild depression, anxiety disorders, compulsive neurosis, marriage and relationship problems, and individuals interested in self-exploration or self-discovery (Ozawa-de Silva 2007: 414).

> In a typical weeklong Naikan session, clients stay at the center, which is usually the house of the Naikan practitioner. Each day, from early morning until night, they review their lives from the perspective of a significant person in their lives, usually starting and ending the week with their mothers. This is punctuated every two hours by a *mensetsu*, or interview, during which they report on their self-examination thus far to the practitioner. The method of recollection is clearly delineated: clients must keep to Naikan's "three themes": (1) what the client received from that person, (2) what he or she gave back to that person, and (3) what trouble he or she caused that person (Ozawa-de Silva 2007: 414).

In recent years, Naikan practice has been made available to some degree in Western countries. The scale is relatively small, however, and as can be seen Naikan is dependent on a period of isolation and sensory deprivation that might be difficult to replicate in a clinically accessible format such as the eight or ten 150-min group sessions typical of MBSR and similar interventions. The focus on introspection and regret for past wrongdoing might also not go down well with many Western clients.

The various practices I have been discussing suggest both some of the possibilities and some of the complexities of introducing further techniques and approaches into the Western context. They are all quite different from the mindfulness-based approaches that derive from MBSR, although as I noted elements of *metta* or LKM have been incorporated into some of the MBSR-derived practices.

A more extensive study might consider a variety of other practices in some detail, but here I shall confine myself to listing a few of the other possibilities that we might consider:

(1) Focus and concentration practices, using the breath or other concentration focuses: Thai and Burmese *samatha*, Tibetan *shiné* (Singh et al. 2016; Maclean et al. 2010).

(2) Open awareness practices such as Ch'an/Zen, Dzogchen, or Mahamudra—again, classically practiced in strongly ritualized contexts and frames and/or preceded by substantial preparatory practice (Mruk and Hartzell 2003; Harrison 2006; Rosch 2007; Amihai and Kozhevnikov 2015).

(3) Mantra practices, ranging from the simple and basic mantra practice of Transcendental Meditation to the much more complex forms used in conjunction with deity yoga in Tibetan practices (Rutledge et al. 2014; Amihai and Kozhevnikov 2015).

(4) 'Subtle body practices,' such as *kriya yoga*, or some forms of Tibetan *trulkor* or Chinese *qigong*. These involve internal visualizations of and exercises with internal 'flows' within the body, often combined with breathing and/or with simple movements (cf. Samuel and Johnston 2013; see also Saraswati 2006; Zope and Zope 2013; Chaoul 2013).

(5) Movement-related techniques (including those which use breathing in more complex ways than simple awareness, and again often involving subtle body concepts)—e.g., more complex forms of Tibetan *trulkor* ('yantra yoga'), *qigong* or *taijiquan* (Tai Chi Chuan), aspects of martial arts training (Lu and Kuo 2014; Wei et al. 2014; Li et al. 2002).

I have deliberately included examples of Hindu and Daoist-associated practices as well as Buddhist. In fact, there has been substantial research on both Hindu yogic practices and Daoist practices such as *qigong*, going back many years, and much of this has been sidelined by the current vogue for mindfulness practice. This is something of a political issue within the research community. Rather than focusing on differences that are primarily significant for advanced religious practitioners, we should be looking across the full range of material available to us.

An important issue here is that, even if we confine ourselves to the Buddhist practices, we cannot and should not assume that these practices are doing the same thing. Here I think the word 'mindfulness,' as deployed by the originators of the main mindfulness-based therapies, has exerted a certain seductive and misleading power. Who, after all, can be opposed to mindfulness? But mindfulness as one way of phrasing the ultimate goal of the teachings is one thing, while mindfulness as a label for a specific set of techniques, such as MBSR and MBCT, is another (cf. Chiesa and Malinowski 2011). It is very easy for the two to become conflated.

One of the few studies that has looked at different cognitive and physiological mechanisms across a range of meditation traditions is by Ido Amihai and Maria Kozhevnikov (Amihai and Kozhevnikov 2014, 2015). This was a comparative study of four modes of meditation practice: *samatha* and *vipassana* on the Theravāda side, Deity yoga and Dzogchen (*rigpa*) on the Tibetan. Amihai and Kozhevnikov report that while both Theravāda practices enhance the activity of the parasympathetic nervous system, indicative of relaxation, the Tibetan practices lead to activation of the sympathetic system, indicative of arousal.

This effect overrides the opposition one might expect to find between the practices involving more focussed styles of awareness (*samatha*, Deity yoga) and those involving more distributed styles (*vipassana*, Dzogchen). It is however consistent with what the Tibetan tradition itself implies:

the conceptualization of meditation as a relaxation response seems to be incongruent with Tibetan views of Vajrayana Tantric practices, which do not presuppose relaxation. Indeed, Vajrayana "generation stage practices, such as "visualization of self-generation-as-Deity", which are to precede the "completion stage"" practices pertaining to realization of emptiness (Rig-pa) are aimed at achieving a wakeful state of enhanced cognition and emotions through the use of visual imagery and the emotional arousal associated with it. (Amihai and Kozhevnikov 2014: 2)

They propose that 'Vajrayana and Theravada styles of meditation are correlated with different neurophysiological substrates' (Amihai and Kozhevnikov 2014: 14). This is of interest, among other reasons, because relatively little research has been done on Tibetan modes of practice.

Another of Kozhevnikov's papers provides a detailed account of the neurocognitive and somatic components of *gtum-mo* meditation, the well-known Tibetan practice that generates heat within the body (Kozhevnikov et al. 2013). Here again we are dealing with something that is certainly a meditative or yogic practice, but is clearly very different in both mechanism and purpose from the familiar Theravadin practices from which MBSR and MBCT were derived.

The Social Context of the New Consciousness Techniques

I have suggested in the last section that we might look at a much wider range of techniques as possibly useful in contemporary society. From this point of view, MBSR, MBCT, and the related techniques represent a significant opening, and I would not want to put too much stress on their limitations. Their true significance may be less in what they are able to achieve in their present versions than in their role in legitimating procedures that work with human consciousness within central areas of contemporary biomedicine.

From another point of view, that of the social context of mindfulness practices, the situation is less promising. A number of recent critiques have suggested that mindfulness practices in their present form can easily be seen as little more than an 'ethically neutral performance enhancement techniques' (Purser and Milillo 2014: 9). Thus, corporate executives can learn to be better at their jobs through mindfulness practices, regardless of the ethical implications of the jobs themselves. Mindfulness practices can also be a way of deflecting discontent and unrest among employees. Thus, Purser and his co-author suggest that:

> corporations have jumped on the mindfulness bandwagon because it conveniently shifts the burden onto the individual employee; stress is framed as a personal problem, and mindfulness-based interventions are offered as means of helping employees cope and work more effectively and calmly within such toxic environments. Cloaked in an aura of care and humanity, this corporate takeover refashions mindfulness as a safety valve, a way to let off steam and as a way of coping and adapting to the stresses and strains of corporate life. What we are left with is an atomized and highly privatized version of mindfulness training […]. Mindfulness training has wide appeal because it can be utilized as a method for subduing employee unrest, promoting a tacit acceptance of the corporate status quo, and as an instrumental tool for keeping attention focused on corporate goals. (Purser and Milillo 2014: 16)

There is something profoundly uncomfortable about the adoption of the mindfulness techniques by large corporations, government agencies, and even military units whose policies in other respects are far away ethically from the core concerns of Buddhism with relieving human suffering. It may be possible that, as one proponent of mindfulness workshops for global leaders and corporate management, Michael Chaskalson, has suggested, that as these people practice mindfulness, they will be become more considerate, more ethical, and more responsible.[12] This is what Purser and Milillo call the 'Trojan Horse' argument; Monsanto and similar corporations will become better global citizens as a result of institutionalizing such individualized forms of mindfulness (Purser and Milillo 2014: 16). It would seem equally or more likely, however, that individualized, de-ethicized, and de-contextualized forms of mindfulness will simply facilitate a more efficient version of business as usual. The promotion of mindfulness-based therapies lower down in the structure can also be easily seen, as Purser and Milillo suggest, in instrumental terms. Their proposed solution is to return to the Buddhist roots of mindfulness, which have 'the potential of calling into question economic materialism […], acquisitiveness and unbridled consumption' (Purser and Milillo 2014: 18).

If we move from the practice of the mindfulness therapies to the research that provides the legitimation for it, we find that many of the same issues turn up again. I suggested in a recent article that the empiricist and individualistic assumptions of much neuroscience and cognitive science allow research in these fields to reinforce the impoverished and decontextualized view of the mindfulness-based techniques (Samuel 2014). I referred there to the British sociologist Steven Rose's comments in a paper called 'The Need for a Critical Neuroscience' (Rose 2012):

> The truth is that in order to approach consciousness as a neuroscientist, one first has to strip the term of any of its richer meanings. […] Consciousness in this neuroscientific sense has been taken out of history and culture; there is no possibility of understanding the extraordinary transitions in consciousness that have occurred through, for instance, the emergence of the women's movement in the 1970s. Instead, consciousness is simply what happens when you are awake, the obverse of being asleep. […T]he essential human meanings embedded in our being conscious have somehow been lost in this reduction. (Rose 2012: 58–59)

Rose suggests a move to an ecological and 'enactive' view of consciousness, in which it is seen as part of an ongoing process in which both 'world' and 'mind' are constituted through mutual interaction. Such an approach, modeled in part on the work of scholars such as Humberto Maturana and Francisco Varela (Maturana and Varela 1980; Varela and Depraz 2003), allows for a variety of different kinds of 'consciousness,' which do not have to be understood simply as a by-product of neuronal activity within the brain, and which can indeed both reflect the wider social world within which we live and generate a critical perspective on how it functions.

---

[12]In a talk at the Transpersonal Psychology section meeting of the British Psychological Society, University of Northampton, September 2014.

The Religious and Transpersonal Context of the New Consciousness Techniques

Rose's critique of science, like the autopoetic model of Maturana and Varela, is just about compatible with an expanded materialism, but it strains a materialist account to its limits. This suggests a need to confront the whole question of the religious and transpersonal dimensions within meditation and related techniques. Buddhism and other Asian traditions we are discussing clearly work with understandings of the world that go beyond the materialist reduction of consciousness to an epiphenomenon of neural and physiological processes within individual bodies.

Buddhism itself does not arise out of a materialist perspective, and several of its core assumptions, for example regarding consciousness and its continuity between lives, are clearly incompatible with any such perspective. Elsewhere I have pointed out the difficulties that this raises for contemporary 'dialogs' between Buddhism and science, such as those within the Mind and Life Institute presided over by His Holiness the Dalai Lama (Samuel 2014). There are a couple of ways in which one can respond to this conflict of worldviews, which of course does not only occur on the boundary between Buddhism and science, but on any number of similar boundaries between pre-modern and contemporary understandings of the universe. One is to insist on maintaining the materialist perspective at all costs and to erect a hermetic *cordon sanitaire* between scientific and other models of the universe. The other is to allow for a mutual interaction and cross-fertilization between scientific and non-scientific models, which might generate new understandings that draw from both sources. As someone who has been working for some years on modes of healing in Tibetan and other societies that seem to be efficacious, in ways that fit poorly into the canons of evidence-based medicine but are nevertheless real enough for those who practice them, I find myself increasingly drawn to the second option.

In support of this option, it is worth considering the nature of Western science a little more closely. To begin with, Western science is not in fact a unitary, mutually consistent body of ideas that has effectively and finally established the validity of the materialist model. It is a collection of separate fields and projects, often based on mutually contradictory assumptions. Science does not deal in facts, despite the tendency of popular magazines and other mass media to present it as if it does, but in theories and theoretical frameworks that always retain a fundamentally hypothetical character. Any conviction that science has finally and irrevocably established the truth of a thorough-going and exclusive materialism is bound to go far beyond the available evidence.

In addition, many areas of science are deeply invested by large-scale financial interests. Evidence-based medicine and its relationship to the pharmaceutical industry is surely a classical example (Healy 2012; Goldacre 2013). This is something of which we are all increasingly aware, hence the frequent requirements for disclosure of relevant financial and other interests in relation to such areas, but it is nevertheless difficult to appreciate the massive scale on which it can affect and distort scientific results. This is not simply a question of the results of individual studies, but the wider issue of the procedures by which studies are carried out, the fields which receive funding, the scholars who are endorsed, rewarded, and employed, and the areas of knowledge which are seen as legitimate. Even where we can see this in relation to fields of which we have some direct knowledge, we may fail to see it in other areas of which we know less.

There is substantial evidence for a more complex relationship between consciousness and material reality than the default versions accepted by workers in fields such as neuroscience and biomedicine. The complex interrelationship between individual, society, culture, and environment alluded to above already points in this direction, as is particularly evident in areas such as healing (see, e.g., Samuel 2015b). Then there are the bodies of experimental evidence that are routinely ignored or marginalized, such as those supporting the continuity of consciousness outside a material basis (see, e.g., Kelly et al. 2009). We might also note theoretical assumptions, such as non-locality in quantum mechanics, that are both central to important areas of contemporary science but incompatible with conventional models of consciousness as a simple epiphenomenon of a material reality that can be fully explained within its own terms.

All these suggest that a degree of modesty might be appropriate in relation to the strident claims that Western science is the fount of all valid knowledge and that it has established the truth of a fully materialist account of the universe. Whether we like it or not, we are all living in an increasingly pluralist world, in which the dominance of specifically Western understandings cannot be guaranteed. All of us, wherever we might be positioned in relation to particular knowledge-claims, might be better advised to accept the limited nature of our partial understandings of the universe.

If, then, we find ourselves employing therapeutic techniques whose underlying postulates conflict with the default assumptions of Western science, it may be that rather than struggling to elide and remove the offending aspects, we might be better off to seek to live with them, and see where that takes us, to 'stay with the open question, and not to seek for resolution or an answer,' as the late Francisco Varela put it in an interview not long before his death (Samuel 2014: 566).

Perhaps I can lay the rest of my cards on the table at this stage by saying that I feel that for all of its massive scale, the 'Mindfulness movement' is something which needs to be seen in a far wider context, the ongoing transformation of Western, and indeed global consciousness as humanity comes to terms with a rapidly changing world, changing both in terms of human populations and in terms of the planet on which we live. The connections between human beings are changing very rapidly in nature and intensity. New technologies are continually changing our sense of who we are and how we relate to each other.

The limits placed upon us by the planetary ecology are beginning to impact on us in ways that we can no longer ignore and push aside. Populations that have previously been relatively insulated from each other by massive obstacles

of distance and difficulty of travel are impacting directly upon each other, as tens of millions of people seek to move from impoverished and politically unstable areas of the world to those where conditions are more tolerable. Reactive nationalisms seek to keep these populations out, causing human disasters on an increasing scale, as with the tens of thousands of people dying while attempting to cross the Mediterranean from Africa to Europe, or to travel from Southeast Asia to Australia. All these problems are likely to get far worse before they improve, and this implies that we are going to have to learn to live together in ways very different from those that served us in the past. We need to develop modes of knowledge that can aid and facilitate these new learnings.

I think that a sense of that wider context is helpful in looking at the whole question of the Mindfulness movement and what it has to teach us. If we regard mindfulness techniques merely as a way of solving our individual problems, we are likely to find that the collective problems we are ignoring will rapidly undercut our individual solutions. If we see the mindfulness techniques more as an opportunity to recognize the presence and significance of collective and social processes within the sphere of consciousness, and to explore their relationship to the world of material reality, we may find ourselves better equipped to create the new forms of knowledge and the new ways of working that might give the human population of this planet a chance of surviving the next fifty or a hundred years.

To put this in slightly different terms: I do not myself see Buddhism as providing a comprehensive answer for how to live in what it describes as the world of *samsara*, of eternally circling around driven by desire, repulsion, and ignorance. In fact, it has never claimed to do so, since its fundamental aim has been to motivate us to reject and transcend that world. However, Buddhism, and I would add other non-Western and pre-modern traditions of knowledge, may indeed be a valuable dialog partner in finding as good a solution as we can to live in the everyday world. As long as we insist, however, that these traditions of knowledge be denatured, impoverished, and rewritten so as to be compatible with our current canons of knowledge, we are unlikely to have much in the way of productive dialog. I think myself that this is the challenge, and the promise, of the new consciousness techniques, and why the issues we are dealing with here are of real importance for the future of our world.

References

American Psychiatric Association. (2013). *Diagnostic and statistical manual of mental disorders* (5th ed.). Washington, DC: American Psychiatric Publishing.

Amihai, I., & Kozhevnikov, M. (2014). Arousal vs. relaxation: A comparison of the neurophysiological and cognitive correlates of Vajrayana and Theravada meditative practices. *PLoS ONE, 9*(7), e102990.

Amihai, I., & Kozhevnikov, M. (2015). The influence of Buddhist meditation traditions on the autonomic system and attention. *Biomed Research International*, Article ID 731579.

Aspen Centre for Living Peace. (2016). [On-line advertisement for forthcoming CBCT training by Geshe Lobsang Tensin, March 6–10, 2016.] Retrieved from http://aspenlivingpeace.org/cbct

Britton, W. (2014). Meditation nation. *Tricycle*, April 25, 2014. Retrieved from http://www.tricycle.com/blog/meditation-nation

Buddhaghosa, B. (2010). *The path of purification* (*Visuddhimagga*) (Pali by B. Ñāṇamoli, Trans.) (4th ed.). Kandy: Buddhist Publication Society

Chaoul, A. (2013). Open channels, healing breath: Research on ancient Tibetan yogic practices for people with cancer. In G. Samuel & J. Johnston (Eds.), *Religion and the subtle body in Asia and the West: Between mind and body* (pp. 100–114). London and New York: Routledge.

Chiesa, M., & Malinowski, P. (2011). Mindfulness-based approaches: are they all the same? *Journal of Clinical Psychology, 67*(4), 404–424.

Chiesa, A., & Serretti, A. (2010). A systematic review of neurobiological and clinical features of mindfulness meditations. *Psychological Medicine, 40*, 1239–1252.

Chilson, C. (2015). Contemplation over compulsion: Naikan as an analytical meditation for treating addiction. Paper given at the 5th Annual Tung Lin Kok Yuen Canada Foundation Conference ('Buddhism and Wellbeing: Therapeutic Approaches to Human Flourishing'), University of British Columbia, May 28–30, 2015.

Cousins, L. S. (1973). Buddhist *jhāna*: Its nature and attainment. *Religion, 3*, 115–131.

Germer, C. K., & Neff, K. D. (2013). Self-compassion in clinical practice. *Journal of Clinical Psychology, 69*, 856–867.

Goldacre, B. (2013). *Bad pharma: How drug companies mislead doctors and harm patients*. New York: Faber and Faber.

Goyal, M., Singh, S., Sibinga, E. M. S., Gould, N. F., Rowland-Seymour, A., Sharma, R., et al. (2014). Meditation programs for psychological stress and well-being: A systematic review and meta-analysis. *JAMA Internal Medicine, 174*(3), 357–368.

Harrison, J. R. (2006). Analytic meditative therapy as the inverse of symbol formation and reification. *Journal of Religion and Health, 45*(1), 73–92.

Hawthorn, U. B. (2013). What is cognitively-based compassion training? [Interview with Lobsang Tenzin Negi.] Retrieved from http://psychcentral.com/blog/archives/2013/09/25/what-is-cognitively-based-compassion-training/

Healy, D. (2012). *Pharmageddon*. Berkeley: University of California Press.

Hofmann, S. G., Grossman, P., & Hinton, D. E. (2011). Loving-kindness and compassion meditation: Potential for psychological interventions. *Clinical Psychology Review, 31*, 1126–1132.

Hutcherson, C. A., Seppala, E. M., & Gross, J. J. (2008). Loving-kindness meditation increases social connectedness. *Emotion, 8*(5), 720–724.

Jung, C. G. (1968). Concerning mandala symbolism. In C. G. Jung (Ed.), *Collected Works of C. G. Jung* (Vol. 9(1), 2nd ed., pp. 355–384). Princeton, NJ: Princeton University Press.

Kabat-Zinn, J. (2003). Mindfulness-based interventions in context: Past, present, and future. *Clinical Psychology: Science and Practice, 10*(2), 144–156.

Kelly, E. F., Kelly, E. W., Crabtree, A., Gauld, A., Grosso, M., & Greyson, B. (2009). *Irreducible mind: Toward a psychology for the 21st century*. Lanham: Rowman & Littlefield.

Kozhevnikov, M., Elliott, J., Shephard, J., & Gramann, K. (2013). Neurocognitive and somatic components of temperature increases during g-tummo meditation: Legend and reality. *PLoS ONE, 8*(3), e58244.

Li, M., Chen, K., & Mo, Z. (2002). Use of qigong therapy in the detoxification of heroin addicts. *Alternative Therapies in Health and Medicine, 8*, 50-4 & 56-9.

Lu, W., & Kuo, C. (2014). Breathing frequency-independent effect of Tai Chi Chuan on autonomic modulation. *Clinical Autonomic Research, 24*, 47–52.

MacLean, K. A., Ferrer, E., Aichele, S. R., Bridwell, D. A., Zanesco, A. P., Jacobs, T. L., et al. (2010). Intensive meditation training improves perceptual discrimination and sustained attention. *Psychological Science, 21*, 829–839.

Maturana, H. R., & Varela, F. J. (1980). *Autopoiesis and cognition: The realization of the living*. Dordrecht, Boston and London: D. Reidel.

Mitchell, C. (2009). *Asperger's syndrome and mindfulness: Taking refuge in the Buddha*. London and Philadelphia: Jessica Kingsley.

Mitchell, C. (2013). *Mindful living with Asperger's syndrome: Everyday mindfulness practices to help you tune into the present moment*. London and Philadelphia: Jessica Kingsley.

Mruk, C. J., & Hartzell, J. (2003). *Zen and psychotherapy: Integrating traditional and nontraditional approaches*. New York: Springer.

Ozawa-de Silva, C. (2007). Demystifying Japanese therapy: An analysis of Naikan and the Ajase complex through Buddhist thought. *Ethos, 35*, 411–446.

Ozawa-de Silva, C. (2015a). Mindfulness of the kindness of others: The contemplative practice of Naikan in cultural context. *Transcultural Psychiatry, 52*(4), 524–542.

Ozawa-de Silva, C. (2015b). *Exploring the wider ethical and cognitive resources of Buddhism for psychotherapy: The case of Japanese Naikan Practice and CBCT (cognitively-based compassion training)*. Paper given at the 5th Annual Tung Lin Kok Yuen Canada Foundation Conference ('Buddhism and Wellbeing: Therapeutic Approaches to Human Flourishing'), University of British Columbia, May 28–30, 2015.

Ozawa-de Silva, B. R. (2015c). *Healing through compassion: The implementation of Buddhism-derived compassion training in schools and other contexts*. Paper given at the 5th Annual Tung Lin Kok Yuen Canada Foundation Conference ('Buddhism and Wellbeing: Therapeutic Approaches to Human Flourishing'), University of British Columbia, May 28–30, 2015.

Ozawa-de Silva, C., & Ozawa-de Silva, B. (2010). Secularizing religious practices: A study of subjectivity and existential transformation in Naikan therapy. *Journal of Scientific Study of Religion, 49*, 147–161.

Ozawa-de Silva, B. R., Dodson-Lavelle, B., Raison, C. L., & Negi, L. T. (2012). Scientific and practical approaches to the cultivation of compassion as a foundation for ethical subjectivity and well-being. *Journal of Healthcare, Science and the Humanities, 2*, 145–161.

Pace, T. W. W., Negi, L. T., Dodson-Lavelle, B., Ozawa-de Silva, B., Reddy, S. D., Cole, S. P., et al. (2013). Engagement with cognitively-based compassion training is associated with reduced salivary C-reactive protein from before to after training in foster care program adolescents. *Psychoneuroendocrinology, 38*, 294–299.

Purser, R. E., & Milillo, J. (2014). Mindfulness revisited: A Buddhist-based conceptualization. *Journal of Management Inquiry, 24*, 3–24.

Rocha, T. (2014). 'Dark knight of the soul'. *The Atlantic*, June 25, 2014. Retrieved from http://www.theatlantic.com/health/archive/2014/06/the-dark-knight-of-the-souls/372766/

Rosch, E. (2007). More than mindfulness: When you have a tiger by the tail, let it eat you. *Psychological Inquiry, 18*, 258–264.

Rose, S. (2012). The need for a critical neuroscience: From neuroideology to neurotechnology. In S. Choudhury & J. Slaby (Eds.), *Critical neuroscience: A handbook of the social and cultural contexts of neuroscience* (pp. 53–66). Oxford: Blackwell.

Rutledge, T., Nidich, S., Schneider, R. H., Mills, P. J., Salerno, J., Heppner, P., et al. (2014). Design and rationale of a comparative effectiveness trial evaluating transcendental meditation against established therapies for PTSD. *Contemporary Clinical Trials, 39*(1), 50–56.

Samuel, G. (2008). Autism and meditation: Some reflections. *Journal of Religion, Disability and Health, 13*, 85–93.

Samuel, G. (2013). Panentheism and the longevity practices of Tibetan Buddhism. In L. Biernacki & P. Clayton (Eds.), *Panentheism across the world's traditions* (pp. 83–99). New York: Oxford University Press.

Samuel, G. (2014). Between Buddhism and science, between mind and body. *Religions, 5*, 560–579.

Samuel, G. (2015a). The contemporary mindfulness movement and the question of non-self. *Transcultural Psychiatry, 52*(4), 485–500.

Samuel, G. (2015b). Healing. In A. J. Strathern & P. J. Stewart (Eds.), *The Ashgate research companion to anthropology*. Farnham, Surrey: Ashgate.

Samuel, G. (2016, in press). Tibetan longevity meditation. In H. Eifring (Ed.), *Asian traditions of meditation*. Honolulu: University of Hawai'i Press.

Samuel, G., & Johnston, J. (Eds.). (2013). *Religion and the subtle body in Asia and the West: Between mind and body*. London and New York: Routledge.

Samuels, J. (1995). Healing sand: A comparative investigation of Navajo sandpaintings and Tibetan sand maṇḍalas. *Journal of Ritual Studies, 9*, 101–126.

Saraswati, S. S. (2006). *A systematic course in the ancient Tantric techniques of Yoga and Kriya*. Munger, Bihar, India: Yoga Publications Trust.

Shonin, E., Van Gordon, W., Compare, A., Zanganeh, M., & Griffiths, M. D. (2015). Buddhist-derived loving-kindness and compassion meditation for the treatment of psychopathology: A systematic review. *Mindfulness, 6*, 1161–1180.

Singh, N. N., Lancioni, G. E., Karazsia, B. T., Felver, J. C., Myers, R. E., & Nugent, K. (2016). Effects of samatha meditation on active academic engagement and math performance of students with attention deficit/hyperactivity disorder. *Mindfulness, 7*, 68–75.

Slegelis, M. H. (1987). A study of Jung's mandala and its relationship to art psychotherapy. *The Arts in Psychotherapy, 14*, 301–311.

Spek, A. A., van Ham, N. C., & Nyklíček, I. (2013). Mindfulness-based therapy in adults with an autism spectrum disorder: A randomized controlled trial. *Research in Developmental Disabilities, 34*, 246–253.

Sweet, M. J. (1996). Mental purification (*blo sbyong*): A native Tibetan genre of religious literature.

In J. I. Cabezon & R. Jackson (Eds.), *Tibetan literature: Studies in genre. Essays in Honor of Geshe Lhundup Sopa* (pp. 244–260). Ithaca, New York: Snow Lion.

Varela, F. J., & Depraz, N. (2003). Imagining: Embodiment, phenomenology, and transformation. In A. Wallace (Ed.), *Buddhism and science: Breaking new ground* (pp. 195–230). New York: Columbia University Press.

Wei, G., Dong, H., Yang, Z., Luo, J., & Zuo, X. (2014). Tai Chi Chuan optimizes the functional organization of the intrinsic human brain architecture in older adults. *Frontiers in Aging Neuroscience, 6*(Article 74), 1–10.

Williams, J. M. G., & Kuyken, W. (2012). Mindfulness-based cognitive therapy: A promising new approach to preventing depressive relapse. *British Journal of Psychiatry, 200*, 359–360.

Zope, S., & Zope, R. (2013). Sudarshan kriya yoga: Breathing for health. *International Journal of Yoga, 6*, 4–10.

Mindfulness: Traditional and Utilitarian

5

David Brazier

The Changing Meaning of a Word

The term mindfulness, as currently in use, derives from translations of Buddhist texts. The choice of the word "mindfulness" to translate the Sanskrit *smṛti* (Pali *sati*) and mindful for the corresponding adjective *smṛitimant* seemed appropriate at the time of the first translations approximately a hundred years ago. Subsequent conceptual development and usage in Western psychology and popular culture have, however, distorted this meaning considerably. Now, therefore, we can distinguish Buddhist mindfulness (*smṛiti*) from the contemporary utilitarian mindfulness. What has become the standard definition of the latter is that it is a form of deliberate non-judgemental attention to phenomena occurring in the here and now. In the Latin languages, this has been translated as "pleine conscience," "consciencia plena," etc., which, in English, comes out as "full consciousness." This latter rendering seems even further from the original Buddhist meaning if we take "consciousness," as we usually do, as indicating those mental functions that are not unconscious.

I propose to unpack some of the difficulties that arise directly from the modern definition in contrast to the original. By this means, I intend to bring out the fuller significance, usefulness, and problems associated with the different usages and also, in passing, make some comments upon the spirit of our times.

Let us begin with the word mindful in the English language. Here, fundamentally, mindfulness refers to the state or act of keeping something in mind. Near synonyms are "to beware," "to consider," "to remember," "to call to mind," "to heed," "to be cognisant of." Thus, one might say, "mindful of the danger posed by the enemy, the general sent for reinforcements," or "although surrounded by filth, he was always mindful of what his mother had taught him about hygiene." We can see from these ordinary usages that being mindful generally implies keeping something in mind in a way that generates a degree of tension that tends toward an action of some kind.

There is clearly a significant difference of nuance between this original meaning and that adopted in current psychological discourse. The significant dimensions of difference are, firstly, that traditionally mindfulness was a form of memory, bringing something from the past into the present, rather than an immediate grasp of what presents in the present; secondly, that traditionally mindfulness was commonly connected with the maintenance of values or standards of some kind, or with wariness—that it was something to take into account that shaped or contrasted with the facts given by the immediate

D. Brazier (✉)
Instituto Terapia Zen International; Head of the
Order of Amida Buddha, Cher, France
e-mail: dharmavidya@fastmail.fm

© Springer International Publishing Switzerland 2016
R.E. Purser et al. (eds.), *Handbook of Mindfulness*,
Mindfulness in Behavioral Health, DOI 10.1007/978-3-319-44019-4_5

situation, whereas in the contemporary psychological usage, mindfulness refers to direct perception of the immediate situation itself rather than to anything that tempers that perception; and thirdly, that being mindful implied the creation of a certain tension or discipline in the mind, whereas contemporary usage equates mindfulness with the relaxation or elimination of stress.

The original English meaning is much closer to the meaning of *smriti* in Buddhism than is the new one. It appears that, at the time of original translation, the word mindfulness represented the intention of Buddha more closely than it does in contemporary usage.

However, the new meaning has permitted mindfulness training and mindfulness practice to penetrate the worlds of medicine, education, mental health, the military, and popular culture generally. Popular imagery of mindfulness on the Internet commonly shows a person with an empty thought bubble. Somehow, the word made up of "mind" and "full" has come to mean a mind that is empty. This is a remarkable transformation. It is tempting to suggest that the contemporary utilitarian version should really be called mind emptiness rather than mindfulness.

The transformation has, however, been for a purpose. It seems as if there may have been an element of deliberate strategy in presenting mindfulness in this way, a way that is in part a redefinition and in part simply a stripping down of original mindfulness to the point where it now only suggests one limited part of its original much fuller range of meaning.

In this transformation, something has been lost, and something has been gained. The gain is primarily in the direction of wider public exposure and acceptability. The loss has been in the shedding of the emotive, ethical and imaginative dimensions. The fact that such a loss brings such a gain can tell us something about the current state of our society, the spirit of our times, and even the danger that we are placing ourselves in by valuing such a reduction over the richness that was originally on offer.

The Importance of Scientific Credentials and the Paradox of Religious

A wide range of the so-called mindfulness-based practices have come into being, and these are said to be scientific. The scientific attitude is, in theory, a phenomenological one, not a moral system. It aims to come at things without the pre-supposition. This aim is questionable, as much philosophy of science shows, scientific conclusions being, in practice, validated substantially by reference to the coherence of findings with the pre-established thinking. Nonetheless, the mind empty idea can be seen as not incompatible with science in a way that an emotive, imaginative, or moral meaning would not.

Associating something with science is a much favored modern method of gaining social approval. Many things that have been well known for centuries can suddenly gain a new fillip of popularity if it gets into the news that science has discovered or confirmed them. Those keen to popularize mindfulness have made full use of this strategy, and it has become common knowledge that there now exist thousands of scientific studies affirming the efficacy of the mindfulness method. No matter that the vast majority of such studies are of very poor quality, the aura of science has given the method credibility.

The importance of paying attention to the here and now in a non-judgmental manner has thus been established as an ideal in the popular mind and in the practices of many individuals, groups, and organisations, not only for the conduct of scientific experiments, but as a way of life, a mode of public activity, a remedy for the ills of modern society, and even the epitome of spiritual development. This popularization and hyperbole have thus been accompanied by a rhetoric that eulogizes living in the here and now as a philosophy of life, not merely as a first-aid technique or psychotherapeutic procedure.

The founders of the mindfulness movement have been skillful in presenting "mindfulness" in a manner that is stripped of all necessary associations with religious practice so that it could become acceptable to a culture that has become secular and utilitarian in its core political and institutional values. We are now left with a rhetoric that, while admitting its origins, stridently denies that utilitarian mindfulness has anything to do with Buddhism. This refocusing, however, has both a direct effect and a paradoxical effect.

The first way that it has been effective is that it has made utilitarian mindfulness acceptable to the many people who want to believe that they are free from religion and religious attitudes. Arguably, such supposed freedom is a self-deception in most cases since our whole Western culture is indelibly imbued with values derived from our religious history and denial of a few key dogmas is nowhere near sufficient to erase these. Our hearts and minds are still full of them, however atheistic or secular the image of ourselves that we try to present. Nonetheless, the first step was to become acceptable to academia and to the medical establishment, and this necessitated a casting off of all liens connecting mindfulness with anything that carried the scent of religion. For mindfulness to be approved, it needed to look as scientifically based as possible.

However, there is a second, more paradoxical, level of effectiveness now becoming apparent. The very frequency and strength of the denial of religious connection can become a powerful form of counter-suggestion. The more times we say that mindfulness has nothing to do with Buddhism, the more times the ordinary member of the public hears the word Buddhism and the more times she or he gathers that something of great value has been derived from that source. The very denial, therefore, has served as a rather effective apologetic carrying a hint of the Buddhist message into corners of Western life and society that it would have found difficult to reach by any other means. Thus, the effort to accommodate secular requirements has, in this case, also served to undermine that secularism to a degree. Buddhism—the religion that is not a religion—has infiltrated Western culture and may modify some of our most fundamental presuppositions without there ever being much conscious exposure of what is really taking place.

A Trojan Horse?

Secularization has required that the new mindfulness has little or no value connotation. However, now that the word and the practice have gained a widespread currency, and some shadows of the original meaning do seem to be beginning to infiltrate the contemporary dialogue to which this popularity gives rise. Does mindfulness really have no connection with ethics, compassion, spiritual wisdom, or even spiritual practice? Could it be that mindfulness is, as some have said, a kind of Trojan horse, smuggling spiritual values into the bastions of secularism, or, to change the metaphor, is it the thin end of a wedge that could, when driven home, disrupt the very certainties that support the secular, scientistic, outlook that those founders have skillfully pandered to?[1] Buddhism, presenting itself in the West as the religion that is not a religion, has given us many riddles to ponder. Be this as it may, we can surely glean some understanding of what has become of our culture by looking into the origins and contemporary necessity of this shift of meaning.

Any collectivity of human beings tends to gain its greatest cohesion from shared values. These values are not necessarily overtly displayed, and they are the hidden ruts which all cultural traffic tends to fall into or find itself compelled to follow. I wish to suggest that the mushrooming growth in the popularity of mindfulness not only reveals some of these ruts but also holds out the tantalizing possibility of changing them. Three of these socio-cultural ruts that here become salient are narcissism, immediate gratification, and an overvaluation of consciousness and rationalism.

---

[1]The first time I heard this suggestion, it was made by Stephen Batchelor.

Narcissism

One foundation of our cultural value system is the assertion of Adam Smith that if each person pursues his own good and interest a "hidden hand" will transform the resulting activities into what is optimally to the benefit of the common-wealth. The modern, psychologized version of this notion has forgotten about "the hidden hand" and has become the idea that it is not possible to love another without loving oneself first. Self-love has become something of a dogma of the popular psychology world.

How is this relevant to the sudden popularity of mindfulness? In the current utilitarian version, we are often told that only the here and now exists. This is an assertion of extreme solipsism. If taken literally, it would mean discarding awareness of whatever is not present to one's immediate awareness. The "here and now" is the time and place where one, oneself, is. The implication is that only oneself is important, and even, only oneself and one's experience—nothing else—is worthy of the status of being considered to exist.

It seems probable that our great-grandparents would have regarded such an assertion as the height of irresponsibility, but now, it seems to speak eloquently to the spirit of the times. Although the "me generation" has, perhaps, passed, something of its solipsistic philosophy has remained entrenched in the foundations of our contemporary cultural life.

Immediate Gratification

We live in an age of speed and consequent stress. Quantity has come to matter more than quality in many areas of life. I have a suitcase that belonged to my mother. It is more than fifty years old, and it serves its purpose very well. I also have several suitcases purchased more recently, none of which I expect to last more than five years, usually less. The amount spent on suitcases in our economy is much greater than it was in Mother's day. Therefore, we are richer—or are we? Multiplied up to a national level, the gross national product keeps rising, and we are told this is the prime indicator of the wealth of nations, but, as in the case of suitcases, it may actually be a measure of how much more junk we throw away. More and quicker is better, we think, but this addiction to the short term does not satisfy the heart. I obtain much deeper satisfaction from my mother's old suitcase. It is redolent with meaning and memory that fills my heart and mind. With my heart-mind thus filled with good things, I feel spiritually nourished. This filling, rather than emptying, is surely a significant dimension of the original meaning of mindfulness, that one's mind be full of the good things that one has remembrance of, and it was in this form that mindfulness was a spiritual enrichment of the highest degree as well as a protection. In this form, mindfulness was a factor of enlightenment: A person who is continually replete with such spiritual riches is kind, wise, and enlightened. This is a great deal more than simply having a technique that brings one back to the here and now. Indeed, the here and now stripped of such associations becomes a rather barren spot. Nonetheless, this is the way our society seems to be going: toward speedy, preferably instant, satisfaction of immediate wants and an attention span that correspondingly narrows down to sound bites, tweets of information, present sensation, and instant benefit.

Into this frenetic milieu arrives a technique that seems to offer peace in the moment and validation of instant attention, while epitomizing the rejection of any concern with the past or the future. It is taken up with alacrity, not only by individuals who find difficulty understanding why the ever increasing speed of their lives seems to yield little or nothing in terms of increased satisfaction, but also by the directors of great corporations who readily understand the benefit of keeping their employees focussed upon the immediate without concern for the longer term as well as of being seen to be benevolent, giving them something in compensation for the ever greater stresses that working in such environments generates.

Those who engage in such practice do indeed experience an immediate benefit. Mindfulness feels good. We can be grateful for this gift.

Knowing how to be mindful, one can more effectively function in the midst of an often frenetic milieu and can discharge the tensions that arise when, instead of allowing oneself to be swept along, one tries, instinctively but unskillfully, to hold onto some stability in the midst of the flux. Mindfulness offers the possibility of simply enjoying being swept along. There are definite and immediately tangible benefits. The patient facing a difficult and frightening medical procedure is well-advised to not think about it, but to redirect attention to something more immediate. The person carrying a heavy load of psychological guilt can achieve some relief by forgetting the past for a minute and concentrating strongly upon the sound of the waves crashing on the rocks as she walks along the beach. Many psychiatric patients are helped considerably by contact with nature whose immediacy and beauty have a capacity to drown out the haunting shadows of a disturbed life. The businessman about to give a crucially important presentation to a potentially hostile audience may steady his nerves by taking a moment to scan his body. Sir Francis Drake, facing the task of battling with the vastly superior Spanish armada, wisely and famously took time to finish his game of bowls before setting sail. There is here certainly a wisdom and a usefulness, and these techniques of immediate awareness should not be under-estimated. It is also wise, however, to realize that they only constitute one part of what was originally intended by the Buddha when he recommended a *smṛitimant* life.

Is Consciousness Overvalued?

Much of the history of Western philosophy, including the social philosophies that it has given rise to, not excepting socialism and capitalism, has revolved around attempts to bring a greater degree of rational organization, justice, and understanding to human affairs. Surely, this is a good motive. However, it seems to necessitate conscious control of many processes that are complex to a degree that baffles human intelligence. In our attempt to improve the world, are we trying to outdo God or Nature? In any project of such a kind, there has to be a serious danger of hubris.

Could it be that the universally acknowledged stress of modern life is, fundamentally, a product of our attempt to control the uncontrollable? We value conscious choice. We believe that more choice is better than less. However, beyond a certain point, as the range of options multiplies, confusion sets in. Conscious rationality thus presents to us as both a blessing and a curse. It has been a blessing for us to enjoy the fruits of modern technology, but a curse to be facing ecological failure and the danger of nuclear war. It is a blessing to be able to sustain a sophisticated society that delivers a vast array of goods, but it often feels like a curse to have to work in such an environment or even to have to make sensible decisions about which brand of domestic gas to buy when faced with a wide range of complicated tariffs.

Mindfulness seems to offer a balm in this situation. On the one hand, it is a form of consciousness. It confirms our belief that being conscious is a good thing. However, at the same time, it relieves us of having to make difficult choices. Simply, we have to pay attention to whatever is happening. One simple act suddenly replaces all the complexity that was driving us crazy. Surely this is a blessing.

Of course, the complexities do not go away, and they go into abeyance for a brief time. Where the original mindfulness in Buddhism was supposed to lead one to a complete change of lifestyle that eliminated the causes of stress, modern mindfulness simply gives one the ability to set them on one side for a short period. When one has completed one's mindfulness exercise, the bills still need to be paid, the report still needs to be written, the deadline is still hanging, and the difficult person that one has to deal with in the finance department is still sending messages marked "Urgent." Nonetheless, one has been able to set all that aside for a moment.

Utilitarian mindfulness thus chimes with one of the prime values of our culture while briefly and intermittently giving us relief from the very difficulties that value tends to generate. The question arises, however, whether we should, in fact, be re-examining that value. Are we trying to be too conscious? Are we trusting nature too little? Mindfulness of both the original and the modern kind does at least hint at an enhanced willingness to let be, to trust in deeper processes than those that we can possibly be consciously aware of.

The Value of Unmindfulness

The sudden popularity of mindfulness does seem to partake of the same taken-for-granted over-valuation of consciousness. The paradox here is that mostly we are unconscious of the extent to which we are overvaluing consciousness. However, it does seem worthwhile to ask ourselves, when evaluating utilitarian mindfulness, what is its contrary. What is it to not be mindful? Not being mindful in the original Buddhist sense meant not having learned anything of value, not being grounded in wisdom, neglecting funda-mental truths gleaned either from listening to teachings or from one's own contemplative exploration. Presumably, not being mindful in the sense of utilitarian mindfulness means not paying attention or paying attention in a non-deliberate fashion or not being mentally present or not being concerned with things of the present moment or simply being unconscious. Yet, are not all of these mental states sometimes valuable?

Attention

Let us take the question of paying attention. Our brains and senses are so constructed that paying full attention to one thing precludes attention to other things. If I am paying full attention to what I am writing at this moment, I might not notice that the cat has entered the room and established itself on the settee. I am unconscious of the cat, even though it is in the here and now, but this is not a problem. I might not even notice that it is getting dark outside the window. I might suddenly wake up to the fact that I need to switch on a light if I

am to continue my work much longer. Was I being unmindful in not having consciousness of the here and now reality of darkness in the room? If so, was it not because I was being mindful of the work before me? It is simply a fact that one cannot be aware of everything at the same time and nature has evolved us this way for good reason. In fact, attention is not only a matter of increasing consciousness, and it is also, always, a matter of increasing unconsciousness at the same time. When I concentrate on one thing, I with-draw attention from other things. If I give intense attention to one thing, then the number of things from which I withdraw attention is much greater than the number of things to which I give atten-tion. Paying attention therefore is, actually, just as much, or more, an exercise in unconsciousness as it is in consciousness.

This is how hypnosis works. Hypnosis is largely a matter of directing attention. If a patient can give attention in a sufficiently total way to something, then this can yield anesthesia in the area that she is not paying attention to. It is a matter of distraction. Much of the benefit of utilitarian mindfulness can be understood in this way. By learning to pay sharp attention to, say, the taste of a grape that I might now be eating, I effectively distract myself from all the other things that might otherwise be taking up brain space. If those other things include worries, fears, guilt feelings, troubling memories, and the like, then, in the short run at least, I will experience an immediately effective sense of relief. When I was a child, I taught myself relaxation exercises not unlike the mindfulness body scan. Starting at my toes, I would systematically work through mus-cle by muscle my relaxation technique. This was very effective in the dentist waiting room or in bed at night in times of insomnia.

Points to note here are, firstly, that the bene-ficial effect is not a function of increased con-sciousness as such, but rather of distraction from consciousness of troubling content and, sec-ondly, that such distraction can be effective whatever the object of attention: It does not have to be something of the present moment neces-sarily. Anything that holds the attention suffi-ciently will suffice.

All this has relevance also to the Buddhist origins of the method. Buddhism offers an extensive range of methods of meditation. The basic definition of meditation is that it is an activity in which one has the mind dwell upon a wholesome object. In traditional Buddhism, much mental cultivation consisted in so training the mind that attention to wholesome objects drove out attention to unwholesome ones. While paying attention to one's mantra or guardian deity, one was not paying attention to the innumerable manifestations of greed, hate, and delusion with which the ordinary untrained mind is mostly populated.

Attention and distraction are intimately related to one another. Systematic distraction can play a valuable role in breaking unhealthy habits of mind. Perhaps we do not think of mindfulness as a form of distraction, but it is quite credible that this is how it often works. Where people in traditional societies might have found visualizing a powerful deity sufficiently riveting to do the trick, modern disillusioned people need something different. They need something that has the prestige of science behind it and that accords with their belief system focussed on the evidence of the senses. The idea of the here-and-now can be such a sacred icon to a person of modern education.

Non-deliberate Attention and Mental Absence. Is mind wandering such a bad thing? When we are asking ourselves what would be non-mindful, one of the things we must consider is mind wandering. However, it is surely apparent that mind wandering is essential for health and is the source of a great deal of creativity. In order to go to sleep, one needs to enter into a mind-wandering process. While the thought train is under conscious, control one will not fall asleep. The reason that the traditional method of "counting sheep" was often effective is that it is a sufficiently boring activity that after a few animals have passed by the mind's eye, the mind wanders off onto something else. Sleep soon follows. For persons who are sufficiently mentally dedicated to endlessly bring the mind back to sheep, the method does not work.

Why does the mind wander? Because it has its own priorities. There is important work for the unconscious mind to do, and it tries to do it in the interstices between conscious thought. When there are insufficient of these, sleep may take over. If this does not happen, a person may become ill or go crazy because the basic work of the mind is not getting done. We need unconscious periods.

Clearly, there is some value in training the mind and bringing it under conscious control some of the time, but this can be over-done. The unconscious mind attends to all the regular things that are vital to our health and survival. The conscious mind is peripheral to this. The things it does are valuable, but they could never replace the more basic functions that go on almost entirely out of awareness.

Nor does the conscious mind even have a monopoly on problem solving. We are all familiar with the phenomenon of having a problem that seems insoluble in the daytime, going to bed and sleeping, and finding, the following morning, that the solution seems obvious. How did we fail to see the matter clearly the night before? Quite clearly, important processing has gone on during the night. Recent studies have also shown that the unconscious mind can even handle mathematics. Experiments have been done in which a mathematical problem is flashed onto a screen while a subject's mind is distracted by a powerful stimulus. Those subjects asked about the mathematical problem have no awareness of having seen it. Yet when subsequently given number recognition tests, the same subjects have much quicker reaction times when the number that is the solution to the mathematical problem appears than they do to other numbers. This has to indicate that the mind did, in fact, process the number problem even though the subject had no conscious knowledge of ever having seen it.

The moral of this appears to be that we undervalue the unconscious mind at our peril. If mindfulness is conceived as a means of increasing consciousness at the expense of the unconscious, then we may be making a serious mistake. It is vital for the mind to be allowed time in which to wander. Being unmindful may be the source of creativity.

In any case, it is certainly important to sleep. Any philosophy that set up the idea of awareness 24/7 would be an invitation to madness. Even as His Holiness the Dalai Lama admits, a good night's sleep is a great boon. Consciousness and unconsciousness need each other, yet, of the two, the unconscious is the more vital and fundamental. A genuine mindfulness needs to take this into account.

Original Buddhist mindfulness did so. Mindfulness meant internalizing wholesome wisdom to the point where it was part of one. It did not cease when one ceased to be consciously aware of it, just as one does not cease to love somebody when one is not thinking of them. The things that one was mindful of were accessible—they could be brought to mind and placed before the mind—but this was because they were internalized to the degree that they no longer themselves depended upon such attention. A skilled craftsman knows his craft and can bring necessary know-how to bear when appropriate, but that know-how does not disappear as soon as he stops paying attention to it. The Buddha often used this kind of analogy.

Problems with Here-and-Now-ISM

Even a cursory review of Internet sites promoting mindfulness will throw up reference to the idea that "only the here and now exists." This idea is presented as a justification for that aspect of utilitarian mindfulness that takes things of the here and now to be the only things worthy of attention. This rhetoric can be seen as an essential part of the hypnotic process by which the contemporary person is induced to take the objects of attention presented in mindfulness training to be of sufficient importance to rivet the attention. We pay attention to things because we think they are important or sacred in some way. The converse is also true: The things that we think are important are the ones that we habitually pay attention to. If I wish you to adopt a

habit of attention, then I have to convince you that the object is important.

We have already seen how modern people are susceptible to this rhetoric because it fits in with ideas about the prestige of science and the culture of rationalism and sensory evidence. The assertion that only the here and now exists, therefore, finds a ready audience and does not meet with an ingrained resistance.

It is, however, worth our while pausing long enough to ask whether this philosophy is a satisfactory one. We have already raised some doubts when, under the heading of Narcissism, we asked whether the giving of primary importance to the here and now is not an extreme form of solipsism. I would like now to take this line of argument a bit further.

Firstly, let us ask what counts as "here" and what counts as "now." In contemporary usage, here and now seems to imply only what is registering in my senses in this very moment. In the Buddhist texts, the phrase more commonly has the implication "in this very lifetime." These are not the same thing.

If "here" means where I am, how extensive is that? Here in this room? Here on planet earth (still a tiny drop in the cosmos)? Does anybody really think that when things are not impacting on one's own senses they have ceased to exist? I am in the sitting room. Has the kitchen ceased to exist? If so, where is my dinner going to come from? These may seem trivial questions, but I think they do illustrate that no sane person actually lives his or her life as though only what is "here" exists. It is not true in experience.

The same is true in regard to "now." Do we mean now this minute, this day, this year? As the philosopher Bergson pointed out, time is actually a matter of duration. Moments are purely conceptual. If we take "now" to refer to a moment only, then it is perhaps more true to say that it is just such a supposed now that does *not* exist, while the flow of durations of time is reality as we experience it. Similarly, if, as many scientifically minded people might be inclined to do, we make the senses into a criterion, then it is apparent that there is a time lag and that what the brain registers and forms into image has already passed by the time that one has got the image formed. Thus, the "now" that one experiences does not exist at the time that one does so. "Now" thus becomes a very slippery concept because time is not actually made up of a series of now moments, it is duration and flow.

In any case, just as with the disappearance of the kitchen, no sane person lives without an awareness of the past and the future. Sanity involves an ability to flow through duration and make sense of experience in temporal terms. Without such an extended sense of time, there is no sense at all. Some philosophies, including some Buddhist philosophies, have tried to envisage existence as if time does not exist, but this does not appear to be how Buddha understood it, nor does it accord with common experience.

Although some support can be found in some Buddhist texts for the efficacy of paying attention to the present moment as a practice that is sometimes useful, the general drift of the mainstream of Buddhist philosophy is concerned with causation and consequence unfolding over not merely short, but even vast, extensions of time. It is in the Buddha's ability to see the rising and falling of beings according to their deeds that we recognize his wisdom and ability to teach and impart to us something of immense value.

Leaving these more philosophical considerations aside, at a practical everyday level, here-and-now-ism is not a very good guide to the good life. My friend is in hospital; therefore, she is not here. Should I, therefore, consider her to no longer exist and pay her no further attention? My mother is dead, ten years since. Should I consider her non-existent and expunge her memory from my mind? I have a project in hand to renovate a barn adjacent to my house; the work all lies in the future—how does thinking that only the present exists help me to achieve the desired and? Indeed, does such a philosophy permit me to have "ends" at all? I think it is fairly apparent that bringing attention to the here-and-now, however defined, is a technique that can be intermittently useful, but does not constitute a satisfactory life philosophy and that the idea that only the present should be classified as "existent" is a self-deception.

McMindfulness

Further, the times when paying attention to what is immediately present is most useful are times when doing so should, in a healthy person, be naturally compelling anyway. If we need a special technique to tell us how to do something that should be happening quite naturally, then it is presumably because our lives have become so unnatural that we no longer trust our nature to perform as it should. The real remedy for such a situation would not be to add a technique, but to change the lifestyle. A serious critique of the mindfulness fashion is made by those who see that business corporations have taken to offering mindfulness courses to their staff in order to help the staff to cope with stress inflicted upon them by work in the very same corporations. The suggestion is that what is happening is that employees are being given a "benefit" the nature of which is to make them even more exploitable than before and that this can serve as a means for the corporation to avoid looking at the harm that it is inflicting or the unnaturalness of the lives that it is requiring of its servants.

In general, our society has a tendency to seek remedies for ills rather than their elimination. When a person suffers from insomnia, the natural recourse would be to find the cause and do something about it. If the insomnia is due to worry, then there is a need to solve the problem that is being worried about. If to a lack of exercise in the daytime, then there is a need to take more exercise. However, our modern way is, often enough, to take a sleeping tablet and never do anything about the root problem. Is the application of mindfulness becoming a kind of "treatment" of this kind? Others have written much more extensively about this issue. Here, I merely mention it for its relevance to the here-and-now issue. The sleeping tablet is a supposedly instant remedy, but it involves a deliberate ignorance of the cause and effect operating in the longer duration.

Mindfulness Old and New

Let me try and summarize the main points made so far and relate them to mindfulness in its original Buddhist form.

A New Popular Word

There is currently a thriving fashion for "mindfulness." Most of us know a little of how this contemporary idea was developed by the American John Kabat-Zinn for use in medical settings and has since spread to self-help psychology and even to business and the military. As now defined, mindfulness means a form of deliberate, non-judgemental attention to what is immediately present, either in one's body or in one's immediate surroundings. If a person bumps his head on a low doorframe or spills his tea because he was lost in thought, we say that he was not being sufficiently mindful. This kind of mindfulness has been found to be effective in interrupting trains of thought and so can be usefully therapeutic for people whose thought train is full of anxiety or morbidity. Mindfulness has thus become a "treatment," or prophylactic, for stress, depression, anxiety, and other forms of negative rumination. Essentially, it involves developing a skill and habit of periodically interrupting such a thought train by bringing the mind into direct awareness of something immediately present, such as the blueness of the sky, the taste quality of what one is eating, the temperature difference between one's two hands, or the quality of sensations spontaneously occurring in the body. This is an effective technique. When done well, it empties the mind, at least momentarily, of troubling obsessions. Since modern life is replete with such obsessions, the arrival of a simple method for reducing their pernicious effect is to be welcomed. It does not, however, deal with the underlying causes.

Buddhist Origins

The idea of mindfulness comes from Buddhism, but do we then think that this means that Buddhism teaches that one should discard the past and future and live in the present moment all the time? How is it that Buddha himself told so many gripping stories about the past lives or about his own night of enlightenment long in the past? Why, when confronted with Kisagotami whose baby had recently died, did he send her round the village to hear the reminiscences of grief of all the other people whose near and dear relatives had died? Why did he not tell her to interrupt her grief and just dwell in the present moment? Why did he not teach her some immediate attention techniques that would get rid of the unpleasant feeling—a kind of mindfulness pill to relieve the symptoms? In his wisdom, he sent her into a series of encounters with the past and with the great depth of feeling that can flow from it when one focuses the mind on things no longer present that have been momentous in one's life. He did not tell her to let go of such things, but to hold them and learn. He did not take away her grief, he intensified it. By doing so, he taught her great compassion and enabled her to go beyond herself. He filled her heart and mind with a painful yet highly beneficial remedy. He thus taught her a different kind of mindfulness.

The Past and Future Are Important

In fact, the word "mindful" is a translation of the Sanskrit word *smṛti* (Pali *sati*) which comes from the word for "remember." In Buddhism, mindfulness principally refers to a kind of remembering. It is a remembering that serves one well in the present and helps to sow good seeds for the future. Thus, it is not just any remembering, but "right remembering." This correct way of keeping something sane and helpful in mind is something that the Buddha considered very important. He made it one of the limbs of the "Eightfold Path" and one of the elements in the "Seven Factors of Enlightenment." In fact, we can say that it is one of the ways of understanding what enlightenment is. Enlightenment is to have the mind and heart always full of good things, always full of such things as will lead to a good, noble, wise, and compassionate life. Buddhism is not concerned only with the present moment, but with the long term, the very long term. Another way of seeing the basic Buddhist message is to say that Buddha taught that all the important things in life flow from the heart; when a person acts with a good heart, good follows and when with a bitter heart, trouble follows. This happens over a period of time. A good heart is none other than a heart filled with good things, or, we can say, a mind full of sane objects. Here, "heart" and "mind" refer to the same thing. When we bear in mind good inspiration from the past, our actions in the present build a good future. This is generally true in the short term, and it is absolutely true in the very long term. This is the Buddhist understanding.

The Paradox of the Present Moment

So, it is valuable to understand the Dharma correctly and not get hold of the wrong idea. Buddha said that getting hold of the Dharma in the wrong way is like getting hold of the wrong end of a snake. If you get hold of the wrong end it may bite you. If you get hold of the right end, you can get it to give you good medicine. If we think that Buddhism is just having an empty mind in the present moment, we could get bitten by a mistaken idea. Buddha wants us to have hearts that are full, full of good things, full of sane and healthy things that will stand us in good stead: full of the inspiration that we have received from good teachers; full of what we have learnt from experience, both inspiring and difficult; full of good principles and understandings. When our mind is filled with something good in this way, then we have a protection. Further, this is a protection against what may happen in the present moment. In the present moment, one may be in danger of being overwhelmed, by fear, by greed, by envy, by bitterness. How is one to deal with these things that

sometimes catch one unawares? By having some longer-term good in one's heart to return to. So living fully in the present does not mean cutting off the past or ignoring the future, and it does not mean having an empty mind. Rather it means having a fullness of life that comes from a heart full of good inspiration. In Buddhism, this is called "taking refuge." When difficulties come along, we have something in which to take refuge, something that will sustain us through the time of darkness.

The Proper Function of Consciousness

Even though one of the most important developments in psychology in the past hundred years has been the work of those who have tried to understand the importance of the unconscious mind, modern psychology still has a tendency to overvalue consciousness. Why do we have a conscious mind? Could we not get on without it? Many animals seem to do pretty well with very little of it, responding always to instinct. We have instincts too. Why not just rely on them? The times when we find our conscious mind most useful is when instinct does not do the job, for some reason. When my instincts are happily digesting my dinner, I am not conscious of it. When something goes wrong, I become conscious of my stomach. Would it be better for me to have full consciousness of my stomach process all the time? Certainly not. It is much better to let instinct get on. Consciousness can interfere too much. It is possible to have too much consciousness. Consciousness is for solving problems, and in this proper function, it is immensely valuable. However, there are many things that are not problematic, and it is better to keep it that way. Making total consciousness a goal would be a serious mistake and lead to a very unnatural life. The translation of "mindfulness" into "full consciousness" is rather unfortunate. It can easily give the wrong idea. The aim of Buddhist practice was not that one makes everything conscious. That would, in any case, be impossible. Buddha said that he was somebody who slept well. We spend a third of our life in the unconscious state of sleep, and at least half of the rest of the time the mind is wandering. This is important. We should give the unconscious mind a chance and room to do its proper job. So the idea that mindfulness means being conscious of everything all of the time is a mistake. It is all right to reminisce, to daydream sometimes, to sleep, and to live life in a natural way; just do it with a good heart.

Conclusion

So, I hope that from this analysis we can see that there is a value in the modern utilitarian mindfulness, but that we should be careful not to take the idea too literally, nor take it to too much of an extreme, since there is a deeper meaning in the original Buddhist mindfulness. The modern utilitarian mindfulness is non-judgemental and so is the original, but the original mindfulness is intrinsically full of good values where the modern form can be empty of them. The modern form is simply a matter of attention, whereas the original involves, in addition, the whole heart-mind, including reminiscence, thought, imagination, emotion, bodily feeling, foresight, and so on. The modern kind of mindfulness values the present moment. Buddhism does too, but in a way that is informed by the past and future and has room for dreams, visions, and all that is inspiring and uplifting. Buddhist mindfulness is full of love, compassion, joy, and peace, and these wonderful qualities are real for each of us because, in the Buddhist perspective, our life stream unfolds over the course of eternity and is not just limited to the immediate present. Let your mind be full of Dharma, and the present moment will look after itself is closer to the Buddhist message.

Distinguishing between modern utilitarian mindfulness as a mind-emptying technique for treating minds that are full of noxious content and original Buddhist mindfulness as a life path of filling the mind with wisdom and compassion enables us to reflect more deeply upon what is happening to our culture. It throws an interesting light upon the present vogue for mindfulness classes. It also reveals what may be the hidden potential inherent in this new interest. The contemporary idea is that normal is good and that anything bad is abnormal and needs fixing with a remedy. The Buddhist idea is that there is much more to play for than a return to normality, that it is not a matter of fixing abnormalities, but of reaching far beyond toward the highest human potential, which is the potential of a heart-mind truly full of wisdom and compassion.

There is, surely, a possibility that the wide dissemination of this technique, known to be of Buddhist origin, may create a new openness to some of the deeper values that were inherent in the original, but which have had to be set on one side for the moment in order to get the modern stripped-down version established in those parts of the body-social that are resistant to any such intrusion of wholesome values. The future development is bound to be interesting.

Can "Secular" Mindfulness Be Separated from Religion?

Candy Gunther Brown

Introduction

Mindfulness has become mainstream. Hospitals and prisons offer "Mindfulness-Based Stress Reduction" (MBSR), public schools teach students to put their "MindUP," and Google trains employees to "Search Inside Yourself." Mindfulness entered the American cultural mainstream as marketers tactically muted religious-sounding Buddhist terminology by foregrounding secular-sounding scientific and commercial linguistic frames (Woodhead 2014, p. 15). Thus, many Americans have embraced mindfulness as a secular, scientific, fee-for-service technique to reduce stress, support health, and perhaps even cultivate universal ethical norms. Asking "Can 'secular' mindfulness be separated from religion?" suggests several related questions: What does it mean for a practice to be religious, spiritual, and/or secular? Is Buddhism a religion? What *is* mindfulness? Is mindfulness inherently Buddhist and/or religious? If mindfulness can theoretically be separated from religion, have particular "secular" programs disentangled mindfulness from religion?

This chapter questions the supposition that "secular" mindfulness programs teach a purely secular, universal technique. It argues that nominally secular programs instill culturally and religiously specific and contested worldviews, epistemologies, and values. To be clear, this chapter does *not* argue (or deny) that mindfulness is "inherently" religious (or secular) or that it has some intractable "essence." Such claims, though made by some (including both mindfulness advocates and detractors), tend to be analytically flattening and embedded in metaphysical ideas that are empirically unfalsifiable. The more fruitful question may be how, in particular cultural contexts, "mindfulness" might be conceptualized, communicated, and practiced in ways that explicitly or implicitly convey religious meanings and/or facilitate religious and spiritual experiences. After defining key terms, this chapter explores three common patterns: (1) Code-Switching, (2) Unintentional Indoctrination, and (3) Religious and Spiritual Effects.

Defining Practices as Religious, Spiritual, Buddhist, or Secular

There is no single, universally accepted, historically stable, politically neutral definition of "religious," "spiritual," or "secular." Many people assume that they "know it when they see it," but often common-sense definitions obscure cultural blind spots and charged agendas. Historian Jeff Wilson aptly notes that such terms are not "mere statements of fact," but "markers of value employed strategically by agents" who, in speaking of mindfulness as secular or religious, are

C.G. Brown (✉)
Indiana University, Bloomington, IN, USA
e-mail: browncg@indiana.edu

© Springer International Publishing Switzerland 2016
R.E. Purser et al. (eds.), *Handbook of Mindfulness*,
Mindfulness in Behavioral Health, DOI 10.1007/978-3-319-44019-4_6

"making an argument" to serve particular projects (Wilson 2014, p. 9). One need not throw up one's hands at recognizing that every concept—including the religious and the secular—is an "arbitrary construct" that "never corresponds fully with reality," because such concepts can nevertheless be useful in facilitating the classification of "real phenomena" and finding out "empirically where the classifications break down" (Berger 2014, p. 17). An important caveat is that the classificatory project itself reflects a distinctively modern, and in a certain sense metaphysical, assumption that the religious and the secular can be objectively identified, distinguished, and potentially disentangled (Taylor 2007, p. 13). Particular individuals or groups may, regardless of such classifications, retain deep convictions about the inherent or essential nature of practices in question that cannot be negated by analytical fiat.

This chapter understands religion as encompassing beliefs and practices perceived as connecting individuals or communities with transcendent realities, aspiring toward salvation from ultimate problems, or cultivating spiritual awareness and virtues (Durkheim 1984, p. 131; Smith 2004, pp. 179–196; Tweed 2006, p. 73). Religion may be identified by the presence of "creeds" (explanations of the meaning of human life or nature of reality), "codes" (rules for moral and ethical behavior), "cultuses" (rituals or repeated actions that instill or reinforce creeds and codes), and "communities" (formal or informal groups that share creeds, codes, and cultuses), or by "ultimate ideas," "metaphysical beliefs," "moral or ethical system," "comprehensiveness of beliefs," and "external signs" such as an enlightened founder, sacred writings, gathering places, keepers of knowledge, and proselytizing (Albanese 2013, pp. 2–9; Adams in *Malnak v. Yogi,* 1979, at 208–210; U.S. v. Meyers 1996, at 1483).

Such definitions do not sharply distinguish religion from spirituality, both of which make metaphysical (more-than-physical) assumptions about the nature of reality. Confusion arises because, in recent decades, the term religion has accrued negative cultural connotations that induce many people to substitute euphemisms such as "spirituality" or "scientific" to deny that practices have a religious nature. Identifying as "spiritual, but not religious" signals one's rejection of religious (and especially Christian) dogmas and institutions, and may be tactically employed to overcome cultural resistance, gain access to state-funded institutions from which religion has been legally barred, or qualify for health insurance coverage. Definitions that differentiate religion and spirituality tend to associate the former with bureaucracy and the latter with individual quests for ultimate reality, while noting that overlaps are so extensive that they are difficult to disentangle (Shapiro 1992, p. 24; Stratton 2015, p. 101).

Modern classifications of world religions—Buddhism included—have a complex history tied to European colonialism. Europeans invented the term Buddhism in the nineteenth century, though religious traditions now identified as Buddhist have a much longer history. Modern Buddhisms, out of which contemporary mindfulness practices developed, took shape through encounters between European orientalists and Asian reformers (McMahan 2008, p. 20). Buddhist modernizers have often found it useful to deny that Buddhism is a religion at all—preferring the language of science, universal spirituality, or philosophy (Lopez 2008, p. 32).

Historically, the word "secular" emerged in the context of Roman Catholic Canon Law to differentiate a priest who lived in the world (*saeculum*) from a priest who lived in a religious cloister. The term sometimes separated the religious from the secular world, and sometimes distinguished between the observable and unseen worlds (Casanova 1994, pp. 13–14). Today, the secular most often gets defined in relation to religion and spirituality: "either the absence of it, the control over it, the equal treatment of its various forms, or its replacement by the social values common to a secular way of life" (Calhoun et al. 2011, p. 5). Although people imagine the secular as the opposite of the religious—and assume that a practice can either be one or the other, not both—concepts of the secular, the religious, and the spiritual have often intermingled and co-constituted one another (Asad 2003; Taylor 2007; Jakobsen and Pellegrini 2008).

It is important that "secularization" may denote not the disappearance of religion, but the relabeling of religion to scaffold religious perspectives on ultimate reality while addressing practical concerns with health and commerce. Often the *same* individuals oscillate between secular and religious language in talking about the *same* practices depending upon audience or purpose at the time. Mindfulness marketers may employ religious and secular discourses simultaneously: describing religious concepts with language of science and spirituality; through self-censorship, selecting certain concepts or practices to omit disclosing while emphasizing others; and by means of camouflage, or concealing followed by carefully timed, gradual introduction of spiritual nuggets as perceived benefits win over cautious novices (Bender 2010, p. 42; Zaidman et al. 2009, pp. 605–606).

Assertions that mindfulness is secular beg the question of what it means to secularize a Buddhist practice. Marketers insist that mindfulness has been secularized, without defining the terms religion or secularity, explaining how mindfulness has been secularized, or exploring the corollary that a secularized practice presumably started off religious. Promoters employ one or more of six linguistic tactics: (1) Mindfulness passes as "purely secular" through circular speech acts of linguistic substitution; this is not religion, it is secular. (2) Spokespersons may avoid the terms Buddhism, religion, spirituality, or meditation altogether, or disavow that mindfulness is Buddhist, New Age, or religious. (3) Some concede that Buddhists have practiced mindfulness for millennia. This serves a two-fold function of, first, authenticating mindfulness as empirically validated, and, second, communicating that modern mindfulness has been unmoored from ancient religion. (4) Promotional texts signal that advocates are knowledgeable about religion and undeserving of the criticism that mindfulness is backdoor Buddhism. Analogies characterize such worries as irrational: Misperceiving mindfulness as making one

Buddhist is akin to worrying that eating pizza will make one Italian or drinking coffee will make one Ethiopian. In addition to making fears of religious contamination seem ridiculous, such analogizing associates Buddhism with foreign ethnicity and implies that Americanized mindfulness is free from Buddhist cultural and religious "baggage." (5) Rhetoric categorizes mindfulness as a scientific technique rather than a religious ritual. It does so by referencing brain anatomy and fMRI studies showing changes in brain structure and function, and by employing terms with scientific cachet, such as neuroplasticity, awareness, stress reduction, cognitive skills, and social and emotional learning. (6) Marketers assert that mindfulness cultivates universal virtues, such as compassion, and can be practiced by Christians, Jews, Muslims, and atheists without religious conflict. As this chapter explores, none of these tactics fully disentangles mindfulness from religion.

Defining Mindfulness

The etymology of the term "mindfulness," as commonly employed in twenty-first-century American culture, can be traced to Pali language Buddhist sacred texts, especially the *Satipaṭṭhāna Sutta*, or "The Discourse on the Establishing of Mindfulness." *Sammā sati*, often translated as "right mindfulness," comprises the seventh aspect of what is frequently translated as the "Eightfold Noble Path" to liberation from suffering, the fourth of the "Four Noble Truths" of Buddhism (Wilson 2014, p. 16).

The best-known contemporary definition of mindfulness, coined by Jewish-American molecular biology Ph.D. Jon Kabat-Zinn, is "paying attention in a particular way: on purpose, in the present moment, and non-judgmentally" (1994b, p. 4). Kabat-Zinn privileges the term mindfulness precisely because it is capacious enough to carry "multiple meanings," both seeming to denote a universal human capacity and also functioning as "place-holder for the entire dharma," an "umbrella term" that "subsumes all of the other elements of the Eightfold

Noble Path" (2009, pp. xxviii–xxiv; 2011, p. 290). He authenticates his decision to feature mindfulness with "the words of the Buddha in his most explicit teaching on mindfulness, found in the *Mahasattipathana Sutra*, or great sutra on mindfulness." It is the "direct path for the purification of beings, for the surmounting of sorrow and lamentation, for the disappearance of pain and grief, for the attainment of the true way, for the realization of liberation [Nirvana]— namely, the four foundations of mindfulness" (2009, p. xxix). For Kabat-Zinn, mindfulness is a Zen Buddhist "koan" that invites deep questioning (2015, para. 3).

The term mindfulness does "double-duty," signifying a stripped-down, therapeutic technique for "regulation of attention" and potentially evoking a "comprehensive" Buddhist worldview and way of life—*Buddhadharma* (Kabat-Zinn 2009, pp. xxviii–xxix; Stratton 2015, p. 103; Winston in Wilks et al. 2015, p. 48). It is significant that secularized mindfulness programs purge much identifiably Buddhist terminology, yet retain mindfulness. For instance, savoring a single raisin is not particularly meaningful—until framed as mindfulness. The term makes room for certain teachers to introduce normative frameworks and metaphysics that they imported from Buddhism whether consciously or culturally, and directs initiates to where they can find resources to go "deeper." As one secular mindfulness teacher admitted, "we can't hide" the Buddhist roots, since novices "only need to Google 'mindfulness' to find out!" (Wilks 2014b, December 8).

Pattern #1: Code-Switching

Certain of the foremost promoters of mindfulness switch back and forth between describing the practice as "completely secular" and embodying the "essence" of *Buddhadharma*. They do so not only to offer therapeutic benefits to a culture resistant to non-Christian religion, but because they are confident that mindfulness—even stripped of Buddhist vocabulary—is inherently transformative. When speaking to Buddhist audiences, promoters describe their tactics as "skillful means," "stealth Buddhism," a "Trojan horse," or a "script." These spokespersons exhibit what linguists term "Code-Switching" and sociologists call "frontstage/backstage" behavior —moving between vocabularies of multiple cultures to achieve complex goals (Chloros 2009; Goffman 1959; Laird and Barnes 2014, pp. 12, 19). As psychologist Daniel Goleman boasts of his own efforts, "the Dharma is so disguised that it could never be proven in court" (1985, p. 7).

Skillful Means

No individual leader or program model better illuminates the skillful means tactic than Jon Kabat-Zinn's promotion of "secular" Mindfulness-Based Stress Reduction (MBSR). Indeed, the unsurpassed influence of Kabat-Zinn and MBSR merits extended discussion. Founded in 1979 as the Stress Reduction and Relaxation Clinic, as of 2015, the University of Massachusetts Center for Mindfulness in Medicine, Health Care, and Society (CfM) had enrolled 22,000 patients, certified 1000 instructors, spawned more than 700 MBSR programs in medical settings across more than thirty countries, and become a model for innumerable mindfulness-based interventions (MBIs) in hospitals, prisons, public schools, government, media, professional sports, and businesses (CfM 2014b, para. 1; Wylie 2015, p. 19). MBSR has, moreover, been cited by legal and policy analysts as a primary example of "secular meditation techniques" that "seem not to contain any spiritual or religious teachings" (Masters 2014, p. 260), "do not make any metaphysical or religious assumptions," and are "not committed to substantive ethical standards about what is good, bad, right or wrong" (Schmidt 2016, pp. 451–452).

In an article for *Contemporary Buddhism*, Kabat-Zinn frames MBSR as skillful means for mainstreaming *Buddhadharma*. He developed MBSR "as one of a possibly infinite number of skillful means for bringing the dharma into mainstream settings. It has never been about

MBSR for its own sake" (2011, p. 281). In an interview with Buddhist monk Edo Shonin, Kabat-Zinn says of MBSR that "what it is—now I have to use some Buddhist terminology—it is the movement of the Dharma into the mainstream of society. Buddhism really is about the Dharma—it's about the teachings of the Buddha." Denying that MBSR is "McMindfulness" (one of a number of critiques Kabat-Zinn's rhetoric has provoked among Buddhist scholars, e.g., Purser and Loy 2013), he insists that "what is practiced in Buddhist monasteries is essentially no different from what is taught in MBSR" (2015, para. 6). MBIs are "secular Dharma-based portals" opening to those who would be deterred by a "more traditional Buddhist framework or vocabulary" (Williams and Kabat-Zinn 2011, pp. 12, 14). An "example of 'skill in means' (*upāya-kauśalya*): it provides a way of giving beings the opportunity to make a first and important initial step on the path that leads to the cessation of suffering" (Gethin 2011, p. 268). Merely stripping what Kabat-Zinn summarily dismisses as "unnecessary historical and cultural baggage," MBSR preserves what is "essential" of the "universal dharma that is co-extensive, if not identical, with the teachings of the Buddha, the Buddhadharma" (Williams and Kabat-Zinn 2011, p. 14; Kabat-Zinn 2011, p. 290). As he put it in another interview, "what we're really trying to do is to create an American Dharma, an American Zen" (1993, p. 36). Kabat-Zinn felt comfortable "glossing over important elements of Buddhist psychology (as outlined in the Abbidharma, and in Zen and Vajrayana teachings)," reasoning that these "could be differentiated and clarified later" once the practical benefits of mindfulness had been demonstrated (2009, pp. xxviii–xxix).

MBSR focuses on "stress" as a catch-all malady to which most people can relate, yet can also be presented as "authentically" Buddhist since it "has the element of *dukkha* embedded within it" (Kabat-Zinn 2011, p. 288). According to Kabat-Zinn, MBSR aims to "elevate humanity" by instilling "fundamental teachings of the Buddha about the nature of suffering and the possibility of the sort of transformation and liberation from suffering" (2015, para. 7). The "invitational framework" of "stress reduction" encourages MBSR participants to:

> dive right into the experience of *dukkha* in all its manifestations without ever mentioning *dukkha*; dive right into the ultimate sources of *dukkha* without ever mentioning the classical etiology, and yet able to investigate craving and clinging first-hand, propose investigating the possibility for alleviating if not extinguishing that distress or suffering (cessation), and explore, empirically, a possible pathway for doing so (the practice of mindfulness meditation writ large, inclusive of the ethical stance of *śīla*, the foundation of *samadhi*, and, of course, *prajñā*, wisdom—the eightfold noble path) without ever having to mention the Four Noble Truths, the Eightfold Noble Path, or *śīla*, *samadhi*, or *prajñā*. /In this fashion, the Dharma can be self-revealing through skillful and ardent cultivation (2011, p. 299).

In Kabat-Zinn's formulation, although framed as "secular" therapy, MBSR reveals each and every one of Buddhism's Four Noble Truths and cultivates the Eightfold Noble Path to the cessation of suffering. The "particular techniques" taught in MBSR are "merely launching platforms or particular kinds of scaffolding to invite cultivation and sustaining of attention in particular ways" that bring one to "ultimate understanding" that "transcends even conventional subject object duality" (Kabat-Zinn 2003, pp. 147–48). Kabat-Zinn aims at nothing less than "direct experience of the noumenous, the sacred, the Tao, God, the divine, Nature, silence, in all aspects of life," ushering in a "flourishing on this planet akin to a second, and this time global, Renaissance, for the benefit of all sentient beings and our world" (1994a, p. 4; 2011, p. 281).

In the early years, Kabat-Zinn "bent over backward" (in his words) to select vocabulary that prevented both patients and hospital staff from recognizing MBSR as the "essence of the Buddha's teachings" (2011, p. 282). In addressing the public, Kabat-Zinn has steadfastly insisted that "you don't have to be a Buddhist to practice" mindfulness (1994b, p. 6). Over time, as scientific publications (which Kabat-Zinn pioneered in publishing) lent credibility to mindfulness, he felt it was safe to begin to "articulate its origins and its essence" to health professionals, yet "not so much to the patients," whom he has intentionally continued to leave uninformed about the "dharma that underlies the curriculum" (2011, pp. 282–83).

Despite secular posturing, Buddhism pervades MBSR and many offshoot MBIs. This can be seen through a closer examination of: (1) program concept, (2) systematic communication of core Buddhist beliefs, (3) teacher prerequisites, training, and continuing education requirements, and (4) resources suggested to MBI graduates.

MBSR Program Concept

Kabat-Zinn first trained as a Dharma teacher with Korean Zen Master Seung Sahn. Eclectically inclined, in developing MBSR Kabat-Zinn drew from Soto Zen, Rinzai Zen, Tibetan Mahamudra and Dzogchen; a modernist version of Vipassana, or insight meditation, modeled after Burmese Theravada teacher Mahasi Sayadaw; as well as hatha yoga, Hindu Vedanta, and other non-Buddhist spiritual teachers (Kabat-Zinn 2011, pp. 286, 289; Dodson-Lavelle 2015, pp. 4, 47, 50; Harrington and Dunne 2015, p. 627). Although he still trains with Buddhist teachers, Kabat-Zinn stopped identifying as a Buddhist when he realized that he "would [not] have been able to do what I did in quite the same way if I was actually identifying myself as a Buddhist." Kabat-Zinn also insists that "the Buddha himself wasn't a Buddhist," since "the term Buddhism is an invention of Europeans." Yet, Kabat-Zinn views his "patients as Buddhas," since "literally everything and everybody is already the Buddha" (2010, para. 4; 2011, p. 300).

In the "origins" story narrated by Kabat-Zinn for Buddhist audiences, while on a spiritual retreat at the Buddhist Insight Meditation Society in 1979, he had a flash of insight to "take the heart of something as meaningful, as sacred if you will, as Buddhadharma and bring it into the world in a way that doesn't dilute, profane or distort it, but at the same time is not locked into a culturally and tradition-bound framework that would make it absolutely impenetrable to the vast majority of people" (2000, p. 227). Kabat-Zinn refers to his development of MBSR

as his "karmic assignment" and "personal koan" (2011, p. 286).

Systematic Communication of Core Buddhist Beliefs

MBSR consists of eight 2.5–3.5 h classes, plus a 7.5 h retreat and 45 min daily of personal practice. Classes foreground instruction in three easy-to-learn techniques: hatha yoga, body scan, and sitting meditation. On February 28, 2015, Bob Stahl, Adjunct Senior Teacher for the CfM's Oasis Institute and co-author of *A Mindfulness-Based Stress Reduction Workbook* (2010), posted a two-page unpublished document (probably written much earlier) to a secure online CfM forum for MBSR teachers, under the topic "MBSR Underpinnings." The document details session by session how the MBSR class sequence provides a "full expression" of "the essence of the dhamma," including the "4 noble truths, 4 foundations of mindfulness, and 3 marks of existence." Page one, "The Heart of the Dhamma," enumerates key Buddhist doctrines, citing their sources in Buddhist sacred texts: "1. Four Noble Truths (Dhammacakkappavattana Sutta)," beginning with "Suffering/Stress" and culminating with the "8-fold Path to freedom," "II. Three Marks of Existence (Anattalakkhana Sutta) … Suffering, Impermanence, No Self," and "III. Four Foundations of Mindfulness," including mindfulness of the "Breath," "postures of the Body," "Teachings (Dharmas)," and the "7 Factors of Awakening." The second page, "Central Elements of MBSR: The Essence of the Dhamma," begins with an explanatory note:

> Without explicitly naming the 4 noble truths, 4 foundations of mindfulness, and 3 marks of existence, these teachings are embedded within MBSR classes and held within a field of loving-kindness. MBSR is a full expression of the 4 noble truths: suffering, its causes, and the path to freedom.

For instance, "Class 1 contains the 1st noble truth and marks of existence … suffering, impermanence, and the selfless nature evoked by body scan … Class 4 begins to investigate the causes of stress/suffering (2nd Noble truth) … Class 5 points to the 3rd noble truth. …. Classes

6–8 draw from the 4th Noble Truth, the 8-fold path." Stahl presumably circulated this insider document to remind MBSR teachers of the principles that they should be communicating in each class session and to respond to potential criticisms that MBSR is a dilution of the Dharma.

Margaret Cullen, one of the first ten CfM-certified MBSR instructors, confirms many of the details of Stahl's unpublished summary in an article published in *Mindfulness* (2011). In Cullen's account:

> The intention of MBSR is much greater than simple stress reduction. Through systematic instruction in the four foundations [as defined by the *Satipaṭṭhāna Sutta*] and applications in daily life, as well as through daily meditation practice over an 8-week period, many participants taste moments of freedom that profoundly impact their lives.

For example, the body scan is "designed to systematically, region by region, cultivate awareness of the body—the first foundation of mindfulness." Sitting meditation begins with "awareness of the breath," proceeding to "systematic widening of the field of awareness to include all four foundations of mindfulness" (p. 188). This promotes "insights into no-self, impermanence and the reality of suffering," dispels "greed, hatred, and delusion," and leads "automatically" to "enlightenment" (p. 192). MBSR also has "elements of all of the *brahma vihāras* [loving-kindness, compassion, sympathetic joy, equanimity] seamlessly integrated into it" (p. 189). Cullen concludes that MBSR represents a new "lineage" of Buddhism, a distinctively "American," though no less Buddhist, formulation of the Dharma (p. 191).

Melissa Myozen Blacker, who spent twenty years as a teacher and director of programs at CfM, corroborates key statements by Stahl and Cullen. Blacker recalls that "the MBSR course was partly based on the teachings of the four foundations of mindfulness found in the *Satipatthana Sutta* ... and we included this and other traditional Buddhist teachings in our teacher training." Yet, "for the longest time, we didn't say it was Buddhism at all. There was never any reference to Buddhism in the standard eight-week

MBSR class; only in teacher training did we require retreats and learning about Buddhist psychology" (Wilks et al. 2015, p. 48). Given the need to present MBSR as secular in order to achieve its mainstreaming, Kabat-Zinn and other movement leaders have had to rely heavily upon MBI teachers to "embody" the Dharma.

Teachers do so, in part, by reading poems that "evoke particular feelings and moods"—favoring spiritual poets such as thirteenth-century Sufi mystic Rumi and American metaphysical authors Walt Whitman and Mary Oliver. An MBSR teacher interviewed by this author in 2015 observed that the choice of whether to use poems seems to determine if participants report spiritual experiences. MBSR and MBCT teacher Jenny Wilks recalls that one participant objected to a poem read in a colleague's class as "New Agey" and "brainwashing" (Wilks in Cheung 2015, p. 7). Wilks acknowledges that "key Dharma teachings and practices are implicit ... even if not explicit" in secular classes, which present "more of a distillation than a dilution"—a form of "highly accessible Dharma" (2014a, September 8, Sect. 4–6). By contrast, Wilks worries that "explicitly Buddhist ethics could potentially offend participants who are atheist, Christian, [or] Muslim" (2015, p. 7).

The MBSR model relies on teachers to convey not only techniques, but also worldviews to students. MBI teacher training includes "cultivation of a particular attitudinal framework" and "assimilating a particular view of the nature of human suffering" (Crane et al. 2010, p. 82). The CfM "Standards of Practice" guidelines specify that MBSR teachers cultivate "foundational attitudes" of "non-judging, patience, a beginner's mind, non-striving, acceptance or acknowledgement, and letting go or letting be" (Santorelli 2014, p. 10). These attitudes are, according to psychologist Steven Stanley, "related to core virtues found in early Buddhist texts, such as generosity, loving-kindness, empathetic joy and compassion" (2015, p. 99). Instilling these attitudes is important because the "insights that arise for MBI participants" do so within the "scaffolding created by the teacher and the curriculum. It is the job of the skillful teacher to engage directly with the students, challenging beliefs, inviting deeper exploration, suggesting where and how to pay attention, all within the framework of a secular articulation of the four foundations of mindfulness" (Cullen 2011, p. 190). As one MBSR teacher interviewed by this author in 2015 explains, MBSR class discussions are not "open-ended." If a participant shares an experience at odds with MBSR assumptions, that person gets "corrected," encouraged to "look again," or "given a different answer." Participants are "very, very carefully guided to land at a certain answer" and "given a way of understanding experiences" that "set up what people are supposed to value and change."

Teacher Training

MBSR has rigorous—and specifically Buddhist—prerequisites, training, and continuing education requirements for teachers. *Teaching Mindfulness: A Practical Guide for Clinicians and Educators*, with a foreword by Kabat-Zinn, lists as prerequisites for MBSR teachers a "3-year history of daily meditation practice; participation in two 5-day or longer mindfulness retreats in the Theravada or Zen traditions; [and] three years of body-centered practice, such as Hatha Yoga" (McCown et al. 2010, p. 15). Although there are variations in requirements for the growing number and range of MBIs, many urge personal mindfulness practice, retreat experience, and ongoing supervision. This is true, for instance, of Mindful Schools and UCLA's Mindfulness Awareness Research Center (Mindful Schools 2016, Sect. 7; Winston in Wilks et al. 2015, p. 50). Requirements for "continuing professional development" of MBSR/MBCT and other MBI teachers in the UK Network for Mindfulness-Based Teacher Trainers include annual residential retreats and "an ongoing and regular process of supervision/peer supervision of teaching, and inquiry into personal practice by an experienced teacher," with the goal of integrating insights gained through personal practice into teaching (Crane et al. 2010, p. 81).

Movement leaders emphasize the personal mindfulness practice of instructors because they envision mindfulness as "not simply a method,"

but "a way of being" (2003, p. 149). Specifically, Kabat-Zinn insists, "all MBIs are based on … Buddhadharma" (2011 p. 296). The MBI instructor must "translate" meditation into a "vernacular idiom," but "without denaturing the dharma dimension. This requires some understanding of that dimension, which can come about only through exposure and personal engagement in practice—learned or deepened either through meditation retreats at Buddhist centers or through professional training programs in MBSR with teachers who have themselves trained in that way, or, ideally, both" (2003, p. 9). The CfM "Principles and Standards" require that the MBSR teacher be a "committed student of the dharma, as it is expressed both within the Buddhist meditation traditions and in more mainstream and universal contexts exemplified by MBSR" (Kabat-Zinn and Santorelli, n. d.) In other words, Buddhist training constitutes a necessary qualification for teaching secular mindfulness in the MBSR model.

Retreats at Buddhist centers play a prominent role in training MBSR teachers. Kabat-Zinn describes the "periodic sitting of relatively long (at least 7–10 days and occasionally much longer)" retreats as an "absolute necessity" and "laboratory requirement" for MBSR teachers (2011, p. 296). He recommends retreats in the "Buddhist Theravada tradition (vipassana)," such as those offered by the Insight Meditation Society: http://www.dharma.org (2003, p. 154). The CfM—the fountainhead of MBI training—requires silent, residential, teacher-led "vipassana retreats (or an equivalent)" as a prerequisite for interns (Kabat-Zinn 2003, p. 154). To be considered for MBSR Oasis Teacher Certification, two of at least four retreats must be nine days or more. The CfM's Oasis Institute for Professional Education and Training lists fourteen acceptable retreat centers—all of which are Buddhist (CfM 2014c, Sect. 3).

Certain Buddhist retreat centers, such as Spirit Rock, California and the Insight Meditation Society, Massachusetts, host 9-day retreats specifically for MBI professionals. According to Margaret Cullen, MBI retreats share with other retreats offered at these centers "the same reliance on the original teachings of the Buddha, and Dharma talks are offered to illuminate essential components of Buddhist philosophy. Sitting and walking practice are taught much as they would be at any other vipassana retreat." Participants may be asked to observe the Five Buddhist Precepts (abstaining from killing, stealing, sexual misconduct, lying, or intoxication) for the duration of the retreat (Hickey 2010, p. 174). The primary distinctive of MBI retreats is that they draw more international participants, and the leaders are more actively involved in modeling practices such as walking meditation and silent meals (Cullen 2011, p. 192).

As an example of an MBI retreat, a CfM listserv advertised "Convergence: An Insight Meditation Retreat with Saki Santorelli, Carolyn West and Bob Stahl, May 8–15, 2015," in West Hartford, Connecticut. The retreat invites "anyone teaching or aspiring to teach mindfulness in healthcare, psychology, education, science, government, or in the business and corporate sectors" to explore how the "Insight (Vipassana) Meditation Tradition influences MBSR and all MBIs." The ad promises that "through the direct practices of the four foundations of mindfulness you will learn how the essence of these wisdom teachings (Dharma): the four noble truths and the three characteristics of existence intersects and informs all MBIs." Perhaps reflecting worry about mission drift, the retreat focuses on instructing secular mindfulness teachers in the Buddhist foundations of "all" MBIs.

Graduate Resources

MBSR offers graduates resources to maintain and deepen their meditation practice. The program ends with an invitation to join an ongoing meditation community such as an Insight Meditation Society, which Kabat-Zinn describes as having "a slightly Buddhist orientation" (1990, p. 436). The CfM's MBSR "Authorized Curriculum Guide" directs teachers in week eight to encourage participants in "keeping up the momentum and discipline developed over the past 7 weeks" through "books, recordings, graduate programs, free all day sessions for all

graduates 4 times per year; mention retreat centers" (Blacker et al. 2015, p. 28). The CfM FAQs webpage recommends that graduates "expand" their understanding by reading; suggested books explain Buddhist and metaphysical doctrines (2014a, para. 28). For example, Kabat-Zinn's *Wherever You Go, There You Are* (1994b) devotes chapters to "ahimsa" (non-harming) and "karma" (consequences), suggests placing the hands into "mudras" that are "associated with subtle or not-so-subtle energies" or turning the palms up in receptivity to the "energy of the heavens," and alludes to the chakra system by explaining that the "solar plexus" helps contact "vitality" (pp. 113–14, 154, 217–19, 220–25).

Guided meditation audio recordings supplement the MBSR course. The CfM website links to Kabat-Zinn's website, which sells three four-CD (or MP3) series. The first series, designed for use during MBSR, suggests particular ways of framing and interpreting meditation experiences. Listeners hear repeatedly that "judgmental and critical thoughts" are "afflictive," whereas "non-conceptual" meditation on this-moment bodily sensations offers access to a "realm of oneness" that is "awareness itself" (1.1, 1.2).

As one progresses from the first through the third series of meditations, the content becomes progressively more explicit in its Buddhist references. Third series recordings teach foundational Buddhist beliefs, for instance that suffering is caused by "greed or aversion or delusion or ignorance," and mindfulness offers a path for "freeing ourselves from all our conditioning of mind and heart and the suffering it brings with it" (3.1, 3.2, 3.8). Guided meditation 3.4 reveals that what MBSR terms "Choiceless Awareness" is "known in the Chinese Zen tradition, in the Chan tradition, as silent illumination or the method of no method. In Japanese Zen it is sometimes called *Shikantaza*, which translates literally as just sitting nothing more. In the Tibetan tradition it is often called *Dzogchen* or mind essence or the great natural perfection. ... The Tibetans refer to this as self-liberation." Thus, Kabat-Zinn insists that MBSR teaches the same authentically Buddhist practice, "secularized" only by renaming.

Kabat-Zinn's "Heartscape" meditation opens by framing loving-kindness meditation as an authentic Buddhist practice: "Loving-kindness or *metta* in the Pali language is one of four foundational practices taught by the Buddha known collectively as the heavenly abodes or the divine abodes: loving-kindness, compassion, sympathetic joy, and equanimity … used for the most part to cultivate Samadhi, or one-pointed concentrated attention" (3.2). The "evoked qualities emerge"—both from the formal practice of *metta* and from "all the mindfulness practices" since they contain the same "essence"—with a "power" for "transfiguring the heart," resulting in the "heart's liberation." Because all beings are interconnected in the "lattice structure of reality," the "world benefits and is purified from even one individual's offering of such intentions." Individuals "literally and metaphorically" possess a "capacity for love" that is "limitless," and thus "the web of all life" may be "shifted" through one person's practice.

This *metta* meditation begins by speaking blessings over oneself: "May I be safe and protected from inner and outer harm. May I be happy and contented. May I be healthy and whole to whatever degree possible. May I experience ease of wellbeing." The "field of loving-kindness" expands first to loved ones and ultimately to "our state," "our country," "the entire world," "all animal life," "all plant life," "the entire biosphere," and "all sentient beings." "May all beings near and far … our planet and the whole universe" be "safe and protected and free from inner and outer harm," "happy and contented," "healthy and whole," and "experience ease of wellbeing." It is worth noting the similarity in phrasing between this guided meditation and those used in a growing number of secular programs, for instance Mindful Schools and Inner Kids. Such programs typically "secularize" the "May I/you be" blessings by labeling them "heartfulness" or "friendly wishes" instead of *metta* (Bahnsen 2013; Greenland 2013, Sect. 4).

Stealth Buddhism

Certain of Kabat-Zinn's numerous disciples refer to their favored tactic as "stealth Buddhism" (and some Buddhist commentators credit Kabat-Zinn with coining the phrase). Trudy Goodman, founder of Insight LA, in California, describes her approach as "Stealth Buddhism" in a podcast interview by that title aired on Buddhist Geeks.com (2014). Goodman teaches secular mindfulness classes in "hospitals, and universities, and schools, and places where as Buddhists we might not be so welcome especially state places," given the "separation of church and state." Although advertised as secular, such classes, in Goodman's view, "aren't that different from our Buddhist classes. They just use a different vocabulary," but "anyone who practices sincerely, whether they want it or not," is going to experience "healing from the delusion that we have about who we are, this fundamental illusion that we carry, about the 'I' as being permanent and existing in a real way … I think it's inevitable." Interviewer Vincent Horn concurs that the effects are "independent of whether one is trained in a Buddhist context, or in a new, non-Buddhist Buddhist context," what Emily Horn describes as the "new American religion." The interview ends with all three laughing aloud at their promotion of "stealth Buddhism."

Trojan Horse

Stephen Batchelor, meditation teacher and advocate of "Secular Buddhism," popularized the phrase "Buddhist Trojan horse." Once mindfulness has been "implanted into the mind/brain of a sympathetic host; dharmic memes are able to spread virally, rapidly and unpredictably" (2012, p. 89). In a Buddhist Geeks.com podcast titled "The Trojan Horse of Meditation," Kenneth Folk identifies his teaching of meditation in the Silicon Valley as a "stealth move" in which he "sneak[s]" into mindfulness training his own

Buddhist "value systems" of "compassion and empathy" (2013, para. 13–18). Buddhist Geeks producer Kelly Sosan Bearer adds that just getting elites on the cushion is enough because meditation is an "inherent process" that leads to awakening (para. 33).

Scripting

Actress and movie producer Goldie Hawn boasts of writing a "script" to sneak Buddhist meditation "into the classroom under a different name because obviously people that say 'oh meditation' they think oh this is 'Buddhist'." Hawn's script is *The MindUP Curriculum* for K-8 classrooms published through Scholastic Books. In an address to Buddhist insiders at the Heart-Mind Conference of The Dalai Lama Center for Peace-Education, Hawn says that MindUP "all started" with "His Holiness" (who "gave me my mantra") and the Dalai Lama Center ("it's karma"). Hawn explains: "I'm a producer, I'm gonna put this show on the road … and I got the script written, and I call it a script because it is, it's one step of how the story gets told of how you're able to facilitate the best part of you" (2013). The MindUP script replaces the terms "Buddhism" and "meditation" with "neuroscience" and "Core Practice." Hawn's goal is to see MindUP "absolutely mandated in every state … that's our mission" (2011, para. 67–68).

The Hawn Foundation hired a team of educators, neuroscientists, and psychologists to work with Buddhist meditators in constructing the MindUP curriculum. The result might be described as "bricolage"—a loose assemblage of cultural symbols and rituals, some secular and others religious (Hatton 1989, p. 75). The bulk of the content has little to do with the "signature" Core Practice of meditation or broader understandings of mindfulness, despite the frequent peppering of lessons with this term. Simplified instruction in brain anatomy ("reflective, thinking prefrontal cortex" = good, "reflexive, reactive amygdala" = bad) and exhortations to be kind to others (pause for a moment before hitting another kid back), oneself (if you actually try your vegetables you might like them), and the earth (recycle instead of littering) may produce educational and social effects regardless of whether students do or do not engage in the Core Practice, conceive of what they are doing as "mindful," or are even introduced to this term.

The MindUP curriculum promises that what makes it distinctive is three-times-daily "brain breaks" of "deep belly breathing and attentive listening" that instill "empathy, compassion, patience, and generosity," virtues derived from though not credited to Buddhist ethics (Hawn Foundation 2011, pp. 11–12, 40–43, 57). The curriculum emphasizes that "to get the full benefit of MindUP lessons, children will need to know a specific vocabulary," chiefly the term mindfulness—circularly defined as the opposite of "unmindfulness"—and repeated multiple times per lesson, suggesting that mindfulness is the key to any positive attitude or behavior. Lessons encourage children to think of role models who act in mindful ways—the custodian who picks up trash, a doctor who keeps calm in emergencies, or an imaginary dinosaur who eats its vegetables—though none of these role models may ever have meditated. But defining mindfulness as synonymous with virtue makes it seem urgent to use the Core Practice.

Frequent repetition of the term mindfulness also points children and parents to where they can find resources to deepen their practice. Following links from The Hawn Foundation website leads to Buddhism. The website includes a "Science Research Advisory Board" page that, as of August 2016, lists exactly one board member: Kimberly Schonert-Reichl, Associate Professor of Human Development Learning and Culture and Special Education at the University of British Columbia (Hawn Foundation 2016). This page links to biographical sketches describing Schonert-Reichl as a "long-time partner with the Dalai Lama Center for Peace-Education" and to videos of Schonert-Reichel reassuring Buddhist audiences at the Vancouver Peace Summit and at the Garrison Institute that "secularized" classroom mindfulness effectively advances "Buddhist Contemplative Care" (2009, at 1:11:28; 2011, at 38:35; 2012, para. 2).

Buddhist Critiques of Deception as Wrong or Unskillful Speech

Certain Buddhist religious leaders and scholars have sharply criticized the "secular" mindfulness movement as self-contradictory or deceptive (Shonin et al. 2013; Purser 2015). Thupten Jinpa Langri, translator and interpreter for the Dalai Lama, has "often told" movement leaders that they "cannot have it both ways. It is either secular, or you want to say it's the essence of Buddhism, therefore it's a Buddhist practice" (2013, quoted in Purser 2015, p. 26). Brooke Dodson-Lavelle, director of the Mind and Life Institute's Ethics, Education, and Human Development Initiative, calls attention to a Mindfulness in Education Network e-mail list on which "regular postings appear that either blatantly or suggestively describe ways in which program developers and implementers have 'masked' or 'hidden' the Buddhist roots of their mindfulness-based education programs." Dodson-Lavelle elaborates that "The sense is that one needs to employ a secular rhetoric to gain access into educational institutions, and once one's 'foot is in the door,' so to speak, one is then free to teach whatever Buddhist teachings they deem appropriate." Threads also imply that "discussions concerning secularization are merely semantic games designed for 'them,' because 'we' all really know what is going on here" (2015, p. 132). Such critiques indicate that not all Buddhists consider stealth approaches to be right or skillful speech.

Pattern #2: Unintentional Indoctrination

Despite semantic games self-consciously played by some, many secular mindfulness promoters are convinced that the mindfulness technique is non-religious, produces scientifically validated health benefits, and instills universal values. They may nevertheless unintentionally communicate more than a religiously neutral technique. This is because suppositions about the nature of reality can become so naturalized and believed so thoroughly that it is easy to infer that they are simply true and universal, rather than recognizing ideas as culturally conditioned and potentially conflicting with other worldviews. Stephen Batchelor observes that "although doctors and therapists who employ mindfulness in a medical setting deliberately avoid any reference to Buddhism, you do not have to be a rocket scientist to figure out where it comes from. A Google search will tell you that mindfulness is a form of Buddhist meditation." Thus, Batchelor continues, an "unintended consequence" of even an eight-week secular MBSR course can be that it opens for participants "unexpected doors into other areas of their life, some of which might be regarded as the traditional domains of religion" (2012, pp. 88–89).

Calling attention to the "fallacy of values-neutral therapy," Buddhist mindfulness teacher Lynette Monteiro argues that "regardless of the intention to not impose extraneous values," it is problematic to define MBIs as secular because Buddhist values are "ever-present and exert a subtle influence on actions, speech and thoughts," potentially disrespecting client values (2015). Monteiro is not alone among Buddhist commentators in worrying that purportedly secular mindfulness programs fail to present "truly belief neutral" programming that respects the religious diversity of participants (Oman 2015, p. 52; Warnock 2009, p. 477; Farias and Wikholm 2015). As psychologist Stephen Stratton concludes, the presumed "distinction between the secular and the religious and/or spiritual when it comes to meditation in general and mindfulness in particular" may be "simplistic. A more culturally aware perspective might suggest that religious-spiritual dimensions are always potentially present, even in overtly secular processes," an observation that calls for ethical reflection (2015, p. 113).

Buddhist and Christian Assumptions Compared

Secular mindfulness programs instill a religious worldview that clashes with other worldviews.

This can be illustrated by a comparison with historic Christian teachings: Christian scriptures encourage meditation not on the breath or body, but the Bible's revelation of God as Creator of breath, body, and everything else. Rather than non-judgmental, accepting awareness of the present, Christian teachings encourage rejection of certain thoughts and feelings as wrong; repentance of past sins and grateful remembrance of God's redemptive work in history; faith in God's future promise of eternal life and striving to live a holy life. Instead of envisioning life as suffering or seeking to extinguish attachments or escape the cycle of death and rebirth, Christians view life as a good gift from God and anticipate that God will grant individuals the desires of their heart as they delight themselves in God. Contrary to waking up to realize that everything is impermanent, there is no self, or that awareness itself is the ultimate reality, Christians affirm that a personal God created each individual as a unique, enduring self for the purpose of eternal relationship with God. For Christians, the source of suffering is sin, or disobedience to God, and the only path to end suffering was paved by God's love for humanity, demonstrated through Jesus's atoning death and resurrection, and which can only be appropriated through repentance and faith in Jesus as one's personal Savior. In place of locating the source of compassion in the non-dual realization that everyone is part of the same Buddha nature, Christians adopt a dualist belief that a transcendent God is love and the source of human compassion.

Although many Buddhists and Christians share certain vocabulary, such as "compassion" and "loving-kindness," they may define these terms so differently that they aspire toward competing ideals. For example, Christians place a high value on sacrificial love—purportedly demonstrated by Jesus's willingness to sacrifice his life for the sake of fundamentally other "selves." Christians view their own highest calling as to love others—even when doing so means sacrificing one's own needs for those who give nothing in return. To imply that compassion and loving-kindness relieves one's own suffering and promotes one's own happiness because everyone shares the same nature may be perceived as conflicting with central Christian values.

Differences Among Buddhist Schools

The universality of assumptions and values communicated by secular mindfulness is further belied by disagreements among Western convert Buddhists. For instance, Buddhists differ about whether the goal of mindfulness should be sensory enhancement or detachment (Purser 2015, p. 30); stress reduction or induction (Lopez 2012, para. 14); non-judgmental acceptance or ethical discernment (Dreyfus 2011, p. 51); happiness or dissatisfaction (Heuman 2012, para. 1). Brooke Dodson-Lavelle systematically compares three major "secular" meditation programs: MBSR, Cognitively-Based Compassion Training (CBCT), and Innate Compassion Training (ICT) in her Emory University dissertation (2015). Although all three employ secular, universalist "rhetoric," they stem from "competing," yet all "very Buddhist," understandings of the causes and solutions of stress and suffering, and rival expectations about the innate capacities for compassion in human nature. None of these programs are morally or ethically neutral, but rather "tell people, at least implicitly, stories about what they *ought* to be thinking, feeling, or doing" (Dodson-Lavelle 2015, pp. 7–10, 21, 95, 161, 163).

The Dalai Lama's interpretation of "secular ethics," used to validate such programs as secular, accepts as self-evident that all people share goals and values such as avoidance of suffering and compassion—and, in so doing, may "dangerously overlook the natural capacity humans possess for violence and evil" (Ozawa-de Silva 2015, p. 1; Dodson-Lavelle 2015, p. 168). In Dodson-Lavelle's experience teaching all three programs, the notion that "all beings want to be happy and avoid suffering" has "failed to resonate" with many participants, further calling into question the universality, and hence the presumed secularity, of the values communicated (Dodson-Lavelle 2015, pp. 17, 96–99, 162).

What Kabat-Zinn seems to mean in asserting that "the dharma" is "universal," thus non-religious, is that dharmic assumptions are universally true (Davis 2015, p. 47). This claim may, however, be undercut by his choice of an "untranslated, Buddhist-associated Sanskrit word" (Helderman 2016, p. 16). Jeff Wilson argues that "Dharma is itself a religious term, and even to define it as a universal thing is a theological statement" (2015). Stephen Batchelor suggests that each of the Four Noble Truths is a "metaphysical statement" that can neither be proven nor refuted (2012, p. 93). As professor of counseling David Forbes (2015) explains, the "myth of the given" is that reality can be objectively presented and directly perceived. Exhortations to "wake up" and "see things as they are" gloss hidden cultural constructs and the favoring of one set of lenses with which to view and interpret reality over another. For instance, seeking to attenuate desire and cultivate equanimity reflects a culturally specific ideal affect that values "low-arousal emotions like calm" (Lindahl 2015, p. 58). Universalist rhetoric privileges the perspectives of mindfulness promoters, many of whom are white and economically privileged, as "objective and representative of reality," "standing outside of culture, and as the universal model of humans" (DiAngelo 2011, p. 59; Ng and Purser 2015, para. 4). This is not only a culturally arrogant position; it is precisely a *religious* attitude—a claim to special insight into the cause and solution for the ultimate problems that plague humanity.

Pattern #3: Religious and Spiritual Effects

Promoters of secular mindfulness cite scientific research to support their claim that mindfulness is an empirically validated technique rather than a religious ritual. Regardless of the strength of the scientific evidence, appeals to science are beside the point of this chapter. The same practice can exert both secular and religious effects simultaneously. Abundant scientific research demonstrates that religious and spiritual practices promote physical and mental health (Koenig et al. 2012; Aldwin et al. 2014). Historian Jeff Wilson observes that Buddhism has repeatedly gained access to new cultures by offering "this-worldly or practical benefits" that make Buddhist religion seem relevant and appealing (2014, p. 4).

Anecdotal Reports

Anecdotes can be cited of individuals finding their way into Buddhism after being introduced to mindfulness through a secular course. As one MBSR graduate testified, "I took an 8 week Mindfulness Based Stress Reduction Course two years ago without knowing anything about Buddhism. …That program spurred my curiosity and here I am learning all about the Four Noble Truths" (JKH 2015).

Secular mindfulness teachers often attest that secular classes provide a doorway into Buddhism. Dharma teacher Janette Taylor reflects that "there are different levels of the Dharma to be taught," and that "taking the more secular approach at first, gives more people a doorway that they can enter easily. Then, once they gain their footing, they become more willing to explore the more transcendent aspects of the Dharma" (2013). Trudy Goodman says the "really interesting question" is what people "do after" they take a secular class, answering that some "sort of migrate into Buddhism" (2014). Melissa Myozen Blacker recalls of her experience at CfM that "after eight weeks," MBSR participants were "transformed," further noting that the Boundless Way Zen Temple where she now serves as abbot attracts people who "even ten years ago, wouldn't have come to a Zen temple" (Wilks et al. 2015, p. 54). Jenny Wilks likewise recounts that Buddhist retreat centers have "seen an increase in the numbers of people coming on retreats and many of them have started with a secular eight-week course" (2014a, Sect. 4). Stephen Batchelor observes that "on every Buddhist meditation course I lead these days, there will usually be one or two participants who have been drawn to the retreat because they want to deepen their practice of 'secular mindfulness'" (2012, p. 88). Neuroscience researcher and MBSR teacher Willoughby Britton reports that a number of students who have been introduced to mindfulness through college courses have subsequently taken off time to go on long retreats, often in Asia, and/or ordain as Buddhist monks/nuns (2011, para. 37). In both her research on "the Varieties of Contemplative Experience" and her MBSR/CT clinic, she has seen a number of individuals who came to meditation through MBSR describe a meditation-induced loss in sense of self that was accompanied by significant levels of distress and impairment of functioning (2014, para. 30). Kabat-Zinn seeks to soften his admission that "a lot of patients do go deeper into Buddhism and do retreat practice" by adding that some "also go to Catholic and Jewish retreat centers" (2010, para. 32). Similarly, Margaret Cullen suggests that many MBI graduates "report a deeper connection to their own faith tradition, and its attendant moral code" (2011, p. 189). Although intended to distance MBIs from Buddhism, these latter statements undermine the assertion that MBIs are fully secular or non-*religious*.

Research Studies

Research studies confirm anecdotal reports of an association between secular mindfulness and increased religiosity. Psychologists Tim Lomas and colleagues conducted in-depth narrative interviews with thirty Buddhist meditators. Most had first tried meditation for secular reasons, such as stress management, but for many of them, "meditation became their gateway to subsequent interest in Buddhism" (2014, p. 201).

Quantitative survey research by psychologist Jeffrey Greeson and colleagues of participants in MBSR classes (2011, $n = 279$; 2015, $n = 322$) taught by CfM-trained instructors found a significant correlation between increased mindfulness and spirituality. Most participants in the 2011 study enrolled wanting improved mental health (90 %), help managing stress (89 %), and improved physical health (61 %); half (50 %)

agreed that "exploring or deepening my sense of spirituality" motivated enrollment. After eight weeks, 54 % reported that the course had deepened their spirituality, including personal faith, meaning, and sense of engagement and closeness with some form of higher power or interconnectedness with all things. The study concludes that mental health benefits of secular mindfulness can be attributed to increases in daily spiritual experiences. In 2015, Greeson's team replicated the finding that "increases in both mindfulness and daily spiritual experiences uniquely explained improvement in depressive symptoms" (p. 166). Smaller studies (Astin 1997, $n = 28$; Carmody and Kristeller 2008, $n = 44$) also report associations between MBSR and increased spirituality scale scores.

In a study of Vipassana retreat participants ($n = 27$), Deane Shapiro found that practitioner intentions shifted over time along a continuum from self-regulation, to self-exploration, to self-liberation (1992, pp. 33–34). Shapiro also found a statistically significant relationship between religious orientation and length of practice. Longer-term meditators were less likely to be religious "Nones" or monotheists and more likely to identify as Buddhist or with "All" religions.

Reaching an apparently conflicting conclusion, a study of prisoners ($n = 57$) participating in a 10-day Vipassana retreat found no significant change in post-intervention scores on the Religious Background and Behavior Questionnaire (RBBQ). From this, the authors conclude that "mindfulness meditation, even when taught in a traditional Buddhist context, may be attractive and acceptable to those of other religious faiths, and involvement in such practices does not threaten engagement in non-Buddhist religious practices" (Bowen et al. 2015, p. 1461). It is important that the authors measured religious affiliation before the retreat, but not afterward, so it is possible that participation induced unrecognized changes in religious affiliation. The authors excluded the meditation item from their analysis (because it would have increased the post-intervention scores), reporting only a composite score for the other five items: "thought about God," "prayed," "attended religious services," "studied holy writings," and "had direct experiences with God" (p. 1458). It is therefore possible that decreases in certain measures (for instance of Christian spirituality) offset increases in others (such as Buddhist spirituality), or that the contents within measures shifted (for instance from study of Christian to Buddhist holy writings). Such shifts would reconcile the study's findings with other research suggesting that even secularized Buddhist meditation increases reported spiritual experiences.

Research on mantra meditation offers additional insight. Psychologists Amy Wachholtz and Kenneth Pargament (2005, 2008) compared groups focusing on a spiritual phrase (e.g., "God is good") with "internal" secular (e.g., "I am good"), "external" secular (e.g., "Grass is green"), and progressive muscle relaxation groups. Although the spiritual meditation groups reported "significantly more daily experiences of a spiritual nature," the authors were surprised to note that the other groups also reported increased daily spiritual experiences. The authors ponder that "if secular meditation was truly devoid of spirituality, then the number of spiritual experiences should not have been affected by this ostensibly secular meditation technique." They infer that "secular meditation tasks represent *less*-spiritually oriented, rather than *non*-spiritually oriented, meditation tasks" (2005, p. 382). The authors suggest that because "historically, meditation has been embedded in a larger spiritual matrix … it may be impossible to disconnect meditative practices fully from this larger context. Thus, the distinction between 'secular' and 'spiritual' meditation may be overdrawn" (2008, p. 363). These findings suggest difficulties with the project of secularizing meditation.

Conclusion

Although it may be theoretically possible to separate mindfulness from religion, and specifically Buddhism, this has often not occurred despite the use of secularizing rhetoric. Upon closer examination, the asserted boundaries

between Buddhist and secular mindfulness in many instances dissolve. A basic difficulty is that the term mindfulness, in the contemporary American cultural context, does double-duty—opening onto a comprehensive Buddhist worldview and way of life even when introduced as a mere therapeutic technique. The problem is made worse because secular mindfulness movement leaders have intentionally engaged in Code-Switching tactics (skillful speech, stealth Buddhism, Trojan horse, scripting). Jon Kabat-Zinn's MBSR model is a prime example of an MBI infused at every level—concept, structure, teacher training, and graduate resources—with carefully camouflaged Buddhist content. It is not enough, however, to eschew deception. Advocates may truly believe that mindfulness is secular because its values seem to them self-evidently universal and science validates its practical benefits. But it is easy to confuse culturally and religiously specific diagnoses and prescriptions for the ultimate problems that plague humanity with universally shared goals, values, and human capacities. Given the pervasiveness of explicit and implicit Buddhist content in many MBIs, it should come as no surprise that research suggests that even nominally secular mindfulness programs produce religious and spiritual effects.

Returning to definitions of the religious, spiritual, and secular that opened this chapter, the secular mindfulness movement may provide a potent illustration of the difficulty, if not impossibility, of disentangling these co-constructed categories. Mindfulness is steeped in transcendent beliefs and enacted through practices that purportedly connect individuals with ultimate reality, trace a path to salvation from suffering, and cultivate spiritual awareness and virtues. The mindfulness movement has its own creeds or compelling explanations of what is real; implies codes of moral and ethical behavior; reinforces its creeds and codes through cultuses or repeated words and actions; and is practiced through formal and informal communities. Ultimate ideas, metaphysical beliefs, a comprehensive worldview, and external signs of religion and spirituality can all be identified. Mindfulness might be understood as secular if one reduces religion to rhetoric and secularity to this-worldly effects. However, if one means by secular the absence of religious and spiritual beliefs and practices, this is a harder case to make. Like the Catholic priest who is secular because he lives in the world rather than a cloister, the secular mindfulness movement continues to carry Buddhist religious influences into the mainstream. Whether or not mindfulness *can* be separated from religion, in today's cultural milieu secular and religious mindfulness seem conjoined twins.

References

Adams, A. (1979). Concurrence in *Malnak v. Yogi*. 592 F.2d 197 (3rd. Cir.). Retrieved January 12, 2016, from https://casetext.com/case/malnak-v-yogi

Albanese, C. (2013). *America, religions and religion* (5th ed.). Boston: Wadsworth.

Aldwin, C. M., Park, C. L., Jeong, Y. J., & Nath, R. (2014). Differing pathways between religiousness, spirituality, and health: A self-regulation perspective. *Psychology of Religion and Spirituality, 6*(1), 9–21.

Asad, T. (2003). *Formations of the secular: Christianity, Islam, modernity*. Stanford, CA: Stanford University Press.

Astin, J. A. (1997). Stress reduction through mindfulness meditation: Effects on psychological symptomatology, sense of control, and spiritual experiences. *Psychotherapy and Psychosomatics, 66*, 97–106.

Bahnsen, M. (2013). *Healthy habits of mind*. Persona Film. Retrieved January 2, 2016, from http://www.mindfulschools.org/resources/healthy-habits-of-mind/

Batchelor, S. (2012). A secular Buddhism. *Journal of Global Buddhism, 13*, 88–89.

Bender, C. (2010). *The new metaphysicals: Spirituality and the American religious imagination*. Chicago: University of Chicago Press.

Berger, P. L. (2014). *The many altars of modernity: Toward a paradigm for religion in a pluralist age*. Boston: De Gruyter.

Blacker, M., Meleo-Meyer, F., Kabat-Zinn, J., Koerbel, L., & Santorelli, S. (2015). *Authorized curriculum guide for Mindfulness-Based Stress Reduction 2015*. Worcester, MA: CfM.

Bowen, S., Bergman, A. L., & Witkiewitz, K. (2015). Engagement in Buddhist meditation practices among non-Buddhists: Associations with religious identity and practice. *Mindfulness, 6*(6), 1456–1461.

Britton, W. (2011). The dark night project. Interview by V. Horn. BG232 Retrieved January 2, 2016, from http://www.buddhistgeeks.com/2011/09/bg-232-the-dark-night-project/

Britton, W. (2014, April 25). Meditation nation. Interview by L. Heuman. *Tricycle*. Retrieved January 2, 2016, from http://www.tricycle.com/blog/meditation-nation

Calhoun, C., Juergensmeyer, M., & Van Antwerpen, J. (2011). *Rethinking secularism*. New York: Oxford University Press.

Carmody, R., & Kristeller, M. (2008). Mindfulness, spirituality, and health-related symptoms. *Journal of Psychosomatic Research, 64*(4), 393–403.

Casanova, J. (1994). *Public religions in the modern world*. Chicago: University of Chicago Press.

CfM [Center for Mindfulness in Medicine, Health Care, and Society]. (2014a). FAQs. Retrieved April 25, 2016, from http://www.umassmed.edu/cfm/stress-reduction/faqs/

CfM [Center for Mindfulness in Medicine, Health Care, and Society]. (2014b). History of MBSR. Retrieved April 25, 2016, from http://www.umassmed.edu/cfm/stress-reduction/history-of-mbsr/

CfM [Center for Mindfulness in Medicine, Health Care, and Society]. (2014c). Mindfulness meditation retreats. Retrieved April 25, 2016, from http://www.umassmed.edu/cfm/training/detailed-training-information/meditation-retreats/

Cheung, K. (2015, November). *To teach or not to teach explicit ethics in mindfulness programs: Right question but we need to ask the right audience*. Paper presented at the American Academy of Religion, Atlanta, GA.

Chloros, P. G. (2009). *Code-switching*. New York: Cambridge University Press.

Crane, R. S., Hastings, R. P., & Williams, J. M. G. (2010). Training teachers to deliver mindfulness-based interventions: Learning from the UK experience. *Mindfulness, 1*, 74–86.

Cullen, M. (2011). Mindfulness-based interventions: An emerging phenomenon. *Mindfulness, 2*(3), 186–193.

Davis, J. H. (2015). Facing up to the question of ethics in mindfulness-based interventions. *Mindfulness, 6*(1), 46–48.

DiAngelo, R. (2011). White fragility. *International Journal of Critical Pedagogy, 3*(3), 54–70.

Dodson-Lavelle, B. (2015). *Against one method: Toward a critical-constructive approach to the adaptation and implementation of Buddhist-based contemplative programs in the United States*. Ph.D. dissertation, Emory University.

Dreyfus, G. (2011). Is mindfulness present-centered and non-judgmental? A discussion of the cognitive dimensions of mindfulness. *Contemporary Buddhism, 12*(1), 41–54.

Durkheim, É. (1984). *The division of labor in society* (reprint ed.). New York: Simon & Schuster.

Farias, M., & Wikholm, C. (2015). *The Buddha pill: Can meditation change you?*. London: Watkins.

Folk, K. (2013). The Trojan horse of meditation. Interview by V. Horn, E. Horn, & K. S. Bearer. BG296. Retrieved January 2, 2016, from www.buddhistgeeks.com/2013/09/bg-296-the-trojan-horse-of-meditation/

Forbes, D. (2015, June). *Critical integral urban education: From neoliberal to transformational?* Paper presented at Mindfulness & Compassion: The Art and Science of Contemplative Practice Conference, San Francisco State University, CA.

Gethin, R. (2011). On some definitions of mindfulness. *Contemporary Buddhism, 12*(1), 263–279.

Goffman, E. (1959). *The presentation of self in everyday life.* Garden City, NY: Doubleday.

Goleman, D. (1985, Summer). *Inquiring Mind, 2*(1), 7.

Goodman, T. (2014). Stealth Buddhism. Interview by V. Horn & E. Horn. BG331. Retrieved January 2, 2016, from www.buddhistgeeks.com/2014/08/bg-331-stealth-buddhism/

Greenland, S. K. (2013). The inner kids program. Retrieved April 26, 2016, from http://www.susankaisergreenland.com/inner-kids-program.html

Greeson, J. M., Smoski, M. J., Suarez, E. C., Brantley, J. G., Ekblad, A. G., Lynch, T. R., et al. (2015). Decreased symptoms of depression after Mindfulness-Based Stress Reduction: Potential moderating effects of religiosity, spirituality, trait mindfulness, sex, and age. *The Journal of Alternative and Complementary Medicine, 2*(3), 166–174.

Greeson, J. M., Webber, D. M., Smoski, M. J., Brantley, J. G., Ekblad, A. G., Suarez, E. C., et al. (2011). Changes in spirituality partly explain health-related quality of life outcomes after Mindfulness-Based Stress Reduction. *Journal of Behavioral Medicine, 34*(6), 508–518.

Harrington, A., & Dunne, J. D. (2015). When mindfulness is therapy: Ethical qualms, historical perspectives. *American Psychologist, 70*(7), 621–631.

Hatton, E. (1989). Lévi-Strauss's "bricolage" and theorizing teachers' work. *Anthropology & Education Quarterly, 20*(2), 74–96.

Hawn, G. (2011, April 20). Goldie Hawn talks "MindUP" and her mission to bring children happiness. Interview by M. Schnall. *Huffington Post.* Retrieved January 4, 2016, from http://www.huffingtonpost.com/marianne-schnall/goldie-hawn-mindup_b_850226.html

Hawn, G. (2013). Address for heart-mind 2013. The Dalai Lama Center for Peace-Education. Retrieved January 2, 2016, from https://www.youtube.com/watch?v=7pLhwGLYvJU

Hawn Foundation. (2011). *MindUP curriculum: Brain-focused strategies for learning—and living, Grades Pre-K-2.* New York: Scholastic.

Hawn Foundation. (2016). Scientific research advisory board. Retrieved August 6, 2016, from http://thehawnfoundation.org/research/scientific-research-advisory-board/

Helderman, I. P. (2016). Drawing the boundaries between "religion" and "secular" in psychotherapists' approaches to Buddhist traditions in the United States. *Journal of the American Academy of Religion.* doi:10.1093/jaarel/lfw003

Heuman, L. (2012, Fall). What's at stake as the Dharma goes modern? *Tricycle.* Retrieved January 2, 2016, from http://www.tricycle.com/feature/whats-stake-dharma-goes-modern?page=0,1

Hickey, W. S. (2010, June). Meditation as medicine. *Crosscurrents,* 168–184.

Jakobsen, J. R., & Pellegrini, A. (Eds.). (2008). *Secularisms.* Durham: Duke University Press.

JKH. (2015, May 15). Comment on R. K. Payne, What's ethics got to do with it? The misguided debate about mindfulness and morality (May 14, 2015). *Tricycle.* Retrieved January 4, 2016, from http://www.tricycle.com/blog/whats-ethics-got-do-it

Kabat-Zinn, J. (1990). *Full catastrophe living: Using the wisdom of your body and mind to face stress, pain, and illness.* New York: Delacorte.

Kabat-Zinn, J. (1993). Bringing mindfulness into mainstream America. Interview by B. Gates & W. Nisker (updated December 2007). *Inquiring Mind, 10*(1), 34–42.

Kabat-Zinn, J. (1994a). *Catalyzing movement towards a more contemplative/sacred-appreciating/non-dualistic society.* The Contemplative Mind in Society Meeting of the Working Group, Sponsored by The Nathan Cummings Foundation & Fetzer Institute (September 29–October 2). Retrieved April 26, 2016, from http://www.contemplativemind.org/admin/wp-content/uploads/2012/09/kabat-zinn.pdf

Kabat-Zinn, J. (1994b). *Wherever you go, there you are: Mindfulness meditation in everyday life.* New York: Hyperion.

Kabat-Zinn, J. (2000). Indra's net at work: The mainstreaming of Dharma practice in society. In G. Watson, S. Batchelor, & G. Claxton (Eds.), *The psychology of awakening: Buddhism, science, and our day-to-day lives* (pp. 225–249). York Beach, ME: Weiser.

Kabat-Zinn, J. (2003). Mindfulness-based interventions in context: Past, present, and future. *Clinical Psychology: Science and Practice, 10*(2), 144–156.

Kabat-Zinn, J. (2009). Forward. In F. Didonna (Ed.), *Clinical handbook of mindfulness* (pp. xxv–xxxii). New York: Springer.

Kabat-Zinn, J. (2000, October 7). Mindfulness and the cessation of suffering. Interview by D. Fisher. Retrieved January 4, 2016, from http://www.lionsroar.com/mindfulness-and-the-cessation-of-suffering-an-exclusive-new-interview-with-mindfulness-pioneer-jon-kabat-zinn/

Kabat-Zinn, J. (2011). Some reflections on the origins of MBSR, skillful means and the trouble with maps. *Contemporary Buddhism, 12*(1), 281–306.

Kabat-Zinn, J. (2015, May 18). This is not McMindfulness by any stretch of the imagination. Interview by E. Shonin. *The Psychologist.* Retrieved January 4, 2016, from https://thepsychologist.bps.org.uk/not-mcmindfulness-any-stretch-imagination

Kabat-Zinn, J. (n.d.) *Guided mindfulness meditation.* Series 1–3. Retrieved January 4, 2016, from http://www.mindfulnesscds.com/

Kabat-Zinn, J., & Santorelli, S. F., with Blacker, M., Brantley, J., Meleo-Meyer, F., Grossman, P., Kesper-Grossman, U., Reibel, D., et al. (n.d.). Training teachers to deliver Mindfulness-Based Stress Reduction: Principles and standards. Retrieved January 2, 2016, from http://www.umassmed.edu/cfm/trainingteachers/index.aspx

Koenig, H., King, D., & Carson, V. B. (Eds.). (2012). *Handbook of religion and health* (2nd ed.). New York: Oxford University Press.

Laird, L. D., & Barnes, L. L. (2014, October). *Stealth religion in the borderland: Undercover healers in the hospital.* Paper presented at Conference on the Hospital: Interface between Secularity and Religion, Boston University, MA.

Langri, T. J. (2013). McGill University's Advanced Study Institute, Mindfulness in Cultural Context Conference.

Lindahl, J. (2015). Why right mindfulness might not be right for mindfulness. *Mindfulness, 6*(1), 57–62.

Lomas, T., Cartwright, T., Edginton, T., & Ridge, D. (2014). A religion of wellbeing? The appeal of Buddhism to men in London, United Kingdom. *Psychology of Religion and Spirituality, 6*(3), 198–207.

Lopez, D. S., Jr. (2008). *Buddhism & science: A guide for the perplexed.* Chicago: University of Chicago Press.

Lopez, D. S., Jr. (2012, Winter). The scientific Buddha: Why do we ask that Buddhism be compatible with science? *Tricycle.* Retrieved January 13, 2016, from http://www.utne.com/mind-and-body/the-scientific-buddha-zm0z13sozlin.aspx?PageId=4

Masters, B. (2014). The (f)law of karma: In light of *Sedlock v. Baird*, would meditation classes in public schools survive a first amendment establishment clause challenge? *California Legal History, 9*, 255–295.

McCown, D., Reibel, D., & Micozzi, M. S. (2010). *Teaching mindfulness: A practical guide for clinicians and educators.* New York: Springer.

McMahan, D. L. (2008). *The making of Buddhist modernism.* New York: Oxford University Press.

Mindful Schools. (2016). Year-long certification. Retrieved January 4, 2016, from http://www.mindfulschools.org/training/mindful-schools-certification/

Monteiro, L. (2015, November). *Ethics and secular mindfulness programs: Sila as victim of the fallacy of values-neutral therapy.* Paper presented at the American Academy of Religion, Atlanta, GA.

Ng, E. & Purser, R. (2015, October 2). White privilege and the mindfulness movement. *Buddhist Peace Fellowship.* Retrieved January 4, 2016, from http://www.buddhistpeacefellowship.org/white-privilege-the-mindfulness-movement/

Oman, D. (2015). Cultivating compassion through holistic mindfulness: Evidence for effective intervention. In T. G. Plante (Ed.), *The psychology of compassion and cruelty: Understanding the emotional, spiritual, and religious influences.* Santa Barbara, CA: Praeger.

Ozawa-de Silva, B. (2015, November). *Contemplative science, secular ethics and the Lojong tradition: A case study.* Paper presented at the American Academy of Religion, Atlanta, GA.

Purser, R. (2015). Clearing the muddled path of traditional and contemporary mindfulness: A response to Monteiro, Musten, and Compson. *Mindfulness, 6*(1), 23–45.

Purser, R., & Loy, D. (2013, July 1). Beyond McMindfulness. *Huffington Post.* Retrieved January 5, 2016, from http://www.huffingtonpost.com/ron-purser/beyond-mcmindfulness_b_3519289.html

Santorelli, S. F. (Ed.). (2014). *Mindfulness-Based Stress Reduction (MBSR): Standards of practice.* Worcester, MA: CfM.

Schmidt, A. T. (2016). The ethics and politics of mindfulness-based interventions. *Journal of Medical Ethics 0*, 1–5. doi:10.1136/medethics-2015-102942

Schonert-Reichel, K. (2009). Vancouver peace summit 2009: Educating the heart and mind (September 27–29). Retrieved January 4, 2016, from https://www.youtube.com/watch?v=suojNzKZ8ew&feature=youtu.be&t=1h11m28s

Schonert-Reichel, K. (2011, November). Buddhist contemplative care symposium. Retrieved January 2, 2016, from http://www.garrisoninstitute.org/cae-videoaudio/2011-symposium-videos/1091-susan-kaiser-greenland-kim-schonert-reichl-and-patricia-broderick

Schonert-Reichel, K. (2012, September 25). Interview by Dalai Lama Center. Retrieved January 4, 2016, from http://dalailamacenter.org/blog-post/interview-dr-kimberly-schonert-reichl

Shapiro, D. H. (1992). A preliminary study of long-term meditators: Goals, effects, religious orientation, cognitions. *The Journal of Transpersonal Psychology, 24*(1), 23–39.

Shonin, E., Van Gordon, W., & Griffths, M. (2013, April 18). Mindfulness-based interventions: Towards mindful clinical integration. *Frontiers in Psychology*, 1–4.

Smith, J. Z. (2004). *Relating religion: Essays in the study of religion.* Chicago: University of Chicago Press.

Stanley, S. (2015). Sīla and Sati: An exploration of ethics and mindfulness in Pāli Buddhism and their implications for secular mindfulness-based applications. In E. Shonin, W. Van Gordon, & N. N. Singh (Eds.), *Buddhist foundations of mindfulness* (pp. 89–113). Cham, Switzerland: Springer.

Stratton, S. P. (2015). Mindfulness and contemplation: Secular and religious traditions in Western context. *Counseling and Values, 60*(1), 100–118.

Taylor, C. (2007). *A secular age.* Cambridge, MA: Harvard University Press.

Taylor, J. (2013, January 14). Comment on Heuman, 2012.

Tweed, T. A. (2006). *Crossing and dwelling: A theory of religion.* Cambridge, MA: Harvard University Press.

U.S. v. Meyers (1996). 95 F.3d 1475 (10th Cir.) Retrieved January 12, 2016, from http://www.leagle.com/decision/1996157095F3d1475_11392/U.S.%20v.%20MEYERS#

Wachholtz, A., & Pargament, K. I. (2005). Is spirituality a critical ingredient of meditation? Comparing the effects of spiritual meditation, secular meditation, and relaxation on spiritual, psychological, cardiac, and pain outcomes. *Journal of Behavioral Medicine, 28*(4), 369–384.

Wachholtz, A., & Pargament, K. I. (2008). Migraines and meditation: Does spirituality matter? *Journal of Behavioral Medicine, 31*(4), 351–366.

Warnock, C. J. P. (2009). Who pays for providing spiritual care in healthcare settings? The ethical dilemma of taxpayers funding holistic healthcare and the First Amendment requirement for separation of church and state. *Journal of Religion and Health, 48*, 468–481.

Wilks, J. (2014a, September 8). Secular mindfulness: Potential & pitfalls. *Insight Journal*. Retrieved January 4, 2016, from http://www.bcbsdharma.org/2014-10-8-insight-journal/

Wilks, J. (2014b, December 8). Comment on C. B. Brown, Mindfulness: Stealth Buddhist strategy for mainstreaming meditation? (December 2, 2014). Retrieved January 4, 2016, from http://www.huffingtonpost.com/candy-gunther-brown-phd/mindfulness-stealth-buddh_b_6243036.html

Wilks, J., Blacker, M. M., Boyce, B., Winston, D., & Goodman, T. (2015, Spring). The mindfulness movement: What does it mean for buddhism? *Buddhadharma: The Practitioner's Quarterly*, 46–55.

Williams, J. M. G., & Kabat-Zinn, J. (2011). Mindfulness: Diverse perspectives on its meaning, origins, and multiple applications at the intersection of science and Dharma. *Contemporary Buddhism, 12*(1), 1–18.

Wilson, J. (2014). *Mindful America: Meditation and the mutual transformation of Buddhism and American culture*. New York: Oxford University Press.

Wilson, J. (2015, December 4). Mindfulness Inc.: Buddhist practice beyond religion. Lecture at Smith College. Northampton, MA. Retrieved January 4, 2016, from http://www.smith.edu/buddhism/videos.php

Woodhead, L. (2014). Tactical and strategic religion. In N. M. Dessing, N. Jeldtoft, J. S. Nielsen, & L. Woodhead (Eds.), *Everyday lived Islam in Europe* (pp. 9–22). Farnham, U.K.: Ashgate.

Wylie, M. S. (2015). The mindfulness explosion. *Psychotherapy Networker, 39*(1), 19–45.

Zaidman, N., Goldstein-Gidoni, O., & Nehemya, I. (2009). From temples to organizations: The introduction and packaging of spirituality. *Organization, 16*(4), 605–606.

The Mindful Self in Space and Time

Jack Petranker

Perhaps the immobility of the things that surround us is forced upon them by our conviction that they are themselves, and not anything else, and by the immobility of our conceptions of them. *Marcel Proust*, Swann's Way

When someone sits down to practice mindfulness, he or she is usually operating within a basic framework taken over from ordinary experience. Within that framework, there is the self, or subject, who is practicing mindfulness, and there is the object of which one is mindful (for instance, the breath). This basic framework, which situates the self as knower or perceiver in a world that is known, is so fundamental, so thoroughly taken for granted, that it is mostly invisible (Dreyfus and Taylor 2015).

There are reasons for thinking it would be helpful to take a closer look at this situating framework—to strip away its cloak of invisibility. After all, the Buddha taught as one of the three hallmarks of reality that no matter where one looks, no self can be found. This teaching of no-self (*anatta*, Skt. *anātman*), found in most if not all Buddhist traditions (Gethin 1998), is often considered one of the distinguishing characteristics of the Buddha's teachings. If that is so, does it really make sense to take the subject/object framework as a given in the practice of mindfulness?[1]

This is not an abstract question. Most Buddhist traditions agree that the commitment human beings make to the existence of the self, or to being guided by its wants, its hopes, and its fears, is the source of profound suffering and unhappiness.[2] If the practice of mindfulness does not directly call this commitment into question, such suffering will continue unchecked. The result will be to limit in advance the benefits that the practice of mindfulness offers.

The modern mindfulness movement teaches practitioners to cultivate awareness in the present moment; as a representative definition puts it, mindfulness consists of 'non-elaborative and non-judgmental present-centered awareness' (Dreyfus 2011, p. 42). If we look to the Theravāda tradition, however we find that such 'present-centered' mindfulness (hereafter, mindfulness[pc]) is only a first step. As the practice unfolds, mindfulness[pc] leads directly to investigation of the claims of the self. In the systematic teachings known as the *Abhidhamma* (Sanskrit

---

[1]Challenging the existence of the self is not necessarily the same as questioning the subject/object or self/world framework, a concern not universal among different Buddhist traditions. For my purposes, it is enough to acknowledge that the two questions are connected. My thanks to Linda Heuman for emphasizing to me the importance of this distinction.

J. Petranker (✉)
Mangalam Research Center for Buddhist Languages, Center for Creative Inquiry, Berkeley, CA, USA
e-mail: petranker@att.net

[2]A few examples: Śāntideva, in the classic Mahāyāna text known as the *Bodhicāryāvatāra* (Padmakara Translation Group, 1997, ch, 8, verse 134, writes, 'All the harm with which the world is rife,/ All fear and suffering that there is,/ Clinging to the 'I' has caused it!' In the Pāli Canon, the Buddha tells his followers, 'Nothing whatsoever is to be clung to as 'I' or 'mine'' (Goldstein 2003, p. 134). Tulku Urgyen Rinpoche (1999, p. 15), a twentieth-century Tibetan master, writes, 'An ordinary person's attention strays according to any movement of mind. Suddenly there is the confusion of believing in self and other, subject and object, and this situation goes on and on repeating itself endlessly. This is samsaric existence.'

© Springer International Publishing Switzerland 2016
R.E. Purser et al. (eds.), *Handbook of Mindfulness*,
Mindfulness in Behavioral Health, DOI 10.1007/978-3-319-44019-4_7

Abhidharma), this happens through analyzing appearance into constituent parts: the *khandhas* (Sanskrit *skandhas*), *āyatanas*, or *dhātus*; see also the *Satipaṭṭhāna Sutta* (MN, 10). Such analysis is undertaken for the express purpose of challenging the commitment to a self.

Two citations from the Pāli Suttas help make this developmental approach to mindfulness practice clear. In the *Arittha Sutta* (SN 54,6), the Buddha tells a disciple who describes mindfulness of the breath as simple awareness of breathing in and breathing out that this is only the beginning; that bringing mindfulness practice to its 'culmination' requires moving on to mindfulness of a host of other factors, including mental fabrication, impermanence, and relinquishment.[3] In the *Bhaddekharatta Sutta* (MN, 131), the Buddha explains that it is not enough for 'the ideal lover of solitude' to cut off attachment to past and future if he is 'drawn into present things.' This happens, the Buddha explains, if he views the constituent factors of reality (the *khandhas*) as being the self, as belonging to the self, or as otherwise involved with the self.

While mindfulness in its traditional context thus seems to evolve quite naturally into practice that challenges the subject/object framework of experience, mindfulness[pc] lacks the resources to turn practice in this direction. The aspects of mindfulness practice that question the self and its central role in experience are simply not part of the training found in such popular mindfulness[pc] programs such as Mindfulness-Based Stress-Reduction (MBSR) and Mindfulness-Based Cognitive Therapy (MBCT) (Rosch 2015).

Whether the simpler form of practice introduced in mindfulness[pc] does violence to traditional Buddhism or extracts its essence is a debate I do not wish to enter. Instead, my intention is to explore a different approach to mindfulness: one that builds on mindfulness[pc] but takes it in a different direction. The basic shift I want to make is this: Instead of understanding mindfulness in terms of time, I will look

at it in terms of space. As I intend to show below, this space-centered approach to mindfulness gives a way of questioning the subject/object framework without having to rely on traditional Buddhist teachings.

Through Thoughts to Space

The focus on present experience in the modern mindfulness movement is closely associated with a critique of our overreliance in thinking. Programs such as MBSR and MBCT tell us that one important reason we fail to focus on the present moment is that we are caught up in our thoughts, which generally center on the past and the future, and in the judgments and emotions that those thoughts stir up (Puhakka 2015; Rosch 2015). Mindfulness[pc] trains practitioners to focus instead on the present moment so that they can get 'out of their heads' and back into sensory, embodied experience (Kabat-Zinn 2006).

One way to think of this shift is in spatial terms. Often when we have a 'thought' (using the term broadly to include, for instance, daydreams and memories), we find ourselves inhabiting a realm that the thought itself sets up. This is why we speak of being 'lost in thought': We have shifted our attention from the shared world of embodied experience to the private realm of the thought. Described in this way, the goal of mindfulness[pc] is to help us shift from 'thought space' to embodied space.

Does this way of speaking rely too heavily on a metaphor? While we certainly do speak of thoughts using space imagery ('deep in thought,' 'caught up in an idea,' 'falling into a reverie'), perhaps this is just a way of speaking, not to be taken all that seriously (cf. Lakoff and Johnson 1980). Usually, we think of space as inherently and exclusively physical—the container for physical objects. If that is so, any talk of thought space can only involve a kind of analogy.

Still, why insist that embodied space—the space of sensory experience and physical measurements—is the only space there is? Thoughts cannot exist in physical space, but perhaps they can 'exist' in another kind of space. The human

---

[3] I am grateful to David McMahan for calling my attention to this Sutta.

understanding of space has varied considerably across cultures and time (Cornford 1936). There seems no reason to rule out adopting an expanded understanding of space if it seems useful to do so.

From Existence in Space to Appearance in Space

To say that an object 'exists' amounts to saying that it appears in physical space. By this definition, thoughts—or perhaps more accurately, the contents of thoughts—do not exist. Still, the contents of thoughts, dreams, fantasies, etc., do *appear*, just like existing physical objects. So as between the two concepts 'appearance' and 'existence,' appearance seems more fundamental.[4] To put it differently, physical objects have one way of appearing; thoughts and daydreams have another. We do not have to privilege one over the other.

Once we shift our focus from existence to appearance, we are ready to speak of different kinds of space. Following Tarthang Tulku (2015), I will sometimes speak of such different spaces as fields or space fields. In the field of physical space, whatever objects appear also exist, but in other kinds of space, the link between appearance and existence may not hold. This seems to be the case for thoughts. 'Non-existent' thoughts and other mental events appear and are 'real' in terms of the space within which they appear, even though they do not exist.

To clarify this distinction, try this simple experiment: Right now, think of your car. Most likely, when you do this an image of your car arises in your mind. That imagined car does not exist, unlike the 'actual' car. Still, the image has appeared. Where does it appear? One answer is to say that it appears in the mind, but this does

little to clarify what is happening. We could also say, however, that it appears in mental space, a kind of space that allows for just such appearances. If we are ready to allow for these two kinds of space—mental and physical—we need a more expansive understanding of space. We could say that space is simply what allows appearance—in all possible modes—to appear (Tarthang Tulku 1977).

If space is what allows appearance, and if different space fields allow different ways of appearing, then the usual framework that takes as a given the subject/object, self/world structure, is no longer quite so self-evident.[5] In fact, it starts to seem problematic. Subject and object both appear. Do they both appear in the same space? If the answer is yes, why can we take hold of objects but not the subject? If the answer is no, does this mean we inhabit two kinds of space at once (as Descartes maintained)? These are not just philosophical conundrums. As we shall see, they can fruitfully be brought into the practice of mindfulness.

The Space of Subject and Object, Self and World

The Buddha taught that a mistaken belief in the existence of the self is universal (Gethin 1998), so questioning it goes against the grain. It is especially difficult in the culture of modernity and post-modernity, which puts the self—its wants, its needs, its judgments, its possessions, and its experiences—at the heart of reality in ways that may be unprecedented (Taylor 1992, 2007). No longer living in a meaningful cosmos, we have come to think of human fulfillment as identical to self-fulfillment, and we have made personal experience the key to seeking such fulfillment.

The practice of mindfulness[pc] is wholly consistent with this emphasis on personal experience

---

[4]Compare what the discipline of phenomenology calls the 'phenomenological reduction': the decision to suspend claims about what does or does not exist in favor of an inquiry into how things appear. Interestingly, Edmund Husserl, the founder of phenomenology, spoke of carrying out the phenomenological reduction as a rigorous meditative practice that is transformative for those who thoroughly engage it (Cogan 2016).

[5]The subject/object framework is not the same as the self/world framework, but there is considerable overlap. I shall use both descriptions, depending on the context.

and personal well-being. Often, it is presented as a way to shift the nature of experience in the direction of greater psychological well-being (e.g., through the reduction in stress or the relief of physical or mental symptoms). McMahan (2008) speaks in this context of the psychologization of Buddhism, while Taylor (2007), speaking more generally of spirituality, follows Rieff in calling it the triumph of the therapeutic. Carette and King (2005, p. 101) refer to it as the privatization of Asian wisdom traditions; Huntington (2015), more polemically, calls it narcissism.

When we look at the privatization of mindfulness practice in spatial terms, we can say that the space we inhabit has been interiorized. Although our bodies inhabit physical space, the realm of meaning and of our fundamental concerns is found in our hearts, our minds, or our souls. As Taylor writes (2007, p. 540), 'the depths which were previously located in the cosmos, the enchanted world, are now more readily placed within.' The result is a self that he describes as 'buffered': locked away in its own space, separate from the world it encounters. Cf. Tarthang Tulku (1987, ch. 5).

In encouraging a shift from thought space to embodied space, mindfulness[pc] stays squarely within this prevailing framework. When we make the fundamental move of turning from the 'fabricated' space of thoughts and fantasies, from past and future toward the immediacy of the present, we still find ourselves inhabiting the interiorized space of experience. Mindful in the present moment, we operate in a space where objects appear to a knower, and 'I' am that knower. All possible experiences are those of a self inhabiting its world.[6]

As long as we dwell in a space that structures experience in terms of this self/world framework, we will have a hard time making sense of the Buddha's teachings on no-self. And absent a link to more traditional Buddhist teachings (such as the teaching on the three marks of existence,

mentioned above), mindfulness[pc] is unlikely to be of much help. Here is where a different approach to mindfulness, centered on space as field, can be of benefit.

The Fields of Space

Considered as the field within which appearances arise, space proves multiple, for different kinds of appearances arise within different fields. When we are lost in thoughts, we inhabit one field; when we focus on the breath, we inhabit another; when we dream, we inhabit still another. The space of each field allows certain kinds of events and experiences that are not allowed in other spaces. Walking down the street, I inhabit a field that I share with houses, gardens, and passerby. But if, as I walk, I fall to thinking of an earlier conversation, I enter and inhabit an entirely different field. The physical world that I was in a moment earlier disappears, and a new world makes itself available.[7]

In putting matters this way, I am following Tibetan lama Tarthang Tulku (1977, 2015), who several decades ago introduced to the West a 'vision of reality' in which space is posited as an active 'force' that has the property of allowing appearance. But I could also point to other sources. For instance, phenomenological thinkers similarly speak of space as shifting in accord with the various worlds we inhabit, based on our interests, concerns, and ways of understanding (Husserl 2002; Merleau-Ponty 1962; Dreyfus and Taylor 2015). Sartre (1958, p. 42) offers an illuminating example: If I enter a café expecting to see Pierre and do not find him, 'Pierre is absent

---

[6]This is not to say that physical objects are 'only' our experience of them, like objects in a dream. Nothing I am saying here speaks to issues of ontological status. In this sense, the approach I am taking is broadly phenomenological.

[7]This point is made quite clearly in *The Questions of King Milinda* (Rhys Davids 1963, vol. I, p. 127), an important text in the Theravādin tradition. King Milinda asks how it is possible that an advanced practitioner can transport himself instantly to the Brahma world, one of the highest heavens. In reply, the sage Nagasena asks the king where he was born and if he remembers some activity there. When the king tells him he was born at a place about 200 leagues distant and that he does remember doing something there, Nagasena replies, 'So quickly, great king, have you gone about two hundred leagues.'

from the *whole* café. . . . Pierre absent haunts [the space of] this café.' [emphasis in original][8]

Within a given field, space has characteristics that determine what it does or does not allow. For instance, for a physical object to be present before me, both I and the object must inhabit the same physical space, and that space allows for certain relationships. Some obvious examples are location, distance, and separation. The vase is 'over there' and I am 'here': that is, somewhere else, but within the same field.

Once we have attuned ourselves to such spatial relationships, we can make them the focus of mindful awareness. For instance, instead of being aware of the vase, I can be aware of the distance between me and the vase. When I do, my sense of that distance may shift. Mindful in this way, I have begun to engage the space of the field I currently inhabit.

Let us call this approach 'field-centered mindfulness,' or mindfulness^{fc}. As we shall soon see, practicing mindfulness^{fc} can lead us to engage experience differently. Ultimately, it can call into question the framework for experience within which we ordinarily operate. Here, an analogy with fields as they are understood in quantum physics can be helpful. Particle physicists tell us that at the subatomic level, the field within which particles occur is more fundamental (more real) than the particles themselves, which are better understood as excitations of the field (Jepsen 2013; Healey 2008). Similarly, once we engage the field through mindfulness^{fc}, entities such as 'subject' and 'object' may be better understood as expressions of the field.

Attuned to the field, we are no longer bystanders disengaged from a world that we observe at a distance (Tarthang Tulku1987). Nor can we think of that world as 'objective' in the usual sense, for the field we inhabit allows not only for physical objects, but for the meanings we assign those objects, as well as the meanings that guide our actions. As we learn to be mindful of such aspects of experience in a field-centered way, we naturally grow more attuned to the field itself. The basic self/world framework for experience can come under scrutiny.

As I began work on this chapter late in 2015, the Golden State Warriors, a professional basketball team, had captured the attention of millions of sports fans by a run of consecutive wins, mostly just overpowering its opponents. Sports writers and fans began comparing them to the 1996 Chicago Bulls, another basketball team often considered the best of all time. As it happens, both the Bulls and the Warriors were trained by their coaches in practicing mindfulness (Jackson and Delehanty 2006; Kawakami 2015). The Bulls were pioneers in the use of MBSR—what I have been calling mindfulness^{pc}. They were taught to use it in ways familiar in the literature—to reduce identification with fleeting thoughts and feelings, to practice not thinking, to deal more effectively with their emotions, and to cultivate inner harmony. The Warriors were trained in a different understanding of mindfulness, which their coach described as follows: 'It's thinking the game. It's not just trying [to] out-talent people; it's not trying to go for your individuals stats [statistics]. It's being mindful of the right way to do things.' In other words—at least up to a point—the Warriors were asked to practice mindfulness of the field, the whole. The common description of a skilled basketball player as being able to 'see the whole court' gives a sense of what this might be about.[9]

The comparison between 'Bulls mindfulness' and 'Warriors mindfulness' should not be taken too seriously. It does suggest, however, what it might mean to practice mindfulness^{fc} and also why one might want to do so. If mindfulness practice asks us to be attentive to experience, why not practice it in a way that engages the field

---

[8]Phenomenological approaches maintain that the world of things and events as we encounter it can only be described as 'objectively real' from the viewpoint of the physical sciences. The world or worlds we inhabit, in contrast, are constructed by the meaning we assign them. They maintain that separating out 'subjective' experience from the 'objective' reality that manifests in physical space is the mark of a discredited Cartesian dualism (Dreyfus and Taylor 2015; Husserl 2002).

---

[9]Another way to put this, one that was in fact regularly used to describe Warrior's style of play, is that they were unselfish (Strauss 2015).

as a whole, the field within which we live and act, the field that makes our lives what they are?

Consider again the *Bhaddekharatta Sutta*, discussed above. The Buddha's teaching there seems to me to support a field-centered understanding. The one 'drawn into present things' inhabits a world with the self at its center, engaging appearance on the basis of the self's concerns and desires. In the space of that world, the self is everywhere; to repurpose a phrase made well known by Kabat-Zinn, 'Wherever you go, there you are.'[10] It is this unthinking acceptance of a field understood in terms of a self/world framework that mindfulness[fc] calls into question. It does so by investigating the possibility of experience centered in the field as a whole.

The Field as a Whole

Physical space could be said to 'operate' in terms of three characteristics: location, distance, and separation. In the self/world framework, the self is subject to these characteristics as well: I am located 'here,' I encounter things that are located 'there,' and I am aware of the distance that separates 'me' here from 'that thing' over there.

If different kinds of experiences take place in different space fields, however, this description of what space allows will be much too limiting. For instance, can we really speak of thoughts or desires as being located in one place? And what would it mean to say we are distant from our own intentions? In each of these cases, it seems to make more sense to say that mental events *pervade* the field within which they arise. Here again, quantum physics provides an useful analogy. We are told that in the field of subatomic particles (our conventional physical world, theorized at a vastly different scale), it is wrong to speak of a

particle being located in one place or even as having its own distinct identity. Instead, the principle of nonlocality applies: Nothing is located anywhere; put differently, everything is located everywhere (Musser 2015).

What would it mean to put this kind of field-centered view into practice? Consider a story told by Joseph Goldstein (2016), a well-known Vipassana teacher. He describes meditating on his breath, with 'nothing special going on.' At a certain point, he asks himself about his mental attitude. Having asked this question, he immediately notices a subtle 'wanting' in operation, and in the instance he notices it, it releases.

One way to think about the shift Goldstein describes is that he shifts from a breath-centered field to an attitude-centered field. This would be consistent with what I said above about getting lost in a thought. When it comes to attitudes, however, the shift is more subtle. You do not necessarily fall into an attitude, though you can; rather, the field that you inhabit expands, and you become aware of what had gone unnoticed before. Following Tarthang Tulku (2015), we could speak here of becoming aware of the 'feel' of the field. Goldstein becomes aware of the attitude that pervades the field of his awareness, along with the activity of breathing. Does his awareness of breathing also pervade the field?

Let us look a bit more closely, building on Goldstein's brief description. Seen in a field-centered way, there is wholeness to breathing. The ribcage expands, the diaphragm rises and falls, and air enters the nostrils. But that is not all. I am the one who is breathing, and my sense of 'I-ness,' of being the owner of the breathing, is a part of the field as well. And there is more. I breathe, and if my eyes are open, I also perceive. My seeing the table before me is part of the field, together with the breathing, the feel of the field, and the narrative that tells me that I am the one doing the practice.

In the field-centered experience, then, a wholeness is available that seems at odds with the usual self/world framework. This is not a matter of shaping experience to suit some particular teaching—for instance, the Buddhist

---

[10]What this might mean is suggestively presented in the film *Being John Malkovich* (1999). For reasons never explained, the characters in the film have the ability to use a kind of 'chute' located at a particular place in physical space to enter the mental space that constitutes the mind of John Malkovich. At a certain point, Malkovich, himself one of the film's characters, enters his own mind. When this happens, he encounters nothing but himself.

analysis of experience into five *skandhas* (*khandhas*), or a perspective grounded in non-duality. When we let ourselves see it, the wholeness of the field is simply there.

Mindfulness[fc] invites engaging the whole of experience in this integrated way. When we do, the identity and status of entities within the field—the ways they interact and the qualities they exhibit—become more fluid and less positioned. The self/world structure begins to lose its grip on our awareness. This happens not because we have some special kind of experience, but because we see the world in a field-centered way.

cI do not mean to draw too sharp a distinction. Mindfulness[pc] and mindfulness[fc] both challenge conventional forms of awareness, and the experiences they invite will overlap. Even if mindfulness[pc] generally operates within the self/world framework, it can also move beyond it. Experienced meditators, for example, may report awareness of the breath in which the subject/object structure drops away, and there is 'just breathing.'

Still, a field-centered approach to mindfulness makes a more radical questioning of the self/world framework more likely. Here is an example: I said above that mental events, in contrast to objects of perception, tend to pervade their fields. But this way of putting the matter actually commits us to separating the field of physical experience from the field of the mental, and that may not go far enough. If the field of experience is truly integrated, we cannot say that one kind of space operates in the mental realm and another in the physical realm. Experience is both physical *and* mental, subjective *and* objective. As Dreyfus and Taylor (2015) argue (proceeding from very different assumptions and with very different concerns), the self is inseparable from its world.

Suggestions for Exploration

In the new way of seeing that mindfulness[fc] invites, new questions arise. For instance, if the objects of awareness and the content of thoughts have no specific location, can they be possessed? Can they have an owner? Do their claims to be

real operate in the same way?[11] Such questions can readily be multiplied. Here are a few examples, starting with some of the characteristics we ordinarily take for granted when we limit our conception of space to the field of the physical.

Locatedness

Suppose I hear the sound of tires on the road. I immediately put things in their place: 'car going by, outside to my left.' The fog of familiarity descends, and the possibility for immediate awareness is lost. The practice of mindfulness[pc] offers new possibilities. The sound arises, and I note this arising. If I also label it as 'car going by,' that is either a further mental event or a refinement inherent in the initial perception. Less bound to familiar labels, I have the opportunity to notice other particulars, such as the tonal qualities of the sound and the bodily sensations that the sound triggers. Other elements of the experience may present themselves as well; for instance, the judgments and associations that follow the initial perception, the sense of myself as the one who is hearing, and even the operation of a variety of mental events that let me label the source of the sound and its location. Through all this, however, the self/world framework continues to operate.

Mindfulness[fc] builds on what mindfulness[pc] introduces. It questions the self/world framework by engaging the field within which the experience arises. For instance, do I actually *experience* the sound of the car going by as being located in a particular place ('to my left')? If I attend to the hearing with a sense of the field as a whole, I may find that my experience of hearing is not local at all: Although I attach a location to it, the sound is everywhere in the field of experience.

---

[11]These questions can also be put in more traditional Buddhist terms. In discussing how Buddhism challenges our commitment to the self, Ganeri (2007, p. 174) writes: 'We are not in error when we think of the world in a person-involving way; it is just that we could do better ... by thinking of it in some other way altogether, by standing in a different cognitive relation with the world.' Compare, in a non-Buddhist context, Dennett (1986). Similar issues arise when we consider our propensity to frame the world in terms of narrative (Bruner 1987; Tarthang Tulku, 1987).

This is not to deny the phenomenon of located-ness, but rather to inquire into what it entails.

Distance and Separation

Suppose I am looking at the vase on the table in front of me. In ordinary experience, I would automatically be aware of the distance that separates me from the vase, without really noticing it. In practicing mindfulness[pc], I might focus on the vase with much greater precision and immediacy, but the fact of a separating distance between me and the vase would remain self-evident and thus beyond the range of awareness. For mindfulness[fc], however, 'distance' and 'separation' are themselves part of the field in operation, available to investigate. For instance, 'distance' is ordinarily 'distance from me.' If I am sensitive to my own location in relation to what I see—if both subject and object arise within an unitary field—'distance' might resolve into something quite different: perhaps a sense of connection, or even non-separation.

Thoughts

I pointed out earlier that mindfulness[pc] emphasizes freeing ourselves from thoughts, especially those that center on the present and the future. For mindfulness[fc], however, thought space is just another field. Although we usually say that thoughts happen 'in' the head, experientially there is little or no foundation for this. As with other mental events, thoughts—at least those thoughts that frame the field we inhabit in the moment—pervade the field in which they arise. And even for thoughts that just seem to pass by, like clouds in the sky, there is no easy way to assign them a 'where.' Finally, is it even really accurate to say that a thought appears, rather than the content of thoughts?

I have noted already a special quality of thoughts (I include here dreams, fantasies, and the like): We can get lost in the field they manifest. When we do, awareness simply disappears. Only afterward do we notice that we have been gone—seemingly nowhere at all. How does this happen? Perhaps it is not enough to say that thoughts pervade the field within they arise; perhaps thoughts actually give birth to that field. If we understood the field of thinking differently, if we had ways to explore it, could we free ourselves from thoughts' anesthetizing effect?

Meaning

The usual view that perceives the world in terms of 'things located in physical space' is most readily maintained when we encounter objects with which we have no special relationship: not the blue willow dish that my mother gave me, but a circular blue and white patch. Objects that have meaning for us, on the other hand, quite naturally encourage field awareness. Looking at a photograph of someone I care about who is not present, I could simply engage it as a physical object, next to the stapler and in front of the clock. But I could also let the image in the photograph evoke the presence of the missing person. Immediately, the whole of the field is pervaded with meaning. The situation is just the reverse of Sartre's example of 'Pierre who is not at the café': Now it is the presence of the missing person that is everywhere.[12]

The Self at the Center

In the field of what appears, the sense of self arises. If we look more closely, the sense of the self at the center manifests in a variety of ways. There is the feel of owning experience, enjoying experience, seeking experience, reacting to experience; there is the sense of the one who decides or judges, who makes associations. Such aspects of the field stand on their own. We do not have to engage them as pointers confirming the

---

[12]MBSR, the best known version of mindfulness[pc], includes in the training it gives students the practice of *metta* (Sanskrit *maitri*) or loving kindness, in which one wishes for the happiness of others. Rosch (2015) and others have pointed out that while this practice is well known in traditional Buddhism, it is not usually considered a mindfulness practice. However, it fits quite well with the practice of mindfulness[fc]. The meaning that pervades a field is one that we ourselves can activate, and that is just what the practice of loving kindness does. Interestingly, flooding the field of experience in this way is precisely how the practice of loving kindness, and the other 'immeasurable states', is presented in the Canon. See the *Tevijja Sutta*, DN 13.

self as a separately existing entity with its own fixed identity. When we let the sense of self pervade the field, inseparable from whatever else appears, the self/world framework becomes problematic in a new way.

The Presence of Others

If mindfulness[fc] lets us experience in ways that do not rely on the self/world framework, how does this affect our engagement with others? When we adopt and hold on to the self's position, we tend to assign a fixed position and identity to the other as well. One way to think of positions is as distortions in the field, in somewhat the same way that gravity is said to amount to a distortion of physical space. Would a position-free, field-centered way of engaging experience remove certain kinds of distortions? Would the result be to change our relationship with others?

The possibility for inhabiting space fields pervaded with meaning suggests that this may be so. In such fields, space becomes in a literal sense heartfelt. In such a heartfelt space, could each encounter enact intimacy? It is useful to recognize that we are free to explore these possibilities, for mindfulness[fc] is not inherently passive: One way to explore the field is to vary what arises within it and observe the results.

A New Vision for the Practice of Mindfulness

We learn in the course of growing up that we inhabit a space shaped by the self-world framework. While it may lead to experiences in which this framework vanishes, mindfulness[pc] gives us no direct way to question this 'self-evident' truth or even to realize that there is a question to be investigated. Mindfulness[fc], on the other hand, gives us access to field potentialities not shaped in accord with the self/world framework. When we put questions of distance, separation, positioning, identity, and so forth at the center of practice, we gain the ability to investigate basic blind spots in our experience.

Reflecting on the relation of MBSR (for my purposes, a stand-in for mindfulness[pc]) to traditional Buddhist teachings, Kabat-Zinn (2011) describes it as a skillful means (a technical Buddhist term that he adapts for his own purposes). Easily taught and practiced, making no overtly religious demands on those who take it up, it seems to offer relief from suffering of many kinds, from stress, anxiety, and depression to various forms of mental and physical pain. If embedding it in its Buddhist context would drive away many of those who could otherwise benefit, he argues, why insist on doing so?[13]

This may seem like common sense, but there is an assumption at work that many critics have questioned. Do we really know what it means to 'strip away' the Buddhist context within which mindfulness has always been practiced in the past? Perhaps those who decide to set aside traditional Dharma teachings and focus on present-centered awareness are taking too narrow a view of what the Buddha taught. Perhaps, in carefully carving away aspects of traditional Buddhism that we moderns are unable to digest, advocates of mindfulness[pc] are blocking access to teachings that could offer a different set of benefits for our fractious times.

These two views represent the two sides in the current debate over mindfulness. As I said at the outset, I do not want to enter that debate. Rather, in introducing mindfulness[fc], I have hoped to engage what might be called a more comprehensive Buddhist vision of reality.

A vision of reality is not the same as a doctrine or philosophy that tells us what is real. Traditionally, mindfulness practice has been used to deepen insight into impermanence, no-self, and the three poisons (greed, hatred, and delusion). These are fundamental truths on which the Buddhist tradition could be said to rest (Rosch 2015), though they receive little overt attention from those who teach mindfulness[pc].[14] When I

---

[13]As Sharf (2014) points out, a similar calculus has been made at other moments in history when Buddhism was entering cultures where it was not previously known.

[14]It is always possible that at least a few people who are introduced to mindfulness[pc] will go on to more traditional forms of Buddhist practice. But my own casual discussions with teachers of MBSR and similar programs suggest that this does not often happen. In a more systematic study, Rosch (2015) has written about her experience in attending

speak of the Buddhist vision of reality, however, it is not such truths and teachings I have in mind. Vision is not about theory and explanation.

We can clarify this distinction by looking back to an earlier time in Western thought. Writing around the time that Mahāyāna Buddhism was taking form, the Hellenistic thinkers of Greece and Rome presented philosophy as *therapeia* (Hadot 1995; Ganeri and Carlisle 2010): a therapy for the soul or simply for the condition of being human. In modern times, we place therapy in the realm of psychology, distinct from what we consider to be the abstract truths of philosophy. But the notion of philosophy as *therapeia* challenges this split. It suggests that if we see the world differently, that may be all the therapy we need. Such seeing is not 'theoretical,' at least in our modern sense. Instead, it involves activating a new vision.

Perhaps it is as *therapeia* in this sense that mindfulness[fc] can extend the range of benefits that mindfulness[pc] makes available. With its active commitment to phenomenological inquiry and a readiness to question what everyone takes for granted, mindfulness[fc] seems well suited to bringing new dimensions to our present circumstances and ways of knowing.[15]

(Footnote 14 continued)
three MBSR trainings and conducting interviews with a number of participants. She concludes that most participants in MBSR training do not develop much sense of what mindfulness (i.e., mindfulness[pc]) is, nor do they actually practice it very much. Instead, they are more likely to engage practices that they find familiar or easy to understand (e.g., relaxation practices, generating loving kindness, and hatha yoga, all of which are a part of the training). If true, this makes it quite unlikely that they will go on to engage more traditional Buddhist teachings. Of course, there will always be exceptions.

[15]I can speak to this point in terms of my own experience. As a practicing Buddhist, I find that the writings of contemporary Buddhist teachers and writers often offer valuable insights, images, and practices. Yet I have my doubts that new insights and new practices are really what I need. I am not lacking in good Dharma advice and sound instruction; it is just that I do not follow it. Perhaps this is due to my own psychology—a personal failing. But it seems more fruitful to trace it to my commitment to a limited vision of the way things are. Expanding that vision seems to me the *therapeia* that will serve me—and others who share my situation—best.

In his discussion of the current mindfulness boom, Wilson (2014) has offered a helpful refinement of the oft-repeated observation that Buddhism adapts to each new culture it enters. This process, he suggests, comes about when Buddhist teachers are able to find a niche in their new surroundings that no one else is adequately filling, a way of addressing 'local concerns and desires.' The niche that mindfulness[pc] has found is related to the psychological needs and feelings of alienation, negativity, and burn-out that seem endemic to our times. The parts of Buddhism that speak to these needs—more effectively, in some cases, than other approaches—are taken up and developed; the parts that do not are peeled off and discarded.

This way of putting the matter suggests to me is that there is more than one niche in which Buddhism could make a home. Those who worry that mindfulness[pc] threatens to sweep away essential aspects of the Buddha's teachings might do well to look for such a niche. As I see it, they do not have look far. As countless thinkers have insisted for over two centuries (Taylor 2007), we moderns live in a disenchanted world, a world where we are not at home, where traditional sources of meaning are no longer available and where the various alternatives on offer seem largely ineffective.

By engaging a different aspect of Buddhism, the practice of mindfulness[fc] might help fill this gap, this wound in the heart of modernity. It draws on those aspects of the Buddhist tradition that invite us to question vigorously all our ordinary assumptions about what is real—not through sophisticated conceptual analysis, but simply through learning to see differently. It offers a way to engage a vision of reality more consistent with the Buddha's realization.

Perhaps we can no longer find 'fullness' (Taylor 2007) through the sense that we are embedded in a meaningful cosmos, a vision that has guided most civilizations throughout time. Then could fullness appear within the immediacy of experience itself? Mindfulness[fc] explores this possibility. By turning from what appears within the field to the field itself, from the identities of situated entities to a non-localized engagement with the ever-shifting whole of what is so, it invites us to activate a different dimension of experience.

For many people, mindfulness[pc] offers a valuable opportunity to find balance in their lives. But because it largely moves within the governing framework of self and world, it easily supports the tendency to live out our lives in a radically psychologized space, in which 'fullness' is reinterpreted as authenticity, and the truths toward which religion has always aimed are understood entirely in terms of personal, subjective experience (Sharf 1998; Tarthang Tulku 1987). Mindfulness[fc] may help expand and ultimately challenge this psychologized understanding.

In engaging a new vision of what is so, mindfulness[fc] starts from the commitment to question what common sense holds to be true, a commitment it shares with science. Perhaps this commonality is significant. It may be that the continuing dedication by generations of scientists to a radical empiricism that stakes out no positions in advance has now prepared the way for something new. Perhaps the time is right for an approach to inquiry that lets us recognize, precisely in the gaps left unexplored by the scientific view, a niche where Buddhist-inspired modes of questioning can find a home.

Toward the end of his life, Husserl (1970) concluded that the task of a scientific philosophy must be to understand the lifeworld, the everyday world of experience. The Buddha taught that we can do more than understand the lifeworld—the world of samsara. Through cultivating our own capacities, we can change that world, for ourselves and perhaps for others as well. If a field-centered way of being mindful can contribute to this possibility, opening new dimensions of our human being to inquiry, we have every reason to investigate it more fully.[16]

---

[16]My thanks to Linda Heuman, Hayward Fox, and Michael Gray for helpful comments on earlier drafts of this article.

References

Canonical Texts

Arittha Sutta (SN 54,6) Trans. Thanissaro Bhikkhu. http://www.accesstoinsight.org/tipitaka/sn/sn54/sn54.006.than.html. Accessed December 26, 2015.

Bhaddekaratta Sutta, MN 131. Trans. Bhikkhu Ñanananda. http://www.accesstoinsight.org/tipitaka/mn/mn.131.nana.html. Accessed February 6, 2016.

Tevijja Sutta, DN 13. Trans. Nyanaponika Thera. http://www.buddhanet.net/ss02.htm. Accessed December 26, 2015.

Secondary Sources

Bruner, J. (1987). *Actual minds, possible worlds.* Cambridge, MA: Harvard University Press.

Carette, J., & King, R. (2005). *Selling spirituality: The silent takeover of religion.* London: Routledge.

Cogan, J. (2016). The phenomenological reduction. The Internet Encyclopedia of Philosophy. www.iep.utm.edu/phen-red/. Accessed February 14, 2016.

Cornford, F. M. (1936). The invention of space. In G. Murray & H. A. L. Fischer (Eds.), *Essays in honour of Gilbert Murray* (pp. 215–235). London: George Allen & Unwin.

Dennett, D. (1986). The self as narrative center of gravity. In F. Kessel, P. Cole, & D. Johnson (Eds.), *Self and consciousness: Multiple perspectives.* Hillsdale: Erlbaum.

Dreyfus, G. (2011). Is mindfulness present-centred and non-judgmental? A discussion of the cognitive dimensions of mindfulness. *Contemporary Buddhism, 12*(1), 41–54.

Dreyfus, H., & Taylor, C. (2015). *Retrieving realism.* Cambridge, MA: Harvard University Press.

Ganeri, J. (2007). *The concealed art of the soul: Theories of self and practices of truth in Indian ethics and epistemology.* New York: Oxford University Press.

Ganeri, J., & Carlisle, C. (2010). *Philosophy as therapeia. Royal Institute of Philosophy Supplement* (Vol. 66, pp. 187–218). Cambridge: Cambridge University Press.

Gethin, R. (1998). *The foundations of Buddhism.* Oxford: Oxford University Press.

Goldstein, J. (2003). *One Dharma: The emerging western Buddhism.* New York: Harper Collins.

Goldstein, J. (2016). Who knows: An interview with Joseph Goldstein. *Tricycle.* Spring 2016.

Hadot, P. (1995). *Philosophy as a way of life* (M. Chase, Trans.) Oxford: Blackwell. (Original work published 1981)

Healy, R. (2008). Holism and nonseparability in physics. *Stanford Encyclopedia of Philosophy*, www.plato.stanford.edu/entries/physics-holism/#OHQM. Accessed December 27, 2015.

Huntington, C. W, Jr. (2015). The triumph of narcissism: Theravāda Buddhist meditation in the marketplace. *Journal of the American Academy of Religion, 83*(3), 624–648.

Husserl, E. (1970). *The crisis of European sciences and the transcendental phenomenology* (D. Carr, Trans.). Evanston: Northwestern University Press. (Original work published 1936)

Husserl, E. (2002). Foundational investigations of the phenomenological origin of the spatiality of nature: The originary ark, the earth, does not move (F. Kersten Trans., L. Lawlor Rev.). In E. Husserl & M. Merleau-Ponty (Eds.), *Husserl at the limits of phenomenology: Including texts by Edmund Husserl*. Evanston, IL: Northwestern University Press.

Jackson, P., & Delehanty, H. (2006). *Sacred hoops: Spiritual lessons of a hardwood warrior* (p. 2006). New York: Hyperion.

Jepsen, K. (2013). Real talk: Everything is made of fields. *Symmetry: Dimensions of particle physics.* Retrieved from http://www.symmetrymagazine.org/article/july-2013/real-talk-everything-is-made-of-fields

Kabat-Zinn, J. (2006). *Coming to our senses: Healing ourselves and the world through mindfulness.* New York: Hyperion.

Kabat-Zinn, J. (2011). Some reflections on the origins of MBSR, skillful means, and the trouble with maps. *Contemporary Buddhism, 12*(1), 281–306.

Kawakami, T. (2015). Luke Walton, Steve Kerr and the Warrior's four core values: joy, mindfulness, compassion and competition. *San Jose Mercury News.* November 24, 2015.

Lakoff, G., & Johnson, M. (1980). *Metaphors we live by.* Chicago: University of Chicago Press.

McMahan, D. L. (2008). *The making of Buddhist modernism.* New York: Oxford University Press.

Merleau-Ponty, M. (1962). *Phenomenology of perception* (C. Smith, Trans.). London: Routledge & Kegan Paul. (Original work published 1945)

Musser, G. (2015). *Spooky action at a distance: The phenomenon that reimagines space and time—and what it means for black holes, the big bang, and theories of everything.* New York: Scientific American/Farrar, Straus, and Giroux.

Puhakka, K. (2015). Encountering the psychological research paradigm: How buddhist practice has fared in the most recent phase of its Western migration. In E. Y. Shonin, W. Van Gordon, & N. N. Singh (Eds.), *Buddhist foundations of mindfulness.* Cham: Springer.

Rhys Davids, T. W. (1963). *The Questions of king Milinda.* New York: Dover Publications.

Rosch, E. (2015). The emperor's clothes: A look behind the Western mindfulness mystique. In B. D. Ostafin, M. D. Robinson, & B. P. Meier (Eds.), *Handbook of mindfulness and self-regulation* (pp. 271–292). New York: Springer.

Sartre, J. P. (1958). *Being and nothingness: An essay on phenomenological ontology* (H. E. Barnes, Trans.). London: Methuen. (Original work published 1943)

Sharf, R. (1998). Experience. In M. C. Taylor (Ed.), *Critical terms in religious studies* (pp. 94–116). Chicago: University of Chicago Press.

Sharf, R. (2014). Mindfulness and mindlessness in early Chan. *Philosophy East & West, 64*, 933–964.

Strauss, E. S. (2015) Big three dominates: No streak, but Warriors still at the peak. www.espn.go.com/blog/golden-state-warriors/post/_/id. Retrieved February 13, 2016.

Tulku, Tarthang. (1977). *Time, space, and knowledge: A new vision of reality.* Emeryville: Dharma Publishing.

Tulku, Tarthang. (1987). *Love of knowledge.* Berkeley: Dharma Publishing.

Tarthang Tulku. (2015). Space field. In J. Petranker (Ed.), *Inside knowledge: How to activate the radical new vision of reality presented to the world by Tibetan lama Tarthang Tulku.* Berkeley: Dharma Publishing. [Originally published 1990]

Taylor, C. (1992). *Sources of the self: The making of the modern identity.* Cambridge: Harvard University Press.

Taylor, C. (2007). *The secular age.* Cambridge: Harvard University Press.

Tulku Urgyen Rinpoche. (1999). *As it is.* Kathmandu: Rangjung Yeshe Publications.

Wilson, J. (2014). *Mindful America: The mutual transformation of Buddhist meditation and American culture.* New York: Oxford University Press.

Selling Mindfulness: Commodity Lineages and the Marketing of Mindful Products

8

Jeff Wilson

U.S. Army veteran Jose Arana smiles out from the February 2015 cover of *Mindful*, relaxed but well put together in his camouflage uniform and short haircut. A gold ring on his left ring finger unobtrusively announces his married status, while the sea foam green cover background and yellow/blue/pale green color scheme of the text (no angry reds or energetic oranges here) exude a calming tone. The magazine's cover text suggests that the contents inside mainly relate to health ("Healing Our Vets," Rewire Your Fearful Brain," "5 Steps to Sleep Soundly Through the Night") and self-care ("Taking Time for What Matters," "Not Getting What You Want May be *Just* What You Need," "Savoring the Complex Flavors of Chocolate," "Caring for Others, A Deeper Kind of Love," "How Meditation and a Whitewater Adventure Lead to Newfound Strength") (Mindful 2015b, i).

There is little here to suggest a connection between the magazine and religion. Sharp-eyed readers might wonder why Arana is seated cross-legged on the floor, rather than standing or sitting on a chair, the common American cultural practice. Others could speculate about that word "meditation" that briefly pops up, or even interrogate the title *Mindful*, though many would not automatically accord them a religious status.

Readers might be surprised, therefore, to learn that *Mindful* is devoted to promoting the Buddhist-derived practice of mindfulness meditation and that its editor-in-chief, editorial director, publisher, most of the board of advisors, and many other staff persons are Buddhist or were trained to meditate in explicitly Buddhist settings. Seeking to promote this Buddhist practice, the makers of *Mindful* have strategically left out nearly all overt references to Buddhism in order to appeal to the widest possible consumer base. This eliding of Buddhism is the result of a long series of changes and choices performed in and on modern Buddhism. And just as all results then themselves become causes, the production of de-Buddhified mindfulness results in the mindfulness movement becoming ever more commodified, diversified, and competitive in turn.

Sometimes to find Buddhism in the mindfulness movement, however, we just need to look slightly below the surface. Opening the front cover of the February 2015 issue of *Mindful*, we discover the inside cover and first page occupied by a huge display ad. On the left, a stack of jewel-encrusted gold bracelets hovers in the middle of the page; on the right, a well-manicured white woman's hand reaches into the picture, her wrist adorned by more golden bracelets, and a glass Buddha statue resting in her upturned palm. Inset pictures show pendants and other jewelry, with Buddhist-related imagery such as Buddhas and lotuses. In the upper left-hand corner is the

J. Wilson (✉)
Renison University College, University of Waterloo, Waterloo, ON, Canada
e-mail: jeff.wilson@uwaterloo.ca

© Springer International Publishing Switzerland 2016
R.E. Purser et al. (eds.), *Handbook of Mindfulness*,
Mindfulness in Behavioral Health, DOI 10.1007/978-3-319-44019-4_8

company's name and tagline: BuDhaGirl, Mind-ful Glamour. This is mindfulness and Buddhism for sale, or rather, it is jewelry for sale, with mindfulness and Buddhism as associations designed to intrigue and please the buyer.

These, then, are two of the most important modes of the mindfulness movement: on the one hand, the selling of Buddhist meditation practice through the use of secular images and rhetoric; on the other hand, the selling of secular products through the use of Buddhist images and rhetoric. Rather than antagonistic or opposite phenomena, these two trajectories represent entwined pro-cesses that result from the ongoing encounter of Buddhism and contemporary Western culture. Both are enabled through the auspices of the ascendant "secular" mindfulness movement and its promotionary vehicles. Since *Mindful* is one of the most important and representative of such vehicles, this chapter presents an analysis of the February 2015 issue in order to excavate the forces at work in the marketing of mindful products—including mindfulness, which has been turned into a product by a new professional class of non-monastic meditation instructors, health gurus, and scientists. In the process, lin-eages of production and representation that mir-ror or influence the mindfulness movement—such as Veggie Tales, Ben and Jerry's Ice Cream, and Nike—are explored as contributing elements in the commodification of mindfulness.

Buddhism in the Economy

What is going on in these situations, where Buddhism is sublimated, appropriated, and sold to make revenue (even if for a nonprofit maga-zine)? In part, these are outcomes from the market-driven mediascape that we all inhabit. Writing in 2007, Mara Einstein notes,

> In a culture where we spend more time with media than any other activity other than working and sleeping, and that media is supported by advertis-ing and marketing, it should not be surprising that religion would need to take on aspects of the market in order to stay relevant within the culture. It is at its base a product, competing against an

overwhelming number of other products in the consumer marketplace. (Einstein 2007, pp. 93–94)

Einstein's comments can be refined further, as the advent of ubiquitous personal mobile devices connected to the Internet means that media (and advertisements) are never more than a quick glance away, such that even such divides as "work" and "viewing media" collapse and media consumption penetrates nearly all facets of life. Churches and meditation groups now make "turn off your cellphone" announcements just like movie theaters, suggesting that congregants (that is to say, consumers) face similar media-checking temptations, perhaps because attendance at reli-gious services is conceptualized as an entertain-ment choice similar to going to the cinema. Certainly, religions' greatest competitors these days are not other religions, but the vast universe of (often media-based) distraction and entertain-ment options that aggressively compete for peo-ple's attention, disposal income, and allegiance.

From a certain angle of view, Buddhist interactions with markets and economics are hardly new. As Lionel Obadia notes in "Is Buddhism Like a Hamburger? Buddhism and the Market Economy in a Globalized World:"

> important historical and ethnographical data, col-lected in very different national and cultural con-texts, has established that Buddhist monasticism has always been concerned (and associated) with economic activity in the societies surrounding and hosting it. The works of Gregory Schopen, for instance, clearly exhibit the presence of an intense reflection on economic activities within Buddhist monasticism, since these institutions needed to be involved in 'business matters,' even in the ancient forms of scriptural Buddhism (2004). Further, in the *Digha Nikaya* (sacred Buddhist texts) the foundations of happiness lie unambiguously on an economic foundation: Buddhist lay believers might live in economic security (Pali: *atthi-sukha*), enjoy wealth (*bhoga-sukha*), and want to be free from debts (*anana-sukha*). (Obadia 2011, p. 103)

As I explain in the introduction to *Mindful America*, adjusting Buddhist practice and pre-sentations so as to offer new modes of exchange is part of what allowed Buddhism to flourish across multiple cultural borders and through many different time periods (Wilson 2014, pp. 2–

6). Buddhism has never been static, has never been apart from economics, has continually sought new adherents and patrons, and has always been in the process of adapting to evolving cultural environments.

But we have to note some important differences between earlier adaptations and the current mindfulness movement. First, earlier economic exchanges typically involved Buddhist patrons paying for the production of desired goods, tangible or intangible, but in either case often not consumed in a straightforward fashion by the laity. Monks were given donations of money, food, or material items (such as robes) so that they might generate merit that was then circulated ritually to laypeople or their ancestors; in other cases donors sponsored the creation of Buddhist images and scriptures, which were usually owned and used by the monastery, not the donor. Thaumaturgic items, such as charms, amulets, talismans, or fortunes, were consumed by purchasers, but mindfulness-based meditation practices such as vipassana were not part of this system of Buddhist magic. However, the prevailing pattern of paying for the production rather than the consumption of Buddhist products is generally unsatisfying to the individualistic, materialistic, capitalistic, and consumeristic culture that has abetted the wider mindfulness movement. Therefore, new models of consumption have taken hold in the dissemination of mindfulness and its associated benefits in the West.

Second, the pace of adaptation and transformation is greatly increased compared to earlier eras of Buddhist history. Rapid travel, instantaneous communication across distances, access to a sea of information about previously unknown peoples and practices, the saturation of everyday life with media, and the constant competition between sellers for finite consumer dollars in an overcrowded marketplace all combine to push an ever-accelerating rate of change in modern life. Subjected to these forces, change in the presentation, understanding, and selling of Buddhist meditation likewise accelerates. Whereas it took approximately 150 years for the Buddhist foundations of the mindfulness movement to be laid in the West, it took only a few decades for it to go from experimental stress relief practice to omnipresent panacea, and change is so rapid at this point that it is no longer possible to stay abreast of all the ways in which mindfulness is being promoted or used to promote other products.

Alternate Ancestors: Evangelicals

One reliable way to narrate the rise of the mindfulness movement is to follow its movement from Asian monks to North American lay teachers to medical and scientific practitioners and finally to non-Buddhist life coaches, self-help authors, and entrepreneurs. This is the trajectory that I trace in chapter one of *Mindful America*. But there are other channels of influence and change at work in the mindfulness movement that could be productively explored. For example, in *Selling God* R. Lawrence Moore chronicles the process whereby Protestant Christians pioneered the commodification of religion in America, with Bible-based diet fads, evangelical sex manuals, Christian rock music, godly exercise books, and more (Moore 1994, pp. 253–255). Via the genre of products called Christian Living, Christians worked out many practices only recently experimented with by proponents of Buddhist-derived mindfulness. As Moore explains, "Christian Living covers a broad spectrum of titles, but fundamentally it is about making Christianity widely accessible by relating religious and spiritual themes in a practical way to life and relationships" (Moore 1994, p. 42). This same impulse lies at the heart of the mindfulness movement, which applies meditation to mundane, practical matters in an accessible manner.

One example of this strategy among Christians is the animated children's series Veggie Tales, which successfully markets Christian morality and biblically derived stories to kids by not dressing them in the trappings of Christianity. This allows Veggie Tales to out-compete other evangelical kids' products as it both satisfies conservative Christian viewers and reaches much larger audiences who would avoid overt religious messaging. Indeed, Veggie Tales went from the evangelical fringe to the mainstream through the use of this tactic, as demonstrated by its direct-to-video origins and later adoption into NBC's Saturday morning cartoon lineup.

A main representative of this approach to selling Christianity is Joel Osteen, senior pastor of Lakewood Church in Houston, America's largest megachurch. Osteen's best-selling books include such seemingly non-religious titles as *Your Best Life Now: 7 Steps to Living at Your Full Potential* (2004) and *Become a Better You: 7 Keys to Improving Your Life Every Day* (2007). Once drawn to open the book the reader discovers that the content is Christian, but also aggressively non-denominational: Osteen's message is geared toward using Christian practices to receive earthly wealth and success, revealing his connection to the theological interpretation of Christianity known as "prosperity gospel." His presentation is also meant to be accessible to non-Christians, in the hope that they will find Christian-based practices to improve their lives, and will perhaps commit to Christianity as a result. Chapters echo themes subsequently found in mindfulness publications like *Mindful,* such as "Reprogramming Your Mental Computer" (*Mindful*: "Rewire Your Fearful Brain"), "Choosing the Right Thoughts," "Be Happy With Who You Are," and "Keep Your Heart of Compassion Open (Osteen 2004, vii–viii)." We could locate Christian Living products such as Veggie Tales and best-selling prosperity gospel preachers such as Joel Osteen as alternate ancestors of the mindfulness movement—first to be subjected to the economic and social forces that have now come to bear on mindfulness practice, evangelicals responded with a host of changes to the style and marketing of Christianity that have influenced a generation of self-help books, happiness coaches, and spiritual entrepreneurs.

The media kit for *Mindful* asks the question, "Why is mindfulness hitting the mainstream?

(Mindful 2015a, n.p.)." Their multi-pointed answer points to the positive nature of mindfulness: "Anyone can do it. It helps. It's evidence-based. It's a way of living. It offers hope for the future (Mindful 2015a, n.p.)." Left undiscussed is the careful scrubbing of mindfulness's Buddhist nature, which is at least as important an explanation for how it has been able to enter the mainstream. *Mindful*'s hiding of its Buddhist roots parallels efforts to sell Christianity to a wider audience. Overt Buddhism is essentially banished from the editorial content of *Mindful*, replaced by thinly disguised Buddhist content. For example, Pema Chödrön, the most famous Buddhist nun in North America, is referred to simply as a "meditation teacher (Halliwell 2015, 38)." Traditional Theravada metta (lovingkindness) meditation practice is just called "offering kindness," without reference to its Buddhist origins (Furtado 2015, 56). Carolyn Gimian's article "The Perks of Disappointment" advises readers on the three types of disappointment: "Not Getting What You Want," "Getting What You Want," and "Not Knowing What You Want (Gimian 2015, 72, 74075)." These are presented as non-religious insights that the author has reached based on her life experience. Nothing tips off the reader that these are in fact central teachings given to her by Chogyam Trungpa, whose Buddhist lineage Gimian is one of the primary holders of—they are, of course, commentaries on dukkha. And mindfulness itself is a translation of sati, the Pali term for meditative awareness, attention, and remembrance. The media kit quotes best-selling mindfulness promoter and TV personality Dan Harris for support: "*Mindful* magazine is a fantastic resource for people looking to learn about mindfulness in a smart, secular, and science-oriented way (Mindful 2015a, n.p.)." Yet given the thoroughly (though unacknowledged) Buddhist nature of *Mindful*'s content, this assertion of "secular" demands analysis. In the context of *Mindful*, and perhaps for the entirety of the mindfulness movement, "secular" appears to mean "Buddhism packaged as if it isn't Buddhism."

Alternate Ancestors: LOHAS

Another alternate lineage can be established between *Mindful* and the LOHAS movement. As Mara Einstein explains:

> For some Mind/Body/Spirit seekers, products themselves become the conduit for expressing their spiritual beliefs. This market segment—and it is a *market* segment and not a spiritual practice per se —which embodies both the more traditional 'save the earth' philosophies and the market orientation of the New Age, is the Lifestyles of Health and Sustainability (LOHAS) movement. LOHAS is defined as 'a market segment focused on health and fitness, the environment, personal development, sustainable living, and social justice' (LOHAS, n.d.). This movement is a defining example of the marriage between belief system and the market. Called Cultural Creatives or Lohasians, consumers of LOHAS products claim that it is through their product choices—responsible investing, organic products, eco-tourism, green products, and so on—that they can change the world and themselves. Building on the work and spirituality movement of the 1990s, the purveyors of LOHAS products believe that business and spirituality can work in tandem, that being socially and environmentally responsible is not mutually exclusive from business practices. The appeal of these products is evident in that they represent a market of almost $230 billion made up of close to 50 million U.S. adults. (Einstein 2007, p. 206)

Ben and Jerry's ice cream is a good example of LOHAS products: It is more expensive than other ice creams and has the same effect on your waistline, but careful marketing has associated Ben and Jerry's with "healthy" ingredients, concern for the environment, and progressive values.

LOHAS is the latter-day market instantiation of what used to be termed the New Age. In fact, the influential magazine *New Age Journal* was rebranded as *Body + Soul*, then rebranded as *Whole Living*, and now has ceased production as an independent magazine—instead, it has been absorbed into *Martha Stewart Living*, signaling the completion of the process whereby the New Age (and its consumers) moved from the fringes to the mainstream of modern North American culture (Einstein 2007, 199). We can note in passing how it progressively shed its overtly religious connotations, yet its content largely remained the same, as did the imagery used to sell that content—even as the name shifted away from alternative spiritual connections, the magazine's new incarnations continued to offer cover images of fit women sitting in a modified lotus position, a visual trope that Buddhologist Scott Mitchell labels the "Tranquil Meditator." As Mitchell describes, "the Tranquil Meditator represents a particular kind of this-worldly nirvana. She represents a practical way for individuals to attain a relaxed and healthful state of mind as a way to alleviate the stresses caused by the modern world. She is not concerned with any metaphysical or existential issues that might be brought up by doing meditation—she wants to be happy. In this way, meditation becomes, as Iwamura would say, safe for cultural consumption, nothing more harmless than a trip to a spa or a vacation." (Mitchell 2014, p. 86). The Tranquil Meditator is the most common cover image for *Mindful,* either in direct representations of in implied versions such as Arana's seated pose for the February 2015 issue.

LOHAS products abound in the current market, and *Mindful*, like most mindfulness purveyors, attempts to position itself as catering to LOHAS consumers, the better to draw LOHAS advertising dollars. The media kit beckons, "Why advertise with *Mindful* media? To reach a fresh and largely untapped audience of early adopters who prioritize a well-balanced approach to physical and mental health; gravitate toward products embedded with health benefits, thoughtfully designed, joyful, tasty, interesting or storied; look to purchase products that are environmentally friendly and socially responsible (Mindful 2015a, n.p.)." These are precisely the up-market base of LOHAS consumers, and the media kit assures potential advertisers that "*Mindful* offers a highly committed readership looking for brands, products, and services that speak to their values and reflect the mindful world (Mindful 2015a, n.p.)." According to the magazine's own research, 77 % of their readers "are willing to pay a premium for Natural/Organic products," and *Mindful* employs a "LOHAS Ad Director" as one of its two primary staff people assigned to advertising management (Mindful 2015a, n.p.).

Mindful's appeal to LOHAS sellers is successful, as evidenced by the appearance of LOHAS product advertisements throughout the magazine. Page 5 of the February 2015 issue—opposite the second part of the issue's table of contents—is a full-page display ad for Eden Chili. Since chili has no direct link to mindfulness, Eden tries to sell its product by connecting it to health, purity, and environmentalism. The ad copy crows that it is a "PURE and PURIFYING chili of beans, whole grains, vegetables, and healthy mushrooms…cooked in SAVORY sauce at our certified organic cannery Mindful (2015b, 5)." Eden Chili sells at $53 per dozen cans.[1] Page 22 is another full-page display ad, this time for Toe Talk socks. Socks too have no natural connection to mindfulness, so the seller tries to tie them to yoga and meditation. "From the yoga or Pilates studio to mindfulness practice or meditation, Toe Talk's stylish line of socks provides a personal mantra to help you focus your mind and invigorate your soul (Mindful 2015b, 22)." The fit female model is seated, but this time in a yoga pose rather than a meditation one, while above floats the tagline "Be mindful from head to toe (Mindful 2015b, 22)." The "personal mantras" are essentially imperatives stitched across the top of the socks: one orders the wearer to "Be Mindful," while another tells the wearer to "Seek Balance," and a third pair sports the command "Inhale" on one sock and "Exhale" on the other (Mindful 2015b, 22). Toe Talk socks retail for $10 per pair.

Alternate Ancestors: iPhone

One of the contributing factors to the diversification of mindful commodities is the growth of the tech sector for communication and entertainment products. The proliferation of technological gadgets and their role in contemporary life provides an ever-expanding set of platforms for which mindful products can be designed. Furthermore, since science and technology carry positive associations as varied as "secular," "modern," "cool," and "relevant," when mindfulness enters the tech realm it participates in the aura of wonder, chic, and fandom that exist around successful electronic products lines, such as Apple's iPhone, iPod, and iPad. This serves to further disassociate mindfulness from Asia, Buddhism, religion, the past, and tradition, although these can be creatively re-appropriated at strategic moments if they provide additional selling power in specific contexts.

Advertisements for tech-related mindful products are common in *Mindful*. Page 17 of the February 2015 issue is a full-page display ad for eMindful, an online service that claims "Since 2007, our members have realized exceptional results" via their platform for delivery of meditation instruction, brain games, mindful apps, and chat functions (Mindful 2015b, 17). The page is dominated by four ascending columns that resemble the signal strength bars on a personal mobile device, while connected text reveals them as measurements on a bar graph with the explanation that eMindful users reported "27 % improved sleep, 37 % less stress, 40 % stopped smoking, 59 % reversed metabolic syndrome (Mindful 2015b, 17)." The silhouette of a fit female figure sprints along the top of the columns, like a runner dashing toward the finish line. The statistics evoke a world of metrics, scientific assessment, and rigor, while the tagline "Improving life one moment at a time" suggests that mindfulness makes life better and can be tackled by brief meditative pauses, rather than being a long-term, primarily monastic or religious discipline (Mindful 2015b, 17).

Developed by founder and CEO Kelly McCabe Ruff, a long-time senior executive at Salomon Brothers, Lehman Brothers, and Citicorp, eMindful sells itself to individuals as "stress reduction" while marketing to employers with claims to "reduce your employee healthcare costs and improve productivity (eMindful 2016, n.p.)." Access to basic content costs $10 per month,

---

[1]Prices in this chapter were obtained by my research assistant, Laura Morlock, in mid-2015, by checking the vendors' websites or appropriate online stores, such as Amazon.com. Because these prices fluctuate frequently, those listed here are for illustrative purposes, and may or may not correspond precisely to those current when the February 2015 issue of *Mindful* first debuted.

while premium access costs $16 per month. eMindful also runs promotions, such as the 1 % Challenge, which points out that since a day has 1440 min in it, being mindful for just fourteen minutes would equal only 1 % of the day yet will "help build resilience, reduce stress, improve relationships, and create more joy in every area of your life (eMindful 2016, n.p.)." Users who commit to fourteen minutes of mindfulness practice per day for thirty days are given access to eMindful content; for every 25,000 min that 1 % Challenge participants perform eMindful donates $1000 to MindUP, the mindfulness-in-schools program created by actress Goldie Hawn; particularly diligent users can win a year's subscription to *Mindful*, VIP membership in eMindful, and even a "brain sensing Muse headband" valued at $500 (eMindful 2016, n.p.).

Lower down the tech scale, yet still a part of the phenomenon, is the meaning to pause® bracelet, which appears in a half-page advertisement on page 76 of the February 2015 issue of *Mindful*. The bracelet is a string of beads visually identical to a Buddhist mala, except that it includes a beige oval with the word "pause" and an inset button. As the ad relates, the bracelet will vibrate every sixty or ninety minutes, thus "prompting you to pause, ...reflect on your intentions and reframe your thoughts (Mindful 2015b, 76)." This mild shock treatment will "create a ripple of mindfulness" and "inspire mindful change," as well as providing adornment that signals one's commitment to mindful values (Mindful 2015b, 76). Essentially, this is the wedding of old-school Buddhist technology (mala beads) with contemporary wearable tech such as the Fitbit. Potential customers are told they can personalize their bracelets with photos, phrases, and a range of different beads.

Also in the tech camp are online mindfulness training programs designed to teach the student how to become a mindfulness instructor themselves. For example, page 14 of *Mindful*'s February 2015 issue sports a half-page ad for the Mindful Life Program's teacher certification, while page 19 is a full-page display ad for Mindful Schools' certification program (Mindful 2015b, 14, 19). Both are one-year online

programs with a limited in-person component (two five-day residential workshops for Mindful Life Program, two-week-long summer retreats for Mindful Schools). Those who complete the Mindful Life Program will be certified to teach Mindful Life Program to others, while the Mindful Schools program bills itself as "in-depth training for professionals interested in integrating mindfulness into their work with children and adolescents (Mindful 2015b, 10)." Mindfulness instruction in schools, as exemplified by MindUP and similar programs, is a rapidly growing phenomenon, and thus certification in this field potentially provides enhanced employment opportunities—after working online with mindfulness teachers, a student could then use their certificate to obtain a job training children to do mindfulness meditation. The Mindful Life Program costs $4800, while the year-long certification of Mindful Schools costs $4950.

Insights from Rachel Wagner and Christopher Accardo's "Buddhist Apps: Skill Means or Dharma Dilution?" are useful in discussing these mindful products. As they note:

> How does the medium of the smartphone itself shape the teachings received through it? One of the most powerful implicit messages that the iPhone sends for those utilizing Buddhist apps is that the Dharma can be learned alone, most likely with headphones on, and possibly even while working out at the gym or commuting to the office, even if the app isn't explicitly designed for that purpose. iPhones are deeply personalized devices: the user determines how and when an app will be used— not one's community. The Dharma, says the medium of the iPhone app, is more inclined to be seen as something you squeeze in between other things you should be doing. It is a personal, private thing. (Wagner and Accardo 2014, p. 140)

In the specific cases just examined above, we can see that *Mindful* readers are taught by these ads that mindfulness is a personalized service one buys like other apps in online markets, something that does not require face-to-face instruction or community support. You can learn all you need for a mindfulness practice from eMindful's app-based portal; you can learn to teach others what they need via online courses, without close work with a teacher. Invariably, the instructors for these products are presented as educated, upper-middle-to-upper class professionals, not Buddhist monks (the sole teachers of mindfulness for the previous 2500 or so years). Mindfulness's benefits can be gained in brief pauses, fourteen minute chunks, and one moment at a time. It can be done not only at work but while working, not only in school but while learning, not only while observing the world but also while immersing oneself in virtual realities. We should note that *Mindful* itself hosts a robust suite of online services, including mindful.org which receives hundreds of thousands of unique visitors, the *Mindful* iPad Edition, a weekly electronic newsletter, and the video platform MindfulDirect, all of which are available for ad placement purchase by sellers. The magazine also sports popular Facebook and Twitter accounts.

Alternate Ancestors: Nike

So far we have examined some of the products that appear in *Mindful*—but what of the magazine's policies around advertisements? And who are the consumers that these ads are targeting? According to *Mindful*'s media kit, the print magazine has a readership of 210,000, 79 % of whom are female (compared to 50.8 % of the American population) (Mindful 2015a, n.p.). The readers' median household income is $80,000, and 73 % are employed—the majority of those in professional, executive, or managerial positions (Mindful 2015a, n.p.). Their median age is fifty-five years of age and 85 % have a college degree, while 55 % also have a postgraduate degree (Mindful 2015a, n.p.). In class terms, they are well above the American average: According to the U.S. Census Bureau, the median household income is $53,482 and only 29.3 % of persons over twenty-four years old have a bachelor's degree or higher, while only 58.7 % of the civilian female population over fifteen years of age is employed (Unites States Census Bureau 2015).[2]

Mindful's surveys indicate that large majorities of its readers travel for leisure, love to read, are the primary grocery shoppers for their homes, and have active exercise regimens (Mindful 2015a, n. p.). A majority also travels for education or personal development, and half practice yoga (Mindful 2015a, n.p.). These characteristics closely correspond to those of LOHAS consumers, who are an especially desirable consumer demographic due to their high levels of disposal income and belief that purchasing products is itself a form of spiritual or holistic practice, as well as a way of creating and exhibiting responsible identities.

Advertisements in *Mindful* are subject to certain limitations. As their copy and contract requirements state, "We reserve the right to reject advertisements which, based on our judgment, are not consistent with our publication's objectives, standards, and editorial convictions, as well as ads which in our estimation will not achieve the advertiser's aims. Because *Mindful* endeavors to offer a view and voice for the application of secular mindfulness practices, from time to time we may suggest changes to copy and/or imagery in keeping with this goal (Mindful 2015a, n.p.)." In other words, advertisements that appear in *Mindful* have been approved on ideological grounds by the chief editorial and publishing leaders at the magazine and thus can be considered as endorsed by *Mindful*. For the print edition, those ads fall into two broad categories. First are full-color display ads. Full-page ads cost $2995, half-page ads cost $1995, the inside front cover costs $3650, the inside back cover costs $3350, and the back cover of the magazine costs $3900 (Mindful 2015a, n.p.).[3] All of the products discussed thus far fall into this category.

The second category is the Marketplace section, consisting of several pages of smaller product advertisements at the back of the magazine.

---

[2]*Mindful* is a joint American–Canadian production, and it has readers in countries outside the United States. We should be careful, then, about too easily conflating *Mindful* readers with Americans. Nevertheless, the large majority of *Mindful* readers are in fact American.

[3]Ad prices come from the 2016 media kit of *Mindful*, obtained in Fall 2015. They may not, therefore, precisely correspond to the actual cost of the ads in the February 2015 issue, but should nonetheless be reasonably close. They are used here for illustrative purposes, since actual costs may have differed for any given advertisement due to bulk buying, changing ad rates, delays in delivery, and other factors.

1/4-page ads cost $750 in the Marketplace, while 1/8-page ads cost $500 (Mindful 2015a, n.p.). There are discounts available for multiple-issue purchases for both categories. Marketplace products are sometimes repeats of the items or services featured in the full-color ads sprinkled throughout the magazine: For instance, Toe Talk and meaning to pause® bought Marketplace space in addition to their display ads in the February 2015 issue. Other products are only advertised in the relatively cheap Marketplace, such as Mindfulicious granola bars, "the first bar ever to include mindful eating tips for savoring every bite!" ($8.95 per bar) or Passion and Presence's mindful sexuality couples retreats (starting at $995) (Mindful 2015b, 79).

With all of this in mind, it is possible to return to our opening advertising example, BuDhaGirl, for a deeper analysis. The BuDhaGirl ad on the inside front cover and first full page is dominated by images, with minimal, carefully chosen, text. The primary ad copy is just four sentences long:

> a day begins
> BuDhaGirl is a story of awareness and new beginnings—
> beautiful objects and jewelry for bringing the mind back into
> balance and focus. Our story and our legacy is, simply,
>   MINDFUL GLAMOUR.
>   (Mindful 2015b, ii, 1)

The most important aspect for analysis of BuDhaGirl's ad is its branding approach. Modern branding is primarily about narrative: The product is embedded in a carefully crafted story that gives it a "biography," elicits emotional resonance, and invites the consumer to include herself in the narrative. As Mara Einstein notes,

> Branding is about making meaning—taking the individual aspects of a product and turning them into more than the sum of their parts. It is about giving consumers something to think and feel about a product or service beyond its physical attributes. (Einstein 2007, p. 70)

In this way, branding actually mirrors processes found within religions, which are also about meaning making. BuDhaGirl tells the reader twice that their product is a story, one about being mindful, starting fresh, and being beautiful.

Their jewelry is superior to others' because when you don a BuDhaGirl product in the morning, it stimulates you to be momentarily mindful in a way that putting on another company's bracelet does not.

The BuDhaGirl ad informs consumers about how to use its products through pithy instructions. Floating over the glass Buddha statue are the words "turning routines into rituals," implying that adorning the body can be transformed from a rote action into a meaningful or even sacred practice (Mindful 2015b, 1). The ad also includes the free-floating tagline "pause. be aware. be it (Mindful 2015b, ii)." At the top of the ad is the BuDhaGirl logo, a circle with the initials B and G (for Buddha and Girl) repeated in a stylized form that causes them to resemble an abstract rendering of the Buddha's head, complete with his famous tiered hairstyle; the logo is pink, connecting it to femininity according to contemporary genderings of color. The logo and brief imperative tagline are clear latter-day descendants of one of the most successful branding campaigns of all time: the Nike "swoop" and slogan "Just do it." Just as Nike managed to associate its sportswear products with personal achievement and confident, powerful self-identity through advertisements that crafted narratives around its iconic logo and slogan, BuDhaGirl seeks to communicate that its product carries Buddhist-type values of mindfulness and simplicity, mixed with valuation of style, beauty, glamor, and conspicuous (yet conscious) consumption. Even on a barely noticeable level, the ad strives to frame its product: For instance, the woman's hand supporting the Buddha statue is wet, as if she had just stepped out of the shower, suggesting cleanliness and purity.

Is all of this effort at crafting an image and story for mindfulness-associated jewelry worth it? From a purely economic standpoint, the answer would appear to be "likely." The various products featured in BuDhaGirl's *Mindful* ad range from $95 to $220, with an average cost of $136. If we use the basic rate for an inside cover and full-page ad of $6645, then BuDhaGirl would need to sell forty-nine units in order to recoup their advertising cost (49 × $136 = $6664). Thus, BuDhaGirl would have to sell a unit to only one out of every 4286 readers (210,000 ÷ 49 = 4285.71) to earn back its $6645; if one in every thousand readers bought one unit, that would generate $28,560, for a profit of $21,915 after advertising costs.[4]

But do readers actually buy products advertised in *Mindful*? According to the magazine's media kit, after reading *Mindful* 50 % bought a book by a featured writer, and 25 % purchased something advertised in the Marketplace (Mindful 2015a, n.p.). If we imagine that 25 % of readers also buy products from the magazine's full-page ads (a conservative estimate, given the far greater prominence of such ads), then BuDhaGirl could make a very tidy profit indeed. Just 1 % of *Mindful*'s 210,000 readers (2100) buying a single $136 unit would amount to an intake of $285,600. With such profits potentially available in the commodified world of mindful products, why *would not* sellers seek to associate their jewelry, granola bars, apps, and chili with mindfulness? And surely they would give strong consideration to precisely how to utilize Buddhism and mindfulness in the branding and marketing of their product, as well as whether to splurge on more expensive advertisements in *Mindful*.

Conclusions

Mindfulness is a valuable commodity in today's competitive marketplace. Because it has been wrested from the control of Asian monks and their Buddhist followers, it is now available to three interrelated business types: meditation instructors, who sell guidance that used to be disseminated via face-to-face apprenticeships under Buddhist monks that largely fell outside the economic realm; material product vendors who sell items alleged to enhance mindfulness or

---

[4]These calculations are intentionally simplistic, as they leave out additional considerations such as the cost of raw materials, employee wages, and other costs of production. They are intended only to broadly demonstrate the level of revenue and profit involved.

who use mindfulness to imbue their wares with an aura of healthfulness, authenticity, spirituality, environmentalism, and/or chic; and mindfulness promotion vehicles such as magazines and websites that make money off of the advertising dollars of the first two groups. All three types claim to be motivated by compassionate concern for overworked, stressed-out, unhealthy modern people; all three also contribute to the ongoing commercialization of mindfulness meditation.

The relationships with Buddhism that these three groups display vary according to the interests of the individual seller and the current situation. *Mindful* is largely run by Buddhists but is committed to presenting mindfulness as secular, and thus it buries Buddhism beneath a veneer of secularity. Given North America's overwhelmingly non-Buddhist population, this decoupling of mindfulness from its traditional constituency allows meditation to be marketed to a far larger audience. Meanwhile, Buddhism re-appears in the advertisements that make up a third of the magazine's content, such as page 15's ad for DharmaCrafts meditation cushions, Insight L.A.'s page 74 ad for "secular and Buddhist classes," the Monastery Store's page 78 ad for "Buddhist Statues," and of course BuDhaGirl's two-page spread, the largest advertisement in the entire February 2015 issue (Mindful 2015b, 15, 74, 78, ii, 1). Buddhism is now a resource for the commodified mindfulness movement, to be deployed strategically when it serves a purpose. It can be a branding tactic, provide "exotic" aesthetics, serve as a symbolic code for values and ideals, and also be a convenient foil for secularists who use it to stand in for larger issues or a target for biases toward "religion" as a whole. Mindfulness thus becomes a colonizer of Buddhism, mining the tradition for convenient resources as needed, rather than mindfulness being an integral part of the religious culture of particular Asian societies.

These developments are not without consequences. For instance, when non-religious online companies such as Amazon penetrated the religious market through sales of Christian books and products, they negatively affected local Christian bookstores, both by taking away immediate dollars and by training new generations to shop online instead of developing a habit of visiting such stores (Einstein 2007, p. 50). In a similar way, the secularization of mindfulness redirects potential adherents from Buddhist temples and meditation centers, channeling them into a professionalized, fee-for-service model that decreases the relevancy of mindfulness's traditional purveyors at the very moment that mindfulness comes to have its greatest impact on the wider culture.

A further consequence is the heightened competition around mindfulness: Because money can be made off mindfulness, it invites competition and consumption, as increased variety raises the likelihood that a consumer will encounter mindfulness products and also that they will find a product that interests them, which injects more dollars into the system and thereby heightens the incentive for competitive innovation, in an unending cycle. Quality control becomes ever harder to enforce as profits accrue to those who can best gain customers through savvy marketing and product design—these do not necessarily mean that the products are of low quality, but customers flock to products that have the slickest ads or apps with the nicest interface, which are not guaranteed to be those with the most reliable instructors or deepest understanding of meditation. The logic of the market takes over and all sellers must devote an increasing percentage of their budget to advertising, market research, and employees devoted to ad design and publicity, lest they be crowded out by their competitors who are already spending on such things. Those who cannot afford to do so command a shrinking share of the market, regardless of the superiority or inferiority of their product.

None of this should be surprising. A commercialized, media-driven, capitalistic system such as that seen in the United States inevitably generates transformations in religion and culture, and Buddhism has always been reinvented by newer generations looking to promote those aspects they find valuable and to profit financially or materially on the usefulness of their religion. While the relatively rapid proliferation of mindfulness sellers and products is remarkable, it makes sense when we consider that the same forces that gave us Veggie Tales, Ben and Jerry's, iPhone, and Nike are at work upon Buddhism as well.

References

Einstein, M. (2007). *Brands of faith: Marketing religion in a commercial age*. New York: Routledge.

eMindful. (2016). eMindful.com. https://emindful.com/. Accessed January 22, 2016.

Furtado, T. (2015). Getting started: Kindness. *Mindful, 2* (6), 52–59.

Gimian, C. (2015). The perks of disappointment. *Mindful, 2*(6), 70–76.

Halliwell, E. (2015). Lean into fear. *Mindful, 2*(6), 34–39.

Mindful. (2015a). *Media kit 2016*. New York: The Foundation for a Mindful Society.

Mindful. (2015b). *Mindful, 2*(6).

Mitchell, S. A. (2014). The tranquil meditator: Representing Buddhism and Buddhists in U.S. popular media. *Religion Compass, 8*(3), 81–89. doi:10.111/rec3. 12104

Moore, R. L. (1994). *Selling god: American religion in the marketplace of culture*. Oxford, England: Oxford University Press.

Obadia, L. (2011). Is Buddhism like a hamburger? Buddhism and the market economy in a globalized world. *The Economics of Religion: Anthropological Approaches, 31*, 99–120. doi:10.1108/S0190-1281 (2011)0000031008

Osteen, J. (2004). *Your best life now: 7 steps to living at your full potential*. New York: Hachette.

Osteen, J. (2007). *Become a better you: 7 keys to improving your life every day*. New York: Free Press.

Unites States Census Bureau. (2015). QuickFacts. http://www.census.gov/quickfacts/table/PST045215/00. Accessed January 22, 2016.

Wagner, R., & Accardo, C. (2014). Buddhist apps: Skillful means or Dharma Dilution? In G. P. Grieve & D. Veidlinger (Eds.), *Buddhism, the internet, and digital media: The pixel in the lotus*. New York: Routledge.

Wilson, J. (2014). *Mindful America: The mutual transformation of Buddhist meditation and American culture*. New York: Oxford University Press.

Mindfulness and the Moral Imperative for the Self to Improve the Self

Richard K. Payne

Introduction

The idea of self-improvement, of bettering one-self, is central to American values. In any number of different expressions, it is a basic theme of our social discourse. The hero narrative propagated in our culture is most commonly structured around an individual's efforts to achieve a better life. This is, for example, the "official" ancestral story of America, "they came as immigrants to these shores seeking merely the opportunity to make a better life for themselves." The national myth of America as a land of new beginnings, where everyone can start over, requires ignoring Native Americans, those brought as slaves, and those fleeing from rather than aspiring to. The national trope of new beginnings is intimately linked to expectations of self-improvement.

Dan P. McAdams has abstracted the narrative structure of what he calls the "redemptive self," a pattern he finds in "highly generative American adults," people who

> shape their lives into a narrative about how a gifted hero encounters the suffering of others as a child, develops strong moral convictions as an adolescent, and moves steadily upward and onward in the adult years, confident that negative experiences will ultimately be redeemed. Redemption may take the form of atonement for past wrongs, upward social mobility, political or emotional emancipa-

tion, recovery and healing, personal enlightenment, or the progressive development and fulfillment of the good inner self....more than other kinds of life stories, the redemptive self underscores the narrator's belief that bad things can be overcome and affirms the narrator's commitment to building a better world. (McAdams 2006, p. 241).

While not the only life story narrative found in American culture, it is a dominant one and is a variant version of the founding myth of new beginnings in the quest toward self-improvement, which it serves to amplify.

Although the two are often used synonymously, it is important to take a moment here at the beginning to distinguish between self-help and self-improvement as we will be using the terms here. Self-help emphasizes the abilities and attitudes needed to acquire the skills and competencies to allow one's autonomy in the pursuit of any number of specific goals. One might take a class or purchase a manual that will show you how to do your own income tax return, for example. Self-improvement is a subset of self-help, that in which the project or the object is not something like preparing one's tax return, but rather improving the self itself.

The Myth of the Frontier: DIY or Die

One of the origin myths of American individualism is the frontier where, we are told, survival depended on the ability to "do it yourself." The myth is built up of images of at first itinerant

R.K. Payne (✉)
Institute of Buddhist Studies, Berkeley, CA, USA
e-mail: rkpayne1@mac.com

© Springer International Publishing Switzerland 2016
R.E. Purser et al. (eds.), *Handbook of Mindfulness*,
Mindfulness in Behavioral Health, DOI 10.1007/978-3-319-44019-4_9

visitors to the frontier (hunters, explorers, and mountain men) and then later permanent settlers (ranchers and farmers) living in an often hostile environment, isolated from one another, and with no skilled craftsmen to call upon. American individualism is presumed by much of the self-help literature generally, and the origin myth tells us that this is a "natural" characteristic that follows from the frontier experience of rugged individuals responsible for themselves and only for themselves. Kevin M. Kruse has discussed the mythology of the frontier with its "promises of a fresh start" as having "long attracted Americans looking to reinvent both themselves and their nation." (Kruse 2015, p. 8).

The actuality of life on the frontier seems, however, to have been rather different. Survival may have meant the ability to do many things oneself, but it also required shared collective action. The era following the Civil War, for example, saw the rise of the Grange movement together with other rural cooperatives ("coops"). Whether intentional or not, it seems likely that the collectivist actualities of the frontier experience were displaced by an idealized image of individualism, as that version of the self was developing in the late nineteenth and early twentieth centuries under the influence of capitalism and industrialization. The individualistic ethic of "do it yourself" is reflected in self-help literature and in turn in self-improvement programs, such as mindfulness training programs.

A Brief and Impressionistic History

Purpose

The purpose of this historical sketch is to emphasize the historicality and socially constructed nature of the ethos of self-improvement —its values, beliefs, and presumptions. Like much in religion generally, the values, beliefs, and presumptions of self-improvement are presented by its proponents as ahistorical, eternal truths. Examining the historical background of the self-improvement culture also makes it possible to the question whether it corresponds or conflicts with the culture of Buddhism as the latter has developed in its own, separate historical fashion. Were we instead to simply presume that the ethos of self-improvement is universal, the effects of placing Buddhism in that culture would either be obscured to the point of invisibility, or minimalized. Frequently, the strategy for obscuring or minimalizing the differences is for proponents to assert that what they are promoting is the eternal or true essence of Buddhism, the true heart of the Dharma, while whatever does not accord with that representation is dismissed as "merely cultural accretions." The claim goes on to conclude that such supposed and inconvenient accretions can be freely dispensed with, not only without loss, but actually to the benefit of freeing the truth from the merely conventional, the merely cultural. This serves, however, to conceal the sociohistorical locatedness of present-day interpretations, facilitating the representation of those interpretations as if they are the true and eternal significance of the Buddha's teachings.

The Sketch

The culture of self-improvement has a very long history in Euro-American society, one that reaches back to ascetic traditions of the Mediterranean world during Greco-Roman and Judeo-Christian times. This is the early end of the trajectory of moral philosophy that runs through the Renaissance and Reformation to colonial America. The religious themes of that trajectory created the moral imperative to self-improvement found in Puritan religious thought, continues to inform American popular religious culture, and are perhaps particularly evident in the culture of self-improvement.

Pierre Hadot has claimed that the goal for Greek philosophy is self-transformation. In Hadot's terminology, it was "a way of life, both in its exercise and effort to achieve wisdom, and in its goal, wisdom itself. For real wisdom does not merely cause us to know: it makes us 'be' in a different way." (cited in Storhoff and

Whalen-Bridge 2010, p. 2). Such conceptions appear to have contributed a ground for the development of the ascetic impulse in the Greco-Roman world. Gavin Flood has described asceticism as "the discovery or opening out of an interior world." (Flood 2004, p. ix). And the corollary is the self as a private inner space —"This sense of subjectivity is closely linked to the idea of interiority or inwardness and…the modern notion of the self is constituted by a sense of inwardness—that we are beings with inner depths and inner resources." (Flood 2004, p. 17). This "discovery" has carried forward all the way to present-day American religious culture. That culture is psychologized, that is, presumes the concepts, concerns, and categories of psychology, and it is this private inner space which is conceived to be the location in which religious experience is to be sought, for example, mystical union with God, enlightenment, or awakening. Prior to this psychologized form of popular religious culture coming to hegemonic dominance, the understanding of "hearing voices" was not always understood to be an internal event (Schmidt 2000). Today, however, some prominent contemporary forms of Evangelical religion do conceive God's communication as taking place internally—though hearing God speak is a learned skill. According to T.M. Luhman, "Newcomers [to the Vineyard, an Evangelical church] soon learn that God is understood to speak to congregants inside their own minds. They learn that someone who worships at the Vineyard must develop the ability to recognize thoughts in their own mind that are not in fact their thoughts, but God's" (2012, p. 39).

This conceptualization of the mind as an interior space was further refined by Augustine. According to Phillip Cary, "The new and specifically Augustinian contribution to the notion of the inner space—the thing that distinguishes it from previous forms of Platonist inwardness—is precisely that Augustine's inner space is actually private" (Cary 2000, p. 5). However, while the privacy of interior space is for Augustine a consequence of our

"estrangement from the one eternal Truth and Wisdom that is common to all" (ibid.), modern conceptions of interiority as found not only in the Evangelical understanding just mentioned, but also in the ethos of self-improvement tends to see that interior space as the location of divine wisdom, inner sources of strength, one's true nature, and so on.

Renaissance

The conception of the self as malleable, that is, as something that one can create or alter as one desires, takes on much of its familiar modern form in the Renaissance and has been referred to by Stephen Greenblatt as "self-fashioning." Self-fashioning required that the self be understood as malleable, the reshaping of which could be undertaken as an intentional project, fundamentally no different from other kinds of projects directed toward fashioning the world. Greenblatt notes that "Perhaps the simplest observation we can make is that in the sixteenth century there appears to be an increased self-consciousness about the fashioning of human identity as a manipulable, artful process" (Greenblatt 1980, p. 2). One of the important factors in creating this sense that it is possible to fashion oneself is mobility. While modernity is often decried for its lack of stability, and the move from community (*Gemeinschaft*) to society (*Gesellschaft*) as the irremediable loss of a sense of integrity of being and belonging, Greenblatt notes the centrality of mobility—social, economic and geographic—in creating the freedom to refashion oneself (ibid., p. 7). In this sense, it is ironic that the characteristic of modernity that anti-modernists blame for what they see as a devastating loss of rootedness is at the same time the source of personal freedom to be self-responsible for one's own identity in the world.

Reformation

While it was a theme of Renaissance thought, during the Reformation self-fashioning seems to have merged with the tradition of religious asceticism and came to inform Protestant, particularly Puritan thought. The moral imperative for self-improvement as found in American popular religious culture itself grows out of the Puritan tradition, which promoted religious self-improvement. This has been described by Sacvan Berkovitch as a "war of the self on the self" or "automachia" (Bercovitch 1975).

This Puritan blending of religious asceticism and self-fashioning is the basis of what Max Weber called "inner-worldly asceticism," that is, not the separating from society by living in some isolated retreat or monastery, but rather a disciplined lifestyle leading to salvation while still actively involved in society and commerce. In his cultural history of psychotherapy, Philip Cushman describes the understanding of the self in the early modern era:

> Most importantly, it was a self that was instrumental; that is, it could manipulate the material world and transform it. Miraculously, the self could also manipulate and transform *itself*: it was pure, independent, instrumental consciousness... (Cushman 1995, p. 381).

Cushman goes on to discuss the historical trajectory of the religious values that were given form by the English Puritans, entered America in the colonial era, and have continued to mold American popular religious culture through the late twentieth century. Drawing on the characteristics of the trajectory that he describes, we can see the stages of social development in interaction with those values.

Colonial America Forward

The beginning of Puritan colonial America was communal, with an emphasis on piety, hard work, and compliance to God's will through a hierarchical social order (ibid., 36). The religious traditions of the time contain within themselves an ambivalence that continues to mark popular religious conceptions of the self. "Puritanism and to a lesser extent other Protestant religions at this time developed a notion of the self that stressed an increased self-consciousness and a deep self-suspicion....[T]he recognition of self-deception in Puritan ideology added to the increasingly problematic nature of the early modern self"

(ibid., pp. 381–382). Faith was understood to provide the sole access to salvation, and

> indeed, saving faith was the miracle of the incarnation reenacted in the individual soul. Hence the experience of saving faith was a special experience, a discreet experience, which had its own unique quality (Michaelsen 1976, p. 253).

Eliciting and discerning this transformative experience became a central concern in American popular religious culture in the face of the anxieties posed by eternal damnation.

In the first half of the eighteenth century, the First Great Awakening (ca. 1730 to ca. 1745) contributed to a societal shift from a hierarchical orientation in both society and religion to an egalitarian, democratic modality. In response to this shift some religious leaders attempted to encourage a return to compliance with social authority, but the new religiosity of the period also promoted a highly emotional expression, and a sense of motivating God by preparing oneself for sanctification (ibid., pp. 37–38). There was also created what Cushman calls a "dominationist sense," that is, not only of changing oneself but also changing others (ibid., p. 92).

Lastly, Cushman notes a liberationist mode in the twentieth century. This mode is evident in the mid-century development of humanistic psychology, which embraced without question "the post–World War II configuration of the self." This self was conceptualized as "subjective, often antitraditional, ahistorical, and preoccupied with individualistic concerns such as personal choice, self-realization, and the apolitical development of personal potential" (ibid., p. 243).

Despite these many vacillations in the dominant religiosity of the US as described by Cushman, and the transformation—and in some cases reversal—of values and beliefs, the enduring heritage from the Puritans includes the moral imperative of self-improvement. This is discussed by Cushman as the Puritan combination of salvation with work (ibid., pp. 382–383), or one might say, salvation through work, including in the culture of self-improvement, labor on the self. And since the self is always inherently inadequate, "depraved" in the terminology of Puritan theology, such labor is never-ending.

In American society, the value placed on self-improvement encountered capitalism in the era following the Civil War, and an initial commodification of self-improvement marked by Samuel Smiles's *Self-Help* (discussed more fully below) led to the culture of self-improvement as we know it today. The combination of religious values and economic themes is an enduring and pervasive part of this culture. "Protestantism strongly emphasized voluntary action, and in the American context self-help strategies for living have been notoriously combined with popular religions, making spirituality and self-help a central aspect of American culture" (Illouz 2008, p. 157). This is the framework within which Buddhism, understood as a form of self-improvement, and more recently secular mindfulness have become commodified. One characteristic of self-improvement Buddhism is its emphasis on individualism, a characteristic that expresses the typically American valuing of the autonomous individual—to the extent that individualism as a form of subjectivity is often conflated with democracy as a political philosophy. The emphasis on individualism also then folds neatly into the conservative themes of neoliberalism. This is evident for example in the disjunction between working conditions that produce and may even glorify stress, and some recent corporate efforts to reduce stress and improve productivity—not by making structural changes to the conditions of employment, but by placing responsibility for overcoming that stress on the individual and coopting Buddhism along the way (Purser 2015).

Individualism and economics are entangled at the very birth of self-improvement. In 1859, Samuel Smiles established self-improvement as a cultural concept when he published his treatise *Self-Help, With Illustrations of Character, Conduct, and Perseverance*. His intended audience was members of the *petit bourgeoisie*, known as the "uneasy class" of the Victorian era, precursors to what Jayne Raisborough, drawing on the work

of Anthony Giddens, calls the "jostled self" (2011, p. 28). Smiles's work sought to motivate people to self-discipline—particularly discipline of one's body and bodily desires—in the service of the "virtues of industry, frugality, temperance, and honesty" (Smiles 1859, Chap. X). Self-improvement through disciplining the self was promoted as the means of achieving economic stability and well-being in the face of the social and economic disruptions of the nineteenth century—industrialization, urbanization, mechanization, speedier transportation and communication, as well as modern warfare, and modern imperialism. The similarity with contemporary self-improvement discourse in the face of economic displacement and social instability is not accidental, but rather a repetition of the same dynamic—emphasizing that it is the individual who needs to learn to adapt to changing social conditions. As with other forms of neoliberal ideology, any notion of collective action or community formation is absent, or is denigrated either implicitly or explicitly as ineffectual. This kind of implicit denigration of social action is found in the trope often repeated in self-improvement culture that the only real change comes from within. This trope reinforces neoliberal individualism by effectively denying the role of institutionalized inequity in inhibiting social change or even the betterment of individuals.

Even when secularized asceticism is bound to what Charles Taylor calls "human flourishing" (2007) it retains its moral imperative. Whether wearing a hair shirt or running miles daily, there is still the drive imposed by the need to improve oneself. Now instead of asceticism with the goal of forcing the will into submission to God, it is for the sake of longevity or good health—but at the same time the secular asceticism of self-improvement transcends these pragmatic goals and carries the moral imperative that one *should* undertake some such regimen in order to improve oneself. The moral imperative toward self-improvement motivates our modern asceticism, and the therapeutic culture gives it shape and direction.

Buddhism as Self-improvement

From its very earliest introduction into Euro-American society, the Buddhist tradition has been integrated as a form of self-improvement. Although Schopenhauer's quasi-Stoic interpretation of Buddhism (App 2014) is no longer in fashion, it is perhaps the most widely formative introduction of Buddhism to Euro-American society. Although Schopenhauer and his followers interpreted the goal of Buddhism as the suppression or conquest of the will, rather than happiness or self-fulfillment, his view promoted the idea of self-control, that is self-improvement in the Stoic mold (App 2011). In this sense it is not the case that there ever was an originally pure form of Buddhism available to Euro-American audiences—teachers from Asia themselves actively molded the presentation of the teachings to accord with Western sensibilities (McMahan 2008). The reinterpretation of Buddhism as a form of self-improvement is not, therefore, a creation of late twentieth century culture, and the pervasion of the understanding of Buddhism as a kind of self-improvement from such an early period overdetermines this interpretation.

The culture of self-improvement has appropriated Buddhism as a part of a century and a half long fascination with the exotic in general, and the "Mystic East" and its "ancient wisdom" in particular. And, Euro-American proponents of Buddhism have themselves made use of the self-improvement culture as a ready-made vehicle for promoting Buddhism. Some might justify this by claiming that since the origins in the sangha that formed around Śākyamuni, Buddhism has been a self-improvement program, or perhaps even the original self-improvement program. This, however, obscures the fundamental ways in which self-improvement has its own ethos, that is, set of values, presumptions, beliefs, rationales, and technologies. By interpreting Buddhism as a program of self-improvement, the entire ethos of the self-improvement culture is imposed onto Buddhism. It is, therefore, important for us to turn our attention specifically to understanding that ethos.

The Ethos of Self-improvement

In American society today, self-improvement exists in the overlap between general self-help and popular religious culture. The ethos of the self-improvement culture provides the sense of moral imperative that motivates people to pursue the various regimens and strategies of self-improvement. One of the values of that ethos which we can utilize as an organizing principle for discussing the ethos is the emphasis on individualism, and as mentioned above this emphasis aligns self-improvement with neoliberal economics.

This alignment is demonstrated in a study by Lars Ahlin. His research examined over 800 articles appearing in the Swedish newspaper *Svenska Dagbladet*'s business and women's sections between 1974 and 1995. The two sections of the newspaper carried articles with themes emphasizing the "freedom of the individual" (Ahlin 2013, p. 188). Those in the business section were oriented toward neoliberal political and economic theories that emphasized the freedom of the individual from societal constraints, whether imposed by government or unions. The competitive world of business was presented as the realm in which the individual could fully manifest him/herself, but only if freed from the legal and moral limitations imposed by Swedish society.

The articles in the women's section, carrying the same theme of the freedom of the individual, took the tone of New Age spirituality. Here society was also presented as a force constraining the individual from full personal expression and fulfillment. Social institutions that failed to recognize, much less encourage, the "primacy of the individual" were characterized as oppressive. In this representation "all sorts of oppressive institutions and collectives" (ibid., p. 182) were characterized as needing to be overcome in order to create the New Age in which full individual freedom would be possible.

These two ideologies, neoliberal and New Age, shared a basic tenet—the primacy of the individual—and employed similar terminologies. Beyond these superficial similarities, however, the two systems of thought—which Ahlin referred to as "competitive individualism" and "growth individualism," respectively—actually diverged greatly from one another. In Ahlin's analysis, neoliberal agents, such as the Swedish Employer's Association which sought to turn public opinion against governmental and union controls of business—actively promoted the rhetoric of growth individualism in order to achieve their ends. Ahlin's concluding paragraph summarizes this dynamic and is worth quoting in full.

> Clearly New Age had something to offer the corporate world. Firstly as an ally, promoting a view of the individual and society converged with views essential for neoliberalism. Secondly, as something that would promote efficiency and growth. This was not, however, the spiritual world upon with New Age was fundamentally based. What the corporate world wanted were the techniques used in New Age to stimulate the individual's spiritual development. Specifically, these techniques could also serve as useful tools in the quest for greater efficiency, increased performance, and hence increased [corporate] growth in the highly competitive world of which the corporate world was and is a part (ibid., p. 188).

While Ahlin's study reveals a case of intentional manipulation of popular religious culture in promoting neoliberal ideology, there is an affinity even in the absence of such intent. Jayne Raisborough calls attention to the comments of "philosopher Axel Honneth [who] warns against seeing self-transformation as a tool strategically developed by a neoliberal progressive agenda. Rather than the result of a 'deliberate strategy,' self-transformation and self-realisation have a longer and quite diverse history which has been gradually appropriated or 'transmuted' to become an ideology of neoliberalism" (Raisborough 2011, p. 13).

Although individuality is often naturalized in the ethos of the self-improvement culture as the human norm that has been distorted or repressed by society, it is a value that is part of a system of values and preconceptions regarding the nature of human being in the world. While the value may be absolutized as either natural or God-given, the "ideal of a self-interested, rational, and calculating subject" (McGee 2005, p. 4) is in the ethos of self-improvement only to be actualized through hard work on oneself, or as Raisborough puts it, the self becoming a site of labor (2011). This individualized self is also theorized as autonomous, and supposedly contains within itself, hidden like a golden treasure, resources that will become increasingly effective in one's quest for self-improvement through practice.

This theme of "inner wisdom" is proximately sourced for American religious culture in the ideas of "mental science" in the late nineteenth century. An early formulation of this idea is found in the work of Phineas P. Quimby (1802–1866), who built on the ideas of a mental substance popularized by Mesmerism, developing on the theme of healing. Likened to soil,

> the spiritual substance of the mind was a place where beliefs germinated and subsequently affected both the conscious and unconscious aspects of the body. Hidden in the mind, this substance was accessible to those spiritual healers who could discern its contents, bring those contents to the attention of the patient, and make them manifest. This was possible because there was a portion of every soul, regardless of illness, that was not sick. It was this portion that the healer summoned to action (Haller 2012, p. 57).

The similarity between ideas of inner wisdom and the conception of the unconscious being popularized by Freudian psychologists is not incidental, as the latter was an important basis upon which New Thought and other related religious forms claimed a scientific basis. Sydney Ahlstrom has used the term "harmonial religions" to identify what he considers to be the most prominent unifying belief of these forms. Harmonial religions are described as encompassing:

> those forms of piety and belief in which spiritual composure, physical health, and even economic well-being are understood to flow from a person's rapport with the cosmos. Human beatitude and immortality are believed to depend to a great degree on one's being "in tune with the infinite" (Ahlstrom 1972, p. 1019).

The philosophical idealism of inner wisdom and harmony are known in the self-improvement culture of the present under a variety of names, one of the more commonly deployed being the "law of attraction." Jessica Lamb-Shapiro has pointed out that while there is certain common sense understanding that believing in one's own abilities does contribute to the likelihood of success, and conversely that believing that one is incapable fosters both beliefs and instances of failure. She indicates, however, that "people who believe in the law of attraction interpret the relationship between thoughts and action more literally. They believe one's internal reality impacts not just the way one subjectively experiences an external reality, but the objective external reality itself" (Lamb-Shapiro 2014, p. 122). Along with individualism, inner wisdom and harmony constitute important values and beliefs in the ethos of self-improvement.

Another element in the ethos of self-improvement is the ideal of wholeness, which is extended by some as an explanatory device for diagnosing the human condition. In many instances this is based on the Stoic or Cynic belief that society is responsible for fragmenting an originally whole person, leaving one with only partial aspects of oneself. As a consequence, one is never authentically, or fully oneself. The different roles one fulfills—employee or employer, spouse, child, parent, church member, and so on —impose a particular way of being in the world that does not reflect the authentic inward person who is a whole. The social imposition of role-specific behaviors not only leads one away from one's inner wholeness, but also the diverse and often diverging concerns of these different roles lead to being out of the present moment.

In addition to wholeness and authenticity, spontaneity is also valued. Preplanning, rational reflection, and calculation are seen as barriers to immediate and direct contact with one's true innermost self, since they require thinking about the past or future rather than being purely in the present moment. It is sometimes claimed that this is the goal of Buddhist practice, a misunderstanding perhaps perpetuated by the terminology of "no mind" (wu xin, 無心) used by some Zen teachers, perhaps most influentially, Suzuki (1969). The misunderstanding arises when *wu xin*

is treated as the goal of practice. In some usages, such as in the Chan tradition, it refers to the absence of defiled thought, while as a translation for the Yogācāra *acitta*, it refers to the five states of consciousness that a practitioner can enter, and which may be of benefit in the development of one's practice, but which are definitely not theorized as the goal of Buddhist practice (Digital Dictionary of Buddhism 2010, "wu xin (無xin)" and "wu wei wu xin (五位無心)").

Under a program of present-minded spontaneity, one is advised to remove layer by layer all pretensions, social conventions, false self-representations, linguistic formulations, and thus come finally and truly to know who one is, and be able to spontaneously and freely express that personal truth, the reality of that inner self. But a systematic, planned program of self-improvement is inherently, that is, logically, contradictory to these ideals of authenticity and spontaneity.

Premodern ascetics exercised the will to overcome the "temptations of the flesh" as a means to experiencing the transcendent, the divine. By objectifying one aspect of oneself as something that is to be improved by another aspect of oneself, such a medieval asceticism of self-control is inherently similar to a modern asceticism of self-fashioning. Thus, modern secular ascetics exercise the will to overcome the temptations of the flesh as a means of benefitting oneself—goals such as longevity, health, an end of suffering, and true happiness. In both cases, there is a division of the self from itself, a self-fragmenting. While such self-objectification may have benefits in some situations, it is an artifice, a mental construct of who we are. Without the moderating effects of self-compassion, self-fragmentation in pursuit of self-improvement is at odds with the development of an integrated sense of self as an ongoing, changing process—the wholeness otherwise held up as a value in much of self-improvement rhetoric. While it may seem contradictory to say so, that ongoing, changing process that is the self is also self-directing—to use an image of non-duality found in many Buddhist descriptions of the mind, it is like the river that carves out its own channel by flowing.

Anxieties of the Malleable Self

Self-improvement culture integrates two strains of thought from two distinct sources, both of which are sources of anxiety for the malleable self. On one hand the moral imperative to self-improvement derives from Protestant, particularly Puritan sources. On the other, the goal of personal perfection derives from the strains of harmonial religion, or what J. Stillson Judah has called "metaphysical religion" (1967). Judah's work makes clear that it is the syncretic character of the metaphysical movements that is responsible for the integration of Eastern religious traditions into American popular religious culture. Because this was the vehicle through which both Hindu and Buddhist religious praxis was introduced, the syncretic form in which only select elements of these traditions were reconfigured and reinterpreted as resources for the metaphysical movements has always been and remains the only form widely familiar in the West. In contrast immigrant Buddhist religious traditions have largely been marginalized in American religious culture (Payne 2005).

The Ever-Receding Horizon of Perfectionism

Perhaps fundamental to self-reflective conscious awareness is the awareness of being able to have some control over oneself. Intuitively, this makes sense as it would be from the realization of self-control of one's actions that one learns that one is in control of oneself, and that reflexive arc is what instigates self-reflective conscious awareness. At some point, the ability to control one's actions takes a further self-reflexive turn and becomes the ability to control one's thoughts and feelings, that is, to control one's self. However,

> The self is never really finished improving…. Self-help directs us to specific goals, and we may reach those goals; but everything we achieve is temporary. A constant sense of renewal is what keeps the self-help industry alive, year after year. No matter how satisfied you are with your life, the specter of failure always looms ahead, and there will always be a book to help you through it (Lamb-Shapiro 2014, p. 163).

The very ethos of self-help not only builds upon a sense of lack, inadequacy, incompleteness, but instills and reinforces those feelings as the self is placed against an ever-receding horizon of perfection. In this way the self-improvement model creates unending consumption. The process necessitates an oppositional relation that binds the lack, which is diagnosed by a self-improvement system, to the treatment, which that system then offers as fulfilling the lack it has defined. In her study of what she calls "the makeover culture," Jayne Raisborough comments on the "'after' stage of the journey of self-transformation" portrayed in self-improvement media, noting that

> The "after" appears as a solid and unproblematic moment of success where dreams come true, where the inner self beats the external self into compliance, bodies are sculpted, esteem is supercharged and people get the look/home/confidence —the self—they always wanted. Yet, this success is only short-lived because the "after" is temporally fragile; the ceaseless momentum of the makeover culture rolls these moments into new beginnings and new projects of the self (Raisborough 2011, p. 142).

Meredith Jones summarizes this aspect of the ethos of self-improvement saying that "in the makeover culture the process of *becoming something better* is more important than achieving a static point of completion" (as cited in Raisborough 2011, p. 142). From this Buddhist's perspective the very idea of a static point of completion is an illusion.

Perfectionism Versus Human Depravity

Several authors have noted the open-ended character of self-improvement, particularly under the burden of perfectionism. However, one of the ironies of the self-help culture is that its religious background generally denies the ideal of perfectionism per se. According to Luther, for example, salvation is not the consequence of our own actions. Rather, "God 'imputes' the merits of the crucified and risen Christ through grace to a fallen human being, who remains without inherent merit and who, without this 'imputation', would remain unrighteous" (MacCulloch 2003, p. 119). Similarly, Calvin held that "None of our talents or capacities can lift us from this abyss in our fallen state, only an act of free grace from God" (ibid., p. 195).

Though this message may have created a sense of hopelessness, futility, or apathy among some believers, the general societal effect was rather the opposite. American religious culture of the nineteenth century:

> was formed by Puritan dissidents whose depictions of God and humanity laid the groundwork for the nation's rigorous morality. Central to their beliefs was the conviction that humans were lost in sin, saved only by the grace of God and not by any merit or effort on their own part. Nevertheless, the elect and damned were each accountable for their condition.... Even though the doctrine of salvation decreed that God's grace rather than human efforts opened the gates of heaven, and that nothing an individual did could in any way "earn" that access, the doctrine had the effect of unleashing an intense activism among believers who concluded that grace, once bestowed, must result in a thoroughgoing change—both internal and external—in the sinner's now-sanctioned life (Haller 2012, p. 214).

What we have referred to here as the moral imperative to self-improvement is therefore based on the ambivalence between an inherent inadequacy and the need to demonstrate a state of grace—or in contemporary terms, exhibiting oneself as happy, successful, popular, and self-confident.

The Rhetorically Bifurcated Self

Self-improvement structures subjectivity into two poles—the self as agent and the self as object. Despite the grounding of mindfulness in Buddhist teachings regarding the absence of a permanent self, the ethos of self-improvement imposes its own conception of the nature of the self (Rakow 2013). This bifurcation of the subject has been described by Eva Illouz in summarizing the effect of psychotherapeutic theory on management, which brought about a management culture based on ideas of communication and cooperation. She says:

> the precondition for "communication" or "cooperation" is, paradoxically, the *suspension of one's emotional entanglements in a social relationship*. To the extent that emotions point to the entanglement of the self in a social relation, they also point

to one's dependence on others. Emotional [self-] control thus points to a model of sociability in which one must display the ability to remove oneself from the reach of others in order to better cooperate with them. The emotional control of the type propounded by the therapeutic persuasion is at once the mark of a *disengaged self* (busy with self-mastery and control) and of a *sociable self*—bracketing emotions for the sake of entering into relation with others (2008, p. 104).

The two terms that Illouz uses to describe the self-contradictory bifurcation she identifies, "disengaged self" and "sociable self," constitute the same oppositional relation as that between the self who improves (agent self) and the self who is improved (object self)—the structure of subjectivity essential to the ethos of self-improvement.

Beyond Stylistics: From Rhetorical Bifurcation to Crypto-Ātman

Today the culture of self-improvement interfaces with both psychotherapeutics and religion, and along with those float on a broad ocean of Christian and Cartesian presumptions basic to popular religious culture. The attribution of agency and autonomy to the self who improves (agent self) means that it is easily misunderstood as a permanent self—an *ātman* that is eternal and unchanging, the core of consciousness and the will. The ease with which this mistaken conception arises and has come to permeate popular religious understandings of Buddhist teachings, including those in the self-improvement culture, follows from those Christian and Cartesian presumptions basic to Western popular religious culture. That this conception of the agent self as transcendent to the fallibilities of social, embodied, emotional, sexual, and cultural existence means that the common rhetorics depending on such a conception are at variance with the Buddhist teaching of *anātman/anatta*—the utter absence of any permanent, eternal, absolute, unchanging essence, self, or soul.

Although interacting with both psychotherapy and religion, self-improvement operates more explicitly in the realm of commodity capitalism than do either psychotherapy or religion, both of which tend more toward a service industry model, or at least embrace that model even if struggling with the market realities of consumer capitalism. The effects of this commodification can be seen in certain stylistics of marketing and commodification. The by now familiar weekend workshops or handbooks for changing your life in eight weeks or training programs offering certification are evident ways in which the culture of self-improvement has influenced the presentation of Buddhism and mindfulness practice in popular culture.

There is, however, a more foundational—and indeed more problematic—way in which it has influenced Buddhist teachings. There is a fundamental conflict between understanding of the self presumed by the ethos of self-improvement, particularly its psycho-technology, and an understanding of the self that is more explicitly grounded in Buddhist thought. In brief, the self of self-improvement is fragmented so one part can control the others, while Buddhist understandings more generally draw on an understanding of the self as a process involving a non-dual integrity between consciousness (Skt. *citta*) and its objects (Skt. *caitta*). Buddhist teachings regarding the nature of the self—a non-dual integrity of consciousness and its objects that in no way constitutes a permanent, eternal, and absolute or unchanging self (*ātman*) —diverge so greatly from Western cultural presumptions about the self as an autonomous agent, that the latter exert a consistent "gravitational pull" on the interpretation of Buddhist teachings in the context of popular religious culture, such as mindfulness. Importantly, where mindfulness is presented as a "secular" teaching, a psycho-technology informed only by "scientific" understandings of the self, this gravitational pull of interpretation is simply made invisible, and the idea of an autonomous agent self is naturalized, considered the norm and original condition of the self.

The culture of self-improvement is located within the larger, and largely invisible, therapeutic culture. While the therapeutic culture includes psychotherapy, it refers more broadly to a pattern of diagnosis and prescription. In its self-improvement manifestation, however, the diagnostic process defines the self as needing exactly the kind of cure on sale. In other words it is the cure that defines the diagnosis. And, the cure on offer by self-improvement requires a person to fragment themselves into the agent–self who controls and directs the process of self-improvement, and the object–self that needs to be improved.

We might then call this the dualistic ontology of self-improvement, it is the self controlling the self, that is, the exercise of one's own "will-power" to change oneself. The idea that the will has power is based on nineteenth century ideas of energy, including psychic energy. This is another model of the self or the mind that reifies what should more accurately be treated as processes.

The dual nature of the self as both agent and object is more than simply a grammatical matter of self-reference, a harmless sort of objectification of oneself as a grammatical object, but one carrying no practical consequence. Language matters, and the language of self-improvement programmatically divides the self against itself. This then not only creates the rhetorically bifurcated self, but creates the conditions by which notions of the self as autonomous agent existing independently of its objects are integrated as part of the teaching of mindfulness practice.

Developmentally, this ability to look at oneself objectively, that is, to treat oneself as an object, can be valuable. In decision-making, there are often different values at play: Should I do this? or should I do that? Thinking through the possible consequences of one's actions or imagining different possible outcomes requires just such a division and such an act of self-objectification. Indeed, the person who has no doubts because they have not developed the ability to think about themselves as an object in this way may be dangerous in their relations with others. Yet, as useful as such an ability is, its value is heuristic—it is valuable for certain projects. As the metaphor familiar from a variety of Buddhist teachings would have it, this is a kind of medicine that is taken to cure a specific disease. Once the disease is cured, the medicine is no longer needed. Self-fragmentation should neither be established as a permanent condition, nor does its heuristic value justify its reification as a metaphysical essence, an *ātman*.

Programs of self-improvement are, however, based on just this idea that one can enforce a regimen of self-improvement on oneself. I am not questioning the benefits of programs of exercise or weight loss or regular meditation. However, given the ideas inherent in American popular religious culture, the programmatic self-fragmentation of self-improvement as such can reinforce the image of a self separate and independent from all other things—an understanding of oneself that most Buddhist traditions consider to be the key to creating suffering for oneself and for others. The self that is to be changed is not the only self that is objectified, but so also is the self that is to do the changing. This fragmentation of the self supports thinking of one fragment—the one to which agency is attributed —as a permanent, eternal, absolute, and unchanging self, that is, as an *ātman*. As noted above, self-improvement authors have long attempted to justify their claims as scientific. The late twentieth century focus on brain sciences led to a new understanding of how the brain operates, one that has become something of a standby in contemporary explanations of the efficacy of self-improvement technologies being offered. The way in which this has been integrated into self-improvement demonstrates the strength of the "gravitational pull" of the common cultural presumptions regarding the self prevalent in the self-improvement ethos. Both the bifurcation of the self and the identification of one part as an autonomous agent (crypto-*ātman*) are central to the theory of the self-requisite to the supposed efficacy of self-improvement.

Misreading Neural Plasticity

Contrary to long accepted understandings of the brain as static, it is now understood that neural connections do change throughout life, a phenomenon known as neural plasticity.

A manifestation of reifying one fragment of the self as an autonomous agent is the way that neural plasticity is frequently interpreted as scientific proof of the efficacy of mindfulness and other types of meditation. In this interpretation, by concentrating my mind on something like my breath or compassion I cause my neural connections to change—the reorganization of neural connections is interpreted as the physical effect of a mental cause. Other versions employ will as a third term, attributing autonomous efficacy to the will as the agency by which neural connections are realigned in new ways. The problematics of this attribution are evident when one considers that the will of the theory of mind employed by self-improvement programs is what Schopenauer calls the "empirical will" (App 2014, 78), which as a part of a person does not have the ability to leverage or bootstrap change in the entirety of the person.

In a delightfully naïve instance of circular reasoning, neural plasticity is then taken as evidence of the real existence of will as a spiritual or mental, but nonphysical, power able to effect physical change (for example, Schwartz and Begley 2002; Begley 2007; and Hanson and Mendius 2009). This idea of will as a mental force capable of effecting physical change to neural connections seems as if it is motivated by the desire to establish the validity of belief in a soul—a nonphysical source of agency. In Buddhist jargon, we can understand such interpretations as introducing a crypto-*ātman*—an essence of some kind that exists autonomously and while it can effect the material world, is not effected by it.

The bifurcation and control of the object self by the agent self actually stands in contradiction to the values of authenticity and spontaneity that are a common element in the rhetoric of self-improvement. Authenticity, the expression of who one truly is in their innermost essence, is held as a moral value, and often presented as the source of creativity. Yet at the same time, the ethos of self-improvement is that one's innermost essence is malleable, it can be formed and re-formed—refashioned as an effort of will and intent. Authenticity presumes some true innermost self, while self-improvement presumes that the innermost self is malleable. This is the contradiction identified by Illouz discussed above—one must learn to control oneself in order to act spontaneously.

Conclusion: Marketing the Unattainable

Over the course of the twentieth century, religion has become increasingly optional, a matter of individual choice rather than societal mandate. As such it is a leisure time activity, and therefore is marketed as a leisure product. During this same time period, the rise of the "therapeutic culture" created a focus on the diagnosis and cure of individual problems. Self-improvement partakes of the commodification and marketing of both religious and therapeutic culture. Having been integrated into the culture of self-improvement, mindfulness competes with

similar products on the market that combine self-help, positive thinking, and the pursuit of happiness and self-realization with spirituality. They all place happiness and contentment solely within the agency of an individual and thereby dovetail a neoliberal discourse that naturalizes the idea of individual autonomy and simultaneously conceals the supra-individual forces of the social and material world (Rakow 2013, p. 486).

Contrary to the analysis that traditional Protestantism has been largely displaced, what we find is that a core value of the Puritan roots of American religious culture, the moral imperative to self-improvement, remains central to the culture of self-improvement. The enduring sense of inadequacy deriving from seventeenth century Puritanism creates open-ended consumption of self-improvement products by twenty-first century Americans. Part of the marketing is the appeal of individualism, that while so much else may be outside your control, you have the power to at least take charge and mold oneself into the successful and happy person you know you want to be, you know you can be.

References

Ahlin, L. (2013). Mutual interests? Neoliberalism and new age during the 1980s. In F. Gauthier & T. Martkainen (Eds.), *Religion in consumer society: Brands, consumers and markets.* Farnham, England and Burlington, Vermont: Ashgate.

Ahlstrom, S. E. (1972). *A religious history of the American people.* New Haven: Yale University Press.

App, U. (2011). *Richard Wagner and Buddhism.* Rorschach and Kyoto: UniversityMedia.

App, U. (2014). *Schopenhauer's compass: An introduction to Schopenhauer's philosophy and its origins.* Wil, Switzerland: UniversityMedia.

Begley, S. (2007). *Train your mind, change your brain: How a new science reveals our extraordinary potential to transform ourselves.* NY: Random House, Inc.

Bercovitch, S. (1975). *The Puritan origins of the American self.* New Haven: Yale University Press.

Cary, P. (2000). *Augustine's invention of the inner self: The legacy of a Christian Platonist.* Oxford: Oxford University Press.

Cushman, P. (1995). *Constructing the self, constructing America: A cultural history of psychotherapy.* Boston: Addison Wesley Publishing.

Digital Dictionary of Buddhism. (2010). wuwei, wuxin (五位無心). Retrieved from http://www.buddhism-dict.net/cgi-bin/xpr-ddb.pl?4e.xml+id%28%27b4e94-4f4d-7121-5fc3%27%29

Greenblatt, S. (1980). *Renaissance self-fashioning: From more to Shakespeare.* Chicago: University of Chicago Press.

Flood, G. (2004). *The ascetic self: Subjectivity, memory and tradition.* Cambridge: Cambridge University Press.

Haller, J. S, Jr. (2012). *The history of new thought: From mind cure to positive thinking and the prosperity gospel.* West Chester, Pennsylvania: Swedenborg Foundation Press.

Hanson, R., & Mendius, R. (2009). *Buddha's brain: The practical neuroscience of happiness, love and wisdom.* Oakland: New Harbinger.

Illouz, E. (2008). *Saving the modern soul: Therapy, emotions and the culture of self-help.* Berkeley: University of California Press.

Judah, J. S. (1967). *The history and philosophy of the metaphysical movements in America.* Philadelphia: The Westminster Press.

Kruse, K. M. (2015). *One nation under God: How corporate America invented Christian America.* New York: Basic Books.

Lamb-Shapiro, J. (2014). *Promise land: My journey through America's self-help culture.* New York: Simon and Schuster.

Luhman, T. M. (2012). *When God talks back: Understanding the American evangelical relationship with God.* New York: Vintage Books.

MacCulloch, D. (2003). *The reformation: A history*. New York: Penguin Books.

McAdams, D. P. (2006). *The redemptive self: Stories Americans live by*. Oxford: Oxford University Press.

McGee, M. (2005). *Self-help Inc.: Makeover culture in American life*. Oxford: Oxford University Press.

McMahan, D. (2008). *The making of Buddhist modernism*. Oxford: Oxford University Press.

Michaelsen, R. (1976). The Beecher family: Microcosm of a chapter in the evolution of religious sensibility in America. In F. E. Reynolds & D. Capps (Eds.), *The biographical process: Studies in the history and psychology of religion*. The Hague: Mouton.

Payne, R. (2005). Hiding in plain sight: The invisibility of the Shingon mission to the United States. In L. Learman (Ed.), *Buddhist missionaries in the era of globalization*. Honolulu: University of Hawai'i Press.

Purser, R. E. (2015) Trickle down mindfulness: Examining and questioning core assumptions in the corporate mindfulness movement. Paper presented at the American Academy of Religion conference, Atlanta, Georgia, in the Economics and Capitalism in the Study of Religion seminar.

Raisborough, J. (2011). *Lifestyle media and the formation of the self*. Houndmills, Basingstoke, Hampshire: Palgrave Macmillan.

Rakow, K. (2013). Therapeutic culture and religion in America. *Religion Compass, 7*(11), 485–497.

Schmidt, E. L. (2000). *Hearing things: Religion, illusion, and the American enlightenment*. Cambridge: Harvard University Press.

Schwartz, J. M., & Begley, S. (2002). *The mind and the brain: Neuroplasticity and the power of mental force*. New York: HarperCollins.

Smiles, S. (1859). Self-help, with illustrations of character, conduct, and perseverance. Reprint, 2002. In Peter W. Sinnema (Ed.) *Self-help*. New York: Oxford University Press.

Storhoff, G., & Whalen-Bridge, J. (2010). *American Buddhism as a way of life*. Albany: State University of New York University Press.

Suzuki, D. T. (1969). *The Zen doctrine of no-mind*. London: Rider.

Taylor, C. (2007). *A Secular Age*. Cambridge: Harvard University Press.

The Critique of Mindfulness and the Mindfulness of Critique: Paying Attention to the Politics of Our Selves with Foucault's Analytic of Governmentality

Edwin Ng

Introduction

This chapter offers an account of a style of thought accompanying a mode of critical practice. This style of thought has enabled for me insight into how the adaptation of mindfulness across multiple domains has to negotiate the dominant logics of the present neoliberal capitalist order of things. It has also helped me to explore how mindfulness might function as a disruptive technology of the self within and against these dominant logics. I have found in the work of Michel Foucault, an incisive set of analytical tools and conceptual schemas with which to investigate the microphysics of power channeling through the contemporary mindfulness trend. I will demonstrate how Foucault's analytic of governmentality has helped me to work through the ethico-political stakes in my coterminous personal and professional practice of mindfulness. I consider myself an Engaged Buddhist and I am developing a livelihood within the institutional space of the university, which is a key site for the production of knowledge on contemporary mindfulness. But because the analytic of governmentality is not strictly speaking a theoretical program but a style

of thought, the critical purchase of this account of Foucault's work is not restricted to Engaged Buddhist concerns. I hope this chapter will arouse curiosity among others who are exploring mindfulness in other areas of personal, professional, private, or public concerns.

The chapter first unpacks Foucault's genealogical analysis of neoliberalism as a historically contingent regime of truth about human nature and reality 'as it is.' According to this regime of truth, an array of knowledges, procedures, techniques, and expertise play out as social or institutional mechanisms and as individualizing practices. These mechanisms and practices facilitate the experience of 'free choice' in the production of a subject of interest: the neoliberal subjectivity of *homo economicus*. This will shed light on why mindfulness is so malleable and adaptable across diverse settings. The chapter then connects Foucault's account of neoliberalism with his reevaluation of the ethical practices of spiritual self-constitution in antiquity. This will clarify how an analytic of governmentality hinges on a mindful interrogation of the potential consonance or dissonance that may be generated between, on the one hand, being 'subject to someone else by control and dependence,' and on the other, the cultivation of 'identity by a conscience and self-knowledge' (Foucault 1982: 781). With this dual understanding of subjectivity, we have to investigate the meanings and uses of mindfulness as the

E. Ng (✉)
Faculty of Arts and Education, Deakin University, Melbourne, VIC, Australia
e-mail: Edwin.a.ng@gmail.com

© Springer International Publishing Switzerland 2016
R.E. Purser et al. (eds.), *Handbook of Mindfulness*,
Mindfulness in Behavioral Health, DOI 10.1007/978-3-319-44019-4_10

emergent and contested outcome of techniques of domination and techniques of the self.

A Foucauldian understanding of the subject as constituted by historically contingent practices is accompanied by an understanding of ethics as critical practice, where the task of critique is performed with an ethos or attentive 'limit attitude' toward the dominant logics of the present juncture. By connecting Foucault's curiosity about neoliberalism with his curiosity about ethical self-cultivation, the chapter will propose some ways by which the practices of critique and mindfulness might reciprocally nourish one another as the critique of mindfulness and the mindfulness of critique. The chapter concludes with some reflections on how this ethos of critical mindfulness might relate to the challenges facing scholars and researchers of mindfulness working within the increasingly corporatized institution of the university, which is a key site and relay point for the production of knowledge on mindfulness in the contemporary world.

Governmentality, Neoliberalism, and the Production of Subjectivity

One way to begin to unpack the analytic of governmentality is to read it as a portmanteau concept, governmentality. Governmentality does not simply refer to the processes of the state. Rather, Foucault evokes the way 'government' was used in sixteenth century Europe, where it referred more generally to the conduct of persons and to the types of knowledge and practices by which individuals may conduct themselves responsibly as moral subjects. This broader understanding of government persisted into the eighteenth century, where the term appeared in political tracts and also in philosophical, religious, medical, and pedagogic works that addressed matters like self-control, guidance for family, advice for the soul, and so forth (Foucault 1982: 790). Describing it variously as the government of self and others or the art of government, the object and objective of government are not only 'the legitimately constituted forms of political or economic subjection but also *modes of action*, which [are]

destined to act upon the possibilities of action of other people. To govern, in this sense, is to structure the *possible field of action* of others' (Foucault 1982: 790; emphasis added). The analytic of governmentality is thus concerned with the multifaceted processes—the interplay between fields of knowledge, modes of practices, types of expertise, systems of norms and values, sites of private and public activities—which steer 'the conduct of conduct.'

For Foucault, the Christian pastorate was a prelude to governmentality because it provided the individualizing logics for the constitution of the subject (Foucault 2007a, b:147–148). According to Foucault's coterminous account of the 'genealogy of the modern state' and a 'history of the subject,' the modern (Western) state emerged out of the complex interplay between 'political' and 'pastoral' power. The notion of political power is traced to the Greek *polis* and it is relayed through procedures concerning rights, universality, public space, and so forth. Pastoral power, on the other hand, derives from Christian understandings about the relationship between the shepherd and his flock, and it is relayed through procedures concerning the comprehensive guidance of individuals. Pastoral power is tied with the production of truth, as exemplified by the methods of analysis and techniques of reflection of the Christian confessional apparatus, which was designed to secure knowledge of the 'inner truth' of the individual—in confessing a truth about herself or himself, the individual is constituted by an objectifying act of knowing as a subject of that truth, who may be guided on a path to salvation. On Foucault's account, the governmental logics of pastoralism overflowed their ecclesiastical confines and began to take shape in secularized forms with the formation of the modern state, which saw the proliferation of rational knowledge and expertise across the natural and human sciences about the individual and the population as a whole (Foucault 2007a, b: 236–237). With the rise of liberalism and the attendant shift in understanding the economy as a conceptually and practically distinguished space with its own intrinsic laws, a specific art of government emerged which aimed neither at salvation in an afterworld nor at increasing the welfare of the sovereign state. Rather, liberal governmentality took civil society as the starting point and the freedom of the individual as the critical yardstick for governmental action.

Neoliberalism as Political Ontology

This brief contextual overview of governmentality provides a backdrop for us to broach the question of neoliberalism, and more specifically, to understand how neoliberal governmentality turns on the production of a particular subjectivity. In critical discourses about neoliberalism, there is general consensus that neoliberalism is not simply a shift in ideology but a transformation of how ideology functions (and as we shall see, it becomes questionable if the classical understanding of ideology is still adequate). Under neoliberalism, it is not the state or a dominant class that provides the generative conditions of ideology, but the everyday experience of market logics which are taken as a general matrix of society. Neoliberalism in this sense refers not only to the programs of the political or economic realm or an ideal of the state but to human existence as a whole. Fredric Jameson has captured neoliberal conditionings with the pithy statement: 'The market is in human nature' (1991: 263). And we find in the opening pages of David Harvey's A Brief History of Neoliberalism an observation about its everydayness: 'Neoliberalism… has pervasive effects on ways of thought to the point where it has become incorporated into the common-sense way many of us interpret, live in, and understand the world' (2007: 3). In Harvey's influential analysis, neoliberalism is regarded as a deliberate and effective attempt to restore power and wealth to the dominant class by using received ideals like liberty, choice, and rights to deflect attention away from the grim realities of 'free market fundamentalism' at local and transnational levels (Harvey 2007: 7). Resistance against neoliberalism must therefore expose the workings of this move to restore class privilege (Harvey 2007: 202–203).

Without going so far as to claim that Harvey's Marxist-inflected assessment of the 'false consciousness' of neoliberalism is wrong, I want to draw attention instead to how a Foucauldian approach offers an alternative perspective for understanding how neoliberalism shapes the ground of social reality today. With the analytic of governmentality, neoliberalism is not treated as an ideological lie that hides or distorts truth, but as a regime of truth that provides the conditions for making reasonable judgments with regard to social and political actions and relations. Neoliberalism, in other words, is treated as a political ontology (Oksala 2013). Neoliberal governmentality draws on and modifies liberal governmentality, where economic rationality centered on what Adam Smith had described as humankind's tendency to barter, truck, and exchange. Examining the work of the proponents of the Freiburg School of economics like Walter Eucken and Wilhelm Röpke (also known as 'Ordoliberals) as well as the Chicago School of economics (focusing in particular on Gary Becker), Foucault shows that neoliberalism extends on classical liberalism's process of making economic rationality a general matrix of society by *shifting the focus from exchange to competition* (Foucault 2008: 12). The transformation between 'classical' and 'neo' forms of liberalism unfolds through a shared 'anthropology' of the figure of *homo economicus*.

Homo Economicus and Human Capital

The transformation in *homo economicus* from a creature of exchange to a competitive creature entailed a redefinition of the 'worker' and 'labor.' For the neoliberals, classical economics has only conceived of labor as something purchased on a market or as something tied to the production of a commodity, but not from the perspective of the worker as a *subjective choice*. As Foucault observes, the neoliberals wanted to 'ensure that the worker is not present in the economic analysis as an object… but as an active economic subject.' And to do this, 'we must put ourselves in the position of the person who works; we will have to study work as economic conduct practiced, implemented, rationalized, and calculated by the person who works.' (2008: 223). Labor, according to this logic, is regarded as one among other 'substitutable choices' (2008: 222). The rationale for engaging in labor would be to ostensibly secure a wage. But as Dilts (2011: 136) has noted, 'wage' is the market term referring to the price paid for a unit of labor power from the point of view of *exchange*. If the grounding point of view is to be anchored on the worker's subjective choice, the wage should not be understood as the price paid in the market but as an *income stream*, which is a return on an investment, or more precisely, a return on 'human capital' (Schultz 1972).

What distinguishes human capital from other forms of capital is that it must necessarily be tied to *a body with certain abilities, attributes, and qualities*—and this is pivotal for understanding neoliberalism as the production of a particular mode of subjectivity, and why mindfulness is so malleable and adaptable under neoliberal conditions (Dilts 2011: 136). The abilities, attributes, and qualities of any given body are necessarily constrained to some degree by one's biological makeup and the social circumstances that one is born into (e.g., ethnicity and class). But with the two analytical shifts mentioned above—that is, the reconceptualization of labor as one subjective choice among substitutes, and the redefinition of wages as an income stream—these constraints are no longer simply impediments. Rather, they become opportunities for the individual to work on and transform their initial investment of human capital with different *technologies of the self*. Foucault defines them as techniques 'which permit individuals to effect by their own means or with the help of others a certain number of operations on their own bodies and souls, thoughts, conduct and way of being, so as to transform themselves in order to attain a certain state of happiness, purity, wisdom, perfection, or immortality' (Foucault 1993: 203). Contemporary technologies of the self include cosmetic surgery, therapy, life-coaching, exercise, and of course, mindfulness training. Labor conceptualized in terms of human capital is thus not confined to paid work, but any activity

that maximizes an individual's potential to secure any form of material or immaterial future return: 'Any activity that increases the capacity to earn income, to achieve satisfaction, even migration, the crossing of borders from one country to another, is an investment in human capital' (Read 2009: 28). In this manner, homo economicus shifts radically from being a 'partner of exchange' to an 'entrepreneur of himself [sic]' (Foucault 2008: 226). Dilts paints an arresting portrait of the neoliberal view of homo economicus.

> The neo-liberal analysts look out at the world and do not see discrete and identifiable firms, producers, households, consumers, fathers, mothers, criminals, immigrants, natives, adults, children, or any other 'fixed' category of human subjectivity. They see heterogeneous human capital, distinct in their specific attributes, abilities, natural endowments, skills. They see entrepreneurs of the self. They see homini œconomici, responsive agents to the reality of costs and benefits attached to activities, each of which are productive of satisfaction (Dilts 2011: 138).

The figure of homo economicus is not an anthropological self, but a minimal and theoretically 'empty' subject that is simply an array of activities which can be objectively known and managed. If there are no discrete entities but only entrepreneurial subjects engaged in self-optimizing conduct, then the object and objective of governmentality become principally about dealing with responsive subjects of a reality that already exists: 'Now, all that matters for questions of who one is, for the 'truth' of a subject, are the activities of that subject, the behaviors, the conducts, and the accumulation of skills and qualities that allow for the self to arrive at a self-understanding of those activities as producing some benefit. All that matters, in the end, is identifying the truth of this reality' (Dilts 2011: 139). This is the dominant logic driving the mindfulness trend.

The Governmentality of the Mindfulness Trend

There have been some recurring points of contention regarding the widespread uptake of an individualistic and therapeutic approach to mindfulness popularized by the Mindfulness-Based Stress Reduction (MBSR) model. In the opinion piece for Salon.com I co-authored with Purser and Ng (2015), we tried to take stock of the debates that have ensued since the publication of Purser and Loy's now viral article 'Beyond McMindfulness' (2013). We reiterated that the dominance and privileging of an individualistic, therapeutic mindfulness needs to be interrogated, especially in institutional or corporate contexts where a narrow focus on helping employees manage stress levels may deflect attention away from the systemic or structural conditions that are inducing stress in the first place. Without collective critical attention to these conditions, an individualistic and therapeutically oriented mindfulness risks becoming an alibi for what critical psychologist Smail (2005) has described as 'magical volunteerism,' a form of victim blaming that privatizes stress by placing the responsibility squarely on the individual. We highlighted David Gelles advice in *Mindful Work* as an example. Gelles asserts that stress 'isn't something imposed on us' but is something 'we imposed on ourselves.' He also claims that '[w]e live in a capitalist economy and mindfulness can't change that,' thus implying that the best course of action is to allow individuals the free choice to use mindfulness to foster greater well-being for themselves amidst the vicissitudes of this social reality (Gelles quoted in Pinsker 2015).

On such accounts, afflictive experiences of stress are not regarded as inevitable, but curiously the capitalist economy and its excesses are. It is as if the market logics of competitive self-interest simply reflect the ontological truth of social and political relations 'as it is,' and thus the responsible thing to do is to conduct ourselves (with the aid of mindfulness or other means) as responsive subjects of this reality. In our commentary, we observed that advocates of corporate mindfulness tend to evoke a 'Trojan horse' hypothesis to argue that an individualistic, therapeutic mindfulness can act as a disruptive technology that would in time generate change from within even the most dysfunctional systems. Against such a vision for corporate mindfulness, the critique of McMindfulness is dismissed by Gelles as a 'seductive nefarious

vision' that paints a false picture of covert 'brainwashing' which denies the therapeutic benefits of mindfulness to individuals and potential organizational change. In an interview, Kabat-Zinn (2015a) has even dismissed the critique of McMindfulness as merely a imaginary vision conjured up by a sole individual which misrepresents the reality on the ground. Though curiously, in his commentary in *The Guardian* on the policy paper released in the UK, *Mindful Nation*, Kabat-Zinn (2015b) acknowledges the challenges of McMindfulness. However, he reduces these challenges to the proper accreditation of mindfulness teachers, and the further development of evidence-based research to keep up with market interest in mindfulness. The dominant logics of neoliberalism governing existing programs and research remain unquestioned.

An analytic of governmentality can help to clarify these points of contention and overcome the critical impasses that have stalled around them. To critique contemporary mindfulness with an analytic of governmentality is not to bemoan the victory of capitalist ideology, whereby the 'ruling ideas' of the dominant class have been accepted by all. It is not about revealing the ideological lie of a political or economic program which has hidden the truth, nor is it about exposing some conspiracy. As Hamann writes, 'Governmentality is not a matter of a dominant force having direct control over the conduct of individuals; rather, it is a matter of trying to determine the conditions within or out of which individuals are able to freely conduct themselves' (2009: 55). The task is to investigate the *ontological conditions and effects* of the varied ways by which a historically contingent arrangement of present reality is subjectivized through the array of knowledges, procedures, techniques, and expertise that are diffused across the multiple domains of private and public lives.

Neoliberal governmentality turns on the self-constitution of 'responsibilized' individuals who exercise free choice to work on specific dimensions of their conduct—whether it be the maintenance of a fitness regimen with the help of a personal trainer or a reality TV program like *The Biggest Loser*, or the enhancement of creativity with the guidance of a professional consultant or brain training videogame, or the cultivation of focus, concentration, and composure with a daily mindfulness or yoga routine at home or in schools or workplaces. Under neoliberal governmentality, these practices become equivalent practices of the self within a flattened rationality of investment. According to this economic rationality, freedom is experienced as the subjective choice individuals make to tweak those specific attributes, abilities, and qualities of the body that make them viable as self-optimizing subjects.

Neoliberal Subjectivity

We catch a glimpse of the everydayness of neoliberal governmentality in the story behind MNDFL, the self-styled 'premier drop-in meditation studio' in New York City.[1] MNDFL was conceived during a conversation between its founders Ellie Burrows and Lodro Rinzler on why there was not a drop-in studio for meditation just like there are drop-in studios for beauty care. In Schulson's (2016) commentary on MNDFL, he opines that millennia-old practices of meditation are not necessarily equivalent to practices of beauty care, and discusses the extent to which practices like mindfulness have become thoroughly commercialized. While a Marxist-inflected analysis of the commodity fetishism shaping the contemporary mindfulness trend would be a productive way to further unpack Schulson's observation, this is not my aim here. I want to interrogate the rationale behind MNDFL from a different angle. I am more curious about the implications when diverse practices of the self—whether they be meditation or beauty care practices—become substitutable alternatives. The key here is to recall that neoliberal subjectivity may be constituted with diverse knowledges, procedures, techniques, and expertise across multiple domains of everyday life. For it shows that *homo economicus* 'is not a natural being with predictable forms of conduct and ways of behaving, but is instead a form of

subjectivity that must be brought into being and maintained through *social mechanisms of subjectification*…. [and] produced by way of forms of knowledge and relations of power aimed at encouraging and reinforcing *individual practices of subjectivation*' (Hamann 2009: 42; emphasis added).

Nikolas Rose's canonical work on the 'psy disciplines' has shown that they have played a pivotal role in normalizing the social mechanisms of subjectification and the individual practices of subjectivation enabling the production of neoliberal subjectivity (Rose 1989, 1996). This occurs not by way of coercion but by way of the freedom to choose. Rose contends that the knowledge practices of Western psychology are the contemporary successors of the spiritual *askēsis* of Greek philosophers and the Christian confessional apparatus and that they channeled the pastoral power of neoliberal governmentality in two key ways. Firstly, the psy disciplines provided the terms for human subjectivity to be translated into the governmental language of schools, prisons, workplaces, and other social sites. Secondly, they enabled subjectivity and intersubjectivity to become objects of intervention by providing the conceptual and practical tools for addressing problems concerning intelligence, development, maladjustment, group dynamics, and the like. The psy disciplines thus 'made it possible to think of achieving desired objectives—contentment, productivity, sanity, intellectual ability—through the systematic government of the psychological domain' (Rose 1996: 70). Jeremy Carrette and King's (2005) much cited study on contemporary spirituality has further developed the insights offered by Rose's work. They chart the ways in which the psychologization of religion enabled the development of individualist and corporate forms of spirituality. Refracted through the language of the New Age, self-help, and body/mind/spirit genres, practices like meditation and yoga have been extracted from their historical, cultural and/or religious contexts and are rebranded as secular technologies of the self. The dominance and privileging of an individualistic and therapeutic approach to mindfulness has developed under these conditions.

---

[1]MNDFL http://mndflmeditation.com.

Importantly, the purpose of these Foucauldian-informed analyses is not simply to bemoan the totalizing reach of neoliberal governmentality. Rather, it is to expose the historical contingency of its arrangements, and to show that if neoliberal subjectivity must be constantly produced and maintained, then the very process of self-constitution by which neoliberal subjectivity is produced also holds the potential for its resistance and refusal. For Foucault, '[t]here is no first or final point of resistance to political power other than in the relationship of self to self.' (2005: 252). If an analytic of governmentality understands subjectivity in its dual sense—that is, to be subjectified by a dominant influence *and* to be subjectivated by one's own effort and conscience—it is in the process of subjectivation where the relationship of self to self may be cultivated differently, so that one may govern oneself differently within and against the dominant logics of neoliberalism.

Political Spirituality, the Care of the Self, and Ethics as Critical Practice

Foucault's 'Ethical Turn' and Zen Encounter

Foucault's account of neoliberal governmentality was delivered in a series of lectures in 1979 entitled, *The Birth of Biopolitics* (2008), which were part of the annual program that was held at the Collège de France. It was the only series in the program where he located his analysis within the twentieth century, and systematically addressed the topic of neoliberalism. From this point, his work would take what has been described as an 'ethical turn' as he shifted his attention to the aesthetic and ascetic dimensions of spiritual cultivation in the Greco-Roman world (Milchman and Rosenberg 2007). This shift in perspective was a way for him to fine tune the question of 'how individuals are able, are obliged, to recognize themselves as subjects,' a problematic which he admitted was not clearly isolated in his earlier work on the history of

madness, the clinic, and the prison where the focus was on techniques of domination rather than techniques of the self. It is worth noting briefly here some of the events leading to the lectures of 1979 and the emergent ideas Foucault was developing at the time before he took the long journey back to antiquity, as it were.

In April of 1978, Foucault traveled to Japan and stayed at a Zen temple where he experimented briefly with meditation. When asked about his thoughts on the practice, he said: 'With so little experience, if I have been able to feel something through the body's posture in Zen meditation, namely the correct position of the body, then that something has been new relationships which can exist between the mind and the body, and moreover, new relationships between the body and the external world' (Foucault 1999: 112–113). Upon his return to Paris in May, Foucault participated in a roundtable discussion where he proposed the idea of 'political spirituality,' describing it as 'the will to discover a different way of governing oneself through a different way of dividing up true and false' (Foucault 1991: 82). Then in September and November, he traveled to Iran to write a series of journalistic essays on the revolution that had begun in January that year. In one of the essays, Foucault again raised the question of political spirituality, posing it in relation to the historical conditions set in motion by 'the great crisis of Christianity' (Foucault quoted in Afary and Anderson 2005): 209—that is, the historical moment when the logics of pastoralism overflowed into the secularized knowledges, procedures, and techniques of indivdualization necessary for the conduct of conduct within the modern state.

This brief survey of the events and emergent ideas leading up to the *The Birth of Biopolitics* reveals that Foucault developed his account of neoliberalism as part of a larger critico-political itinerary of understanding the resistive or transformative potential immanent in the process of subject formation. And more curiously, it reveals the interest he showed toward Zen meditation as a technology of the self which might actualize such a resistive and transformative potential.

Foucault did not elaborate on his Buddhist interest in his subsequent works, and my point here is not to insinuate that there is some hidden Buddhist commitment in his work. Rather, it is simply to show that his intuitions about Zen meditation invite us to consider how contemporary mindfulness (Buddhist oriented or otherwise) might be understood, through the connection between his account of neoliberal governmentality and his history of the care of self, as a technology of the self that could perform the function of political spirituality (Carrette 2000).

Reevaluating the Care of Self

Foucault addresses the theme of spirituality in his lecture series of 1982, *The Hermeneutics of the Subject*, which opens with a reevaluation of the Ancient Greek precept of 'the care of self' (*epimeleia heautou*) and its relation to the more famous maxim of 'know thyself' (*gnōthi seauton*). Taking Socrates as a point of departure, Foucault claims that throughout antiquity, the philosophical question of 'how to have access to truth' was always twinned with, and even subordinated under, the spiritual question of 'what transformations in the being of the subject are necessary for access to truth.' On his reading, there were three requisites of the *epimeleia heautou* which oriented the *gnōthi seauton*. Firstly, the care of self begins with the taking of a 'general standpoint, of a certain way of considering things, of behaving in the world, undertaking actions, and having relations with others.' *Epimeleia heautou* thus expressed 'an attitude toward the self, others, and the world' based on care and concern (Foucault 2005: 10). Secondly, the care of self required a form of attention turned toward 'oneself.' *Epimeleia heautou* thus entailed 'a certain way of attending to what we think and what takes place in our thought' (Foucault 2005: 11). Thirdly, the care of self did not merely refer to an attitude or a form of attention turned on the self, but also designated 'a number of actions exercised on the self by the self, actions by which one takes responsibility for oneself and by which one changes, purifies,

transforms, and transfigures oneself.' *Epimeleia heautou* must therefore be performed and cultivated with certain practices and exercises, such as 'techniques of meditation, of memorization of the past, of examination of conscience, of checking representations which appear in the mind, and so on' (Foucault 2005: 11). But why, Foucault asks, has the care of self been largely forgotten in the history of Western philosophy?

> How did it come about that we accorded so much privilege, value, and intensity to the "know yourself" and omitted [and] left in the shadow, this notion of care of the self, that, in actual fact, historically [...] seems to have supported an extremely rich and dense set of notions, practices, ways of being, forms of existence, and so on? (2005: 12)

While the displacement of the care of self occurred gradually over the centuries, Foucault identifies the 'Cartesian moment' as pivotal. With the Cartesian approach, the self-evidence of the subject's existence was placed at the source of its own being, and thus came to figure as the origin and point of departure of the philosophical method. Where knowledge of oneself formerly required the testing of the self-evidences of experience in order to transform the subject's ground of existence, the Cartesian approach proceeds by first establishing the impossibility of doubting the ground of one's existence as a self. The *gnōthi seauton* now takes objective knowledge of oneself as a fundamental means to access truth, and by the same gesture precludes the *epimeleia heautou* from the field of philosophical thought. Henceforth, Foucault (2005: 15) argues that a line could be drawn between 'philosophy' and 'spirituality.' Philosophy becomes 'the form of thought that asks what it is that enables the subject to have access to the truth and which attempts to determine the conditions and limits of the subject's access to truth.' Knowledge of truth, in this mode of philosophical thought, is obtained by way of an objectifying process. What gets relegated by this dominant form of philosophical thought is an understanding of spirituality, which entails 'the search, practice, and experience through which the subject carries out the necessary transformations on himself [sic] in order to have access to the truth,' including practices involving 'purification, ascetic exercises, renunciations, conversions of looking, modifications of existence, etc.' (Foucault 2005: 15). The knowledge of truth, in this mode of spirituality, is cultivated by way of a subjectivizing process.

A Fourfold Analysis of Ethics-Based Morality

The care of self provided the ancient Greeks with an ethos of spiritual self-cultivation by which they may pursue philosophy as a way of life, a pursuit that was oriented by the praxis-ideal of an art of living (Foucault 1988a). Among the domains of activity which they attended to with the ethos of the care of self were the erotic relations between men. On Foucault's account, the Greco-Romans employed an ethics-based rather than code-based approach to morality in their problematization of erotic relations. Moral code refers to the 'prescriptive ensemble' gathering together 'a set of values and rules of action that are recommended to individuals through the intermediary of various prescriptive agencies.' On Foucault's account, values and rules may be articulated in terms of a coherent doctrine or explicit teaching, or 'transmitted in a diffused manner, so that far from constituting a systematic ensemble, they form a complex interplay of elements that counterbalance and correct one another, and cancel each other out on certain points, thus providing for compromise or loopholes' (Foucault 1990: 25). Ethics, on the other hand, refers to 'the manner in which one ought to form oneself as an ethical subject acting in reference to the prescriptive elements that make up the code' (Foucault 1990: 26). Ethics, in this sense, describes the *rapport à soi* or relation of self to self involving four interrelated dimensions: (1) the ethical substance, the part of oneself that serves as the prime material for the cultivation of moral conduct; (2) the mode of subjection, the rationale or justification according to which a person chooses to meet the obligation to conduct themselves as an ethical subject; (3) the ethical work, the activity a person engages

into transform oneself into the ethical subject of one's actions; and (4) the *telos*, or the mode of being that is the aim of one's ethical work. These four dimensions could be summarized as the 'what,' 'why,' 'how,' and 'goal' of ethics.

For Foucault, moral codes remain relatively stable from one historical and cultural context to another, while ethics are more susceptible to modification and rearticulation. Consider, for instance, how a similar moral code 'One shall not sleep with boys' could be found in Classical Greece and a later period when Christianity took hold in Europe; but the ways in which a person constitutes herself or himself as an ethical subject in relation to this moral code differ. For the Greeks, the *ethical substance* was pleasure, because a lack of restraint and failure to exercise moderation in erotic activities with boys present the dangers of hubris and disrepute, while for the Christians, the *ethical substance* was desire or concupiscence, because any such erotic yearning, for boys or otherwise, was a sign of the first sin. Accordingly, the *mode of subjection* for the Christians was divine law, a juridico-religious injunction which they had to obey, while for the Greeks, it was an aesthetico-political decision to transform one's life into a work, the expression of beautiful conduct. The *ethical work* for the Greeks, then, entailed personal and social duties and techniques of contemplation and body exercises that would allow one to cultivate a healthy asceticism toward oneself and greater responsibility in the use of pleasure. For the Christians, on the other hand, it entailed the constant examination of hidden desires and of bringing them to light with self-renouncing confessional practices. The *telos* for the Christians was thus the self-renunciation, while for the Greeks, it was the goal of self-mastery.

What interests Foucault about the Greco-Roman world is not the specific content of their sexual ethics but the orienting praxis-ideal of an art of living. His goal is neither to pursue a history of ancient Greek philosophy nor to discover its 'true' meaning. Nor is he implying that we can simply transplant the solutions from the past to solve present problems. Rather, the purpose is to describe how a particular domain of experience becomes a 'problem,' and in revealing the historical contingencies by which a domain of experience is problematized, to allow for new vantage points and tactics for a critical ontology of ourselves by turning what is 'given' about a present problem into a question. When asked about the relation between modern practices of self and those in antiquity, Foucault (1988b: 247) says that it is important 'to point out the proximity and the difference, and, through their interplay, to show how the same advice given by ancient morality can function differently in a contemporary style of morality.' In so doing, Foucault's research into antiquity opens up the question about the extent to which the 'what,' 'why,' 'how,' and 'goal' of ethical practices of self-constitution at any given juncture may be modified and rearticulated in relation to the prevailing norms of the day.

Ethical Self-Cultivation

O'Leary (2002: 37) has argued that Foucault reads in the ancient Greek praxis-ideal of an art of living 'a critical indictment of our modern modes of self-relation: In the comparison between the two, it is the ancient model which, for all its faults, triumphs.' Yates (2010: 81) has similarly claimed that 'the self-governance of the self ultimately furnishes a model that may, in effect, restore even to modern subjects an impetus for vigilance and action in an age otherwise caught up in the coercive structures of power and governmentality.' But what exactly is the difference between the ancient model of self-relation and neoliberal model of self-relation which might allow for such a possibility? Because if we consider Foucault's account on the care of self alongside his account of the neoliberal governmentality, his emphasis on spiritual self-cultivation does, at first blush, appear to lend 'itself quite nicely to neoliberalism's aim of producing free and autonomous individuals concerned with cultivating themselves in accord with various practices of the self' (Hamann 2009: 48).

To clarify this, we need to keep in mind that Foucault's history of the care of self or genealogy of the subject begins from the standpoint *that*

there is no transcendental subjectivity, no such thing as a 'sovereign, founding subject, a universal form of subject one could find everywhere.' For Foucault, 'the subject is constituted through practices of subjection, or, in a more anonymous way, through practices of liberation, of freedom, as in Antiquity, starting of course from a number of rules, styles and conventions that are found in the culture' (1988c: 50). The neoliberal subjectivity of homo economicus is enabled by a theory of human capital, which universalizes the entrepreneurial logics of competitive self-interest as a general matrix of all social relations. Foucault, it appears, is inviting us to see that neoliberal governmentality treats *the subject as a subject* and not simply as an object of knowledge–power, when it conceives of freedom as the choice that an individual makes to invest in future returns as an entrepreneur of the self. But his genealogical analysis also reminds us that *homo economicus* is a subject constituted by practices, which are performed in relation to the rules of a particular 'game of truth' or 'regime of veridiction' organized around economic rationality, *about which a history can be given*. With an account of the history of the care of self, Foucault further invites us to see that—if practices of the self must necessarily be performed in relation to the rules, styles, and conventions of their constitutive culture—the truth of practices of the self as 'free' requires a self-conscious account of how those rules, styles, and conventions are at play. Or more precisely, the practice of freedom requires that we be mindful of how we conduct ourselves under the rules of the historically contingent game of truth, within which practices of the self may be performed as free practices. As Dilts writes:

> ...if we must accept some degree of the neo-liberal understanding of the subject, then we must think very seriously about the care of the self, about the kinds of individuals that we form ourselves into— never forgetting, however, that we are constrained, that we are already governable, or that we can succumb to something that forms and reforms us. We must take part in that work ethically rather than satisfactorily (2011: 143).

An analytic of governmentality allows us to take seriously the shift in perspective enabled by the theory of human capital: The shift in self-relation experienced as the ongoing work and outcome of practices involving technologies of the self. But this shift in perspective also provides the opening for us to see—through a reevaluation of the care of self as the reorienting shift in attitude necessary for a transformation in subjectivity—that because practices of the self experienced as choices 'are already taken as practices of freedom' (Dilts 2011: 145), this is also the moment where they may become part of an ethical project. This is an opening for political spirituality. The pivot around which this possibility turns is an understanding of ethics as critical practice, or the task of critique as the cultivation of an ethos or 'limit attitude' toward the dominant logics shaping our individuality and experience of present reality.

The Limit Attitude of Critique

Foucault's coterminous genealogy of the subject and of the art of government, was a way for him to understand the emergence of 'the art of not being governed like that and at that cost,' or a history of the 'critical attitude' in the West (Foucault 2007a, b: 45). The task of critique, he says, is not a matter of decrying that things are not right as they are, but 'a matter of pointing out on what kinds of assumptions, what kinds of familiar, unchallenged, unconsidered modes of thought the practices that we accept rest' (Foucault 1988d: 154); the task of critique is 'to show that things are not as self-evident as one believed, to see that what is accepted as self-evident will no longer be accepted as such' (Foucault 1988d: 155). In the essay 'What is Enlightenment?,' Foucault engages with Kant's essay of the same title, reconceiving modernity not as a historical epoch but as an attitude. He describes this attitude as 'a mode of relating to contemporary reality; a voluntary choice made by certain people, a way of thinking and feeling; a way, too, of acting and behaving that at one and the same time marks a relation of belonging and presents itself as a task' (1984: 39). Describing it as the

critical ontology of ourselves, Foucault writes that it:

> …has to be considered not, certainly, as a theory, a doctrine, nor even as a permanent body of knowledge that is accumulating; it has to be conceived as an attitude, an ethos, a philosophical life in which the critique of what we are is at one and the same time the historical analysis of the limits that are imposed on us and an experiment with the possibility of going beyond them (1984: 50).

This understanding of critique is not restricted to professional intellectual work, but is rather an activity that may be located across the diverse domains of everyday life. It bears repetition that the task of critique is 'neither a form of abstract theoretical judgment nor a matter of outright rejection or condemnation of specific forms of governance. Rather it is a practical and agonistic engagement, reengagement, or disengagement with the rationalities and practices that have led one to become a certain kind of subject' (Hamann 2009: 57). O'Leary (2002) has illustrated this 'limit attitude' by taking up Foucault's invitation to reconsider how 'what,' 'why,' 'how,' and 'goal' ethics may be rearticulated in relation to prevailing norms. He proposes that the self—or more precisely, the forces of becoming aggregating as the specific attributes, abilities, and qualities of one's body by which the experience of selfhood take shape—be taken up as the ethical substance ('what') for a contemporary praxis-ideal of an art of living. The rationale or mode of subjection ('why') for doing so is the recognition that if there is no such thing as a transcendental self, the open-endedness of subjectivity invites an attitude that treats life as an ongoing work of critical reflection, experimentation, and crafting. The ethical work ('how') for a contemporary art of living could thus be performed with techniques of the self that allow one to work through the normative influence of social mechanisms by which 'our individuality is given to us in advance through ordered practices and forms of knowledge that determine the truth about us.' Accordingly, the *telos* ('goal') would the pursuit of freedom, understood not as some historical constant or final state of being, but as the ongoing search for and invention of diverse social relations and ways of becoming.

I have identified this approach to ethics as critical practice in the erotic activities of the participants of a qualitative study on bisexual and genderqueer lives (Ng and Watson 2013; Watson 2012). The study revealed that practices like cross-dressing, anonymous or polyamorous sexual encounters, and BDSM allowed the participants to engage in the critical, experimental, and transformative work of refusing and detaching themselves from the normative influence, the symbolic and actual violence of fixed binary gender identities and sexual essentialisms. They were able to use the body's capacity for desire and pleasure as the ethical substance, performing ethical work with techniques of the self related to erotic activities, to cultivate the interpersonal and social relations and ways of becoming that may otherwise be denied to them in a heteronormative world. While the practices constituting their subjectivities as bisexual or genderqueer individuals may involve consumerist and even entrepreneurial self-optimizing activities like beauty care and fashion, they are not simply consumptive but are also ethico-political attempts to perform the task of critique, in that Foucauldian sense. Others have located the critical attitude in the ways in which female snowboarders negotiate media discourses of femininity in snowboarding culture (Thorpe 2008); in the ways in which negative stereotypes about disability and essentialist disability identity are contested with technologies of the self that facilitate the process of 'coming out' as a disabled person (Reeve 2002); in the ways in which narratives of illness oriented around the ethos of the care of self allow ill people to reclaim their experiences from the potentially dehumanizing medical appropriation of illness (Frank 1998); and in the ways in which reflexive writing on Facebook may become a tool for self-transformation (Sauter 2014).

Experience and Critical Resistance

To be sure, any such attempts at critical resistance are not free from ambiguities or contradictions; there is no guarantee that they would effectively disrupt or defuse the normative operations of power. But this is precisely why the care of self presents us with an invitation to ongoing experimentation. Foucault's use of the French word *expérience* is instructive here. Firstly, it plays on the dual meanings of 'experience' and 'experiment.' Secondly, it points to the relational and collective dimensions of critique, since Foucault does not evoke the notion of 'experience' to presuppose or secure a transcendental subject, but to investigate the mutualizing dynamics between 'fields of knowledge, types of normativity, and forms of subjectivity' in a particular culture. Thus, as Lemke argues, 'the ethos of critique Foucault envisions is not a solitary attitude or a mode of individual self-fashioning; it is closely connected to existing forms of government' (Lemke 2012: 65). To be clear, I am not suggesting that the people in the aforementioned studies are Foucauldian specialists; I am not expecting people who are not professional intellectuals to be able to articulate their experience in the language of academic discourse. But precisely on this score, I am trying to show that we can nevertheless take seriously the possibility that people (professional intellectuals or not) have the capacity to engage in the task of critique by way of ethics or ethical self-cultivation as critical practice.[2] The critical attitude Foucault speaks of can be cultivated by 'specific intellectuals' across the varied domains of everyday life:

> Within these different forms of activity, I believe it is quite possible … to do one's job as a psychiatrist, lawyer, engineer, or technician, and, on the other hand, to carry out in that specific area work that may properly be called intellectual, *an essentially critical work. When I say "critical", I don't mean a demolition job, one of rejection or refusal, but a work of examination that consists of suspending as far as possible the system of values to which one refers when … assessing it.* In other words: *What am I doing at the moment I'm doing it?* (Foucault 1988e: 107; emphasis added)

---

[2] I have attempted to illustrate this in a media commentary, 'Mindfulness and Self-Care: Why Should I Care?' http://patheos.com/blogs/americanbuddhist/2016/04/mindfulness-and-self-care-why-should-i-care.html.

'What am I doing at the moment I'm doing it?' evokes a questioning attitude that dovetails with the general principles of mindfulness. If the potential for a critical ethos or 'limit attitude' may be actualized across the varied domains of private and public lives—and if mindfulness today has well and truly entered these domains, such that we may now find people cultivating mindfulness in relation to work, education, healthcare, sports, and even sex—my argument is that a key task confronting all of us invested in the ongoing development of mindfulness is to take seriously the possibility that mindfulness may function as a disruptive technology of the self. Thus, I am in a qualified sense agreeing with advocates of corporate or institutional mindfulness when they claim that mindfulness could potentially help to bring about change and transformation from within prevailing systems. This is why I have explicated in some detail the critico-political itinerary in Foucault's analytic of governmentality. Importantly, I hope I have shown that the potential of mindfulness (or other practices of the self) as a disruptive technology *within and against* prevailing systems has to co-dependently arise with interventions into the political ontology of the present milieu. The rules of the neoliberal game of truth are not to be taken for granted as an inevitable part of reality 'as it is' which has existed all along. Rather, they are the unquestioned givens of present reality which are to be problematized as questions by subjecting them to the task of critique—or more precisely, a two-pronged task of the critique of mindfulness and the mindfulness of critique.

The Critique of Mindfulness and the Mindfulness of Critique

Predicated on a dual understanding of subjectivity, the analytic of governmentality investigates, on the one hand, 'the points where the technologies of domination of individuals over one another have recourse to processes by which the individual acts upon himself [sic],' and on the other, 'the points where the techniques of the self are integrated into structures of coercion or domination'

(Foucault 1993: 203). More importantly, an analytic of governmentality tracks the different ways by which the voluntary conducts of individuals function as the 'contact point,' where these two techniques feedback into or displace one another. The idea of a two-pronged task of the critique of mindfulness and the mindfulness of critique is thus a proposal that mindfulness practice and research include within its purview, the multi-perspectival inquiries of governmentality.

The Critique of Mindfulness

With the critique of mindfulness, the objective is to interrogate the normative operations of power shaping the adaptation and use of mindfulness across different settings. That is, the critique of mindfulness attends to the points where the technologies of domination of individuals over one another have recourse to processes by which individuals act upon themselves. Some possible ways to do this would be to extend on the analyses of Rose and others like Carrette and King. For instance, the development of the mindfulness trend could be mapped against analyses of the governmentality of the happiness or positivity psychology movement (Binkley 2014). Will Davies's *The Happiness Industry: How the Government and Big Business Sold Us Well-Being* (2015) provides a detailed genealogical analysis of the 'science of happiness.' He shows that attempts to make the relationship between mind and world, the subjective experience of individuals and especially of workers, amendable to instrumental analysis and intervention have been developing since at least the early 1800s. Davies locates on this historical continuum, the proliferation of studies on the neuroscience and psychological benefits of mindfulness and other techniques for well-being, the rapid growth of the market for apps and devices related to meditation and fitness, and the interest and support shown by financial elites at the World Economic Forum toward these trends—all of which intensified after the 2008 global financial crisis, which left in its wake a challenge of restoring moral authority and trust in capitalism.

The critique of mindfulness could investigate the extent to which the moral injunction to be happy, along with techniques and devices being marketed, is being used to further the agendas of the powerful. The critique of mindfulness could also be performed within a specific domain like education. For example, Forbes (2015) has observed that a fixation on using mindfulness for the 'emotional-regulation' of students leaves the discriminatory, inequitable, anti-critical, and depoliticizing structural arrangements in the education system unaddressed. This issue is especially pertinent in the push to implement mindfulness programs in impoverished inner-city schools attended by students of color; the difficulties they face in school cannot be reduced to a individualized matter of poor emotional management skills, but should rather be investigated as a symptom of larger sociopolitical problems related to race, class, and poverty. The critique of mindfulness could bring more nuance to such debates by considering, for example, the governmentality of youth 'at risk' (Besley 2010), or by considering discourses on the neoliberal governmentality of education in general (Peters et al. 2009).

The Mindfulness of Critique

With the mindfulness of critique, the objective is to experiment with ways to better articulate and actualize the potential of mindfulness as a critically, ethically, and politically enabling practice grounded in the care of self. That is, the mindfulness of critique attends to the points where the techniques of the self are integrated into structures of coercion or domination, and more precisely, the points where techniques of the self might disrupt or defuse the normative operations of these structures. Some possible ways to do this would be to extend on the emergent discourses on contemplative education and activism, as exemplified by the itinerary of The Center for Contemplative Mind in Society. Building on the inroads paved by therapeutic mindfulness, this emergent field of studies recognizes the importance of attending to the somatic and emotional

dimensions of experience—not just for the purpose of stress reduction, but for the fostering of critical sensibilities and the political will to address the sociopolitical conditionings behind interpersonal and institutional acts of discrimination, exploitation, and other forms of injustice.

The mindfulness of critique could establish new lines of dialogical inquiry by engaging with Foucauldian understandings of the role of the body in the care of self. The mindfulness of critique could also bridge the concerns of contemplative education and activism with humanities and social science scholarship on the affective and emotional registers of sociocultural politics (Seigworth and Gregg 2010; Thrift 2004). This body of scholarship is premised on the understanding that the bodily, affective forces of everyday encounters are conduits for the exercise of power (think of the rhetoric surrounding the 'War on Terror'). But because these forces shaping public moods and collective sentiments (of fear or hope, for example) are never fully containable or controllable, they may be harnessed for the fostering of new relational capacities and social actions. These lines of inquiry dovetail with current neuroscientific interest on the impacts of mindfulness on empathy and compassion. The mindfulness of critique could thus broaden the purview of mindfulness research by establishing new channels of dialogue between different academic disciplines. With a richer and wider range of conceptual resources and methodological approaches, empirical studies of new models of mindfulness interventions could conceivably be developed which are not just individualistically and therapeutically oriented but also critically and civically oriented.

Future Promises

These preliminary suggestions for some possible ways forward are inspired by my ongoing curiosity about the reciprocity between Buddhist understandings and critical cultural theory (Ng 2016), and by the new inquiries I'm developing on the possible ways to cross-fertilize critical pedagogy with the principles of mindfulness. I have limited my suggestions to the educational context and to the humanities and social science discourses I am familiar with. But other ways forward could conceivably be developed with the frameworks and methods of other disciplines, like critical psychology, critical management studies, or the emergent field of critical neuroscience (Stanley 2012; Spicer et al. 2009; Choudhury and Slaby 2012). Keeping in mind the way Foucault evokes *expérience*, the critique of mindfulness and the mindfulness of critique can be regarded as *an experiential experiment with the experience of present reality*, a task which can be pursued via different vectors and assemblages of dialogical exchange and experimentation. And because the analytic of governmentality requires attentiveness toward the 'contact point' of voluntary conduct where techniques of domination and techniques of the self feedback into or displace one another, the critique of mindfulness and the mindfulness of critique also place an ethical demand on those of us engaging with mindfulness as part of an academic livelihood: Are we paying attention to the role of mindfulness, and more specifically to our role in producing knowledge on mindfulness, within the institutional environment of the university?

Mindfulness is becoming commonplace in the health and/or student services of university campuses. Commenting on this trend in the UK, social psychologist Steven Stanley suggests that it is not just a concern with student well-being that has prompted the uptake of the practice, but also a desire to boost attainment. He acknowledges that there is 'limited evidence' to indicate that mindfulness could help students perform better in terms of grades, but also raises a cautionary note: 'Most of the distresses and challenges happening in universities are to do with broader institutional goals and objectives rather than an innate problem with students' mental health.' Important as it may be to assist university students or staff with their mental health struggles, a reductive interpretation and use of mindfulness that privatizes the generative conditions of stress, risks becoming 'a sticking plaster that prevents us seeing the roots of the problems' (Stanley quoted in Swain 2016). As we have seen, this is the same line of argument that others have raised with regard to corporate mindfulness, and with regard to the push to implement mindfulness in schools to manage student behavior. Let me reiterate that to raise this cautionary note is not to trivialize the emotional or mental difficulties faced by individuals, nor is it to dismiss the possible therapeutic benefits of mindfulness. This is simply to recognize that the prevailing individualistic and therapeutic approach to mindfulness popularized by the MBSR model (or for the matter, a traditional Buddhist approach to mindfulness) is limited in its capacity (because it is not their primary focus and aim) to address the broader systemic and structural generative conditions of stress or anxiety in the contemporary world. We thus face a collective search for a more critically and civically oriented discourse of mindfulness. This search does not obviate the task of helping individuals with their personal well-being but supplements it with more expansive diagnoses of the dominant cultural logics and the precarious sociopolitical conditions and effects of this historical moment.

Conclusion: What Are We Doing or not Doing with Mindfulness?

By way of conclusion, what if we take Stanley's observation about the problems facing the university as a kind of mirror, angling it in such a way so that it does not just cast a reflection on student life and learning but also on academic life and labor? I trust that academic readers of this book have encountered the growing body of critical discourses, or at least the commentaries in higher education and/or mainstream news publications, on some of the key problems troubling the university under neoliberal conditions. I trust I am not the only one who is struggling to maintain conviction and hope in the direction in which academia is moving, not just in Australia where I presently reside but also in the USA and the UK. As I write this, I have been working as a casually employed academic (or an adjunct, as it is known in the USA) for almost ten years. I have no savings and live with constant anxiety about my ability to meet basic material needs. Because of the decline in continuing or tenured appointments and decreasing state funding to higher education, I am part of a growing population that is laboring at the center of the system and delivering the bulk of classroom activities to secure benchmarks of 'teaching excellence,' while also marginalized at the periphery without the professional support or security or any clear prospects of ongoing employment. I observe myself and others—doctoral students, new Ph.D. holders, research fellows, senior professors, along with support staff—being subject to managerial surveillance at every turn, all having to meet the endless demands of technologies of audit, feedback, performance, and risk management. Under this regime, research quality is measured according to the income it secures for the university as business, and teaching is evaluated according to student retention numbers and 'client satisfaction.' Aspirations of collegiality, collaboration, altruism, or activism are suffocating in this climate of competitive individualism and precarity; though some have been able to surf the wave of managerialism to carve a career in the business of knowledge brokering.

I have examined elsewhere how an analytic of governmentality can help us contest the micropolitics of the vicissitudes of academic life and labor, by shedding light on the ways in which apparatuses of audit, feedback, performance, and risk management function to produce the 'responsibilized' subject of the neoliberal academic regime (Ng 2015). To connect this with the present discussion, I wonder whether those of us who are working in academia need some form of mindfulness-based intervention. But intervention against what? To be sure, mindfulness has helped me to manage the stresses of academic life and labor. But is it our fault that we are feeling stressed out, feeling like we are a fraud all the time because we have no time to read or think, or tragically enough, feeling like we are being 'unproductive' when we do find time to read and think? Is this an innate problem with our mental

health? Or is it a symptom of something larger than this constricting question of 'me'? In which case, what are we doing or not doing with mindfulness as both a life practice and object of study?

What if we collectively interrogate how the apparatuses of audit, feedback, performance, and risk management are impacting on perceptions of the relative value and merits of the different fields of human knowledge, and on the viability of cross-disciplinary exchange and reciprocal learning? How might this impact on the scope and objective of mindfulness research, which at present reflects the symbolic and financial privileges that the psy-medical-cognitive disciplines command in the neoliberal knowledge economy? More curiously, how might we factor in our personal commitment to maintain mindfulness in everyday activities (including the vicissitudes of academic life and labor), to also investigate its mutual influence on our professional research of mindfulness? Why has not this question about the role of the subject in its own discourse received sustained attention, given that it may illuminate the theoretical or methodological oversights that arise with the unavoidable inbuilt limitations of our respective disciplinary training and preferred approach to mindfulness?

I invite practitioners and scholars/researchers engaging with Buddhist or secularized approaches to mindfulness to explore the usefulness of an analytic of governmentality in shedding light on these questions, the ethico-political implications of which extend beyond the immediate institutional space of the university to texture the critical ontology of ourselves. The analytic of governmentality reminds us that the exercise of *power over* people co-dependently arises with people's *power to* act. A collective search for a civically and critically oriented mindfulness is a way to probe the limits of the present: a task for ongoing experimentations with the care of self, and for the cultivation of different ways of becoming and new capacities for social action, relations, and freedoms. I dedicate these aspirations to the conversations to come on *the critique of mindfulness and the mindfulness of critique*, as we collectively pay attention to the workings of power flowing through the politics of our selves today. The workings of power are at once the obstacles and openings by which the promises made in the name of mindfulness may be fulfilled for a more promising future.

References

Afary, J., & Anderson, K. (2005). *Foucault and the Iranian revolution: Gender and the seductions of Islam*. Chicago: University of Chicago Press.

Besley, T. (2010). Governmentality of youth. *Policy Futures in Education, 8*(5), 528–547.

Binkley, S. (2014). *Happiness as enterprise: An essay on neoliberal life*. Albany: SUNY Press.

Carrette, J. (2000). *Foucault and religion: Spiritual corporality and political spirituality*. London, New York: Routledge.

Carrette, J., & King, R. (2005). *Selling spirituality: The silent takeover of religion*. London and New York: Routledge.

Choudhury, S., & Slaby, J. (Eds.). (2012). *Critical neuroscience: A handbook of the social and cultural contexts of neuroscience*. Wiley-Blackwell, Chichester: West Sussex.

Davies, W. (2015). *The happiness industry: How the government and big business sold us well-being*. London: Verso.

Dilts, A. (2011). From 'entrepreneur of the self' to 'care of the self': neo-liberal governmentality and foucault's ethics. *Foucault Studies, 12*, 130–146.

Forbes, D. (2015, November 9). They want kids to be robots: Meet the new education craze designed to distract you from overtesting. *Salon*. Retrieved from http://www.salon.com/2015/11/08/they_want_kids_to_be_robots_meet_the_new_education_craze_designed_to_distract_you_from_overtesting/.

Foucault, M. (1982). The subject and power. *Critical Inquiry, 10*(4), 777–795.

Foucault, M. (1984). What is enlightenment? In P. Rabinow (Ed.), *The foucault reader* (pp. 32–50). New York: Pantheon Books.

Foucault, M. (1988a). *The care of the self*. Translated by R. Hurley. New York: Vintage Books.

Foucault, M. (1988b). The return of morality. In L. Kritzman (Ed.), *Michel foucault: Politics, philosophy, culture* (pp. 242–254). London: Routledge.

Foucault, M. (1988c). An aesthetics of existence. In L. Kritzman (Ed.), *Michel foucault: Politics, philosophy, culture* (pp. 47–53). London: Routledge.

Foucault, M. (1988d). Practicing criticism. In L. Kritzman (Ed.), *Michel foucault: Politics, philosophy, culture* (pp. 152–156). London: Routledge.

Foucault, M. (1988e). On power. In L. Kritzman (Ed.), *Michel foucault: Politics, philosophy, culture* (pp. 96–109). London: Routledge.

Foucault, M. (1990). *The use of pleasure*. Translated by R. Hurley. New York: Vintage Books.

Foucault, M. (1991). Questions of method. In G. Burchell, C. Gordon, & P. Miller (Eds.), *The foucault effect: Studies in governmentality* (pp. 73–86). Hemel Hempstead: Harvester Wheatsheaf.

Foucault, M. (1993). About the beginnings of the hermeneutics of the self: Two lectures at dartmouth. *Political Theory, 21*(2), 198–227.

Foucault, M. (1999). *Religion and culture*. Selected and edited by J. Carrette. London: Routledge

Foucault, M. (2005). *The hermeneutics of the subject*. Edited by F. Gros and translated by G. Burchell. New York: Picador.

Foucault, M. (2007a). What is critique? In S. Lotringer (Ed.). *The politics of truth*. With an introduction by J. Rajchman (pp. 41–75). California: Semiotex (e).

Foucault, M. (2007b). *Security, territory, population: Lectures at the collège de France 1977–1978*. Edited by M. Senellart and translated by G. Houndmills, Basingstoke, Hampshire; New York: Burchell, Palgrave Macmillan.

Foucault, M. (2008). *The birth of biopolitics: Lectures at the collège de France 1978–1979* Edited by M. Senellart and translated by G. Burchell. Houndmills, Basingstoke, Hampshire; New York: Palgrave Macmillan.

Frank, A. (1998). Stories of illness as care of the self: A foucauldian dialogue. *Health, 2*(3), 329–348.

Hamann, T. (2009). Neoliberalism, governmentality, and ethics. *Foucault Studies, 6*, 37–59.

Harvey, D. (2007). *A brief history of neoliberalism*. Oxford, New York: Oxford University Press.

Jameson, F. (1991). *Postmodernism; or, the cultural logic of late capitalism*. Durham, NC: Duke University Press.

Kabat-Zinn, J. (2015a). This is not McMindfulness by any stretch of the imagination. *The Psychologist*. Retrieved from https://thepsychologist.bps.org.uk/not-mcmindfulness-any-stretch-imagination

Kabat-Zinn, J. (2015b, October 20). Mindfulness has huge potential—but McMindfulness is no panacea. *The Guardian*. Retrieved from http://www.theguardian.com/commentisfree/2015/oct/20/mindfulness-mental-health-potential-benefits-uk

Lemke, T. (2012). *Foucault, governmentality, and critique*. Boulder, Colorado: Paradigm Publishers.

Milchman, A., & Rosenberg, A. (2007). The aesthetic and ascetic dimensions of an ethics of self-fashioning. *Parrhesia, 2*, 44–65.

Ng, E. (2015). The question of faith in a micropolitics of the Neoliberal University. *Sites: A Journal of Social Anthropology and Cultural Studies, 12*(1), 153–177.

Ng, E. (2016). *Buddhism and cultural studies: A profession of faith*. Palgrave Mcmillan: New York, London.

Ng, E., & Watson, J. (2013). A foucauldian and deleuzian reading of autopoetic bisexual lives. *Writing From Below, 1*(2), 73–90.

O'Leary, T. (2002). *Foucault and the art of ethics*. New York, London: Continuum.

Oksala, J. (2013). Neoliberalism and biopolitical governmentality. In J. Nilsson & S. Wallenstein (Eds.). *Foucault, biopolitcs, and governmentality*, Södertörn Philosophical Studies 14. StockholmL Södertörn University.

Peters, M., Besley, T., Olssen, M., Maurer, S., & Weber, S. (Eds.). (2009). *Governmentality studies in education*. Rotterdam: Sense Publishers.

Pinsker, J. (2015). Corporations' newest hack: Meditation hack. *The Atlantic*. Retrieved from http://www.theatlantic.com/business/archive/2015/03/corporations-newest-productivity-hack-meditation/387286/

Purser, R., & Loy, D. (2013, July 1). Beyond McMindfulness. *The Huffington Post*. Retrieved from http://www.huffingtonpost.com/ron-purser/beyond-mcmindfulness_b_3519289.html

Purser, R., & Ng, E. (2015, September 27). Corporate mindfulness is bullsh*t: Zen or no Zen, you're working harder and being paid less. *Salon*. Retrieved from http://www.salon.com/2015/09/27/corporate_mindfulness_is_bullsht_zen_or_no_zen_youre_working_harder_and_being_paid_less/

Read, J. (2009). A genealogy of homo-economicus: Neoliberalism and the production of subjectivity. *Foucault Studies, 6*, 25–36.

Reeve, D. (2002). Negotiating psycho-emotional dimensions of disability and their influence on identity constructions. *Disability and Society, 17*(5), 493–500.

Rose, N. (1989). *Governing the soul: The shaping of the private self*. New York, London: Routledge.

Rose, N. (1996). *Inventing our selves: Psychology, power, and personhood*. Cambridge, New York: Cambridge University Press.

Sauter, T. (2014). 'What's on your mind?' Writing on facebook as a tool for self-formation. *New Media and Society, 16*(5), 823–839.

Schulson, M. (2016). *Put your money where your mind is: a for profit meditation studio opens in New York*. Retrieved from http://religiondispatches.org/put-your-money-where-your-mind-is-a-for-profit-meditation-studio-opens-in-new-york/

Schultz, T. (1972). Human capital: Policy issues and research opportunities. *Human Resources*.

Seigworth, G., & Gregg, M. (Eds.). (2010). *The affect theory reader*. Durham, London: Duke University Press.

Smail, D. (2005). *Power, interest and psychology: elements of a social materialist understanding of distress*. Ross-on-Wyde: PCCS Books.

Spicer, A., Alvesson, M., & Kärreman, D. (2009). Critical performativity: The unfinished business of management studies. *Human Relations, 62*(4), 537–560.

Stanley, S. (2012). Mindfulness: Towards a critical relational perspective. *Social and Personality Psychology Compass, 6*(9), 631–641.

Swain, H. (2016, January 26). Mindfulness: The craze sweeping through schools in now at a university near you. *The Guardian*. Retrieved from http://www.theguardian.com/education/2016/jan/26/mindfulness-craze-schools-university-near-you-cambridge

Thorpe, H. (2008). Foucault, technologies of self, and the media: Discourses of feminity in snowboarding culture. *Journal of Sport and Social Issues, 32*(2), 199–229.

Thrift, N. (2004). Intensities of feelings: Towards a spatial politics of affect. *Geografiska Annaler, 86*(1), 57–78.

Watson, J. (2012). *Re-visioning bisexuality: Rhizomatic cartographies of sex, gender and sexuality (unpublished doctoral dissertation)*. School of Humanities and Social Sciences: Deakin University, Melbourne.

Yates, C. (2010). Stations of self: Aesthetics and ascetic in foucault's conversion narrative. *Foucault Studies, 8*, 78–79.

A Meta-Critique of Mindfulness Critiques: From McMindfulness to Critical Mindfulness

Zack Walsh

Meta-Critique or: A Critique of Ideological Critiques

Critiques of mindfulness have now become so popular that they compete for the public's attention alongside regular reports of mindfulness' purported benefits. In just the last two years, commentators declared 2014 the year of mindfulness (Robb 2015; Gregoire 2014), then a popular backlash emerged (North 2014), and now, commentators seem poised to critique the critique (Delaney 2015; Gregoire 2015; Drougge 2016). However, as Mary Sykes Wylie (2015) argues in her historical account of these trends, critics who employ Buddhist ethics to critique secular mindfulness assume a reactionary position that is fated to produce its own antithesis. Religiously based ethical critiques produce deeper ideological trenches between critics and apologists, without advancing a process for their reconciliation, because by imposing an interpretive frame from outside, these critiques produce nothing but endless cycles of future critique between contrary religious and secular perspectives.

Rather than engage a tired debate over the potential benefits and drawbacks of mainstream adaptations of mindfulness, this chapter will outline the terms of that debate in an attempt to curtail the proliferation of online commentaries that lack self-reflexivity and suffer from a poor understanding of opposing viewpoints. By offering a critical summary of online critiques, this chapter will analyze how secular, scientific, religious, economic, and political ideologies attribute certain characteristics and prescribe certain values to mindfulness, in order to produce particular representations that are somehow more authoritative and valuable than their alternatives.

The guiding assumption of this meta-critique is that neither secular mindfulness nor critiques of mindfulness are value-free. The semiotics of mindfulness reflects particular ideologies and their associated values. As Payne (2014) argues, mindfulness, like all tools, "are ideologies—they exercise the values of their makers and instantiate those values in their users" (para. 13). Using mindfulness in schools (Forbes 2015), the military (Purser 2014), or Occupy Wall Street (Rowe 2015) and marketing it to stock traders (Dayton 2011) or people who want mind-blowing sex (Marter 2014) each affirm particular ideologies and sets of values that inform mindfulness practices, whether that includes an ethic of caregiving, a sensitivity to economic injustice, a drive for profit, or a desire for satisfaction.

One assumption underlying many online critiques is that as Western culture, secularism, and science transform meditation into mindfulness, it

Z. Walsh (✉)
Claremont School of Theology,
1325 N College Ave, Claremont, 91711 CA, USA
e-mail: walsh_zack@yahoo.com

becomes uncritical of how mindfulness is refashioned into a tool for ideology. Though this line of critique is often assumed in some critical circles, it is often a non-starter for apologists who remain largely unaware or unconcerned by the impact of ideology on mindfulness—a dividing line which is largely responsible for polarizing online debate. Despite their prevalence and power, there has yet to be an extensive critical examination of how mindfulness practices are shaped by these implicit ideologies and values. In fact, what is unique about the mindfulness revolution may be the way in which such an absence of critical inquiry has propelled the growth of mindfulness and its institutionalization. Secular and scientific communities have largely represented mindfulness as a value-free practice with universal benefit, which disguises how particular ideologies and values shape mindfulness to serve particular interests, as opposed to the general public interest. This guise of universality has allowed mindfulness to be marketed as a panacea, even though it is represented and practiced in ways that satisfy specific interests.

Critics who resort to Buddhist philosophy or accounts of individual experience mask the historical and social relations conditioning the mindfulness revolution, and critics who impose their own religious perspective or who debunk the science of mindfulness distract from a critical approach to the larger sociocultural phenomenon. If one wants to analyze how and why particular representations of mindfulness are generated to satisfy specific interests, then critiquing the specific content of debates is less important than critiquing how ideology informs them. This meta-critique analyzes the conditions under which the mindfulness revolution emerges to satisfy a narrower set of interests than what is explicitly claimed or desired. It cross-examines how power and interest shape mindfulness and how its investments are supported by people's uncritical enthusiasm for mindfulness, the ideologies and values underlying them, and the conditions supporting them.

Mindfulness and Universalism

As pragmatic religious modernizers from Asia transformed meditation into mindfulness with the help of modern psychology, mindfulness was decontextualized, separated from its association to traditional objects of meditation (the eightfold path), and shaped by new desires and demands. The fact that "the word meditation is not acceptable but mindfulness is" (Pradhan 2016, para. 18) reflects the West's underlying insecurity with what meditation represents and a rebranding of the term to allay those anxieties (Patterson 2015).

Some online critiques have recognized the emergence of the mindfulness revolution in social and historical contexts (Ng 2014; Goldberg 2015a), which scholars have documented more extensively elsewhere (McMahan 2008; Braun 2013; Wilson 2014), and in some cases, they have also recognized that mindfulness has been transformed through a process of cross-cultural exchange that discredits the search for cultural purity (Goldberg 2015b). But, while these starting points seem noncontroversial, in fact, the cultural identity of mindfulness has been a key site of contention, contesting how mindfulness is represented and how an emergent identity politics is coalescing to resist its formalization and institutionalization.

Religious and scientific communities recognize that mindfulness means many things to many different people. In Buddhism, the definition of mindfulness varies across different traditions and includes "eighteen elements or factors of mind that support mindfulness" (*Lion's Roar* 2015, para. 14). In the psychological literature, it can refer to a state, a trait, or a process which changes meaning across varied historical, cultural, and scientific contexts, all of which are inherently difficult to study and compare (Vago n.d.). "There are at least nine different questionnaires that claim to define and measure mindfulness, but no standard of reference exists which can be used to evaluate such questionnaires" (Flores 2015, para. 5). Robert Sharf's survey of traditional and modern Buddhist critiques illustrates that it is a challenge for scientific research to establish causal correspondences between traditional practices and the outcomes science expects to find, because traditional practitioners do not model modern, scientific understandings of mental health (McGill's Division of Social and Transcultural Psychiatry 2013). In the *Handbook of Mindfulness: Theory, Research, and Practice*, Rupert Gethin (2015) states that "it is not clear what standard we might use to judge any given account of mindfulness as either wanting or fitting" (p. 9).

Amidst this panoply of meanings, the public has not been discriminating and mindfulness has become a catchall term, referring to an entire movement, a basic human capacity, and several different practices that cultivate that capacity in relation to various different "'outcome qualities', such as compassion, patience, and equanimity" (*Lion's Roar* 2015, para. 18). The progenitor of modern mindfulness, Jon Kabat-Zinn, has often contributed to this general confusion about what mindfulness is. He defines it not as a technique, but as "a way of being… a way of seeing, a way of knowing, even a way of loving" (2005, p. 58). He maintains an ambiguous stance toward both the cultural rootedness and universal value of mindfulness, considering it to be "a universal dharma that is co-extensive, if not identical, with the teachings of the Buddha." He says, "[Mindfulness is] a place-holder for the entire dharma… meant to carry multiple meanings and traditions simultaneously." On the other hand, his working definition defines mindfulness as a universal and innate human capacity to cultivate "moment-to-moment, non-judgmental awareness," leading secular mindfulness proponents like Barry Boyce, editor of *Mindful* magazine, to claim that "the fundamental mindfulness that we all have… is obviously not an invention of Buddhism" (*Lion's Roar* 2015, para. 56).

By absorbing cultural particularism in universal rhetoric, Kabat-Zinn maintains that mindfulness is "one of seven factors of enlightenment," according to the Abidharma, and yet also "a kind of umbrella term for the Dharma in some much larger and universal sense" (*The Psychologist* 2015, para. 24). He situates mindfulness in Buddhist contexts. "As has been richly documented, the MBIs (mindfulness-based interventions) are in themselves outgrowths of Buddhism" (Knickelbine 2013b). And yet at the same time, he says elsewhere that the essential difference between Buddhist teachings and practices and the meditation practices that underlie MBSR (mindfulness-based stress reduction) and MBCT (mindfulness-based cognitive therapy) might be zero depending on the quality of the teacher (*The Psychologist* 2015). Presumably, good secular mindfulness instructors provide the same essential teachings on the nature of mind and self that Buddhist meditation and ethics provide.

Absorbing cultural particularism in universal rhetoric is, in the view of critics like Candy Gunther Brown, a strategic move to market mindfulness. Andy Puddicombe, a former monk turned CEO of the popular meditation app, Headspace, which "recently landed $30 million in new funding" (Morford 2015, para. 4), said, "I always teach View, Meditation, and Action," even if I never mention Buddhism (Widdicombe 2015, para. 45). To critics like Brown, mindfulness advocates like Puddicombe strategically replace religious language with scientific language to reframe Buddhist meditation as a secular practice. Science is used as the common idiom for economic and cultural capital to bring together religious and secular communities around common interests. Though Brown's critique wrongly assumes that mindfulness is essentially religious (Davis 2013), she reveals the logical fallacy committed by apologists who claim secular mindfulness cultivates virtue when such a claim cannot be made on the basis of current science, but only "as a tenet of the eightfold path of Buddhist awakening" (Brown 2015, para. 11). Either secular mindfulness advocates are making faith-based claims on the basis of science that does not exist or they are harboring unclaimed religious beliefs. In either case, there is no rational or empirical basis to justify universal claims about the benefits of mindfulness and its ethical foundations, except on the basis of implicit ideology. This is why critiques of mindfulness should not focus on the religious/secular divide and its ethical implications, but rather, on why these claims are made, by whom, for whom, and to what effect.

One way to illustrate how universalist claims function ideologically is to examine the Trojan horse hypothesis (Purser and Ng 2015a). This hypothesis posits that secular mindfulness contains implicit ethics which do not require a priori empirical support, because they are a universal and essential aspect of the practice itself. Kabat-Zinn claims there is an intrinsic social dimension to mindfulness and ethics are built into the practices (Williams and Kabat-Zinn 2015, p. 294; Kabat-Zinn 2015). When examined rationally, however, this perennialist claim that mindfulness possesses universal ethics could only be justified, Payne (2014) argues, if one maintains an "a priori conception of the subject as an isolated individual with private access to a pre- or trans-cultural and universal cognitive ground of consciousness" (para. 6). Universalist claims effectively ignore the social and historical dimensions that shape mindfulness practice, and as Ed Ng (Purser and Ng 2015b) argues, they position the dominant white male perspective as the invisible subject at the center of discourse. This implicit perspective was especially visible in *Time* magazine's special issue on "The Mindful Revolution" (Pickert 2014), which featured a beautiful, white, blond woman on the cover (Piacenza 2014). While universalist claims imply that everyone benefits from mindfulness, they occlude how mindfulness is ideologically framed and employed to serve particular interests.

In public discourse, apologists frequently use the Trojan horse hypothesis as a rhetorical strategy to deflect critiques implicating mindfulness practices in injustice. Payne (2015) argues that using the Trojan horse hypothesis in this way is "a means of marketing mindfulness programs while simultaneously blunting upper middles class liberal sensitivities to social inequity" (para. 11). By positing an intrinsic relationship between mindfulness and ethics, apologists can make unjustified ethical claims that escape critique. This strategy is most often employed against

critics who argue that "offering mindfulness to individuals in corporations will, at best, offer stress relief or create what Kevin Healy has described as 'integrity bubbles' for select individuals, while systemic corporate dysfunction continues unabated." Purser and Ng (2015a) have called this the corporate quietism hypothesis. There is no empirical evidence to suggest that either the Trojan horse hypothesis or the corporate quietism hypothesis is true, though apologists and critics often assume one or the other position and offer anecdotal evidence to support it.

The validity of either position is not what matters for this study, since each hypothesis positions itself as more authoritative, despite a lack of evidence to support its claims. What is important is the way in which universal, asocial, and ahistorical representations of mindfulness which support the Trojan horse hypothesis mask the enormous influence that current social, political, and economic interests exercise over mindfulness. Modern mindfulness practices that present themselves as universal practices for individual stress reduction and self-improvement are popular among people and institutions in large part because they internalize neoliberalism and offer practices for discipline and control.

Mindfulness and Neoliberalism

In general, critiques of McMindfulness contest precisely this tendency of mindfulness to serve neoliberalism. Ron Purser is one of the most vocal critics and his *Huffington Post* article co-authored with David Loy (2013), called "Beyond McMindfulness," may be the most widely circulated critique to date. Its relative success is due in part to its clear critique of how neoliberal ideology shapes mindfulness. It argues that corporate mindfulness "conveniently shifts the burden onto the individual employee: stress is framed as a personal problem, and mindfulness is offered as just the right medicine to help employees work more efficiently and calmly within toxic environments" (para. 14). Bhikkhu Bodhi and Slavoj Žižek join this line of critique by claiming that mindfulness is in danger of becoming the perfect ideological supplement to capitalism (Eaton 2013; Žižek 2001), and since 2013, there has been a marked acceleration of publications about how mindfulness is used to alleviate symptoms of stress without addressing how stress is generated by social systems and environmental problems. A blog post on the *Contemplative Pedagogy Network* summarizes the chorus of ongoing critique, saying that mindfulness blames the individual for suffering and encourages psychological adjustment, "rather than addressing the external cause of stress" (Barratt 2015, para. 1).

McMindfulness can be viewed as an expression of a more widespread tendency for neoliberalism to shape spiritual practices, as Honey (2014) documented in "Self-Help Groups in Post-Soviet Moscow: Neoliberal Discourses of the Self and Their Social Critique." She argues that spiritual practices have evolved to emphasize "the centrality of the self in attainment of wellbeing, practices of self-realization and self-control, and the sale of practices and ideas of the self in the marketplace" (para. 3). Through ethnographic research, she identifies several "core concepts within the self-help sphere which have been linked to the production of a neoliberal self: personal responsibility, self-control and development, self-blame, commodification, and depoliticization" (para. 8). These core concepts have appeared frequently in online critiques of mindfulness, demonstrating the degree to which mindfulness is in fact being shaped by neoliberalism.

Despite the growth of McMindfulness critiques, however, there has been a parallel growth of denialism that has polarized debate. For instance, Kabat-Zinn has completely dismissed the idea of McMindfulness, arguing that the social critiques that Purser and Loy initiated "throw grenades at something that is at least 99 % healthy for people" and these critiques are not worthy of our attention because they "just came out of one person's mouth" (*The Psychologist* 2015, para. 9). In response to critics, mindfulness advocate Gelles (2015) similarly retorts, "rarely, if ever, does exposure to meditation make someone a worse person" (p. 203).

Yet McMindfulness critiques have enjoyed broad public appeal, suggesting that they have tapped a cultural nerve and should not be dismissed as the invention of a few cranks. When Kabat-Zinn blames Purser and Loy for giving voice to a much larger social issue and when Gelles (2015) addresses critiques by appealing to anecdotal evidence just to "put them to rest" (p. 203), they effectively silence public discourse and erase the concerns of a much broader public. Could these denials be informed by ideology, since ideological biases are already evident in how advocates present mindfulness? For instance, in *Mindful Work*, Gelles takes Honey's core concepts of the neoliberal self for granted and uses them to frame mindfulness (Horton 2015). Gelles (2015) writes, "Stress isn't something imposed on us. It's something we impose on ourselves" (p. 85). In *The New Yorker*, Purser cites a Stanford study showing that on the contrary, "most workplace stress is caused by things like corporate dysfunction and job insecurity—not by 'unmindful employees'" (Widdicombe 2015, para. 43). But in spite of this, Gelles discounts the impact of structural forces by framing mindfulness around the neoliberal self. Whereas Purser argues, "Corporations like mindfulness… because it 'keeps us within the fences of the neoliberal capitalist paradigm'" (Widdicombe 2015, para. 43), apologists for secular mindfulness vehemently deny this claim. But their denial may be the result of ideological biases against critiques of capitalism. In an interview for *The Atlantic*, Gelles retorts, "We live in a capitalist economy, and mindfulness can't change that… The focus, I hope, is on the employees themselves" (Pinsker 2015, para. 26).

By framing mindfulness around the neoliberal self, people reduce mindfulness to a private practice without social impact, used primarily for daily maintenance, emotional regulation, and self-improvement. These apologists deny the possibility for mindfulness to structurally transform society. They dismiss sociological issues and refocus debate on psychological questions, like "What does it feel like?" (Heuman 2014, para. 42) and "Is mindfulness, as currently construed, useful or not?" (Segall 2013, para. 2)

Ultimately, this detracts from understanding who benefits, how they benefit, and why. In *The Guardian*, Moore (2014) warns that "This neutered, apolitical approach is to help us personally —it has nothing to say on the structural difficulties that we live with. It lets go of the idea that we can change the world; it merely helps us function better in it" (para. 10).

Honey claims the depoliticization of spiritual practices like mindfulness is an effect of neoliberalism. On the other hand, in Seth Segall's (2015) blog post on "The Politics of Mindfulness," he claims, "Teaching the Dharma… transcends politics" (para. 13). Segall's view represents apologists' claims that mindfulness is both a placeholder for the universal dharma and a secular practice with universal ethics. Yet this universalist rhetoric ignores the material and social relations that constitute how mindfulness is represented and practiced to serve specific interests. One of the leading popularizers of mindfulness, Thich Nhat Hanh, expresses a similarly apolitical view:

> …as long as business leaders practice 'true' mindfulness, it does not matter if the original intention is triggered by wanting to be more effective at work or to make bigger profits. That is because the practice will fundamentally change their perspective… We need not fear that mindfulness might become only a means and not an end because in mindfulness the means and the end are the same thing (Confino 2014, para. 5).

By equating the means and the end with "true" mindfulness, Hanh ignores any critical investigation into the power dynamics informing how mindfulness is practiced, by who, and for what purpose, while at the same time asserting that Hanh's particular understanding of mindfulness is universal.

If apologists recognize that mindfulness is embedded in political and economic relations, but if they dismiss structural critiques out of hand, because this embeddedness represents a historical continuity with the past, then they are also effectively depoliticizing the practice (Wylie 2015, para. 48). Acceptance of the political and economic status-quo is common among apologists who deflect critiques of mindfulness' implication in ethically questionable institutional practices. According to (Purser and Ng 2015b), "corporate mindfulness apologists ardently believe that structural and transformative change comes by working within the system" (para. 11). They fail to view mindfulness challenging the current system, because transformation is restricted to within that system, or as congressman Tim Ryan (Ball 2014) tells critics, "To transform the process, you've got to be part of the process" (para. 22).

Debates about the application of mindfulness in the military provide good illustrations of how apologists ignore critics' ethical concerns by framing mindfulness in exclusionary ways. Secular Buddhist Mark Knickelbine (2013a) argues that "the battleground soldier finds him or herself in a vast matrix of social conditions which he or she has little power to control," so "those of us who object to warfare should strive to make the outcomes of mindfulness more widespread in our society" (para. 21), rather than question the social conditions of the soldier. Conversely, in *Salon*, Stone (2014) argues that apologists are "omitting entirely the option of not putting soldiers in traumatic situations to begin with as a stress-reduction strategy" (para. 8). Instead, apologists take standard military institutions and procedures for granted, which restricts the locus of change to the individuals working within the system, because as (Purser and Ng 2015b) points out, "Ethical behavior and stress are insourced to individuals; social structures and systems of power are simply viewed as a given" (para. 12).

Amishi Jha, who received $4.3 million to develop mindfulness-based mind fitness training (MMFT) for the military, openly assumes this apolitical stance (Purser 2014). In an interview with *Inquiring Mind*, she (Gates and Senauke 2014) says, "That's the starting point. I'm not debating, 'Should there be a military? Should there be war?'… [soldiers'] stress is not so much about the nature of the conflict or whether they should be engaged in it, it's about whether they themselves did something they didn't feel was right" (para. 13, 31). In a 2011 white paper, fellow researchers Elizabeth Stanley and John Schaldach invoked the Trojan horse hypothesis, claiming that MMFT "could provide greater cognitive and psychological resources for troops to act ethically and effectively in today's morally-ambiguous and emotionally-challenging operational environment" (p. 8). This brief rejoinder to critics' concerns at the conclusion of their report not only attempts to defer judgment, but it relegates ethics once again to the individual. Similarly, when Kabat-Zinn (2015) says that the all-party parliamentary group on mindfulness in the U.K. "will be addressing some of the most pressing problems of society at their very root—at the level of the human mind and heart" (para. 13), he is also effectively reducing questions pertaining to the social ethics of mindfulness to a matter of individual ethics.

The social imaginary around mindfulness seemingly collapses whenever it confronts social issues, largely due to a lack of critical thinking that interrogates power. At the International Symposium of Contemplative Studies (2014), Purser notes that "corporate mindfulness trainers are constrained by their dependency under corporate sponsors to ensure that such programs do avoid disruption of social harmony." Yet many apologists seem unconcerned by this, arguing that "it is not within the remit of mindfulness programmes to question the modus operandi of the corporations who employ the services of mindfulness consultants" (Whitaker 2013, para. 12). Titmuss (2016) writes:

> It is unfair to expect mindfulness coaches to address deep issues. We should not think for a moment that mindfulness courses will change the underlying ideology of people in power who seek to maximise gain and control… mindfulness does not appear to offer more than [what is presently conceived] nor should we make demands that it should offer more (para. 11, 22–23).

Mindfulness apologists seem to echo Margaret Thatcher's neoliberal dictum: There is no alternative. Rubin (2014) writes in *The New Yorker*

that "to expect it to be otherwise seems to me either to overstate the power of meditation or to understate that of capitalist ideology" (para. 12). To the apologists, people are stuck with what they have. Mindfulness is not for social change.

Apart from these rhetorical dismissals, individualistic and depoliticized forms of mindfulness are also used to police attention away from social issues, allowing "the conditions of our neoliberal political economic situation [to be] unquestioned and accepted as inevitable" (Ng 2015, para. 20). Apologists who deflect critiques are also policing public discourse. Purser and Ng (2015b) argues, "When confronted with engaged Buddhist criticisms, mindfulness advocates seem to lack the psychosocial stamina to extend intellectual hospitality to views that question the limitations of neoliberal, individualized mindfulness programs" (para. 6). This general dismissal of criticism by "mindfulness advocates [who] seem unwilling to engage with the issues at hand [displays] a kind of 'bad faith'" (para. 18).

In many ways, this bad faith is the cause for the increased polarization of online debates. On his teaching blog, the meditation instructor and apologist, Kenneth Folk (2013), discredits the entire body of critique as "strident moralism and impotent hand-wringing" and warns students that "Every moment of making love to ideas is one you could have spent paying attention to your experience" (para. 7, 9). In addition, mindfulness advocates who reduce the complexity of issues raised by critiques to the inadequacy of particular instructors (*Lion's Roar* 2015; Olendzki 2015), in effect redirect the public's attention back to individual responsibility, ignoring that they unconsciously validate critics concerns by doing so. Advocates who emphasize better education, better instruments, and higher standards for quality control similarly project neoliberal ideology by refocusing a structural problem on individuals (Sherwood 2015), and those who propose greater access and increased funding as the preferred solutions have already accepted current forms of mindfulness, simply sidestepping critique (Kabat-Zinn 2015; Knickelbine 2013b).

Mindfulness©

The conceptualization of mindfulness as a depoliticized self-help technique has another major and profitable consequence. By disassociating meditation from historical and social contexts and by adapting it to fulfill new needs, mindfulness develops into a vast array of profitable commodities. Representing mindfulness as universal allows for it to be shaped by an enormous diversity of possible representations recontextualized in the dominant ideology of the new culture. Kabat-Zinn's (2005) various contradictory definitions leave open the questions of what to be mindful of (p. 108). In "Elixir of Mindfulness," the critic Glenn Wallis (2011) argues that mindfulness can be directed toward any object and assume almost any form, because it has become a floating signifier "empty of any determinate and demonstrable object of signification" (para. 4).

The slipperiness of the concept allows mindfulness to be easily molded as a tool for ideology. Quoting Lewis Carroll's Humpty Dumpty, Segall (2013) concedes that mindfulness has come to mean "just what we choose it to mean, neither more nor less" (para. 2). When mindfulness means nothing in particular, it can mean anything in general. According to Wallis (2011), mindfulness is the new mana. It "can be filled with any sense desired by the user" (para. 4). What it means and how it is used now largely depend on how mindfulness is marketed (Holloway 2015). Mindfulness is tailored to meet consumer demands, and marketing strategists cherry-pick science to improve its marketability. Its variety of commodity forms are vast, including "Mindful Parenting, Mindful Eating, Mindful Teaching, Mindful Politics, Mindful Therapy, Mindful Leadership, Mindful Recovery"—the list goes on (Wallis 2011, para. 15). No matter your social position or means, there is a brand of mindfulness to satisfy your needs (Krupka 2015). Mindfulness is not just a product. It is also a brand marketed to enhance social capital. As some critics point out, it has become so trendy, mindfulness is "a badge of enlightened and self-satisfied consumerism" (Hefferman 2015,

para. 10) and "a class signifier," especially among a subset of Caucasians in Silicon Valley (Ehrenreich 2015, para. 4).

According to NIH statistics, "Americans spent some $4 billion on mindfulness-related alternative medicine in 2007" (Pickert 2014, para. 13). Something free is now repackaged and sold in countless books, magazine, CDs, studio lessons, therapy sessions, courses, wearable technologies, and online apps. People's minds have become colonized by private interests. They can no longer enjoy a few free moments of silence. Pervasive noise pollution, digital technologies, notifications, and nudges have invaded people's mental space, which they are forced to buy back to cultivate their attention and cognitive resources. Practices and materials that induce states of mind-wandering are even sold as supplements to mindfulness, so that whether people are working or taking a break, they are still being productive (Manthorpe 2015; Korda 2015; Biswas-Diener and Kashdan 2015).

In none of these commodified forms of mindfulness are the ideologies of neoliberalism or the value of productivity ever questioned. Placing the focus of mindfulness squarely on attentional training is a task-oriented approach that leaves the values around which the task was framed unquestioned. This narrow approach to mindfulness polices people's attention by regulating thoughts and behaviors that violate social norms, and it normalizes the conditions under which one practices (Krupka 2015). Whatever is viewed as distracting is directed away from one's attention, without recognizing that distraction is not only experienced phenomenologically, but also socially mediated through what society conceives as the primary focus (namely work). Judgments about what is distracting are left unquestioned and people's attention is supposed to return to the task at hand, as if it were the only thing of value.

By forestalling critical inquiry, advocates reduce mindfulness to a set of practices that support dominant ideologies and values. They promote practices like mindfulness-based stress reduction without necessarily questioning whether people's needs are best served by the values instantiated in these practices. They direct people's practice toward enhancing functionality and productivity and they sacrifice the opportunity to conceive and practice mindfulness according to alternative values oriented toward different goals, such as ecological sensitivity, social and economic justice, voluntary simplicity, esthetic enjoyment, creativity, or spontaneity.

Exclusionary practices that define mindfulness according to dominant social norms are not only prevalent in the marketplace, but also in the contemplative sciences. In *The Atlantic*, Tomas Rocha quotes contemplative scientist Willoughby Britton, explaining how dominant economic and cultural values shape the science of mindfulness. Britton (North 2014) argues that funding agencies are more interested in studies that "develop hypotheses around the effects of meditation… that promise to deliver the answers we want to hear" (para. 5). Timothy Caulfield (2015) claims that mindfulness research falls prey to the white hat bias—"a bias leading to the distortion of information in the service of what may be perceived to be righteous ends" (Cope and Allison 2010, p. 83). Furthermore, Purser and Cooper (2014) argue that "The appeal to science for legitimacy and validation is based largely on faith in promises about science, not in science itself" (para. 15). It is widely recognized even in the scientific community that common myths about mindfulness have propelled public enthusiasm, but far outpaced the development of scientific evidence (Wikholm 2015; Miller 2014; Hart 2015).

By privileging the sciences over humanities, contemplative studies effectively reduce meditation to an individual technique with psychological and neurobiological effects, while discounting the historical and social ecology of a contemplative life and worldview (Mind and Life Europe 2015). Science that abstracts mindfulness practice from its context and defines it operationally within a field of established social norms is partially responsible for reducing mindfulness to certain prescribed myths (Walsh 2016). Many critics are quick to point out that people "confuse co-relationships with causal factors" (Pradhan 2016, para. 16) and they need to examine a

person's entire life, rather than just a brain scan, to determine meditation's effects (Bieber 2014; Salzberg 2015). But to reverse these trends, the ideologies and values underlying public demand for mindfulness and the select interests they represent must be critically interrogated. Critics need to engage a much broader public discussion on the value of mindfulness and how it can serve broader coalitions of interest.

Critical Mindfulness

What is called for is not just more diverse representations of mindfulness that respond to the needs of marginalized people, or alternative forms of practice that engage different ways of knowing. As important as these may be, what is also needed is for mindfulness practitioners to engage critical inquiry, so that they interrogate the ideologies and values around which mindfulness is framed, and so they challenge the concentrations of power and interest that give rise to commodified forms of mindfulness.

Commodity forms of mindfulness are one of the primary targets of critique, because the sale and marketing of mindfulness advance particular practices and ideas about mindfulness which do not represent the interests of everyone. Commodified mindfulness empowers privilege and prevents broader awareness of the social and historical conditions, many of which are unjust, that allowed for the formation of these forms of mindfulness to profit some and exclude others. Forms of mindfulness which are less impacted by market demands and more focused on palliative care undoubtedly serve an important role in alleviating stress and trauma, but they do not address their underlying social causes.

On the contrary, critical mindfulness exposes how mindfulness is commodified and how non-instrumental approaches to mindfulness subvert that commodification process by cultivating it in the context of nonattachment. Ironically, the mindfulness instructor and activist, Jesse Maceo Vega-Frey (2015) argues that "this tendency of commodities to wrap themselves in the illusion of a separate *selfness* that exist outside the conditions of their creation is precisely the kind of delusion that mindfulness is designed to destroy" (para. 7). The commodification of mindfulness requires continuous commodification in the future to resolve the new needs generated by an instrumental approach to practice. Unlike their commodified counterparts, noninstrumental approaches to mindfulness reveal the connection between people's perceptions of mindfulness, their desire for commodified forms of mindfulness, and the conditions which generate those perceptions and desires (Scalora 2015; Crouch 2011; Burkeman 2015; Morford 2015). Although everyone comes to meditation practice for the wrong reasons, Barry Magid (Bieber 2014) argues, "real practice is subversive and deconstructive of all the reasons that initially brought us to it" (para. 13).

Now apologists of secular mindfulness and social critics should move past the polarizing debate in which each opposing camp dismisses the other, based on anecdotal evidence or unwarranted claims that support either the Trojan horse hypothesis or corporate quietism hypothesis. It is a mistake for apologists to confuse critique with criticism, as Richard Payne (2015) says, because critiques are not denying the important role that mindfulness can play in alleviating suffering. But apologists need to stop dismissing critiques. They need to take them seriously for debate to move forward. Critiques do not argue that mindfulness is inherently dangerous or that access to mindfulness must be limited. Rather, they argue that context and intention matter, and mindfulness should not be used to reinforce an implicit ideology or structure of power without question. Mindfulness practices need to represent a wider range of social interests, they need to probe deeply into practitioner's context and intentions, and they need to incorporate social ethics into a critical awareness of contemporary issues in ways that support positive transformation.

The sine qua non for incorporating critiques into current practice is an incorporation of critical inquiry, which "entails a mindful questioning of the habits and forces of 'attention policing' and 'border control'—the critique of mindfulness and the mindfulness of critique" (Ng 2015, para. 43). Nothing should be outside the purview of collective critical inquiry—not neoliberalism, Buddhism, capitalism, or the military. Mindfulness practitioners need to reflect on social and historical contexts and situate them within an identity politics, rather than claiming mindfulness to be a universal practice occluding the neoliberal, Buddhist monk, or dominant, white male as the model individual. When Kabat-Zinn (2011) says that he "sees the current interest in mindfulness and its applications as signaling a multi-dimensional emergence of great transformative and liberative promise… akin to a second, and this time global, Renaissance" (p. 290), practitioners need to be skeptical and ask who is doing the framing, why and to what effect. They should consider "how might the dominant frames surrounding mindfulness be reassembled to direct attention differently," and they should consider whether it is possible "to [direct] attention towards a particular view, without bracketing things outside the border of the frame?" (Ng 2015, para. 30)

Critical approaches to mindfulness politicize mindfulness. Whether or not people are aware, mindfulness has always been political. It is inextricably linked to how one leads one's life in relation to others. Spiritual activists already realize the intrinsic connection between awareness and action, theory and praxis. They meditate to support social action, and their social action is part of their meditation. They also recognize that, "If the problem is systemic, the solution needs to be a change in the character of the system," not an internalization of the problem (Vishvapani 2014, para. 7). To his credit, Kabat-Zinn says he does not reduce mindfulness to a psychological intervention or an instrumental way of practicing, and he distinguishes between nonjudgmental awareness and discernment (Scalora 2015; Genju 2015). But rhetoric aside, MBIs are not presented around a prescribed ethical frame (Pradhan 2016), and instead, they assume the ethical frame they are provided. Using mindfulness to reduce stress without questioning how the stress is generated tacitly reinforces the social system within which one practices. To address this problem, Bhikkhu Bodhi (Duerr 2015) argues that social, economic, and environmental concerns are not "the domain of mindfulness but of its companion, *sampajañña*, 'clear comprehension'" (para. 17). Although, mindfulness may increase sensitivity and responsiveness to collective suffering, it requires critical reason and social awareness of present injustices to effectively broaden one's circle of concern. In response to critiques of McMindfulness, the mindfulness movement should replace universal, asocial, and ahistorical views of mindfulness with critical, socially aware and engaged forms of mindfulness.

References

Ball, M. (2014, September). Congressman moonbeam: Can representative tim Ryan teach Washington to meditate? *The Atlantic.* Retrieved from http://m.theatlantic.com/magazine/archive/2014/09/congressman-moonbeam/375065/

Barratt, C. (2015, November 26). *Mindfulness, social change and the 'neoliberal self'* [blog post]. Retrieved from http://contemplativepedagogynetwork.com/2015/11/26/mindfulness-social-change-and-the-neoliberal-self/

Bieber, M. (2014, March 3). *Everyone comes to meditation practice for the wrong reason: A conversation with psychoanalyst barry magid* [blog post]. Retrieved from http://www.mattbieber.net/magid/

Biswas-Diener, R., & Kashdan, T. B. (2015, September 16). Why mindfulness is overrated. *Fast Company.* Retrieved from http://www.fastcompany.com/3051122/know-it-all/why-mindfulness-is-overrated

Braun, E. (2013). *The birth of insight: Meditation, modern buddhism, and the burmese monk ledi sayadaw.* Chicago: University of Chicago Press.

Brown, C. G. (2015, February 4). Mindfulness meditation in public schools: Side-stepping supreme court religion rulings. *Huffington Post.* Retrieved from http://www.huffingtonpost.com/candy-gunther-brown-phd/mindfulness-meditation-in_b_6276968.html

Burkeman, O. (2015, April 7). Meditation sweeps corporate america, but it's for their health. Not yours. *The Guardian.* Retrieved from http://www.theguardian.com/commentisfree/oliver-burkeman-column/2015/apr/07/mediation-sweeps-corporate-america

Caulfield, T. (2015, May 4). Be mindful, not mindless. *Policy Options.* Retrieved from http://policyoptions.irpp.org/issues/is-it-the-best-of-times-or-the-worst/caulfield/

Confino, J. (2014, March 28). Thich nhat hanh: Is mindfulness being corrupted by business and finance? *The Guardian.* Retrieved from http://www.theguardian.com/sustainable-business/thich-nhat-hanh-mindfulness-google-tech

Cope, M. B., & Allison, D. B. (2010). White hat bias: Examples of its presence in obesity research and a call for renewed commitment to faithfulness in research reporting. *International Journal of Obesity, 34,* 84–88. doi:10.1038/ijo.2009.239.

Crouch, R. (2011, March 31). *The myth of mindfulness* [blog post]. Retrieved at http://alohadharma.com/2011/03/31/the-myth-of-mindfulness/

Davis, E. (2013, May 3). Is yoga a religion? Evangelical Christians in California tried to ban yoga in schools. So where is the line between the body and the soul? *Aeon.* Retrieved from http://aeon.co/magazine/society/erik-davis-is-yoga-a-religion/

Dayton, G. (2011, April 12). Mindfulness—The most important trading psychology skill for traders. *Daily FX.* Retrieved from https://www4.dailyfx.com/forex/fundamental/article/guest_commentary/2011/04/12/Guest_Commentary_Mindfulness_The_Most_Important_Trading_Psychology_Skill_for_Traders.html

Delaney, B. (2015, October 18). If 2014 was the year of mindfulness, 2015 was the year of fruitlessly trying to debunk it. *The Guardian.* Retrieved from http://www.theguardian.com/commentisfree/2015/oct/19/if-2014-was-the-year-of-mindfulness-2015-was-the-year-of-fruitlessly-trying-to-debunk-it

Drougge, P. (2016, March 4). *Notes toward a coming backlash: Mindfulne$$ as an opiate of the middle class* [blog post]. Retrieved from http://speculativenonbuddhism.com/2016/03/04/notes-towards-a-coming-backlash-mindfulne-as-an-opium-of-the-middle-classes/

Duerr, M. (2015, May 16). *Toward a socially responsible mindfulness* [blog post]. Retrieved from http://maiaduerr.com/toward-a-socially-responsible-mindfulness/

Eaton, J. (2013, February 20). American buddhism: Beyond the search for inner peace. *Religion Dispatches.* Retrieved from http://religiondispatches.org/american-buddhism-beyond-the-search-for-inner-peace/

Ehrenreich, B. (2015). Mind your own business. *The Baffler, 27.* Retrieved from http://thebaffler.com/salvos/mind-your-business

Flores, N. (2015, September 11). Here's why you need to question mindfulness in classrooms. *Huffington Post.* Retrieved from http://www.huffingtonpost.com/natalie-flroes/heres-why-you-need-to-question-mindfulness_b_8112090.html

Folk, K. (2013, July 1). *Not a productivity tool* [blog post]. Retrieved from http://kennethfolkdharma.com/2013/07/why-meditation-is-not-a-productivity-tool/

Forbes, D. (2015, November 8). They want kids to be robots: Meet the new education craze designed to distract you from overtesting. *Salon*. Retrieved from http://www.salon.com/2015/11/08/they_want_kids_to_be_robots_meet_the_new_education_craze_designed_to_distract_you_from_overtesting/

Gates, B., & Senauke, A. (2014, Spring). Mental armor: An interview with neuroscientist amisha jha. *Inquiring Mind*. Retrieved from http://www.inquiringmind.com/Articles/MentalArmor.html

Gelles, D. (2015). *Mindful work: How meditation is changing business from the inside out*. New York: Houghton Mifflin Harcourt.

Genju. (2015, May 18). *Mindfulness, ethics & the baffling debate* [blog post]. Retrieved from http://108zenbooks.com/2015/05/18/mindfulness-ethics-the-baffling-debate/

Gethin, R. (2015). Buddhist conceptualizations of mindfulness. In K. W. Brown, J. D. Creswell, & R. M. Ryan (Eds.), *The handbook of mindfulness: Theory, research, and practice* (pp. 9–41). New York: Guilford Publications.

Goldberg, M. (2015a, April 18). The long marriage of mindfulness and money. *The New Yorker*. Retrieved from http://www.newyorker.com/business/currency/the-long-marriage-of-mindfulness-and-money

Goldberg, M. (2015b, November 23). 'Where the whole world meets in a single nest:' The history behind a misguided campus debate over yoga and 'cultural appropriation.' *Slate*. Retrieved from http://www.slate.com/articles/double_x/doublex/2015/11/university_canceled_yoga_class_no_it_s_not_cultural_appropriation_to_practice.html

Gregoire, C. (2014, January 2). Why 2014 will be the year of mindful living. *Huffington Post*. Retrieved from http://www.huffingtonpost.com/2014/01/02/will-2014-be-the-year-of-_0_n_4523975.html

Gregoire, C. (2015, March 16). Why mindfulness will survive the backlash. *Huffington Post*. Retrieved from http://www.huffingtonpost.com/2015/03/16/mindfulness-backlash_n_6800924.html

Hart, A. (2015, October 24); Mindfulness backlash: Could meditation be bad for your health? *The Telegraph*. Retrieved from http://www.telegraph.co.uk/women/womens-life/11942320/Mindfulness-backlash-Meditation-bad-for-your-health.html

Heffernan, V. (2015, April 14). The muddied meaning of 'mindfulness.' *New York Times*. Retrieved from http://www.nytimes.com/2015/04/19/magazine/the-muddied-meaning-of-mindfulness.html

Heuman, L. (2014, October 1). Don't believe the hype. *Tricycle*. Retrieved from http://www.tricycle.com/blog/don%E2%80%99t-believe-hype

Holloway, K. (2015, July 11). Mindfulness: Capitalism's new favorite tool for maintaining the status quo. *AlterNet*. Retrieved from http://www.alternet.org/personal-health/mindfulness-capitalisms-new-favorite-tool-maintaining-status-quo

Honey, L. (2014). Self-help groups in post-soviet moscow: Neoliberal discourses of the self and their social critique. *Laboratorium: Russian Review of Social Research, 1*, 5–29. Retrieved from http://www.soclabo.org/index.php/laboratorium/article/view/330/1028

Horton, C. (2015, June 25). *'Mindful work': Drowning social ethics in a sea of neoliberal niceness* [book review]. Retrieved from http://carolhortonphd.com/mindful-work-drowning-social-ethics-in-a-sea-of-neoliberal-niceness-book-review/

International Symposium for Contemplative Studies. (2014, November 2). *Search inside and outside yourself: Challenging current conceptions of corporate mindfulness*. Retrieved from https://www.youtube.com/watch?v=9JfUKiZc8aA

Kabat-Zinn, J. (2005). *Coming to our senses: Healing ourselves and the world through mindfulness*. New York: Hyperion.

Kabat-Zinn, J. (2011, June 14). Some reflections on the origins of MBSR, skillful means, and the trouble with maps. *Contemporary Buddhism: An Interdisciplinary Journal, 12*(1), 281–306. doi:10.1080/14639947.2011.564844

Kabat-Zinn, J. (2015, October 20). Mindfulness has huge health potential– but McMindfulness is no panacea. *The Guardian*. Retrieved from http://www.theguardian.com/commentisfree/2015/oct/20/mindfulness-mental-health-potential-benefits-uk

Knickelbine, M. (2013a, August 2). *Mindfulness on the battlefield* [blog post]. Retrieved from http://presentmomentmindfulness.com/2013/08/mindfulness-on-the-battlefield/

Knickelbine, M. (2013b, August 12). *From both sides: Secular buddhism and the 'mcmindfulness' question* [blog post]. Retrieved from http://secularbuddhism.org/2013/08/12/from-both-sides-secular-buddhism-and-the-mcmindfulness-question/

Korda, J. (2015, November 20). The lost factor in the buddha's path to happiness. *Huffington Post*. Retrieved from http://www.huffingtonpost.com/josh-korda/the-lost-factor-in-the-bu_b_8605052.html

Krupka, Z. (2015, September 22). How corporates co-opted the art of mindfulness to make us bear the unbearable. *The Conversation*. Retrieved from https://theconversation.com/how-corporates-co-opted-the-art-of-mindfulness-to-make-us-bear-the-unbearable-47768

Lion's Roar. (2015, May 5). *Forum: What does mindfulness mean for buddhism?* Retrieved from http://www.lionsroar.com/forum-what-does-mindfulness-mean-for-buddhism/

Manthorpe, R. (2015, December 10). Mind-wandering: The rise of an anti-mindfulness movement. *The Long and Short*. Retrieved from http://thelongandshort.org/society/is-mind-wandering-an-anti-mindfulness-movement

Marter, J. (2014, July 28). Mindfulness for mind-blowing sex: 5 practices. *Huffington Post*. Retrieved from http://www.huffingtonpost.com/joyce-marter-/mindfulness-for-mindblowi_b_5608649.html

McGill's Division of Social and Transcultural Psychiatry. (Producer). (2013, September 23). *Mindfulness or mindlessness: Traditional and modern buddhist* critiques of 'bare awareness.' [Video file]. Retrieved from https://www.youtube.com/watch?v=c6Avs5iwACs&feature=youtu.be

McMahan, D. L. (2008). *The making of buddhist modernism*. New York: Oxford University Press.

Miller, L. D. (2014, April 30). *Recent meta-analysis finds varied efficacy of group meditation interventions* [blog post]. Retrieved from http://effortlessmindfulness.wordpress.com/2014/04/30/recent-meta-analysis-finds-varied-efficacy-of-group-meditation-interventions/

Mind & Life Europe. (2015). *Interview with prof. martijn van beek - esri 2015*. Retrieved from http://www.mindandlife-europe.org/interview-with-prof-martijn-van-beek-esri-2015/

Moore, S. (2014, August 6). Mindfulness is all about self-help. It does nothing to change an unjust world. *The Guardian*. Retrieved from http://www.theguardian.com/commentisfree/2014/aug/06/mindfulness-is-self-help-nothing-to-change-unjust-world

Morford, M. (2015, September 22). *Meditation jumps the shark: Headspace sells you nothingness* [blog post]. Retrieved from http://blog.sfgate.com/morford/2015/09/22/meditation-jumps-the-shark/#photo-688818

Ng, E. (2014, December 11). Who gets buddhism 'right'? Reflections of a postcolonial 'western buddhist' convert. *ABC Religion and Ethics*. Retrieved from http://www.abc.net.au/religion/articles/2014/12/11/4146841.htm

Ng, E. (2015, March 12). Who gets mindfulness 'right'? An engaged buddhist perspective. *ABC Religion and Ethics*. Retrieved from http://www.abc.net.au/religion/articles/2015/03/05/4191695.htm

North, A. (2014, June 30). The mindfulness backlash. *The New York Times*. Retrieved from http://op-talk.blogs.nytimes.com/2014/06/30/the-mindfulness-backlash/

Olendzki, A. (2015, Spring). The mindfulness solution. *Tricycle*. Retrieved from http://www.tricycle.com/new-buddhism/mental-discipline/mindfulness-solution

Patterson, C. (2015, July 16). Mindfulness for mental health? Don't hold your breath. *The Guardian*. Retrieved from http://www.theguardian.com/commentisfree/2015/jul/16/mindfulness-mental-health-breath

Payne, R. K. (2014, February 18). *Corporatist spirituality* [blog post]. Retrieved from https://rkpayne.wordpress.com/2014/02/18/corporatist-spirituality/

Payne, R. K. (2015, October 27). *Sheep's clothing? Marketing mindfulness as socially transforming* [blog post]. Retrieved from https://rkpayne.wordpress.com/2015/10/27/sheeps-clothing-marketing-mindfulness-as-socially-transforming/

Piacenza, J. (2014, March 31). Time's beautiful, white, blonde, 'mindfulness revolution.' *Huffington Post*. Retrieved from http://www.huffingtonpost.com/joanna-piacenza/time-mindfulness-revolution_b_4687696.html

Pickert, K. (2014, January 23). The mindful revolution. *Time*. Retrieved from http://time.com/1556/the-mindful-revolution/.

Pinsker, J. (2015, March 10). Corporations' newest productivity hack: Meditation. *The Atlantic*. Retrieved from http://www.theatlantic.com/business/archive/2015/03/corporations-newest-productivity-hack-meditation/387286/

Pradhan, R. (2016, January 1). *Mindfulness and growing pains* [blog post]. Retrieved from http://levekunst.com/mindfulness-and-growing-pains/

Purser, R. (2014, Spring). The militarization of mindfulness. *Inquiring Mind*. Retrieved from http://www.inquiringmind.com/Articles/MilitarizationOfMindfulness.html

Purser, R., & Cooper, A. (2014, December 6). Mindfulness' 'truthiness' problem: Sam harris, science and the truth about buddhist tradition. *Salon*. Retrieved from http://www.salon.com/2014/12/06/mindfulness_truthiness_problem_sam_harris_science_and_the_truth_about_buddhist_tradition/

Purser R., & Ng, E. (2015a, September 27). Corporate mindfulness is bullsh*t: Zen or no Zen, you're working harder and being paid less. *Salon*. Retrieved from http://www.salon.com/2015/09/27/corporate_mindfulness_is_bullsht_zen_or_no_zen_youre_working_harder_and_being_paid_less/

Purser, R., & Ng, E. (2015b, October 2). White privilege and the mindfulness movement. *Buddhist Peace Fellowship*. Retrieved from http://www.buddhistpeacefellowship.org/white-privilege-the-mindfulness-movement/

Purser, R., & Loy, D. (July 1, 2013). Beyond mcmindfulness. *Huffington Post*. Retrieved from http://www.huffingtonpost.com/ron-purser/beyond-mcmindfulness_b_3519289.html

Robb, A. (2015, December 31). How 2014 became the year of mindfulness. *New Republic*. Retrieved from https://newrepublic.com/article/120669/2014-year-mindfulness-religion-rich

Rowe, J. K. (2015, March 21). Zen and the art of social movement maintenance. *Waging Nonviolence*. Retrieved from http://wagingnonviolence.org/feature/mindfulness-and-the-art-of-social-movement-maintenance/

Rubin, J. (2014, July 10). Meditation for strivers. *The New Yorker*. Retrieved from http://www.newyorker.com/books/page-turner/meditation-for-strivers

Salzberg, S. (2015, April 26). The challenges of seeing meditation only through a scientific lens. *On Being*. Retrieved from https://www.onbeing.org/blog/the-challenges-of-seeing-meditation-only-through-a-scientific-lens/7492

Scalora, S. (2015, March 20). Mindfulness-based stress reduction: An interview with jon kabat-zinn. *Huffington Post*. Retrieved from http://www.huffingtonpost.com/suza-scalora/mindfulnessbased-stress-r_b_6909426.html

Segall, S. (2013, December 19). *In defense of mindfulness* [blog post]. Retrieved from http://www.existentialbuddhist.com/2013/12/in-defense-of-mindfulness/

Segall, S. (2015, April 23). *The politics of mindfulness* [blog post]. http://www.existentialbuddhist.com/2015/04/the-politics-of-mindfulness/

Sherwood, H. (2015, October 28). *Mindfulness at risk of being 'turned into a free market commodity.'* Retrieved from http://www.theguardian.com/lifeandstyle/2015/oct/28/mindfulness-free-market-commodity-risk

Stanley, E. A., & Schaldach, J. M. (2011). *Mindfulness-based mind fitness training (MMFT.)* Mind Fitness Training Institute.

Stone, M. (2014, March 17). Abusing the buddha: How the u.s. army and google co-opt mindfulness. *Salon*. Retrieved from http://www.salon.com/2014/03/17/abusing_the_buddha_how_the_u_s_army_and_google_co_opt_mindfulness/

The Psychologist. (2015, May 18). *'This is not McMindfulness by any stretch of the imagination.'* Retrieved from https://thepsychologist.bps.org.uk/not-mcmindfulness-any-stretch-imagination

Titmuss, C. (2016, January 1). The buddha of mindfulness. *The politics of mindfulness* [blog post]. Retrieved from http://christophertitmussblog.org/the-buddha-of-mindfulness-the-politics-of-mindfulness-2

Vago, D. (n.d.) What is mindfulness? *We are amidst the mindful revolution, but do we have a handle on what mindfulness even is?* [blog post]. Retrieved from https://contemplativemind.wordpress.com/what-is-mindfulness/, http://www.rsablogs.org.uk/2014/socialbrain/mindfulness/

Vega-Frey, J. M. (2015, November 7). The buddeoisie blues: The price and peril of genetically modified dharma. *Medium*. Retrieved from https://medium.com/@dolessforpeace/the-buddeoisie-blues-160400a3adbe#.ln5q4ssju

Vishvapani. (2014, February 20). Mindfulness is political [blog post]. Retrieved from http://www.wiseattention.org/blog/2014/02/20/mindfulness-is-political/

Wallis, G. (2011, July 3). *Elixir of mindfulness* [blog post]. Retrieved from http://speculativenonbuddhism.com/2011/07/03/elixir-of-mindfulness/

Walsh, Z. (2016). A critical theory-praxis for contemplative studies. *Journal of the International Association of Buddhist Universities*, VII. Ayutthaya: Mahachulalongkornrajavidyalaya University Press. Retrieved from http://www.iabu.org/JIABU2016v7.

Whitaker, J. (2013, December 21). *Mindfulness: Critics and defenders* [blog post]. Retrieved from http://www.patheos.com/blogs/americanbuddhist/2013/12/2013-as-the-year-of-mindfulness-critics-and-defenders.html

Widdicombe, L. (2015, July 6). The higher life: A mindfulness guru for the tech set. *The New Yorker*. Retrieved from http://www.newyorker.com/magazine/2015/07/06/the-higher-life

Wikholm, C. (2015, May 22). Seven common myths about meditation. *The Guardian*. Retrieved from http://www.theguardian.com/commentisfree/2015/may/22/seven-myths-about-meditation?CMP=fb_gu

Williams, J. M. G., & Kabat-Zinn, J. (2015). *Mindfulness: Diverse perspectives on its meaning, origins and applications*. New York: Routledge.

Wilson, J. (2014). *Mindful america: The mutual transformation of buddhist meditation and american culture*. New York: Oxford University Press.

Wylie, M. S. (2015, January 29). How the mindfulness movement went mainstream—And the backlash that came with it. *AlterNet*. Retrieved from http://www.alternet.org/personal-health/how-mindfulness-movement-went-mainstream-and-backlash-came-it

Žižek, S. (2001, Spring). From western Marxism to western Buddhism. *Cabinet Magazine, 2*. Retrieved from http://www.cabinetmagazine.org/issues/2/western.php

Notes Toward a Coming Backlash Mindfulness as an Opiate of the Middle Classes

Per Drougge

> The "Western Buddhist" stance is arguably the most effective way for us to fully participate in capitalist dynamics while retaining the appearance of mental sanity.
> *(Žižek 2001a: 13).*

> This is what we are obliged to posit here: The historical tendency of late capitalism—what we have called the reduction to the gift and the reduction to the body—is in any case unrealizable. Human beings cannot revert to the immediacy of the animal kingdom (assuming indeed the animals enjoy themselves such phenomenological immediacy).
> *Jameson (2003: 717).*

Introduction

An earlier version of this chapter (Drougge 2014) first appeared in a multi-disciplinary, Swedish anthology on mindfulness a couple of years ago. The situation in Sweden shows many similarities to that in the USA and other Western countries, as mindfulness-based interventions have become an integral part of the clinical/therapeutic mainstream during the past decade, while various forms of, more or less well-defined, "mindfulness" are also a staple of the expansive management, coaching, and self-help industries. But whereas critiques of mindfulness and the mindfulness industry have received considerable attention and sparked lively debates in the Anglophone world, such critical discussions have largely been absent in Sweden.

Both the international success of mindfulness and local differences in its reception and application are important topics that remain under-studied, but they will only be touched upon tangentially here. My main purpose with collecting these notes and commentaries was simply to introduce a few critical perspectives on mindfulness and the mindfulness movement to a Swedish audience. Most of the material will likely be familiar to readers of this volume, but the text can hopefully still function as a kind of compendium for a reader who wants a quick overview of current debates, especially as they have played out in popular media.

Some of the arguments presented here are familiar from related contexts, such as discussions about the role of psychotherapy and psychoanalysis under capitalism (e.g., Zaretsky 2013), or the debates about psychiatry and anti-psychiatry during the 1960s and 1970s (Ohlsson 2008). Other objections to the mindfulness industry can be seen as variations or specifications of a more general criticism of commodified "spirituality" (e.g., Carrette and King 2005; Hornborg 2012; Webster 2012). A different set of objections have been raised by

P. Drougge (✉)
Department of Social Anthropology, Stockholm University, Stockholm, Sweden
e-mail: per@drougge.eu

© Springer International Publishing Switzerland 2016
R.E. Purser et al. (eds.), *Handbook of Mindfulness*,
Mindfulness in Behavioral Health, DOI 10.1007/978-3-319-44019-4_12

Buddhist scholars and practitioners. Central to their critique have been the interpretation and use of the Pali term *sati*, considered to have lost important aspects of its original meaning when translated as "mindfulness" and grafted onto a new and very different context (cf. Bodhi 2011; Gethin 2011). Linked to this, is a concern that the mindfulness movement is doing a disservice to the Buddhist tradition by misrepresenting and trivializing it. More recently, there have also been some methodological objections to the numerous studies cited as proof that mindfulness actually works as intended, as well as a growing interest in possible, unwanted effects of meditation practice.

I have chosen a few, what I consider particularly interesting and pertinent topics, but I also want to emphasize that what follows is not an attempt to formulate or synthesize a coherent mindfulness critique. Taken together, however, these objections represent a potentially devastating critique of conceptual fuzziness, grandiose claims, dubious self-presentation, cynical appropriation and (mis-)use of Buddhist concepts and practices, anti-intellectualism, and— not least—how mindfulness functions as a control mechanism and ideological lubricant in an increasingly harsh, neoliberal order (or "the opiate of the downward-mobile middle classes," as someone put it.[1])

Until recently, most of the critical discussions about mindfulness occurred in the shadow of uncritical media hype and rarely seemed to involve committed proponents. Quite suddenly, something seemed to change, however. There has been an upsurge in critical engagement with both mindfulness and the mindfulness industry, and the number—and intensity—of debates triggered by articles published in places such as *The New York Times, Huffington Post,* or *Salon.com* indicates both a growing need and a new willingness to think critically about the issues mentioned above. This is an interesting and promising phenomenon, not least since it

shows how online debates allow for discussions across disciplinary and professional borders, sometimes blurring the distinction between academic, professional, and popular discourses.

For some time now, I have been interested in the way Web sites, blogs, discussion boards, and various "social media" have opened up new venues for critical discussions, not only of mindfulness but of Western Buddhism as a whole. I have therefore made a point of mainly quoting material which is freely available on the Internet (Source volatility is an unavoidable problem with this kind of material. All of the web pages that I am referring to were freely available as of April 2016, but links sometimes go bad, and web pages can be locked up behind pay walls or disappear without warning.).

There are other reasons why it can be problematic to use web publications, including blogs (and their comment fields), this way. Online discussions rarely allow for subtle arguments, and the opinions expressed are sometimes hasty and ill-informed. It should be noted, therefore, that the selected examples were all written by highly qualified debaters. And even if the tone is sometimes sharply polemical or bantering, the material is also of considerable substance.

For reasons of space, I have abbreviated and simplified some rather complex arguments, and I have made no attempt to summarize the long and winding discussions that followed some of the quoted articles. This omitted material would make for a very interesting netnography, though, as it provides valuable examples of the way mindfulness practitioners think about their profession, and how they respond to criticism.

As for the predicted backlash, I guess it is still too early to determine whether it has arrived, and what was the impact of these debates. Given the faddish nature of the arenas where mindfulness has been most successful, and the incessant demand for new products and services, it seems reasonable to believe that the popularity of mindfulness has reached its peak, however. For what the observation is worth, around the time this article was first published, the word "backlash" itself begun appearing in many discussions

---

[1]Tom Pepper: https://speculativenonbuddhism.com/2013/07/10/buddhism-as-the-opiate-of-the-downwardly-mobile-middle-class-the-case-of-thanissaro-bhikkhu/.

about mindfulness.[2] The important question, of course, is what will come next, what will be the response to it, and to what extent the tools and arguments used for a critique of mindfulness will be applicable.

Mindfulness and Buddhaphilia

As a social anthropologist, my main research interest concerns the formation of Western Buddhism (more specifically the role of monasticism as an ideal and lived experience). Over the past few years, however, I have probably spent more time thinking and writing about "secular" mindfulness, and the rise of a global mindfulness industry, than "religious" Buddhism. In a way, this trajectory (which is not uncommon) makes perfect sense: Mindfulness has become a mass phenomenon, and while its connection with Buddhism is generally acknowledged, this is also a puzzling topic, surrounded by questions.

Whether secular/therapeutic mindfulness should be understood as a form of (crypto) Buddhism or something else is a question which has been the subject of some controversy. Many (but far from all) practitioners are quick to point out that what they are teaching is not Buddhism. Like the vast majority of their clients, and the consumers of mindfulness literature, most of them also do not self-identify as Buddhists. A scholar like Wilson (2014), on the other hand, treats the proliferation of mindfulness-labeled products and services as a paradigmatic example of how Buddhism adapts to and gains mass appeal in a new host-culture by offering practical or worldly benefits. As will be suggested below, Western Buddhism and the mindfulness movement share so many key features that the latter could arguably be understood as an extreme form of "Buddhist modernism." It also seems significant that MBSR-style mindfulness has such a prominent place within the nascent movement of "Secular Buddhism."[3]

As we know, the practice of mindfulness has often been marketed as a kind of "neutral" technique, stripped both of religious beliefs and cultural specifics. Paradoxically, many of its proponents also seem convinced that "mindfulness," as taught today, constitutes the very essence not only of Buddhism but of *all* major "wisdom traditions." This perennialist—and, one is tempted to add, chauvinist—assertion has often also been part of what is undeniably a very successful sales pitch.

To the critical observer, on the other hand, this may seem as a naïve (or very smart) attempt at having it both ways: The dubious invocation of a "2500-year-old, unbroken tradition" gives one kind of legitimization to the mindfulness project, while a meditation practice wrapped in the language of popular psychology, neuropsychiatry, and management-speak will be much more appealing to both clinicians and a mass audience than one that comes with the bells and smells, foreign terminology and ethical demands of conventional Buddhism.[4]

This jani-form nature of mindfulness raises several interesting questions about the role of religion (specifically Buddhism) in today's "post-secular" society, but also about distinctions such as religious/secular and soteriological/therapeutic. The ambiguous connection to a religious tradition also makes the significant impact mindfulness has had on ostensibly secular contexts such as medicine, social work, education, the penitentiary system, human resource management, and the military, into something quite remarkable, and something which itself deserves a closer study.

---

[2](E.g. http://op-talk.blogs.nytimes.com/2014/06/30/the-mindfulness-backlash/; http://www.telegraph.co.uk/women/womens-life/11942320/Mindfulness-backlash-Meditation-bad-for-your-health.html; http://www.huffingtonpost.com/2015/03/16/mindfulness-backlash_n_68009-24.html).

[3](E.g., http://secularbuddhism.org/forums/topic/mbsr-as-secular-buddhist-practice/).

[4]To make things *really* confusing, the mindfulness movement's undisputed front-figure, Jon Kabat-Zinn, has been quite open with how he sees MBSR as a form of *upaya*, or "skillful means," in the service of Buddhist mission.

In part, the enthusiastic and uncritical reception of mindfulness can probably be explained by the common but questionable idea that Buddhism is less a religion than a kind of (proto) science which just happens to always resonate with current paradigms, be it quantum mechanics, cognitive neuroscience, or something else. (cf. Lopez 2008, 2012).

Using some terminology borrowed from the Speculative Non-Buddhism movement (Wallis et al. 2013), I would also suggest that this could be seen as a symptom of a widespread tendency toward uncritical admiration of all things Buddhist, or *buddhaphilia*.[5] And I do not think it would be too far-fetched to suggest that both the inflated claims of the mindfulness industry and the readiness with which substantial parts of the Western Buddhist community has embraced it have something to do with the *principle of sufficient Buddhism*.[6]

Mindfulness in Wonderland

The first example is a blog entry, *Elixir of Mindfulness*, written by Glenn Wallis, first published on his Web site speculativenonbuddhism.com[7] and later in the e-journal *non+x* (Wallis 2012). Although the text precedes the debates of the last few years, it nevertheless raises a few important topics that do not seem to have received the kind of attention they deserve. The author holds a doctorate in Buddhist Studies, has published a number of translations and commentaries on Buddhist texts (e.g., Wallis 2002, 2004, 2007), and is now the chair of a program in applied meditation at the Won Institute of Graduate Studies in Philadelphia. As founder of the

intellectual movement known as Speculative Non-Buddhism, Wallis has also gained a reputation (or, in some circles, notoriety) as a sharp critic of both Western Buddhism and secular mindfulness (Wallis et al. 2013).

Elixir of Mindfulness begins with the observation that today's mindfulness industry has successfully established itself on the competitive market for naïve utopias that was previously dominated by healers and preachers, therapists, hypnotists, Theosophists, self-help gurus and actual gurus, and so on. Like those enticers, mindfulness also comes with the promise of a universal aid, or an *elixir*, against all human suffering. This assertion may seem over the top, too drastic, or applicable only in cases of the most vulgar abuse. Wallis, however, cites for evidence a popular Web site, mindful.org, which could be seen as representative of the business mainstream.[8] The Web site proudly proclaims that a dose of mindfulness will both enhance our enjoyment and appreciation of everyday life and help us deal with life's most difficult changes— in a way which makes the critical reader wonder (in the words of Wallis) if the copy was written by some latter-day Dale Carnegie who forgot to take his Adderall.

Indeed, there appears to be no limit to what can be achieved by means of mindfulness. According to mindful.org, mindfulness is helpful in such diverse contexts as nursing, death and dying, parenting, healing and health, intimate relationships and sex, consumerism, finances, cooking and diet, entrepreneurship, creativity, sports, activism, education, environmental protection, prison advocacy, and so on, ad nauseam (As it has become quite common to see mindfulness advocates state that "mindfulness is no panacea," it is interesting, perhaps revealing, to contrast this admission with the marketing hype of the mindfulness industry.).

---

[5]See Iwamura (2011) for a discussion of the fetishization of the "Oriental monk."

[6]A parallel to French Philosopher François Laruelle's "principle of sufficient philosophy": a "pretension that all things under the sun are matters for x-buddhism's oracular pronouncements, and that the totality of pronouncements … constitutes an adequate account—a unitary vision—of reality … (Wallis et al. 2013: 138)."

[7]https://speculativenonbuddhism.com/2011/07/03/elixir-of-mindfulness/.

[8]Behind mindful.org is "The Foundation for a Mindful Society," which later also started publishing the bi-monthly magazine *Mindful* which could be described as a kind of equivalent to publications like *Yoga Journal* or *Runner's World* and has close links with popular Buddhist glossies like *Lion's Roar* (formerly *Shambhala Sun*) and *BuddhaDharma*.

This cheerful sanctity is not a North American phenomenon, but seems to be a feature inherent to the global mindfulness industry, something which can be illustrated by a few examples of Swedish books and CDs: *Mindfulness Exercises for Children 4–7 Years* (Lundgren Öhman and Schenström 2011); *Mindfulness for Parents* (Andersen Cerwall and Stawreberg 2009); *Mindfulness in School* (Terjestam 2010); *Mindful Eating: Eat Well, Feel Good, Lose Weight with Mindfulness, Compassion, and CBT* (Palmkron Ragnar et al. 2016); *Mindfulness in Life: Guided Meditations for Men* (Engström 2009a) ("Become a more consciously aware, present husband, lover, father and manager. Learn to observe and manage your thoughts so that they do not constitute an obstacle for you. Train your capacity for attention and mindfulness and to have the patience, perseverance and acceptance to grow. Treat every relationship, love meeting or golf swing as the unique moment it is. Increased desire and joy of life comes for free."); *Mindfulness in life: Guided Meditations for Women* (Engström 2009b) ("Mindfulness is a practice that can enrich your life by learning to listen to your body, embrace your femininity, and manage stress. Mindfulness means to have contact with the present, to the present moment as it is—pleasant or unpleasant, good or bad—because that's what right now is right here."). Would a certain over-satiety appear after such a hefty dose of locally produced mindfulness literature, there is also *Heartfulness: Your Way to Happiness in the Present. The New Wave of Mindfulness* (Åkesdotter 2011).

The Anglophone market is filled with an almost incalculable number of similar titles. Alongside more conventional books on mindfulness and depression, anxiety, pain, obesity, anorexia, addiction, love, sex, childbirth, parenting, aging, and death, one can also find difficult to categorize but evocative works with titles such as *Mindfulness for Law Students: Using the Power of Mindfulness to Achieve Balance and Success in Law School* (Rogers 2009) and *The Mindful Dog Owner: What Your Dog is Teaching You About Living Enlightenment* (Stephens 2012). Jeff Wilson's *Mindful America* (2014) contains long lists of mindfulness publications with similarly intriguing titles.

In addition, of course, are the scores of books on Buddhism and Buddhist meditation with the word "mindfulness" in their titles, and a book like *Savor: Mindful Eating, Mindful Life* (Hanh and Cheung 2010) appears to be a hybrid of Buddhist and self-help discourses. According to the blurb: "With the scientific expertise of Dr. Lilian Cheung in nutrition and Thich Nhat Hanh's experience in teaching mindfulness the world over, *Savor* not only helps us achieve the healthy weight and well-being we seek, but also brings to the surface the rich abundance of life available to us in every moment."

How is it possible that mindfulness can accomplish all this? And what do all books, CDs, apps flooding the market even mean by "mindfulness"? Both researchers and practitioners with theoretical interests have long pointed out that Jon Kabat-Zinn's oft-quoted definition: "Mindfulness can be thought of as moment-to-moment, non-judgmental awareness, cultivated by paying attention in a specific way, that is, in the present moment, and as non-reactively, as non-judgmentally, and as openheartedly as possible" (Kabat-Zinn 2005: 108), is unsatisfactory. Wallis presents several examples from mindful.org and identifies four broad categories:

Mental operations: attention, concentration, change in focus, value-free observation of consciousness content, etc.

1. Behavior: kindness, compassion, common consideration, love, deep listening, slowing down, niceness, being good.
2. Traditional practices: various forms of Buddhist meditation, generalized/unspecified meditation, contemplation, various styles of yoga.

3. Indistinct something-or-others: openness, relating more effectively to thoughts and feelings, lovingkindness,[9] going inward, letting go, acceptance, truly seeing someone, minding mind, being *in* the moment, being moment to moment, just being.

The list of more or less diffuse descriptions ends with a recent formulation, attributed to Jon Kabat-Zinn, which simultaneously confirms the concept's elusive nature and shows how mindfulness appears to be yet another form of "spirituality" (with or without scientific claims):

> Mindfulness is not a technology. It is a way of being, a way of seeing, a way of knowing.

The lack of a clear definition and the baffling diversity of phenomena that fall within the concept lead Wallis to describe "mindfulness" as a textbook example of what is known as an "empty" or "floating" signifier.[10] Wallis points to the similarities between the usage of the term "mindfulness" and the Melanesian word *mana*, anthropologist Claude Lévi-Strauss's famous example of a floating signifier. Wallis then asks whether "mindfulness" can be said to function in the same way as *mana*: An amorphous concept that, in a Humpty-Dumpty-like fashion, can mean whatever the user wants. The only consistency, perhaps, is that the various usages all circle around the notion of some kind of life-giving *elixir*. Seen in this way, it becomes easier to understand the (unreasonable) expectations and (grandiose) claims linked to

mindfulness, as well as the futility of the search for clear definitions of the term.

Corporate Mindfulness and Its Discontents

The next example is an article/blog post titled *Beyond McMindfulness*, written by David Loy and Ron Purser, published on the *Huffington Post's* "Huffpost Religion Blog" in the summer of (2013). The article became the starting point for a debate that lasted for the rest of the year,[11] and "McMindfulness" has since become a recurrent trope in many discussions about mindfulness.

Loy is a philosopher, a Zen Buddhist teacher, and the author of several books that can perhaps be described as an attempt to formulate a critical theory with Buddhist overtones (Loy 1996, 2002, 2003, 2008). Purser is also a Zen Buddhist teacher, as well as a professor of management, and a business consultant. While this combination of professions might strike an outsider as somewhat peculiar, it makes perfect sense within the context of North American Zen, where clergy often hold day jobs, and other teachers have published books with titles like *Zen and Creative Management* (Low 1976) or *Zen at Work: 50 Years in Corporate America* (Kaye 1997). What

---

[9]"Lovingkindness" is a common English translation of the Buddhist concept of *metta*, encompassing a generalized outlook characterized by compassion and kindness, and specific meditation practices with the aim of cultivating such qualities. In this context, one could note the emergence of "compassion-focused therapy," an "integrated and multimodal approach that draws from evolutionary, social, developmental and Buddhist psychology, and neuro science" (Gilbert 2009: 199).

[10]"An "empty" or "floating signifier" is variously defined as a signifier with a vague, highly variable, unspecifiable or non-existent signified. Such signifiers mean different things to different people: They may stand for many or even *any* signifieds; they may mean whatever their interpreters want them to mean" (Chandler n.d.).

[11]Discussions about the article has played out on several blogs and Web sites, with the participation of both "religious" and "secular" Buddhists as well as mindfulness practitioners. Here is a selection: *American Buddhist Perspective* (Justin Whitaker): http://www.patheos.com/blogs/americanbuddhist/2013/12/2013-as-the-year-of-mindfulness-critics-and-defenders.html
108 Zen Books (Genju): https://108zenbooks.com/2013/08/02/on-mindfulness-muggles-crying-wolf/
Mindfulness Matters (Arnie Kozak): http://www.beliefnet.com/columnists/mindfulnessmatters/2013/09/mcmindfulness-revisted.html
Off the Cushion (Rev. Danny Fisher): http://www.patheos.com/blogs/dannyfisher/2013/07/your-practice-is-not-all-about-you/
Secular Buddhist Association (Mark Knicklebine): http://secularbuddhism.org/2013/08/12/from-both-sides-secular-buddhism-and-the-mcmindfulness-question/
The Existential Buddhist (Seth Zuihō Segall): http://www.existentialbuddhist.com/2013/12/in-defense-of-mindfulness/.

makes *Beyond McMindfulness* stand out, is rather that it turned out to be a scathing critique of what Buddhist Studies scholar Payne (2014) calls "corporatist spirituality," i.e., the use of spirituality for corporate ends.

Beyond McMindfulness is a short article of only a few pages. It touches, however, on several problematic aspects of today's ostensibly unadorned and secular mindfulness, especially its selective appropriation of Buddhist thought and practice, as well as the way in which mindfulness is increasingly used by a cynical and manipulative corporate culture. Purser has since published other articles in the same vein (e.g., Purser and Ng 2015, 2016a, b, c) which has made him and his co-authors somewhat controversial figures within the mindfulness movement.

Loy and Purser note, initially with appreciation, that mindfulness has become a part of the North American mainstream, and that this kind of meditation today is commonplace in large corporations and government agencies, schools, prisons, and even the military. "Millions of people are receiving tangible benefits from their mindfulness practice: less stress, better concentration, perhaps a little more empathy," write the authors, before adding that there is a darker side to the "mindfulness boom." What they particularly oppose is the secularization of mindfulness which, paradoxically, arguably has been a necessary condition for its success and widespread application. Decoupled from its ethical and soteriological context, it is argued, this Buddhist-derived, contemplative practice loses its radical, emancipatory potential, and what remains is not much more than a self-help technique for dealing with psychosomatic disorders and foster more focused (and thus productive) middle managers.

That recontextualized, therapeutic or medicalized mindfulness is a watered-down version of "real" Buddhist practice has been a fairly common (although not undisputed) form of criticism, coming mainly from "traditional" Buddhists.[12] In

mindfulness practice, it is claimed, the actual purpose of Buddhist practice (namely liberation or enlightenment) is substituted for comparatively trivial goals, such as well-being and stress reduction.[13] The argument has also often been made that traditional (or "religious") Buddhist practice is based on a triad of meditative absorption (*Samadhi*), ethical conduct (*sila*), and insight (*prajna*), and, furthermore, that a meditation practice lacking the latter two aspects easily degenerates into an unproductive, narcissistic pursuit.

Loy and Purser go a step further in their criticism when they point out that today's mindfulness is often used in a way that is not only ineffective in accessing the deeper causes of human misery (according to Buddhism the three "poisons" of greed, anger, and delusion), but rather *strengthens* these causes. How? In situating concentration, relaxation, well-being, and *gelassenheit* within the same economic system that is dependent on—and indeed can even be said to produce—these very poisons.

Many mindfulness enthusiasts seem to assume that the cultivation of "mindfulness" through meditation practice in itself is either a value-neutral training, or something that will automatically produce positive ethical consequences. To this point, Loy and Purser mention that an important distinction is made in classical Buddhism between "right attention" (*samma sati*) and "wrong attention" (*miccha sati*).

That mindfulness has become so popular in the corporate world can, obviously, be explained by the fact that the practice is not only marketed as a way to increase employees' concentration and thus their productivity. Mindfulness is also advertised as a kind of respite from the modern world of work insecurity and competition. When a worker's unhappiness and stress persist despite

---

(Footnote 12 continued)
more "traditional" varieties to a great extent is shaped by "modernist" ideals (McMahan 2008).

[13]Lopez (2012) argues, for example, that stress reduction is not a traditional goal of Buddhist meditation. Many of its forms seem rather intended to evoke a kind of existential crisis and should rather be described as a way to create stress than a means of relieving it.

---

[12]Terms such as "real" or "traditional" Buddhism (or Buddhists) should be used with great care, if at all. Not least because much contemporary Buddhism, including

mindful breathing exercises, and despite attentively chewing raisins, it is now understood that the responsibility lies with the individual, specifically with her lack of mindfulness. Here, one could also add that the often-repeated encouragement to assume an "accepting," "non-judgmental," and "non-reactive" attitude, of course, fits like a glove for the employers who want their employees to passively accept the social and economic *status quo* of the workplace.

As was mentioned above, *Beyond McMindfulness* attracted considerable attention and sparked several lively debates. Although many commentators expressed agreement with the author's argument, one can easily get the impression that the larger debate that followed its publication confirms what Loy and Purser write at the end of the article: That many mindfulness practitioners and advocates consider "a more ethical and socially responsible view of mindfulness" as an irrelevant and unnecessarily politicized criticism of a "personal journey of self-transformation."

Even if sympathetic to the authors' analysis, one could argue that they make a mistake by injecting Buddhist *ideals* into a late-capitalist *reality*. It has been argued, for instance, that today's Western Buddhism often serves a very similar, if not identical, ideological function as do secular mindfulness (Wallis et al. 2013; Pepper 2014a). A critique similar to that presented in *Beyond McMindfulness* could also be directed against certain features of contemporary, Asian Buddhism. One example would be the Japanese Zen establishment; even though it no longer actively supports brutal militarism (Victoria 2006), it is still fairly common for Japanese companies to send their employees to Zen temples in order to cultivate self-discipline, endurance, conformity, and obedience (Victoria 1997).

Depression, Perennialism, Anti-Intellectualism, and the Specter of *Atman*

The third example of online mindfulness critique is neither an article nor a blog post, but a short paper by Robert H Sharf with the title *Mindfulness or Mindlessness: Traditional and Modern Critiques of "Bare Awareness."* Presented at the Division of Social and Transcultural Psychiatry's Advanced Study Institute at McGill University in 2013, it was recorded on video and later posted on YouTube.[14] As Sharf's paper challenges several common notions regarding both mindfulness and Buddhism in general, it has generated some debate although the video clip has not received the same kind of attention as, say, *Beyond McMindfulness*. Nevertheless, this is also an excellent and accessible introduction for anyone interested in situating today's mindfulness in an historical context and that of Buddhist doctrine and practice.

Sharf, professor of Buddhist Studies at Berkeley University, is perhaps most well known as the author of a few, oft-quoted articles critiquing popular views on Zen and Buddhism in general (e.g., Sharf 1993, 1995a, b). *Mindfulness and Mindlessness* can be seen as a continuation of that work, and—just like Loy and Purser—he problematizes the relation between the Buddhist tradition and mindfulness. Sharf does this from a different perspective, however, historically informed and highly critical of the modernist understanding of the Buddhist tradition which has been so important for the formation of today's mindfulness discourse. It is pointed out, for example, that today's mindfulness, contrary to a common claim, is *not* the essence of a 2500-year-old, unbroken tradition. Rather, today's mindfulness is a development of the assemblage of ideas and practices known as "modernist" or "Protestant" Buddhism—a reform movement born out of the meeting of Asian Buddhism and Western colonialism and missionary activities during the 19th century.

Sharf's exposition covers a lot of ground in a somewhat rhapsodic fashion and I will focus on two issues of particular interest. The first concerns the relation between Buddhist practice, mental health, and happiness. The second deals

---

[14]https://www.youtube.com/watch?v=c6Avs5iwACs. A written version was later published in *Transcultural Psychiatry* (Sharf 2015). See also Sharf (2014).

with the concept *sati* or "mindfulness" itself, and how it has been connected with ideas of unmediated, or "bare" awareness.

Mindfulness and Mindlessness opens with a discussion of Buddhism and depression, taking as its starting point an essay by Sri Lankan anthropologist Gananath Obeyesekere (Obeyesekere 1985), where a contemporary, Western description of depression which places a "generalization of hopelessness" at the core of the disorder, is juxtaposed with an orthodox (Theravada) Buddhist outlook. The similarities are striking, but Obeyesekere's (and Sharf's) point is obviously not that depression (at least in a modern, clinical sense) would be particularly common among Sri Lankan Buddhists, or that their religion aims at to bring on clinical depression. Rather, Obeyesekere suggests the experience of hopelessness and loss in Western society exists in a free-floating manner, while in different social contexts (in this case a traditional Theravada Buddhist one), it can be anchored to a shared ideology. This is not the place to go into the cross-cultural implications of this, but rather use it as a starting point for a short discussion of how the popular Western image of Buddhism has changed over time and thus paved the way for the idea of it as a "science of happiness" which fits so well with the mindfulness ethos.[15]

It is not that long ago since Western textbooks would describe Buddhism as a life-denying, pessimistic, or nihilistic religion. While it is easy to dismiss such descriptions today, we should perhaps ask ourselves if our current image of Buddhism has not gone too far in the opposite direction. In recent decades, Buddhism has often been presented as a kind of "happiness project," symbolized by laughing monks rather than stern, emaciated ascetics. Among other things, this version of Buddhism allows affluent sympathizers to enjoy its privileges while, at the same time, upholding a detached, cynical distance toward the vicissitudes of samsaric existence. This is why philosopher Žižek (2001b) has described

Western Buddhism as the "perfect ideological supplement" to capitalist dynamics. Hardly surprising, it is also this kind of Buddhism which has inspired the mindfulness movement.

Meditation, which in most Buddhist traditions was an activity engaged in by only a small elite of religious specialists and ascetics with the explicit purpose of cutting all ties with the world, is here presented as a method for improving health, well-being, and professional performance (as well as our sex life and golf swings). Sharf reminds us that the orthodox, Theravada Buddhist, outlook is rather dark. To be alive means that we are suffering, and the only way out is liberation from *samsara*, which demands that we abandon all hope of finding lasting happiness in worldly existence. As a contrast to current, sanguine ideas of meditation, he goes on to quote a passage from Buddhaghosa's classic *Visuddhimagga* with its descriptions of the fearful stages (so-called *dukkha nana*) advanced yogis traverse before attaining final liberation.[16]

Even though one should be careful not to take classic meditation manuals at face value (cf. Sharf 1995b), it is worth considering that the canonical literature often describes the Buddhist path as one filled with fear and loathing, and that the idea of Buddhist meditation as remedy for depression, chronic pain, substance abuse, personality disorders and whatnot is an entirely new phenomenon.

Like many other critics of mindfulness, Sharf admits that it may have some therapeutic value, and he mentions the "substantial body of empirical (if contested) data, that suggest it does." He adds, though, that many years' contact with experienced meditators has made him

---

[15]For an insightful and thought-provoking elaboration on Sharf's paper and Obeyesekere's essay, see Tom Pepper's *Nirvana and Depression* (in Pepper 2014a).

[16]Interestingly enough, there seems to be a growing interest in these stages within certain groups of (predominantly younger, North American) convert Buddhists. Considered unavoidable, these *dukkha nana* are here referred to as "the dark night of the soul," a concept borrowed from the Spanish 16-th Century mystic St. John of the Cross, and are often described in terms reminiscent of clinical descriptions of depression, anxiety, and depersonalization.(cf. Ingram 2008). Some meditation researchers have also showed an interest in this "dark side of meditation" and its implications for intensive or prolonged mindfulness practice.

skeptical; not only do they exhibit behaviors at odds with common notions of what constitutes mental health—even more important, perhaps, is that they likely "do not aspire to our model of mental health in the first place." And this, Sharf concludes, is a real challenge when we want to understand the connection between Buddhist meditation and its desired outcome.[17]

When the continuity between the Buddhist tradition and today's mindfulness is emphasized, we are often reminded that "mindfulness" is a translation of the Pali term *sati* (Skt. *Smṛti*), a central concept within canonical Buddhism. The English word has become the standard translation, even though *sati* literally means "memory" or "remembrance" (cf. Gethin 2011). As this memory often has to do with remembering one's goals as a Buddhist practitioner, the term could possibly also be translated as "alignment" or something similar. Within "modernist" Buddhism and among mindfulness practitioners, however, there has been a strong tendency to interpret *sati*/mindfulness as "bare awareness." One example would be the descriptions of mindfulness as a kind of "pure witnessing" that is something radically different than thinking itself.

Sharf points out that this approach to meditation as a "non-judgmental, non-discursive attending to the moment-to-moment flow of consciousness" has a long history in Buddhism; it can be found in the Chan/Zen[18] and Dzogchen traditions, and it was prominent in the "modernist" interpretation of the Theravada school which is the foundation of contemporary Vipassana practice. It is important to know, however,

that this tendency always has been controversial and that it, contrary to what is commonly assumed today, cannot be said to be representative of the entire Buddhist tradition.

Sharf also suggests that the underlying ideology of mindfulness could be seen as an example of what scholars of religion call "perennialism." (The idea that mystics in all times and places have had access to a common experience which is "unconstructed" and not conditioned by social, cultural, historical, and linguistic influences.) Today's mindfulness seems to be particularly influenced by a version of perennialism which Sharf calls the "filter theory"—an almost logophobic idea that our normal, conditioned discursive processes do not connect us with reality but rather function as a filter, locking us out from it. The purpose of a contemplative practice, then, is understood as a kind of radical *de-conditioning*, rather than the *re-conditioning* or gradual change in perspective and outlook that characterizes more traditional, monastic forms of Buddhist practice.

Here, I would suggest, we find a clue to the anti-intellectualism found within both Western Buddhism and the mindfulness movement, and which has been the target of much criticism from the Speculative Non-Buddhism movement (e.g., Pepper n.d.). Both revolve around the notion that the roots of human suffering are to be found in destructive, individual patterns of thought, but have curiously little to say about the art of thinking *better*. The solution, rather, seems to be to create a distance to one's own thinking process, or even to think *less*. Thinking itself is seen (thought of!) as a hindrance and the major cause of suffering.

It should come as no surprise, then, that brain scientist Taylor's enormously popular book (2008) and TED video, where she describes the debilitating effects of a stroke as a taste of Buddhist nirvana is so often mentioned approvingly by Western Buddhists and proponents of mindfulness.

In a comment on *Mindfulness or Mindlessness*, Pepper (2014b) also argues that the ideal of "bare awareness" presupposes some kind of uncreated, pure, and transcendent mind or

---

[17]*Pace* Sharf, I would add that the hope of improving psychological health and emotional well-being often seems to be an important motivation for taking up intensive meditation practice, even within a monastic regimen.

[18]See Hori (2000), however, for a problematizing discussion about the notion of "pure" or "prediscursive" awareness in the context of orthodox Rinzai Zen. Here, I could also add that I have noticed that the word "mindfulness" is used fairly often in the context of Zen practice, but that it has less to do with an inward focus on thoughts and feelings than what could be seen as its opposite: paying attention to the task at hand, be it *zazen* meditation or everyday activities.

consciousness, in other words, a soul or *atman*. For anyone not subscribing to such beliefs, the promise of attaining any kind of "mindful" awareness *sub specie aeternitatis* must appear fraudulent, or at least misleading. A person engaging in this kind of futile exercise will likely experience herself as a failure. Alternatively, she will "succeed"—but only by mistaking the "observing self" for some kind of transcendental awareness.

The perennialist and quietist ideology shared by the mindfulness movement and much Western Buddhism is not simply a question of highly abstract, metaphysical, sometimes mystified, assumptions. It has practical and political consequences as well, "ethically dubious and politically reactionary," Sharf suggests, with references to Arendt, Levinas, and the Japanese "Critical Buddhism" movement. As an example he mentions the North American, Buddhist quarterly *Tricycle* with its advertisements for all kinds of "dharmic" commodities, and the similarly entrepreneurial and commercial spirit with which mindfulness programs are marketed.

But It Works, Doesn't It?

After presenting a number of critical perspectives on the mindfulness phenomenon, it seems only fair to end with a few words about the counter-arguments put forward by representatives of the mindfulness movement.

The most common response to the kind of critiques summarized above seems to be the assertion that, in the final analysis, mindfulness actually *works*. Suffering individuals really *are* being helped by mindfulness, we are being told. Never mind that it is less than clear as how or why this is, or even what, exactly, is meant by "mindfulness." Mindfulness "works," even though (or perhaps just because) it helps shaping exactly the kind of obedient, quietist, detached subjects demanded by the market. It "works," even though expensive mindfulness retreats and courses are marketed in the same vulgar and hyper-active way as any other commodity. Mindfulness "works," even though it is not what

we are told ("The essence
"2500-year-old technique
and joy," and so on). It
"non-judgmental aware
social, cultural, and ling
impossibility or would re
or *atman*. And so on.

Relatively few critics of mindfulness and the mindfulness industry have challenged this claim that mindfulness "works." Indeed, the idea that "meditation is good for you" has become so axiomatic that it would seem absurd to question it. Which is obviously a good reason to do exactly that.

The "growing body of evidence" to the efficacy of mindfulness mentioned by Sharf is often invoked by representatives of the mindfulness movement. But although there exists an abundance of studies which seems confirm many of its claims, there are also good reasons to take them with a grain of salt. Several meta-studies (e.g., Ospina et al. 2007; Goyal et al. 2014) mention several methodological flaws common to studies of meditation and mindfulness, including research bias, a lack of active reference groups, and insufficient attention to placebo effects. In a recent interview (Heuman 2014), Willoughby Britton (meditation researcher at Brown University Medical School) acknowledges the problematic nature of many such studies and also mentions several adverse effects of meditation practice, such as depression, confusion, and depersonalization, which until recently have received only scant attention in the scientific literature.

Another, naïve but surprisingly common, response from its proponents is that mindfulness, in itself, is a "pure" or neutral technique, but that its critics are motivated by some sinister agenda or "ideology." An obvious (and very Zizekian) reply would be that this is a good example of "ideology at its purest." It is exactly when we posit our beliefs and practices outside of ideology that ideology functions most effectively. One could also argue that there simply is no thing such as mindfulness-in-itself, but that these practices, ideals, and approaches labeled always are embedded in a social and cultural context

178

...akes them meaningful and comprehen-... Mindfulness has a very specific history, ...en hidden behind layers of mystification,[19] and there are good reasons why we should investigate its ideological functioning.

Let me end on a personal note. I am obviously very skeptical toward the mindfulness phenomenon, and I find some aspect of the mindfulness industry quite repulsive. Even so, I have been a bit hesitant to attack (if by proxy) an activity which, besides the obvious charlatans and peddlers of snake oil, engages many well-meaning and sometimes idealistic individuals. But, as sociologist Roland Paulsen (2008) writes about a similar phenomenon, it is an important task to "critically analyze their frauds and castles in the air and call them by their right names." I hope this contribution can serve to ignite a critical discussion—If mindfulness really has something of value to offer, its proponents won't have anything to lose, except a number of cherished illusions.

References

Åkesdotter, C. (2011). *Heartfulness: din väg till glädje i nuet*. Linköping: Heartfulness Inspiration.

Andersen Cerwall, H., & Stawreberg, A-M. (2009). M*indfulness för föräldrar*. Stockholm: Bonniers.

Bodhi, B. (2011). What does mindfulness really mean? A canonical perspective. *Contemporary Buddhism: An Interdisciplinary Journal, 12*(01), 19–39.

Carrette, J., & King, R. (2005). *Selling Spirituality: The Silent Takeover of Religion*. London: Routledge.

Chandler, D. (n.d.). *Semiotics for beginners* (Online version of the author's *Semiotics: The Basics*). http://visual-memory.co.uk/daniel/Documents/S4B/

Drougge, P. (2014). Anteckningar inför en stundande backlash. Mindfulness som ett medelklassens opium. In K. Plank (Ed.), *Mindfulness: Tradition, tolkning och tillämpning*. Lund: Nordic Academic Press.

Engström, M. (2009a). *Mindfulness i livet för mannen: guidade meditationer*. Acora: CD. n.p.

Engström, Maria. (2009b). *Mindfulness i livet för kvinnan: guidade meditationer*. Acora: CD. n.p.

Gethin, R. (2011). On some definitions of mindfulness. *Contemporary Buddhism: An Interdisciplinary Journal, 12*(1), 263–279.

Gilbert, P. (2009). Introducing compassion-focused therapy. *Advances in Psychiatric Treatment, 15*, 199–208.

Goyal, M. et al. (2014). Meditation programs for psychological stress and well-being: a systematic review and meta-analysis. *JAMA Internal Medicine 174*(3): 357–368. Also: http://archinte.jamanetwork.com/article.aspx?articleid=1809754

Hanh, T. N., & Cheung, L. (2010). *Savor: Mindful eating, mindful life*. New York: HarperCollins.

Heuman, L. (2014). Meditation nation. *Tricycle blog*, April 25. http://tricycle.org/trikedaily/meditation-nation/

Hori, G. Victor Sōgen (2000). Koan and Kensho in the Rinzai Zen curriculum. In S. Heine & D.S. Wright (Eds.), *The Koan: ÎÊTexts and contexts in Zen Buddhism* (pp. 280–315). New York: OXford University Press.

Hornborg, A-C. (2012). *Coaching och lekmannaterapi: en modern väckelse?* Stockholm: Dialogos.

Ingram, D. M. (2008). *Mastering the core teachings of the Buddha: An unusually hardcore dharma book*. London: Aeon. Pdf-version available at: http://static.squarespace.com/static/5037f52d84ae1e87f694cfda/t/5055915f84aedaeee9181119/1347785055665/

Iwamura, J. N. (2011). *Virtual orientalism: Asian religion and American popular culture*. Oxford: Oxford University Press.

Jameson, F. (2003). The end of temporality. *Critical Inquiry, 29*(4).

Kabat-Zinn, J. (2005). *Coming to our senses: Healing ourselves and the world through mindfulness*. New York: Hyperion.

Kaye, L. (1997). *Zen at work: A Zen teacher's 50-year journey in corporate America*. New York: Random House.

Lopez, D. S. (2008). *Buddhism & science: A guide for the perplexed*. Chicago: University of Chicago Press.

Lopez, D. S. (2012). *The scientific Budha: His short and happy life*. New Haven: Yale University Press.

Low, A. (1976). *Zen and creative management*. Garden City: Anchor Press.

Loy, D. (1996). *Lack and transcendence: The problem of death and life in psychotherapy, existentialism, and buddhism*. Atlantic Highlands: Humanities Press.

Loy, D. (2002). *A buddhist history of the West: Studies in lack*. Albany: SUNY Press.

Loy, D. (2003). *The great awakening: A buddhist social theory*. Boston: Wisdom Publications.

Loy, D. (2008). *Money, sex, war, karma: Notes for a buddhist revolution*. Boston: Wisdom Publications.

Lundgren Öhman, C., & Schenström, O. (2011). *Mindfulnessövningar 4—7 år*. CD. Luleå: Mindfulness Center.

McMahan, D. L. (2008). *The making of buddhist modernism*. Oxford: Oxford University Press.

Obeyesekere, G. (1985). Depression, buddhism, and the work of culture in Sri Lanka. In: A. Kleinman & B. Good (Eds.), *Culture and depression. studies in the anthropology and cross-cultural psychiatry of affect and disorder*. Berkeley: University of California Press.

Ohlsson, A. (2008). *Myt och manipulation. Radikal psykiatrikritik i svensk offentlig idédebatt 1968—*

---

[19]On the issue of mystification of mindfulness, see Wilson (2014).

1973. Diss. Stockholm: Acta Universitatis Stockholmiensis.

Ospina, M. B. et al. (2007). *Meditation practices of health: State of the research*. Evidence report/technology assessment no. 155. AHRQ Publication no. 07-E010. Rockville: Agency for Healthcare Research and Quality.

Palmkron Ragnar, Å., & Lundblad, K. (2016) *Mindful eating* (sic)*: Ät bra, må bra, gå ner i vikt med mindfulness, compassion och KBT* n.p: Argos.

Paulsen, R. (2008). Via Negativa. En antiauktoritär läsning av Kaj Håkanson. In S. Wide, F. Palm & V. Misheva (Eds.), *Om kunskap, kärlek och ingenting särskiljt. En vänbok till Kaj Håkanson*. Uppsala: Sociologiska institutionen, Uppsala universitet.

Payne, R. K. (2014). *Corporatist spirituality*. https://rkpayne.wordpress.com/2014/02/18/corporatist-spirituality/

Pepper, T. (n.d.). On buddhist anti-intellectualism. *Non +x*. (3). http://www.nonplusx.com/app/download/503005004/On+Buddhist+Anti-Intellectualism.pdf

Pepper, T. (2014a). *The faithful buddhist*. n.p. (Self-published e-book.).

Pepper, T. (Ed.). (2014b). Mindfulness or mindlessness. *The faithful buddhist*.

Purser, R., & Loy, D. (2013). Beyond McMindfulness. *Huffington post*. http://www.huffingtonpost.com/ron-purser/beyond-mcmindfulness_b_3519289.html

Purser, R., & Ng, E. (2015). Corporate mindfulness is bullsh*t: Zen or no Zen, you're working harder and being paid less. *Salon.com*. http://www.salon.com/2015/09/27/corporate_mindfulness_is_bullsht_zen_or_no_zen_youre_working_harder_and_being_paid_less/

Purser, R., & Ng, E. (2016a). Cutting through the corporate mindfulness hype. *Huffington post*. http://www.huffingtonpost.com/ron-purser/cutting-through-the-corporate-mindfulness-hype_b_9512998.html

Purser, R., & Ng, E. (2016b). Mindfulness and self-care: Why should i care? Part 1. http://www.huffingtonpost.com/edwin-ng/mindfulness-and-self-care-why-should-i-care_b_9613036.html

Purser, R., & Ng, E. (2016c). Mindfulness and self-care: Why should i care? Part 2. http://www.patheos.com/blogs/americanbuddhist/2016/04/mindfulness-and-self-care-why-should-i-care.html

Rogers, S. L. (2009). *Mindfulness for law students: Using the power of mindfulness to achieve balance and success in law school*. Mindful Living Press.

Sharf, R. H. (1993). The Zen of Japanese nationalism. *History of religions 33*(1), 1–43. (Reprinted from *Curators of the Buddha: The study of buddhism under colonialism*, by D. S. Lopez, Ed., Chicago: University of Chicago Press).

Sharf, R. H. (1995a). Sanbōkyōdan. Zen and the way of the new religions. *Japanese Journal of Religious Studies, 22*(3–4), 417–458.

Sharf, R. H. (1995b). Buddhist modernism and the rhetoric of meditative experience. *Numen, 42*, 228–283.

Sharf, R. H. (2014). Mindfulness and mindlessness in early chan. *Philosophy East and West, 64*(4), 933–964.

Sharf, R. H. (2015). Is mindfulness buddhist? (and why it matters). *Transcultural Psychiatry, 52*(4), 470–484.

Stephens, L. (2012). *The mindful dog owner: What your dog is teaching you about living enlightenment*. Verbatim.

Taylor, J. B. (2008). *My stroke of insight: A brain scientist's personal journey*. New York: Viking.

Terjestam, Y. (2010). *Mindfulness i skolan: om hälsa och lärande bland barn och unga*. Lund: Studentlitteratur.

Victoria, B. (Daizen). (1997). Japanese corporate Zen. In J. Moore (Ed.) *The other Japan: Conflict, compromise, and resistance since 1945*. Armonk: M.E. Sharpe. Bulletin of concerned Asian scholars.

Victoria, B. (Daizen). (2006). *Zen at war*. Lanham: Rowman & Littlefield.

Wallis, G. (2002). *Mediating the power of Buddhas: Ritual in the Mañjuśrīmūlakalpa*. Albany: SUNY Press.

Wallis, G. (2004). *Basic teachings of the Buddha*. New York: Modern Library.

Wallis, G. (2007). *The Dhammapada. Verses on the Way*. New York: Modern Library.

Wallis, G. (2012). Elixir of mindfulness. *Non+x. Issue 2*. http://www.nonplusx.com/issues-1-4/

Wallis, G., Pepper, T., & Steingass, M. (2013). *Cruel theory—sublime practice: Toward a revaluation of buddhism*. Roskilde: Eyecorner Press.

Webster, D. (2012). *Dispirited: How contemporary spirituality makes us stupid, selfish and unhappy*. Winchester: Zero Books.

Wilson, J. (2014). *Mindful America: The mutual transformation of buddhist meditation and American culture*. Oxford: Oxford University Press.

Zaretsky, E. (2013). *Psykoanalysen och kapitalismens anda. Fronesis, 44–45*, 53–79.

Žižek, S. (2001a). *On belief*. London: Routledge.

Žižek, S. (2001b). From Western Marxism to Western Buddhism. *Cabinet Magazine*, (2). http://www.cabinetmagazine.org/issues/2/western.php

Is There a Corporate Takeover of the Mindfulness Industry? An Exploration of Western Mindfulness in the Public and Private Sector

Christopher Titmuss

A friend in the USA kindly provided me with a copy of the February, 2014 issue of *Time* magazine, the North American weekly news publication. This particular issue highlighted the explosion of interest in mindfulness to the degree that mindfulness has become an industry.

The cover of the magazine showed a beautiful woman, probably a model, displaying a mindful presence. The main story of that issue addressed what the magazine referred to as the *Mindful Revolution*. The arrival of mindfulness on the cover of this particular widely read publication confirmed the penetration of mindfulness into the public and private sectors of society.

While reading the article, my mind flashed back to 1979. Jon Kabat-Zinn, a microbiologist and dedicated Buddhist meditator, who lived in Boston, had participated in a number of retreats with myself and other Dharma teachers at the Insight Meditation Society in Boston, Massachusetts, USA. During the retreat, he had a sudden insight that he could adapt these teachings and practices for patients in hospital dealing with physical pain and suffering stress and anxiety. In a one to one interview on the retreat, after this insight, Jon outlined his vision and asked for my response. I gave him full support. I felt he was ready to take on such a challenge. A mutual friend and fellow meditator, Brian Tucker in Boston, joined Jon, along with other meditators, to establish his first administrative team.

Jon provided the standard definition for secular mindfulness in the West: *'Mindfulness means paying attention in a particular way; on purpose, in the present moment, and nonjudgmentally.'*

In *The Psychology of Awakening* (1999), Jon Kabat-Zinn wrote a chapter for the book on the early developments of his Mindfulness-Based Stress Reduction (MBSR) program that he created. In his chapter, he explained:

> The way I propose to do this is to tell you something about my own life trajectory and work. In school, at least in the United States at that time, you were constantly being evaluated and judge the how you performed, and hardly ever acknowledged as being a whole person. At gatherings of professionals and at parties, the common way for male intellectuals and academics to reach out in conversation was to say 'Where are you? Translated, that meant.' 'What recognisable institution are you affiliated with?' That, instead of perhaps' How are you? Or "Who are you? It was a kind of discourse, a way of relating, that I always had a great deal of trouble with. I often have the impulse, which I usually recognised as hostile and kept in check, to say Why I'm standing in front of you! Where do you think I am? When I came across meditation and the consciousness disciplines, they meant an enormous amount to me, in part because they emphasise so much a clear seeing an acceptance of the present moment rather than being so caught up in one's head that one literally live there full-time. I dropped into meditation (that's another story), and started practising as much as I could. It was love at first sight."

C. Titmuss (✉)
Totnes, Devon, England, UK
e-mail: christopher@insightmeditation.org

© Springer International Publishing Switzerland 2016
R.E. Purser et al. (eds.), *Handbook of Mindfulness*,
Mindfulness in Behavioral Health, DOI 10.1007/978-3-319-44019-4_13

A Flash on a Retreat in 1979 of 'Jon Kabat-Zinn'

Much of the direction our work at the hospital has taken over the past seventeen years came to me in a flash, maybe lasting 15 seconds on a retreat in the spring of 1979 at the Insight Meditation Society in Barre, Massachusetts. The retreat was led by Christopher Titmuss and Christina Feldman, who are guiding teachers at Gaia House in Devon. The flash had to do with the question of how to take the heart of something as meaningful, as sacred if you will, as Buddha Dharma and bring it into the world in a way that doesn't dilute, profane or distorted, but at the same time is not locked into a culturally and tradition bound framework that would make it absolutely imperishable to the vast majority of people who are nevertheless suffering and who might find it extraordinarily useful and liberative.

Jon then continues to explore further in the chapter the remarkable developments of his mindfulness programs. During the 1980s, I recall following with much interest the remarkable benefits of his 10-week mindfulness training course for patients in the basement of Worcester hospital. On visits to Boston, I would stay for a couple of days at the home of Jon and Myla, his wife. Gwanwyn (my daughter's mother) and I drew inspiration for the name of our daughter, Nshorna, from Naushon, the name of the daughter of Jon and Myla. Naushon, an American Indian word that means 'Spring' or 'Planting Season' and also the name of a small island near Martha's Vineyard. Gwanwyn is the Welsh word for 'Spring.' So we gave our baby daughter (born July 3, 1981) the same name with a slightly different spelling, and added an 'a' to the end of the name. Mother and daughter then shared the same name in different languages.

Teachers in the Buddhist tradition recognize that new teachers emerge out the depth of exploration of certain meditators, who wish to share their understanding and experiences with others. A person steps into the role of a Dharma teacher through the support and sanction of a teacher. The tradition considers it questionable to launch oneself into such a teaching role without the endorsement from a senior teacher.

Sometimes, the new teacher requires the guidance of his or her teacher(s) to develop in the role. A senior teacher then acts as a mentor and ongoing support for the new teacher. To his credit, Jon had years of practice in both the insight meditation tradition and Zen tradition. He could stand on his own two feet to develop the forms, methods, and techniques suitable for patients in the medical center. He did not need a mentor. If a senior teacher expresses reservations about the readiness of a retreatant to be a teacher, he or she must express those reservations. I had no reservation in giving Jon the strongest encouragement. Of course, he might well have gone ahead in any case or, if necessary, received acknowledgment elsewhere. I recall that the vast majority of the teachers at the Insight Meditation Society (IMS) were supportive of his initiative.

During the 1980s, Jon and his team engaged in thorough scientific studies to show the direct benefits to patients of the fledgling Mindfulness-Based Stress Reduction (MBSR) program. There was growing appreciation for Jon's tremendous work to reduce significantly the pain level of patients and also offer practices to cut their stress about their present and future situation.

Jon told me that he gave patients a comprehensive questionnaire before they embarked on his stress reduction course. From his Buddhist training, he knew the importance of motivation and the necessity to sustain it, especially through the difficult times. It meant that once he had established that his patients were fully committed to MBSR, he invited them to join the course with practices in the basement of the hospital as well as at home. This initial preparation in mindfulness training sets the tone for the years ahead.

I remember having a conversation with another mutual Dharma friend, Larry Rosenberg, the founder of the Insight Meditation Center in Cambridge, Massachusetts about the remarkable work of Jon, a friend of Jon for years. In the early 1980s, Larry said to me that one day Jon would appear on the cover of *Time* magazine. *Time* might have chosen Jon as a foremost pioneer of the mindfulness secular movement in the USA for its cover but instead opted for a beautiful, slim blonde as a mindfulness pinup.

Mindfulness in the Public Sector

Throughout the 1980s and 1990s, various mindfulness programs, including MBSR, gathered momentum in the public sector, especially in hospitals, clinics, schools, and prisons. A growing number of practitioners deepened their skills in mindfulness in order to facilitate regular classes in public institutions. The benefits of mindfulness practice began spreading throughout North America and other Western countries. Management and professionals in the health service appreciated the benefits for their patients and clients. They also saw mindfulness training as a real resource for themselves as well. The numerous demands in public service often have its toll on staff.

Medical staff took part in classes and watched and listened to video/audio teachings and guided meditations. More and more people took part in the training setting aside time daily for mindfulness/meditation practices from a few minutes to 30 min or more. These mindfulness exercises also included movement as a therapy to stretch out the various limbs of the body to release tension, dissolve contraction, and experience waves of emotions coming and going.

I recall a mindfulness teacher in Israel, a former Buddhist nun based in Thailand, offering mindfulness classes to young children in primary school. The children would hold a flower at arm's length to breathe mindfully in and out the fragrance. The child would lie on the yoga mat on the floor with four children mindfully holding one of the hands or one of the feet of the child. Child would chew on a raisin to experience the different flavors of the taste. With these classroom exercises every few days, the sense of friendship and connection developed between the children. Teachers reported that the children not only enjoyed the mindfulness practices but found they could concentrate more in the class. Best of all the mindfulness classes reduced bullying and fear within the class, as the children learned to trust each other through getting closer to each other.

These stories of the power of mindfulness spread to other schools in Israel with requests for such classes. The same kind of development took place in many cities and towns around the world. Mindfulness then became another string in the violin to contribute to a holistic education with connection between the pupils, and between the pupils and teacher.

University students began reading newspaper reports, articles and essays on the power of mindfulness with its roots in the Buddhist tradition and its secular application. These university students began to take up mindfulness as a subject for their thesis, while more and more school teachers, psychologists, psychotherapists, and social workers were attending Buddhist retreats, especially in the Theravada tradition, to experience mindfulness training alongside strict insight meditation practice.

Some of the thesis for a BA, MA, or PhD became available on the Internet. The range of thesis around mindfulness and its widespread application gave extra authority and credence to mindfulness as a legitimate branch of psychology. The tools for the application of mindfulness marked a significant shift away from the reliance upon talking therapy. People could practice mindfulness so they find and maintain calmness, clear comprehension, and a genuine sense of well-being.

Whether sitting, walking, standing, or reclining, these mindfulness practices empowered practitioners so they could give attention to issues of past, present, and future without increasing their levels of stress. Stress, fears, and worry plague the lives of many people. Some people who were dependent on medication, such as anti-depressants, sleeping tablets, and medication for migraines were able to gradually wean their way off medication.

As a result, people begin to feel stronger in themselves with a greater sense of self-worth. A small number of people, whether employees in public institutions or those in need of support in public institutions, were inspired to deepen their experience and understanding of mindfulness and meditation. This led some to participate in weekend, weeklong, or longer residential Buddhist retreats. These retreats provide the opportunity for sustained mindfulness/meditation practices from early morning to late evening. Many hours of silence throughout the day, formal meditation along with teachings, practices, and one-to-one meetings with the teacher triggered a greater range of experiences, conventional, and spiritual.

Mindfulness and the Brain

The exploration of mindfulness and meditation neatly coincided with the wide range of experiments and observations taking place in neuroscience. The word 'plasticity' entered into the language of neuroscience referring to the adaptability of the brain to changes in its environment, including the inner life. The brain makes an impact on the mind and the mind impacts upon the brain. Various trials continue to be undertaken with beginners of mindfulness and meditation and experienced mindfulness meditation practitioners. Neuroscientists endeavored to measure the way mindfulness practices impact upon the brain.

A team of Harvard University researchers (2011) engaged in a study of a group of participants in an eight-week mindfulness meditation program to measurable changes in brain regions. The study showed that meditation-produced changes over time in the brain's gray matter.

'Although the practice of meditation is associated with a sense of peacefulness and physical relaxation, practitioners have long claimed that meditation also provides cognitive and psychological benefits that persist throughout the day,' said study senior author Sara Lazar, a teacher of psychology at Harvard Medical School.

Some of the pioneers of mindfulness programs regularly make reference to the findings of neuroscience to help justify the benefits of mindfulness and meditation. The tradition has relied upon the direct experience of the meditator rather than science. Generally, Western science continues to believe that matter is the basis of the mind rather numerous causes and conditions serve as the basis for the arising of both mind and matter.

While research in neuroscience generates interesting results, neuroscience relies upon scientific materialism as the authority to determine the benefits of mindfulness/meditation. The Buddhist tradition adopts a very different view. The mindfulness teacher engages in a dialogue with the meditator to find out the depth of experience, the insights, and the enduring value. We do not have to measure the brain for that and then offer scientific and largely abstract papers.

A belief in scientific research can carry more weight than 2600 years of mindfulness and meditation in the Buddhist tradition. There is a forgetfulness that neuroscience has become the new kid on the block with very little real depth of experience of meditation and of the power of mindfulness. Science overlooks the abiding connection of mindfulness/meditation with ethics, spiritual experiences, and the dynamics of the inner and outer. The teachings and practices embrace profound realizations, an understanding of environmental issues, wisdom, and an authentic awakening. Attaching wires to the heads of meditators cannot measure the depth of experience, lasting insights, and liberation.

Rather than spending months, and substantial cost, in scientific research, it would be far more beneficial for neuroscientists to spend the equivalent time in sustained mindfulness/meditation practice through a variety of meditation retreats. The researchers could touch an inner depth unavailable and inaccessible to congested mind of scientific terminology, proliferation of thought, and attachment to notions of measurement. Any measurement of the brain/mind/mindfulness/meditation will change slowly or quickly according to conditions. These measurements will always be impermanent and unreliable.

It is important to reiterate again the immense benefits for individuals, patients, clients, school children, prisoners, and the range of professionals in the public sector. Society needs an army of wise and skilled mindfulness mentors to offer with modest charges their services for the inner well-being of stressed out and unhappy citizens enduring various kinds of pain in mind and body. The increased dependency on medication makes its impact on people with the ongoing side effects through the digestion of powerful chemicals into the organic cells of the body and the sensitivities of heart and mind.

Mindfulness practitioners welcome these ancient practices that contribute in a healthy and natural way to inner well-being, calmness of perception, and empathy with others through meditation and loving-kindness practices. Emotional and psychological health for adults and children need to take priority rather than belief in the quick fix of addictive drugs. To the delight of the pharmaceutical industry, doctors and psychiatrists widely prescribe drugs for painkillers, anti-depressants, and so-called attention deficiency. The numbers who take such drugs grow year by year. Far too much of science, including neuroscience, serves the vested interests of the pharmaceutical industry and the military.

Mindfulness of the Power of Corporations

Along with some other Western nations, the USA has lost its way as a democratic institution with its replacement with an elite to govern a country rather than a democratic will. With their incredible wealth, this elite exercises exceptional political influence. They run powerful corporations, the major banks, and many become influential figures in public office.

With the regular ritual of the vote, citizens continue to believe that they exercise influence in terms of who they vote into high office. Whether for the Republican or Democratic Party, the votes of citizens barely make any real difference to policy. Differences between the two major political parties appear wafer thin. Influential lobby groups and huge corporations set the agenda—regardless of who sits in the White House or who is elected as a senator or congressman. Elected politicians depend upon the wealth and influence of the elite, especially the oligarchy that runs much of the media.

Decade after decade, the phenomenal gap between the super-rich and the rest of society grows bigger due to the excessive profit on consumer goods, services, and large-scale investment in the stock market. Research in Princeton University shows that the privileged elite grabbed 95 % of all income gains while the rest of citizens experience a 12 % drop in their income.

In the USA, several major corporations have received a total taxpayer bailout running into trillions of dollars from the Federal Reserve and the Treasury Department. These staggering sums of money are gleaned from the middle classes and poor to support powerful corporations. These companies continue to employ every loophole available to evade paying taxes through tax havens in certain overseas countries such as Luxemburg and the Bahamas and private off-shore accounts. These unpaid taxes should go to the poor, the sick, the elderly, the young, schools, hospitals, the infrastructure, environmental needs, and overseas aid. Full payment of taxes shows corporate ethics, social responsibility, empathy, and compassion for poor members of society.

The application of mindfulness in corporations has failed to address widespread public concerns about the massive sums involved in tax avoidance. There is a long list of elite and powerful businesses who pay little or no tax. They include Amazon, Apple, Bank of America, Google, Goldman Sachs, JP Morgan Chase, General Electric, Boeing, Microsoft, Monsanto, Starbucks, Google, McDonald's, Caesars entertainment, Exxon, and numerous others.

Many of these major corporations bring in mindfulness trainers for management to reduce their stress levels and develop the power of attention in order to focus more clearly on the aims of the corporation to achieve targets in a focussed and relaxed environment as possible. These corporations have welcomed with open arms the explosion of interest in mindfulness. They approve wholeheartedly of the standard Western definition of mindfulness namely to 'be in the present moment, non-judgmentally.' This definition absolves the company from any criticism from the staff.

The CEOs and their boards of directors offer mindfulness tools to various people in the company since these workshops exclude enquiry into the causes for the widespread stress in office life, the factory or warehouse. The corporations, who offer mindfulness programs, know they do not have to face the coupling of mindfulness with inquiry and ethics which contribute to wisdom. Mindfulness workshops in corporations do not address stress due to the outer demands upon the staff. There is no exploration in the mindfulness industry of the pressure upon white and blue collared workers at home or abroad to achieve corporate goals.

The current definition of mindfulness must sound like sweet music to the bosses, who find themselves free from taking any responsibility for stress they impose on those beneath them in the hierarchy. These powerful global businesses have become involved in the maximization of profit regardless of the abuse of workers, the willful destruction of the environment, and the rapid depletion of resources. Millions associate mindfulness exclusively and narrowly with the reduction of stress rather than a depth of mindfulness about what causes suffering to arise owing to outer and inner circumstances.

We only have to take a single product, such as the mobile phone, to see the long interconnected chain that starts with poorly paid workers for the mining industry and ends up as a fancy, addictive, and very expensive gadget in our pocket. For the mobile phone to find its way into our pocket or handbag, we become involved in the mining industry, production industry, transportation industry, banking industry, computer industry, satellite industry, advertising industry, fashion industry, and marketing industry. Every industry pursues to maximize profit in the course of development of the product. A wise approach acknowledges the importance of profit alongside ethics, social responsibility, payment of taxes, and respect for people and environment.

All of these industries, which feeds our addiction to the latest technology, work on the principle of the maximization of profit through the maximization of management/worker and minimal cost per unit. There is a cynical disregard for the environmental consequences impacting on land, minerals, air, and water. We have perhaps two to three generations to bring about a dramatic transformation of these global corporations and their abuse of people and environment. We need to develop small-scale sustainable businesses working to support local people. The mindfulness industry remains mute on such global issue and, instead, has opted to being in the present, non-judgmentally rather than bring mindfulness to the conditions that perpetuate suffering for millions in the present and future. Despite the claims of *Time* magazine, a mindful revolution has not yet begun in the corporate world.

Obsessed with increasing their market share, this powerful elite has become oblivious to the lives of workers, refugees, the poor, and the marginalized. Governments and corporations have identified themselves with religious conviction with the notion of progress, a concept that has come to take deep root in the westernized mind, as if progress marked an inviolable truth, self-evident that only the blind and naive would deny.

Teachings from other parts of the world have never grasped on to such a distorted view of global reality. The scientific evolutionists cherish the notion of progress while living in blatant denial of unfolding circumstances of past, present, and future. These evolutionists will boast of the progress that we have made in the past 5000 years and, more specifically, in the past 200–300 years since the dawn of the so-called era of enlightenment. Western mindfulness organizations generally prefer to ignore the view from the East and identify with progress as some kind of fundamental truth.

Mindfulness teaching from the Buddhist tradition have a radically different world view. The teachings in the East never took on board notions of evolutionary progress and belief in the movement of humanity toward a higher level of consciousness based on science, reason the power of technology. Like religion, these metaphysical assumptions obscure the opportunity to recognize the harsh reality of what's on the ground.

The East has adopted a much more realistic viewpoint, especially in the Buddhist tradition. Instead of evolution as a central scientific concept, the Buddhist tradition uses the word 'becoming' to signify an ongoing process. Darwin used the word 'evolution' to refer to the adaption in biology of plant life according the environment. Scientists and atheists hijacked the word 'evolution' from its original content and made a grand metaphysic out of the concept. The same fate fell on yoga which became detached from ethics, spiritual disciplines, simple lifestyle, and liberated way of being and found itself reduce to a form of exercise. And now a similar fate has fallen upon mindfulness with its restriction to person stress reduction rather than serving as a factor in a body of teachings for inner–outer change.

Instead of the Western concept of 'progress,' the long-standing teachings in the East make reference frequently instead to 'change,' to 'impermanence.' Meditation in impermanence refers to a clear comprehension to what arises, to what stays, and to what passes. There are no assumptions whatsoever that we, as a species, continue to make progress into eternity or achieve an ideal state, nor assume that we evolve in the right direction.

Rather than restricting mindfulness to a series of meditative relaxation exercises and sharing of these experiences, mindfulness workshops need to explore the deep issues as well. We need mentors in corporations who have investigated the soul and shadows of the corporation, engaged in much online research beforehand, and have the capacity to offer a workshop that looks deeply into the impact of outer corporate demands upon the inner life of all the workers and the environment. Mindfulness leaders in the corporate world need to ask deep questions to participants.

A sampling of such deep questions would include: *Who am I? What are the causes, inwardly and outwardly, for stress? What are we doing here? What is most troubling about this company? Does this company bring any harm to people or environment? What is more important than making money? What is more important than working hard to achieve company targets? What contributes to peace of mind, empathy with customers and society? What would show empathy and kindness to the paid workers in the company?*

What am I missing out on through working so many hours a week at work and at home? Are we afraid to speak up? What wise changes, inwardly and outwardly, am I willing to explore?

Another key word in the Buddhist tradition is 'development.' The concept of development, whether through the individual, group, or organization, refers specifically to the reduction of greed, violence, and delusion. As human beings, we develop through generosity, kindness, and clarity. A large business may grow extensively becoming more and more influential, as it gains a bigger hold of the market share. Growth of the market share does not indicate any development of the business. The rapacious desire for growth, the violence of the strategies, the delusion of self-importance, and egotistical hunger for power indicates a lack of development from the CEO downwards.

We witness the growing evidence of the wisdom of the East that recognizes the principle of what arises, stays for a while, will pass. This principle is as sure and the certain as the ball thrown into the air, stays for a while in air, and falls to the ground. The start, stay and end, birth, living, and death applies to whatever is seen, heard or sensed, inwardly or outwardly. Progress cannot go on and on and on. We witness the impact of the industrial, technological culture with its obsessive desire for growth, progress, and evolution. We are also witnessing the consequences. Far too many corporations express a pathological madness through attachments to expansions and profit rather than living in tune with what arises, stays, and passes.

We witness climate change including intense heat waves, melting ice caps, shrinking glaciers, widespread destruction of rainforests, severely polluted environments, dying rivers and lakes, oceans infected with plastics and pollution, freezing temperatures, and dangerously fundamental beliefs in science and religion. Our corporations, governments, and military impose their will on people, land, water, air, and resources. All sentient life has become threatened through the greed, violence, and delusion of powerful institutions. We find the concentration of these poisonous states of mind in our major institutions.

As citizens, we also have to address these issues in our personal lives and at the local level. Mindfulness leaders locally and globally have the potential to make an important contribution but within their present remit.

The desire and the craving for more creates a corresponding reduction elsewhere. More here means less elsewhere. A simple science of reality. The teachings in the East of impermanence, change, an unfolding process, of arising, staying, and passing meet with first-hand experience, insight, and realization. Concepts of Western enlightenment, evolution, and progress see more and more absurd as we look with honest eyes at the vulnerable state of this world.

Selfie Mindfulness in Corporations

Powerful corporations have offered their staff regular mindfulness classes. These corporations have opened the doors to mindfulness. These mindfulness teachers have the opportunity to expand their current limited remit. They have to dig deep into themselves and learn from others about the methods, values, beliefs, and strategic aims of corporations. The Buddhist tradition places much emphasis on 'upaya'—skillful means. The mentors have to find a skillful means in language and the tools to address the corporate inner–outer dynamics. They need to address the personal, the interpersonal, and the ideological structure of the corporation with regard to workers, customers, and the environment.

For example, the current belief system claims that we create our own stress. The mentors seem to take for granted that stress belongs to the self, created exclusively by the self and one needs a few helpful practices for the self to reduce the stress. Corporate leaders know that any reduction in stress will mean less time off work, greater cooperation in the office, sustainable attention at meetings and more available energy to meet company targets.

We can describe this as 'selfie mindfulness.' This mindfulness finds its limits wrapped around the individual working in the corporation without any extension of interest beyond the personal. Selfie mindfulness focuses exclusively upon the personal self for the reduction of stress without any enquiry into the totality of the work environment.

It is not unusual for the range of office workers to hate their job. They hate the toxic atmosphere and hate spending nine or 10 h per day or more working at their desks. Hostile, aggressive, and intense demands running through any business, great or small has an emotional, psychological, and physical impact upon the staff. Negativity and tension in the office then extend itself into personal, family and social life.

It is not surprising that staff resent the upper echelons of management and everything that the corporation represents. The staff arrive on a Monday morning and finds themselves counting off every hour until Friday evening in order to have a break from a tightly controlled regime, regardless of any perks, promotions, or bonuses. Employees regularly go to the media to detail the pressure and demands imposed upon them. These whistle-blowers have experienced the futility of endeavoring to get the bosses to listen to their needs. With little or no representation in terms of unions or other kinds of counsel, the staff have to pursue their own initiatives knowing their the independent voice may result in the loss of their job, loss of promotion, demotion, or a general disregard for their complaint.

The *New York Times* reported in detail the intensity of the workplace of Amazon employees, who find it harder and harder to sustain the demands made upon them. The subheading in the *New York Times* article in August 2015 read 'The Company (Amazon) is conducting an experiment in how far it can push white-collar workers to get them to achieve ever expanding ambitions.'

The *New York Times* said that Amazon proudly boasted that it had set standards that were 'unreasonably high.' When Amazon workers, whether in the office or on the floor of the workhouse, were unable to match the demands imposed upon them, they were likely to be fired. Staff reported that they would regularly observe stressed out staff putting their face in their hands and crying. They were unable to cope with the demands of Amazon bosses. Amazon constantly measure how well or otherwise employees perform. They push office, warehouse, and factory workers to their limit to maximize efficiency at the lowest possible costs per shipment, per person.

Amazon keeps a check on those who achieve their tasks and find out those who do not meet their demands. Such relentless striving underlines the Darwinian principle of survival of the fittest. The *New York Times* article reported the conflicts that take place in committee meetings with the thrashing out of various ideas without the necessary considerations for the feelings and thoughts of others at the same meeting.

While other major corporations adopt a softer policy, they simply employ different strategies to reach the same kind of goal, namely maximization of profit through minimization of cost. Google, Facebook, and other big businesses provide benefits in the office building, such as gyms, meditation rooms, cafeterias offering nourishing food, and substantial bonuses. Google has become the best know multi-national for its endorsement of mindfulness Google established its Wisdom 2.0 conferences to promote mindfulness and meditation with a variety of workshops led by prominent mindfulness teachers, some of whom have a long training in the disciplines of the Buddhist tradition.

Google have offered around 1700 members of their mindfulness courses to establish a more friendly, happy, and cohesive climate in the workplace. Plenty, but not all, report they enjoy working for Google since they experience less stress than elsewhere.

Despite years of mindfulness programs, hosting major international conferences, and regular invitations to spiritual/religious leaders, there is no evidence of mindfulness practices penetrating deeply into the body of Google or any other corporation. Performance matters most to these companies, whether they adopt the hard approach of Amazon or softer approach of Google. Corporate soul searching is not on the mindfulness agenda.

Critics of Google point out:

- In 2013, Google made £3 billion in revenue in the UK and only paid £11.6 m in taxes to support the people of Britain. Corporate tax avoidance has become widely known as Google Tax. A Parliamentary committee referred to Google's contrived tax avoidance schemes as 'calculated and unethical.'

  – Widespread reports of Google case misuse of its power to promote its business.
  – Google unfairly uses its products to oust smaller competitors
  – European Union accused Google of cheating competitors by distorting Internet search results in favor of Google
  – Top results on Google searches are not solely based on relevance but manipulated according to which company paid the most.
  – Google has been criticized for censoring its search results in compliance with the laws of various countries in order to maximize revenue.
  – Critics doubt the validity of Google's 'Don't be evil' motto.

No doubt, some of the 1700 Google workers in Silicon Valley in the USA, who attend mindfulness courses, have their dirty fingerprints on their computer keyboards in terms of some of the corporate behavior of Google. I cannot find any reports through Google Search of mindfulness leaders teaching computer programmers to become mindful of their feelings, thoughts, and intentions in such forms of anti-trust behavior and their disregard for ethical responsibility.

Corporate employees in various companies will offer praise or find fault with their colleagues which they report to their bosses. Bosses will weed out those at the bottom of the scale, who have been outperformed by others. In the corporate world, the values of loyalty and commitment, along with performance rank high. The staff may work 50–80 h or more per week. They often advocate the strengths of their corporation and detail the perceived weaknesses of other businesses. This shows to bosses and colleagues an expression of loyalty.

A member of staff may have to reduce their hours of work owing to stress, health issues, parenting, sickness in the family, or the simple wish to give more time to life outside the office. Bosses may interpret time off work as a sign of weakness and a lack of commitment to the company. Far too often, bosses have little real interest in the ongoing welfare of the staff outside of the company; work takes priority over everything else. As people get older, they can struggle with the long working hours. They fear they will not meet company targets. The workers in the office, warehouse, and on factory floor find themselves looking over their shoulder at the new employees 10 or 20 years younger with a vigor and determination greater than their own. This pressure leads to frequent turnover of the staff unable to cope with the demands made upon them, as well as the demands that they make upon themselves.

With competition, progress, and struggle as the persistent thread in corporate life, corporations have adopted with unquestioning obedience the Darwinian model of 'survival of the fittest.' A Gallup opinion poll revealed that 70 % of employees hate their job and feel disengaged from it. Many experience thoughts and fantasies of switching to a more thoughtful and sane career. More and more resign. Quite often, a large mortgage, children's education, addiction to a certain lifestyle keeps many trapped in the daily boredom/demands of the workplace.

They find themselves as much a prisoner of opulence as the poor and marginalized prisoners of poverty. Alcohol and drugs become the flight from the drudgery and demands of the desk-bound job and endless repetition of various committee meetings. Gossip around others becomes a welcome diversion from the intransigence of office life.

Mindfulness teachers have their work cut out to change the ethics, values, and demands on the staff of a corporation and provide a harmonious and congenial workplace with mindfulness and enquiry addressing every aspect of an entire business. This will require a comprehensive training for mindfulness teachers to develop a noble path to transformation. This would indicate the first step to an authentic 'mindful revolution.'

We can hardly expect the Buddhist tradition, which points directly to the emptiness of the ego, of the self, to settle for selfie mindfulness.

The Voices of Concern About Western Mindfulness

We currently witness a sectarian divide, troubling and unnecessary, between the religious aspects of Buddhism and secular mindfulness. Some mindfulness teachers imagine that orthodox religious Buddhists resent secular mindfulness because these Buddhists have moved away from the religious traditions of monasteries, monks, nuns, taking refuge in the Buddha, Dharma and Sangha, chanting and devotion, as well as beliefs in rebirth and Nirvana. Some religious Buddhists, both ordained and lay people, feel the religion protects the full body of the Buddha's teachings and the tradition while the secular Buddhists only wish to preserve part of the practice, such as mindfulness.

We witness a plethora of essays, articles, critiques, and polemics on the direction of mindfulness. This essay serves as another contribution. I believe it is time for the pioneers of Western mindfulness to expand in far reaching ways the application of mindfulness into every area of corporate life rather than selected aspects. We cannot expect religious Buddhists to influence the boardroom and management of corporations who show a lack of wisdom and compassion in their pursuit of power and wealth. Mindfulness teachers and secular Buddhists have that responsibility.

Bhikkhu Nanaponika the German monk and author of *The Four Foundations of Mindfulness*, Ajahn Buddhassa of Thailand, Venerable Thich Nhat Hahn of Vietnam, Buddhist Monks and nuns residing in the West, Insight Meditation centers, Jon Kabat-Zinn, and others have made an important and significant contribution toward establishing mindfulness as one of the practices the Buddha strongly emphasized in his teachings.

It is important to note, however, that if the practitioner goes one degree off the noble path, she or he will get lost in the undergrowth and cutoff from the exploration of the depth of the teachings and practices. Jon Kabat-Zinn and other mindfulness teachers acknowledge the importance of the discipline of regular meditation retreats with its threefold training on ethics, mindfulness meditation, and insight/wisdom to experience a comprehensive understanding of the Dharma. Yet, there is an immense gap between the threefold training and what most mindfulness leaders offer to corporations.

I know people in the religious tradition who express appreciative joy to dedicated mindfulness leaders working to reduce suffering of individuals. Through mindfulness training, participants learn how to handle daily life situations with clarity. Participants develop much more friendship and loving-kindness. Such training certainly constitute some important preparatory steps along the great way to an authentically liberated and awakened life.

We can only applaud the immense benefits of mindfulness, whether mindfulness training for a woman giving birth, a person dealing with a painful injury or sickness or a mindfulness practice to reduce anxiety levels. Mindfulness can contribute to the uplifting of the spirit through deepening of connection with the ordinary from watering a flower or looking up at the night sky.

We, the seniors in the Dharma, have to ask whether corporations have hijacked mindfulness to use it primarily for its own selfish ends. If so, then the corporate takeover of mindfulness sweeps away indispensable features of the noble path, which addresses inner and outer conditions that trigger stress, suffering and harm to people, animals, and resources.

In the November 13, 2013 issue, *The Economist*, the UK business magazine, naively boasts: 'Western capitalism seems to be doing rather more to change eastern religion than eastern religion is doing to change Western capitalism.' Unlike the Buddhist tradition, Western capitalism will not last 2500 years. The Earth simply lacks the resources to sustain the psychotic and rapacious capitalist greed to exploit the wealth of the Earth. Climate change slowly but surely brings more and more environmental destruction and unbearable pressure on more and more people.

Mindfulness mentors working in large corporations can make a significant contribution to a spiritual renaissance in businesses. I have searched the Internet and cannot see a shred of evidence to show mindfulness in the workplace has brought about any radical change in corporate thinking for the present and future generations. To my knowledge, I do not know of any mindfulness trainers in the corporate world who can report any deep change in the values of the company.

The Buddha on Mindfulness

The Buddha recognized the deep significance of mindfulness for human existence. Various spiritual/mystical/religious teachings of that era, and prior to it, make occasional reference to mindfulness. Historical evidence shows that the Buddha expounded on mindfulness on various occasions as a factor on the path to awakening. He said the wise, who live fully liberated lives, also live mindful lives.

In one of his talks, the Buddha gave a comprehensive summary on mindfulness. The summary should serve as the basis for every mindfulness practitioner, whether a pioneer, teacher, leader, mentor, mindfulness practitioner, meditator, employer, or employee in the public or private sector. Below readers will see a comprehensive summary of the Buddha's discourse that shows the way to a genuine liberated way of life. Readers could read slowly and mindfully through this relatively short discourse to develop a sense of its significance.

Great Discourse on the Applications of Mindfulness (*Digha Nikaya. DN.22*)

Thus have I heard: once the Buddha was staying among the Kurus in the market town of Kammasadhamma. He said: 'There is this one way to the purification of beings, for the overcoming of grief and distress, for the disappearance of pain, for gaining the right path and for the realization of Nirvana—that is to say the four applications of mindfulness.'

One abides contemplating body as body. One sits down, holding his body erect, having established mindfulness around him. Mindfully, he breathes in and out, knowing a long breath and a short breath. One trains oneself to breathe in and out and calm the whole bodily process. He contemplates the body internally and externally and the arising and passing of phenomena. Mindfulness is established to the extent necessary for knowing. One abides not clinging to anything.

In whatever way his body is disposed—sitting, walking, standing, and reclining—one knows how it is disposed. One is clearly mindful of whatever one is doing—eating, drinking, passing urine and excrement, waking up, falling asleep, speaking, or silent. One abides, not clinging to anything. He reflects on all the parts of the body, internally and externally. He reflects on the body as elements—earth, air, heat, and water. He reflects on the body as a corpse. 'It will become like that. It is not exempt from that fate.'

One contemplates feelings as feelings. One knows when one feels a pleasant feeling, a painful feeling and a feeling that is neither painful nor pleasant, a spiritual feeling and a worldly feeling. One contemplates feelings inwardly and outwardly and their passing nature. There are feelings present so mindfulness is established to the extent necessary for knowing. One abides not clinging to anything in the world.

One contemplates states of mind as states of mind. One knows the desirous state of mind as that and a mind not in such desire as that, an angry mind state as that and non-angry state of mind as that. One knows confusion, and its absence, contraction and non-contraction, depth of meditation and absence, surpassed and not surpassed, free and not free, a developed mind and one that is not. One abides knowing the mind internally and externally. One abides knowing arising and passing states of mind. Mindfulness of states of mind is present just to the extent

necessary for knowing. One abides not clinging to anything.

One contemplates the Dharma. One contemplates the presence and absence of any of the five hindrances, the Four Truths of the Noble Ones, the relationship of sense doors to the sense objects, and factors for awakening. One knows how anything comes to arise and pass. One knows suffering, the conditions for it, the cessation of it and the way to the cessation. Mindfulness of the Dharma is established to the extent necessary for knowing. One abides not clinging to anything.

Whoever practices these four applications of mindfulness for seven years down to seven days can expect one of two results. One is fully realized and liberated. Or, if any substrate is left, there is no more returning to a mundane way of life. The practitioners rejoiced and delighted at his words.

Having read through the discourse, readers will notice the Buddha referred to four applications of mindfulness namely body, feelings, states of mind and the Dharma (teachings and practices, inner and outer).

Readers should take complete full notice of the importance the application of mindfulness, internally AND externally, for the resolution of suffering. One can only conclude that there has been a huge error of judgment by mindfulness leaders in identifying the personal inner self as the cause of any stress. This warped and one-sided view shows a conspicuous misunderstanding of the application of mindfulness. It leads to a neglect of inquiry for the conditions for stress and the catastrophic assumption that we 'create our own reality.' As a legal 'person,' corporations have to take responsibility for their behavior.

Mindfulness teachers in the corporate world need to apply mindfulness and inquiry equally to both the inner conditions and external factors that generate stress, anxiety, and fear among the workforce from the lowest paid to those pursuing big bonuses. We live in a culture of the self. We refer to the self-made billionaire, self-created problems, self-acceptance, self-improvement, self-knowledge, self-interest, self-compassion, and self-development. The persistent use of the language of 'self,' of 'I' and 'my' brings about a gap between self and other, between the inner and the outer. Owing to suffering, we easily blame ourselves and make ourselves into a victim or we can attack others to avoid any responsibility.

This 'self' language places all the responsibility upon the individual in the office and not upon the perpetuation of daily demands of the bosses. Employees can end up adopting the view that criticism of the policies of the corporation shows denial, an unconscious psychological strategy to avoid taking responsibility for one's own stress. Mindfulness teachers can feed into this view rather than hold a corporation accountable.

Staff often point out that insecurity contributes significantly to their stress and sleepless nights. Poor financial management, changes in the money markets, new technology and up-and-down cycles in the markets have nothing to do directly with the life of employees. They find themselves experiencing the consequences of events outside their control. Employees experience fear of reduction of income, loss of benefits, demotion, or loss of employment. Staff experience stress due to the demands made upon them, the toxic atmosphere, the absence of unions or other forms of organizational protest, long days, intensity of environment, feelings of lack appreciation, and the day-to-day powerlessness.

Mindfulness facilitators rarely address the unhealthy external circumstances of corporate life. The instructors are more likely to suggest mindful meditations on the breath, or taking mindful in and out breaths before picking up the phone or mindful exercises in listening to another without the mind wandering. Such practices will never change the culture of greed, aggressive policies, and desire for self-aggrandizement. After 10–20 years of mindfulness in the corporate world, it is time for the mentors to take bold steps and open up the exploration into the very soul of the corporation.

Acting with a fearless wisdom, mindfulness mentors have the opportunity to be a real force for meaningful change for the welfare of all

sentient beings in the present and future. Pioneers and leaders in the current mindfulness industry need to develop an understanding of the external conditions in the corporate world that contribute to suffering on a global level.

Four Noble Truths

Rather than mindfulness teachers adopting glib rhetoric about enlightened self-interest and the mindfulness revolution, they could draw upon the Four Noble Truths, the hub of the teachings of the Buddha (*Four Foundations of Mindfulness. Middle Length Sayings. MN 10.44*).

Mindfulness experts can evolve to show confidence in a deep inquiry into corporate suffering instead of offering only a handful of palliative techniques to reduce stress. The Four Noble Truths are:

1. There is suffering. Suffering arises through not getting what we want, losing what we have, being separated from who and what we love and through inflaming body/forms, feelings, perceptions, thoughts, and consciousness.
2. Suffering arises due to causes and conditions
3. There is the resolution of suffering
4. There is a way to resolve suffering.

These Four Noble Truths (or more precisely The Four Truths of the Noble Ones) apply to the individual and the collective, such as corporations.

The Four Noble Truths Applied to the Corporate World

First Noble Truth: There is suffering. What suffering arises in the company from top to bottom in the hierarchy? What suffering does the corporation cause when corporations avoid moral responsibility? What suffering arises when the corporation engages in exploitation of loopholes in the law and taxation? What suffering arises when corporations inflame their products through advertising and marketing?

Second Noble Truth: What are the causes and conditions for suffering in the corporation? Drawing on the Buddha's teachings, Nagarjuna, the 2nd century sage, named the four conditions for what arises. The four conditions have immense significance influence giving rise to policies, strategies, and everything from stress to peace of mind and onto every event. The four conditions are as follows:

a. A strong condition(s) from the past.
b. Surrounding conditions in the present.
c. Variety of conditions leading up to the suffering in the present.
d. All the conditions, major and minor, near and far, for what arises.

Third Noble Truth: There is the resolution of suffering in the corporation. This may require the application of any one or more of the four conditions. It requires commitment. If a corporation shows no willingness to change its behavior, then it requires the voice of the public upon the corporations and government. Third noble truth confirms wisdom, love, and liberation from suffering.

Fourth Noble Truth: The way to the resolution of suffering includes the willingness to address the issues of suffering. This requires commitment, integrity with employees taking risks to show corporate malpractice and change the culture.

If, as human beings, we are going to develop, then we must be willing to look into all four conditions at the personal and institutional level. The purpose of looking deeply into causality is to take the suffering out of events. We apply links of the noble path including right understanding, right intention, right speech, right action, right livelihood, right creative effort, right mindfulness, and right concentration. Every one of the links serves the deepest interests of all those in the private and public sector and all consumers.

Seasoned in knowing and exploring the four conditions, people at work can understand what arises, endures, and passes and respond with wisdom to events.

The networks of caring and compassionate organizations remain determined to change capitalism and its crude ideology of relentless competition, maximization of profit and power at no matter what the cost. Mindfulness teachers need to go much more deeply into the nature of mindfulness and supportive conditions to transform the persistent abuse of corporate power.

We may be living in the last two or three generations of life on Earth. Thoughtful networks for meaningful change show no interest in corporate mindfulness since the captains of mindfulness exclude the external realities.

Mindfulness teachers might jeopardize their future workshops with corporations if they start to question, even in a gentle and respectful way, any unethical policies, demands and goals of the company. Mindfulness in the West requires a new definition to replace the outdated one, namely 'mindfulness is being in the present moment, non-judgmentally.'

The new definition would be 'mindfulness addresses the inner and outer, with clear judgments.'

Corporate Mindfulness and the Pathologization of Workplace Stress

14

Alex Caring-Lobel

Introduction

In the late 19th and early 20th centuries, as the first nonmonastic meditation movement and its modern construal of "mindfulness" was sweeping across colonial Burma, the management philosophy of Frederick Winslow Taylor, the world's first management consultant, was in the process of transforming the American corporation. The imminent exhaustion of rich new geographical markets (like the Burmese Kingdom once was) that fed capitalism's growth for most of its history has led to changes in its structure and with it new forms of labor that make fewer demands on the human body and far more on the human mind in the form of cognitive and emotional engagement. New technologies of management have emerged to bolster such engagement and mitigate the most harmful aspects of the psychological dispositions that competition in our current, highly financialized iteration of capitalism necessitates. Behaviorist interventions like mindfulness meditation, derived from the Buddhist tradition, represent one such technology.

Modern mindfulness has its roots in the third Anglo-Burmese War. At its end, the British Empire consolidated its soft power in Burma by abolishing the monarchy, the most sacred role of

which was support of the *sasana*, the Buddhist monastic community (Braun 2013). Once the site of education, the monasteries gave way to the secular schools introduced by the British. As Burmese religious and educational institutions were drastically undermined, Buddhist reformers set out to preserve the religious tradition in the best way they knew how: currying material support directly from the laity and educating them in Buddhist philosophy. This included teaching meditation—a practice the Buddha did not consider fit for those who had not renounced worldly desire—to laypeople, and en masse. It was the first time this had ever occurred in the history of Buddhism. The forms of Vipassana meditation these reformers taught would be rebranded in the late 20th century as "mindfulness."

Meanwhile in America, F. W. Taylor confronted the greatest threat facing capitalist production: labor, the fount of capitalists' wealth, was being worn down by the physical demands of industrial work. The discontent this incited among workers jeopardized their ongoing cooperation with managers and owners, a pact on which the American capitalist project delicately rested. Taylor's solution was to bring science to bear upon the organization and engineering of work, thereby ending industrial conflict once and for all. Scientific management, or "Taylorism," promised a "complete change in the mental attitude of both sides," "the substitution of hearty brotherly

14

A. Caring-Lobel (✉)
Mexico City, Mexico
e-mail: helloalexcl@gmail.com

© Springer International Publishing Switzerland 2016
R.E. Purser et al. (eds.), *Handbook of Mindfulness*,
Mindfulness in Behavioral Health, DOI 10.1007/978-3-319-44019-4_14

cooperation for contention" (U.S. Government Printing Office 1912, p. 1389). Following Taylor's engineering of the use of workers' bodies, his successors would apply those theories to crack the worker's psyche, the final frontier of work design.

These thinkers would take Taylor's insights and turn management theory into a "soft" science, one concerned with the feelings and emotions of workers as complete human beings, *social* beings, with hopes and aspirations. Psychological insights, however, would be administered only to secure workers' acquiescence. Elton Mayo, the foremost "industrial psychologist," found that by bringing human concerns into the design of work, work could provide people with a source of meaning and purpose, thus quelling industrial unrest without conceding material improvements in the form of compensation or work conditions. Gestures of interest on the behalf of the worker by the corporation, however empty, would improve morale, which Mayo claimed increased productivity.

Over the past decade, these two modernist narratives—that of mindfulness (in particular, forms of mindfulness-based stress reduction [MBSR]) and that of corporate management—have intersected in the contemporary office, where mindfulness techniques and technologies have quickly proliferated. The reasons for this are overdetermined and involve both practical and ideological considerations, but we can safely say they spring from the specific needs of capital in central, highly developed labor markets, where the rapid expansion of postindustrial productive forces increasingly marshal the emotional, psychological, and cognitive faculties of workers to the point of strain. The science of mindfulness promises to address the worker discontent that these forms of labor engender without confronting the social and economic causes of this discontent. In this chapter, I will argue for the repoliticization of the forms of worker stress and discontent that workplace mindfulness rhetoric and praxis obfuscate by framing them in purely psychological terms.

Why Corporations Have Taken Interest in Mindfulness

Mindfulness first entered the corporate world as a luxury relaxation technique and status symbol for those running the show. It is now being disseminated to employees with promises of improved employee health (i.e., lowered healthcare costs for employers) and increased productivity (i.e., greater value extracted from employees).

Of course, mindfulness appeals to employees as well as executives. Who would not want to test the purported benefits of the practice for him or herself? Mindfulness owes its workplace success, however, to its specific promise to address worker discontent without calling into question the current distribution of power. It is no coincidence that power is concentrated among the business executives, managers, and administrative elites who theorize and oversee workplace behaviorist interventions involving meditation and emotional intelligence. The depoliticization of the stress and discontent of working people, the historical foundation for organizing against capital, depends on the psychological reductionism proffered by the latter. Viewing themselves as the guardians of the global capitalist economy, management elites have historically viewed themselves as fulfilling a duty (today, as "leaders," as managers are now commonly glossed) to redress the harmful psychological dispositions and social challenges of the day, despite the fact that these challenges invariably stem from the particular organization of society on which they depend for their position.

Many of the recent challenges to the global capitalist economy—and even humanity's future survival—consist of what economists benignly call "negative externalities," the costs and consequences of business that neither producers nor consumers but the public end up paying for, such as air and water pollution. Misery could be considered one of these externalities if not for the fact that the contemporary economy, which increasingly depends on our passion, sustained attention, and emotional engagement, does pay the price. Having done away with much of the trade union movement and the organized left— the traditional voices of worker discontent— capitalism now faces an unprecedented crisis. Labor, the source of its wealth, is stressed, depressed, deflated, and generally unhappy, and its discontent is costing the world's most robust economies hundreds of billions of dollars a year in the form of lost productivity (Gallup 2013). This crisis has sparked in business a renewed interest in the thoughts, feelings, and general contentment of workers. It is in this context that the stewards of capital have struck up a partnership with Buddhist elites.

In addition to daily meditation instructions from Jon Kabat-Zinn, this year's World Economic Forum in Davos included sessions exploring how to build "mindful organizations" and harness stress for creativity, taught by "leading experts on neuroscience and mindful leadership." There was even a "mindfulness dinner" hosted in collaboration with the *Harvard Business Review*.

Talk of mindful executives and managers (usually glossed "mindful leaders") serves as but one example of the unbridled enthusiasm around the mindfulness movement, which identifies the dissemination of mindfulness practices as a means to correct all manner of civilizational ills. Mindfulness's rapid ingress into public institutions like clinics, prisons, and schools has been mirrored in the private sector, where proponents have found some of their most generous benefactors among the world's business elite. Corporate implementation of mindfulness, commonly referred to as "corporate mindfulness," has stoked hopes for a better future among business executives, human resources departments, and liberals alike.

Many of the largest corporations in the USA, such as Aetna, Google, and Target, have developed their own mindfulness programs for staff. In the USA, mindful business has its own annual conference, Wisdom 2.0, where a panoply of figures, from Buddhist monks, to hero venture capitalists, to at least one African dictator (Kagame 2014), give business and "leadership" advice to economic and political elites, along with anyone else willing to pay the steep cost of admission.

The sincere fascination among managers, CEOs, and economists with all things meditation might seem contradictory given mindfulness's liberal, New Age, and Buddhist associations. But mindfulness might be just what business needs in order to keep humming along: its pseudo-leftist critique of social relations, unwavering focus on individual well-being, and adoption of therapeutic language only serve to obscure the reasons it's so lauded by management consultants and has been successfully deployed in a diversity of workplaces.

The simple reason for the proliferation of these programs is that these they address worker happiness directly, which has proven critical to the bottom line in the form of productivity and efficiency. But more essential might be their role in producing specific affective valences that can obscure and thus help maintain unequal power relations in the workplace. In fulfilling these functions, meditation and mindfulness in the workplace can be understood within the history of industrial management, serving as only the latest tool in a long line of management techniques and technologies that have, throughout the history of management theory, been substituted for workplace democracy.

Following the failure of state socialism and the weakening of trade union movements, elites now worry that the most imminent threat to capitalism might be some combination of lack of engagement and general apathy (Gelles 2015, p. 105). Resistance is cropping up in new, uncoordinated ways that nonetheless cut deeply into corporate profits.

Gallup estimates that only 13 % of the global workforce is now "engaged," with nearly one in five workers in the USA and Europe "actively disengaged." The cost of active disengagement to the US economy alone is estimated at around $550 billion a year (Gallup 2013). As the trade union movement continues to be forced to make huge concessions across the developed world, forms of unorganized resistance, which are hence difficult to recognize as such, have arisen. Resistance can be as simple as calling in sick to work due to depression or burnout or coming into work and accomplishing little by failing to muster the will to be a "good," productive employee that day. A Canadian insurance company's research found that a third of absenteeism cases—or "calling in sick"—were due to stress and burnout (Murphy 2014). In the modern firm, where management is ideally warm and friendly, sickness becomes one of the only viable means to resist managerial prerogatives or refuse work. Yet sickness, especially in the form of chronic mental illness, has become a very real ailment of postindustrial workers. In the USA, the leading cause of absenteeism is now "depression" (Folger 2013); in the UK, the leading cause of absence from work is "stress" (Gelles 2015, p. 133).

Resistance can express itself in chronic mental health issues that can threaten lives, or just a simmering general apathy. And while these forms of lateral resistance cannot stand in for worker collectives or a larger worker movement, they have nonetheless succeeded in scuttling capital to the tune of billions of dollars a year—figures not lost on the Davos crowd. The extent of these losses, as ubiquitous as employee discontent, is difficult to overstate. Labor is all but bleeding capital in the form of generalized unhappiness.

With the postindustrial workforce beginning to show signs of mental and affective depletion, burnout is the new catchword, and its cause—stress—is on the tip of everyone's tongue. As commerce in "The Fourth Industrial Revolution," the theme of this year's World Economic Forum gathering, comes to demand more and more in the way of emotional and psychological engagement, the management of stress becomes key to sustaining it.

The epidemic of worker stress reflects the current needs of capital and the corresponding changes in work design (to say nothing of the demands of the realization of value—i.e., consumption). Yet capitalist interest in various aspects of worker well-being is anything but

new. Indeed, at the turn of the century, a similar fear fixated around "fatigue," the wearing down of laborers' bodies through repetitive physical tasks. In the late 19th and early 20th centuries, fatigue stood as the greatest threat to labor's cooperation with capital, and thus, the economy in general. Industrialists' response consisted of the erecting of modern management theory, specifically "scientific management," which would establish the fundaments of work design that form its bedrock to this day (Braverman 1998, p. 60). Whenever management brings in an outside consultant or contractor—whether to "restructure" the company or teach mindfulness meditation—they are following in the tradition established by Fredrick Winslow Taylor, the world's first management consultant.

Scientific Management

The scion of a wealthy Quaker family in Philadelphia, F.W. Taylor could trace his line back to the original Mayflower pilgrims. His impressive pedigree and sterling academic record earned him admission to Harvard, but following a nervous breakdown of sorts, Taylor instead took up an apprenticeship at a machine shop, where he eventually became a foreman. There he noticed that the shop workers were not pushing themselves nearly as hard as they could (what he called "soldiering"), and productivity and—for his employers—profits were diminished as a result. His solution was to institute what he understood to be the first "scientific" approach to the control of humans in the production process, which he would eventually bring to Bethlehem Steel, one of the biggest corporations at the time. Science was to be put to work in service of productivity and efficiency.

On the heels of the robber barons, the Heroic Age of American business resulted in corporations of unprecedented size and complexity and a proliferation of monopolies not unlike what we are witnessing today. Progressives feared these great corporations, perceiving them to be lacking in accountability, a suspicion the earliest muckraking journalists—such as Ida Tarbell, who exposed the corrupt business practices of industrialist John D. Rockefeller—would confirm.

By the early 1900s, following brave activist and journalists' exposure of reprehensible working conditions, companies began to take a deeper interest in the well-being of their employees in order to present a good face to the public, discourage unionization and bargaining rights, and preclude the most undesirable result of discontented labor—the strike (Anteby and Khurana 2012).

Liberals likewise saw in scientific management an amicable solution to the so-called labor problem, a means to create harmony out of the class antagonisms that threatened the social order in a time of global labor unrest when the threat of communism loomed large. Tarbell praised Taylor for his contribution to "juster human relations" (Kanigel 1997, p. 505). For US liberals and progressives in government, scientific management could provide "objective" rules with which to manage these corporate megaliths.

It was, in fact, the famous labor activist Louis Brandeis, dubbed "the people's attorney" and appointed to the Supreme Court by Woodrow Wilson, who coined the discipline "scientific management" in 1910. "Of all the social and economic movements with which I have been connected none seems to me to be equal to this in its importance and hopefulness," wrote Brandeis et al. (1971, p. 385). A friend of Taylor, Brandeis is largely responsible for his popularity.

But Taylor relied more on rhetoric and showmanship than reproducible findings. Through repetition, Taylor fashioned a kind of mythology of scientificity where it did not exist. Taylor's most lasting contribution was thus not a science of management but a simple idea. In bringing a scientific attitude toward management, he did not create a true science of management, but something quite different: A management technique supposedly vetted by the objective discipline of science. In fact, his is likely the first articulation of the *idea* of management as a field in general. Scientific management, he said, was not an efficiency device, but "a complete mental revolution" (Taylor 1911). At bottom,

Taylor's was the theory of "the one best way" to perform work in the interest of the control of labor that is bought and sold—the purpose of management and truly the cornerstone of Taylor's thought (Braverman 1998, p. 62). In spite of his supporters' occasional allusions to its reformative potential with vague platitudes of social justice, cooperation, and democracy, it was only this.

Taylor had convinced his audiences that "there has never been a strike of men working under scientific management" (Montgomery 1989, p. 254). To progressives, for which the proliferation of the ideas of scientific management depended, its purpose was this—to reconcile labor and capital—and according to Taylor, it worked. Through the implementation of the scientific management of labor, the overall cost of labor power could be reduced, and with it, the price of raw materials. Wages could increase and the overall standard of living would improve. The private corporation, driven by the accumulation of private wealth, progressives hoped, would be brought to serve the public good.

In contrast to the grandiose rhetoric surrounding scientific management, its actual content was quite modest, consisting of tweaks like economizing the discrete number of movements performed by laborers during the production process and adding rest breaks for them to recover from fatigue. Though these changes yielded increased productivity, most savings were likely cannibalized by Taylor's exorbitant consultant fees (Stewart 2009, p. 47).

While it may have concerned itself exclusively with the engineering of work, Taylorism was undergirded by a specific utopian vision, one made by and for the ruling class. All management techniques and theories, from the steel factory floor to the Googleplex, must inspire, and inspire Taylor did. His vision survives in contemporary management theory and the techniques and technologies it deploys, including corporate mindfulness programs. Through teaching meditation, liberal and progressive teachers brought in by management believe they can transform corporations into ethical actors from the inside out. But in order to understand how such an idea has gained purchase, it is necessary to look at the development other disciplines that grew out of scientific management.

The Mind This Time

Ultimately unable to quell worker discontent, Taylorism gave way to the so-called human relations management theories. Taylor's focus was on the work itself, especially the careful engineering of manual labor. Once employee well-being had been subjected to a purely economic rationality, the mind of the worker—the locus of discontent—naturally became management theory's next target.

Taylor's successors, some of whom openly attacked his methods, sought to bring the "human element" of labor to the fore through applied psychology and sociology (Stewart 218). Their mission, however, was wholly consonant with Taylor's: to bring the worker into cooperation with a naturalized scheme of work. After all, the goal of scientific management, according to Taylor, was to bring about "a complete mental revolution." Industrial psychology, which would eventually evolve into the Human Relations movement, would complete this revolution. The new engineers of labor only addressed this "human dimension" in order to more directly target the discontent of the worker, the greatest threat to industrial relations, and never to improve the worker's lot, as Taylor's liberal supporters had hoped. Psychology would be recruited to preserve industrial power relations through the engineering of sentiment.

The first author to schematize industrial psychology, the Prussian-born German Hugo Münsterberg, explained his project thus: "the psychological experiment is systematically to be placed at the service of commerce and industry" (1913, p. 3). His object of study was "all variations of will and feeling, of attention and emotion, of memory and imagination" (1913, p. 28). All in the interest of controlling the laborer in order to interpolate him or her into that scheme of work designed by the industrial engineer. "Whether these ends are the best ones," wrote Münsterberg, "is not a care

with which the psychologist has to be burdened" (1913, p. 19).

In the end, not even executives would be exempt. Münsterberg put particular emphasis on what he called "mental fitness," a concept that would become a fixture of business-speak in the white-collar world of executive coaches, especially among today's mindfulness-peddling hucksters in Silicon Valley.

Yet Münsterberg focused on forming new habits among workers in particular. "We ask … under what psychological conditions we can secure the greatest and most satisfactory output of work from every man; and finally, how we can produce most completely the influence on human minds which are desired in the interests of business" (1913, pp. 23–24). With such statements, we can already see how these consultants made demands on workers that reached beyond their workplaces. These predecessors to the Human Relations movement would largely concern themselves with hiring practices, searching for workers whose general "mental qualities" made them loyal, suitable to perform work effectively, and who were disinclined to stir up industrial conflict. Once psychology was put in service of industry, the specific scheme of domination it desired would not be confined to it.

Industrial psychology took hold in America after Münsterberg's death, with the arrival in of Mayo (1924). Departing from the at times severe unfeeling of Taylor, the Australian psychologist busied himself to understand the worker's emotional life to design a more "humane" approach to management. He nevertheless strictly belonged to Taylor's lineage, extending the maven's theories to the role of the individual and group psyches of workers within the industrial process, for which the industrial psychologist and later the human resources expert would serve as "the maintenance crew for the human machinery" (Braverman 1998, p. 60).

Dumbfounded by strikes under Taylor's scientific management in his home country, Mayo dismissed radicalized workers as "irrational" and their revolutionary theories and general discontent as psychopathological in origin. "Industrial unrest is not caused by mere dissatisfaction with wages and working conditions," he wrote, "but by the fact that a conscious dissatisfaction serves to 'light up' as it were the hidden fires of mental uncontrol" (Mayo 1922, p. 64).

Mayo advocated business to take "account of human nature and social motives." If action was not taken, he warned, strikes and sabotage would continue to hobble industry (Gillespie 1991, p. 99). Research into and implementation of work-design interventions would be entrusted to Mayo's ilk of administrative elites. "This is not one way," he wrote, "but the only way to save society" (Mayo 1924, p. 597). If workers lacked, in Stewart's words, "basic rational capacity to act in their own self-interest" (2009, p. 135), they would have to be educated by these elites. In doing so, they would define self-interest (as understood by psychologists) as consonant with the corporate body, placing it at the center of their project.

Mayo, among other consultants associated with the humanistic management movement, considered man's dissatisfaction with his work life to be the most imminent threat to the control of labor and production and thus to the entire social order, which appeared increasingly precarious. Mayo's sense of this precarity, however, verged on the apocalyptic. He seemed to believe that industrial conflict—specifically the actions of trade unionists and socialists—would lead to the complete collapse of Western civilization. Underpinning Mayo's opinion of unions was a general disdain for democracy, which, he said, "as at present constituted, exaggerates the irrational in man and is therefore an antisocial and decivilizing force," an "ignorant ochlocracy" (Trahair and Zaleznik 2005, p. 221). In order to discourage unionization and socialism, the biggest threats to civilization in Mayo's estimation, elites would have to find a way to simultaneously discourage unionization and dispense with industrial conflict altogether, all without altering the balance of power. His end goal was thus simply successful domination, a state of lasting peace attributable to the acquiescence of the worker to his own subordination.

To accomplish this, the psychological handicaps of workers, as defined and enumerated by the discipline, were figured as the primary impediment to the control of labor. According to Mayo's professional opinion, workers were suffering from something akin to mental illness. Mayo went as far as to dismiss socialism, anarchism, and trade unionism as symptoms of some combination psychiatric illness and severe physical fatigue, conditions that required treatment.

In seeking to recover what he understood to be the humanity of the worker, Mayo did just the opposite. He reduced workers to unstable bundles of emotions inscrutable to themselves. Material conditions, compensation, control of production as well as the fruits of labor—nothing within the capitalist economy need be modified. Such factors took backseat to the psychological dispositions of the worker, which could be manipulated through appealing to his social and existential needs. "A happy worker is a good worker" emerged as the central tenet of the humanistic philosophy of management (Stewart 2009, p. 115). Psychotechnics and self-betterment would thus be substituted for democracy.

"Whenever we now hear that managers must focus on the 'whole person', and not just the 'employee', or that employee happiness is critical to the bottom line, or that we must 'love what we do' or bring an 'authentic' version of ourselves to work," writes Davies, "we are witnessing Mayo's influence" (2015, p. 123). Mayo was the first to prioritize the morale of the industrial worker as it relates to productivity, thereby bringing the worker's entire emotional life—previously excluded from work matters—into the realm of economic rationality. Breaking from the utilitarian theorists before him, Mayo instrumentalized happiness as a means to stability rather than an end in and of itself.

Inspired by his specific allegiances (and a healthy income from his clients), Mayo willfully misapprehended social and political issues as personal and pathological. While his approach might seem antiquated and crass, such reduction of social malaise to psychological dysfunction continues to dominate organizational theory. Indeed, industrial psychology provides the theoretical foundation for the deployment of corporate mindfulness programs today.

The Social Vision of Corporate Mindfulness

It is no mystery why the fundamental principles of scientific management and its successive theories—industrial psychology in particular—have seen a recent resurgence. In 2014, levels of inequality reached their most extreme point to just before the Great Depression (Saez and Zucman 2014), the heyday of Mayo's strain of management theory. With their dedicated lobbyists and large philanthropic organizations, today's massive corporations of unprecedented size, scope, and reach, threaten not just civil liberties like privacy but even the foundations of democracy itself. The concerns of the age of scientific management have returned with a vengeance.

As we have seen, the vision of ushering in a better world by reorienting business to serve the greater good has served as the social basis for management theory from its Taylorist beginnings. Since then, the demotic religious character of management movements has only increased (Stewart 2009, p. 262). Mindfulness, with its widespread, if not slightly hippieish, appeal to liberals and progressives, represents its culmination.

Just as progressive supporters of Taylorism like Brandeis and Tarbell championed the capacity of scientific management to reign in big business at the turn of the century, liberals today champion corporate mindfulness as a solution to the new Robber Baron economy. But for reasons that have already been indicated, this support is fundamentally misguided. Like past management interventions, corporate mindfulness is recruited to alleviate problems of labor without acknowledging, much less pursuing, questions of power or political economy. As Chade-Meng Tan, the creator of Google's program celebrates, "mindfulness can increase my happiness without changing anything else" (Tan 2012, p. 134).

Facing corporations accountable to neither the state nor the public, liberals have resigned themselves to behaviorist interventions, focusing on personal responsibility and launching appeals to the morality of the individuals who belong to business and economic institutions. But such emphasis commits the same categorical error that Mayo willfully makes by framing questions formerly understood in the lexicon of political economy as matters of health and psychological well-being. Furthermore, this emphasis is entirely political. Targeting worker habits and attitudes as the exclusive objects of reform, management theory's program presents no social vision beyond the extension of the status quo, albeit under more harmonious circumstances— the simultaneous absence of tension despite the absence of justice, or what Martin Luther King, Jr. called "negative peace" (1994).

It is precisely due to this lack of social imagination that an alternate vision of social transformation, one that directly contradicts the very purpose of management theory to control alienated labor, must be yoked to corporate mindfulness. It is this false hope that sustains its implementation, much like in Taylor's time liberals' hope in scientific management ensured its proliferation, which only served to further disempower workers. Within this vision, meditation is understood as not just a tool for inner-development—or, as in the Buddhist context, for liberation—but also for creating a better society, person by person, from the inside out. Comportment, however, has never improved the lot of the disempowered.

The Mindfulness Ethic and the Spirit of Capitalism

Unable to hold large corporations or global financial institutions accountable, liberals have retreated to a kind of moral sentimentalism, reframing social and political problems as matters of individual choice and personal responsibility. This attitude is exemplified in reactions of several corporate mindfulness instructors to the financial crisis of 2007–2008. Both Kabat-Zinn (Boyce 2011, p. 59) and Norman Fischer, one of the principal meditation teacher of Google's "Search Inside Yourself" (SIY) program for employees (Boyce 2010), attribute the tanking of the US housing market and global economy to a lack of mindfulness. Fischer in particular holds that workplace mindfulness programs "make the world a better place through the 'technology' of meditation":

> For the people who take the [Google] course, it makes a difference in how they operate, how they communicate. They learn that they do not have to leave their emotions at the door when they come to work. That is big. If Wall Street traders, for example, had had more emotional intelligence, they might have realized the crazy derivatives they created were wrong (Boyce 2010).

Such a diagnosis should be immediately rejected, as it functions as a broad exoneration of the economic system as well as of the banks and loan companies that precipitated the crisis. Moralistic understanding of systemic problems is naturally favored by elites, who by definition benefit from current political, economic, and social relations. By persuading the public that economic crises are moral issues, matters of the human heart to be addressed by looking inward, elites curb political action and democratic deliberation regarding what is collectively to be done (Johnson 2013). With complex social issues reduced to moral rectitude, business-as-usual becomes further entrenched as "the one best way."

"The root cause of our current economic and civilizational crisis is not Wall Street…not infinite growth…not Big Business or Big Government," echoes management author and *Huffington Post* contributor Otto Scharmer. It is "between our ears." Scharmer's solution to our civilizational crises?: "A new type of *awareness-based collective action leadership school*" that incorporates mindfulness meditation instruction for more effective management (Scharmer 2013). "Leader" managers and CEOs, who take home on average about 330 times more than the typical worker, will deliver us (AFL-CIO 2014).

Chade-Meng Tan, Google's "guru" who heads their SIY program, goes as far as to identify Google's program as a means to world peace. "Social or political structures…tried to create peace from the outside in. My idea is to do the reverse, to create world peace from the inside out" (Boyce 2010). All this as Google spends record amounts on lobbying to have its say in politics: well over $60 million since 2009, just $15 million short of Lockheed Martin, the world's largest defense contractor (Solomon 2014). The hypocrisy is hard to ignore.

For consultants like Chade-Meng Tan and Norman Fischer, one does not have to choose between being moral and being successful. "The holy grail is that everybody in the organization—especially the leaders—everybody is wise and compassionate, thereby creating the conditions for world peace," says Meng (Tan et al. 2012). While science might be able to account for some of the health benefits of meditation, more than a little magical thinking is required to validate the social vision behind its workplace implementation. Meng's intellectual move consists of an extrapolation of Mayo's views of industrial conflict to global conflict of all kinds. Just as Mayo suggested "that conflicts were not a matter of competition over scarce resources but rather resulted from tangled emotions, personality factors, and unresolved psychological conflicts" (Illouz 2008, p. 14), Meng suggests that wars between nations do not occur over competition over resources but rather lack of what he calls "emotional intelligence," a kind of measurement developed by corporate consultants and popularized by psychologist Daniel Goleman that measures what might we might call the emotional capital of employees in much the same way as IQ might serve as a measurement for cognitive intelligence.

When personal solutions are prescribed for political problems and the reductive force of moral sentimentalism reigns, wars occur because people's minds are not at peace. Capitalist firms strive for profits because they are "greedy." In the words of sociologist Ulrich Beck, "[s]ocial crises appears as individual crises" (Beck 1992, p. 100)

While mindfulness promises to further its own baked-in, implicit ethics, one purportedly backed by science, the studies cited to back these claims (Gelles 2015, pp. 133–134); DeSteno 2013) conflate ethics with social niceties. The primary study cited, for example, is built around a contrived situation in which people choose whether or not to give up their seat for someone with crutches.

The research of Harvard Business School professor Michel Anteby suggests that such allusions to some kind of implicit ethics in business instruction without elaboration might be worse than no mention of ethics whatsoever. In *Manufacturing Morals: The Values of Silence in Business School Education* (2015), Anteby argues that under-specifying morality in business education amounts to an endorsement of a normative viewpoint—one that benefits the established order and elites' interests. Silence, in this sense, becomes a way to condone glaring social inequities.

Moreover, being guided by an ethic does not necessarily make one ethical. The good of the scorpion, after all, is not that of the frog. Likewise, the good of capitalism is the good of neither labor nor democracy—it is simply the good of capital.

Yet proponents and defenders of corporate mindfulness regularly mistake social relations based on economic realities as a fault of culture, as if corporate culture existed independently from corporations' economic context and mandate to generate profits.

"Corporate culture, and its values, has a big effect on all our other instructions," Mirabai Bush, who gained notoriety by teaching mindfulness courses at Monsanto, told *Tricycle* magazine in an interview (2001). Bill Duane, senior manager at Google, considers SIY a "sort of organization WD-40, a necessary lubricant between driven, ambitious employees and Google's demanding corporate culture. Helping employees handle stress and defuse emotion helps everyone work more effectively" (Kelly 2012). In other words, mindfulness helps employees cope with the culture that a competitive workplace operating within a competitive

economy necessitates. It is what makes the long hours and unreasonable performance demands tolerable. Mindfulness, then, is recruited to mitigate the most harmful cultural effects that the company depends upon, not reform that culture, which would require a redistribution of power.

The corporation's prerogatives are a problem of neither individual comportment nor corporate culture. We cannot say whether or not the corporation's ends are ethical because they are the same for any; it is rather a question of the nature of capitalism and the capitalist firm. Milton Friedman hit the nail on the head when he wrote that "[t]here is one and only one social responsibility of business—to use its resources and engage in activities designed to increase profits." Who are workers—let alone CEOs, who are obligated to increase profits for shareholders or face termination by the board, appointed by the shareholders—to change the nature of the capitalist firm? Corporations function as they do not because they are immoral, but because they are *amoral*. Only in a society accepting of the absurd premise of corporate personhood could a corporation be said to be mindful.

Mindfulness meditation has not made its way into the modern corporation because it increases creativity or "leadership" skills. Nor has it because it makes for more ethical employees, and hence a more ethical corporation. Management has prescribed meditation neither to improve employee health nor well-being. Though it might do all these things, the reason the highest grossing corporations have adopted mindfulness meditation courses is far simpler, and almost too obvious to state, but bears repeating: It increases profits.

This is no secret. It is the first thing the corporate consultants who sell these programs will tell you, and they say the same to the media. But this is not an inspiring vision, a world with even greater profits for the 1 %. Instead, mindfulness meditation is sold as a tool to lower stress, increase individual well-being, and "optimize… impact and influence" (Google's "Search Inside Yourself" program). Furthermore, it is sold as a panacea to the worst excesses of capitalism. These programs, whose purpose it is to further the inalienable drive of capitalism to privatize wealth, will also mitigate the detrimental effects that very drive creates in terms of emotional and psychological health. Our economic and social system demands a fundamental dishonesty on our part about what it is and what it is not in order to function. Oriented in this way, a cultural critique of capitalism and its pretense of reform actually help to sustain this dishonesty.

No less is demanded of those who enter businesses to instruct staff in mindfulness exercises. They never hear the profit-boosting, absenteeism-busting spiel. Many supporters of workplace mindfulness interventions—especially teachers of meditation, insofar as they themselves are not management consultants but trained administrators of mindfulness technologies—remain insulated from the logic of their implementation. Employees, having as their only contact the administrator of mindfulness instruction, remain doubly insulated from this logic. This is an essential feature not only of corporate mindfulness but also of the management consultancy field in general. Taylor was able to do what he did precisely because he entered the workplace as an outsider. Management consultants and administrators can eschew labor tensions—that is, the conflicts of interest between workers and management/ownership—in spite of the fact that they are always brought in by the latter. This fact maintains regardless of whether instructors are elite consultants or simply mindfulness teachers, whether they are arch-capitalists or socialists. Indeed, the more insulated (or, we could say, "Buddhist") these outside mindfulness technicians are from their purpose in that role, the more successful they and the interventions they present are likely to be in fulfilling that purpose. Entering the firm as neutral parties interested only in the health and well-being of employees, they nevertheless serve the prerogatives of the management that pays them and to whom they are held accountable.

Unable to surmount the conflict of interest between labor and capital (of which management is surrogate) in the workplace, management theory and its executive acolytes attempt to say it away by convincing workers that their interests and those of business are one in the same. This represents a logical extension of industrial psychology, which seeks to inculcate the specific worker habits and attitudes most amenable to business and the needs of the market without confronting questions of political economy. Indeed, today's popular psychology is largely reflective of trends in institutional psychology and its fixation on optimization.

Arianna Huffington, the founder and CEO of *Huffington Post*, which is more responsible than any other media outlet for popularizing mindfulness, is one such aforementioned executive acolyte. "There is a growing body of scientific evidence that shows that these two worlds [of spirituality and capitalism] are, in fact, very much aligned—or at least that they can, and should, be," she writes in 2013 blog column. "I do want to talk about maximizing profits and beating expectations—by emphasizing the notion that what's good for us as individuals is also good for corporate America's bottom line" (Huffington 2013).

The vision of a harmonious alignment of interests between worker and firm is furthered by its promotion by mindfulness instructors. Again, Norman Fischer is the best representative of this trend, reciting the company line of his employer on command:

> [Google's] main value is not the hard-nosed, hard-edged, profit-seeking mind. It's the creative mind, the altruistic mind. They really believe that if you give room to and foster the creative altruistic mind, you will make money and you'll also be able to do good things. . . . my goal of personal integrity and the goals of the corporation and the participants seem to line up. (Boyce 2010)

He goes on to describe Google's "20 %" policy (discontinued in 2013), a common practice among large technology companies in which employees are encouraged to use one-fifth of their time to work on company-related pet projects, benevolently "saving the world through technology," in Fischer's words. But this is a gross misrepresentation, as the creation of products like Gmail, Google News, Google's autocomplete system, and AdSense, the advertising engine that accounts for around a quarter of the company's $60 billion yearly revenue, has its origins in the 20 % program (Tate 2013).

The contention that the interests of the worker and firm are one and the same is furthered not just through rhetoric but also through the very staging of forms of welfare capitalism (or "industrial paternalism," to distinguish the term from its other meaning referring to social democratic policies), including mindfulness interventions. Mayo's signature was to demonstrate that showing attention to employees could be a greater boon to productivity than improving working conditions or offering material incentives, at least in the short run. This accounts for one of the subtler reasons why mindfulness teachings in the workplace are almost always accompanied by the enumeration of the practice's scientific health benefits: the worker needs to be made aware that the company is taking an interest in his or her well-being in order for the intervention to perform its primary ideological function, one that has real, demonstrable consequences in terms of increasing employee loyalty and productivity—what is known as the "Hawthorne Effect" from Mayo's signature study at the eponymous company. Advising others in management and human resources at the Wisdom 2.0 conference, Meng himself notes how important studies of the neuroscientific benefits of meditation are for administrators of workplace interventions (Tan et al. 2012), who should be expected to impart this knowledge in their presentations.

The language in which consultants and their firms sell their services to companies lays bare this specific rationale. SIY advertises its services with client testimonials, one of which states the program's benefits thus: "It says to the employees that you care about them." Or, as Kabat-Zinn told the audience at Wisdom 2.0, mindfulness enables managers and executives navigate their companies "in ways that really help them [employees] *feel* like they belong, and their lives meaningful" (emphasis mine) (Kabat-Zinn 2013). In a short article in *The Guardian*, "leadership guru and practicing Buddhist" Sander Tideman tells the author how "it makes a real difference if you go to your office in the morning

thinking, 'wow, I'm helping provide hygiene for kids in India', to arriving at your office and thinking 'I sell soap.'" For SIY, what is important is not caring about employees but making employees *feel* that you care about them. For Kabat-Zinn, the point of mindfulness is not to provide meaningful work or actually value employees but to convince them that this is the case. For Tideman, it is not purposeful work that improves morale but the feeling that the work is purposeful, regardless of the reality. All of these examples espouse a particular instrumental approach to the feelings and emotions of employees that originated with Elton Mayo. It is the final expression of labor as commodity, which reduces people to a means for generating profit for others. The discontents of alienated labor are refigured as personal inadequacies and questions of personal responsibility to and for oneself.

Management's show of interest in the employee's well-being also serves to obscure the actual relationship of power. Given the employer's show of beneficence, to shed light on or question this balance of power would seem inappropriate or discourteous.

In these ways, corporate mindfulness retains the raison d'etre of scientific and psychological management interventions: to neutralize tension between workers pushed to their limits and management, who engineer work and adjust the worker according to the demands of capital.

Recovering the Artistic Critique

Part of what obscures the social functions and ideological underpinnings of management interventions is management theory's recovery of what French sociologists Eve Chiapello and Luc Boltanski term the "artistic critique" of capitalism (Boltanski and Chiapello 2005). This critique venerates the realization of human potential and bemoans the erosion of social bonds and ethical behavior under capitalism in much the same style as more radical commentaries. Chester Barnard, one of Mayo's most prominent successors, in particular emphasized "self-realization" over profits, a shift in values most would sympathize with. The genius of management theorists like Barnard was to instrumentalize this cultural critique in the service of business, specifically to increase engagement and productivity. Managers and theorists viewed themselves as managing neither "men nor work" but "administering a social system" (cited by Wolin 2004, p. 361). In other words, they sought to repair the social bonds eroded by the capitalist mode of production without challenging that mode of production (or consumption, for that matter) but rather accepting it as an inevitable given, much like improving "race relations" might take precedence over eradicating racism. Acknowledging that the competition on which the capitalist economy depends contributes to social divisiveness, management theorists nevertheless skirt questions of ownership and democracy under capitalism, viewing "the basic problem as one of restoring communal solidarity in the industrial age" (Wolin 2004, p. 364).

For mindfulness in particular, special emphasis is put on health and holistic well-being. Once mindfulness becomes a part of the job, the job becomes a source of whatever health and well-being mindfulness contributes to.

Buddhist-derived meditation techniques in particular have benefited from recent overlaps with trends in management theory. As "networks" and "the permanence of change" became *the* two key tropes of management literature in the 90s (Budgen 2000), so did MBSR—which claimed to provide insight, however rudimentary, into the impermanence of mental events and the interconnection of all things—attract a newfound corporate audience.

What makes these "new spirit" programs so seductive, and to baby boomers especially, is that they recuperate the romantic and libertarian undercurrents of '68 that value radical individualism, the primacy of individual well-being and growth, and horizontal structures of management over hierarchical control (Budgen 2000). "Leaders" and "coordinators"—terms for the managers of the "liberated firm"—would gain authority by "acceptance" through superior "communication skills." Such thinking found a

comfortable home in the new corporation, which appropriated late-60s tropes while reinventing new systems of control. "Enlightened" employers—enamored with hot terms like "networks," "diversity," and the symbol of the "visionary" leader, and so revolted by hierarchy and top-down control—embraced the radical individualism of the previous generation's counterculture and rendered it compatible with the new spirit of capitalism. The "network," no longer associated with organized crime, became the ultimate symbol of forward-thinking and the financialization of markets (Budgen 2000). Indra's Net suddenly seemed to coincide with the underpinnings of business and finance. Interconnectedness and interrelatedness—what was at least generally understood in early Buddhism as *bondage*—found a new, albeit vague, significance in management-speak.

In the corporation, mindfulness is lauded on its own terms at the same time as it is instrumentalized as a productivity device. Businesses can make appeals to spiritual values to increase loyalty and efficiency without offering more worldly rewards like pay, healthcare plans, vacation time, and reasonable hours, to say nothing of having a real say in their own work and the missions of their companies.

While meditation is associated with an escape from economic and disciplinary rationalities, it is deployed in the corporate setting to further precisely these rationalities. A scientific study, for example, found that meditation retreats can produce the same affective targets, or levels of relaxation, in less time than vacations to, say, the beach, thereby reducing the amount of time employers need to give their employees to achieve the same economic outcome (Chang and Pichlhoefer 2015).

Work—what ties the individual to society and endows his or her life with meaning—has always, in some sense, been the most significant source of self-actualization and self-realization. Thus, while the more recent humanist iterations of scientific management that embrace the follow-your-bliss attitude would at first seem to undermine the core of management—the refinement of methods of control—they instead involve an internalization of those very mechanisms. As Meng has said regarding workplace mindfulness, "your work will become a source of your happiness" (2012, p. 301). The genius of recent humanist iterations of management theory, however, lies in the way this significance can be retained without examining the purpose of work, which can, admittedly, be a depressing prospect, considering a great part of postindustrial work does not provide a socially useful function.

Moreover, mindfulness's particular emphasis on self-management and self-control meshes seamlessly with the horizontal management structures favored among today's forward-thinking employers and employees alike. "The whole point of mindfulness-based stress reduction," writes Kabat-Zinn, "is to challenge and encourage people to become their own authorities, to take more responsibility for their own lives, their own bodies, their own health" (Kabat-Zinn 1994, pp. 191–192). The most important shift in management over the past half-century has been the transition from top-down control to *self*-control. Mindfulness techniques have become its ideal expression, encouraging self-management in place of formal supervision. If your most important manager is yourself, constraints become exclusively personal, the result of your personal dispositions, and the only thing holding you back from self-actualization. "We're looking for alignment, finding our deepest values, envisioning how they'll take us to our destination and the resilience we need to achieve that," as Meng told the *New York Times* (Kelly 2012).

What is perhaps most striking about the corporate mindfulness movement in contrast to scientific management or industrial psychology is its emphasis on targeting the "leaders" of firms—senior managers and executives. On the one hand, this simply mirrors recent trends in management theory that emphasize the leader's vision and the culture and values with which they imbue the organization as most essential to encouraging "the convergence of forms of individual self-control, since the controls voluntarily exercised by everyone over themselves are more likely to remain consistent with one another of

their original source of inspiration is identical" (Boltanski and Chiapello 2005, pp. 80–81). But this emphasis also points to an essential feature of management theory in general. Braverman argued that "no part of capitalist employment is exempt from the methods which were first applied on the shop floor," (1998, p. 88) meaning that even the executive is not exempt from the encouragement of certain attitudes and habits necessitated by economic competition and profit-seeking. Elite mindfulness consultant Janice Marturano, founder of the Institute for Mindful Leadership, agrees that in our current age, even business executives have come to "feel disconnected from their own values" (Marturano 2014).

Perhaps the good news is that Marturano's prescription to introduce the practice of mindfulness to rise to "the challenges at hand," echoed by so many consultants, cannot actually provide deliverance. As much as they level a critique of the unbridled greed, rampant individualism, and runaway egocentricity under capitalism—a critique even echoed by the economic elites of the World Economic Forum and *Harvard Business Review*—management solutions will continue to fail to solve the underlying problem, as they reproduce the categorical error of identifying individual choice and responsibility as the only worthwhile vector for change. Corporate mindfulness offers a technical solution to a social and political problem. Like other management trends, it also regularly misattributes prosperity to the innovation, creativity, and leadership skills of those who profit directly from the real source of our prosperity: the hard work of the many.

This meritocratic view only functions because it also appeals to workers. Scientific management met great success because it "multiplied the loci of control" within the expanded firm, shifting power from traditional capitalists to technocrats who established their authority using "the rhetoric of science, rationality, and general welfare" (Illouz 2013). Innate privilege would be displaced by new forms of rationality, which, far from given, would become articulated and popularized through decades of work on the part of industrial psychologists. Mindfulness represents a recent culmination of the meritocratic ideas of industrial psychologists in which one's status in a company is legitimized by one's emotional skills or level of mindfulness or inner-development rather than one's class privilege or pedigree. This idea is appealing, but its actual implementation only serves to retrospectively authorize the power of the few over the many, meritocracy be damned.

It's You

With stress alone costing US businesses as much as $300 billion a year (Gelles 2015, p. 84), largely in the form of indirect healthcare costs related to reduced productivity and absenteeism, companies seek to "mount an aggressive approach to wellness, prevention, screening, and active management of chronic conditions," (ibid.) a whole area of health concerns heretofore excluded from the purview of human resources. But "for all this talk of stress," Gelles writes, "we rarely examine its root causes." One cannot help but agree. However, Gelles identifies the root cause of stress as the individual's mental and behavioral habits:

> If stress results from out-of-control thinking, the solution, it stands to reason, is learning how to, if not control our thoughts, at least not let them control us. . . . That's where mindfulness comes in. Stress isn't something imposed on us. It's something we impose on ourselves. As a popular saying in mindfulness circles goes, 'Stress isn't what's happening. It's your reaction to it.'. . . That is, stress emanates from a mismatch between our expectation of how things should be and the way things are. It is the result of us not being able to control our own thoughts. (2015, pp. 84–85)

We can dismiss this as a kind of solipsism. After all, Kabat-Zinn first developed MBSR in the 1970s specifically to aid those struggling with chronic pain and illness against the unnecessary psychological suffering accompanying their diseases or disorders that they could do little about. Workplace mindfulness, on the other hand, is specifically meant to address the far from natural stress that postindustrial work produces.

Stress is not something that comes from nowhere. We should understand these statements from Gelles as political in nature, as they assume that any causes of stress other than those springing from the deficient self-management of one's own thoughts to be illusory. It also dismisses that glorious gap between "how things should be and the way there are" as a kind of impediment to flourishing. This is, after all, the space of the social imagination. It is also the space in which democracy operates. Here, it is swiftly and thoughtlessly pathologized as a lack of mental control. (Recall Mayo: "Industrial unrest is. . . caused. . . by the fact that a conscious dissatisfaction serves to 'light up'. . . mental uncontrol.") This is an incredibly effective way to depoliticize stress and neuter any kind of collective, critical response.

The most radical challenge to such pathologization would be to ask the worker *why* he is stressed, to empower him to speak for himself. William Davies puts it well in *The Happiness Industry*:

> human beings may have their own considered reasons to be happy or unhappy, which may be just as important as the feelings themselves. . . . we have to recognize that they possess authority to speak for their own thoughts and bodies. . . . Were, for instance, someone to describe themselves as "angry," a response focused on making them feel better might entirely miss the point of what they were saying. . . . In a monistic world [i.e., one in which all pleasures and pains can be located on a single scale], there is merely sentiment, experiences of pleasure and pain that fluctuate silently in the head. (2015, pp. 33–34)

For treating stress, mindfulness's ingression into the workplace represents, in the words of sociologist Kristin Barker, an "expansion in the terrain of the pathological" (Barker 2014). Within such a pathological framework, the kinds of stress and strain that contemporary postindustrial work induces are abstracted from the very working conditions that cause them, implicitly blaming stress on the one who experiences it. Stress is refigured as a free-floating, inevitable problem to which mindfulness is the solution. In this way, mindfulness recasts workplace stress and discontent within a narrow etiology particular to psychological behaviorism, which sublimates an earlier, political vocabulary of worker discontent.

The refashioning of social and political issues as evidence of personal shortcomings and psychological dispositions goes hand-in-hand with this specific understanding of workers as people in need of help, even charity. Technocratic capitalist control depends on such pathologization, on the manipulation not only of thoughts through ideology but also that of affect.

Once again, these attitudes have their origin with Mayo, who dismissed the labor agitator who challenged inequality in the workplace and beyond as "disoriented" and "usually a genuine neurotic" and his theories of alternate political economic systems as "very largely [neurotic] phantasy constructions" (Quoted in Gelles 2015, p. 121). Inequality is what makes this attitude—indeed, the managerial attitude in general—possible. Strategies of manipulation only make sense within such a framework.

What is particularly disturbing about this state of things is how the very same managerial elite whose existence depends upon the subjugation of the worker offers her sanative programs to manage her discontent. This creates a kind of double bind for workers. If work is the primary mode that ties an individual to society, then social control, as Herbert Marcuse observed, is "anchored in the new needs which it has produced" (1971). For the engineer of work, what he or she defines as "good" habits, which management "leaders" gently suggest with a beneficent smile, fall into the scheme of domination, yet at the same time make the situation of the worker more tolerable. Long before Taylor, after all, management theory first cropped up as a way to secure the cooperation of slave labor and discourage uprisings. In this sense, managers who represent the interest of the ruling class always manage precisely the condition of non-autonomy, non-self-determination, and non-freedom of others.

This pairing of sanative concerns regarding "well-being" with profit-making represents one of the more insidious developments of capitalism in the 20th century, especially because it comes to inform the way people think about themselves, subjectivities that follow them beyond the private sphere of profit-making to the public and civic spheres.

Davies posits that a "single ideal" lies behind any managerial initiative, and argues that an inverse relationship exists between democracy and the behaviorist solutions put forward by psychologists and engineers of work:

> that individual activity might be diverted towards goals selected by elite powers, but without either naked coercion or democratic deliberation. . . . When we put our faith in "behavioral" solutions, we withdraw it from democratic ones to an equal and opposite extent. (2015, p. 88)

Insofar as it seeks to develop certain skills among workers, corporate mindfulness—or any technology that requires the modification of habits—fosters specific attitudes, and therefore falls under what Foucault called technologies of domination. Davies identifies such modifications, based on behaviorist understandings that view people as clusters of irrational sentiment, as fundamentally undemocratic. Kabat-Zinn unknowingly spoke a profound truth on mindfulness in the workplace when he conceded that, when mindfulness is brought to managers and executives, "the applications unfold on their own" (Kabat-Zinn 2013).

The defining quality of corporate mindfulness is that it always begins, "despite occasional protestations to the contrary," as economist Harry Braverman wrote of scientific management,

> not from the human point of view but from the capitalist point of view. . . . It does not attempt to discover and confront the cause of this condition, but accepts it as an inexorable given, a 'natural' condition. . . . It enters the workplace not as the representative of science, but as the representative of management masquerading in the trappings of science. (1998, p. 59)

In the words of senior Google manager Bill Duane, "business is a machine made out of people" (Kelly 2012). And when these people are not viewed as a means, they are simply a problem.

"If you have people," he says, "you have problems."

A Shift in Power

Democracy is not in reality what it is in name, until it is industrial as well as civil and political.
—John Dewey (1886)

The purpose of Taylorism—or "scientific management"—and its offshoots of industrial psychology and industrial sociology has always been the application of science to the control of labor and its adjustment to the needs of capital. The recent trend of corporate mindfulness is firmly rooted in these needs, continuing this long tradition.

Although corporate mindfulness techniques have been presented as a radical departure from the past—as business tends to regard any of its products, old or new—they merely represent the most recent tool in an established tradition of business management. The movement shares a telos not with Buddhism or even Buddhist mindfulness practice, but of scientific management and industrial psychology. It is deployed to adjust work and worker and manager and executive—a transformation on all sides—to the needs of capital, but does not pause to question whether the needs of capital in any way represent or fulfill human needs. As the newest iteration of industrial psychology, corporate mindfulness has participated in corporate capitalism's commodification of the mind, of the "optimization" of people to generate ever-greater profits.

The biggest issue with scientific management and its newest iteration, corporate mindfulness, is that it offers psychotechnical solutions to what are fundamentally political problems. In doing so, it not only neglects to redress those problems but also obscures them, rendering them all the more impenetrable. It seeks to modify man within the scheme of work that is further concentrating wealth among the few and is destroying our planet.

The hidden brutality of work lies in the way it incorporates the human mind, with its whole range of aspirations and emotional potentialities, into the profit-creating machine. Corporate mindfulness's pretensions to change the world through management techniques are at times least as earnest as they are misguided. They stem from a hope that the capitalist corporation can be other than what it is, and that social reformation can be brought about by self-improvement.

The message is clear: Don't try to change the world. Manage it.

The individual but shared discontent of so many Western workers, traditionally voiced through organizations like labor unions, could potentially serve as fertile ground for organizing mass movements. Instead, attention has been redirected toward the purportedly inexorable proliferation digital technologies that are said to make undue solicitations upon attention that lead to stress, anxiety, and depression, and the myriad fixes, presented in the workplace and beyond, for coping with these demands. To counter this, we must attend to the social and ethical frameworks that orient mindfulness meditation programs and practices, as well as the social relations that shape our day-to-day lives at work.

The promise of management theory is to dissolve all antagonisms between ownership/management and labor, and between the corporation and society. But it will fail, as it always has. Business cannot succeed in this because the work conditions it depends upon militate against its resolution. Yet the question remains what will be made of this failure.

In the past, worker discontent—however diffuse, however general—has at times become politicized as demands for better working conditions and higher wages. Movement-building has always depended on a combination of discontent and moral outrage. It is the management theorist's duty to defuse both.

But we must take the generalized discontent endemic to postindustrial society far more seriously than the manager or technocrat. It is not work itself, which gives purpose and meaning, which connects us to one another, but its dominative aspect that leads to discontent. Other organizations of work might create real meaning and purpose among workers, and not just the false feeling of it.

Disempowered workers who have little control over the organization of their work are naturally unhappy. It is a recipe for anxiety and depression. The repoliticization of stress and emotional ineptitude (as defined in contradistinction to the "Emotional Intelligence" that mindfulness is supposed to engender) might involve establishing correlations between disempowerment and anxiety and depression. Research has shown that the countries with the highest levels of income inequality also report the highest incidences of stress, worry, and anger among its populations (De Neve and Powdthavee 2016). De Neve and Powdthavee have found that it is in particular wealth inequality, and not wealth or lack thereof, that seems to have the most detrimental effect on well-being: "a 1 % increase in the share of taxable income held by the top 1 % hurts life satisfaction as much as a 1.4 % increase in the country-level unemployment rate" (2016). In general, we still know very little about how inequality affects psychological well-being. Still, it is clear that "unhappiness and depression are concentrated in highly unequal societies" like the USA and UK (Gelles 2015, p. 9). "Among wealthy nations," writes Davies, "the rate of mental illness correlates very closely to the level of economic inequality across society as a whole, with the USA at the top."

To challenge inequality on a societal scale, the workplace is an ideal place to start. Throughout the 20th century, corporations served as an incubator for the industrial psychological and managerial attitudes, both of which are predicated on inequality, that would eventually proliferate throughout popular psychology and culture. Workplaces, cooperatives in particular, can potentially incubate more democratic movements. But first workers must demand the one thing they have not been given: democratic control of the workplace.

Concluding Remarks

I have only sought out to understand how mindfulness functions within the workplace. To that end, perhaps the most notable thing about mindfulness meditation in the workplace in particular is how its implementation achieves effects that have little to do with mindfulness itself, which is simply a calming relaxation technique that involves the focusing or reorientation of attention. Instead, I have strived to lay bare the ideological goals that underpin mindfulness interventions in the workplace.

The most salient feature of the new mindfulness rhetoric is the restriction of discourse to talk of psychological effects to the exclusion of all else. The more directly and reliably these effects are observed—such as is believed to be the case with neuroscience's psychosomatic surveillance tools, despite the fact that specific psychological states must always be extrapolated and abstracted from a combination of interviews and neuroscientific data—the better. While psychology and the production of specific psychological effects in particular must not be excluded from the analysis of mindfulness meditation in the workplace or elsewhere, neither should it be limited by it. Neuroscience and psychology both have their own ways of speaking about experience. Both participate in producing the subjects of study they purport to observe. We can see, for example, how Elton Mayo's construal of workers as unstable bundles of emotions inscrutable to themselves still has purchase today, and how such a construal discredits various forms of worker radicalization and political awakening. It is high time we repoliticized stress and articulated our discontent, together.

References

Anteby, M. (2015). *Manufacturing morals: The values of silence in business school education*. University of Chicago Press.

Anteby, M., & Khurana, R. (2012). Employee Welfare. Retrieved from http://www.library.hbs.edu/hc/hawthorne/02.html

Barker, K. K. (2014). Mindfulness meditation: Do-it-yourself medicalization of every moment. *Social Science and Medicine, 106*, 168–176. doi:10.1016/j.socscimed.2014.01.024.

Beck, U. (1992). *Risk society: Towards a new modernity*. SAGE Publications.

Boltanski, L., & Chiapello, E. (2005). *The new spirit of capitalism*. Verso.

Boyce, B. (2010, August 26). Google Searches—Mindful. Retrieved from http://www.mindful.org/google-searches/

Boyce, B. C. (2011). *The mindfulness revolution*. Shambhala.

Brandeis, L. D., Urofsky, M. I., & Levy, D. W. (1971). *Letters of Louis D. Brandeis: Volume II, 1907–1912: People's Attorney*. State University of New York Press.

Braun, E. (2013). *The birth of insight: Meditation, modern Buddhism, and the Burmese monk Ledi Sayadaw*. University of Chicago Press.

Braverman, H. (1998). *Labor and monopoly capital: The degradation of work in the twentieth century*. Monthly Review Press.

Budgen, S. (2000, January). A New "Spirit of Capitalism" [Review of Luc Boltanski and Ève Chiapello, Le Nouvel esprit du capitalisme]. *New Left Review*, pp. 149–156. Retrieved 2016, March 2 from https://newleftreview.org/II/1/sebastian-budgen-a-new-spirit-of-capitalism

Chang, A., & Pichlhoefer, O. (2015, June 4). *Recovery from work: A comparison of the effects of meditaiton retreat programs and leisure holiday vacataions in working European and American adults*. Lecture presented at Mindfulness & Compassion Conference in San Francisco State University, San Francisco.

Contemplating Corporate Culture [Interview by H. Tworkov]. (2001). *Tricycle: The Buddhist review 10*(4).

De Neve, J., & Powdthavee, N. (2016, February 16). As the rich get richer everyone else gets less happy. Retrieved from http://www.theguardian.com/sustainable-business/2016/feb/16/inequality-happiness-rich-getting-richer-poor-wellbeing-life-satisfaction

Desteno, D. (2013, July 06). The morality of meditation. Retrieved from http://www.nytimes.com/2013/07/07/opinion/sunday/the-morality-of-meditation.html

Dewey, J. (1886). *The Rise of great industries*. Speech to the Political Science Association, cited in Jay Martin, *The education of John Dewey* (p. 109). New York: Columbia University Press, 2002.

Folger, J. (2013, July 10). The causes and costs of absenteeism in the workplace. Retrieved from http://www.forbes.com/sites/investopedia/2013/07/10/the-causes-and-costs-of-absenteeism-in-the-workplace/

Gallup. Report: State of the American Workplace. (2013). Retrieved from http://www.gallup.com/services/176708/state-american-workplace.aspx

Gelles, D. (2015). *Mindful work: How meditation is changing business from the inside out*. Houghton Mifflin Harcourt.

Gillespie, R. (1991). *Manufacturing knowledge: A history of the Hawthorne experiments*. Cambridge: Cambridge University Press.

Huffington, A. (2013, March 18). Mindfulness, meditation, wellness and their connection to corporate America's bottom line. Retrieved 2015, April 1 from http://www.huffingtonpost.com/arianna-huffington/corporate-wellness_b_2903222.html

Illouz, E. (2008). *Saving the modern soul: Therapy, emotions, and the culture of self-help*. University of California Press.

Illouz, E. (2013). *Cold intimacies: The making of emotional capitalism*. Wiley.

Johnson, D. V. (2013, July 22). Revolutionizing ethics. Retrieved from https://www.jacobinmag.com/2013/07/revolutionizing-ethics/

Kabat-Zinn, J. (1994). *Wherever you go, there you are: Mindfulness meditation in everyday life*. New York: Hyperion.

Kabat-Zinn, J. (2013, March 28). Applied mindfulness in business and life: Jon Kabat-Zinn, Melissa Daimler. Retrieved from https://www.youtube.com/watch?v=AjqDbJR6mo0

Kanigel, R. (1997). *The one best way: Frederick Winslow Taylor and the Enigma of Efficiency*. New York: Viking.

Kelly, C. (2012, April 28). O.K., Google, Take a deep breath. Retrieved from http://www.nytimes.com/2012/04/29/technology/google-course-asks-employees-to-take-a-deep-breath.html

King, M. L. (1994). *Letter from the Birmingham jail*. Harper San Francisco.

Marcuse, H. (1971). *An essay on liberation*. Beacon Press.

Marturano, J. (2014). *Finding the space to lead: A practical guide to mindful leadership*. Bloomsbury Publishing.

Mayo, E. (1922). Industrial peace and psychological research. *Industrial Australian Mining Standard, 5*, 16.

Mayo, E. (1924, October). Civilization—the perilous adventure. *Harper's Monthly Magazine, 149*, 590–597.

Montgomery, D. (1989). *The fall of the house of labor: The workplace, the state, and American labor activism, 1865–1925*. Cambridge University Press.

Münsterberg, H. (1913). *Psychology and industrial efficiency*. Houghton Mifflin.

Murphy, F. (2014, June 26). Employee burnout behind a third of absenteeism cases. Retrieved from http://www.covermagazine.co.uk/cover/news/2352243/employee-burnout-behind-a-third-of-absenteeism-cases

Paywatch 2014. (n.d.). Retrieved March 2, 2016, from http://www.aflcio.org/Corporate-Watch/Paywatch-2014

President Kagame at Wisdom 2.0 Conference- San Francisco, 14 February 2014. (2014). Retrieved from https://www.youtube.com/watch?v=fPefVu4btgA

Saez, E., & Zucman, G. (2014). Wealth inequality in the United States since 1913: Evidence from capitalized income tax data. doi:10.3386/w20625

Scharmer, O. (2013, February 5). Collective mindfulness: The leader's new work. Retrieved 2014, April 7 from http://www.huffingtonpost.com/otto-scharmer/collective-mindfulness-th_b_4732429.html

Solomon, J. (2014, October 1). Top 10 companies lobbying Washington [Web log post]. Retrieved from http://money.cnn.com/2014/10/01/investing/companies-lobbying-10-biggest-spenders/

Stewart, M. (2009). *The management myth: Debunking modern business philosophy*. W. W. Norton.

Tan, M. (2014, March 20). 3 steps to build corporate mindfulness the Google way: Karen May, Meng Tan, and Bill Duane. Retrieved from https://www.youtube.com/watch?v=zVlXE1UtJKY

Tan, C. M., Goleman, D., & Kabat-Zinn, J. (2012). *Search inside yourself: The unexpected path to achieving success, happiness (and world peace)*. HarperCollins.

Tate, R. (2013, August 21). Google couldn't kill 20 percent time even if it wanted to. Retrieved from http://www.wired.com/2013/08/20-percent-time-will-never-die/

Taylor, F. W. (1911). *The principles of scientific management*. Harper & Brothers.

Taylor, U. S. C. H. S. C. t. I. t., Management, O. S. o. S., Taylor, F. W., & management, U. S. C. H. F. (1912). *The Taylor and other systems of shop management: Hearings before special committee of the house of representatives to investigate the Taylor and other systems of shop management under authority of H. Res. 90. [Oct. 4, 1911-Feb. 12, 1912]*. U.S. Government Printing Office.

Trahair, R. C. S., & Zaleznik, A. (2005). *Elton mayo: The humanist temper*. Transaction Publishers.

Wolin, S. S. (2004). *Politics and vision: Cotinuity and innovation in western political thought* (expanded ed.). Princeton University Press.

Mindfulness in the Working Life. Beyond the "Corporate" View, in Search for New Spaces of Awareness and Equanimity

Massimo Tomassini

The Advent of Organizational Mindfulness

Among the media images associated with mindfulness, those of corporate managers engaged in various types of supposedly stress-reducing postures are just a little less common than those of closed-eyed, softly-smiling, white charming women. In the last decade mindfulness in the business world has become a relevant topic for media coverage and a symptom of how Buddhist meditation has extended its appeal to the most canonical areas of neo-liberal capitalism within the mutual transformation process between Buddhism and Western culture (Wilson 2014). Prompted by the media coverage and other relevant factors, a number of mindfulness-based interventions within large companies, corporations, and business schools have burgeoned (Chaskalson 2011; Tan 2012; McKenzie 2015; Marturano 2013; Hunter 2013; Gelles 2015), giving rise to a social phenomenon usually defined as *corporate mindfulness*.

The grand majority of such interventions—with few exceptions—for instance, those within the "Corporate-Based Mindfulness Training" approach (Hougaard 2015)—have been conducted along the guidelines of the MBSR protocol established by Jon Kabat-Zinn and colleagues (2003; Santorelli 2000; Segal et al. 2002; Williams et al. 2013). All such interventions—carried out by consultants, counselors, corporate trainers or by the emerging category of licensed MBSR instructors—have been developed for the sake of enhancing employees' well-being and fostering growth in the organizations' overall levels of effectiveness and productivity.

Conceptually the protocol implemented in business contexts is no different from the one used in clinical applications—as reported in a vast scientific literature (Chiesa and Serretti 2009; Chiesa and Malinowski 2011)—and in an extremely wide range of other applications: from parenting, education and elderly care to prison inmates' treatment and troopers' before-fire training (Ergas 2015; Kabat-Zinn and Kabat-Zinn 2010; Stanley et al. 2011). As a whole, such applications can be placed within the category of "mindfulness based interventions" (Cullen 2011); but, despite the common roots established by MBSR, they seem quite distinct in terms of goals and styles. On one extreme end of such differentiation, in clinical environments, the protocol has reached the highest peaks of acknowledgment by scientific and health care communities—after initial suspicion—and has gained the status of a commonly used therapeutic tool (Kabat-Zinn 2011). On the other extreme, in business

M. Tomassini (✉)
Department of Education Science, Roma Tre University, Rome, Italy
e-mail: massimo.tomassini@uniroma3.it

contexts, MBSR has only rarely been tested according to scientific procedures, following formalized experiments (Davidson et al. 2003; Nielsen and Kaszniak 2006; Levy et al. 2012). In these contexts, in fact, scientific assessment activities have to face many more difficulties than in clinical settings, mostly due to the lack of standard parameters (e.g. regarding the reduction of pain) and to environmental conditions (e.g. little time to be spent in trials and tests).

The fashion factor, prompted by the above mentioned media coverage, has played an important role in the diffusion of MBSR in business contexts, but other factors have also been relevant, like the soft appeal of mindfulness' Buddhist roots. This factor was left in the background to avoid the generation of misunderstandings and rejections, (Kabat-Zinn 2011) but it was undoubtedly capable of adding a certain allure of mystical exoticism to otherwise overly clinical and dry practices. Good results have been reached in terms of the satisfaction and acceptance of the involved audiences: a significant phenomenon like corporate mindfulness couldn't have lasted and proliferated as it has without producing significant impacts on its clients. Keeping in mind, however, that "clients" in this case are both the involved company populations and the management staffs that demanded, and paid for, these interventions.

Many beneficial effects of mindfulness are illustrated in a dedicated literature (Chaskalson 2011; McKenzie 2015; Gelles 2015) which—mostly in a journalistic style—deals with multitasking, decision-making, leadership, busyness, social interaction and many other relevant aspects of individual and organizational life. The prevailing tone of this literature is encomiastic: more than informing about the topic, the aim seems to be convincing the reader of the benefits of the product. In many ways these types of books look like extended versions of the material that has abounded in the last few years in print and online magazines, such as *Huffington Post* and *Time* (e.g. Pickert 2014), as well as in outstanding economic newspapers (e.g. Gardiner 2012).

Inevitably, such a large wave of praise for mindfulness has attracted many criticisms. For many—especially those who retain a specific concern about its Buddhist roots, mindfulness should not be used as a managerial tool for the joint improvement of employees' well-being and organizational effectiveness. Several representatives of official Buddhist traditions, and lay scholars as well, have underlined the contradiction between the idea of mindfulness as a tool for getting better performance, happiness and well-being, on one side, and the original meanings of a consciousness which is intrinsically refractory to any instrumental use and makes sense only in terms of aimlessness and selflessness, on the other side (Bodhi 2011). Moreover the lack of ethical commitment has been widely assumed to be a fundamental obstacle to acknowledging MBSR's relationship to the Buddhist tradition. With even more intensity, corporate mindfulness has been attacked by representatives of an area of critics (not necessarily coinciding with the typical "engaged Buddhism") that tend to couple Buddhism and social engagement. The term "McMindfulness", gone viral on the web, represents well the positions of those who denounce both the commodification of mindfulness, a fundamental tenet of Buddhist practices and spiritual horizon, and its subjugation to the rules of a typically neo-capitalistic industry—the "happiness industry"—whose main raw materials are human emotions.

This chapter tackles some aspects of the, until now, briefly reported situation. It aims to define the starting point of a future debate about the feasibility of new interventions, not necessarily labeled as "mindfulness-based", but in which mindfulness practice could play some kind of role without being misinterpreted or abused. To this end, it seems necessary to attain a better understanding of the phenomenon known as *corporate mindfulness*, whose implications go far beyond the corporate world. They also have very much to do, in general, with the interplay between contemplative activities and life, especially working life, dynamics. The main hypothesis is that it would be possible to carry on interventions in which mindfulness practices could be paralleled by reflective practices hinging on challenges that arise in working life.

Both the more recent wave of organizational mindfulness-based approaches that have risen up through the introduction of MBSR and the previously used "cognitive mindfulness in organizations" approach—developed by Karl Weick and his school—are taken into consideration in this chapter. Subsequently, the criticisms that have been raised against dominant organizational MBIs are analyzed; both as they emerge from the Buddhist field and are related to fundamental aspects of the Teachings and from the "McMindfulness" opponents (also related to more general criticisms of the neo-liberal society) as well. Some hypotheses about new kinds of awareness-oriented interventions that rely on working life experiences are presented in the last section and in the concluding remarks.

Cognitive Mindfulness

A variety of different practices have been developed over time in organizational contexts to introduce mindfulness for productive purposes, according to principles and criteria which have very little to do with the corporate mindfulness wave. In particular the approach to organizational mindfulness developed by Weick and colleagues in the 1990s seems to represent an interesting attempt at creatively identifying the role of mindfulness within organizational processes (Weick et al. 1999; Weick and Putnam 2006; Weick 2009). Weick's approach to mindfulness (still followed by a number of scholars and practitioners), is overtly instrumental: mindfulness is considered a cognitive/behavioral skill (or quality, or capability), largely synonymous with purposeful attention and care in performance of activities. Therefore mindfulness is considered highly valuable when continuously applied and developed in organizations, in particular within areas such as operations, organizational design and management. High Reliability Organizations (HROs, i.e., air traffic control systems, nuclear-powered aircraft carriers, wildland firefighting teams), where avoiding accidents is a crucial imperative, are the best fields for recognizing the importance of cognitive

mindfulness. The latter tends to warrant closer attention to every detail of organizational flow in the workplace, and to foster individual and collective processes which may suppress the "tendencies toward inertia" implicit in every organization (Weick et al. 1999).

An important reference for such an approach has been provided by Langer (2000); she asserts that mindfulness is "a flexible state of mind in which we are actively engaged in the present, noticing new things and sensitive to context" (p. 220), allowing different action improvement opportunities, such as "the differentiation and refinement of existing distinctions; the creation of new categories out of the continuous streams of events that flow through activities; a nuanced appreciation of the action context and of alternative ways to deal with it". As noted by Baer (2003), in this view mindfulness is more concerned with awareness of external events than inner experiences (thoughts and emotions), and more concerned with goal-oriented cognitive tasks rather than nonjudgmental observation. A reconciliation of the external and internal sides of mindfulness is attempted by Dane (2010) within a definitively "Western" context. It is based on the correlation between *task performance* (in which attention goes towards internal phenomena and is related to the individual's degree of expertise) and *task environment* (that draws attention to external phenomena).

Weick's approach seems to actually be much more subtle than those of other advocates of organizational mindfulness. It is based on a sort of homology between meditation practice and organizational action, converging on the key phenomenon of attention, which allows for the seizing of important aspects of the effective organizational action, such as the suspension of judgment, a wariness of routine solutions, or the search for weak signals. Mindfulness—in this cognitive and instrumental version—is the engine for subtle, almost invisible, but nonetheless crucial activities in HRO, machinations which hinge on the interpretation of everything that influences the functioning or failure of specific plant components. It suggests a continuous reframing of established knowledge in relation to unexpected deviations that need to be corrected through individual action and inter-subjective sharing (Weick 2009). Key, in this regard, is the concept of *organizing* that Weick put forward in contraposition to *organization* in order to underline the nature of the organizational phenomenon, which is primarily based both on dynamic exchanges between people and on individual and collective attempts to establish viable and acceptable forms of sense-making (Weick 1995). Mindfulness is at the heart of five fundamental forms of organizational/interpretative action: "preoccupation with failure", "reluctance to simplify interpretations", "sensitivity to operations", "commitment to resilience", and "underspecified structuring" (Weick et al. 1999). These types of action take on a peculiar form within the dialogue between Western "organizational" mindfulness and "Eastern" mindfulness, whereby the former is mainly centered on individual cognition and collectively shared, written or non-written, rules while the latter is centered on inner and conceptually amorphous states (Weick and Putnam 2006). The mindful organizational agent, from this perspective, becomes increasingly knowledgeable and reliable about the procedures and practices he or she is involved in through full participation in organizing processes. The logic that drives organizing relies implicitly on inherent qualities or principles—such as concentration and mindfulness—that are typically objectives of meditative practices. The five previously quoted organizing processes associated with HROs incorporate properties which are intrinsically homologous to what meditation is deemed to induce or enhance. "Preoccupation with failure", for instance, can induce stable concentration and potentially vivid insights. "Reluctance to simplify" and "sensitivity to operations" can involve a kind of awareness to detail which is akin to concentration and mindfulness, and "commitment to resilience" can generate insights for future actions (Weick and Putnam 2006).

From a viewpoint which is highly coherent with Weick's approach, several correlations can be identified between mindfulness and reflection in organizations in order to reinforce the cognitive dimension of organizational behaviors. Mindfulness, from this perspective, is seen as both a state of mind and a mode of practice. It permits the "questioning of expectations, knowledge and the adequacy of routines in complex and not fully predictable social, technological, and physical settings" (p. 468). As such, it represents a prerequisite for reflection-in-action and allows practitioners to reflect on their actions as they go along. In this way the instrumentality of mindfulness is fully confirmed but, once again, it is put in the service of reaching goals (e.g. mutual questioning, job rotation; strategy review), which are nested in the social cooperative texture of the organization.

Organizational MBIs Guided Through MBSR

The organizational mindfulness-based interventions guided, or just inspired by, the MBSR model differ significantly in comparison with Weick's cognitive mindfulness. While the latter is highly "organizational", with an emphasis on forms of interpretation and communication which can be fostered through *mindfulness*, the theoretical-epistemological model which underpins MBSR organizational implementations is definitively "individual" and closely linked to its clinical origin, which lies under the all-encompassing label of *stress*. On one hand cognitive mindfulness (also in relation to its typical environments, continuous cycle processes and HRO, mostly populated by workers and technicians) is largely equatable to the *attention* needed for maintaining balance in complex and delicate organizational/production processes. Across the spectrum is the kind of mindfulness fostered through MBSR-guided interventions that largely coincides with the attention that people in large companies (mostly the higher-level managers and professionals, and their support staffs) have to address in their own mental processes, attitudes and behaviors. It is taken for granted that these individuals are challenged by intense and contradictory conditions of continuous technological innovation, professional creativity, chaos and hierarchal destabilization (Malone 2004). A capacity for "muddling through" and resilience towards stress are therefore fundamental criteria for surviving and thriving in such contexts. People have to face operational conditions which are increasingly volatile while traditional levers (structure, hierarchy role system, procedures) are significantly reduced in comparison to more stable organizational forms.

While Weick's cognitive mindfulness—underpinned by a continuous collective search for sense-making and intimately shared forms of a social construction of reality—fits well within an organizational model which has been dubbed *interpretative-symbolic* (Hatch and Cunliffe 2006), mindfulness of the corporate kind takes place within a rather different model which mainstream *modern* management approaches belong to. It stems from a positivistic epistemology whereby organizations are considered "systems of decision and action driven by norms of rationality, efficiency and effectiveness for stated purposes" (Hatch and Cunliffe 2006, p. 42). Mindfulness seems to introduce a significant variant into this model insofar as it contributes to summoning up the individuals' resources and to fostering their well-being within a bottom-up strategy that complements the formal rationality of the organization; "When practiced diligently mindfulness can help reduce stress, make us more productive, and boost happiness. It can transform not only the way we do our work, but the very work we do".

In many ways the implementation of mindfulness as a tool for personal and organizational development is primarily aimed at reinforcing the "individualized" nature of the organization (Goshall and Bartlett 1997; Coffield 2000; Prochaska et al. 2001). This shifts the burden of initiative and responsibilities from the organization to the individual organizational actors, as bearers of cognitive, emotional, and social competencies and of positive intentions regarding their implementation (Boyatzis 2007). Mindfulness, in this sense, looks like a contribution to reaching kinds of balance which are

organizational and, to an even greater degree, simultaneously *human* and *emotional*, irrespective of given formal requisites and positions in the organizational flow of everyone involved in the practices. Such an "emotionalization" of the working life yields different manifestations, from the simplest ones, in terms of emotional control [like accepting the nervousness that can arise before a video-link presentation or the self-condemnation for a mistake in the use of email, (Chaskalson 2011, p. 20, p. 41)] to more complex cases, like finding the courage to resist the authoritarian attitudes of a dismissive boss (Gelles 2015, p. 91). Most of the literature on MBSR in organizations is ultimately a collection of these kinds of stories, all based on the *re-perceiving* capacity induced by the method (Chaskalson 2011). "Rather than being immersed in the drama of their personal narratives or life stories, participants on MBSR courses learn the skill of standing back a little and witnessing what's going on for them. They learn that the phenomena that arise in practices such as meditation are distinct from the mind contemplating them" (Chaskalson 2011, quoting Goleman 1980, p. 22). On this kind of background the corporate mindfulness movement has built-up a body of techniques that is suitable for managing stress, avoiding burnout, facing the risks of Attention Deficit Hyperactivity Disorders (ADD), and preserving physical health and fitness. This body of techniques is deemed capable of keeping in control the mind's performance and channeling both natural tendencies (like mind-wandering) and technology-induced habits (like multitasking). MIBs thus represent an additional piece of the vast repertoire of well-being techniques made available for a society which has been described as hedonistic and at the same time oppressed by the imperatives of hyper-engaging ways of life (Wallace 2005).

Moreover, the corporate mindfulness movement has generated a panoply of directions and suggestions, repeated and multiplied within countless media messages that hold avoiding multitasking (during desk activities and meetings), being careful about the intensity of emailing, and taking breaks for breathing during the working hours in high regard. A number of recommendations have been also issued regarding how virtually any everyday activity should be carried out in a "mindful" way: eating, walking, sleeping, exercising (mainly through yoga), listening to others, presenting, chatting/talking, making decisions, etc. Such reasonable imperatives are further reinforced when addressed to leaders: they must not only handle their own mental and physical tendencies and attitudes, but also utilize more balanced qualities stemming from mindfulness for the betterment of the overall function they are committed to (i.e. delegating to others). Hence, in the field of leadership mindfulness is directly related to the development of a specific quality of "presence" which underlies the leaders' ability to positively influence their subordinates (Marturano 2014). These almost ineffable personality traits are generically described, as confirmed by experiences in companies from many different cultural systems (from North-America to China), in terms such as humbleness, good listening, effectiveness, courage, and many more. All in all, the fundamental characteristics of leadership excellence have been described through four relevant adjectives: focused, clear, creative and compassionate (Marturano 2014).

Consequently, in the updated (mindfulness-based) version of the *modern* management paradigm the top-down organizational rationality is not overcome but subsumed within forms of leadership that are strongly associated with personal qualities which are deemed as typical of the "resonant" or "primal" leader (Boyatzis and McKee 2005; Boyatzis et al. 2002). "Mindfulness", in this sense—complemented by two other fundamental factors, "hope" and "compassion"— is not necessarily generated by specific practices (like MBSR) but is the product of intentional drives towards self-improvement. Unlike the Weick-type cognitive mindfulness, the focus here is less on attention and more on willingness and achievement.

In many ways, the MBSR-guided organizational interventions represent the updating of traditional HRM/HRD interventions carried out

within well-known streams of organizational culture's development: those inside the stream of the traditional "human resource" approaches (open to workers' needs, including higher level needs such as participation and self-fullfilment: Maslow 1962) and those which presently lie under the hegemony of the *Emotional Intelligence* and *Positive Psychology* approaches. The former has long been aligned with mindfulness principles due to enduring interests of its most successful representative. More recently Goleman has been directly involved, together with Kabat-Zinn, in the development of the field's most coherent effort, the *mindfulness-based emotional intelligence curriculum* at Google, referred to below; and Goleman has definitively included mindfulness within the domain of *focus*, the cardinal virtue of managerial excellence (Goleman 2013). Moreover, within the Positive Psychology approaches, mindfulness is seen as highly homogeneous with proactive behaviors oriented by "learned optimism" (Seligman 2003). Outstanding scholars are engaged by a common effort between mindfulness and positive psychology for opening up new avenues of well-being enhancement. (Brown and Ryan 2003).

However, MBSR-guided organizational interventions aim to introduce new elements to these streams. According to the ambitious statements stemming from the most important field-study of the effects of mindfulness on the employees of a leading bio-tech company, MBSR practice can induce permanent positive alterations in the brain and immune function of participants (Davidson et al. 2003). These kinds of results, reinforced by the bulk of research previously accumulated in clinical implementations, has strongly contributed to the "scientific" appeal of the protocol in business environments. At present the success of MBSR in organizations seems linked to a mix of heterogeneous factors: the hope for relief from the pains of hectic lives is strictly intertwined with the search for more "humanistic" ways of living and working, which leads to a fascination with thousand-year old practices and the search for new horizons for organizational cultures (also including brand-promotion strategies).

Many of these aspects are present in Google's *Search Inside Yourself* program (Tan 2012) whose *mindfulness-based emotional intelligence curriculum* definitively represents the success case in this field. The curriculum, although developed and largely implemented within Google, is intended to address anyone who might be interested in implementing it: Jon Kabat-Zinn underlines that "… it can be used in many ways in many venues… The limits of its usefulness and adaptability are really only the limits of your imagination and embodiment" (p. xiv). The basic idea is to provide opportunities for small groups of a corporation's associates to meet and familiarize themselves with mindfulness. The program is articulated into three main steps: *attention training* (dedicated to fostering calm and clear qualities of mind, as a foundation for emotional intelligence); *self-knowledge and self-mastery* (aimed at becoming able to observe the thought stream and the emotional process with high clarity, "objectively, from a third person perspective"); *creation of useful mental habits,* like "I wish for this person to be happy" type attitudes, creating trust that leads to highly productive collaborations (p. 7). The main aim is the acquisition of a small but significant range of meditative techniques within small-group sessions guided by expert trainers, starting from simple exercises of mindfulness of the breath. Subsequently, following the main MBSR guidelines, participants are accompanied through more complex steps, particularly through exercises regarding the distinction between "focalized" meditation (aimed at reaching states of concentration) and "open" meditation (in which the meditator should try to observe the mental phenomena arising moment-to-moment). In the intermediate steps participants are stimulated towards the parallel exploration of mind and body (also including *body-scanning* and yoga exercises) and reflection about opportunities for including mindfulness in daily life and applying it to problematic situations. Other "reflective" exercises performed in couples or small groups are geared towards; *aware listening* (based on full acceptance of the other person); *aware conversation* (in which listening is reinforced

through specific micro-techniques); *aware emailing* (for a correct use of this fundamental communication tool). Other exercises are geared towards uncovering creative inclinations and personal projects (e.g. "discovering my ideal future").

The most distinguishing aspect of the program is that all of its activities—meditation sessions and exercises—are constituted by the promise of benefits for participants, which are summarized in the motto "optimize thyself"; "The aim of developing emotional intelligence is to help you to optimize yourself and function at an even higher level than what you are already capable of" (p. 17). Under this imperative three goals are established: "stellar work performance", "outstanding leadership" and "the ability to create the conditions for happiness". Moreover, on the company side the benefits are related not only to increased productivity induced by highly effective work behaviors, but also to the promotion of a winning company image, beyond what can be offered by usual marketing expedients. In Google's case, coinciding with its global role, the SIY program is in line with a more than ambitious corporate mission, well represented by the book subtitle "increase productivity, creativity and happiness", (with which *Search Inside Yourself* is offered to broad audiences), and by a self-explicative cover comment signed by the corporate executive chairman: "This book and the course it is based on represent one of the greatest aspects of Google's culture—that one individual with a great idea can really change the world". At its conclusion the book exalts the functionality of matching mindfulness and emotional intelligence, even at the level of world peace, and expresses a committed wish for making the benefits of meditation accessible to all of humanity.

Buddhist Perspectives on Mindfulness in the Working Life

MBSR practice has been exported from clinical to working settings, using the concept of *stress* with few adaptations for the supposedly healthy populations in the latter milieu. It was not too difficult to shift from clinical *stress* equated with "pain", for which mindfulness is a new kind of remedy, to stress as a hindrance to the deployment of "positive attitudes and resources" in arenas in which "…we could hone inner strength and wisdom moment by moment, we would make better decisions, communicate more effectively, be more efficient, and perhaps even leave work happier at the end of the day". Mindfulness can be useful for everybody in these situations, where "…whether you love your job or hate it, you are bringing all your inner resources to bear on your working day" (p. 389). In fact, the possibility of such an export has been noted by Kabat-Zinn since the very first institution of MBSR as a neutral technique, adaptable to widely different situations. It could be implemented in hospitals, which function as "dukkha magnets" pulling in stress, pain of all kinds, disease and illness. It could then be shown to be beneficial in other organizational "magnets" like prisons, schools and work sites, each one producing or attracting its own particular brand of dukkha (Kabat-Zinn 2011).

A substantial reconsideration of the Buddhist roots of mindfulness is integral in this process of design and then implementation of MBSR. A practice which was spoken of as "the heart of Buddhist meditation" had to be relaunched as something having "little or nothing to do with Buddhism per se, and everything with wakefulness, compassion and wisdom… the universal qualities of being human, precisely what the word dharma is pointing to" (Kabat Zinn 2011, p. 283). Important keys for such reconsideration have been both interchangeability in the use of the term *mindfulness*, which can stand for either *meditation* or for *Dharma* or *dharma*, or even for *attention,* as in the well-known definition of mindfulness often quoted as "paying attention in a particular way: on purpose, in the present moment, and non-judgmentally" (Kabat-Zinn 1994, p. 4). Such an understanding of mindfulness, although based on a widely accepted correlation with *bare attention* to internal and external events (e.g. Gunaratana 2002), has been the object of sharp and subtle criticisms from

different sides of the Buddhist world, whose importance goes far beyond the doctrinal field and has direct consequences on the ways in which MBIs can be regarded.

The first important set of criticisms concerns the function of mindfulness itself. Wallace states that mindfulness cannot be equated with bare attention as it is an intentional act, subject to a vast set of determinants, internal and external. Moreover it is not limited by the present-moment as it includes the dimensions of *recollection* and *non-forgetfulness*, and also includes a *retrospective* memory of past events which helps in *prospectively* remembering to do something in the future. Its intended purposes can be fulfilled only when integrated with the parallel function of "clear comprehension" (*sampajanna*).

A second important set of criticisms is about the lack of any ethical dimension in the theoretical underpinnings of MBSR. Contrary to the asserted neutrality of MBSR, mindfulness cannot be seen as neutral: an ethical dimension is omnipresent as it continuously distinguishes between wholesome and unwholesome mental states (Analayo 2003). Mindfulness should be related to the other seven components of the Path: in particular "… the Buddha taught that the foundation for concentration is *sila*, or ethical conduct and that if there is no virtue, the basis for concentration is destroyed" (Goldstein 2012, p. 271).

A tertiary set of criticisms can be identified relating to aspects directly affecting the practice, which cannot be bare because it concerns peculiar personal characteristics linked to a unique biography and personality. These traits are accompanied by a particular historical, social, and cultural context that the personal identity of the practitioner depends on (Bodhi 2011). It should also be recognized that the practice of mindfulness is far from bare, for in no sense is it non-conceptual or non-verbal. On the contrary, as evident in the Mahasi Sayadaw method, it can be supported by precise verbal designations which are helpful relative to the tasks of knowing, shaping and purifying the mind (Bodhi 2011). Finally, it must be acknowledged that mindfulness bears a "receptive" quality, in which full attention is given to the cognized data; this contributes to the training of individual cognition and the reshaping of habitual patterns and worldview (Analayo 2003).

The above criticisms, largely dealing with the equation of mindfulness and attention, go hand in hand with other caveats about modern mindfulness, such as those concerning their reduction to science-based practices ("contemplative practices might be swayed by materialistic premises to explain their efficacy reductively, on the exclusive basis of neurophysiology", Bodhi 2011, p. 35) and more generally their spiritual watering down and subduction under the dominant culture of possessing and well-being (p. 36). However, such Buddhist criticisms and caveats do not imply a definitive ban of modern mindfulness practices: the Teachings are considered open to everybody. Anyone should feel free to take whatever they find useful from the Dharma, even for secular purposes, provided that such uses are in line with generally accepted moral principles and respectful of their unique nature: "…experimenters have entered a sanctuary deemed sacred by Buddhists… they are drawing from an ancient well of sacred wisdom that has nourished countless spirits through the centuries" (Bodhi 2011, p. 38). Those delivering MBIs can access the Buddhist resources, drawing upon them and recommending them to their clients, but must do so carefully (Amaro 2015).

Standpoints such as these mainly refer to clinical MBIs, designed as a healing opportunity and offered to individuals for their own personal development, leaving open the possibility for further steps in terms of morality and wisdom. Concerning their applications outside of clinical settings, the issue seems a bit more muddled, considering that average interventions are aimed at an elementary level of practice development, while instead "…the purpose of meditation is more than just calming ourselves from time to time, getting ourselves out of trouble, but seeing and uprooting the causes which produce trouble and make us not calm to begin with" (Amaro 2015, p. 70). From this viewpoint, mindfulness is seen as shorthand for three distinct psychological qualities, referred to in Pāli by different terms. As

a first rudimentary quality, *sati* can be called *mechanistic mindfulness*, i.e. the simple act of paying attention to an object or action. A second quality, *sati-sampajanna*, results from coupling mindfulness and clear comprehension, and intrinsically includes an appreciation of the practitioner's attitudes, actions, and their consequences. This is therefore described as *right mindfulness* or *informed mindfulness*; it incorporates ethical concerns and the recognition of obstructiveness linked to self-interest. The third quality, that takes form when *sati* encounters wisdom (*panna*) is regarded as that which leads to the full blossoming of human well-being, and it is hence called *noble* or *holistic*, and deemed capable of bringing the meditator into proximity of enlightenment. From this standpoint, according to the kind of phenomenological examination that is typical of *vipassana*, all experiences are interpreted as the mind's representations, and not as the result of fixed and definite external realities. In relation to these terms MBIs in organizations can only be considered to be taking place at the level of *mechanistic mindfulness*, void of ethical content and aspirations for liberation. This brand of mindfulness is very similar to the alertness and focus of a hunting cat, or, even worse, to the sniper's attention.

In an exhaustive review of theoretical standpoints on contemporary mindfulness (Monteiro et al. 2015), mostly dedicated to MBIs in psychotherapy, several of the arguments discussed thus far are presented in a systematic way with a wealth of supporting quotations. A large part of Western mindfulness practices can easily go under the definition of *miccha sati* ("wrong mindfulness"). This is due to a lack of ethical orientation combined with a propensity for forms of absorption which can induce the risk—especially in meditation newcomers—of bypassing experience instead of connecting with it. Such a risk is aggravated by the poor grasp of concepts such as bare awareness, non-judgmental awareness, and non-duality associated with contemporary mindfulness. In average organizational MBIs such a risk is reinforced by the absence of a specific vehicle for the ethical side of meditation. However, Monteiro et al. do not exclude the possibility that such interventions could be carried out effectively with specific attention paid to the individual's development, and towards possible progression towards the higher levels of *sati*.

> ... when designed with sensitivity to the level of psychological safety in the corporate culture and the welfare of the personnel in mind, mindfulness programs can play an important role in training the individual to see the incongruity of values clearly, confront skillfully, and not be frozen by self-blame. In workplace programs, concepts such as being comfortable with uncertainty, taking a nonjudgmental stance to a situation, or cultivating compassionate action are intended to transform emotional reactivity so that the situation can be met with skillful means. It may be naive to think that corporate culture will shift perceptibly even when there are improvements in the individuals' stance to the high-tempo and demands of the workplace. Nevertheless, changes at the ground level can create micro-climates within the work environment that foster support, compassion, and a sense of fellowship (Monteiro et al. 2015, p. 10).

This kind of reasoning seems largely influenced by the authors' experiences in non-profit settings with primary care physicians. Reconciling their view with that of average business cultures seems problematic due to their basic underlying assumptions and values. However, the possibility of carrying on MBIs following those kinds of goals should not be dismissed. Specific case-studies, both in profit and non-profit settings, might be very useful for understanding the depth of practice and attainable results at the individual, group and organizational levels.

Critical Views: The Raising of McMindfulness

The choir of media praises for the virtues of mindfulness has been recently crossed by discordant voices, even from mainstream opinions. An interesting example is offered by the experience of an executive coach and physician who largely supported organizational mindfulness practices in the past. He discovered through his own experience that the introduction of mindfulness practices to organizational

settings posed significant risks, in terms of "avoidance"—when meditation becomes a refuge, more comfortable than thinking, deciding and taking action—and "group-thinking"—when, for instance, group meditation is compelled by pushing requests of a boss who became fond of the practice (Brendel 2015). However, this brand of criticism is definitely inside the logic of mindfulness as a tool for individual well-being and, more or less directly, for organizational effectiveness. In a very different key, the polemic focus of the "McMindfulness" view—now well-known after this self-explanatory phrase went viral on the web (Purser and Loy 2013)—is on the inappropriateness and social dangerousness of modern mindfulness practices, especially when used in corporate contexts. A significant literature has rapidly burgeoned around these kinds of assumptions, both in scientific journals (Purser and Milillo 2014; Purser 2015) and in blogs (Loy 2012; Forbes 2012; Titmuss 2013, White and Cooper 2014, Purser and Ng 2015).

On one side, the rejection is based on the twofold argument that corporate mindfulness betrays the most fundamental tenets of Buddhist Teachings and that it offers substantial support to forms of living and working which are intrinsically based on manipulation and exploitation. The uncoupling from Buddhist roots has generated a substantial de-contextualization and denaturation of the meditation practice, now interpreted as a private practice intended to reduce stress and marketed in ways not far from products advertised to relieve headache or reduce blood pressure. Some of the arguments put forward by traditional Buddhists are borrowed and even reinforced by the McMindfulness advocates. In particular, a mindfulness that is too elementary, unbound by ethical values, in particular those regarding modes of making a living, is seen as *miccha sati*, ready-made for misuses of any kind.

On the other hand, mindfulness training is seen as a commodification phenomenon, i.e. as the nth well-being-inducing product to be sold on an open and flourishing market. "...Rather than applying mindfulness as a means to awaken individuals and organizations from the unwholesome roots of greed, ill will and delusion, it is usually being refashioned into a banal, therapeutic, self-help technique that can actually reinforce those roots" (Purser and Loy 2013, p. 2).

The colonization of mindfulness and its compliance with purposes such as helping executives to become better focused and more productive, is assumed to evince that a substantial diversion from the practice has occurred: the main focus of MBIs is in fact placed on individual factors of dukkha rather than on the social origins of it. Such a diversion is strongly coherent with an "accomodationist" orientation, whose roots can be retraced to the "human relations" movement and the "cow psychology" that was attached to it. Corporations have jumped on the mindfulness bandwagon because it conveniently shifts the burden onto the individual employee: stress is framed as a personal problem, and mindfulness is offered as just the right medicine to help employees work more efficiently and calmly within toxic environments; "Cloaked in an aura of care and humanity, mindfulness is refashioned into a safety valve, as a way to let off steam …a technique for coping with and adapting to the stresses and strains of corporate life" (Purser and Loy 2013, p. 3). In parallel—as noted by Healey (2013, quoted in Purser and Milillo 2014)—the corporate uses of mindfulness, as demonstrated by the Google SIY program, give rise to "integrity bubbles" which tend to reinforce the brand image while keeping the background company aims and the working styles of employees intact. The latter can claim to have satisfaction as practitioners of the corporate mindfulness, but at the same time they are induced to willingly accept much longer working times, even up to eighty-hour per week. Along a similar vein, an important criticism is raised by Titmuss (2013) who claims that corporate mindfulness, due to its lack of moral fiber and its distorted practice of non-judgment, can induce passivity and acquiescence to oppression and violence in working environments.

Different bridges connect within the "McMindfulness" view, generated from inside a rather traditional Buddhist discourse—with critical approaches definitively external to Buddhism but engaged in analyzing its influence on Western cultures and ways of life. Davies (2015), for instance, opens his critical analysis of the increasingly flourishing "happiness industry" by reporting on the participation of an outstanding figure of Buddhist monasticism in the 2014 World Economic Forum in Davos: the symbolic role of such an event is self-evident. This crossover is interpreted as a major example of on-going cultural and political strategies put in place by both governments and big business for subduing human emotions and establishing new regimes based on the selling and buying of emotional well-being. Buddhism, according to this view, is one of the useful reservoirs of ideas and practices (meditation, first of all) from which modern science and technology can grasp what is needed for building up increasingly sophisticated artifacts for controlling individuals and shaping social tendencies. An even more important role is attributed to Western Buddhism in Zizek's analysis of late capitalism, a regime that tends to represents itself as the best of all possible worlds while continuously reproducing its own crises. Of such a regime Buddhism is the paradigmatic ideology: "a pop-cultural phenomenon preaching inner distance and indifference in the face of the frantic pace of market competition". Mindfulness practices are an integral part of "the most efficient way to fully participate in capitalist dynamics while retaining the appearance of mental sanity" (Zizek 2015, p. 9).

Hypotheses for "Mindfulness-Inspired" Initiatives

Keeping in mind the infrastructure which underpins the analyses proposed thus far, "corporate mindfulness" seems to have something to do with strategies based on normalization apparatuses (*dispositifs*). These strategies are aimed at reproducing a societal conformity in accordance with the established power rules, as they allow for peculiar forms of apparent freedom to persist, which include fabricated truths about the self, the body, the psyche, etc. (Foucault 2004; Agamben

2009). At the same time the introduction of mindfulness within modern society's cultural horizons can be seen as an example of societal trends with tendencies towards resistance to the proliferation of power and the search for better ways of life. These tendencies are intrinsically opposed to power forces, according to a perspective not far removed from the *technology of the self* idea that Foucault worked through in his last years (Foucault 1988).

The issue at stake, from this viewpoint, is the feasibility of initiatives in which mindfulness practice plays a role which could be simultaneously respectful of its meanings and intentions and also promote liberation opportunities for individuals and collectivities. From this perspective it seems imperative to overcome the construct and practices of mindfulness as they were developed in both the "cognitive mindfulness" and "corporate mindfulness" movements. Instead of focusing on *stress*—as in the "corporate" approach—or on a functional *sense-making*—as in the "cognitive" approach, the emphasis should be placed on human *experience* developed within working life. Mindfulness practice, then, shouldn't be seen as directly serving organizational action but instead as a non-instrumental background for the natural growth of people engaged in the everyday challenges and opportunities which working life entails. In parallel, in order to profit from the chances that working life can offer for understanding human reality, these "mindfulness-inspired" initiatives should be carried out in a small group dimension (not more than ten–twelve participants) and articulated into two main strands. In one strand mindfulness practice should be preserved as purely an observation of the mind within given, protected, spaces and according to specific instructions. The other one—of a more "reflective" nature—should consist of guided in-group exchanges regarding working life assumed as a significant space for individual experience: participants should find opportunities for both understanding and wisdom (e.g. in terms of self-fulfillment, cooperation and creativity), and for overcoming the causes of stress and uneasiness, which are linked to the "three poisons" of greed, aversion and illusion.

The relationship between the two levels could take advantage of the intrinsic coherence between mindfulness practice and phenomenological approaches to experience analysis (Patrik 1994; Warren Brown and Cordon 2009). However, they should each be characterized as fully autonomous and, at the same time, through reciprocal synergies, related to the positive properties of mindfulness meditation (even beyond the contemplative practice sphere) on one side, and, on the other side, to the benefits that shared reflection on relevant work problems can produce for individual participants.

Typical addressees of the interventions should be identified at the level of *communities of practice* of different kinds, such as:

– communities already formed within specific organizations (e.g., a practical community composed of workers in a department for software development in an IT company, who all share concerns about issues relevant to their own work and organizational life; for instance regarding the pace of work);
– communities based on common professional interests (e.g. a community among program developers working in different companies sharing concerns about relevant issues in their profession; for instance, the ethical conduct of companies in their customer relationships)
– communities of meditators who identify relevant themes of reflection about the working condition in general; for instance, concerning aspirations linked to career development).

Of course, each of these types of communities should be interested in carrying out the suggested mindfulness initiatives, placing full attention on their twofold aim. In each case the implementation of such interventions would require previous activities in order to appropriately promote and clarify the intervention's intents and modalities. It is not inconceivable that specific organizations might be interested in supporting the implementation of such interventions, but the privileged audiences are primarily composed of such people. Therefore these kinds of initiatives should be promoted through bodies and networks—such as: practice centers, social networks and associations of different kinds (cultural, professional, spiritual, etc.) in which people find opportunities for exchanges and may receive specific information.

The average intervention should be addressed to 10–12 people, including at least 10 half-a-day meetings within an overall time of about 6 months, alternating (and keeping well distinct) mindfulness meditation and exercises for reflective awareness; the latter based on group discussions supported by specific tools and techniques. Specific research programs are certainly needed in order to deepen and clarify the lines for the development of "mindfulness-inspired initiatives" briefly mentioned above. Such programs should be aimed at both establishing the theoretical coordinates of the proposed approach and at understanding—through cycles of interviews to appropriate witnesses—the social appeal and practical feasibility of the initiatives. In particular, for the development of the reflective side of mindfulness-inspired initiatives, it would be interesting to develop an intervention model based on the phenomenological method that Varela's school has placed under the heading of "becoming aware" (Depraz et al. 2002), also with regard to organizational contexts (Sharmer 2001). This method (directly stemming from the Husserlian *epoché*) is articulated into three main, cyclically intertwined, processes or "gestures" of *becoming aware*, i.e. of accessing experience. In the first process, *suspension*, the subjects try to break with their "natural attitudes" (or "habitual patterns", or "realist prejudice") that keep them looking at the world—and behaving within it—following established habits. Then, the second process, *redirection*, creates a free space in which various patterns and contents can emerge which are different from (and maybe innovative in relation to) those that the suspension process identified. Redirection shouldn't imply a fixation on what emerged, probably following the most habitual patterns: it should be taken as a form of introduction to the third process, *letting go*, which is aimed at shifting the attention from an active search to an accepting "letting-arrive". The *letting go*, in other terms, is a sort of guarantee that

the *becoming aware* as a whole is kept in motion, and that its recursive function is preserved.

Such a method should be adapted to the functioning of groups in which participants may find ways to both reinforce their own mindfulness practice at the individual level and go through their experience of working life within a collective setting. *Awareness* and *equanimity* could be deemed the guiding principles of such initiatives, in which even the quest for a less stressful way of living and the aspirations for a better use of intelligence at work could be easily integrated, starting from different premises about the meaning of working life and the aspirations of its protagonists.

Final Remarks

This chapter attempts to depict some aspects of the so-called "mindfulness interventions" in work and organizational settings. Most of the attention has been devoted to the "corporate mindfulness" approach, characterized by the use of the MBSR protocol or other similar techniques. Such an approach—typically applied in large companies and institutions of the business world—hinges on the idea that the workplace is the epicenter of highly demanding performances whose stressing effects can be counterbalanced by mindfulness practices. Other positive effects are expected in terms of attention training, self-mastery and creation of useful mental habits. Here "corporate mindfulness" has been assumed to be representative of an updated version of well-established HR policies and activities, reinforced by the contributions of more recent approaches such as those of emotional intelligence and positive psychology. The implementation of mindfulness as a tool for personal and organizational development has been interpreted as being primarily aimed at reinforcing the "individualized" nature of the organization, and at supporting the shift of the initiative and responsibilities burden from the organization to individual organizational actors. Some criticisms have been reported by lay and religious representatives of Buddhism, mostly related to the lack of ethical commitment in MBSR practices and to the instrumental and mechanistic level to which mindfulness is confined within MBSR, along with some potential opportunities for inducing positive changes in the micro-climates and the sense of fellowship in organizations. Even stronger negative remarks have also been expressed by some social critics and activists towards corporate mindfulness. In these cases the concern is mostly about the manipulative nature of the MBSR-led interventions, which can allegedly even induce passivity and acquiescence to oppression in working environments.

Corporate mindfulness practices, however, do not exhaustively cover the entire field of mindfulness in the working life. The "cognitive mindfulness" approach—mostly related to "high reliability organizations"—has also been taken into consideration in this chapter. Within such an approach, instead of emphasizing physical and mental stress, the focus is placed on the risks induced by lack of attention in the execution of work tasks, especially in contexts which require high reliability and interconnection. Mindfulness is therefore deemed to be at the heart of certain forms of organizational/interpretative action that allow for the subtle interpretations of events and a shared sense-making in organizational relations.

A third kind of mindfulness in working life is briefly sketched in the last section of the chapter, in which mindfulness is considered neither an anti-stress remedy nor an attention arouser, but instead a liberating practice, within "mindfulness-inspired initiatives". The keyword in this case—fully hypothetical as no initiative of this kind has been so far implemented—is *experience*. The suggested initiatives, intended to be carried out in small group dimension, should, on one side, support mindfulness practice in a purely meditative stance and, on the other side, stimulate forms of reflective practice among participants. A specific research program is needed for the development of the theoretical premises of such initiatives and the design of field activities for their implementation.

References

Agamben, G. (2009). What is an apparatus. In *What is an apparatus, and other essays*. Stanford: Stanford University Press.

Amaro, A. (2015). A holistic mindfulness. *Mindfulness, 6* (1), 63–73.

Analayo, (2003). *Satipattana: The direct path to realization*. Cambridge: Windhorse.

Baer, R. A. (2003). Mindfulness training as a clinical intervention: A conceptual and empirical review. *Clinical Psychology: Science and Practice, 10*, 125–143.

Bodhi, B. (2011). What does mindfulness really mean? A canonical perspective. *Contemporary Buddhism: An Interdisciplinary Journal, 12*(01), 19–39.

Boyatzis, R. (2007). Competencies as a behavioral approach to emotional intelligence. *Journal of Management Development, 28*(9), 749–770.

Boyatzis, R., Goleman, D., & McKee, A. (2002). *Primal leadership: Realizing the power of emotional intelligence*. Boston: Harvard Business School Press.

Boyatzis, R., & McKee, A. (2005). *Resonant leadership: Renewing yourself and connecting with others through mindfulness, hope and compassion*. Boston: Harvard Business School Press.

Brendel, D. (2015). There are risks to mindfulness at work. *Harvard Business Review*.

Brown, K., & Ryan, R. (2003). The benefits of being present: Mindfulness and its role in psychological well-being. *Journal of Personality and Social Psychology, 84*, 822–848.

Bush, M. (2013a). *Research and practice of mindful techniques in organizations, conversation with Jeremy Hunter*. More than Sound, digital ed.

Bush, M. (2013b). *Research and practice of mindful techniques in organizations, conversation with Daniel Goleman*. More than Sound, digital ed.

Chaskalson, M. (2011). *The mindful workplace. Developing resilient individuals and resonant organizations with MBSR*. Chichester: Wiley.

Chiesa, A., & Malinowski, P. (2011). Mindfulness-based approaches: Are they all the same? *Journal of Clinical Psychology, 67*(4), 404–424.

Chiesa, A., & Serretti, A. (2009). Mindfulness-based stress reduction for stress management in healthy people: a review and meta-analysis. *The Journal of Alternative and Complementary Medicine, 15*(5), 593–600.

Coffield, F. (Ed.). (2000). *Differing visions of a learning society*. Bristol: The Policy Press.

Cullen, M. (2011). Mindfulness-based interventions: An emerging phenomenon. *Mindfulness, 2*(3), 186–193.

Cunliffe, A., & Easterby-Smith, M. (2004). From reflection to practical reflexivity: Experiential learning and lived experience. In M. Reynolds & R. Vince (Eds.), *Organizing reflection*. Aldershot: Ashgate.

Dane, E. (2010). Paying attention to mindfulness and its effects on task performance in the workplace. *Journal of Management,*. doi:10.1177/0149206310367948.

Davidson, R. J., Kabat-Zinn, J., Schumacher, J. M. S., et al. (2003). Alterations in brain and immune function produced by mindfulness meditation. *Psychosomatic Medicine, 65*(4), 564–570.

Davies, W. (2015). *The happiness industry: How the government and big business sold us well-being.* London: Verso.

Depraz, N., Varela, F. J., & Vermersch, P. (2002). *On becoming aware: A pragmatics of experiencing.* Amsterdam: John Benjamin.

Ergas, O. (2015). The deeper teachings of mindfulness-based curricular interventions as a reconstruction of "education". *Journal of Philosophy of Education, 49*(2), 203–220.

Forbes, D. (2012). Occupy mindfulness. *Beams and Struts.*

Foucault, M. (1988). The political technology of individuals. In L.H Martin, J. Gutman, & P.H. Hutton (Eds.), *Technologies of the self. A seminar with michel foucault.* Amherst: The University of Massachusetts Press.

Foucault, M. (2004). *Naissance de la biopolitique. Cours au College de France 1978–79.* Paris: Seuil.

Gardiner, B. (2012). Business skills and Buddhist mindfulness. *The Wall Street Journal.*

Gelles, D. (2015). *Mindful work: How meditation is changing business from the inside out.* London: Profile Books.

Goldstein, J. (2012). *A practical guide to awakening.* Boulder: Sounds True.

Goleman, D. (1988). *The meditative mind. The varieties of meditative experience.* New York: Tarcher.

Goleman, D. (2013). *Focus, the hidden driver of excellence.* London: Bloomsbury.

Goshall, S., & Bartlett, C. A. (1997). *The individualized corporation. A fundamentally new approach to management.* New York: HarperCollins Publishing.

Gunaratana, H. (2002). *Mindfulness in plain english.* Boston: Wisdom Publications.

Hatch, M. J., & Cunliffe, A. L. (2006). *Organization theory: Modern, symbolic and post-modern perspectives.* Oxford: Oxford University Press.

Hougaard, R. (2015). *Bringing dharma into the corporate world: Rasmus Hougaard talks about the potential project.* http://fpmt.org/mandala/online

Hunter, J. (2013, April). Is mindfulness good for business, *Mindful.* http://jeremyhunter.net/wp-content/uploads/2013/02/

Kabat-Zinn, J. (1994). *Wherever you go there you are: Mindfulness meditation in everyday life.* New York: Hyperion.

Kabat-Zinn, J. (2003). Mindfulness-based interventions in context: Past. *Present, and Future, Clinical Psychology: Science and Practice, 10*(2), 144–156.

Kabat-Zinn, J. (2011). Some reflections on the origins of MBSR, skillful means, and the trouble with maps. *Contemporary Buddhism, 12*(1), 281–306.

Kabat-Zinn, M., & Kabat-Zinn, J. (2010). *Everyday blessings: The inner work of mindful parenting.* Kindle books.

Langer, E. J. (2000). Mindful learning. *Current Directions in Psychological Science, 9*(6), 220–223.

Levy, D.M., Wobbrock, J.O., Kaszniak, A.W., & Ostergren, M. (2012). The effects of mindfulness meditation training on multitasking in a high-stress information environment. *Paper given at the Graphics Interface Conference*, May 2012.

Loy, D. (2012, October 16). Can mindfulness change a corporation. In *Buddhist peace fellowship* (online).

Malone, T. W. (2004). *The future of work: How the new order of business will shape your organization, your management style, and your life.* Boston: Harvard Business School Press.

Marturano, J. (2013). *Finding the space to lead. A practical guide to mindful leadership.* New York: Bloomsbury Press.

Maslow, A. H. (1962–1999). *Towards a psychology of being.* New York: Wiley.

McKenzie, S. (2015). *Mindfulness at work. How to avoid stress, achieve more and enjoy life.* New Jersey: Career Press.

Monteiro, L. M., Musten, R. F., & Compson, J. (2015). Traditional and contemporary mindfulness: finding the middle path in the tangle of concerns. *Mindfulness, 6* (1), 1–13.

Purser R., & Ng, E. (2015, Sept. 27). Corporate mindfulness is bullshit: Zen or no Zen, you're working harder and being paid less. *Salon webmagazine.*

Nielsen, L., & Kaszniak, A. W. (2006). Awareness of subtle emotional feelings: A comparison of long-term meditators and nonmeditators. *Emotion, 6*(3), 392–405.

Patrik, L. E. (1994). Phenomenological method and meditation. *The Journal of Transpersonal Psychology, 26*(1), 37–54.

Pickert, G. (2014, February 3). The mindful revolution: The science of finding focus in a stressed multitasking culture. *Time.*

Prochaska, J. M., James, O., Prochaska, J. O., & Levesque, D. A. (2001). A transtheoretical approach to changing organizations. *Administration and Policy In Mental Health, 28*(4), 247–261.

Purser, R. (2015). Clearing the muddled path of traditional and contemporary mindfulness: A response to Monteiro. *Musten, and Compson, Mindfulness, 6*(1), 23–45.

Purser, R., & Loy, D. (2013). Beyond McMindfulness. *The Huffington Post.* http://www.huffingtonpost.com/ron-purser/beyond-mcmindfulness_b_3519289.html

Purser R., & Milillo J., (2014). Mindfulness revisited: A Buddhist-based conceptualization. *Journal of Management Inquiry* 1–22. doi:10.1177/1056492614532315

Salzberg, S. (2014). *Real happiness at work. Meditations for Accomplishment, Achievement and Peace.* New York: Workman.

Santorelli, S. (2000). *Heal thyself. Lessons on mindfulness in medicine.* New York: Random House.

Segal, Z., Williams, M., & Teasdale, J. (2002). *Mindfulness-based cognitive therapy for depression. A new approach to preventing relapse*. New York: The Guildford Press.

Seligman, M. (2003). *Authentic happiness. Using the new positive psychology to realize your potential for lasting fulfillment*. London: Nicolas Brealey Publ.

Sharmer, C. O. (2001, January 2000). The three gestures of becoming aware. Conversation with Francisco Varela. http://www.iwp.jku.at/born/mpwfst/02/www.dialogonleadership.org/Varela.html

Stanley, E. A., Schaldach, M., Kiyonaga, A., & Jha, A. P. (2011). Mindfulness-based mind fitness training: A case study of a high-stress predeployment military cohort. *Mindfulness,*. doi:10.1007/s12671-011-0058-1.

Tan, C.-M. (2012). *Search inside yourself. Increase productivity, creativity and happiness*. London: Harper Collins.

Thanissaro, B. (Geoffrey DeGraff). (1999). *Noble strategy. Essays of the Buddhist path*. San Diego: Metta Forest Monastry publishing.

Titmuss, C. (2013). The Buddha of mindfulness. The politics of mindfulness. http://christophertitmuss.org/blog/

Wallace, B. A. (2005). *The attention revolution, unlocking the power of the focused mind*. Boston: Wisdom Publications.

Wallace, B. A. (2006). *The essence of mindfulness*, translation: L'essenza della mindfulness. In M. Tomassini (Ed.) (2012), *La cura della consapevolezza. Teorie e tecniche della mindfulness*, Milano: Tecniche Nuove.

Warren Brown, K., & Cordon, S. (2009). Toward a phenomenology of mindfulness: Subjective experience and emotional correlates. In F. Didonna (Ed.), *Clinical handbook of mindfulness*. New York: Springer.

Weick, K. E. (1995). *Sensemaking in Organizations*. Thousand Oaks: Sage.

Weick, K. E. (2009). *Making sense of the organization, Vol. 2: The impermanent organization*. New York: Wiley.

Weick, K. E., & Putnam, T. (2006). Organizing for mindfulness: Eastern wisdom and western knowledge. *Journal of Management Inquiry, 15*, 275–287.

Weick, K. E., Sutcliffe, K. M., & Obstfeld, D. (1999). Organizing for high reliability: Processes of collective mindfulness. In R. S. Sutton & B. M. Staw (Eds.), *Research in organizational behavior* (Vol. 1, pp. 81–123). Stanford: Stanford Jai Press.

White, C., & Cooper, A. (2014, March 8). Apple and Amazon big lie. *Salon.*

Williams, J., Mark, G., & Kabat-Zinn, J. (2013). *Mindfulness. Diverse perspectives on its meaning, origins and applications*. London: Routledge.

Wilson, J. (2014). *Mindful America. The mutual transformation of buddhist meditation and american culture*. Oxford: Oxford University Press.

Yanow, D., & Haridimos, T. (2009). What is reflection-in-action? A phenomenological account. *Journal of Management Studies, 46*, 8.

Zizek, S. (2015). *Trouble in paradise: From the end of history to the end of capitalism*. London: Penguin.

Part III
Genealogies of Mindfulness-Based Interventions

Against One Method: Contemplation in Context

16

Brooke D. Lavelle

Introduction

Various secular mindfulness- and compassion-based programs have been developed and implemented in diverse educational, clinical, and other settings in recent years. Many of these programs, including Mindfulness-Based Stress Reduction (MBSR), Cognitively-Based Compassion Training (CBCT), and Sustainable Compassion Training (SCT), have been influenced by diverse Buddhist contemplative traditions that assume different models of mind and methodologies for realizing (possibly distinct models) of enlightenment. MBSR, CBCT, and SCT have been shaped by and in response to their own modern historical–cultural context—which is marked by heightened form of individualism and scientific reductionism—as well as by the ways in which they interpret the category of the secular.

Despite the influences of these Buddhist contemplative and modern cultural frameworks on these contemporary secular programs, MBSR, CBCT, and SCT each claim some form of universal applicability. This underlying assumption—that there is a universal method that can be applied skillfully and effectively in a variety of particular contexts—raises a number of challenges. First, such a perspective assumes there is a universal model of "health" or "well-being." Second, it also assumes that there is a universal cause of stress or suffering that can be overcome through the application of a singular method. Third, as I will suggest below, such universal rhetoric tends to privilege highly individualized descriptions of suffering and health, thereby eschewing social and systemic causes of suffering.

The goal of this chapter was to explore ways in which certain Buddhist contemplative and modern cultural frames both limit and permit different possibilities for health and healing as articulated within contemporary secular programs. The aim is not to determine which modern contemplative program is most authentic or effective, but rather to call attention to the ways in which such frames not only shape or impact practices and programs, but also constitute them. At the same time, this is not purely a critical project. Revealing the dominant frames that shape and inform these programs can highlight our own conditioned and limited biases and thereby help us explore new frames, or new ways of communicating these practices in effective ways to various audiences. In short, this chapter is a call for a more context- and systems-sensitive approach to the design and implementation of secular programs in North America.

B.D. Lavelle (✉)
Greater Good Science Center, University of
California—Berkeley, Berkeley, CA, USA
e-mail: brooke.dodson.lavelle@gmail.com

© Springer International Publishing Switzerland 2016
R.E. Purser et al. (eds.), *Handbook of Mindfulness*,
Mindfulness in Behavioral Health, DOI 10.1007/978-3-319-44019-4_16

Buddhist Contemplative Frames

Mindfulness and compassion are taught and practiced within diverse Buddhist traditions, yet the importance of and methods for cultivating mindfulness and compassion vary across and within these very traditions. These differences are due in large part to the ways in which suffering and its causes are variously defined, conceptualized, and overcome or transformed within the three main Buddhist traditions, namely Theravāda, Mahāyāna, and Vajrayāna. In Theravāda traditions, for example, suffering is understood to arise from the mistaken illusion of a fixed, permanent, and separate sense of self. Practices of mindfulness and compassion are employed to help practitioners gain insight into the selfless or impermanent nature of experience (Gethin 1998). In Mahāyāna traditions, practitioners aim to recognize the emptiness, or lack of intrinsic, independent reality of all phenomena (Pettit 1999). In these traditions, compassion practices in particular are seen not only as supportive, but rather fundamental practices for realizing enlightenment. The Vajrayāna traditions of Tibet, in turn, point toward the innateness or immanence of enlightenment. Practices are thus designed to help reveal the qualities of enlightenment—including compassion—that are present, yet obscured, in the practitioners' mind (Pettit 1999; Makransky 2012).

These diverse Buddhist contemplative traditions variously influenced the design and development of modern, secular programs, including MBSR, CBCT, and SCT. As will become clearer below, these programs not only take their inspiration from these different contemplative practice traditions, but they are also framed by a long-standing traditional debate concerning the nature of mind. To summarize briefly, the debate centers on whether the qualities of awakening or enlightenment are innate to one's mind or whether they need to be constructed or created through cultivation. "innatist" models, which are informed by both Mahāyāna and Vajrayāna traditions, hold that the qualities of awakening are present, yet are obscured by mistaken structures of cognition in the practitioner's mind.

"Constructivist" models, on the other hand, which are informed by other Mahāyāna traditions and Theravāda traditions, hold that practitioners have the potential to awaken, but that the qualities of awakening need be generated. It is important to emphasize that both models require some form of cultivation; the difference lies in whether the qualities of awakening are primarily understood either to be created or made manifest.

These "innatist" and "constructivist" frames have explicitly and implicitly shaped and constrained modern secular interventions. Not only do they shape the style of practices employed, they also impact the starting point of practice itself. In the following section, we will very briefly review the key traditional influences and practices employed within MBSR, CBCT, and SCT. This brief sketch is intended to provide an overview of the programs to orient our discussion and does not presume to capture the range or depth of these modern contemplative programs.

Modern Secular Programs

Mindfulness-Based Stress Reduction (MBSR)

MBSR, developed by Jon Kabat-Zinn, is a participant-centered behavioral medicine program that was originally designed to empower patients whose health needs were not being adequately addressed by standard medical treatments to participate proactively in their own healing (Kabat-Zinn 2000). The MBSR model assumes that stress and suffering arise from an inability to be aware of, present to, and accepting of reality—including pain, illness, challenging life circumstances, and so on—as it is. It is learning to "come to terms" with things as they are, without trying to change them, that healing takes place. The program therefore involves training in mindfulness, which refers to a particular way of paying attention on purpose and without judgment in the present moment, as well as the cultivation of a particular stance or set of attitudes toward the world and one's experiences (Kabat-Zinn 2000). These attitudes include non-judging, patience, "beginner's mind," trust, non-striving, acceptance, and "letting go" (Kabat-Zinn 2000; pp. 31–46).

MBSR assumes that people have a profound, innate capacity for self-healing. The program was influenced by various strands of Buddhism, including certain non-dual strands of Tibetan Buddhism and the American Zen tradition, as well as by the modern Theravāda tradition (for more, see Kabat-Zinn 2011). On Kabat-Zinn's view, "Buddhism is fundamentally about being in touch with your own deepest nature and letting it flow out of you unimpeded. It has to do with waking up and seeing things as they are" (Kabat-Zinn 1994; p. 2). This rhetorical emphasis on naturalness, simplicity, and non-doing reflects the spirit of the so-called "innatist" camp described above. This frame shapes not only the rhetorical style but also the non-analytical and inquiry-driven practices contained within the program itself. As we will see below, this style stands in contract to cognitive or analytical practices employed in CBCT and other "constructivist" programs.

Cognitively-Based Compassion Training (CBCT)

CBCT, developed by Lobsang Tenzin Negi, was originally designed as a means of addressing the rising rate of depression among undergraduate students at Emory University (Negi 2009). Like MBSR, the program has since been adapted for use in a variety of clinical, non-clinical, and educational settings (Ozawa-de Silva and Dodson-Lavelle 2011; Reddy et al. 2013).

CBCT assumes that suffering arises from obsessive self-concern. Compassion, which is defined as the heartfelt wish to alleviate others' suffering—and is thus incompatible with maladaptive self-focus—is framed as the antidote to stress and suffering. CBCT assumes that people have a natural capacity for compassion, but that this capacity typically only extends to one's so-called "in-group." In order to learn to extend compassion to others, one needs to cultivate affection, insight, and empathy. This is typically done through systematic analytic reflection, contemplation, and cultivation of the following capacities: (1) attention; (2) insight into mental experience; (3) self-compassion; (4) impartiality or equanimity; (5) gratitude; (6) affection and empathy; and (7) aspirational compassion. It is understood that through cultivating these capacities compassion will emerge (Negi 2009; Ozawa-de Silva and Negi 2013).

CBCT is heavily influenced by the Mahāyāna Buddhist traditions of Tibet. Its format and sequence closely follow the "seven-point cause-and-effect method," which is one of the most well-known methods for cultivating compassion within the Mahāyāna tradition, except that it omits explicit "religious" references to reincarnation or karma. CBCT also incorporates methods from "Equalizing and Exchanging Self with Others," which is another well-known Tibetan Buddhist method for cultivating compassion (Negi 2009).

CBCT may be understood as a kind of response to MBSR. It is distinct from MBSR in that it involves primarily "constructivist" analytical meditations that emphasize the need to cultivate compassion and its related qualities through a process of reasoning and reflection. Unlike MBSR, CBCT is explicitly normative and prescriptive. Rather than discovering one's innate capacity for self-healing as in MBSR, in CBCT, one is given a map and instructions for cultivating one's capacity for healing.

Sustainable Compassion Training (SCT)

SCT, developed by John Makransky, is a contemplative approach designed to help people realize a power of unconditional care from within that is healing and sustaining, and that is not subject to empathy fatigue and burnout. The program is taught in both Buddhist and secular settings and has also been specially adapted for those in a variety of service professions, including education, health care, and social work (Makransky 2011).

SCT assumes that people have an innate, natural capacity for care and compassion

(Makransky 2007). Though it draws on some contemplative reflections similar to those utilized in CBCT, SCT holds that practitioners need to be empowered to access their potential for compassion relationally. In other words, practitioners need to experience themselves as objects of care and compassion in order to strengthen and extend compassion to others. Compassion is thus developed through three interrelated modes of care or compassion, namely receiving care, self-care, and extending care. Receiving care practices help practitioners experience themselves as the recipient of care; self-care practices involve helping practitioners learn to recognize and "let be" into the innate qualities of care available in their awareness; and extending care practices involve supportive analytical practices to help practitioners become aware of and overcome stereotypes, biases, and other obstacles that limit one's natural capacity to care for others.

Though SCT incorporates "constructivist" practices, it is most heavily influenced by various "innatist" tantric and non-dual strands of Vajrayāna and Mahāyāna Buddhism. SCT thus not only problematizes the so-called debate, but it also presents a challenge to modern frames of individualism and certain conceptions of the secular that will be taken up below. In short, SCT argues that the starting point for the cultivation of compassion in Buddhist contemplative traditions has not been the framework of an autonomously separate self, but rather one of deep relationality. On these models, the power of compassion comes not just from the practitioner's own autonomous efforts, but from beyond the practitioner. In other words, practitioners are empowered to realize their innate capacities for care and compassion. We will explore the implications of this relational frame below.

Limits of Buddhist Frames

The Buddhist contemplative frames presented here form a backdrop from and against which MBSR, CBCT, and SCT were developed. As mentioned, such frames—particularly the innatist/constructivist frames—have implicitly

and explicitly constrained the rhetoric and practice styles employed within in each program. MBSR assumes that the qualities of healing are innate. CBCT assumes that the qualities for awakening need to be constructed. SCT assumes the qualities of awakening are innate, yet also employs constructivist practices. These assumptions influence the metaphors and attitudes of "letting go," acceptance, trust, and "letting be" as well as the rhetoric of naturalness employed within innatist programs. Such assumptions also influence the training, developing, and strengthening metaphors found within constructivist models. How might such metaphors constrain and prime participants' experience? How might the different starting points assumed by such approaches influence not only practice but also one's view of what is possible in terms of health and well-being?

Much of the discussion concerning these different approaches has centered on the validity or authenticity of these models (see, for example Dunne 2011). Yet this focus on authenticity or validity has tended to privilege theory over context and has obscured or prevented constructive inquiry into the potential ways in which various practices may be more helpful or effective for different individuals in different contexts or at different points in their personal or spiritual development. Frames have a way of blinding us to other possibilities—by their very nature they emphasize certain aspects of experience while de-emphasizing others. Frames appear to merely capture reality, yet they are situational, perspectival, and subject to varying motivations, positions, and agendas (see Goffman 1986). Thus, it seems worthwhile to inquire more deeply into the ways in which health, healing, and even freedom from stress and suffering are conceptualized, constrained, and even constituted within these programs.

Modern Cultural Frames

MBSR, CBCT, and SCT have all been shaped and constrained by various Buddhist contemplative models, and they have also been influenced by the North American cultural context in which they have been developed and delivered. Buddhism's transmission and assimilation into a North American cultural context have been shaped by its integration and critique of several dominant themes of modernity, including romanticism, scientific rationalism, and monotheism. Together, these "discourses of modernity" provide the implicit frameworks against which MBSR, CBCT, and SCT have been explicitly framed (McMahan 2008).

To focus our discussion, I would like to draw attention to three distinct sub-frameworks related to each broad domain of modernity: (1) the "individualistic frame," which highlights the influence of romantic expressionism—with its emphasis on individual experience and autonomy and a rejection of authority and tradition—on modern conceptions of autonomy and spirituality; (2) the "scientific frame," which reveals both a particular way of relating to rational-scientific, naturalistic, and highly medicalized discourse; and (3) the "secular frame," which reveals the ways in which programs position themselves in relation to the category of "religion" and "spirituality." The analysis of the "secular frame" will take up most of our attention because it reveals some of the most common and contested ways in which modern contemplative programs present themselves. The full scope of such an investigation into the category of the secular and the process of secularization is of course much wider and richer than can be described here. Nevertheless, this approach should offer us sufficient material to consider the potential limits and opportunities that these modern frames place on contemporary mindfulness- and compassion-based programs.

Individualism

We could trace a number of factors and influences that have given rise to modern conceptions of individualism, including capitalism, democracy, deinstitutionalization, the rise of science and the move toward privatized spirituality, and so on (see, for example, Taylor 1989). Within

and in response to this frame, contemplative practice has come to be seen as an individual self-help tool. Not only have contemplative practices in general been presented as separable from their larger religious or cultural framework, but specific non-relational styles of practice also have become characteristic of modern Buddhism (McMahan 2008). In other words, different practices, especially devotional style practices, have been largely ignored or dismissed as cultural baggage in favor of practices that more readily fit this modern, individualistic, self-help framework. This frame has caused us to miss the broader relational ways in which practices were held and transmitted in diverse Buddhist traditions. This frame arguably has had the effect of shifting the starting and ending point for practice, from one of deep relationality and self-transcendence, to one of an autonomous individuality and self-preservation. We will briefly consider some of the limits of this frame below.

Scientific Reductionism/Rationalism

The scientific study of meditation has also had the effect of separating practice from context. The process of operationalizing contemplative processes and attempting to reduce such processes to either their observable or biological causes has similarly encouraged a way of viewing practice as divorceable from context. Researchers are trained to search isolate and identify the so-called "active ingredient," or to search for the "silver bullet" that works for all people regardless of an individual's constitution, development, or context. This frame not only limits inquiry into the ways in which context affects practice, but also what counts as practice itself. Further, attempts to generalize the effects of contemplative training across individuals, communities, and diverse contexts reveal a positivist–objectivist scientific bias. Such a bias assumes that there is a singular, objective, universal truth that can be found and measured. In terms of contemplative practice, such a bias also assumes that there is a universal cause of suffering, and the goal of research is to discover and effectively apply the universal method of healing.

The Secular Frame

MBSR, CBCT, and SCT all openly trace their roots to various Buddhist contemplative traditions, yet they also explicitly emphasize the secular nature of their programs. Although they each consider their programs to be secular, the founders of MBSR, CBCT, and SCT construct and employ the rhetoric of the secular in different ways. As I will suggest below, these diverse frames variously permit and constrain these programs' goals, practices, and possibilities for healing.

The process and legitimacy of "secularizing" Buddhist-based contemplative practices is one of the most debated issues in the field (see, for example, Monteiro et al. 2015: Purser 2015; Brown 2014). On my view, a number of the issues at stake in this debate hinge on the ways in which the category of the secular is understood and employed. There is no singular definition of the secular, in the same way that there is no singular definition of religion. The categories of the religious and secular are mutually constituted, and the (often blurry) lines between them are drawn in different ways in different times and places. In some contexts, the secular signals what is common to people of diverse or of no religious or spiritual traditions. In other contexts, the secular signals a separation or outright rejection of religion or what might be considered the supernatural (Taylor 2007; Calhoun 2010; Lama 2011).

These two distinct conceptions of the secular—what I call "open" and "closed" models of the secular borrowing from Taylor (2007)— shape current thinking and the debate about the secularization of contemplative practice. Although they do not represent the full range of interpretations of the secular, (nor do they necessarily reflect the ways in which the proponents of MBSR, CBCT, and SCT understand themselves to be secular), they are useful heuristic tools for helping to uncover and analyze the secular frames that are at least implicitly employed in these modern contemplative program.

The "open" model of the secular rests on the simple separation thesis. What counts as religious or secular, however, may vary between open models. Programs that employ extremely open models of the secular, for example, would tend to permit, hold, and attempt to negotiate a variety of worldviews and belief systems. Programs that employ moderately open models of the secular would tend to bracket out religious beliefs, but not outright reject or dismiss them. The "closed" model of the secular hinges on the sociological secularization thesis. On this model, secularization is understood to mark a progressive transformation from the so-called "primitive" religious systems to the modern secular worldview (Casanova 2006). The closed secular frame implies a naturalistic framework that is explicitly contrasted with the supernatural and thus most often entails an outright rejection of religion altogether. Closed models tend to define the secular in more universalizing terms, as a space, or set of views and practices that are free from the trappings of particular cultural and religious beliefs, rituals, and institutions. Below we will briefly consider ways in which MBSR, CBCT, and SCT employ these different secular frames.

MBSR's Secular Frame

Although the closed model of the secular has been critiqued by various scholars for its oversimplification and naïveté, it nonetheless shapes much of current thinking about the secular in the West (Casanova 2006; Taylor 2007). A number of modern mindfulness-based programs including MBSR endorse closed models of the secular as evidenced by the ways in which they claim to preserve the essential, universal features of Buddhist practice while getting rid of unnecessary and overtly religious beliefs and rituals (see, for example, Batchelor 1997; and Kabat-Zinn 2011). Such models tend to downplay the role of context and assert a universal applicability to essentialized Buddhist practice. As Kabat-Zinn notes, "mindfulness, often spoken of as 'the heart of Buddhist meditation', has little or nothing to do with Buddhism per se, and everything to do with wakefulness, compassion, and wisdom. These are universal qualities of being human [...]" (2011, p. 283). On my view, programs that endorse closed models of the secular tend to employ the strongest universalistic rhetoric and are most overtly dismissive of or naïve to the role of context—including the context within which they are embedded and constructed. (Of course, it is important to note that universal rhetoric is often employed as a strategy, and is not necessarily a reflection of one's view on context and conditions.).

CBCT and SCT endorse open models of the secular, yet they interpret the secular frame in different ways. Briefly put, CBCT excludes "religious" practices in secular contexts, SCT, in contrast, sees certain religious worldviews, understandings, and practices as both permissible in secular contexts and as theoretically integral to the SCT framework itself.

CBCT's Secular Frame

CBCT employs the category of the secular to refer to a set of basic human values—including kindness and compassion—that are assumed to be universally shared by all people, regardless of one's belief system (Ozawa de-Silva 2014). CBCT's interpretation of the secular is informed by the Dalai Lama's concept of "secular ethics," which does not deny a role for religion or spirituality in people's lives, but rather claims that people have a "basic human spirituality" that is "more fundamental than religion" (Lama 2011). This "basic human spirituality"—which is the first dimension of spirituality—is a disposition toward love, kindness, and affection. The second dimension of spirituality—which is religion-based—is tied to particular cultural beliefs, customs, and practices. Though the Dalai Lama claims that the second dimension of spirituality, which includes religious beliefs and practices, may be more sustaining and motivating

for people's practices, he holds that "basic human spirituality" is more essential for survival.

This two-tiered approach—in which religious customs, practices, and traditions are effectively bracketed but not rejected from secular space—distinguishes this open secular from closed secular models. It also implies that there is something "more" to religious practices than what is available or articulated in their secular adaptations. (This implication raises important considerations about the role of context in shaping the goals of and motivations for contemplative practice.). Yet somewhat paradoxically, this two-tiered approach, in which the second tier is functionally dropped to provide what is assumed to be common ground, fails to notice or appreciate how people of religious traditions feel the need to draw upon their religious worldviews as a primary basis for their training in the so-called "basic human values" of love, kindness, and compassion.

SCT's Secular Frame

SCT also employs an open secular frame. Yet unlike CBCT, which draws heavily on secular–scientific worldviews at the expense of embracing or exploring commonalities among religious traditions, SCT seeks universality or common ground in patterns of diverse religious practice. To be clear, this is not to suggest that SCT seeks a universal secular space in which people put there religious traditions aside in the name of commonality, but rather to identify a common pattern recognizable to a diversity of people who inhabit different worldviews to be accessible to them within their worldviews (Makransky 2015). While the overt rhetoric of the secular employed by MBSR and CBCT programs assumes a predominantly secular audience, SCT assumes that the majority of people that participate in modern contemplative programs are looking for something deeper than what is typically offered in popular, secular programs that are simply aimed at improving one's health (Makransky 2012). Further, SCT assumes that contemplative programs can be more effective when they engage participants' religious and spiritual worldviews, rather than ignoring them (Makransky 2012). SCT recognizes that the suggestion that religious beliefs should be bracketed in secular programs fails to notice how religious people feel the need to continue to draw upon their own religious worldviews as a primary basis for cultivating qualities like compassion and also well-being. SCT therefore employs an extremely "open" model of the secular in which diverse religious and spiritual worldviews are not only permitted, but also considered potentially helpful or even necessary to the process of healing and transformation.

In short, although each program claims to be secular, the ways in which each program understands, interprets, and rhetorically employs the category of the secular limits and permit various possibilities for practice, accessibility, and implementation.

The Limits of Modern Frames

In the same way that Buddhist contemplative frames shape and inform modern contemplative program, various modern frames shape and limit the content and styles of practice that are permissible within modern programs. While MBSR, CBCT, and SCT each presume a certain degree of universal applicability, they are each constructed from a particular vantage point in a certain time and place. Universalizing rhetoric is common in this field, and is due in part to the various discourses of modernity mentioned above, as well as to ahistorical tendencies within the Buddhist tradition itself (McMahan 2008; Makransky 1999). Although we might understand the drive toward universality in the name of sameness or shared humanity, such rhetoric threatens to mask or negate more complicated, contextual concerns. Such rhetoric can also be used to deny or oppress divergent or marginalized perspectives.

To varying degrees, each of the frames presented above have a way of attempting to mask, obscure, or negate context. Yet it is a mistake to think that modern contemplative programs have

abstracted or lifted up practices from diverse Buddhist contexts and placed them in context-neutral modern contexts. Context is everywhere. And context not only influences practices, it constitutes them. Not only that, but contexts and frames shape and limit what we consider to be possible when we undertake these very practices.

The secular frame limits not only what styles of practice may be permitted within programs, but also limits the very goals and possibilities of programs themselves. It is not only the case that modern interventions tend to avoid or overlook seemingly religious, devotional, or other ritualistic practices, but also that they overlook or take for granted the worldviews in which the motivations for approaching such practices may be very different. This oversight puts the potential efficacy of practices at stake. In addition, the ways in which the secular is constructed and employed (as well as the ways in which it is *perceived* to be employed) have the potential to limit the domain of applicability of such interventions in public settings. Thus, it seems critical that the field takes a closer look at the possible interpretations of the secular and the limits and opportunities that each one affords. It also seems worthwhile to explore intentional dialogue with other spiritual and religious traditions as a way of engaging diverse perspectives and methods for conceptualizing and realizing health and well-being in diverse settings.

The modern frames of individualism and scientific reductionism, combined with certain features of our biomedical paradigm that tend to locate illness within the individual, have implicitly shaped the rhetoric of contemporary programs. In short, MBSR and CBCT in particular are biopsychosocially decontextualized. Despite the rhetoric of interdependence in MBSR, for example, the causes of suffering are squarely located within the individual's ways of perceiving or misperceiving the world, and the path of healing involves individuals changing the way they see and relate to the world (Kabat-Zinn). In CBCT, we find a similar suggestion regarding the causes of suffering, and also an emphasis on need for the individual's personal, internal transformation as the path to overcoming that suffering (Negi 2009).

Some have noted the ways in which such individualistic frames—which at first glance may seem empowering—can be disempowering and dismissive of individuals' suffering (see, for example Baker 2014; Purser 2015). Such a frame assumes that individuals, as autonomous agents, have full control and agency over their health and well-being, thereby ignoring social, political, and economic factors that have contributed to or adversely affected their health (Purser 2015). Moreover, this individualistic frame fails to account for ways in which suffering is also interpersonal and social. For example, caring for a loved one with chronic illness or a disability, or experiencing the loss of a loved one, may be considered social or intersubjective forms of suffering. War and cycles of violence and oppression inflict trauma not only on individuals but also on communities, and this trauma can be passed generationally. Even what counts as suffering is socially constructed (Kleinman 1998). Thus, programs that offer means of overcoming suffering therefore must take into account these causes and thus the need for a broader approach for responding to suffering.

There is another limit to the individualistic frame, and that is that it can lead us to adopt a naïve Trojan Horse approach to systemic and cultural transformation (see, for example, Purser and Ng 2015; see also Batchelor 1997, for a discussion of the Trojan Horse ideology of the so-called "secular" mindfulness programs). On this view, it is assumed that individuals within systems, including the military, corporations, schools, and so on, who "wake up" through contemplative practice will be effective in engendering major institutional transformations. Not only is there no evidence for the effectiveness for this strategy, there is evidence which suggests that programs that focus solely on transformation at the individual level are not effective in engendering systems-wide change (see for example Gordon 2015). Thus, although it is complicated to define and assess context, it is critical to do so. Context is not incidental to the practice itself; it also constitutes it.

Toward a (Re)Frame

Deepening our understanding of Buddhist contemplative and modern cultural frames can reveal further implicit assumptions about the causes of suffering and the possibility of health and healing, and thereby potentially open new areas of dialogue and development. Such investigations may also help the field advance its understanding of systems in ways that impact not only to individuals' health and well-being, but also inspire and effect social change.

Fortunately, a number of individuals have strengthened their calls for more context-sensitive approaches to designing and implementing contemplative programs in recent years (see, for example, Germano 2014). Such approaches require that programs address the relational nature of health and healing, not only at the level of interpersonal interaction, but also at the systems level, in which the individual is considered as part of a dynamic system. Rather than simply assessing the validity, authenticity, or compatibility of different theoretical models, the field can grow by considering the implications of the ways in which various contextual frameworks shape, limit, and permit different possibilities for health and healing. This context-sensitive approach facilitates a natural expansion of the conceptualization of the causes of suffering and the methods for overcoming suffering, thereby allowing practitioners, programmers, and researchers to draw upon diverse, community-based, and ecologically sensitive approaches for healing that have been overlooked because of a narrowly imposed Buddhist contemplative or modern frame.

There is not one vision for a healthy, just society, nor is there one method for achieving health and well-being. There are different visions of human potential that embody complex sets of ideals, hopes, values, and goals. On my view, the field would benefit from a deep inquiry into these diverse practices and methodologies. Yet to be clear, the call for a more diverse, context- and systems-sensitive approach does not mean that we cannot be open to the idea of universal approaches or practices (for that would impose another frame!). To privilege culture or difference at the expense of commonality is to undermine, or at the very least significantly limit, the possibility of adapting and implementing practices across contexts. Perhaps there is a way the field can learn to better hold the wisdom of sameness and difference, and to maintain an open inquiry into the relationship between the universal and the particular, the secular and the religious, and even theory and context.

References

Baker, K. K. (2014). Mindfulness meditation: Do-it-yourself medicalization of every moment. *Social Science and Medicine, 106*, 174. doi:10.1016/j.socscimed.2014.01.024.

Batchelor, Stephen. (1997). *Buddhism without beliefs.* New York: Riverhead Books.

Brown, C. B. (2014, December 2). Mindfulness: Stealth Buddhism for mainstreaming meditation. *Huffington Post.* http://www.huffingtonpost.com/candy-gunther-brown-phd/mindfulness-stealth-buddh_b_6243036.html

Calhoun, C. (2010). Rethinking secularism. *The Hedgehog Review*, 35–48.

Casanova, J. (2006). Rethinking secularization: A global comparative perspective. *The Hedgehog Review*, 1–22.

Lama, Dalai. (2011). *Beyond religion: Ethics for a whole world.* New York: Houghton Mifflin Harcourt.

Dunne, J. (2011). Toward an understanding of nondual mindfulness. *Contemporary Buddhism, 12*, 71–88. doi:10.1080/14639947.2011.564820.

Germano, D. (2014). *Contemplation in contexts: Tibetan Buddhist meditation across the boundaries of the humanities and sciences.* Paper presented at the Mind and Life International Symposium for Contemplative Studies, Boston, Massachusetts.

Gethin, R. (1998). *The foundations of Buddhism.* Oxford: Oxford University Press.

Goffman, E. (1986). *Frame analysis: An essay on the organization of experience.* Boston: Northeastern University Press.

Gordon, L. N. (2015). *From power to prejudice: The rise of racial individualism in midcentury America.* Chicago: University of Chicago Press.

Kabat-Zinn, J. (1994). *Wherever you go, there you are: Mindfulness meditation in everyday life.* New York: Hyperion.

Kabat-Zinn, J. (2000). *Full catastrophe living: Using the wisdom of your body and mind to face stress, pain, and illness.* New York: Bantam Dell.

Kabat-Zinn, J. (2011). Some reflections on the origins of MBSR, skillful means, and the trouble with maps.

Contemporary Buddhism, 12, 281–306. doi:10.1080/14639947.2011.564844.

Kleinman, A. (1998). Experience and its moral modes: Culture, human conditions, and disorder. In *The tanner lectures on human values.* Stanford, CA: Stanford University.

Makransky, J. (1999). Historical consciousness as an offering to the transhistorical Buddha. In J. Makransky & R. Jackson (Eds.), *Buddhist theology: Critical reflections by contemporary Buddhist scholars.* New York: Routledge.

Makransky, J. (2011). Compassion beyond fatigue: Contemplative training for people who serve others. In J. Simmer-Brown & F. Grace (Eds.), *Meditation and the classroom* (pp. 85–94). New York: SUNY Press.

Makransky, J. (2012, April–June). How contemporary Buddhist practice meets the secular world in its search for a deeper grounding for service and social action. *Dharma World.* doi: http://www.rkworld.org/dharmaworld/dw_2012aprjunebuddhistpractice.aspx

McMahan, D. L. (2008). *The making of Buddhist modernism.* Oxford: Oxford University Press.

Monteiro, L. M., Musten, R. F., & Compson, J. (2015). Traditional and contemporary mindfulness: Finding the middle path in the tangle of concerns. *Mindfulness, 6*, 1–13.

Negi, L. T. (2009). *Cognitively-based compassion training: A manual.* Unpublished manuscript, Emory University, Atlanta, Georgia.

Ozawa-de Silva, B., & Dodson-Lavelle, B. (2011). An education of heart and mind: Practical and theoretical issues in teaching cognitively-based compassion training to children. *Practical Matters, 4*, 1–28.

Ozawa-de Silva, B., & Negi, L. T. (2013). Cognitively-based compassion training: Protocol and key concepts. In T. Singer & M. Bolz (Eds.), *Compassion: Bridging theory and practice* (pp. 416–437). Leipzig: Max Planck Institute for Human Cognitive and Brain Sciences.

Ozawa-de Silva, B. (2014). Secular ethics, embodied cognitive logics, and education. *The Journal of Contemplative Inquiry, 1*(1).

Purser, R. (2015). Clearing the muddled path of traditional and contemporary mindfulness: A response to Monteiro, Musten, and Compson. *Mindfulness, 6*, 23–45. doi:10.1007/s12671-014-0373-4.

Purser, R., & Ng, E. (2015). Corporate mindfulness is bullsh*t: Zen or no Zen, you're working harder and being paid less. *Salon.* http://www.salon.com/2015/09/27/corporate_mindfulness_is_bullsht_zen_or_no_zen_youre_working_harder_and_being_paid_less/

Reddy, S., Negi, L. T., Dodson-Lavelle, B., Ozawa-de Silva, B., Pace, T. W., Cole, S. P., et al. (2013). Cognitive based compassion training: A promising prevention strategy for at-risk adolescents. *Journal of Child and Family Studies, 22*, 219.

Taylor, C. (1989). *Sources of the self.* Cambridge, MA: Harvard University Press.

Taylor, C. (2007). *A secular age.* Boston: Harvard University Press.

Mindfulness-Based Interventions: Clinical Psychology, Buddhadharma, or Both? A Wisdom Perspective

David J. Lewis and Deborah Rozelle

Introduction

Mindfulness-Based Interventions in Behavioral Medicine and Clinical Psychology

Mindfulness-based interventions (MBIs) are group-based programs of psychoeducation and self-help skills in behavioral medicine and clinical psychology (including psychiatry, psychoanalysis, and related fields; hereafter, just *psychology* or *clinical psychology*), increasingly employed as aftercare and adjunct care to professional medical and psychological treatments (Bowen et al. 2014, p. 548; Center for Mindfulness 2016b; Pedulla, n.d.) and as self-care. Using in part elementary meditation techniques inspired by Buddhist practices, MBIs have helped many people cope with stress, chronic physical pain, grief, headaches, cancer, and so on, and serious mental health disorders such as major depression, borderline personality disorder, and substance abuse. The two best-known and most widely used MBIs are mindfulness-based stress reduction (MBSR) and mindfulness-based cognitive therapy (MBCT). MBSR and to some extent MBCT are promoted and offered not only for specific issues and disorders, but also for general well-being and emotional health, a way to "mobilize your own inner resources for learning, growing, and healing" (Center for Mindfulness 2016b).

MBIs and related "mindfulness" programs are also increasingly developed for and offered in settings outside of health care such as the workplace (Williams 2015), the military (Stanley et al. 2011), and schools (Felver and Jennings 2015). These often stimulate controversy about applying Buddhadharma[1] to ethically incompatible aims and numerous other issues (Monteiro et al. 2014, pp. 8–11; Purser 2015a, pp. 39–42; Purser and Ng 2015; Stanley 2015). Though our analysis does apply generally to such programs, we are not explicitly addressing the controversies.

Extensive but not always high-quality research has found MBIs to be generally effective, with moderate therapeutic effect sizes equal to those of traditional psychotherapeutic or pharmacological aftercare and adjunct care (Khoury et al. 2013). One major systematic review and meta-analysis has called for more rigorous studies (Goyal et al. 2014, p. 357), and another study found significant publication bias

D.J. Lewis (✉)
Center for Trauma and Contemplative Practice,
Lexington, MA, USA
e-mail: djlewis@triadic.com

D. Rozelle
Clinical Psychologist, Lexington, MA, USA
e-mail: drozelle@triadic.com

---

[1]We favor the term *Buddhadharma* over *Buddhism* whenever possible because the former is an ancient and indigenous Sanskrit word for the doctrines and practices based on the teachings of the Buddha (the Pali equivalent is *Buddhadhamma*), while *Buddhism* is a Westernized name dating to the 1830s, with no equivalent in Pali or Sanskrit (Keown 2003, p. 45).

© Springer International Publishing Switzerland 2016
R.E. Purser et al. (eds.), *Handbook of Mindfulness*,
Mindfulness in Behavioral Health, DOI 10.1007/978-3-319-44019-4_17

in reported results (Coronado-Montoya et al. 2016). Interestingly, research evidence is lacking that meditation is a central effective therapeutic ingredient of MBIs (Dimidjian 2016; Dobkin and Zhao 2011; MacCoon et al. 2014).

The major line of MBIs began with the development of MBSR by Jon Kabat-Zinn at the University of Massachusetts Medical School in 1979, initially to "relieve the chronic pain of patients not sufficiently helped by conventional medical treatment" (Ivey 2015, pp. 382–383). With the advent of MBCT in the 1990s (Teasdale 1999) to address relapse of major depression, the MBSR line of MBIs entered the realm of clinical psychology, where they have grown rapidly in number and reach. Mindfulness-based relapse prevention (MBRP) (Bowen et al. 2011) for relapse of addiction actually has meditative roots that predate MBSR (Bowen et al. 2006, pp. 402–403; Marlatt and Marques 1977). In addition, there are now established MBI protocols for borderline personality disorder, obsessive–compulsive disorder, childbirth and parenting, eating problems and disorders, and others, with still more being researched and developed.

There are two other major streams in clinical psychology of methods not directly related to MBSR but widely considered to be MBIs: dialectical behavior therapy (DBT) using meditation inspired by Zen (Linehan 2014), and acceptance and commitment therapy (ACT) (Harris 2009), a general psychotherapy approach that uses meditation very little, but includes psychological principles related to those of the MBSR line. MBSR is still the archetype and actual prototype for the vast majority of MBIs and actual clients served. The MBSR line also dominates popular and professional press coverage to the extent that it is virtually synonymous with mindfulness-based methods in the public mind.

Source and Content of MBIs

The meditational aspects of MBSR were inspired by and derived from a number of Buddhist meditation practices and teachings that

Kabat-Zinn was familiar with from his experience as a student and teacher of Buddhadharma (Kabat-Zinn 2011, pp. 285–294), and he described it from the beginning as "meditation practice as a modality for achieving self-regulation..." (Kabat-Zinn and Burney 1981). The widespread impression that MBIs are intensive meditation programs is inaccurate, as they contain much other material (Blacker et al. 2009; Rosch 2015, pp. 275,278–283,289). Only about half of the roughly 30 h of *class time over* 8 weeks of MBSR is spent on actual meditation, including "mindful yoga," which is not related to Buddhadharma. An additional total of 30–35 h of meditation is assigned as daily homework, but reports of adherence rates vary widely and they probably center somewhere below 50 % (Vettese et al. 2009, pp. 200–201). Other material in MBSR includes didactic presentations on stress and other topics, various journaling exercises, and instructor-led discussion revolving around awareness of everyday habits, judgments, and stressful situations; gaining mastery over typically unreflective emotional reactions; communication exercises; and other lifestyle topics. Other MBIs replace the non-meditation exercises and didactic material and discussions with appropriate topics for their context, often including 1- to 5-min "mini-practices" for specific therapeutic purposes. Evidence suggests that the non-meditative components and mini-practices of MBIs are active ingredients (Eberth and Sedlmeier 2012, pp. 174,180). In fact, the founders of MBRP and MBCT consider specific mini-practices to be the most important single components in their interventions (Bowen et al. 2011, pp. 60–61; Teasdale et al. 2014, p. 99).

Controversy: Are MBIs Also a Form of Buddhadharma?

Notwithstanding their clear character and origins as psychological methods and eclectic content only partly inspired by and derived from Buddhist meditation techniques, MBIs have increasingly acquired an alternate identity *as a form of*

Buddhadharma itself. Kabat-Zinn himself is the most outspoken advocate of the most pronounced form of this claim. Despite initial discomfort over describing MBSR as "dharma" (2005b, p. xiii, 2011, pp. 282–283), Kabat-Zinn has for many years now been declaring that MBIs are actually "universal dharma … not different in any essential way from Buddhadharma" (2011, p. 296) nor "from… what is practiced in Buddhist monasteries" and that they are "the movement of the Dharma (the Buddhist teachings) into the mainstream of society" (Kabat-Zinn and Shonin 2015), recontextualized "within the frameworks of science, medicine (including psychiatry and psychology), and healthcare" (2011, p. 288). This is despite the fact the MBIs are not promoted as Buddhadharma either in advertising or in actual classes, a contradiction that has led some to accuse MBIs, especially MBSR, of sowing confusion with a "chameleon and shape-shifting" public face (Purser 2015a, pp. 24–26).

It is no surprise that the view of MBIs as a form of Buddhadharma has stimulated a great deal of controversy. Prominent cognitive psychologist Eleanor Rosch studied MBSR firsthand and concluded that "Western therapeutic mindfulness, [of which MBSR is the prototype], is not Buddhism. It may use some techniques borrowed from Buddhism, but it uses them in a different manner and towards different goals" (Rosch 2015, p. 272). Ronald Purser, a frequent critic of many applications of MBIs (2015a, b; Purser and Loy 2013) and an editor of this volume, argues that Buddhadharma and MBIs do not "share the same roots," and "there is common grounds but only on the surface" (2015a, p. 33). Both Rosch and Purser, it should be noted, base their opinions at least in part on participation in actual MBSR courses as well as their personal knowledge of Buddhadharma.

The goals of this chapter are to (a) critically evaluate—and reject—the idea that MBIs embody the essence of Buddhadharma in any meaningful way; (b) replace it with an equally profound but more accurate relationship based not on identity and essence but on form and structure; and (c) do so with a novel

methodology that correlates MBIs with deeper strata of Buddhadharma than usual and promotes mutually fruitful dialogue.

Methodological Underpinnings

A View of MBIs from Within Buddhadharma

Because Kabat-Zinn frequently refers to Buddhist theories and practices as the standards and models for MBIs, it is sound and appropriate to analyze his assertions from within Buddhadharma itself. We therefore use foundational ideas from Buddhadharma that are "not constrained by its historical, cultural and religious manifestations associated with its countries of origin and their unique traditions," (Kabat-Zinn 2011, p. 281) to show that MBIs are *not* a form of Buddhadharma, traditional, or recontextualized and *do not* in any meaningful sense capture its essence.

Regarding the relationship between Buddhadharma and modern scholarship, we agree with Wallace (2003, pp. 26–27) that "the way forward… is through mutually respectful dialogue and collaboration in both empirical and theoretical research… entertaining the possibility of learning about the world *from* Buddhism, as opposed to studying this tradition merely as a means to learn *about* Buddhism." In particular, our analysis accepts traditional Buddhadharma, MBIs, and modern clinical psychology each on its own terms, drawing out relationships among them without reducing one to the other or allowing any to colonize or swallow another (Loy 2014).

Functional Analogy

Some may find traditional Buddhadharma difficult to accept on its own terms, because it conflicts with today's zeitgeist of scientific materialism. In part to facilitate such dialogue, we employ a methodology of *functional* or *structural analogy* (Encyclopedia Britannica

2016; Gentner and Bowdle 2008), which we have used elsewhere to relate Buddhadharma with psychological trauma (Lewis 2015a; Rozelle 2015a; Rozelle and Lewis 2014, pp. 104–105; Rozelle et al. 2014), with developmental relational psychology (Rozelle 2015b), as well as with MBIs (Lewis 2015b).

A functional analogy model correlates concepts that operate similarly relative to the different goals and assumptions of the domains in question, not distorting the significant contrasts between them, but conscientiously incorporating differences into the model. It is an alternative to reducing one domain to another, treating one as a subset of the other, or simply declaring them incompatible.

The analogical approach enables us to use deeper strata of Buddhadharma, some of which are problematic in today's zeitgeist, to cast light on psychological practices and mechanisms. For example, in prior work, we correlate PTSD's retraumatization cycle with Buddhadharma's cycle of life, death, and rebirth (Lewis 2015a, pp. 33–35) in a precise, mutually informative way that respects both the single-life frame of modern clinical psychology and the doctrine of rebirth that is central to traditional Buddhadharma.

Mindfulness: The Term and Concept

The word "mindfulness" was originally chosen as a substitute for "meditation" by Kabat-Zinn to obscure MBSR's Buddhist roots and help make the "dharma essence of the Buddha's teachings … accessible to mainstream Americans facing stress, pain, and illness," but who might be put off by Buddhism (2011, p. 282). He intended for scholars and researchers to later resolve any "important issues that may have been confounded and compounded by the early but intentional ignoring or glossing over of potentially important historical, philosophical, and cultural nuances" (2011, p. 290). Unfortunately, as MBIs have grown rapidly, rather than resolve, the confusion has actually increased to the point where the term "mindfulness" has countless

discrete meanings and shades of meanings, often overlapping and conflicting (Grossman and Van Dam 2011). The word "mindfulness" in any environment now requires explicit or implicit qualification to avoid confusion. For example, it is very common to use "mindfulness" without qualification to stand for MBIs and at the same time for any or all Buddhist meditation practices, creating the unwarranted impression that all of those are the same thing (Chiesa and Malinowski 2011).

In this chapter, we shall use "mindfulness" mainly in one way, as the M in MBIs, MBSR, MBCT, and so on, which identify specific programs with well-defined content and methods, which, as we have pointed out, includes much more than what is usually termed "mindfulness meditation" in any form. We occasionally use "Buddhist mindfulness" for the most common traditional usage: a family of practices that train the attention to levels well beyond the ordinary and deploy it to keep specific content firmly in mind for inner observation (Bhikkhu Bodhi 2011a, p. 25).

Buddhadharma: The Noble Eightfold Path

The *Noble Eightfold Path, The Fourth Noble Truth* (Bhikkhu Bodhi 2013, Chapter Preface, II) lays out Buddhadharma's prescription for action, divided into three groups: *wisdom, ethics*, and *meditation* (or *concentration*), also known as the *three higher trainings.* Meditation is by far the most common focus for discussions relating Buddhadharma and MBIs. The general, though far from unanimous, consensus is that MBIs include relatively elementary and limited forms of meditation by comparison with traditional Buddhist contexts and with important variances (Bhikkhu Bodhi 2011a; Dreyfus 2011). Buddhist ethics in MBIs is also discussed extensively in the literature (Monteiro et al. 2014, pp. 6–11; Purser 2015a, p. 26; Stanley 2015), with wide agreement that such ethics are currently lacking in MBIs. The reasons for and implications of that are often hotly debated, and it may be changing.

The wisdom group of the Eightfold Path, which has been treated far less often with respect to MBIs than ethics and meditation, is the main focus of our analysis, though we shall touch on meditation as necessary. Wisdom is in many ways the most distinctive and dispositive aspect of Buddhadharma. Bhikkhu Bodhi calls wisdom "the primary tool for deliverance" (Bhikkhu Bodhi 2013, Chap. V). Wisdom in the Eightfold Path has two components: *right intention*, which concerns motivations and aims, and *right view*, which is about the nature of reality. Bhikkhu Bodhi (2013, Chap. II) says about right view:

> Right view is the forerunner of the entire path, the guide for all the other factors. It enables us to understand our starting point, our destination, and the successive landmarks to pass as practice advances…. In its fullest measure right view involves a correct understanding of the entire Dhamma or teaching of the Buddha.

There are several ways of formulating right view; we work with one that emphasizes the truths of *impermanence, suffering,* and *no-self,* also called the *three marks (or seals) of existence* that uniquely characterize Buddhadharma's understanding of reality (Bhikkhu Bodhi 2005, pp. 27–28). In a recent paper, Kabat-Zinn (2011, pp. 298–299) acknowledges the importance of the three marks in both Buddhadharma and MBIs, and we will be taking up that formulation in some detail. All schools of Buddhadharma accept the three marks; deep, transformative insight into them constitutes total freedom from suffering, most often translated as *liberation* or *enlightenment*. Liberation is often added as a fourth mark of existence. There are many variations and complexities to the concept of liberation across Buddhist traditions, which are beyond the scope of this chapter.

MBIs and Buddhadharma

Subtler Forms of the Claim of Essential Identity

Few others have explicitly defended Kabat-Zinn's claim that MBSR and other MBIs embody the essence of Buddhadharma. It is echoed by Margaret Cullen, one of the first certified MBSR teachers (2011, p. 188), but to our knowledge not by any other past or present senior colleagues of Kabat-Zinn at his institution (Center for Mindfulness 2016a) nor by the founders of any other MBIs. Many teachers and clients of MBIs do acknowledge and honor the Buddhist roots of MBIs, but most seem content to reap the benefits without any notion that they might be an actual variant of Buddhadharma, much less its replacement in the modern world. Furthermore, there is a widespread presumption that MBIs are indeed secular (Brown 2016). Nevertheless, since Kabat-Zinn is widely considered the father of MBIs, and he writes and speaks widely and commands large audiences (Johns 2015), his assertion is influential and needs to be taken seriously.

The idea that MBIs somehow *are* Buddhadharma, however, is actually quite widespread in more restrained terms. Many, though by no means all MBI teachers and proponents and even many researchers (Garland et al. 2015, p. 293; Vago and Silbersweig 2012, p. 1), convey the conflation in some form. Here is an example from an MBSR teacher's Web site:

> MBSR does indeed honor its Buddhist roots (as Kabat-Zinn often publicly remarks), starting with the motivation of compassion for the suffering of others. Beyond this, within its limited time frame MBSR seeks to make skillful use of key elements from the Dharma in order to provide participants with fresh perspectives on how suffering is generated, and practices to use to work with it. In this regard, any informed Buddhist could come on an MBSR course and identify many elements of the Dharma that have been worked into the course, but without their having been named in specifically Dharma terminology: the Four Noble Truths, loving kindness, the Three Marks of Existence, ethics, and more. (Chase 2015)

This passage seems to be saying that the major differences between MBSR and Buddhadharma are MBSR's "limited time frame" and its avoidance of traditional terminology. In fact, essentially every aspect of convergence claimed actually represents a significant divergence between MBIs and Buddhadharma, and we will be taking up several of them.

Suffering: A Crucial Concept

Perhaps the central idea raised in the above citation is suffering and its relief. Kabat-Zinn's claim also revolves around suffering:

> …since the entire raison d'être of the dharma is to elucidate the nature of suffering and its root causes, as well as provide a practical path to liberation from suffering. All this to be undertaken, of course, without ever mentioning the word 'dharma.' (Kabat-Zinn 2011, p. 288)

Suffering is a central concept in Buddhadharma, the subject of *The Four Noble Truths*, and it transcends cultural boundaries. We structure our analysis around it, from which we can unpack many of the important issues in Buddhadharma in the wisdom group of the Eightfold Path.

Suffering in Buddhadharma

Dukkha

The idea of suffering in Buddhadharma is widely misunderstood. *Dukkha* is the Pali term usually translated as "suffering" in English, but also "anxiety," "dissatisfaction," "frustration," "stress," "uneasiness," "unsatisfactoriness," and others (Wikipedia: article on dukkha 2016). Dependence on any single-word rendering is actually part of the problem, however. Like many crucial terms in Buddhadharma (or any complex field), dukkha stands for a multifaceted construct that can be only conveyed as single word in any language by convention. Bhikkhu Bodhi (2005, p. 26) puts it this way:

> Dukkha has a far wider significance, reflective of a comprehensive philosophical vision. While it draws its affective colouring from its connection with pain and suffering, and certainly includes these, it points beyond such restrictive meanings to the inherent unsatisfactoriness of everything conditioned. This unsatisfactoriness of the conditioned is due to its impermanence, its vulnerability to pain, and its inability to provide complete and lasting satisfaction.

The Suffering of Suffering: Overt Dukkha and the Two Arrows

Of the many dimensions to dukkha in Buddhadharma, we consider two. The first is expressed in the metaphor of the two arrows in the *Sallatha Sutta* (Ṭhānissaro Bhikkhu 1997). Physical pain and discomfort, for example, are the first arrow. But we inflict even further suffering on ourselves, a kind of second arrow, with a sense of distress beyond the simple feeling of pain.

Three factors contribute to second-arrow suffering. First, we wishfully hope and fantasize, perhaps subconsciously, that the pain will never happen, and once it arrives, it will soon depart (Soeng 2004, p. 79), and yet are unable to control it in either way. Second, we feel we own the pain; it is such an intimate part of *me* that when it hurts, it is *my* pain, so *I* hurt. Finally, the pain seems real to us, not imaginary; it has substance. The result is more than mere pain; it is resentment, avoidance, fear, anger, anxiety, revulsion, self-pity, and a broad range of other negative emotions. Dukkha thus arises not merely from painful stimuli, but also because of our innate assumptions about the nature of experience and reality. The first arrow can also be emotional pain, such as loss of a loved one, and the same dynamics are at work as with physical pain.

The dukkha associated with physical and mental distress is called *suffering of suffering*, and because it is usually so obvious, we call it *overt suffering*. This two-arrow principle for overt suffering is often expressed as the aphorism, popular in the MBI world (Smalley and Winston 2011), "pain is inevitable, suffering is optional," meaning we cannot avoid painful stimuli, situations, or emotions, but we can mitigate our dysfunctional reactions to them.

Three Levels of Dukkha

The notions of overt suffering and the second arrow, while certainly important to understanding and bettering the human condition, are obviously not a unique contribution of Buddhadharma. In fact, Buddhadharma posits deeper, subtler, farther-reaching levels of dukkha. These are not apparent to our usual, unreflective consciousness, but become perceptible upon deeper contemplation. The deeper forms of dukkha are a major element of both Theravada and Mahayana Buddhism and are treated extensively in Buddhist writings, including the Sallatha Sutta, though that is rarely if ever mentioned explicitly in MBIs. Purser (2015a, pp. 33–34, 2015c, pp. 680–682) shows the multiple strata of suffering in Buddhadharma, of which MBIs address only the surface layers, and our treatment extends and deepens his.

The Suffering of Change: Subtle Dukkha

The second level of dukkha is called the *suffering of change*, the distress implicit in pleasant experiences because they are transient, bound to change, or disappear (Bhikkhu Bodhi 2011b). The suffering of change has its own second arrow. This is the obsessive craving that arises when the reality of pleasure's transience collides with our hope and fantasy, usually only dimly perceived, that the pleasure will last forever (Gyatso1997, pp. 52–54). As with pain, our feeling of intimately owning the pleasure, the sense that I cannot separate myself from it, is also an essential factor, as is the feeling that the pleasure is real, substantial, and not imaginary. The resulting suffering includes obsession, addiction, dysfunctional attachment, and much else. As with overt suffering, therefore, our innate assumptions about reality and experience create and amplify the suffering.

This suffering of change is related to Buddhadharma's foundational notion of impermanence, the fact of inevitable change, working at the level of everyday experience (Tsering 2005, p. 35). Most people recognize the suffering of change operating in their lives to one extent or another, for example, when they become obsessive about something pleasurable and that leads to difficulty such as addiction. But the suffering of change is rarely as apparent as overt suffering, especially in everyday pleasure, so we may call it *subtle suffering*. The suffering of change does not mean that pleasure is illusory or enjoyment impossible, but simply that it is normally contaminated in this way (Tsering 2005, p. 36).

The Suffering of Conditioned Existence: Hidden Dukkha

The third level of dukkha stands at the heart of what Buddhadharma is about. Called *all-pervasive suffering* or *the suffering of conditioned existence* (Purser 2015c, p. 680; Sopa 2004b, p. 222), it arises from having a mind and body that are, like everything else, always changing and subject to the causal influence of countless other beings, events, and things, a fundamental fact that Buddhadharma calls *conditioned existence*.

Unlike the other two levels, the suffering of conditioned existence is not accessible to ordinary experiential perception (Sopa 2004b, p. 248) but can only be seen directly "by sustained attention to experience in its living immediacy" (Bhikkhu Bodhi 2005, p. 26), that is, transconceptually, by deep meditation powered by heightened concentration. We therefore call it *hidden suffering*. Hidden suffering operates at all times, even in the absence of pleasure or pain and even when the second arrow of either of those has been mitigated.

There is a two-arrow structure at this level of dukkha as well, which we describe experientially even though it is actually transconceptual. Conditioned existence itself is the first arrow; it is a source of suffering because it contradicts our innate misapprehensions of permanence, substantiality, independence, and of being in control. The reaction is to grasp onto something that seems permanent and substantial, namely our own body/mind complex (called the *aggregates*), identifying with it, taking it as our self, when it is simply the conditioned basis of the self, the interdependent phenomena that we project the self onto. Because the body/mind is subject to conditionality and is impermanent and perishable, identifying with it is actually a root of hidden suffering. Wallace (2011, p. 41) puts it this way:

> Our closely held identification with the aggregates is the result of the mind's tentacles grasping for I, me, and mine" and latching on. I'm okay. I'm here and not over there." ...The aggregates themselves are not the problem; they are just the body and mind. But holding them closely, identifying with them, clinging to them, and thereby isolating ourselves creates a tear in the very fabric of our existence. ...The Buddha declared that as long as we hold these aggregates closely, we will suffer.

This passage also describes the doctrine of no-self, one of the three marks of existence, which says, "contrary to our most cherished beliefs, that our personality—the five aggregates—cannot be identified as self, as an enduring and substantial ground of personal identity" (Bhikkhu Bodhi 2005, p. 28).

Once again, this level of suffering, indeed all suffering, though pervasive is not inevitable; it is possible to break the dynamic and end it.

As with subtle suffering, impermanence contributes to hidden suffering, but it is a *radical impermanence* (also called *subtle impermanence*) (Bhikkhu Bodhi 2005, p. 26; Fischer 2012; Gelek Rimpoche 1998). The very act of creation entails change and eventual destruction, and "all the constituents of our being, bodily and mental, are in constant process, arising and passing away in rapid succession from moment to moment without any persistent underlying substance." Radical impermanence and hidden suffering manifest as pervasive death anxiety (Chodron 2007, p. 24) that has been the subject of much Western philosophy, literature, and psychological theory (Becker 1973).

Deep Causation: Hidden Suffering Underlies Overt and Subtle Suffering

In what we might call deep causation, Buddhadharma holds that the third level of suffering is the basis for the prior two levels. (Sopa 2004b, p. 248; Gyatso et al. 2006, p. 31); overt and subtle suffering are actually hidden suffering

playing out in everyday experience. "This tight hold on the aggregates—grasping on to my body, my feelings, and my mind—is the root of our deeply ingrained vulnerability to suffering on all its levels" (Wallace 2011, p. 41).

We can separate deep causation into two misapprehensions that manifest in different forms at each level: *reification* and *identification*. *Reify* means to make something concrete or real, in this case to see phenomena as more substantial and permanent than they actually are. Though the term "reification" is usually associated with the Mahayana notions of *emptiness* and *inherent existence*, it applies as well to impermanence (Makransky 2012, p. 63). At the third level of dukkha, radical impermanence entails a complete lack of substantiality, so the deepest misapprehension is seeing any substantiality at all. Reification plays out at the overt and subtle levels in less absolute terms, namely as the expectation of everyday lack of change that drives the second arrow of overt and subtle dukkha.

The other misapprehension, identification, manifests at the third level as identifying with the reified body/mind. That entails a sense of a self at the center of experience, a *me-ness*. That me-ness manifests at the overt and subtle levels as *ownership* of experience, a sense of *mine-ness*. Ownership of pain and pleasure, as we saw above, is a factor leading to the second-arrow reactions. The Sallatha Sutta uses a term translated as "joined," "attached" of "fettered" to describe this ownership of experience (Bhikkhu Bodhi 2000, pp. 1263–1264; Thera 1998; Thānissaro Bhikkhu 1997).

There is a crucial implication of dukkha's deep causation. No matter how much we mitigate the reactive second-arrow dynamic at the experiential overt and subtle levels, unless we address the misapprehensions of reification and identification at the deepest level, the experiential dynamic will inevitably reassert itself. In effect, there is an endless supply of pleasures, pains, and two-arrow dynamics to create suffering, with no hope of exhausting them one by one. For that reason, Buddhadharma primarily targets seeing through and dissolving the underlying dynamic, the suffering of conditionality. As we shall see, this is a dispositive differentiation between Buddhadharma and MBIs.

Two Realms of Buddhadharma: Everyday/Psychological and Radical/Transcendental

The first two levels of dukkha are, from the Buddhist standpoint, the domain of everyday experience and hence include what the modern world calls psychology. We therefore call this the *everyday/psychological realm*. It deals directly with overt, clearly recognizable mental dysfunction and distress and the dysfunctional nature of pleasure, namely maladaptive craving and addiction.

The main target of Buddhadharma, however, is the basis of all suffering, the third level, which we call the *radical/transcendental realm*. The word "radical" echoes Bhikkhu Bodhi's epithet for deep impermanence (Bhikkhu Bodhi 2005, p. 26). We use "transcendental" not in any mystical or supernatural sense, but to denote the extraordinary high goal of Buddhadharma: to permanently relieve all suffering.

These two realms are our own formulation, intended as simplified, skeletal models of the psychological and transcendental, useful for drawing out major features and, later, relationships with MBIs. They are reminiscent of, but not intended to be the same as two traditional divisions of Buddhadharma, *conventional* (or *relative*) *reality* versus *ultimate reality* (Tsering 2008) and the *mundane* versus the *supramundane path* (Zahler 2009, pp. 5–7), both elaborate constructs with many variations and nuances.

We are omitting, but not denying, Buddhadharma's dimension of rebirth and *karma*. Rebirth is implicit in conditioned existence, also known as *samsara*, the cycle of rebirth. Karma is the source of first-arrow suffering, according to Buddhadharma, because even our external circumstances, including the so-called chance, are the result of actions in prior lives. As we discussed in prior work (Lewis 2015a, pp. 33–35), cyclic phenomena in psychology can be viewed

as analogous to rebirth, but since we are not treating psychological cycles, we need not address rebirth and karma directly.

Right Intention

Right Intention in Buddhadharma

Right intention is the element of the Eightfold Path that addresses motivation and goals; without right intention, one is not practicing Buddhadharma (Bhikkhu Bodhi 2013, Chap. III). The first item in right intention is typically translated as *renunciation*, which is often taken to mean renouncing the pleasures of life. Giving up worldly pleasure, however, only addresses the first arrow of the suffering of change. The real problem of pleasure is the second arrow, the craving and clinging to pleasure or lack of pain, and even deeper, the third level of dukkha and grasping at a mistakenly reified self. As Bhikkhu Bodhi (2013, Chap. III) puts it, "real renunciation is not a matter of compelling ourselves to give up things still inwardly cherished, but of changing our perspective on them so that they no longer bind us."

For these reasons, Tibetan teachings use a term for renunciation that translates as *determination to be free*, that is, taking liberation from all suffering as the goal over worldly pleasure or even relief from pain (Palmo 2011, pp. 63–80; Sopa 2004a, p. 3). In traditional teachings, this is formulated in cosmological terms: one is practicing Buddhadharma only if aimed at escaping the endless round of rebirth in samsara, the realm of conditioned existence and dukkha (Pabongka Rinpoche 1991, p. 127). But it is possible in this life, according to Buddhadharma, to end one's suffering at all three levels, so we can keep the discussion in a single-life frame without diminishing the essential role of the determination to attain liberation.

The role of motivation follows in part from dukkha's deep causation. Relief of overt and subtle suffering is ultimately ineffective without dissolving the dynamic at the third level. Does that mean that a Buddhist practitioner ignores their everyday suffering to focus solely on the deepest level? No, and for good reason. Even to pursue the path one needs a certain amount of leisure and freedom from suffering and material want (Inagaki and Stewart 2003, p. 6). Traditional Tibetan teachings address this fact to motivate practitioners to seize the opportunity of their favorable circumstances (Sopa 2004a, pp. 247–248). It can also be read in more modern terms as motivation to reduce ones basic suffering to enable practice of Buddhadharma. A Buddhist practitioner, therefore, does not neglect their material needs, health, or everyday suffering, but pursues them in the context of motivation toward the higher goal.

Keeping the deeper motivation in mind as much as possible not only infuses intermediate stages with inspiration and determination, but also addresses the paradox of success: incremental progress at any level can actually impede overall progress if accomplished with too strong a sense of agency, that "I did this." If one is not vigilant to keep the no-self goal firmly in mind, therefore, that sense of agency can strengthen the very reified self one is seeking ultimately to dissolve. Conversely, any action done with a motivation to reach liberation that reduces suffering becomes progress toward that end.

All this is why the motivation to be free of suffering is not merely invoked periodically in Buddhadharma; it is intimately integrated into the whole path, including meditation:

> ...in the specific context in which the practice of mindfulness is envisaged by ancient Buddhist texts, in remembering that one should remember the breath, one is remembering that one should be doing a meditation practice; in remembering that one should be doing a meditation practice, one is remembering that one is a Buddhist monk; in remembering that one is a Buddhist monk, one is remembering that one should be trying to root out greed, hatred and delusion. (Gethin 2011, pp. 270–271)

Finally, modern research psychology confirms the common sense in the value of setting high and specific goals:

> Specific, high (hard) goals lead to a higher level of task performance than do easy goals or vague, abstract goals such as the exhortation to "do one's best." So long as a person is committed to the goal, has the requisite ability to attain it, and does not have conflicting goals, there is a positive, linear relationship between goal difficulty and task performance. (Locke and Latham 2006, p. 265)

Right Intention and MBIs: Critique

Kabat-Zinn (2005b, pp. 45–46) endorses the crucial role of motivation in MBIs:

> I used to think that meditation practice was so powerful in itself and so healing that as long as you did it at all, you would see growth and change. But time has taught me that some kind of personal vision is also necessary. Perhaps it could be a vision of what or who you might be if you were to let go of the fetters of your own mind and the limitations of your own body. This image or ideal will help carry you through the inevitable periods of low motivation and give continuity to your practice.
>
> For some, that vision might be one of vibrancy and health, and for others, it might be one of relaxations or kindness or peacefulness or harmony or wisdom. Your vision should be what is most important to you and what you believe is most fundamental to your ability to be your best self, to be at peace with yourself, and to be whole.

This nicely states part of the case for motivation's central role in Buddhadharma, though without mentioning Buddhadharma by name. The goals mentioned, however, vibrancy, health, kindness, peacefulness, harmony, and wisdom, are too nonspecific to serve as useful motivation in any concrete plan of action, let alone serve as right intention's determination to be free.

More to the point, MBIs are designed and promoted for relief of medical and psychological problems, and that is why people take them. A recent study showed reducing negative experiences and general well-being to be by far the predominant reasons for starting and continuing mindfulness meditation, with religion or spirituality registering around 5 % (Pepping et al. 2016,

p. 544). In particular, MBIs target Buddhad-harma's first level of suffering in the form of stress, chronic pain, depression, trauma, and other psychological or behavioral issues (Purser 2015c, p. 681; Teasdale and Chaskalson 2011, p. 92) and to some extent the second level by addressing addiction and promoting general well-being (Garland et al. 2015). MBIs are not focused on the deepest level of dukkha and therefore cannot be meaningfully considered Buddhadharma by Buddhadharma's own standards, across all traditions.

Renunciation and MBIs: Analogical Perspective

We can now begin to replace the idea that MBIs are a form of Buddhadharma with a functional analogy between them. In this model, renunciation, the determination to attain liberation, correlates with commitment to the target of the particular MBI such as preventing depression relapse for MBCT or coming to terms with chronic pain in MBSR. Each specific MBI program is thus addressed by a separate analogical model, constructed using the methodology, with each element of the program treated relative to its target. Some MBI cohorts are relatively homogeneous with respect to goal, and others are more heterogeneous, but we can still analyze most elements of a particular MBI program using the same overall analogy, with some exceptions.

Relational Right Intention: Analogical Perspective

Besides renunciation, the other two elements of Buddhadharma's right intention are focused on one's relationship to others, to develop: good will and harmlessness, or more evocatively loving-kindness toward others; and compassion for their suffering (Bhikkhu Bodhi 2013, Chap. III). These are important in all Buddhist traditions but are particularly salient in the Tibetan Mahayana.

We may also treat these aims analogically. In Buddhadharma, both are developed in intensive meditation, beginning with love and compassion for oneself, then those near and dear, and progressively moving outward to ultimately include all sentient beings, even enemies. MBSR includes a single meditation session on loving-kindness toward others (Blacker et al. 2009, p. 16; Kabat-Zinn 2005b, pp. 182–184). This is unlikely to reach anywhere near the scope, depth, or impact of the equivalent Buddhist practice, but we may view it simply as the MBSR analog of the full Buddhist practice. MBCT contains only loving-kindness and compassion practice toward oneself (Teasdale et al. 2014, pp. 170–193), which is an appropriate analog for a program aimed at depression, whose participants are contending with debilitating self-negativity.

Are MBIs a Gateway to Buddhadharma?

People can and do develop in motivation and sometimes progress from the psychological aims of MBIs into some form of actual Buddhadharma, though one recent survey suggests that the number is low (Pepping et al. 2016, p. 544). Some authors have cited this possibility to justify the lack of true Buddhadharma intention in MBIs (Monteiro et al. 2014, p. 11), but that observation applies to individuals, not to the program. In fact, several other aspects of MBIs belie considering them to be any kind of formal gateway to Buddhadharma. One is the stealth nature of Buddhist principles in MBIs, the fact that Buddhadharma is mentioned in the curricula only as a historical source, if at all (Purser 2015a, pp. 24–26). For example, Buddhism is not mentioned in the official MBSR Curriculum Guide (Blacker et al. 2009) or MBCT workbook (Teasdale et al. 2014) and almost always with the qualifier "ancient" in the MBSR workbook by Stahl and Goldstein (2010). Another is that MBI programs do not appear to explicitly encourage a transition to Buddhadharma, though participants are presumably directed to Buddhist centers if they express interest. In fact, Kabat-Zinn's widely circulated claim that MBIs embody the essence of Buddhadharma could lead participants to conclude that they need seek no further.

A Model of Dukkha

As a foundation for analyzing MBIs with regard to Buddhadharma's right view, we integrate the levels of dukkha and the four marks into a single model, summarized in Table 1.

The "visibility" factor at each level represents the subtlety of the dynamic, the difficulty in seeing it operate and the depth of meditative attention and insight needed to see through and dissolve it.

The dynamic common across levels constitutes what we are calling a functional analogy between Buddhadharma's psychological and transcendental realms indicated in the leftmost column. The two realms have different aims or functions, reflected in different scopes of activity, for example, relief of overt suffering versus all suffering. Relative to those different functions and scopes, however, the levels operate analogously. Another way of looking at this analogical model is that the everyday/psychological realm exhibits the same *form* as the radical/transcendental realm but at a different *ontological level*, a kind of "as above, so below" principle (Braverman 2016).

The model at this point displays a functional analogy between two aspects or realms of Buddhadharma. In the right view section below, we will bring MBIs into the model by showing that they embody many of the same principles as Buddhadharma's psychological realm.

Columns of the Model

The "first-arrow" column represents the conditioned nature of reality at the various levels of

Table 1 The two-arrow dynamic across levels

Realm	Level of dukkha (visibility)	First arrow (reality)	Misapprehension		Second arrow (reactions)
			Reification	Identification	
Everyday /Psychological	Suffering of suffering (overt)	Painful stimulus, thought or emotion	Never pain; pain is real	Ownership; *my* pain	Distress, avoidance, fear, anger, etc.
	Suffering of change (subtle)	Impermanent pleasurable stimulus, thought or emotion	Always pleasure; pleasure is real	Ownership; *my* pleasure	Craving, attachment, addiction, etc.
Radical/Transcendental	Suffering of conditionality (hidden)	Conditioned existence; radical impermanence	Permanence; substantiality	Identification; body/mind is *me*	Suffering of manifestation; death anxiety
	Four marks/realizations	Liberation	Impermanence	No-self	Suffering

The four columns of this table with white background show elements of the two-arrow dynamic at each level, with the third level as the source and prototype for the other two:

(a) First arrow: the reality of conditioned existence induces fear, pain, etc., leading to…;

(b) Reification of conditioned existence by attributing the permanence, control, solidity, that we would prefer, and…;

(c) Identification with the reified, solid, permanent entity, which leads to…;

(d) Second arrow: emotional reactions

permanence and substantiality, from radical impermanence and absence of substantiality to everyday fluctuating experience and apparent physical solidity. The "second-arrow" column is the actual suffering that arises from the misapprehensions, varying in scope and subtlety from all suffering to everyday distress. The "reification" and "identification" columns represent the misapprehensions that drive suffering. The scope of reification ranges from all of conditioned existence to personal mental content. The scope of identification ranges from the entire self to ownership of particular mental contents.

Four Marks in the Model

We propose connecting Buddhadharma's four marks of existence—liberation, impermanence, no-self, and suffering—to the elements of the dukkha dynamic as shown in the bottom line of the table, with gray background. The marks are both truths about existence and realizations on the way to liberation and thus relate most fundamentally to the third level of dukkha, the heart of Buddhadharma's right view.

Placing impermanence under "reification" is clear in both senses: realization of impermanence dissolves the innate tendency to reify things whose actual nature is purely conditioned and interdependent. No-self is a bit more complex. It entails both non-reification and non-identification, but the latter uniquely characterizes it, so we put realization of no-self under "identification." We place suffering under "second arrow," the painful reactions.

Finally, we place liberation under "first-arrow (reality)," because final relief from suffering is tantamount to deeply, transformatively seeing and being fully at peace with the truths of conditioned existence and radical impermanence. That means no longer reifying anything or identifying with the body/mind and no longer fearing impermanence in any form, including death. In other words, liberation occurs when the third-level dynamic is totally stilled and the practitioner rests in unconditioned reality.

Tibetan teachings place the misapprehension of a substantial, permanent self at the root of suffering (Gelek Rimpoche 2009, pp. 186–187), while Theravada doctrine sees impermanence as the deepest truth. In practice, the marks are intimately intertwined (Smith 2015). A practitioner realizes the four marks more or less simultaneously (Bhikkhu Bodhi 2013, Chap. VIII), so we may think of them as aspects of realization as much as steps on a path.

What might the four marks mean for the other two levels, the everyday/psychological realm? That is the crux of how the model casts light on MBIs and the psychological realm in general, which we take up in the next section.

Right View

MBIs and Right View

We now analyze MBIs from the standpoint of Buddhadharma's right view, using the four marks of existence: impermanence, suffering, no-self, and liberation. Kabat-Zinn claims (2011, pp. 298–299) that these:

> ...can [in MBSR] be self-revealing through skillful and ardent cultivation via formal and informal practice in the supportive context of dialogue, inquiry, and skillful instruction...without any need to reference a Buddhist framework or lens for seeing it.

This assertion is central to the claim that MBIs capture the essence of Buddhadharma. We critique this rhetoric on several grounds and conclude that it does not, in fact, embody the essential nature of the four marks. Additionally, Kabat-Zinn's rhetoric is not matched in actual MBI workbooks and offerings. In effect, Kabat-Zinn and some other MBI observers try to make what is actually a psychological level program conform to the transcendental level of Buddhadharma. In doing so, they sometimes reduce Buddhadharma to the psychological level, inflate MBIs to a transcendental level, or both.

In place of the incorrect paradigm, we offer a more accurate, meaningful, and useful one. Instead of saying MBIs embody the essence of

Buddhadharma's transcendental realm, we show that MBIs are actually a psychological analog of the transcendental realm, with a similar structure but at a very different ontological level. We do this for the four marks in turn: first a critique of Kabat-Zinn's assertion and then the alternate analogical perspective. We also briefly document the prevalence of MBIs' ideas, processes, and techniques in modern non-meditative psychotherapy, adding evidence of the psychological character of MBIs.

Impermanence in MBIs: Critique

Kabat-Zinn says this about impermanence in MBSR (2011, p. 298):

> It doesn't take long for novices to the practice of mindfulness to notice that the thinking mind has a life of its own, and can carry the attention away from both the bare attending to sensation in the body and from any ability to rest in awareness with whatever is arising. But over time, with ongoing practice, dialogue, and instruction, it is not unusual for even novice practitioners to see, either spontaneously for themselves or when it is pointed out, that the mind indeed does have a life of its own, and that when we cultivate and stabilize attention in the body, even a little bit, it often results in apprehending the constantly changing nature of sensations, even highly unpleasant ones, and thus, their impermanence.

The "constantly changing nature" of experience is impermanence at the everyday level, not Buddhadharma's radical impermanence. Kabat-Zinn is thus reducing Buddhadharma's most fundamental notion of impermanence to its obvious manifestation in everyday experience. Awareness at the everyday level of impermanence would only be Buddhadharma if it is directed ultimately at the deepest level (Kin et al. 1999, pp. 59–65); otherwise, it is a truism (Thera 2013).

Evidence from other MBI sources indicates that when impermanence is taught at all, it is of the everyday sort and not aimed at or even informed by radical impermanence. The MBSR workbook by Stahl and Goldstein also mentions shifting thoughts and sound (2010, pp. 87–88),

as well as breath (2010, p. 84), as evidence of impermanence. While the impermanence of thoughts could be an invitation to a deeper treatment, there is no such follow-up in any of these sources. The official MBSR Curriculum Guide (Blacker et al. 2009) does not mention impermanence at all, and perhaps most tellingly, Rosch (2015, p. 288) found in the three MBSR courses she studied in depth that "impermanence was treated only as a comforting reminder that painful situations don't last," which seems to reflect the last point in the above citation. MBCT (Teasdale et al. 2014) appears not to discuss impermanence at all.

In addition, none of these sources mentions death, the most poignant, intimate, and impactful aspect of impermanence in Buddhadharma. Death is the subject of crucial teachings and meditations in the Theravada (Bhikkhu Bodhi 2012), Tibetan (Lama Zopa Rinpoche & McDonald 2010) and Zen traditions (Kapleau 2015). It is understandably omitted from MBSR as incompatible with a psychological stress reduction program.

Impermanence in MBIs: Analogical Perspective

We may replace the mistaken direct correspondence between impermanence in MBIs and that in Buddhadharma with analogical correlations. Observation of everyday change is clearly an everyday/psychological analog of radical impermanence, but we propose two others that are less obvious and therapeutically more crucial in MBIs: "present-moment awareness and acceptance" and "thoughts are just thoughts, not things."

The emphasis on present-moment meditation in MBIs has been widely criticized as inadequately capturing the range and depth of actual Buddhist practice (Purser 2015c, pp. 682–683). Bhikkhu Bodhi (2011a, pp. 24–26), for example, concludes that the meaning of "present" in Buddhist mindfulness is more about making the object vividly *present* to awareness than remaining focused on the temporal present.

"Present moment" in MBIs, however, is actually as much about acceptance of experience as remaining in the present moment in meditation.

> The emphasis was always on awareness of the present moment and acceptance of things as they are, however they are in actuality, rather than a preoccupation with attaining a particular desired outcome at some future time, no matter how desirable it might be (Kabat-Zinn 2005b, p. 290).

Williams and Lynn (2010, p. 9) define acceptance "as the capacity to remain available to present experience, without attempting to terminate the painful or prolong the pleasant." In other words, we may view present-moment awareness and acceptance in MBIs as abiding unwelcome change in the form of present pain and not anticipating future pleasure, that is working with the dynamics of everyday impermanence at the first two levels of dukkha. Present-moment awareness is thus a psychological analog of realizing radical impermanence, playing out in everyday experience in the context of the two-arrow dynamic.

Williams and Lynn (2010) also review both the ancient philosophical and religious roots of acceptance and the recent "swell of interest in acceptance, as evidenced by an increase in acceptance-based therapeutic interventions." MBIs are, of course, a prominent player in that recent interest, but it also has independent origins in cognitive psychotherapy, particularly through ACT:

> Acceptance means opening up and making room for painful feelings, sensations, urges, and emotions. We drop the struggle with them, give them some breathing space, and allow them to be as they are. Instead of fighting them, resisting them, running from them, or getting overwhelmed by them, we open up to them and let them be. (Note: This doesn't mean liking them or wanting them. It simply means making room for them!) (Harris 2009, pp. 9–10)

Another important theme in MBIs, especially those focused on serious psychological disorders such as depression and addiction, is that thoughts are just thought, not things or facts, and that they therefore need not compel us to act or feel in harmful ways. For example:

> [MBSR] …thoughts, as "events" in consciousness, distinguishing the event from the content (Blacker et al. 2009, p. 10).
>
> [MBRP] Thoughts are simply ideas, memories, images, and strings of words that arise in the mind from moment to moment that may or may not be reflective of reality. (Bowen et al. 2011, p. 133)
>
> [MBCT] The thought of a meal is not the meal itself—a thought is just a mental event—very, very different from the reality of the experience…We cultivate the ability to experience thoughts as thoughts—as mental events that enter and leave the mind. With this shift, we rob thoughts of their power to upset us or to control our actions. When we see thoughts for what they are—just thoughts, nothing but passing mental events—we can experience a wonderful sense of freedom and ease. (Teasdale et al. 2014, p. 24)
>
> [MBCT] The crucial thing is to learn a new relationship to thoughts—to relate to them as thoughts, mental events that arise and pass away in the mind—rather than as the truth of "how it really is." (Teasdale et al. 2014, p. 152)

This idea is, at bottom, about the solidity of obsessive thoughts and cultivating the ability to see them as insubstantial rather than reified, compelling things. Radical impermanence, on the other hand, aims at the realization that *everything* in experience is mistakenly reified and utterly without substance, not only *some* thoughts, but all of them, including our body, perceptions, and very self. Thus, MBI's version of insubstantiality is squarely in the everyday/psychological realm, again vastly different than but analogous to the corresponding principle in the transcendental realm.

Strictly speaking, MBI analogs of Buddhist realizations should be formulated with respect to the specific goals of the MBI program in question. For example, present-moment acceptance and "thoughts are not things" in MBCT mean becoming aware of potentially depressive thoughts or emotions as they occur, but seeing them simply as events passing through the mind rather than substantial facts that compel other thoughts, emotions, or behaviors. The *3-minute breathing space* practice (3MBS) (Teasdale et al. 2014, pp. 98–100, 123), "the single most important practice in the MBCT program," used in response to potentially depressive mental

events, seems to be particularly aimed at cultivating such an attitude:

> In using breathing spaces in everyday life, you
> acknowledge that there is strong emotion around
> and take a few moments to bring awareness to it
> (as thoughts, feelings, and body sensations), sim-
> ply allowing it to be there without judging it,
> without trying to chase it away or solve any
> problem (Teasdale et al. 2014, p. 123).

Dukkha in MBIs: Critique

Regarding suffering Kabat-Zinn says:

> When we work with people in a medical or psy-
> chological setting, using 'stress' and the sugges-
> tion that 'stress reduction' might be possible as the
> core invitational framework, we can dive right into
> the experience of dukkha in all its manifestations
> without ever mentioning dukkha; dive right into
> the ultimate sources of dukkha without ever men-
> tioning the classical etiology, and yet able to
> investigate craving and clinging first-hand, pro-
> pose investigating the possibility for alleviating if
> not extinguishing that distress or suffering (cessa-
> tion), and explore, empirically, a possible pathway
> for doing so (the practice of mindfulness medita-
> tion writ large, inclusive of the ethical stance of
> śīla, the foundation of samadhi, and, of course,
> prajñā, wisdom—the eightfold noble path) without
> ever having to mention the Four Noble Truths, the
> Eightfold Noble Path, or śīla, samadhi, or prajñā.

In this passage, Kabat-Zinn does not indicate
what he means by "the ultimate sources of dukkha" and "the classical etiology." If it is simply
the two-arrow reactivity dynamic at the first
and/or second levels of dukkha, then, as we have
discussed, it is reducing Buddhadharma to a
psychological level.

Given the emphasis on meditation being "writ
large," samadhi (a highly concentrated state) and
wisdom, it is possible that Kabat-Zinn does mean
the deeper dynamic of the suffering of conditioned existence, but if so that is a significant
inflation of what MBIs actually cover. Rosch
(2015), for example, found that MBSR does not
come even close to discussing the third level of
dukkha. Stahl and Goldstein (2010) mention

suffering many times, but always equated with
stress, tension, and pain. *Full catastrophe living*
(Kabat-Zinn 2005b) adds illness, loss, grief,
misery and others and the aim of worldly joy and
happiness, only a couple of times even remotely
hinting at the complexities underlying transient
pleasure (Kabat-Zinn 2005b, pp. 24, 59), and
never the third level. Even MBRP,
mindfulness-based relapse prevention (Bowen
et al. 2011), which is aimed at substance abuse
relapse and overcoming craving, does not invoke
Buddhadharma's suffering of change even
though that is a major factor in addiction. The
crucial "urge surfing" exercise (Bowen et al.
2011, pp. 66–67) works with highly compelling
impulses and does not lead to a discussion of
Buddhadharma's second level of dukkha.

Dukkha in MBIs: Analogical Perspective

Again, we propose an analogical correlate in
MBIs to replace the mistaken direct correspondence with Buddhadharma's dukkha. Realizing
dukkha in Buddhadharma means becoming
aware, through deep meditation, that the most
fundamental suffering is bound up with our
response to the very fact of conditioned existence, that is, the transconceptual dynamic at the
third level of dukkha. The MBI analog is
becoming aware, in ordinary experience, of the
two-arrow reactivity dynamic at the first two
levels and being able to work with it.

The reactivity dynamic is indeed a pervasive
MBI theme in the form of stress reactivity, reactions to addictive relapse triggers, running on
automatic pilot, depressive reactions, and so on
(Blacker et al. 2009, pp. 8–11; Bowen et al. 2011,
pp. 7, 22, 44; Kabat-Zinn 2005b, pp. 33, 57, 65,
285; Teasdale et al. 2014, pp. 22–28). For example, the MBCT workbook says "our reactions to
unhappiness can transform what might otherwise
be a brief, passing sadness into persistent dissatisfaction and unhappiness" (Teasdale et al. 2014,

p. 22). MBCT aims for the client to become aware of such potentially depressive thoughts or emotions as they arise during the day in order to deactivate them with the 3-minute breathing space mini-practice.

Reactivity is also a long-standing theme in modern clinical psychology, for example, cognitive behavioral therapy (Beck 1979, p. 26), ACT (Harris 2009, pp. 9–10), and trauma therapy (Briere and Scott 2014, pp. 145, 196) to name only a few.

No-Self in MBIs: Critique

Kabat-Zinn connects MBIs with no-self in this passage (2011, p. 299):

> [no-self] reveals itself without any need to reference a Buddhist framework or lens for seeing it.... although this one is trickier and scarier, and needs to be held very gently and skillfully, letting it emerge out of the participants' own reports of their experience rather than stated as a fact. Often it begins with the realization, not insignificant, that 'I am not my pain,' 'I am not my anxiety,' 'I am not my cancer,' etc. We can easily ask the question, well then, who am I? This is the core practice of Chinese Chan, Korean Zen, Japanese Zen, and also of Ramana Maharshi. Nothing more is needed.... Just the question and the questioning ... the inquiry and investigation into the nature of self, not merely through thought, but through awareness itself.

Kabat-Zinn is here simultaneously deflating Buddhadharma and inflating MBIs, both by significant margins and in several ways. The first way is implicitly interpreting "I am my anxiety" literally, as total identification of the self with the pain or anxiety. In fact, that phrase is not a literal assertion that the "I" is nothing but pain; it is a metaphor for what we are calling ownership of the mind's contents. Everyone instinctively understands that whatever the self is, it constitutes a vast array of mental and physical elements and that there is much more to their "I," than their anxiety or pain, no matter how intense or consuming.

Second, while intuitive realization of non-ownership of anxiety or pain is a significant accomplishment in alleviating suffering, it still entails a reified, observer self. Getting from there to no reified observer self at all is an enormous leap, "a radical transformation of our very being... a fundamental shift in our paradigm... a radical reorientation of our frame of reference" (Jinpa 2000, p. 13). Hölzel et al. (2011, p. 547) also address the difference:

> Whereas more advanced meditation practices are required to experience this drastic disidentification from the static sense of self, a de-identification from some parts of mental content is often experienced even in the earliest stages of meditation practice.

Though the realization of no-self does sometimes happen spontaneously with little or no concerted effort, that is extremely rare. It almost always requires considerable energy and skillful support in a Buddhadharma context over years, including working personally with an experienced teacher, emphasis on intensive meditation with extremely stable attention, and a community culture aimed at liberation. This is especially the case in the vast Zen traditions that Kabat-Zinn cites as models (Pawle 2013, pp. 51–52; Wick and Horn 2015). None of these supports is remotely present or pointed to in MBSR's eclectic curriculum, with its emphasis on relieving everyday distress and reticence about its spiritual roots.

Third, Kabat-Zinn is correct that approaching no-self seriously has the potential to destabilize a participant if not handled with utmost skill (Epstein 2013, pp. 94, 133). Very few MBI instructors, however, are equipped by training or experience to deal with such problems, and even if they were, they are enjoined by MBI protocols from giving personal advice (Blacker et al. 2009, p. 2), as MBSR is not actual psychotherapy.

Fourth, very few people who come to MBIs for relief of overt suffering are interested in and open to this depth of intensive spiritual inquiry (Pepping et al. 2016, p. 544). If no-self were even touched on without extensive groundwork, many participants would likely be skeptical or rebellious.

Finally, Kabat-Zinn does a potential disservice by making no-self sound so easy and routine. Such optimism about actually achieving more than intellectual grasp of no-self could, if taken seriously by participants, result in false confidence that might rob them of the promise for subsequent transformative insight.

Deconstructing the reductive rhetoric on no-self is useful, since it is a common misunderstanding. In the context of MBIs, however, it is moot, because they actually do not seriously engage the subject. Rosch (2015, p. 290) found no such questioning of the self in actual MBSR courses, and there is little if any in the MBSR sources for students and instructors that we consulted (Blacker et al. 2009; Kabat-Zinn 2005b; Stahl and Goldstein 2010). Even if no-self were in the curriculum, it would be unlikely to occupy more than a few minutes of group discussion in a crowded curriculum.

To the contrary, MBIs are exclusively focused on repairing and strengthening the psychological self. As evidence, the following terms are sprinkled liberally throughout the MBI resources we consulted, as qualities to be desired and developed, suggesting a strong emphasis on strengthening the everyday self:

> self-regulation, self-reliance, self-compassion, self-awareness, self-care, self-discipline, self-confidence, self-esteem, self-responsibility, self-respect, self-nourishing, self-efficacy

By contrast, such terms are not found in traditional or modern Buddhadharma teachings. Here, for example, is a representative sample of "self-" terms in two volumes of the collected written works of Chögyam Trungpa Rinpoche (McMahan 2008, pp. 45–46), one of the most prominent Tibetan teachers in the West, who was also considered somewhat of a modernizer, though an extremely creative one:

> self-deception, self-perpetuating, self-hypnosis, self-defeating, self-preservation of ego, self-criticism which is helpful, self-satisfied, self-indulgence, self-sacrificing, not self-centered (Trungpa 2003, 2010)

There is nothing inherently wrong, of course, with strengthening the psychological self, acquiring self-regulation skills, and so on. In fact, it is what people come to MBIs for and what MBIs deliver. For those who move into actual

Buddhadharma practice, work on the psychological self is often an important pre- or co-requisite, in the spirit of Engler's (1984, p. 31, 2003) well-known maxim, "You have to be somebody before you can be nobody." From a Buddhadharma point of view, however, strengthening the psychological self only makes sense for a practitioner when motivated by transcendent goals, which are lacking in MBSR.

No-Self in MBIs: Analogical Perspective

The psychological analog of no-self is disidentification from some particular mental contents, what we are calling non-ownership or an *observer stance* toward the contents; they no longer feel so intimately a part of us that their pain, discomfort, or addictive nature are compelling. Strictly speaking, however, the observer stance addresses the identification aspect of Buddhadharma's misapprehended self, not the reification of body/mind that provides a substantive basis for self, but the former is a prerequisite to seeing the latter.

Though the observer stance has been part of MBSR from early on, especially in conjunction with the body scan and sitting meditation (Kabat-Zinn 2005b, pp. 33–34, 89, 297), it is less a goal in itself and more as a means to a non-judging or letting-go attitude and ultimately "mindfulness":

> Non-judging. Mindfulness is cultivated by assuming the stance of an impartial witness to your own experience. To do this requires that you become aware of the constant stream of judging and reacting to inner and outer experiences that we are all normally caught up in, and learn to step back from it. (Kabat-Zinn 2005b, p. 33)

In fact, the observer stance per se is not among MBSR's seven key "attitudinal factors" (Kabat-Zinn 2005b, pp. 33–40), though it supports two of them, acceptance and letting go. There is no explicit cultivation of the observer stance in the official MBSR Curriculum Guide (Blacker et al. 2009) and only a few scattered instances in Stahl and Goldstein (2010, pp. 37, 70, 85, 87). This may be also due to the generic nature of stress, apart from specific stressors or syndromes, and the heterogeneity of many MBSR cohorts, with no issue or malady shared by all participants to target for non-ownership.

Nevertheless, MBSR participants do tend to develop an observer stance. Kerr et al. (2011, p. 86) studied 8 MBSR participants intensively through daily diaries, and the major conclusion that emerged was that "by the end of the course, all participants developed, to some degree, an *observing stance* [italics in original] toward their experience" regardless of its valence. In the absence of explicit content-observing exercises in MBSR, this may be due simply to experience observing and altering one's own mental processes in meditation practices, non-meditative exercises, and group discussions (Stanley and Longden 2016). In meditation, the very acts of noticing when the mind has wandered from the chosen object and redirecting it back are perforce developing an observer stance toward mental content in general, though not applying it to specific content with therapeutic significance. Lutz et al. (2015, p. 640) call this capacity "background awareness," an aspect of "meta-awareness," which is their term for the observer stance.

The observer stance as a more prominent goal was imported into MBIs from the decentering concept in cognitive psychology (Safran and Segal 1996, p. 117), where it has a long history independent of Buddhadharma. It entered the MBSR line when the MBCT founders, all prominent cognitive psychologists, collaborated with Kabat-Zinn in 1993 to develop MBCT (Kabat-Zinn 2005a, p. 432):

> They felt it [MBSR] might provide an effective framework for teaching their patients what they referred to in their specialized terminology as "decentering skills" (meaning the ability to step back and observe in a less self-identified way one's own thinking as it is occurring…), for training them to recognize when their mood was deteriorating (so that they could initiate the inward stance of decentering).

The 3-minute breathing space (Teasdale et al. 2014, pp. 97–100) is a mini-meditation in MBCT that explicitly cultivates the observer stance as a crucial step. The 3MBS is the only actual practice added to MBSR by MBCT and is considered by the founders of MBCT to be "the single most important practice … the cornerstone of the whole MBCT program" (Teasdale et al. 2014, pp. 99, 124). Here is the observer step of the 3MBS:

> In Step 1 we bring thoughts, feelings, and body sensations into the scope of conscious rather than automatic processing. By deliberately bringing an interested awareness to our inner experience, even if it is difficult or unpleasant, we strengthen the approach tendencies of the mind and weaken the tendency to avoid. We also do our best to see thoughts, feelings, and sensations for what they are —just events passing through the mind, rather than realities or infallible messages that something is wrong (2014, p. 100).
>
> Taking a breathing space will not necessarily mean that unpleasant feelings will no longer be present—the crucial thing is that your mind is now in a position to respond to them mindfully, rather than react to them automatically with aversion. (2014, p. 123)

Explicit non-ownership figures prominently in many current MBI methods for specific issues or maladies. The SOBER breathing space exercise in MBRP is adapted from MBCT's 3-minute breathing space:

> SOBER Breathing Space. This is an exercise that you can do almost anywhere, anytime because it is very brief and quite simple. It can be used … when you are experiencing urges and cravings to use. It can help you step out of automatic pilot, becoming less reactive and more aware and mindful in your response….
>
> O—Observe. Observe the sensations that are happening in your body. Also observe any emotions, moods, or thoughts you are having. Just notice as much as you can about your experience. (Bowen et al. 2011, p. 90)

The official generic version of MBSR does not have specific pain management elements, but here is a pain observer exercise from a non-MBSR pain management program endorsed by Kabat-Zinn (Gardner-Nix and Costin-Hall 2009, p. 165):

> And when you are ready
>   Take your mind's eye over
>   To where the pain is.

Bringing awareness to it.
Observing it.
Imagining walking around it
And seeing it from every angle.

In all these cases (except for generic MBSR), the client cultivates and engages the observer stance for a specific purpose, namely to counter the target syndrome that they came to the MBI for, just as the Buddhist practitioner cultivates complete no-self for their goal of liberation from all suffering. And just as no-self combines identification and reification, MBI practices address both observer stance and "thoughts are not things." Thus, the analogical approach reveals a rich functional correlation between MBIs and Buddhadharma.

The general concept of non-ownership or an observer stance toward mental content is known under a variety of names throughout MBIs as well as modern psychotherapy, though with variations and nuances. These include *meta-awareness* (Hölzel et al. 2011, p. 547; Lutz et al. 2015, pp. 640–642); *observer or observing perspective, self, or attitude* (Deikman 1982; Kerr et al. 2011); *objectifying the mind* (Stanley and Longden 2016); *decentering*; and *cognitive defusion* (Harris 2009, p. 9). Aaron Beck, founder of cognitive therapy (CT), a progenitor of cognitive behavioral therapy (CBT), called this concept *distancing* (1970, p. 189), though CB and CBT emphasize working directly with maladaptive mental content much more than the relationship to that content (Dozois and Beck 2011, pp. 26, 30–32, 37).

Liberation in MBIs: Critique

Kabat-Zinn typically approaches the concept of liberation indirectly, rather than explicitly propose that MBSR embodies the essence of liberation in Buddhadharma as he does for impermanence, suffering, no-self, and other Buddhadharma concepts. For example:

If awareness itself is our true nature, then abiding in awareness liberates us from getting stuck in any state of body or mind, thought or emotion, no matter how bad the circumstances may be or appear to be (Kabat-Zinn 2005a, p. 461).

Kabat-Zinn and Mark Williams used liberation expansively though still colloquially in an introduction to Kabat-Zinn's own article (2011, p. 281):

He [Kabat-Zinn] sees the current interest in mindfulness and its applications as signaling a multi-dimensional emergence of great transformative and liberative promise, one which, if cared for and tended, may give rise to a flourishing on this planet akin to a second, and this time global, Renaissance, for the benefit of all sentient beings and our world.

This indirect approach to liberation in Buddhadharma is probably the most that one can credibly posit, because it is such a manifestly transcendent concept that any attempt to reduce it to a psychological construct would patently lose its essence.

Liberation in MBIs: Analogical Perspective

We propose that the MBI analog of Buddhadharma's liberation is the complete and permanent resolution of the target malady or syndrome: depression for MBCT, addiction for MBRP, distress of chronic pain or stress for MBSR, and so on. This accords both with our approach to right intention above and with Kabat-Zinn's claim that both Buddhadharma and MBSR provide "a practical path to liberation from suffering" (Kabat-Zinn 2011, p. 288), but understanding that Buddhadharma and MBIs address very different levels of suffering.

If that seems too mundane to serve as an analog of Buddhadharma's sublime notion of liberation, we should bear in mind the significant gap between the psychological and transcendent realms. Also, modern clinical psychology essentially never promises total resolution of such maladies, nor is it often achieved. Freud famously led his clients to expect no more than to transform "hysterical misery into common unhappiness" (Breuer et al. 1957, p. 305). Modern research tells us that those who have had at least three episodes of major depression, MBCT's target audience, have a 90 % chance of a fourth (American Psychiatric Association 1998, pp. 341–342), and if minor recurrence is included, the rate probably approaches 100 % (Judd 1997, p. 990). In fact, the very point of many MBIs is entirely relapse prevention and symptom management rather than elimination of the syndrome.

From a psychological standpoint, therefore, complete and permanent remission is a lofty but not unattainable goal, and when accomplished is truly worthy of honor and celebration. In fact, these psychological truths provide a glimpse of the surpassing nature of liberation in Buddhadharma.

Conclusion

We respect and honor the work of Kabat-Zinn and others who have brought MBIs and related psychotherapy systems to their current state in health care. Our comparison of MBIs with Buddhadharma is in no way a critique of MBIs themselves, even from a Buddhadharma standpoint. Given the aims and structure of MBIs, it is entirely appropriate for them not to address the transcendent goals and ideas of Buddhadharma. To do so would risk confusion, disappointment, and philosophical disagreement and possibly compromise the program's beneficial effects.

For MBIs' own benefit, however, it is useful to understand that they are not a form of Buddhadharma, recontextualized or otherwise, despite drawing some of their inspiration and practices from Buddhadharma. MBIs do not aim at the goal of Buddhadharma and do not engage the vital ideas that lead there. They pursue relief of everyday suffering rather than its transcendental roots, and this places them squarely within clinical psychology. These observations are not new, though by delving deeper than usual into both the transcultural wisdom aspect of Buddhadharma and the actual curricula of MBIs, we have perhaps provided more and firmer evidence.

That does not mean, however, that there is only a superficial relationship between Buddhadharma and MBIs. MBIs do embody something fundamental about Buddhadharma: not its *essence* but important aspects of its *form*, translated to a very different ontological level. This has been largely missed thus far because of two factors, both addressed by the analogical methodology: preoccupation with either showing or disproving essential identity; and difficulty in accepting many assumptions and beliefs of traditional Buddhadharma on their own terms.

We conclude with three implications.

Conflation of Buddhadharma and MBIs

The analogical model helps explain why the Buddhadharma and MBIs are so easily conflated, even apart from surface similarities such as meditation. First, they have similar structures relative to their different goals, namely the two-arrow reactive dynamic, and this can be mistaken for identical essence. Second, Buddhadharma does work at the psychological as well as the transcendent level. Any psychological phenomenon therefore has two potential frames, Buddhadharma or a purely psychological context, and it is easy to neglect that or confuse them. For example, during a body scan, am I seeking physical or emotional relief, or am I aimed ultimately at realization of no-self? These will entail very different qualities of engagement from the start. Failing to recognize that will lead to misplaced critiques of MBIs, for example, that they have degraded the body scan from an insight practice in Buddhadharma to a relaxation exercise (Rosch 2015, p. 277), when such relaxation may be very helpful for relief of pain and stress.

MBIs as Buddhadharma: Reality Versus Perception

This chapter is about the reality of the relationship between MBIs and Buddhadharma, which we find relatively clear. When it comes to perception, however, the public face of that relationship is ambiguous, and that is due in no small part to Kabat-Zinn's widely circulated rhetoric as against the decidedly secular reality of MBIs' goals, curricula, and official marketing (Brown 2016). Whether this dual perception, which as we have seen is aided by the analogical relationship between MBIs and Buddhadharma, has helped or impeded the spread or effectiveness of MBIs is hard to say. There are some arenas, however, where the misleading rhetoric could actually impair the "mindfulness" movement itself. One is the adoption of MBI-like programs in schools. In a recent challenge to a school mindfulness program in Massachusetts (Legere 2016), Kabat-Zin's numerous statements asserting the essential identity of MBIs and Buddhadharma, including some of the same ones we have cited, became a significant weapon in the hands of those objecting to the program (Broyles 2016, pp. 4, 8, 12, 14–16). While this particular challenge appears to have failed, it sounds a cautionary note about possible serious negative effects of the misleading rhetoric. It is perhaps ironic that MBIs could come under such an attack when in fact they are so firmly not a form of Buddhadharma.

Dialogue

The analogical methodology facilitates dialogue between Buddhadharma and MBIs. For example, there is considerable depth to the Buddhist principles of reification and identification, especially in Mahayana theory and practice, and likewise for the corresponding MBI and psychotherapy principles of the insubstantiality of dysfunctional thoughts and the observer stance toward distressing mental content. Correlating these as we have suggested might open avenues for further MBI theory, research, and development, addressing questions such as: Are these independent causal factors in MBIs? How should they be structured and sequenced in practice for best effect?

In the reverse direction, Buddhadharma frequently starts with everyday psychological experience to develop deeper insight, for example, using filial love to develop compassion (Gelek Rimpoche

2009, Chap. III) or the shock of false accusation to explore no-self (Gelek Rimpoche 2009, Chap. V). The highly articulated constructs of modern psychology might have much to contribute along these lines if viewed as analogical to Buddhadharma. John Makransky (2007) has already done work along these lines that reflects the resonance of modern psychological attachment theory with Mahayana principles and practices. Another idea might be to use present-moment acceptance and decentering from MBIs and associated psychotherapies as an early step to help develop insight into radical impermanence and no-self, thus implementing Kabat-Zinn's idea but in the context of Buddhadharma's Eightfold Path.

A vital step for progress in both directions is to relinquish reductionism and acknowledge and leverage the significant differences between Buddhadharma and psychological systems. Another is for researchers and program developers to go beyond surface understanding by conscientiously engaging the other side in deep collaboration and direct immersion.

References

American Psychiatric Association. (Ed.). (1998). *Diagnostic and statistical manual of mental disorders: DSM-IV* (4 ed., 7. print). Washington, DC.

Beck, A. T. (1970). Cognitive therapy: Nature and relation to behavior therapy. *Behavior Therapy, 1*(2), 184–200. 10.1016/S0005-7894(70)80030-2

Beck, A. T. (1979). *Cognitive therapy and the emotional disorders*. New York, NY: Penguin.

Becker, E. (1973). *The denial of death*. New York: Free Press.

Bhikkhu Bodhi (Ed.). (2000). *The connected discourses of the Buddha: A new translation of the Saṃyutta Nikāya; translated from the Pāli; original translation by Bhikkhu Bodhi*. Somerville, MA: Wisdom Publications.

Bhikkhu Bodhi (2011a). What does mindfulness really mean? A canonical perspective. *Contemporary Buddhism, 12*(1), 19–39.

Bhikkhu Bodhi (2013, November 13). *The noble eightfold path: The way to the end of suffering*. Retrieved January 28, 2016, from http://www.accesstoinsight.org/lib/authors/bodhi/waytoend.html

Bhikkhu Bodhi (2005). *The middle length discourses of the Buddha: A translation of the Majjhima Nikaya*. New York City: Simon and Schuster.

Bhikkhu Bodhi (2011b, July 3). *Four noble truths— Suffering in depth*. Retrieved February 4, 2016, from http://www.beyondthenet.net/dhamma/unsatisfied.htm

Bhikkhu Bodhi (2012, June 27). *Mindfulness of death*. Retrieved April 3, 2016, from http://www.tricycle.com/blog/mindfulness-death-week-4-bhikkhu-bodhis-retreat

Blacker, M., Kabat-Zinn, J., Santorelli, S., & Meleo-Meyer, F. (2009). *Mindfulness-Based Stress Reduction (MBSR) curriculum guide*. Center for Mindfulness in Medicine, Health Care and Society, University of Massachusetts Medical School.

Bowen, S., Chawla, N., & Marlatt, G. A. (2011). *Mindfulness-based relapse prevention for addictive behaviors: A clinician's guide*. New York: Guilford Press.

Bowen, S., Parks, G. A., Coumar, A., & Marlatt, G. A. (2006). Mindfulness meditation in the prevention and treatment of addictive behaviors. In D. K. Nauriyal, M. Drummond, & Y. B. Lal (Eds.), *Buddhist thought and applied psychological research: Transcending the boundaries* (pp. 393–413). London: Routledge.

Bowen, S., Witkiewitz, K., Clifasefi, S. L., Grow, J., Chawla, N., Hsu, S. H., … Larimer, M. E. (2014). Relative efficacy of mindfulness-based relapse prevention, standard relapse prevention, and treatment as usual for substance use disorders: A randomized clinical trial. *JAMA Psychiatry, 71*(5), 547. 10.1001/jamapsychiatry.2013.4546

Braverman, A. (2016). *As above so below meaning*. Retrieved April 5, 2016, from http://www.arkintime.com/as-above-so-below/

Breuer, J., Freud, S., & Strachey, J. (1957). *Studies on hysteria*. New York: Basic Books.

Briere, J. N., & Scott, C. (2014). *Principles of trauma therapy: A guide to symptoms, evaluation, and treatment*. Thousand Oaks, CA: SAGE Publications.

Brown, C. G. (2016). Can secular mindfulness be separated from Religion? In R. E. Purser, D. Forbes, & A. Burke (Eds.), *Handbook of mindfulness: Culture, context and social engagement*. Berlin: Springer.

Broyles, D. R. (2016). *Legal & practical concerns regarding the district's Calmer choice mindfulness curriculum*. National Center for Law & Policy. Retrieved from http://www.capecodtimes.com/assets/pdf/CC190322.PDF

Center for Mindfulness. (2016a). *People*. Retrieved February 25, 2016, from http://www.umassmed.edu/cfm/about-us/people/

Center for Mindfulness. (2016b). *Stress reduction*. Retrieved February 7, 2016, from http://www.umassmed.edu/cfm/stress-reduction/

Chase, J. (2015, March 27). *MBSR and Buddhist mindfulness: Seeking continuities, not differences*. Retrieved December 22, 2015, from http://www.buddhistdoor.net/features/mbsr-and-buddhist-mindfulness-seeking-continuities-not-differences

Chiesa, A., & Malinowski, P. (2011). Mindfulness-based approaches: Are they all the same? *Journal of Clinical Psychology, 67*(4), 404–424. 10.1002/jclp.20776

Chodron, P. (2007). *The places that scare you: A guide to fearlessness in difficult times*. Boston: Shambhala.

Coronado-Montoya, S., Levis, A. W., Kwakkenbos, L., Steele, R. J., Turner, E. H., & Thombs, B. D. (2016). Reporting of positive results in randomized controlled trials of mindfulness-based mental health interventions. *PLOS ONE, 11*(4), e0153220. 10.1371/journal.pone.0153220

Cullen, M. (2011). Mindfulness-based interventions: An emerging phenomenon. *Mindfulness, 2*(3), 186–193. 10.1007/s12671-011-0058-1

Deikman, A. J. (1982). *The observing self: Mysticism and psychotherapy*. Boston: Beacon Press.

Dimidjian, S. (2016, January 7). *What is it about MBCT that makes it effective?* Retrieved February 7, 2016, from http://www.mindfulnoggin.com/blog/what-is-it-about-mbct-that-makes-it-effective

Dobkin, P. L., & Zhao, Q. (2011). Increased mindfulness—The active component of the mindfulness-based stress reduction program? *Complementary Therapies in Clinical Practice, 17*(1), 22–27. 10.1016/j.ctcp.2010.03.002

Dozois, D. J. A., & Beck, A. T. (2011). Cognitive therapy. In J. D. Herbert & E. M. Forman (Eds.), *Acceptance and mindfulness in cognitive behavior therapy: Understanding and applying the new therapies* (pp. 26–56). Hoboken, NJ: Wiley.

Dreyfus, G. (2011). Is mindfulness present-centered and non-judgmental? A discussion of the cognitive dimensions of mindfulness. *Contemporary Buddhism, 12*(1), 41–54. 10.1080/14639947.2011.564815

Eberth, J., & Sedlmeier, P. (2012). The effects of mindfulness meditation: A meta-analysis. *Mindfulness, 3*(3), 174–189. 10.1007/s12671-012-0101-x

Encyclopedia Britannica. (2016). *Analogy: Reason*. Retrieved April 2, 2016, from http://www.britannica.com/topic/analogy-reason

Engler, J. (1984). Therapeutic aims in psychotherapy and meditation: Developmental stages in the representation of self. *The Journal of Transpersonal Psychology, 1*, 25–61.

Engler, J. (2003). Being somebody and being nobody: A reexamination of the understanding of self in psychoanalysis and Buddhism. In J. D. Safran (Ed.), *Psychoanalysis and Buddhism: An unfolding dialogue* (1st ed., pp. 34–72). Boston: Wisdom Publications.

Epstein, M. (2013). *Thoughts without a thinker: Psychotherapy from a Buddhist perspective* (Revised ed.). New York: Basic Books.

Felver, J. C., & Jennings, P. A. (2015). Applications of mindfulness-based interventions in school settings: An introduction. *Mindfulness, 7*(1), 1–4. 10.1007/s12671-015-0478-4

Fischer, N. (2012, June 21). *Impermanence Is Buddha nature*. Retrieved March 29, 2016, from https://www.eomega.org/article/impermanence-is-buddha-nature

Gardner-Nix, J., & Costin-Hall, L. (2009). *The mindfulness solution to pain: Step-by-step techniques for chronic pain management*. Oakland: New Harbinger Publications.

Garland, E. L., Farb, N. A., Goldin, P. R., & Fredrickson, B. L. (2015). Mindfulness broadens awareness and builds eudaimonic meaning: A process model of mindful positive emotion regulation. *Psychological Inquiry*, *26*(4), 293–314. 10.1080/1047840X.2015.1064294

Gentner, D., & Bowdle, B. (2008). Metaphor as structure-mapping. In R. W. Gibbs (Ed.), *The Cambridge handbook of metaphor and thought* (pp. 109–128). New York: Cambridge University Press.

Gelek Rimpoche. (1998). *Odyssey to freedom*. Ann Arbor, MI: Jewel Heart.

Gelek Rimpoche. (2009). *The four mindfulnesses*. Ann Arbor, MI: Jewel Heart.

Gethin, R. (2011). On some definitions of mindfulness. *Contemporary Buddhism*, *12*(1), 263–279. 10.1080/14639947.2011.564843

Goyal, M., Singh, S., Sibinga, E. M. S., Gould, N. F., Rowland-Seymour, A., Sharma, R., … Haythornthwaite, J. A. (2014). Meditation programs for psychological stress and well-being: A systematic review and meta-analysis. *JAMA Internal Medicine*, *174*(3), 357. 10.1001/jamainternmed.2013.13018

Grossman, P., & Van Dam, N. T. (2011). Mindfulness, by any other name…: Trials and tribulations of sati in western psychology and science. *Contemporary Buddhism*, *12*(1), 219–239. 10.1080/14639947.2011.564841

Gyatso, T. (1997). *The four noble truths: Fundamentals of Buddhist teachings*. London, England: Thorsons.

Gyatso, T., Hopkins, J., & Napper, E. (2006). *Kindness, clarity & insight*. Ithaca, NY: Snow Lion Publications.

Harris, R. (2009). *ACT made simple: An easy-to-read primer on acceptance and commitment therapy*. Oakland, CA: New Harbinger Publications.

Hölzel, B. K., Lazar, S. W., Gard, T., Schuman-Olivier, Z., Vago, D. R., & Ott, U. (2011). How does mindfulness meditation work? Proposing mechanisms of action from a conceptual and neural perspective. *Perspectives on Psychological Science*, *6*(6), 537–559. 10.1177/1745691611419671

Inagaki, H., & Stewart, H. (Eds.). (2003). *The three Pure Land sutras* (Rev. 2nd ed.). Berkeley, Calif: Numata Center for Buddhist Translation and Research.

Ivey, G. (2015). The mindfulness status of psychoanalytic psychotherapy. *Psychoanalytic Psychotherapy*, *29*(4), 382–398. 10.1080/02668734.2015.1081267

Jinpa, G. T. (2000). The foundations of a Buddhist psychology of awakening. In G. Watson, S. Batchelor, & G. Claxton (Eds.), *The psychology of awakening: Buddhism, science, and our day-to-day lives* (pp. 10–22). York Beach, ME: Samuel Weiser.

Johns, S. (2015, October 16). Jon Kabat-Zinn gets present with the audience by… Retrieved January 1, 2016, from http://samueljohns.com/post/131447716544/jon-kabat-zinn-gets-present-with-the-audience-by

Judd, L. L. (1997). The clinical course of unipolar major depressive disorders. *Archives of General Psychiatry*, *54*(11), 989. 10.1001/archpsyc.1997.01830230015002

Kabat-Zinn, J. (2005a). *Coming to our senses: Healing ourselves and the world through mindfulness* (1st ed.). New York: Hyperion.

Kabat-Zinn, J. (2005b). *Full catastrophe living* (Delta trade pbk. reissue). New York, NY: Delta Trade Paperbacks.

Kabat-Zinn, J. (2011). Some reflections on the origins of MBSR, skillful means, and the trouble with maps. *Contemporary Buddhism*, *12*(1), 281–306. 10.1080/14639947.2011.564844

Kabat-Zinn, J., & Burney, R. (1981). The clinical use of awareness meditation in the self-regulation of chronic pain: *Pain*, *11*, S273. 10.1016/0304-3959(81)90541-8

Kabat-Zinn, J., & Shonin, E. (2015, May). This is not McMindfulness by any stretch of the imagination. *The Psychologist, the Monthly Publication of The British Psychological Society*. Retrieved from https://thepsychologist.bps.org.uk/not-mcmindfulness-any-stretch-imagination

Kapleau, P. (2015). *The Zen of living and dying: A practical and spiritual guide*. Boston: Shambhala.

Keown, D. (2003). *A dictionary of Buddhism*. Oxford; New York: Oxford University Press.

Kerr, C. E., Josyula, K., & Littenberg, R. (2011). Developing an observing attitude: An analysis of meditation diaries in an MBSR clinical trial. *Clinical Psychology & Psychotherapy*, *18*(1), 80–93. 10.1002/cpp.700

Khoury, B., Lecomte, T., Fortin, G., Masse, M., Therien, P., Bouchard, V., … Hofmann, S. G. (2013). Mindfulness-based therapy: A comprehensive meta-analysis. *Clinical Psychology Review*, *33*(6), 763–771. 10.1016/j.cpr.2013.05.005

Kin, U. B., Confalonieri, P., & Goenka, S. N. (1999). *The clock of vipassana has struck: The teachings and writings of Sayagyi U Ba Khin with commentary by S. N. Goenka*. Pariyatti.

Lama Zopa Rinpoche, & McDonald, K. (2010). *Wholesome fear: Transforming your anxiety about impermanence and death*. New York City: Simon and Schuster.

Legere, C. (2016, March 3). Calmer choice fight settles down at D-Y—News. *Cape Cod Times*. Hyannis MA. Retrieved from http://www.capecodtimes.com/article/20160303/NEWS/160309768

Lewis, D. J. (2015a, May). *Mahayana Buddhism and trauma*. Presented at the Symposium on Trauma and Contemplative Practice, Harvard University Divinity School. Retrieved from http://www.TraumaAndContemplativePractice.org/video/video.html#Lewis

Lewis, D. J. (2015b, June). *The line between psychology and Buddhism: Where are mindfulness-based interventions*. Presented at the Conference on Mindfulness and Compassion: The Art and Science of Contemplative Practice, San Francisco State University. Retrieved from http://www.cmc-ia.org/mcc2015/

Linehan, M. M. (2014). *DBT skills training manual* (2nd ed.). New York City: Guilford Publications.

Locke, E. A., & Latham, G. P. (2006). New directions in goal-setting theory. *Current Directions in Psychological Science, 15*(5), 265–268. 10.1111/j.1467-8721.2006.00449.x

Loy, D. (2014, July 2). *What Buddhism and psychotherapy are learning from each other.* Retrieved April 3, 2016, from http://www.huffingtonpost.com/david-loy/what-buddhism-and-psychot_b_5549963.html

Lutz, A., Jha, A. P., Dunne, J. D., & Saron, C. D. (2015). Investigating the phenomenological matrix of mindfulness-related practices from a neurocognitive perspective. *American Psychologist, 70*(7), 632–658. 10.1037/a0039585

MacCoon, D. G., MacLean, K. A., Davidson, R. J., Saron, C. D., & Lutz, A. (2014). No sustained attention differences in a longitudinal randomized trial comparing mindfulness based stress reduction versus active control. *PLoS ONE, 9*(6), e97551. 10.1371/journal.pone.0097551

Makransky, J. (2007). *Awakening through love: Unveiling your deepest goodness.* Somerville, MA: Wisdom Publications.

Makransky, J. (2012). Compassion in Buddhist psychology. In C. K. Germer & R. D. Seiegel (Eds.), *Wisdom and compassion in psychotherapy: Deepening mindfulness in clinical practice* (pp. 62–74). New York City: The Guilford Press.

Marlatt, G. A., & Marques, J. K. (1977). Meditation, self-control, and alcohol use. In R. B. Stuart (Ed.), *Behavioral self-management: Strategies, techniques and outcomes.* New York: Brunner/Mazel.

McMahan, D. L. (2008). *The making of Buddhist modernism.* USA: Oxford University Press.

Monteiro, L. M., Musten, R. F., & Compson, J. (2014). Traditional and contemporary mindfulness: Finding the middle path in the tangle of concerns. *Mindfulness, 6*(1), 1–13. 10.1007/s12671-014-0301-7

Pabongka Rinpoche. (1991). *Liberation in the palm of your hand: A concise discourse on the path to enlightenment.* Somerville, MA: Wisdom Publications.

Palmo, J. T. (2011). *Into the heart of life.* Ithaca, NY: Snow Lion Publications.

Pawle, R. (2013). The ego in the psychology of Zen: Understanding reports of Japanese Zen masters on the experience of no-self. In D. Mathers, M. E. Miller, & O. Ando (Eds.), *Self and no-self: Continuing the dialogue Between Buddhism and psychotherapy* (pp. 45–52). London: Routledge.

Pedulla, T. (n.d.). *MBCTBoston: FAQs.* Retrieved April 22, 2016, from http://www.mbctboston.com/faqs.html

Pepping, C. A., Walters, B., Davis, P. J., & O'Donovan, A. (2016). Why do people practice mindfulness? An investigation into reasons for practicing mindfulness meditation. *Mindfulness, 7*(2), 542–547. 10.1007/s12671-016-0490-3

Purser, R. E. (2015b). Confessions of a mind-wandering MBSR student: Remembering social amnesia. *Self & Society, 43*(1), 6–14.

Purser, R. E. (2015a). Clearing the muddled path of traditional and contemporary mindfulness: A response to Monteiro, Musten, and Compson. *Mindfulness, 6* (1), 23–45. 10.1007/s12671-014-0373-4

Purser, R. E. (2015c). The myth of the present moment. *Mindfulness, 6*(3), 680–686. 10.1007/s12671-014-0333-z

Purser, R. E., & Loy, D. (2013, July 1). *Beyond McMindfulness.* Retrieved January 10, 2016, from http://www.huffingtonpost.com/ron-purser/beyond-mcmindfulness_b_3519289.html

Purser, R. E., & Ng, E. (2015, September 27). *Corporate mindfulness is bullsh*t: Zen or no Zen, you're working harder and being paid less.* Retrieved January 26, 2016, from http://www.salon.com/2015/09/27/corporate_mindfulness_is_bullsht_zen_or_no_zen_youre_working_harder_and_being_paid_less/

Rosch, E. (2015). The emperor's clothes: A look behind the Western mindfulness mystique. In B. D. Ostafin, M. D. Robinson, & B. P. Meier (Eds.), *Handbook of mindfulness and self-regulation* (pp. 998–999). New York: Springer.

Rozelle, D., & Lewis, D. J. (2014). Eye movement desensitzation and reprocessing and buddhist practice: A new model of posttraumatic stress disorder treatment. In V. Follette, J. Briere, D. Rozelle, J. Hopper, & D. Rome (Eds.), *Mindfulness-oriented interventions for Trauma: Integrating contemplative practices* (1st ed., pp. 102–124). New York: The Guilford Press.

Rozelle, D. (2015a, May). *EMDR and Buddhist practice: A new model of PTSD treatment.* Presented at the 2015 Symposium On Trauma and Contemplative Practice, Harvard Divinity School, Cambridge, MA. Retrieved from http://TraumaAndContemplativePractice.org/video/video.html#Rozelle

Rozelle, D. (2015b, June). *Relational clinical psychology through a Tibetan Buddhist lens.* Presented at the Conference on Mindfulness and Compassion: The Art and Science of Contemplative Practice, San Francisco State University. Retrieved from http://www.cmc-ia.org/mcc2015/

Rozelle, D., Lewis, D. J., & King, A. P. (2014, October). *PTSD and Buddhism: An analogical mapping model.* Presented at the Mind & Life International Symposium for Contemplative Studies, Boston, MA. Retrieved from http://www.iscs2014.org/

Safran, J., & Segal, Z. V. (1996). *Interpersonal process in cognitive therapy.* Jason Aronson.

Smalley, S., & Winston, D. (2011, January 25). *Suffering is optional.* Retrieved February 26, 2016, from http://www.mindful.org/suffering-is-optional/

Smith, D. (2015, January 5). *A world of impermanence: The three marks.* Retrieved March 31, 2016, from http://secularbuddhism.org/2015/01/05/a-world-of-impermanence-the-three-marks/

Soeng, M. (2004). *Trust in mind: The rebellion of Chinese Zen.* Boston: Wisdom Publications.

Sopa, G. L. (2004a). *Steps on the path to enlightenment: A commentary on Tsongkhapa's Lamrim Chenmo,*

Volume 1: The foundation practices. Somerville, MA: Wisdom Publications.

Sopa, G. L. (2004b). *Steps on the path to enlightenment: A commentary on Tsongkhapa's Lamrim Chenmo, Volume 2: Karma*. Somerville, MA: Wisdom Publications.

Stahl, B., & Goldstein, E. (2010). *A mindfulness-based stress reduction workbook*. Oakland, CA: New Harbinger Publications.

Stanley, S., & Longden, C. (2016). Constructing the mindful subject: Reformulating experience through affective discursive practice in mindfulness-based stress reduction. In R. E. Purser, D. Forbes, & A. Burke (Eds.), *Handbook of mindfulness: Culture, context and social engagement*. Berlin: Springer.

Stanley, S. (2015). Sīla and sati: An exploration of ethics and mindfulness in Pāli Buddhism and their implications for secular mindfulness-based applications. In E. Shonin, W. V. Gordon, & N. N. Singh (Eds.), *Buddhist foundations of mindfulness* (pp. 89–113). Berlin: Springer International Publishing.

Stanley, E. A., Schaldach, J. M., Kiyonaga, A., & Jha, A. P. (2011). Mindfulness-based Mind Fitness Training: A case study of a high-stress predeployment military cohort. *Cognitive and Behavioral Practice, 18*(4), 566–576. 10.1016/j.cbpra.2010.08.002

Teasdale, J. D., Williams, J. M. G., & Segal, Z. (2014). *The mindful way workbook: An 8-week program to free yourself from depression and emotional distress*. New York: The Guilford Press.

Teasdale, J. D. (1999). Metacognition, mindfulness and the modification of mood disorders. *Clinical Psychology & Psychotherapy, 6*(2), 146–155. 10.1002/(SICI)1099-0879(199905)6:2<146::AID-CPP195>3.0.CO;2-E

Teasdale, J. D., & Chaskalson, M. (2011). How does mindfulness transform suffering? I: The nature and origins of dukkha. *Contemporary Buddhism, 12*(1), 89–102. 10.1080/14639947.2011.564824

Ṭhānissaro Bhikkhu (1997). *Sallatha Sutta: The arrow*. Retrieved January 29, 2016, from http://www.accesstoinsight.org/tipitaka/sn/sn36/sn36.006.than.html

Thera, N. (1998). *Sallatha Sutta: The dart*. Retrieved March 16, 2016, from http://www.accesstoinsight.org/tipitaka/sn/sn36/sn36.006.nypo.html

Thera, N. (2013, November 30). *The three basic facts of existence: I. Impermanence (anicca)*. Retrieved April 3, 2016, from http://www.accesstoinsight.org/lib/authors/various/wheel186.html

Trungpa, C. (2003). *The collected works of Chogyam Trungpa: Volume three*. Boston: Shambhala.

Trungpa, C. (2010). *The collected works of Chogyam Trungpa: Volume one*. Boston: Shambhala.

Tsering, G. T. (2005). *The Four Noble Truths*. Somerville, MA: Wisdom Publications.

Tsering, G. T. (2008). *Relative truth, ultimate truth*. Boston: Wisdom Publications.

Vago, D. R., & Silbersweig, D. A. (2012). Self-awareness, self-regulation, and self-transcendence (S-ART): A framework for understanding the neurobiological mechanisms of mindfulness. *Frontiers in Human Neuroscience, 6*, 296. 10.3389/fnhum.2012.00296

Vettese, L. C., Toneatto, T., Stea, J. N., Nguyen, L., & Wang, J. J. (2009). Do mindfulness meditation participants do their homework? And does it make a difference? A review of the empirical evidence. *Journal of Cognitive Psychotherapy, 23*(3), 198–225. 10.1891/0889-8391.23.3.198

Wallace, B. A. (Ed.). (2003). *Buddhism & science: Breaking new ground*. New York: Columbia University Press.

Wallace, B. A. (2011). *Minding closely: The four applications of mindfulness*. Ithaca, NY: Snow Lion Publications.

Wick, G. S., & Horn, V. (2015, November 4). *Koan training and the different styles of Zen*. Retrieved April 4, 2016, from http://www.buddhistgeeks.com/2009/02/bg-109-koan-training-and-the-different-styles-of-zen/

Wikipedia: article on dukkha. (2016, February 27). Translating the term dukkha. In *Wikipedia, the free encyclopedia*. Retrieved from https://en.wikipedia.org/w/index.php?title=Dukkha&oldid=707182721

Williams, K. (2015). Mindfulness-based stress reduction (MBSR) in a worksite wellness program. In R. A. Baer (Ed.), *Mindfulness-based treatment approaches: Clinician's guide to evidence base and applications* (pp. 361–376). Cambridge: Academic Press.

Williams, J. C., & Lynn, S. J. (2010). Acceptance: An historical and conceptual review. *Imagination, Cognition and Personality, 30*(1), 5–56. 10.2190/IC.30.1.c

Zahler, L. (2009). *Study and practice of meditation: Tibetan interpretations of the concentrations and formless absorptions*. Ithaca, NY: Snow Lion Publications.

Mindfulness: The Bottled Water of the Therapy Industry

Paul Moloney

The Rise of Mindfulness-Based Therapy

The set of practices widely known as 'mindfulness' derive from ancient India and the Mahayana and Theravada schools of monastic Buddhism—with their traditional aims of seeking to free the mind and body of desire, aversion, and confusion (Crook 2009). Despite the popularity of the approach, the precise meaning of 'mindfulness' remains elusive: it is an English translation of the term 'Sati, which, in the Pali language, as spoken by the historical Buddha, has many meanings—including 'attention', 'memory of the present', 'clear understanding', and 'awareness' (Stanley 2012). In contemporary usage, especially in the Western world, the word 'mindfulness' has been expanded (or appropriated) to encompass a lifestyle, a system of beliefs, a set of moral prescriptions, a social movement, and, above all, a brand of self-development (Davies 2015; Heffernan 2015). Perhaps the least contentious reading ascribes a dual meaning to the term: as the sustained absorption of one's attention in the present moment, on the one hand, and as the set of practices and ethical teachings designed to cultivate this mode of being, on the other hand

(Bachelor 1983; Claxton 1991; Flanagan 2011). For well over a century, Western imperialists, travelers, mystics, and academics have sought to introduce Buddhist ideas to a wider public. In the field of psychology, Gautama Buddha has sometimes been interpreted as the founder of the first systematic 'school' within the discipline, and almost every contemporary sect has claimed Buddhist thought as the seed of its own—from behaviourism to psychoanalysis (Crook 1980; Stanley 2012). In the present day, practitioners of cognitive behavioural therapy (CBT) and of cognitive neuroscience are among the most ardent (Flanagan 2011; Stanley 2013). They are joined by the American Jon Kabat Zinn—emeritus professor of medicine at the University of Massachusetts and, for his huge global readership, a charismatic writer and teacher (e.g. Kabat-Zinn 1991, 1994, 2001). Kabat Zinn first came across Buddhist doctrine and practice as an undergraduate molecular biologist. Impressed by the clinical potential of these methods to help people trapped in physical pain and emotional turmoil, he devised and then introduced his own programme in 'mindfulness-based stress reduction' (or MBSR), at the University of Massachusetts' Medical Center in 1979. Participants were asked to commit to daily practice of mindfulness exercises, mostly in the form of sitting or vipassana meditation (in which attention is focused upon the breath) and also in regular movement awareness and 'body-scanning' exercises, taught in a series of

P. Moloney (✉)
NHS Adult Learning Disability Service, Shropshire, UK
e-mail: paulmolon@hotmail.com

© Springer International Publishing Switzerland 2016
R.E. Purser et al. (eds.), *Handbook of Mindfulness*,
Mindfulness in Behavioral Health, DOI 10.1007/978-3-319-44019-4_18

weekly sessions. These exercises were intended to dispel physical and mental tension and to foster the relaxed state conducive to the pursuit of meditation.

This classroom-based format—familiar to anyone who has ever taken an introductory adult education course—has become the core of the many efforts to adapt Kabat Zinn's ideas to the treatment of problems such as depression, anxiety, 'borderline personality disorder', psychosis, PTSD, obsessive compulsive disorder, and a host of other psychological maladies (Germer et al. 2013; Gaudiano 2014; Roemer and Orisillo 2009). One of the best known of these therapeutic packages is mindfulness-based cognitive therapy (MBCT), as developed by British psychologist John Teasdale and colleagues in the 1990s (Zindel et al. 2013). This approach is at the forefront of what is known as the 'third wave' of cognitive behavioural therapy (or 'CBT'). In the 'first wave', the practitioners of the 1950s and 1960s sought to inculcate better mental health via relatively straightforward— (and for the critics—simplistic)—techniques of conditioning, derived from experiments with animals; in the second phase, they sought to ease distress by the application of behavioural procedures and of 'rational' arguments, calculated to challenge the presumptively mistaken pessimism of their clients (see, e.g. Beck et al. 1987; Ellis and Dryden 1987). In the 'third wave' of the early twenty-first century, these cognitive behavioural methods have been subsumed within a larger curriculum. This features elements of Freudian and 'systemic' thinking and—especially—mindfulness training: intended to help the sufferer to reconnect with their bodily experience of themselves and of their world and to face their distress with greater equanimity and fortitude (Germer et al. 2013; Fuchs 2013; Michelak et al. 2012). In its claimed basis in 'cognitive science' and quantitative clinical research, in its pragmatic willingness to combine disparate and (arguably) incompatible theories of mind and conduct, and, especially, in its optimistic promise of fundamental personal change—this most recent incarnation of CBT is far more traditional than might appear (see Fancher 1996; Moloney 2013a, b; Smail 1987). It is this continuity, together with the reduced costs that attend group treatment, that helps to explain why MBSR, MBCT, and related approaches—such as dialogical behaviour therapy (or DBT)—have been widely adopted within the British National Health Service since 2004 and are endorsed by the official clinical guideline giving bodies of the UK and the USA (see for example, NICE 2016; Gregoire 2015; NIMH 2016).

More than just a therapy, mindfulness is at the forefront of an official utilitarian 'mental health' movement, sweeping through the health and social sciences. Governments and corporate employers are seeking to use behavioural methods to measure and boost 'happiness', 'nudging' as many of us as possible—and especially the poor and the indigent—into the lifestyle choices deemed to be healthier and more sensible (Davies 2015; Moloney 2013a, b; Midlands Psychology Group 2014; Frawley 2015). Taught increasingly in schools, colleges, universities, and workplaces in Britain, the USA, and many other countries—mindfulness is widely seen as a way to reduce stress and to make people more resilient, productive, creative, and amiable: the natural graces said to accompany a quieter and more open and attentive state of mind (Davies 2015; Frawley 2015). It is reported that 70 % of British general practitioners would like to refer their patients to NHS-funded courses on this subject, if only the public health services could meet the demand (*Mindful Nation*). In the UK, the Centre for Mindfulness Research and Practice in Bangor has trained 2500 teachers, enough to transmit the method to 200,000 people each year, and the one thousand plus fee-paying mindfulness courses that have emerged in the UK sell-out within hours of their announcement (Booth 2014). The uptake of mindfulness in the USA is comparable. More than 200 clinics offer mindfulness training, some of them affiliated with prestigious medical centres and tens of thousands of Americans have signed up for these programmes (Barker 2014; Center for Mindfulness 2010). Academic interest in the subject is intense, with over 500 peer-reviewed academic papers issued every year (Barker 2014).

Mindfulness is also a success story for publishers. In the larger bookstores, on many tens of feet of shelf space, the works of Kabbat Zinn are joined by those of psychologists, life coaches, counsellors, Buddhist monks, New Age therapists, neuropsychologists, and celebrities (with some authors bidding to be all of these things). All of them promise that the path to full awareness and health is attainable as long as we make enough effort in every quarter of our daily round —from office routines, to gardening, to childcare —to mention just a handful of topics featured in the many hundreds of book titles. For those who are too anxious or bored to read, crayoning books with titles like *Colour Your Way to Calm* invite the mindful-infilling of intricate flower drawings, 'groovy mandalas' and 'folk art birds'. In the spoof 1960s, children's publication known as the...'*Ladybird Book of Mindfulness*' ... '*the large clear script, the careful choice of words, the frequent repetition and the thoughtful matching of text with pictures all enable grown ups to think that they have taught themselves to cope*' (Hazely and Morris 2015). Popular satire is a sure sign that a mass movement has arrived. Dedicated Websites and other online resources grow apace, including smartphone apps such as 'Headspace'—designed to help its three-quarter of a million subscribers to meditate. In 2014, the global advertising giant JWT announced 'mindful living' as one of its ten trends to influence the world. Consumers had found '*a quasi-Zen desire to experience everything in a more present, conscious way*' (Booth 2014).

Perhaps then it should be no surprise that in Britain, a cross-party group of MPs has issued an official report, entitled *Mindful Nation*, which indorses the application of these practices on a national scale. Disappointed by the lack of provision of MBCT across the country so far, this group of politicians urge that it be made available on the National Health Service to over half a million adults each year by the third decade of the century, especially to those struggling with anxiety and depression. This document recommends the creation of three national institutes to pioneer mindfulness teaching to children in the classroom and the founding of a million pound 'Challenge Fund', for which schools might bid in hopes of training their best teachers to become

expert disseminators. All public sector workers —from nurses to librarians to police officers— should have the opportunity to become proficient in this art, as should the many criminal offenders who have experienced and inflicted suffering because of their impulsivity.

Whether or not this buoyant blueprint will ever be realized in this time of fiscal 'austerity' is an open question, but the reply might turn out to be 'yes'—since its authors and main supporters are members of an influential political and metropolitan elite. In many cases, such as the economist Richard Layard, they are linked to powerful institutions such as the London School of Economics: one of the main motors behind the rise of the 'new management' and 'market-led' culture that has transformed health care and other public services in the UK, and beyond (see Rogers and Pilgrim 2014; Proctor 2009). Layard has been this way before. Via his co-authorship of '*The Depression Report*' in 2006, he was a key instigator of the massive CBT-based Improving Access to Psychological Therapies programme (or 'IAPT', as it is more commonly known)—still underway in England and Wales (Midlands Psychology Group 2008). Like its predecessor, the *Mindful Nation* document builds its authority on what its authors take to be the solid scientific credentials of CBT. It blends a declared humanitarian commitment with a strong fiscal case for psychological treatment—(in this case, 'mindfulness')—as a means of reducing healthcare bills through the prevention of psychological distress, and by getting the disturbed and disabled back to work and off the state sickness benefits roster. In these and many other ways, the report harks back to IAPT as an unmitigated success (see Layard and Clark 2014). However, it ignores the 'scandalously high' rates of client drop out (McInnes 2011), the questionable methods of data collection and outcome measurement, and the accusations of the coercive use of 'therapy' against the unemployed and debilitated—all of which have pursued the scheme from the outset (Freidly 2013; Midlands Psychology Group 2008; Moloney 2013a, b; Watts 2016). The response of the British media to the growing legitimacy of mindfulness in general and to the *Mindful Nation* report in particular has been generally favourable, sometimes near ecstatic—and from some surprising quarters, including academics and journalists with a reputation for sceptical thinking: some of whom, not so long ago, saw political and social change (rather than meditation) as the most sensible retort to widespread malaise and civil decline (see for instance, Khaneman 2010; and Bunting 2014, respectively). The existence of *Mindful Nation* is a sign of unalloyed official approval for what amounts to a CBT–vipassana hybrid as a preventative and curative programme and on a scale that enthusiasts liken, unblushingly, to previous well-received national public health measures such as the introduction of fluoride to British tap water, sixty years beforehand (Booth 2015).

The Scientific Study of Mindfulness: Insights and Warnings from the Talking Therapy Research Field

With all of this excitement, it is not hard to see the appeal of mindfulness for those who suffer intractable personal torments, or for anyone who seeks refreshment in stillness: the perennial fantasy of the modern era of commerce and industry (Pietikainen 2007; Scull 2015), and even more so, for the globalized and wired-up world of the early twenty-first century (Sim 2004; and see Kabbat-Zin 2005). In the words of one British parliamentarian who took a course on mindfulness at Westminster…'*In today's mad whirl, a few well-earthed, indeed profoundly common sense, contemplative insights are truly valuable.*' (*Mindful Nation*, p. 16). But does any of this hold up as good science? Given the popularity of mindfulness and the self-assurance of its promoters, it might seem odd to even pose this question. However, a clue to its logic can be found in the report from the All Party Parliamentary Mindfulness Group itself, which laments the '*inadequate investment in high quality research needed to strengthen the evidence*' (Mindful Nation, p. 24). To encounter qualms like these in an official policy recommendation document is rather like finding stones in a pudding. It suggests a need to scrutinize the ingredients and how they got there—in this case, beginning with the connections between mindfulness-based therapy and the evidence in favour of other forms of psychological treatment upon which its exponents seek to base its credibility.

Before the middle of the twentieth century, clinical assessments of talking treatment focused on efficacy—its apparent helpfulness in ordinary day-to-day clinical practice, as measured by recovery rates. These early studies spotlighted mainly psychoanalytic treatment and suggested that two-thirds of patients improved. Therapists everywhere drew succour from these findings, until the behaviourist psychologist Hans Eysenck claimed to find exactly the same trajectory in groups of untreated people. This implied the near irrelevance of talking therapy and that patient betterment might be attributed instead to seasonal effects upon mood, the benefits of social support, and the dividends of maturation and experience (Eysenck 1952). Though Eysenck's data and conclusions were later challenged, this phenomenon of 'spontaneous recovery' hovers over the field to this day. In consequence, researchers have sought surety in randomized controlled trails, or RCTs. These are clinical experiments, in which sufferers are allotted blindly to either the treatment of interest or one or more comparison groups—which may comprise people who stay on a treatment waiting list, those who get a genuine alternative therapy or, more rarely, a sham (or placebo) one.

RCTs of this kind are numbered in the thousands, with results that have been positive overall but fickle in magnitude. In pursuit of still greater authority, researchers have used the technique of meta-analysis, in which the data from many dozens of studies are blended and then distilled to capture the main trends. In the decades from 1980, a large number of these procedures imply that talking therapy is a reliable technology for achieving personal change (e.g. Smith et al. 1980; Wampold and Imel 2015). Historically, this has been especially so for CBT: the approach that has most readily embraced psychiatric nosology and quantitative outcome measurement, geared to the audit requirements of managed health care (House and Loewenthal 2008; Moloney 2013a, b; Rogers and Pilgrim 2014). In consequence, CBT dominates official treatment guidelines (see Newnes 2014; House and Loewenthal 2008). For the vast majority of therapy professionals, the question of effectiveness is settled. All that remains is to decide what kind of treatment is suited for what kind of problem—and for whom (see, e.g. Roth and Fonagy 2006).

But there are reasons to doubt this straightforward tale of medical and scientific progress. The technique of meta-analysis has always been hostage to the variable quality of the original studies and to the skill and judgment with which they have been selected and standardized (Charlton 2005; Healy 2013; Prioleau et al. 1983). Moreover, the popular tenet that there are specific treatments and techniques for specific problems—like keys and their locks—cannot be reconciled with what the research literature persists in showing: that for the vast majority of psychological problems, there is no solid evidence that any one type of therapy can consistently outperform another, or indeed a convincing placebo, and that treatment success depends neither upon practitioner qualifications and experience, nor therapeutic orientation (Dawes 1994; Feltham 2013; King-Spooner 2014; Moloney 2013a, b; Norcross and Wampold 2011; Wampold and Imel 2015).

It is much harder to scientifically gauge a talking treatment than most people realize. In part, this is because psychological problems do not lend themselves to objective or quantitative measurement in the same way as many physical disorders (Cromby 2015; Midlands Psychology Group, forthcoming). There are few reasons to think that we have direct access to our mental states in the way that is presupposed by many lay people and too often by talking therapists and those who assess their interventions (see for instance, Kahneman 2013; Moloney 2013a, b; Morgan 2008; Schwitzgebel 2011). Indeed, the bulk of the evidence with which psychologists deal are not *observable facts* but *communications*, which are prone, by definition, to misinterpretation, slippage, and distortion (Rickman 2009; Harre 2002; Shotter 1975).

In the field of psychotherapy research, the rewards and the scope for unconscious dissimulation and exaggeration on the part of the client are very high—perhaps uniquely so. This is because, for most people, success or failure in the task of therapy has become a tacit index of self-worth. In the early twenty-first century, people in Western societies are encouraged to believe more strongly than ever in the individual's power—indeed their moral obligation—to overcome whatever problems life thrusts upon them. This outlook has much to do with globalization and consumer capitalism, which place a large economic premium upon personal flexibility and competitiveness (Aschoff 2015; Cushman 1995; Pietikeinen 2007; Throop 2009), and with the influence of the psychology industry itself: which has encouraged us to gaze anxiously inwards in pursuit of the roots of our unease and has profited from the myth of easy personal change attainable via expert help (Illouz 2008; Rose 1989). For the client sitting under the earnest gaze of their therapist, in a scenario reminiscent of the religious confessional—(and with all that that implies)—there is every reason to exaggerate the benefits of treatment, above all to themselves. It is hard to know how common such self-deception might be—the whole topic is poorly researched. However, accounts of failed or abusive therapy (e.g. Bates 2005; Sands 2000; Zilbegeld 1982), painstaking investigations of how client's seek to present themselves in treatment (Kelly 2000; Illouz 2008), and anthropological insights into the power of cultural myths to shape personal narratives of illness and recovery (Fuchs 2013; Lutz 1985; Throop, ibid) —all imply that such distortion is commonplace. Even more so, perhaps, for mindfulness training which can bear a heavy load of expectation, compounded of the latest 'neuropsychological science' and of popular new age-spirituality (Coward 1991; Davies 2015).

Should these arguments prove hard to digest, it is worth recalling that for decades psychologists have shown that in experiments involving human subjects, researchers must take great care if they are to avoid inadvertently sending out subtle but demanding cues that systematically distort the results. These signals involve more than facial expression, gesture, posture, eye contact, and voice timbre; for example, they are also about the prestige of the researcher and of the institution in which they work. Unconscious messages like these can powerfully shape participant conduct in the direction preferred by the investigator: whether we are considering the answers given to questions on ethical and political issues, for instance, decisions made by juries, or patient's judgments of the potency of inert placebo tablets or of genuine medicines. Even laboratory animals can be systematically swayed by unconscious minute gestures or subtle differences in handling (Sutherland 1992). For humans, these effects are strongest when both parties are unaware of them and when the one directing holds some authority (Caldini 1993; Fisher and Greenberg 1997; Sutherland 1992). Moreover, researchers themselves can get snared into seeing what they want to see when testing a favoured hypothesis and especially where the data are elusive or ambiguous (Rosenthal and Rubin 1978). Expectancy biases, as they are known, may account for the recent failure to replicate many classic experiments and observational studies in psychology and other sciences. The original findings were probably artefacts created by overzealous investigators (Lowe 2011).

If it is not be compromised by such issues, the design and conduct of any clinical investigation into a psychological therapy needs to be of a very high standard. The long list of minimum desiderata begins with participants who fully represent the clinical population of interest, experimental and control (or placebo) treatments that are equally compelling for everyone involved, careful double-blind assessment—in which neither the assessors nor the participants know who has received the genuine or fake remedy, and long-term post treatment follow-up. Unfortunately, as thoughtful observers down the years have pointed out, such conditions have rarely, if ever, been met (Erwin 2000;

Dineen 1998; Holmes 2002; Kline 1987; Mair 1992; Newnes 2014; Moloney 2013a, b; Pietikainen 2007; Smail 2005; Shedler 2015; Zilbegeld 1982).

Perhaps the most comprehensive and detailed of these critiques has come from the American academic William Epstein. In the early 1990s, he scrutinized some of the most reputable research in this field and found it to be badly wanting on methodological grounds (Epstein 1993, 1995). Ten years later, Epstein repeated the procedure for the top three international journals for the scientific assessment of the leading therapies, including CBT and behavioural and psychodynamic treatments. Once more, this literature could not sustain its own claims. Placebo treatments, for instance, were either absent or unconvincing, sample sizes were in many cases too small, and the systematic abdication of clients from key groups was downplayed or ignored. The questionnaires used to assess outcome were of dubious validity or prone to second-guessing by clients, and the statistical methods employed to analyse the data tended to inflate the power of treatment and to smooth over its uneven effects, including the likely deterioration of some participants. Finally, this research was compromised by the doubtful independence of the researchers and by their near total reliance upon what clients *said* about the helpfulness or otherwise of the intervention—as opposed to attempts to observe how it might shape their well-being and conduct, beyond the walls of the consulting room. Epstein concludes that, despite decades of being trumpeted as a success, the leading brands of psychological therapy remain unverified and are probably ineffective. This is especially so for patients who are struggling with harsh social and economic circumstances: the main clientele of the publicly funded health and care services in which most therapists work (Esptein 2006, 2013; and see Moloney 2016). These observations, which echo those of earlier critics such as Zilbegeld (1982), have been largely ignored within the professional literature (see Feltham 2013; Moloney 2013a, b; Newnes 2014), although a minority of psychological healers have persisted in reaching similar judgments, on the grounds of their own clinical experience (Davies

1996; Dineen 1998; Hagan and Donnison 1999; Holmes 2010; Lomas 1998; Moloney and Kelly 2008; Smail 1987). While Epstein's reviews have focused mainly upon second-wave CBT, the supporters of mindfulness-based interventions imply (and sometimes claim) that the latter promise to be more effective than previous approaches, owing to their radically new integration of mind, body, and Eastern psychology.

Mindfulness-Based Therapy for Psychological Problems: A Brief Look at the Evidence

Since the introduction to the West of the Hindu practice of transcendental meditation in the 1960s by the Maharishi Mahesh Yogi (Russell 1976)—apparently for commercial reasons (Wheen 2004)—psychologists have tried to show that such techniques, including mindfulness, reliably yield changes in well-being, mental function, and behaviour superior to standard relaxation methods. The results have been wholly equivocal (Blackmore 1993, 2010; Farias and Wikholm 2015; Holmes 1984). The most recent comprehensive meta-analyses of studies looking at this question do little to challenge the original picture. Sedlmeier et al. (2012) took 163 separate trials involving people deemed to be mentally healthy (or 'non clinical') and assayed them for the reported effects of mindfulness and TM on a range of outcomes—including subjective well-being, intelligence, and negative emotions. Both types of mind training emerged as 'moderately beneficial', however, less than a dozen of these studies used any kind of active control therapy. For these investigations, no extra benefit from meditation emerged.

Of course, this does not imply the irrelevance of techniques of mental concentration to the treatment of clinical problems. In the last two decades, a growing number of reviews have weighed the effectiveness of mindfulness therapy in the reduction of stress and in the management of chronic physical and mental conditions, including autoimmune disorders, persistent pain, anxiety, depression, psychosis, eating disorders, and

'borderline personality disorder', to name but a few (Farias and Wikholm 2015). The results have been favourable on the whole, but with large disparities in the size of the treatment effects. Alongside their optimistic reading of the future of these methods, many researchers acknowledge that little is known about the active ingredients that account for the claimed improvements, how long these effects might last, and how they might be shaped by additional therapeutic procedures and by the circumstances in which the clients live. If caveats like these are not hard to find in the research literature (see, for instance, Baer 2003; Gaynor 2014; Khoury et al. 2013; Piet and Houghard 2011), they seldom survive the journey into the pronouncements of the policy analysts and interest groups and still less into popular works on mindfulness (Davies 2015; Barker 2014).

To take just one instance, the *Mindful Nation* report enthuses about Khoury et al.'s meta-analysis of 209 studies, covering more than 12,000 participants. By the standards of the field, this is a big survey. According to *Mindful Nation*, it showed 'large and clinically significant effects in treating anxiety and depression, and the gains were maintained at follow-up' (16). But this large-scale meta-analysis rests upon a painfully slender column of reliable evidence. The measured helpfulness of contemplative therapy wobbled considerably across the different studies and declined in close step with the level of experimental control (or care) that was exerted.

Such trends are the norm within meta-analyses of this kind (see, for instance, Goyal et al. 2014). They suggest that distortion from expectancy effects and other biases must be common throughout the mindfulness health treatment literature. Most researchers are unable or unwilling to apply the proper controls (Khoury et al., p. 769), perhaps in part because of a widespread belief that mindfulness is clinically effective and no longer requires rigorous testing (Farias and Wikholm 2015). Indeed, only one in nine of the investigations in Khoury et al.'s review collected follow-up data or tried to use even basic blinding procedures: serious omissions, both. It is not necessarily reassuring that the reported benefits were larger for psychological than for physical or medical conditions, since, by their nature, mental 'symptoms' are among the most elusive and, as already noted, the easiest to deform at the reporting stage (see Morgan 2008; Schwitzgebel 2011, for a more detailed discussion of the perils of trying to accurately convey one's subjective experience). In the end, the researchers believed there were grounds for cautious optimism, pending further, and stronger research. However, according to the independent Centre for Reviews and Dissemination, even these modest conclusions '*may be over stated, given the poor quality and wide variation between studies*' (Database Abstract of Reviews of Effects (DARE) 2015, p. 1).

If the standards of the research in this area are often wobbly, then investigations into the application of mindfulness to severe depression tend to be among the more thorough, perhaps because the need for a demonstrable remedy is so high. Down the ages, there have been many accounts of those burdened with sadness and despair (Horwitz and Wakefield 2007; Scull 2015). Modern diagnostic systems attempt to capture this kind of suffering under the heading of 'clinical depression'. There is still much debate about the precise elements that comprise the condition, to what extent they entail bodily as well as mental suffering and how they vary across different times and cultures and overlap with other kinds of distress (see, e.g., Fuchs 2013; Horwitz and Wakefield 2007). Nevertheless, there may be a core experience of dark and impacted misery, impervious to persuasion if not to comprehension—that would be recognizable to most people (Scull 2015; Smail 1996). As described by conventional psychiatric frameworks like the DSM, the lifetime risk of developing severe depression in a country like the USA is said to be almost one in four for women and just over one in ten for men. This form of distress recurs in around half of all sufferers, and it can have huge personal and social costs—including long-term debility and suicide (Horwitz and Wakefield 2007; Massouvi et al. 2007; WHO 2005). There is grudging but growing recognition that conventional therapies, such as antidepressant medication and 'second-wave' CBT, might not be as helpful as

once believed (Healy 2013; Johnsen and and Friborg 2015; Moncrieff 2007). The search is on for new 'adjunctive' treatments to go alongside the psychiatric ones. Half a dozen recent clinical trials focusing upon MBCT as a bulwark against this debilitating form of melancholy have found it to be as good as or better than prescribed medication (e.g. Bondolfi et al. 2010; Kuyken et al. 2008; Ma and Teasdale 2004). The UK National Institute of Health and Clinical Excellence (NICE) has for some time featured MBCT in its list of recommended treatments (NICE 2016), although recent meta-analyses have suggested that it might work only for a subgroup of vulnerable patients who have struggled with exceptional emotional hardship in their formative years (e.g. Piet and Hougaard 2011).

A key problem is the paucity of studies that have sought to compare MBCT with an active psychological treatment, or better still, with a convincing placebo. Without this kind of multi-treatment trial, it is impossible to be confident that the claimed benefits derive from the core contemplative elements, as opposed to the more conventional parts of MBCT: including the teaching of cognitive behavioural 'relapse prevention skills' one the one hand, and the mixture of comfort, support, and hopefulness that comes with joining a therapeutic group, on the other hand (c.f. Frank and Frank 1991).

So far, only one enquiry has sought to address these issues (Williams et al. 2014). It is worth discussing this trial in some detail, because it is likely be seen as a landmark due to its large scale —it drew upon 274 participants—its claims to scientific rigour, and the international standing of its main authors. Both of them are academics at renowned UK universities and leading figures within the worlds of CBT and of mindfulness-based therapy research and practice.

The participants for this study were recruited through referrals from medical practitioners in primary care surgeries and mental health clinics, and via community advertisements. All of them had suffered from recurring attacks of depression, but were deemed to be coping—or 'in remission'—for their worst symptoms. In true experimental fashion, they were randomly assigned to one of three groups, the first of which were given MBCT, while the second, regarded as the 'active control' set, were administered a very similar therapeutic package but with the mindfulness component filleted out. In other words, they were treated with cognitive–psychological education, or 'CPE', aimed at delaying the return of their condition. Finally, a third, 'passive control group' did not have the benefit of either form of psychological intervention, but, in common with the first two groups, were encouraged to continue to access their customary National Health Service outpatient treatment in the form of antidepressant medication, plus whatever advice and encouragement they could find.

Aside from the absence of mindfulness teaching for those in the CPE wing of the trial, Williams et al. strove for parity in everything else that was done with the two psychological therapy groups. To this end, they used their own specially adapted version of Kabat-Zinn's mindfulness instruction manual as the framework for eight once weekly relapse prevention classes, followed by a single review meeting six weeks later, and a further one at six months. Every participant was given regular therapy-based assignments to be completed outside of the clinic, but with the difference that those in the MBCT wing were expected to perform mindfulness at home plus simple cognitive therapy assignments, such as keeping a daily diary of their thoughts and feelings. By contrast, the participants in the 'active control group' were neither instructed nor expected to practise the mindfulness itself, and as far as could be told, they did not do any.

As in the trials already described, this study sought to address two issues. First, how long would it take the people in the different groups to relapse to a state of major depression, once the main treatment had been completed? Second, would there be any subgroups of patients, as defined by the history or the severity of their problems, for whom mindfulness-based instruction might prove especially helpful?

The results were mixed, but encouraging. The more vulnerable participants, as defined by number of previous depressive episodes and disclosure of a troubled childhood, seemed to gain most from the MBCT package. When the relapse data for the two treated groups were blended together and considered as a whole—without regard to differences in psychological vulnerability—then the mindfulness-based package proved no better than cognitive treatment alone; predictably, both of these psychological interventions outdid the standard 'pills and reassurance' variety of outpatient care.

These findings might have justifiably been presented the other way around, given the exuberant claims for mindfulness-based therapies that have circulated in newspapers and other media and, with barely less restraint, in some of the clinical research and teaching literature (see Farias and Wikholm 2015). However that may be, the researchers felt that their study had clear strengths when compared with its forebears: client drop out rates were low and were spread evenly over the three groups. More than 90 % of the participants completed at least four treatment sessions—a respectable rate for this kind of trial. Fidelity to the treatment guidelines seemed to be high; every therapist followed the same tightly prescriptive manual under close supervision, based upon video recordings of each treatment session. The quality of therapist training and commitment appeared to be exceptional. Each practitioner had helped to write and pilot the treatment packages used in the study and held at least three years of experience in MBCT instruction. On the other side of the coin, the patients seemed to be convinced of the validity of both forms of psychological therapy: as confirmed by the results of a questionnaire, administered near the start of the trial.

Furthermore, this investigation boasted a total of six separate assessment interviews for each participant, the first one taking place just before the start of the therapy programme and the final one a year after its completion. These appraisals were done by trained personnel and with standardized questionnaires like the Structured Clinical Interview for the DSM IV (or SCID). To ensure that these assessors stayed blind as to the treatment received by their interviewees, the therapy and the appraisals took place in different buildings, to minimize possible cues. Furthermore, these interviewers were asked to report any treatment disclosures by their respondents. On 'the rare occasions' when this happened, the interviewer was replaced.

The authors assert that their findings '*add to the growing body of evidence that psychological interventions, delivered during remission, may have particular beneficial effects in preventing future episodes of major depression, but may be especially relevant for those of highest risk of relapse*' (Williams et al. 2015, p. 285). Their tone is confident; but is it justified?

Perhaps the first thing to note is coyness in the presentation of the data from this trial. The main outcomes for this study were measured via the mean SCID scores for each treatment group. This information is presented in terms of the percentages of individuals in each group who scored high enough on the SCID to qualify as clinically depressed—as having relapsed. However, the mean SCID scores for each group are not given. A pattern of numbers like this can be statistically significant but far more ambiguous, when it comes to gauging the participants' freedom from distress and their ability (or otherwise) to get on with their lives. In the absence of this basic information, the reader is left to ponder, given the widespread tendency within the field for inconvenient or lackluster findings to be elided or buried within abstruse statistical jargon (Dalal 2015; Epstein 2006; Postle 2007; Shedler 2015).

This study also followed its predecessors, in its exclusive reliance upon the participants' own reports about their improvement, as told to their interviewers. There was no attempt to collect commentaries from relatives, carers or associates, nor was there any attempt to make direct observations of the participants' daily activities—including patterns of sleeping, eating, self-care, physical activity, social contact, leisure, and changes in employment status. Hard as they might be to carry out, detailed real-world assessments of this sort are essential if a psychotherapy study is to hold up as good science (Epstein 1995; Kline 1992).

Indeed, the possibility cannot be discounted that the more vulnerable group of participants—the ones who seemed to gain most from the mindfulness exercises—might also have enjoyed better support and encouragement from family, friends, neighbours, colleagues, and other contacts: perhaps enough to boost their commitment to the otherwise marginally helpful practice of mindfulness therapy to the point where it crossed the threshold of apparent clinical effectiveness—at least as measured via the questionnaires used for this study (see Epstein 1995; 2010). It is impossible to say whether something like this happened for sure, since Williams and collaborators collected minimal data on the social and economic circumstances of their participants.

A further difficulty concerns this study's partial reliance upon self-selected clients, recruited via local adverts, whereas people with a history of more severe or protracted mental health problems—(the official targets of this particular RCT)—are often the least inclined to respond to such appeals (Epstein 1995; 2010). More generally, the participants in this study appear to have been aware of their treatment allocation at the outset. Patients taking part in the control section of a study like this one—and who either learn or infer that they are receiving humdrum regular outpatient care alone—can be tempted to overstate their clinical symptoms, in hopes of being reallocated to the presumptive cutting edge therapies that comprise the focus of the study. By embellishing their distress, the 'waiting list' control clients can make the experimental intervention look more helpful, by comparison, than it really is (Epstein 1995, 2006; Kline 1992).

Still more problematic is the question of the allegiance of the therapists themselves. For this kind of project, it is vital that practitioners providing the treatment under test are not tempted to deliver it with more brio than the standard form of help against which it is being compared:—lest they transmit their enhanced expectations for improvement to their patients (Epstein 1995; Goldacre 2009; Kirsch 2009). As veteran MBCT instructors, the therapists in this trial also helped to create the cogntive therapy packages that were used for both of the treatment groups. Williams and colleagues cite this involvement as proof of the instructors' even-handedness in furnishing the treatments. But just the opposite conclusion seems warranted, since these teachers were so evidently immersed in the practice and theory of mindfulness-based therapy: the latest and 'most advanced' phase in the development of CBT. It stretches credulity to think that they did not therefore have a larger personal investment in the meditation-based element of this study, in comparison with the more mundane 'cognitive relapse prevention' part of the trial. Mindfulness teachers, including Kabat Zinn himself, can sometimes evince a quasi-religious belief in the power of the method, running alongside their professed commitment to scientific rationalism (Barker 2014; Davies 2015; and see Kabat-Zinn 2001). In the context of Williams et al. s' study, it is plausible that when administering the CPE treatment, the instructors were less sanguine and might have unintentionally conveyed their diminished expectations to their patients, however subtly (see Caldini 1993; Epstein 1995; Rosenthal and Rubin 1978). While the rating scales completed by the participants did not suggest any real difference in the credibility of the two types of therapy from their point of view, this grading was done just once near the start of the program: perhaps well before most of them could reflect deeply on what they were being taught.

In this kind of research, the question of allegiance applies equally to those who are trying to gauge the effectiveness of the therapy: they should have as little personal and professional stake in the outcome as is humanly possible. This requirement likewise seems to have been violated. The evaluators—the people upon whose judgments the whole study crucially depends—appear to have had strong connections with the primary research team and might even have belonged to it. In which case, they would have been in a position to unknowingly communicate their hopes and expectations to the patients that they interviewed. The audio recording and verification of a sample of these interviews by a separate team of psychiatrists cannot remove this problem. Such a procedure can only confirm that the interviewer wrote down what the patient said. It cannot answer the key question of unintentional inducement or guidance. Moreover, the impartiality of even these

ancillary assessors might be doubted. If they belonged to the same mental health services in Oxford and Bangor with which the key researchers were professionally associated, as seems likely—then they may well have shared a similar commitment to the promise of mindfulness.

In sum, the most serious methodological problems boil down to sampling issues and poor control of expectancy bias and of demand characteristics, but magnified by the nature of this study as a demonstration project: a clinical trial in which every mental health professional was far more intensively coached, supervised, monitored, accountable, and (probably) inspiring—than would be the case in routine clinical practice (Epstein 1995, 2006; Kline 1992; Zilbegeld 1982). Situations like these are likely to yield superior results, even with the most pedestrian of interventions. The relevance of this highly optimized trial to the conditions encountered by clinicians in the workaday world of over stretched public health services is moot (see Davies 1996). If the research into mindfulness as a clinical treatment is less than encouraging, then it is worth recalling that its advocates see the latter as more than a therapy, and it is a valid means of building strength and 'character' in anyone who practises it, starting with the most vulnerable members of society.

'Are You Paying Attention?' Mindfulness in the Classroom

Interest in the use of mindfulness-based methods on youth and in schools has been growing in the last twenty years and more in the UK especially, under the unofficial banner of what has been described as the therapeutic turn in the education system as a whole. In this new regime, students and educators are encouraged to obsess not merely about their academic performance, but about their emotional lives and vulnerabilities—both real and imagined (Ecclestone and Hayes 2011; Furedi 2004). In contrast to the prescriptive and near hymnal tones of the *Mindful Nation* document, however, the lessons to be derived from the most comprehensive research reviews in this field are speculative at best. For instance, Felver et al. (2015) inspected 28 studies that sought to assess the value of MBIs in school settings—mainly in the USA. Happily, many of these drew upon suitably large samples, but only a third used randomization or a control condition, and of these, a mere four attempted a matched active control. There were no attempts to use fully autonomous evaluators or, oddly enough, the school system's routine administrative data on student performance and conduct, which would have been a small step towards greater objectivity. Neglect of vital demographic information on disability and social and economic status of individuals and their communities was near total, making it difficult to interpret such findings as there were. Soberingly, the effects of the training were measured only during the brief lifetime of each intervention, undercutting one of the main justifications for mindfulness practice within the educational field: as a crucible of self-discipline and lifelong resilience (*Mindful Nation* 2015). The reviewers' warm verdict on the promise of mindfulness for schools sits awkwardly next to their final (but familiar) plea that the standards of the research need to be a lot taller.

In a similar vein, Zenner et al. (2014) present their assay of two-dozen studies into the application of contemplative science to the school arena: half of which were never published—presumably because they were originally deemed to be too small to yield firm conclusions. In contrast to the preceding meta-analysis, this one drew upon publications outside of the USA—some of which looked for gains in students' thinking skills and in other academically relevant markers, months or more after the mindfulness training had ended. The reviewers judged that mindfulness-based applications hold the promise of boosting these capabilities, including emotional resilience. Nevertheless, the flaws within this particular meta-analysis resemble those in the previously discussed one. There was the same unwillingness to review progress in the ensuing months and years, the same absence of convincing placebo control groups, of

independent assessment, and finally, of the demographic data needed to make sense of what made a programme acceptable or otherwise within a particular school or educational district. On top of which, the wide differences among the studies in ethos and methods of training and assessment made it impossible to identify which aspects were the most useful. Zenner and colleagues admit that…'*the precise role that the element of mindfulness plays* [in the reported improvements in student wellbeing] *is unknown, as is the extent of the effect that can be attributed to non-specific intervention factors, such as perceived group support, the speciality, and the novelty of the intervention*' […] (17). The findings of studies such as these—dogged by messy and complex situational variables—are even more inconclusive than the clinical research in this field and indeed are echoed in the extremely poor quality of those studies which purport to show the benefits of mindfulness practice for corporate workplaces and boardrooms (see Davies 2015). Perhaps then investigators should seek more tangible evidence for the transformative power of mindfulness—in the form of unique neurological changes wrought by the practice?

Meditation and Well-being: Experimental and Neuropsychological Studies

Cognitive neuroscience is often presented as a brand new enterprise. In truth, written accounts of the relationship between brain and mind stretch back to far antiquity and have been driven down the centuries by the development of new diagnostic instruments (Rose 2005; Uttall 2016). Since the 1960s, interest in the neuropsychological effects of meditation has been strong, and in the last quarter century, neuroimaging studies based upon fMRI and PET scanning devices have started to yield intriguing results. Some of the key changes in subjective experience that are said to accompany mindfulness practice—including the fading away of the narrative based 'self'—have been tracked within the brains of practitioners, as their meditation unfolds (Tagini and Raffone 2010; Stanley 2012). To take another example, fMRI and PET scans seem to confirm that Buddhist meditation practices designed to foster compassion can do just that: as measured in fronto-temporal brain activation which also predicts improvements in generosity —or at least in the willingness to share a small monetary reward more equitably, in the laboratory (Crook 2009; Flanagan 2011). Similar studies of the effects of noise and other surprises upon long-term meditators suggest that the equanimity for which many Buddhist monks are renowned may not be altogether mythical (Austin 2014; Flanagan 2011). The colourful computer images that convey results like these are often compelling and persuasive for many (Weisberg 2008; Wiseman 2016). Once again, however, we have to be careful: both about the procedures that create these findings and especially how they are interpreted and generalized beyond the compass of the laboratory.

Rather than real-time images of the brain in action, brain scans are colour-coded computer-generated inferences about what might be happening beneath the skull. fMRI images specifically are based upon surges in detected levels of oxygenated blood within the brain, which occur when the iron within blood-borne haemoglobin interacts with the magnetic fields generated by the device, to produce what is known as the BOLD signal. However, the final image comes at the end of a long chain of statistical and logical inference in which there is ample room for best guesses and outright mistakes to be turned into apparent truths. At the most fundamental level, there are persistent doubts about the validity of the BOLD signature itself. Cerebral blood flow does not always straightforwardly match brain activity, and it appears increasingly likely that the spatial and temporal resolution of the resulting images is far too wide to grasp the activities of the subtle and widely distributed neural networks that are considered by many to be the most likely source of our mental activity (Cacioppo, et al. 2003; Noe 2009; Rose and Abi-Rachid 2013; Uttall 2011, 2016; Wiseman 2016). The problems do not end here. In most published

investigations, the pictures are a distillation of data harvested from a group of participants, which are then projected onto a map of a notional 'average human brain'. Since few of us have such a nervous system, these representations can be misleading, and because they draw upon vast amounts of composite data, they are also prone to distortion by chance events and can even suggest areas of apparent neural activation where none took place in the central nervous system of any individual participant (Choudhury and Slaby 2012; Wiseman 2016; Uttall 2011, 2016). For all of their fascination, these images are simulations of cerebral action and are necessarily crude, whereas the brain is intricate, subtle, and vital.

As is the case for the investigation of talking therapy, the overall circumstances in which the research is conducted are far from irrelevant. The fMRI scanning suite is a very singular situation. Participants have to be inducted and managed. They are required to lie prone and isolated within the machine and not everyone can tolerate the procedure, which can be noisy, claustrophobic, and boring (Cromby 2015). All of these factors are suspected of altering the blood flow profiles detected by scanning machines, which perhaps accounts for the finding that the same adults performing the same cognitive assignment can show completely different outcomes in different research centres (Kagan 2012).

Like every other scientific tool, the results yielded by brain-scanning equipment depend heavily upon the assumptions shared by the people who use it. Generations of psychology undergraduates operating EEG equipment were warned that it is possible to get an apparently meaningful EEG signal from a bowl of porridge. Things are no different in the case of neural scanners, as revealed by one celebrated study which purported to find 'emotional activity' in the brain of a salmon exposed to pictures of people arguing, and the fish was dead (Bennett et al. 2010). How often in the neuropsychology field have scanning devices been misused in this way—albeit with far more innocence? Very often, according to Craig Bennett and colleagues, who achieved their notable result by following the minimal standards of software calibration that in the rush to obtain publishable results have been commonplace within the field. Exact replication of findings within this field is a rarity, because of the practical limits upon the reporting of complex experimental protocols, and because researchers often use different statistical procedures to analyse their results (Uttal 2016). As a whole, the area suffers from a dearth of control for demand characteristics and participant expectations which even exceeds that for the study of talking therapy (Moloney 2013b; Sanders 2009; Uttall 2011, 2016; and see Horvath et al. (2015a, b), who find similar doubts in regard to the widely reported results of studies of trans-cranial magnetic stimulation—the application of 'mind altering' electrical fields to the brain).

Besides these technical and procedural obstacles, investigators often take a naïve view of emotion as a natural, biological entity, tightly sealed away inside of the central nervous system. But this is questionable science and poor history (Choudhury and Slaby 2012; Kagan 2007, 2012). Even if stable changes in brain functioning could be shown to clearly flow from cumulative meditational practice, it is not clear what the implications might be when it comes to the promised attainment of more happiness or altruism in daily life. Laboratory-based studies that seek to relate changes in brain functioning to decision-making at work or at play are forced to swap relevance for simplicity. For the people and the situations that count the most, neither our subjective experience nor our conduct can be reduced to predictive models of neural activity (Cromby 2015; Uttal 2016; Wiseman 2016).

This observation applies even to animal studies, devised to uncover the fundamentals of feeling and emotion. In the first half of the twentieth century, for instance, neuroscientists had experimented extensively upon caged laboratory monkeys and other vertebrates, cutting out their amygdalas to produce what became known as Kluver-Bucy syndrome, a state of striking tameness and hypersexuality. Experiments like these implied that this small part of the brain must be the locus of emotions such as fear and rage and warranted a seemingly clear clinical logic: cut out the offending bit of nerve tissue and

the distressing emotions will vanish: a sensible enough conclusion, perhaps, until the effects of these operations were seen in animals living in their natural forest environments and communities, rather than in the solitary confinement of the laboratory cage. In these settings, Kluver-Bucy syndrome melts away. Vivisected rhesus monkeys are neither docile nor oversexed, but can be unusually fearful of their compatriots (Brothers 2002). Brains do not exist in bottles. What goes on inside of them has a lot do with the communal and physical world—*and* with the body in which that brain is situated: whether it be healthy or damaged, animal, or human. For our own kind, the social and material environment has even richer significance, because of the huge role played by language and symbolic thought in our daily lives (Brothers 2002; Pilgrim and Bentall 1999).

The nervous system *enables* our experience and our agency, but this is not the same as claiming that it is the only—or even the best— place to look, if we wish to understand them (Harre 2002; Rose 2005). The attempt to elucidate consciousness and feelings by observing the brain's neural activity is like trying to comprehend dancing by watching only the muscles (Noe 2009). Without the relevant brain structures, it might not be possible to feel afraid, but in the end, it is you who are frightened, not your amygdala.[1] Sociological and historical studies of emotional experience and expression agree that our interior lives are not just matters of biology. They are at also about the weight of our personal biography, of the relationships in which we are enmeshed, and of the differences in power and influence that set the terms of mutual engagement. Our feelings and moods echo our changing position in the world and what we are allowed to say and do. Kings have lots of room for showing and (thus for feeling) anger; slaves, women, and the poor—to take just three examples—have more often been consigned to worry and to

---

[1]Paraphrasing an argument made by Danziger in relation to memory and brain function: 'Without a hippocampus you might not be able to remember where you parked your car, but it is you who remembers, not your hippocampus (Danziger 2008, p. 237).

apathy (Bourdieu et al. 2000; Cromby 2015; Gross 2006; Charlesworth 1999).[2]

For all of its promise, brain scan science is still in its infancy. There are still no reliable criteria which allow brain-scanning methods to confirm consciousness in seemingly 'vegetative' patients, nor to diagnosing any form of so-called mental disorder. It is common for doctors and researchers to disagree about the value of functional imaging. The clinicians, who work every day with non-standardized patients, are frequently the more sceptical (Noe 2009).

Should brain-scanning methods one day show that mindfulness practice yields clear and stable shifts in the activity of the central nervous system, no one should be surprised. There are few grounds for believing that 'emotion', 'self', or 'sati' are non-material, ghostly substances. All that we do—including the act of sitting for long periods of quiet attention—must lead to some kind of neural change (Cromby 2015; Kagan 2012; Rose 2005). The most important question hinges upon the personal and social significance of these alterations. Will they show that we have been freed from our many vanities, fears, and worries once we have shut the door of the brain-scanning suite behind us, with its simple tasks and regimens, and have returned to our day-to-day life, with all of its complexities, ambiguities, and travails?

Monks and other very long-term and intensive meditators are unusual individuals. They may be reaping the accumulated rewards of a lifetime of concentrated practice within highly protected settings: beyond those within the reach of the average (or even above average) people who comprise the main readership of publications like *Mindful Nation* (see Blackmore 2010, for an honest and sometimes gruelling account of the personal challenges entailed in protracted mindfulness practice). To hope that the same results can be achieved by people rehearsing vipassana techniques, several times a week during the school term, say, might be the equivalent of expecting children playing hopscotch to become Olympic gymnasts.

Mindfulness: The Bottled Water of the Therapy Industry

According to their many advocates, mindfulness-based interventions hold great promise for curing distress and for crafting a kinder world. But these methods are unproven. Neither commitment to mindfulness practice nor even to monastic Buddhism itself has ever been able to guarantee compassionate or moral conduct. In Japan and China, techniques of mental and spiritual concentration have long been used to enhance the martial skills of warriors and to overcome their reluctance to kill, up until and including the period of the Second World War (Victoria 2004). More recently, the US armed forces have sought to use the technique to improve the efficiency and combat readiness of their soldiers (Farias and Wikholm 2015). In Tibet, a strong tradition of monastic Buddhism has gone hand in hand with autocratic rule and expropriation of the peasantry (Grimshaw 1992; French 2004). In Sri Lanka, Buddhist monks, as community leaders, have too often been the vanguard for implacable race and class hatred (McGown 1992). Even if it were as effective as its supporters claim, mindfulness could never be a treatment or method that 'works' in a relatively straightforward way, like swallowing a medicinal pill. Its effects, such as they are, depend intimately upon context and the aims and values of the user. In the light of these sobering conclusions, the widespread enthusiasm for the approach in the Western world begins to look more and more like an officially endorsed cult. How has this situation come about, and so quickly?

---

[2] Gross is arguing from the basis of the discipline of rhetoric—the analysis of texts in terms of how they socially and politically position the author, an approach that bears a strong resemblance to Foucauldian discourse analysis. For a similar argument, made from a critical rather than historical/rhetorical perspective, See Hacking (2004) in a review of Damassio's book, *Looking for Spinoza: Joy, Sorrow and the Feeling Brain.* On the other hand, there is an obvious limitation in relying upon written accounts or the words that people use, to describe their feelings: can we really be sure that what people say (or do not say) tells us all that we need to know about their subjective experience?

These questions are complex and have more than one answer. For the therapy professionals, the most edifying account might dwell upon a reluctant (but growing) recognition that conventional psychological therapies—committed to a Cartesian divide between mind, body, and world—have not proven to be as helpful in the treatment of enduring and deep-seated psychological problems as was once hoped. In their promise to unite these mental and somatic domains, mindfulness-based therapies seem to beckon towards a new era of more successful treatments (Fuchs 2013; and see Smail 1985).

Less flatteringly, the world of psychological therapy is an industry that, like any other, serves many needs. In the business environment that increasingly shapes health care in the UK and many other countries, reputation, prestige, and income can be secured or lost on the strength of new products and services (Moloney 2013a, b; Newnes 2014; Pollock 2009; Smail 1989). For its critics, second-wave CBT, the traditional market leader, owed its premier position to assiduous marketing and to its long-standing alliance with biomedical psychiatry (Fancher 1996; Pilgrim 2008), but as so often, ubiquity has given way to banality. Challenged by resurgent rivals such as psychoanalysis and humanistic therapy—which once seemed to be the deadest of ducks (Burkeman 2016; Miller 2012; Shedler 2010); faced with charges of disingenuous data manipulation (Dalal 2015; Shedler 2015; Midlands Psychology Group 2008); confronted with empirical evidence of declining effectiveness in the treatment of depression (Johnsen and and Friborg 2015); and recently deposed from its position as the official therapy of choice for public health services in Sweden (Miller 2012): the authority and mystique of second-wave CBT is starting to fray. The creation of mindfulness-based therapies as the third and latest phase of the cognitive behavioural revolution offers an answer to this unprecedented problem.

A further explanation for the rise of mindfulness may reside in the wider public's ambivalence towards biomedicine and the pharmaceutical industry. Even as more and more of us take the drugs that are advertised and prescribed as remedies for our distress, we grow disenchanted with the impersonality and limitations of these supposed chemical cures (Burstow 2015; Healy 2013; Johnstone 2006). Our appetite for healing methods alternative to mainstream biomedicine is matched by our desire that these remedies enjoy some form of 'scientific endorsement' (Barker 2014; Carrette and King 2005). Mindfulness—with its lingering cachet of mystical and esoteric discipline, on the one hand, and with its seemingly solid foundation in neuroscience, on the other hand—seems to fit this dual requirement rather well.

There is nothing new about our yearning for magic. From at least the seventeenth century and the days of Anton Mesmer, there has been a demand in Europe and the USA for self-improvement and healing methods that supposedly awaken our slumbering will and trounce adversity, often with the promise of religious or personal fulfilment at the end (Thomas 2009; Sladek 1973). East Asian philosophies and spiritual practices—romanticized and homogenized by their Anglo-Saxon interpreters—have held an especially strong allure since the nineteenth century (Buruma 2011; Stanley 2012).

If the fascination with mindfulness is only the most recent episode in this dubious chronicle, then it presents one genuine puzzle in the extent to which, in the Western world, it has been endorsed by elite groups in government, academia, and the media. And like the core of a Russian doll, this conundrum sits insides inside a still bigger one. From the beginning of the twenty-first century, governments, think tanks, and transnational corporations have never been so keen to measure and improve what they take to be the 'happiness' of their subjects and with the help of psychological techniques—of which mindfulness is one of the leaders. And yet these same governments are subjecting their less privileged citizens to stringency and duress on a scale that would have been unthinkable, only a generation or two before.

Since 2008, the global economy has been in the worst economic slump since 'the Dark Valley' of the 1930s (see Brendon 1998). Millions of people in Europe, including the citizens of the

UK, have been subjected to an official policy of 'austerity'. Presented as a way to manage the bad debts incurred by previous social democratic governments—(the alleged main cause of the current crisis)—austerity can be better understood as an official catalyst for the free market or neoliberal polices that have helped to shape the politics of the Western world since the 1980s (Hatherley 2016; Mendoza 2015). By degrees, the democratic state has been relegated to helpmate for business interests, intent upon dismantling most of the barriers to the movement of money and markets: including many of the legal and fiduciary protections that formerly sheltered the rights and livelihoods of millions of ordinary citizens (Clark and Heath 2015; Harvey 2005; Judt 2010; Smail 1993; Stuckler and Basu 2013).

In the UK, services once owned by the public and for the common good—health care, education, disability and unemployment benefits, housing, and transport—have been hollowed out and asset stripped (Harvey 2005; Judt 2010; Meeks 2014). While a small percentage of the population has prospered, poverty and inequalities of wealth and health have rocketed to heights reminiscent of the Gilded Age (Dorling 2014, 2016; Wilkinson and Pickett 2012). For the poor, the sick, and the disabled, austerity boils down to an attack upon the state benefits that help them to survive, increased hardship, and the threat or reality of homelessness and of official vilification (Clark and Heath 2014; Desmond 2016; Taylor 2015).

The consequences for non-executive workplaces especially have been dismal. In recent decades in both the USA and Europe, many of them have become regimes of outsourced employment, subcontracting, franchising, and third-party management. More than one in three American workers are hired by an external agency rather than by the company under whose auspices they labour, and Britain is not far behind. Former notions of an unspoken and joint obligation between employee and corporation have all but vanished, together with predictable careers, mutual trust, and expectations of loyalty (Kuttner 2014; Sennett 1998, 2006; Weil 2014). The places in which many of

us have to earn our living have become more fraught and—thanks to the Kafkaesque power of new information technology—more bureaucratic *and* chaotic, at the same time (Bunting 2005; Donner 2014; Fleming 2015). In the early twenty-first century, work seems for most people to have become increasingly invasive and disruptive of private life through the imposition of flexitime, zero hours contracts, and of mobile communication and tracking devices (Fleming 2015; Schneier 2015). Recent surveys confirm widespread malaise, accounting for over a third of all job-related illness in the UK, for example (TUC 2015). In part, these figures may speak of disenfranchised employees having to resort to the language of psychological symptoms for their grievances to be heard (Newton 1998). However, rising self-harm and suicide rates in professions such as teaching and finance suggest that for many, probably most, the distress is real (Fleming 2015; Fisher 2009).

In the midst of this unsettling landscape, we should not be surprised that the wish for distraction or reprieve is strong. Witness the prevalent recourse to alcohol and recreational drugs, the popularity of escapist holidays, and leisure and national lotteries. The demand for quietude as a commodity is just a subtler instance (Sim 2004). Upmarket religious retreats have become popular in recent years, with their promise to unburden each punter of their dependency upon mobile phones, laptops, and similar devices (Carrette and King 2005; Lipton 2007). So-called artistic siestas invite their participants into a state of easeful drifting attention and then sleep, for which the reading of a poem is merely a backdrop. This movement, most evident on the European continent, echoes attempts to establish the afternoon nap as part of the official working day among the self-consciously creative sectors of the corporate elite, including the employees of Google (Pieiller 2014).

The need felt by so many people for mindfulness practice as a form of escape and self-soothing is consistent with these trends. Mindfulness temporarily cocoons the user against the rootlessness and incessant demands of contemporary life, giving them—literally—a breathing space, in which they can get back in touch with what is happening in their own body and find comfort in convivial group membership: an experience that seems to be growing rarer for many (Cromby et al. 2013; Stivers 2004). To their credit, Kabbat Zinn and his followers sometimes acknowledge that the practice cannot compensate for all of the harmful effects of having to get by inside hostile workplaces (see, for example, *Mindful Nation*, ibid.) and that no one should be blamed for supposedly having created their own illness out of wilful pessimism and other bad mental habits (Kabat-Zinn 2005). In both of these cases and many others within the literature, however, lone cautionary statements like these are overwhelmed by a swarm of research findings and personal testimonials which imply that negative attitudes equate to a lack of moral backbone and predispose to poor health (Barker 2014; Coward 1991; Friedly 2013). In the *Mindful Nation document*, this impression is increased by the authors' failure to acknowledge the extent to which mental health problems and poor educational attainment can reflect impoverished circumstances, and dysfunctional communities and schools (Mombiot 2016; Thomas 2014).

In its relentless focus upon the internal world of the individual as the main answer to all personal and communal ills, mindfulness turns each practitioner into the neoliberal subject incarnate; their personal freedom in the marketplace guaranteed, together with full responsibility and accountability—not merely for their own conduct, but for their health and well-being, too (Aschoff 2015; Davies 2015; Purser 2015; Rose 2007; Smail 2005). No wonder so many employers, government agencies, and mental health experts are keen to promote the practice (see Carrette and King 2005).

The traditional teachings of Mahayana and Theravada schools valued meditation as one path among several that led to insight into the transience and contingency of the 'self', so cherished within Western culture. However, the enlightenment experience gained importance only through

dogged practice—(a rarity even in the Buddhist homelands of South East Asia)—and within a set of institutions, rituals, and teachings intended to nurture the seeker as an ethical being. The canonical texts seldom saw mindfulness as an end unto itself, but rather as bridge towards larger moral purposes, including dedication to a wider community. Equanimity, well-being, and 'happiness' as the warm inward glow beloved of Western psychologists were not necessarily unwelcome, but they could be distractions, in the end (Cohen 2010; Crook 1980, 2009; Flanagan 2011; Purser 2014; Trungpa 1973).

For the more reflective of Western practitioners, mindfulness is a discipline that shows how our thoughts and feelings are inextricably intertwined with our physical embodiment and that mindfulness meditation—like consciousness itself—cannot happen exclusively inside of the meditators' head. Rather, it is a social practice, dependent upon the guidance and support of compassionate people. This standpoint—based in part upon the pragmatist philosophy of William James and upon the insights of Wittgenstein and of various systems theorists—promises to correct some of the misguided optimism of people like Richard Layard (see, e.g., Fuchs 2013 ; Michelak et al. 2012; Crook 1980, 2009; Stanley 2012). However, there is still something missing from this account. Clinical experience and the critical realist tradition within the social sciences have yet more to say about the relationship between self, feelings, and the experience of having to live in a world in which most forms of dignity are undermined or denied (Burkitt 2011; Cromby 2015; Sennett and Cobb 1985; Smail 2005). Indeed, Buddhist teachings aver that the inequity and strife that is central to a neoliberal society will foment envy, greed, mistrust, and anger— the 'poisonous emotions'—and hence personal distress on the widest scale: which is just what epidemiologists have been finding (Crook 1980, 2009; Dorling 2016; Wilkinson and Pickett 2012).

We live in a real world that resists wishful thinking and that is structured by social, economic, and material powers over which most of us have little control, and those with the least—the poorest and the most downtrodden—usually suffer the most (Bourdieu et al. 2000; Moloney 2013a; Smail 1996). Our physical embodiment is about more than the mere kindling of experience. It testifies to our common fragility as creatures of flesh and bone, bearing the ineradicable emotional scars of our passage through life—even as they attune us—exquisitely and vulnerably—to our surroundings (Cromby 2015; Bourdieu 1985; Burkitt 2008; Charlesworth 1999; Sennett and Cobb 1985; Smail 2005). To feel secure, to act with a measure of confidence and compassion, we need to have some assurance of belonging, meaning, and stability. When our world begins to crumble, then we undergo a corresponding personal disintegration (Cromby et al. 2013; Doris 2002; Midlands Psychology Group 2012; Smail 1993, 2005).

Intellectual illusions are often the most seductive and damaging of all. The officially endorsed notion of mindfulness as the answer to societal and personal malaise belongs to this category. The advocates of mass meditation invert the quest of Freud and other psychologists, who wanted to use the lessons of therapy to inform the creation of new institutions for the nurture of future generations more humane and capable than their own (Freud 1895; Smail 2005). Instead, the vision of the good society has been turned upside down and inside out. It has become a collection of individuals, sitting in earnest inward gaze. Techniques of meditation can offer temporary sanctuary from the demands and conflicts of a world that for too many has grown colder, and more frightening. But in the end this respite is a fake. Like the advertised promise of superior refreshment from a commercially supplied bottle of drinking water: an over-packaged and inferior product, the mass consumption of which has helped to undermine the willingness of governments to maintain the quality of communal aquifers (Szasz 2007). Rather than looking to mythical internal cures for our personal ills, we need to look outward. The task is to rebuild a public world in which all of us can find a purpose and a place.

Acknowledgement I give grateful thanks to William Epstein and to Paul Kelly, for their thoughtful comments and editorial advice.

References

Aschoff, N. (2015). *The new prophets of capital*. New York: Jacobin.

Austin, J. H. (2014). *Zen-brain horizons. Toward a living Zen*. Massachusetts: MIT Press.

Bachelor, S. (1983). *Alone with others: An existential approach to Buddhism*. New York: Grove Press.

Barker, K. S. (2014). Mindfulness meditation: Do-it-yourself medicalization of every moment. *Social Science and Medicine*, 168–176.

Bates, Y. (Ed.). (2005). *Shouldn't I be feeling better by now? Client views of therapy*. London: Palgrave MacMillan.

Beck, A. T., Rush, J. A., Shaw, B.F., & Emery, G. (1987). *Cognitive Therapy of Depression*. New York: The Guilford Press.

Bennett, C. M., Baird, A. A., Miller, M. B., & Wolford, G. L. (2010). Neural correlates of interspecies perspective taking in the post-mortem Atlantic Salmon: An argument for proper multiple comparisons correction. *Journal of Serendipitous and Unexpected Results, 1*(1), 1–5.

Blackmore, S. J. (1993). *Dying to Live: Near Death Experiences*. London: Prometheus.

Blackmore, S.J. (2010). *Ten Zen questions*. London: One World.

Booth, R., (2014). Mindfulness therapy comes at a high price for some, say experts. *The Guardian*, August 25, 2014.

Booth, R. (2015). Mindfulness study to track effect of meditation on 7000 teenagers. *The Guardian*, July 15th, 2015.

Bourdieu, P. (1985). *Distinction: A Social Critique of the Judgement of Taste*. London: Routledge.

Bourdieu, P., et al. (2000). *The weight of the world*. London: Polity Press.

Brendon, P. (1998). *Dark valley: A panorama of the 1930s*. London: Duckworth.

Brothers, L. (2002). *Mistaken identity: The mind-brain problem reconsidered*. New York: Suny Press.

Bunting, M. (2005). *Willing slaves: How the overwork culture is ruling our lives*. London: Harper Perennial.

Bunting, M. (2014). Why we will come to see mindfulness as mandatory. *The Guardian*, May 6, 2014. http://www.theguardian.com/commentisfree/2014/may/06/mindfulness-hospitals-schools

Burkeman, O. (2016). Therapy wars: The revenge of Freud. *The Guardian*. Thursday 7th January 2016. http://www.theguardian.cm/science/2016/jan/07/therapy-wars. Downloaded 09/01/16.

Burstow, B. (2015). *Psychiatry and the business of madness. An ethical and epistemological accounting.* London: Palgrave MacMillan.

Buruma, I. (2011). *Taming the gods: Religion and democracy on three continents.* Princeton: Princeton University Press.

Cacioppo, J., Bernston, G., Lorig, T., Norris, C., Rickett, E., & Nusbaum, H. (2003). Just because you're imaging the brain doesn't mean you can stop using your head: A primer and set of first principles. *Journal of Personality and Social Psychology, 85*(4), 650–661.

Caldini, R. (1993). *Influence: The Psychology of Persuasion.* London: Harper Collins.

Carrette, J., & King, R. (2005). *The selling of spirituality: the silent takeover of religion.* London: Routledge.

Charlesworth, S. (1999). *A phenomenology of working class experience.* Cambridge: Cambridge University Press.

Charlton, B. (2005). *Infostat, cargo-cult science and the policy sausage machine: NICE, CHI and the managerial takeover of clinical practice.* In M.J. R. Hampton, & B. Hurwitz (Eds.), *NICE, CHI and the NHS reforms: Enabling excellence or imposing control?* London: Aesculapius Medical Press.

Choudhury, S., & Slaby, J. (2012). Introduction: Critical neuroscience - between lifeworld and laboratory. In S. Choudhury & J Slaby (Eds.), *Critical neuroscience: A handbook of the social and cultural contexts of neuroscience.*

Clark, T., & Heath, A. (2014). *Hard times. The divisive toll of the economic slump.* Yale: Yale University Press.

Claxton, G. (1991). *The heart of Buddhism.* London: Aquarius.

Cohen, E. (2010). From the Bodhi tree, to the analyst's couch, then into the MRI scanner: The psychologisation of Buddhism. *Annual Review of Critical Psychology, 8,* 97–119.

Coward, R. (1991). *The whole truth: The myth of alternative health.* London: Faber and Faber.

Cromby, J. (2015). *Feeling bodies: Embodying psychology.* London: Palgrave.

Cromby, J., Harper, D., & Reavey, P. (2013). *Psychology, mental health and distress.* London: Palgrave MacMillan.

Crook, H. (1980). *The evolution of human consciousness.* Oxford: Oxford University Press.

Crook, H. (2009). *World crisis and Buddhist humanism. End games: Collapse or renewal of civilisation.* New Delhi: New Age Books.

Cushman, (1995). *Constructing the self, constructing America.* Cambridge, Massachusetts: Perseus Publishing.

Dalal, F. (2015). Statistical spin: Linguistic obfuscation— The art of overselling the CBT evidence base. *The Journal of Psychological Therapies in Primary Care, 4,* 1–25.

Danziger, K. (2008). *Marking the mind: A history of memory.* Cambridge: Cambridge University Press.

Davies, D. (1996). *Counselling in psychological services.* Milton Keynes: Open University Press.

Davies, W. (2015). *The happiness industry. How the government and big business sold us well-being.* London and New York: Verso.

Dawes, R. M. (1994). *House of cards: Psychology and psychotherapy built on myth.* New York: The Free Press.

Dineen, T. (1998). *Manufacturing Victims.* London: Constable.

Desmond, M. (2016). *Evicted. Poverty and profit in the American City.* London: Allen Lane.

Donner, B. (2014). The wonder of computers: A threat to health and clinical psychology. *Clinical Psychology Forum, 257,* 36–40.

Doris, M. (2002). *Lack of character. Personality and moral behavior.* New York: Cambridge University Press.

Dorling, D. (2014). *Inequality and the 1 %.* London: Verso.

Dorling, D. (2016). *A better politics. How government can make us happier.* London: London Publishing Partnership.

Ellis, A., & Dryden, W. (1987). *The Practice of Rational Emotive Therapy.* New York: Springer.

Epstein, W. J. (1995). *The illusion of psychotherapy.* New York: Transaction.

Epstein, W. J. (2006). *The civil divine: Psychotherapy as religion in America.* Rheno: University of Nevada Press.

Epstein, W. J. (2010). *Democracy Without Decency: Good Citizenship and the War on Poverty.* Pensylvannia: Penn State University Press.

Epstein, W. J. (2013). *Empowerment as ceremony.* New York: Transaction.

Erwin, E. (2000). *Philosophy and psychotherapy.* London: Sage.

Eysenck, H. J. (1952). The effects of psychotherapy: An evaluation. *Journal of Consulting Psychology, 16,* 319–324.

Fancher, R. F. (1996). *Cultures of healing: Correcting the image of American mental health care.* New York: W. H. Freeman and Co.

Farias, M., & Wikholm, C. (2015). *The Buddha pill. Can meditation change you?.* London: Watkins.

Feltham, C. (2013). *Counselling and Counselling Psychology: A Critical Examination.* Ross-On-Wye: PCCS Books Ltd.

Felver, J.C., Celis-de Hoyos, C. E., Tezanos, K., & Singh, N.N. (2015) A systematic review of mindfulness-based interventions for youth in school settings. *Mindfulness.* doi:10.1007/s12671-015-0389-4

Fisher, M. (2009). *Capitalist realim. Is there no alternative?.* London: Zed Books.

Fisher, S., & Greenberg, R. P. (1997). *From placebo to panacea: Putting psychiatric drugs to the test.* New York: Wiley.

Flanagan, O. (2011). *The Bodhisattva's brain: Buddhism naturalized.* Cambnridge: MIT Press.

Fleming, P. (2015). *The mythology of work. How capitalism persists despite itself*. London: Pluto.

Frank, D., & Frank, J. B. (1991). Persuasion and healing: A comparative study of psychotherapy (3rd ed.) New York: Johns Hopkins.

Freud, S. (1895). Project for a scientific psychology. The Standard Edition of the complete psychological works of Sigmund Freud (Vol. 1, pp. 283–397). London: The Hogarth Press, 1953.

Fuchs, T. (2013). Depression, intercorporeality, and interafectivity. *Journal of Consciousness Studies, 20* (7–8), 219–238.

Gaudiano, B. (Ed.). (2014). *Incorporating acceptance and mindfulness into the treatment of psychosis: Current trends and future directions*. New York: Guildford Press.

Gaynor, K. J. (2014). A critical review of mindfulness-based psychological treatments for worry and rumination. *Open Access Behavioural Medicine, 2*(1), 2.

Gregoire, C. (2015). How mindfulness is revolutionizing mental health care. *The Huffington Post. February 23rd, 2015.*http://www.huffingtonpost.com/2015/01/23/neuroscience-mindfulness_n_6531544.html.. Accessed 08 Aug 2016.

Goldacre, B. (2009). *Bad science*. Oxfordshire: Harper Collins.

Germer, C. K., Seigel, R. D., & Fulton, P. R. (2013). *Mindfulness and Psychotherapy* (2 ed.) New York: Guildford Press.

Grimshaw, A. (1992). *Servants of the Buddha: Winter in a Himalayan convent*. London: Open Letters.

Gross, D. M. (2006). *The secret history of emotion: From Aristotle's rhetoric to modern brain science*. Chicago, IL: University of Chicago Press.

Hagan, T. and Donnison, J. (1999). Social power: some implications for the theory and practice of cognitive behaviour therapy. *Journal of Applied and Community Social Psychology, 9*(2):119–135.

Harre, R. (2002). *Cognitive science: A philosophical introduction*. London: Sage.

Harvey, D. (2005). *A brief history of neoliberalism*. Oxford: Oxford University Press.

Hatherley, O. (2016). *The ministry of Nostalgia*. London: Verso.

Healy, D. (2013). *Pharmageddon*. California: University of California Press.

Heffernan, V. (2015). The muddied meaning of mindfulness. *The New York Times Magazine*, April 14, 2015. http://www.nytimes.com/2015/04/19/magazine/the-muddied-meaning-of-mindfulness.html?_r=0

Holmes, D. S. (1984). Meditation and somatic arousal reduction: A review of the experimental evidence. *American Psychologist, 31*(1), 1–10.

Holmes, G. (2010). *Psychology in the real world: Using community based groups to help people*. Ross-On-Wye: PCCS Ltd.

Holmes, J. (2002). All you need is cognitive behaviour therapy? *British Medical Journal, 324*:288–291.

Horvath, J. C., Carter, J. O., & Forte, J. D. (2015a) Evidence that transcranial direct current stimulation (tDCS) generates little-to-no reliable neuropsychologic effect beyond MEP amplitude modulation in healthy human subjects: A systematic review. *Neurospychologia, 66*, 213–236.

Horvath, J. C., Carter, J. O., & Forte, J. D. (2015b). Quantitative review finds no evidence of cognitive effects in healthy populations from single-session transcranial direct stimulation (tDRS). *Brain Stimulation, 8*(2), 535–550.

Horwitz, A. V., & Wakefield, J. C. (2007). *The loss of sadness: How psychiatry transformed normal sorrow into depressive disorder*. Oxford: Oxford University Press.

House, R., & Loewenthal, D. (Eds) (2008). *Against and for CBT: Towards a constructive dialogue?* Ross-On-Wye: PCCS Books Ltd.

Illouz, E. (2008). *Saving the modern soul: Therapy, emotions and the culture of self-help*. Berkley: University of California Press.

Johnsen, T.J., & Friborg, O. (2015). The effects of cognitive behavioural therapy as an anti-depressive treatment is falling: A meta-analysis. *Psychological Bulletin*.

Johnstone, L. (2006). *The limits of biomedical models of distress*. In D. Double (Ed.), *Critical psychiatry: The limits of madness*. Basingstoke: Palgrave MacMillan.

Judt, T. (2010). *Ill fares the land*. London: Penguin.

Kabat-Zinn, J. (1991). *Full catastrophe living: Using the wisdom of your body and mind to face stress, pain, and illness*. New York: Delta.

Kabat-Zinn, J. (1994). *Wherever you go, there you are: Mindfulness meditation in everyday life*. London: Hyperion Books.

Kabat-Zinn, J. (2001). *Mindfulness meditation for everyday life*. New York: Piatkus.

Kagan, J. (2007). *What is emotion?*. New Haven and London: Yale University Press.

Kagan, J. (2012). *Psychology's ghosts: The crisis in the profession and the way back*. New Haven: Yale University Press.

Kahneman, D. (2013). *Thinking, fast and slow*. London: Allen Lane.

Kelly, A. E. (2000). A self-presentational view of psychotherapy: Reply to Gill, Helso and Mohr (2000) and to Arkin and Hermann (2000). *Psychological Bulletin 126*(4), 505–11.

Khaneman, D. (2010). The riddle of experience versus memory. TED Talks. http://www.bing.com/videos/search?q=Daniel+Kahneman+Ted+Talk&view=detail&mid=4041A90FC53FEE7F01854041A90FC53FEE7F0185&FORM=VIRE. Downloaded May 5th, 2016.

King-Spooner, S. (2014). Psychotherapy and the White Dodo. *The Journal of Critical Psychology, Counselling and Psychotherapy, 14*(3):165–173.

Kirsch, I. (2009). *The emperor's new drugs: Exploding the antidepressant myth*. London: Bodley Head.

Kline, P. (1992). *Problems of methodology in studies of psychotherapy*. In Dryden, W., & C. Feltham (Eds.), *Psychotherapy and Its Discontents*. Buckinghamshire: Open University.

Kline, P. (1987). *Psychology exposed: Or the emperor's new clothes*. London: Routledge and Kegan Paul.

Kuttner, R. (2014).Why is work more and more debased? *New York Review of Books*, 61(16), October 23, 2014. http://www.nybooks.com/articles/2014/10/23/why-work-more-and-more-debased/. Accessed 22 Nov 2015.

Layard, R., & Clark, D. M. (2014). *Thrive: The power of evidence based psychological therapies*. London: Allen lane.

Lipton, S. (2007). Monastery chic: The ascetic retreat in a neoliberal age. In M. Davis & D. Bertrand-Monk (Eds.), Evil paradises. Dreamworlds of neoliberalism. New York and London: The New Press.

Lomas, P. (1998). *Doing good? Psychotherapy out of its depth*. Oxford: Oxford University Press.

Lutz, L. (1985). Depression and the translation of life worlds. In A. Kleinman & B. Good (Eds.), *Culture and depression. Studies in the anthropology and cross-cultural psychiatry of affect and disorder*. Berkley: University of California Press.

Mair, K. (1992). The myth of therapist expertise. In C. Feltham & D. Pilgrim (Eds.), *Psychotherapy and its Discontents*. Milton Keynes: Open University.

Massouvi, S., et al. (2007). Depression, chronic diseases and decrements in health: Results from the world health surveys. *The Lancet*, *370*(9590), 851–858.

McGown, W. (1992). *Only man is vile. The tragedy of Sri Lanka*. London: Picador.

McInnes, B. (2011). Nine out of ten people helped by IAPT? *Therapy Today, Letters Section*, 22(1).

Meeks, (2014). *Private Island: Why Britain now belongs to someone else*. London: Verso.

Mendoza, K. (2015). *Austerity: The demolition of the welfare state and the rise of the zombie economy*. Oxford: New Internationalist Publications Ltd.

Michelak, J., Burg, J. M., & Heidenreich, T. (2012). Mindfulness, embodiment, and depression. In S. Koch, T. Fuchs, M. Summa & C. Muller (Eds.), *Body, memory, metaphor and movement*. Amsterdam: John Benjamins Publishing Company.

Midlands Psychology Group. (2016). Psychology as practical biopolitics (forthcoming).

Midlands Psychology Group. (2008). Our big fat multi-million pound psychology experiment. *Clinical Psychology Forum 181*, 34–37.

Miller, S. D. (2012). Revolution in Swedish mental health practice: The cognitive behavioural therapy monopoly gives way. http://www.scottdmiller.com/feedback-informed-treatment-fit/revolution-in-swedish-mental-health-practice-the-cognitive-behavioral-therapy-monopoly-gives-way/. Downloaded, 5th May, 2016.

Mindfulness All-Party Parliamentary Group. (2015). Mindful Nation UK: Report by the Mindfulness All-Party Parliamentary Group (MAPPG), October 2015.

Morgan, A. (2008). *The Authority of Lived Experience*. In A. Morgan (Ed) *Being Human: reflections on Mental Distress in Society*. Ross-On-Wye: PCCS Books Ltd.

Moloney, P. (2013a). *The therapy industry. The irresistible rise of the talking cure and why it doesn't work*. London: Pluto.

Moloney, P. (2013b). What can brain scans really tell us about the effectiveness of psychological therapy? *Journal of Critical Psychology, Psychotherapy and Counselling, 13*(2), 145–163.

Moloney, P. (2016). If the masses have been brainwashed, they've done the laundry themselves.' An interview with William Epstein. *Self and Society* (forthcoming).

Moloney, P., & Kelly, P. (2008). Beck never lived in Birmingham: Why cognitive behavioural therapy might not be as effective a treatment for psychological distress than is often supposed. In R. House & D. Loewenthal (Eds.), *Against and for CBT. Towards a constructive dialogue?* Ross-On-Wye: PCCS Books, Ltd.

Mombiot, G. (2016). *What went wrong?* London: Jonathan Cape.

Moncrieff, J. (2007). *The myth of the chemical cure: A critique of psychiatric drug treatment*. London: Palgrave MacMillan.

Newnes, C. (2014). *Clinical psychology: a critical examination*. Ross-On-Wye: PCCS Books Ltd.

Newton, T. (1998). Stress discourse and individualisation. In C. Feltham (Ed.), *Controversies in psychotherapy and counselling*. London: Sage.

National Institute for Clinical Excellence (NICE). (2016). *Depression in adults: recognition and management*. https://www.nice.org.uk/guidance/CG90/ifp/chapter/About-this-information. Accessed 08 Aug 2016.

NIMH. (2016). *Psychotherapies*. http://www.nimh.nih.gov/health/topics/psychotherapies/index.shtml. Retrieved Aug 08 2016.

Noe, A. (2009). *Out of our heads: Why you are not your brain and other lessons from the biology of consciousness*. New York: Hill and Wang.

Norcross, J.C., & Wampold, B.E. (2011). Evidence-based therapy relationships: research conclusions and clinical practices.*Psychotherapy*, *48*(1):98–102.

Pieiller, E. (2014). The indolent, beguiling artistic siesta: Keep calm and don't carry on. Le Monde Diplomatique. http://mondediplo.com/2014/02/14siesta Downloaded November 20th, 2015

Pietikainen, P. (2007). *Neurosis and modernity: The age of nervousness in Sweden*. Leiden and Boston: Brill.

Pilgrim, D., & Bentall, R. (1999). The medicalisation of misery: A critical realist analysis of the concept depression. *Journal of Mental Health, 8*(3), 261–274.

Pilgrim, D. (2008). *Reading 'happiness': CBT and the layard thesis*. In R. House & D. Loewenthal (Eds.), *Against and for CBT: towards a constructive dialogue?* Ross-On-Wye: PCCS Books Ltd.

Pollock, A. M. (2009). *NHS Plc: The privatisation of our health care*. London: Verso.

Prioleau, L., Murdock, M., & Brody, N. (1983). An analysis of psychotherapy versus placebo studies. *Behavioural and Brain Sciences, 6,* 275–310.

Postle, D. (2007). *Regulationg the psychological therapies: From taxonomy to taxidermy.* Ross-On-Wye: PCCS Books Ltd.

Rickman, P. (2009). Is psychology science? *Philosophy Now, 74,* 6–7.

Roemer, L., & Orisillo, S. M. (2009). *Mindfulness and acceptance based behavioural therapies in practice.* New York: Guilford Press.

Rogers, A., & Pilgrim, D. (2014). *A sociology of mental health and illness* (Fifth Ed.). Berkshire: Open University Press.

Rose, S. (2005). *The 21st century brain.* London: Jonathon Cape.

Rose, N. (2007). *The politics of life itself: Biomedicine, power, and subjectivity in the twenty-first century.* Princeton: Princeton University Press.

Rose, N., & Abi-Rachid, J. (2013). *Neuro.* Princeton, New Jersey: Princeton University Press.

Rosenthal, R., & Rubin, D.B. (1978). Interpersonal expectancy effects: The first 345 studies. *Behavioural and Brain Sciences, 3*: 377–415.

Roth, A., & Fonagy, P. (2006). *What works for whom?: A critical review of psychotherapy research.* New York: The Guilford Press.

Russell, P. (1976). *The TM technique: An introduction to transcendental meditation and the teachings of the maharishi mahesh yogi.* London: Routledge and Kegan Paul.

Sanders, L. (2009). Trawling the brain. New findings raise questions about reliability of FMRI as gauge of neural activity. *Science News, 176*(13), 16.

Sands, A. (2000). *Falling for therapy: Psychotherapy from a client's point of view.* London: MacMillan.

Schneier, B. (2015). *Data and goliath: The hidden battles to collect your data and control your world.* New York: W.W. Norton and Co.

Schooler, J. (2011). Unpublished results hide the decline effect. *Nature, 40*(7335), 437.

Schwitzgebel, E. (2011). *Perplexities of consciousness.* Massachusetts: MIT Press.

Scull, A. (2015). *Madness in civilization.* London: Thames and Hudson Ltd.

Segal, Z., Williams, J. M. G., & Teasdale, J. D. (2013). *Mindfulness based cognitive therapy for depression* (2nd ed.). New York: The Guildford Press.

Senett, R. (1998). *The corrosion of character.* New Haven and Connecticut: Yale University Press.

Sennett, R. (2006). *The culture of the new capitalism.* New Haven and Connecticut: Yale University Press.

Shedler, (2010). The efficacy of psychodynamic psychotherapy. *American Psychologist, 65*(2), 98–109.

Shedler, J. (2015). Where is the evidence for "evidence-based" therapy? *The Journal of Psychological Therapies in Primary Care, 4,* 47–59

Shotter, S. (1975). *Images of man in psychological research.* London: Methuen.

Sim, S. (2004). *Silence: A manifesto.* Oxford: Oxford University Press.

Sladek, J. (1973). *The new apocrypha.* London: Jonathan Cape.

Smail, D. (1985). *Illusion and reality: The meaning of anxiety.* London: Dent.

Smail, D. (1989). Managerialism and power. *Clinical Psychology Forum, 20,* 10–12.

Smail, D. (1993). *The origins of unhappiness: A new understanding of personal distress.* London: Constable.

Smail, D. (2005). Power, interest and psychology. Elements of a social-materialist approach to understanding distress. Ross-On-Wye: PCCS Books Ltd.

Stanley, S. (2012). Mindfulness: Towards a critical relational perspective. *Social Psychology and Personality Compass, 6*(9) 631–641.

Stivers, R. (2004). *Shades of loneliness: Pathologies of a technological society.* Lanam and New York: Rowan and Littlefield.

Szasz, A. (2007). *Shopping our way to safety. How we changed from protecting the environment to protecting ourselves.* Minneapolis: The University of Minnesota Press.

Sutherland, S. (1992). *Irrationality: The enemy within.* London: Constable.

Taylor, D. (2015). 'Conscious cruelty': Ken Loach's shock at benefit sanctions and food banks. *The Guardian.* November 23rd, 2015.

Thomas, K. (2009). *The ends of life: Roads to fulfillment in early modern England.* Oxford: Oxford University Press.

Thomas, G. (2014). Inequality and the Next Generation. *The Psychologist, 27*(4), 2–4.

Throop, E. A. (2009). *Psychotherapy, American culture, and social policy: Amoral individualism.* London: Palgrave MacMillan.

Trades Union Congress. (2015). Risks, E-newsletter on health and safety issues. Number 725, October 24th, 2015. https://www.tuc.org.uk/workplace-issues/health-and-safety/risks-newsletter/risks-2015/tuc-risks-725-24-october-2015

Trungpa, C. (1973). *Cutting through spiritual materialism.* Boston and London: Shambala.

Uttall, W. N. (2011). *Mind and brain: A critical appraisal of cognitive neuroscience.* Massachusetts: The MIT Press.

Uttall, W.J. (2016). *Macronueral theories in congitive neuroscience.* Psychology Press: New York and London: Taylor and Francis.

Wampold, B., & Imel, Z. E. (2015). *The great psychotherapy debate. The evidence for what makes psychotherapy work* (2nd ed.). London: Routledge.

Watts, J. (2016). IAPT and the ideal image. In J. Lees (Ed.), *The future of psychological therapy: From managed care to transformational practice* (pp. 84–101). London: Routledge.

Weil, D. (2014). *The fissured workplace. Why work became so bad for so many and what can be done to improve it.* Cambridge, Massachusetts: Harvard University Press.

Weisberg, D. H. (2008). Caveat lector: The presentation of neuroscience information in the popular media. *The Scientific Review of Mental Health Practice, 6*(1), 51–56.

Wheen, F. (2004). *How Mumbo-Jumbo Conquered the World: A short history of modern delusions.* London: Harper Collins.

Wilkinson, R., & Pickett, K. (2012). *The spirit level* (2nd ed.). London: Penguin.

Williams, J. M. G., Crane, C., Barnhofer, T., Brennan, K., Duggan, D. S., Fennell, M. J. V., et al. (2014). Mindfulness-based cognitive therapy for preventing relapse in recurrent depression: a randomised controlled trial. *Journal of Clinical and Consulting Psychology, 82*(2), 275–286.

Wiseman, H. (2016). *The myth of the moral brain: The limits of moral enhancement.* Massachusetts: MIT.

Zenner, C., Hernleben-Kurz, S., & Walach, H. (2014). Mindfulness-based interventions in schools—A systematic review and meta-analysis. *Frontiers in Psychology, 5*, 1–20.

Zilbegeld, B. (1982). *The shrinking of America. Myths of psychological change.* Boston: Little, Brown and Company.

The Fourth Treasure: Psychotherapy's Contribution to the Dharma

Manu Bazzano

Introduction

A Buddhist practitioner is said to take refuge in the three treasures (Buddha, Dharma and Sangha), potent source of inspiration and support on the path. In doing so, she is reminded of the Buddha's own example, of his teachings (the Dharma) and of the encouraging presence of a community of fellow travellers (Sangha). One interpretation of the three treasures, unofficially attributed, among others, to Taizan Maezumi (1931–1995), is *secular* in the literal sense of the word, i.e. pertaining to history (*saeculum* means generation, age as well as century). This interpretation sees the three treasures manifested in different ages of history. The first one would be the age of the historical Buddha Gautama. The second, the age of the Dharma, of consolidation (some would say institutionalization) of 'Buddhism' as a religious doctrine. The third would be the modern age of Sangha or community. The emphasis in the latter is on ethics, the social dimension and the actualization of the teachings in the crucible of everyday life with others. When I first heard this interpretation, I instinctively linked it to the unorthodox views of Joachim of Fiore (1135–1202), the twelfth century Calabrian mystic, theologian and founder of a Christian monastic order who lived as a hermit in the Sila Mountains near Cosenza. Despite the widely different contexts (Joachim's view was centred on a theology of revelation), there are intriguing similarities between Joachim's reading of the Christian trinity (Father, Son and Holy Spirit) and Maezumi's interpretation of the three treasures. For Joachim, the first name of the trinity corresponds to the age of the father, i.e. the age of the Old Testament; the second (between the birth of Christ and the thirteenth century) to the age of the son and the third, from the thirteenth century onwards, to the age of the spirit. The latter would be characterized by humankind's potentially unmediated contact (via spirit) with God—an optimistic if heretical view of history and theology that some scholars saw as anticipating of several centuries Hegel's (and Marx's) *theodicy*, i.e. the belief in the presence of God's providence (or its secular equivalent, gradual progress towards justice and equality) in the midst of history's evils. This is not the place to discuss in depth whether the notion of a divinely or humanly inspired evolutionary 'progress' is defensible. My hunch is that it is not, yet both notions are evocative. That ordinary people may have access to the divine subverts in one sweep the clergy's millenarian authority and privilege. That community may acquire prominence over other concerns in the current propagation of the Dharma has, likewise, far-reaching implications.

M. Bazzano (✉)
Metanoia Institute, London, UK
e-mail: manubazzano@onetel.com

© Springer International Publishing Switzerland 2016
R.E. Purser et al. (eds.), *Handbook of Mindfulness*,
Mindfulness in Behavioral Health, DOI 10.1007/978-3-319-44019-4_19

Active and Passive Adaptation

That our era is the age of community could mean that community is less a *given* than a *task*, i.e. an area we need to focus on, something to be practiced and cultivated because it has been neglected, rather than factual description of our zeitgeist. Sangha may need to be valued over and above the two other 'treasures' which were dominant in other eras. In the case of the first treasure, this could mean going beyond the deification *and* the secular idealization of *Buddha*—beyond the otherworldly, enlightened archetype *and* the mindful, stress-reducing and allegedly objective physician. It could mean that in order to contribute to the contemporary world, the Dharma would need to be a little more than an exotic system of feudal religiosity at the margins of society or a pragmatic assortment of rescue remedies for frazzled high-achievers.

Similarly, in relation to the second treasure (Dharma), the systematization of the Buddha's teachings and institutionalization of Buddhism as an established religion would take second place—a form of *active adaptation* of its views and precepts to the world we find ourselves in. The notion of active adaptation is borrowed from Adler (2006), who believed that human beings strive for a constructive adaptation to life's challenges and demands, and that striving itself must be appreciated above an idealized and often damaging notion of perfection and a customary fixation with goals and targets. Adler's notion was in turn inspired by the German poet and essayist Lessing (1729–1781) who maintained that if God gave him the choice between truth and the *striving* for truth, he would opt for the latter.

There is of course a world of difference between *no* adaptation, *passive* adaptation and *active* adaptation. In the first instance, one holds on to Dharma teachings as an eternal body of truth impervious to the contingencies of history. This stance possesses a certain appeal, but cannot be said to directly contribute to the struggles and anxieties of the modern world. *Passive* adaptation, on the other hand, entails obeisance to the dominant ideological frame to which the sheer *otherness* of the Buddha's teachings is co-opted,

their existential edge smoothed out and their core made more palatable as yet another product on the shelves of the self-help superstore. This is clearly a case of throwing the baby Buddha out with the bathwater, which is what I believe has happened, on the whole, with the 'mindfulness' phenomenon, employed in some instances by 'those in power as a technology for their own self-serving purposes, unmoored from its ethical meaning' (Forbes 2012).

By adopting a stance of active adaptation, on the other hand, we potentially partake of the world, become implicated in the dust and noise of the marketplace. I have called this approach to Buddhism *mundane* (Bazzano 2013a, b, c), literally *of the world* ('mondo'): earthy, earthly, with no remnants of the cloister and of the 'monkish virtues' derided by Hume (2004), those very same unnatural virtues that adhere so tightly to supposedly secular forms of Buddhism. The term 'mundane' links the Dharma to the best phenomenological tradition, which sees human experience situated in the world and as such inherently *ambiguous* (Merleau-Ponty 1964). It smokes it out of the holier-than-thou dwellings to which Buddhism has been consigned and links it to what is normally considered 'samsaric'. A *mondana* is a sex worker in Italian slang, and *vita mondana* refers to an 'unedifying' existence dedicated to a Fellinian *dolce vita* of glamour and gossip and a celebration of what is conventionally seen as the world of 'mere' appearance.

A mundane conception of the Dharma also affirms that a practitioner can be *in* the world and yet hold a watchful eye and a dignified bearing in its midst—the meditative posture being an embodied expression of *dignity*. By maintaining a dignified stance, I would perhaps feel less compelled to bow to ideological pressures or having to acquire the latest gleaming product on offer. Equally, I'd feel less inclined to join the religious and moralistic chorus of sanctimonious disapproval of the world and be more critical of dominant values and agendas—which in this day and age are those of neo-liberalism. At the same time, I'd recognize that the way in which we experience *dukkha* is not one and the same as the way in which it was experienced by people at the time of the

Buddha. Of course death is certain (and tomorrow as uncertain) now as it was two thousand and five hundred years ago. Of course, craving and reactivity (*tanha*) cause disappointment now as they did then. But to disregard history altogether would be a serious mistake. It would also be an act of indolence. Every era needs to describe human experience anew. This is what the historical Buddha did in relation to the dominant world view of his time. He did not particularly oppose or endorse the latter but was decidedly non-committal with regard to metaphysical questions.

Actively adapting the Dharma could then means retranslating it, re-interpreting it while maintaining its otherness and its existential valence—a difficult task, and for that very reason worth pursuing.

From Secularism to the Mainstream

It would be naïve to assume that a healthy dose of psychotherapeutic knowledge and expertise could 'cure' or prevent the inevitable pitfalls we face in trying to actively adapt the Dharma to the contemporary world. This is because the very same reductionism now in vogue and forcefully endorsed by vested interests is at work in both psychotherapy and Buddhism. For example, humanistic psychology and psychotherapy trainees in the UK are increasingly being taught 'evidence-based', quantitative methods of research, study and practice that are often at variance with the humanistic ethos and with more heuristic, exploratory modes historically associated with therapeutic practice. This mode now in vogue may be alternatively seen as a pure and simple sell-out or, more charitably, as a well-intentioned but misguided attempt to appease and be acknowledged by the 'Father' and the powers-that-be so that funding will not be stopped, our survival ensured and our way of being in the world validated. This way of thinking flies in the face of the wisdom acquired in many decades of social and political activism, something pithily expressed by Audre Lorde: there is no way you can pull apart the master's house by using the master's tools (Lorde 2007).

Another way of registering this very same puzzling inclination to bend over backward in order to please Big Daddy is the widespread use of left-brain type of language to promote right-brain ideas. This can be seen at play in the contemporary neuroscience literature (Schore 2011), although it is fair to concede that this may well be a translation of progressive principles into a language that has wider appeal and applications (Voller 2013, p. 77).

More worryingly, progressive psychology and psychotherapy appear to have engendered, like hippy parents to an 'ultra-square' progeny, a new breed of neo-conservatives—writers and practitioners who officially advocate progressive psychological principles yet show unequivocal signs of having contracted, in a perverse Foucauldian twist, a ravenous hunger for power. They will extol the virtues of randomized controlled trials,[1] refurbish psychopathology and diagnosis, colonize space left for debate at conferences with streams of data and blanket use of *PowerPoint*. The assumptions behind these words and deeds appear to be that progress can be measured by how often state, government and governing bodies adopt a progressive jargon, or by how far we manage to go in convincing the ruling elites that our utterances and practices are legitimate. What is strange is that the most vocal exponents of this pervasive compulsion-to-compromise wish to maintain their 'radical kudos' intact, seeing no contradiction at all between what they preach and their nominal progressive affiliations. We witness something similar with the mindfulness brand, now cheerfully adapted to the corporate world and the military, no (ethical) questions asked. This tendency has been building up for some time and is now being hailed as a welcome development from a once hopelessly peripheral, exotic practice of meditation into the coveted territory of the *mainstream*, a term praised by Kabat-Zinn (2015) as a more preferable goal for the mindfulness movement than the

---

[1] A randomized controlled trial (RCT) is a type of medical test, now extended to counselling and psychotherapy, where people being studied are randomly allocated one or other of the different treatments under study.

outdated secularist paradigm. Both 'mainstream' and 'secularism', however, have become hallmarks of dominant Western 'non-ideological' ideology: the former is characterized by the amount of corporate power it takes to propel a product; the latter, as the recent emergence of militant *laïcité* in France testifies, is increasingly a synonym, at least in Europe, for time-honoured hatred of otherness and suspicion of foreign religious beliefs. This represents a rather perverse turn of events if one considers that originally *laïcité* (its English rendition 'secularism' is at best an approximation) meant avowed neutrality of the state towards religious beliefs and, conversely, affirmation of non-interference by any religion in the running of government.

Upheld in the first article of the French constitution, at its inception this notion of secularism vividly expressed freedom of thought in the religious sphere. The tragic developments of 2015, from the *Charlie Hebdo* killings to the *Bataclan* massacre at the hands of fundamentalists and the polarization that followed, have, however, brought about a sharp transformation in the very meaning of laïcité, to the point where it could be arguably be understood as a liberal form of Islamophobia. This is all the more problematic if we consider the shadowy backdrop of colonial France and the Algerian war of independence. Even when removed from laïcité, the notion of secularism in the English-speaking world carries its own shadows. It will suffice to mention here the 'god delusion' industry, after the book by the same title (Dawkins 2006) which saw a series of authors lining up in hasty and largely uninformed condemnation of religion and religious thought per se. My own criticism is mild criticism compared to Terry Eagleton's amusing lampooning of this reactionary brand of secularism:

> Imagine someone holding forth on biology whose only knowledge of the subject is the *Book of British Birds*, and you have a rough idea of what it feels like to read Richard Dawkins on theology (Eagleton 2006, p. 32).

It is perplexing to find some Buddhist writers directly or indirectly aligned to the ethos of the god delusion brigade. The cover of the renowned Buddhist scholar Stephen Batchelor's *Confessions of a Buddhist Atheist* (Batchelor 2010) is embellished by an endorsement written by none other than Christopher Hitchens, the late, great British essayist notorious for his *volte face* from progressive thinking to right-wing war-mongering. Much more worrying is the presence among the new secularists of Sam Harris, who features prominently among the authors and speakers of the recent 31-day *Mindfulness Summit* event. Harris is well known for his advocacy of torture against Muslim fundamentalists and for his rejection of pacifism on the grounds that its widespread influence would create a situation where thugs inherit the earth (Harris 2005). What he perhaps forgets is that thugs *have* inherited the earth a while back, and it wouldn't surprise me to learn that they signed up for a handful of 'mindfulness' sessions in order to reduce the stress of running a misshapen world.

This is why the term 'secularism' begins to sound almost obsolete here, as it does not quite describe the scope and breadth of the current mindfulness movement's ambition. As it can be inferred by the various contributions to the online Mindfulness Summit in October 2015, a more adequate word for this is 'mainstream'. If *Newscorp* Chairman and CEO Rupert Murdoch practice meditation, to name one among 'outrageously successful people' (Gregoire 2013, Internet File), this is surely a sign that meditation and mindfulness have gone mainstream. Whether this is something to be welcome and be excited about is another matter.

I remember reading several years ago an interview in a renowned Buddhist magazine with people in the Pentagon who regularly practiced mindfulness. Wasn't it wonderful, the journalist pondered, that these people meditated? I found this question deeply disturbing, for I thought, as I still do, that one of the effects of a meditation practice that goes beyond solipsistic concentration and self-absorbed relaxation would in this context be a critical examination of the very notion of war, of the *raison d'être* of the military and of a department of defence. Unless of course one seriously thinks it is wonderful to drop bombs mindfully.

Remembering and Forgetting

There have been considerable efforts recently from the mindfulness movement to offer itself to mainstream culture. Highly symbolic among these is Mark Williams' 'explanation' in the Attlee Room in the British parliament of the 'basics of [a] 2400-year-old tradition' boiled down to 'how to control and measure your breath, thoughts and feelings'. This preceded a call by a cross-party group of MPs to bring mindfulness-based meditation into the public sector 'in a bid to improve the nation's mental health, education and criminal justice system'. The financing of these 'secular meditation courses' is done with the intention of 'reduc[ing] misbehaviour and … improve GCSE results', and in prisons, it will help reducing re-offending (Booth 2015). The above example is only one among many. While one cannot deny the expediency and sincerity of this effort, seemingly aimed at putting to good use the millenarian tradition of the Dharma, the emphasis on reducing misbehaviour and re-offending seems perniciously close to connivance with societal injustice and discrimination if not paired with a critique of current educational and 'corrective' systems. I was faced with similar conundrum a while back when I considered applying for Buddhist chaplaincy work in prisons. Although keen to do this at first, I faced serious doubts after talking to people who had worked as chaplains and qualified prison counsellors. They all seemed to agree on the fact that their effectiveness was at best minimal, at worst counterproductive. The reason for the latter was that they were being perceived by inmates as complicit with the institution. Many emphasized that there was too wide a gap between the compassionate ethos they strived to personify and a structural approach based almost exclusively on punishment and retribution. In this context, they argued, meditation and counselling offered mild consolation and even a deluded, complicit

encouragement that it is possible to lead a good life when surrounded all around by a 'bad life'.

This disquieting image, borrowed from Adorno (2005), is not confined to life in prison but can be extended to our alienated, commodified existence in a contemporary world marred by injustice, suffering and exploitation on a grand scale. In this context, achieving personal meditative serenity and private 'integration' is positively trivial. This type of 'mindfulness' is effectively a form of *forgetfulness*.

In his original discourse on mindfulness, the Buddha encouraged us to *remember* a number of things worth remembering (Thera, Internet file). Among these are the certainty of death and the uncertainty of the time of its occurrence. Being mindful in a world besieged by suffering must also mean *not forgetting*—for instance, not forgetting Auschwitz, Amritsar, Sabra and Shatila, or the Armenian genocide. Not forgetting that our fêted capitalist democracies are founded on colonial abuse, wars, wage labour, unemployment, the violent repression of strikes, anti-Semitism and racism (Merleau-Ponty 2000). This type of remembrance/mindfulness 'travels way beyond the narcissism of personal liberation, the self-absorbing dream of individual psychological integration' (Bazzano 2013a, b, c, p. 70). It goes well beyond the fantasy of *ataraxia*, the imperturbability we love to project on the ancient Greeks and the early Buddhists in India. It goes beyond the fairy tale vision of an innocent and wise humanity at the dawn of civilization.

Entering the Stream

Can 'going mainstream' as the mindfulness movement is enthusiastically doing at present be reconciled with 'entering the stream'? The latter expression is used by the Buddha to describe those who enter the path. Among the synonyms of mainstream we find: conventional, middle-of-the-road, (pertaining to) the majority, standard, ordinary and run of the mill. With the exception of ordinary (as in the Zen turn of phrase 'ordinary mind is the way'), all of these illustrate an altogether different stance to that of entering the stream. A *sotāpanna*, one who entered (*āpanna*) the stream (*sota*), is described in the Dhammapada as a person who has gained spontaneous, intuitive understanding of the Dharma. The fruit of stream entry is said by the Buddha to excel 'sole dominion over the earth, going to heaven [and] lordship over all worlds' (Thanissaro 2013). It also excels, I believe, the gaining of greater cultural currency and status, especially when these are bought at a high price.

Entering the stream could be understood as going *against* the stream—against the cultural, social and political conformities that keep us lulled in tranquilizing complicity. It could be understood as the development of greater awareness, hence greater involvement with the river of the world, that river of becoming which cannot be entered twice. Going mainstream can, on the other hand, be apprehended as forfeiting the investigative, counter-traditional elements that have made of the Dharma a vibrant practice across the centuries. These latter aspects are arguably harder to embrace because they require of us the courage to stand apart from the normalizing institutional truths aimed at the manufacturing of pseudo-individuals. I'm obliquely reminded of Kierkegaard (1980, 1987), a thinker and a poet of religion for whom religion is as far removed from pandering to the mainstream as one can imagine. Paradoxically, only one who is able to stand alone and be a true individual can see through the non-substantive nature of individuality itself. This standing alone often opens up as a result of a crisis in life. It is in attempting to alleviate or travel through the suffering produced by a crisis that most of us come across the Buddha's teachings and the yearning to practice. At this crucial juncture, we can embark on a path of curative therapy that lowers our cholesterol and reduces stress before going back to the traffic jam, or back to 'enjoy[ing] the routine and monotony of the assembly-line' (Purser 2015, p. 8). *Or* we can embrace a path that may turn crisis into opportunity for greater freedom, potential breakdown into breakthrough. For an individual who, sustained by faith in the three treasures, has the courage to attempt the latter another important dimension opens up, one that is sorely missing from the stress reduction, low-cholesterol brand of mindfulness: the *communal* element of Dharma practice.

Communal Feeling and Imperceptible Mutual Assistance

'Entering the stream' is a powerful image that strongly reverberates in western psychology and philosophy. The metaphor of the river is common to significant strands of humanistic psychology. We find it in Carl Rogers (1961), who equates therapeutic progress with the active acceptance of self as *process*, a 'flowing river of change' (p. 122). We find it in Rosenzweig (1999) who compares opinions and concepts—the whole array of what makes up a point of view—to a bowl filled with stream water, which the observer takes home to study, thinking all along to be 'studying' the river. This is impossible, Rosenzweig warns; it is futile to try to comprehend the startling poetry of experience; to believe that in gazing at the water in a bowl we are gazing at the river is a delusion. Both Rogers and Rosenzweig refer to the acceptance of the natural fluidity of the self and of existence as something desirable. Indeed, what is variously referred to as incongruence, neurosis and mental distress in all its manifestations can be compared to an unrealistic desire to stand aside from the stream, from an existence that inevitably flows into the ocean of death. This knowledge often brings about a defensive shrinkage of experience and an almost exclusive focus on self-preservation (Bazzano 2016), something already highlighted in the 1930s by a precursor of humanistic psychology Kurt Goldstein (Goldstein 2000).

An important aspect of this newly acquired fluidity, away from the fear of death and the fear of life is *life with others*, life in society and community. Alone among the pioneering psychologists and psychoanalysts of the early years of the twentieth century, Adler placed *Gemeinschaftsgefühl*—communal feeling or social interest—as the very centre of psychological development, as the very yardstick by which mental health can be assessed (Ansbacher and Ansbacher 1964; Bazzano 2005).

There is a link here between this notion of psychological maturity and the propagation of Buddhism in the West. With the latter, it was the third treasure, Sangha, that became the crucial element, and this in spite of the arguably exotic and 'Orientalist' escapism of the early years. If early Indian Buddhism had as its main objective individual liberation from the wheel of birth and death, later traditions underlined the communal dimension. A bodhisattva operates for the benefit of all beings. She is what Nietzsche (1978) called 'a genius of the heart': a person 'from whose touch everyone goes richer' (p. 200). In Buddhist lore, the bodhisattva is said to vow *not* to enter nirvana until the last being on earth achieves freedom from suffering. A bodhisattva is prepared to work endlessly for the welfare of others, whether she finds herself in heavenly or hellish realms. Zen practice is a collective endeavour. The Buddha Way is realized *together*. 'Leaping beyond the confines of … personal enlightenment—Dōgen says,

> [The Buddhas] sit erect beneath the kingly tree of enlightenment, turn simultaneously the great and utterly incomparable Dharma wheel, and expound the ultimate and profound *prajna* free from all human agency … They in their turn enter directly into the way of imperceptible mutual assistance" (Dōgen 2002, p. 12)

Imperceptible mutual assistance: by sitting together in silence, with no utilitarian aim in mind, we sustain one another. We do so without even trying. Initially, we may take our shoes off and enter the meditation hall with an overriding sense of having to solve a personal problem. We may sit on the cushion feeling alone and isolated, wanting our self-generated concentration and absorption to illumine a way out of our own private suffering. Gradually, we find ourselves becoming more sensitive to the presence of others; they too bring their own private burden of hope, grievance and anxiety. At times, these individual burdens seem lifted in the pervasiveness of our common unspoken intent. Being able to sit together in silence, for long period of time, is an ordinary experience—yet also remarkable. *Imperceptible mutual assistance*: these three words admirably encompass all there is to say on

the subject; they are a form of *Dichtung*, a poetic condensation able to convey deep meaning with a minimal amount of words.

The Existential Unconscious

The conventional view among Buddhists and psychotherapists is that at the very heart of our endeavour is an attempt to make the unconscious conscious. This is based on the belief that becoming more aware will reduce the detrimental influence our biases, aversions and cravings exert on us. Making aspects of the unconscious conscious sounds not only legitimate but crucial when dealing with instances such as trauma, dissociative disorder, several kinds of self-destructive behaviour and so forth.

At the same time, to seriously think that all there is to know about human motivation and emotion can be brought to the surface from the depths, scrutinized under the floodlight of consciousness, understood and duly modified is downright naïve. To think that, given time, the unknown will become eventually known is a form of hubris. It is also the dominant view at present. Asserting *imperceptibility*, as Dōgen does, goes counter the current *Zeitgeist*. That the 'real work' should go on unobserved goes against the current 'embarrassing soap opera romancing of consciousness theory in psychoanalysis' (Bollas 2001, p. 236). Similarly, the recognition that Dharma practice is not only subtle but *undetectable* is placed at the opposite end of the mindfulness project as it has been developed and apprehended thus far.

The fact that a great number of psychoanalysts are now beginning to doubt whether the key tenet at the origins of their discipline, i.e. the unconscious, exists at all is a confirmation of our times of 'hypertrophied consciousness' (Bollas 2007, p 81).

This state of affairs is nothing new—nor is the corresponding critique of the unlimited power attributed to consciousness. Already in 1930, in the midst of his lecture tour in the USA, Otto Rank—by far one of the most gifted and creative early psychoanalysts—spoke of an important cultural battle taking place, not so much between different 'schools' of psychology as between two world views (Rank 1996, pp. 221–27). One view, roughly associated with a scientistic stance, could be called *Promethean*. Acting strongly in response to what he perceived as the rigid determinism of mainstream psychoanalysis, Rank saw the attempt to build a 'scientific' psychology as a failure, for psychology is 'necessarily insufficient' (ibid., p. 222) in explaining the mysteries and vagaries of human nature. According to Rank,

> The error lies in the scientific glorification of consciousness, of intellectual knowledge, which even psychoanalysis worships as its highest god – although it calls itself a psychology of the unconscious. But this means only an attempt to rationalize the unconscious and to intellectualize it scientifically (ibid.).

We find a parallel critique in the writings of another great psychoanalyst, Matte-Blanco (1975, 1999). It is possible, Matte-Blanco persuasively argues, for crucial facets of the *repressed* unconscious to enter consciousness 'once [the] prohibition is cancelled' (Matte-Blanco 1999, p. 87). In contrast, however, the unrepressed unconscious 'cannot enter consciousness owing to its own nature' (ibid.). The reasons for this differ widely in Rank and Matte-Blanco. I will not open here what is a complex, tantalizing discussion in trying to articulate their different ways of amending Freud's notion of the unconscious. What can be said is that common to their stance is the assertion that the unrepressed unconscious cannot by definition made conscious.

The above assertion may in turn be adapted as follows: what is 'wholly other' (Otto 1950), the *mysterium tremendum* of existence or *existential unconscious* is simply beyond the grasp of our consciousness. All that a sincere meditation practice can do is make us aware of the *unknowability* of our being in the world; it may help us realize the impossibility of knowing the real.

What may stem from this is *humility*: through meditative practice, I may begin to see a little more clearly into the absurdity of the claims of

the conscious mind and the ego. This sense of humility constitutes the very ground, according to the great Scottish philosopher David Hume, for the cultivation of a healthy scepticism (Hume 2004) that, if applied, would turn our meditation practice from stress reduction into a practice of open inquiry. At first, the latter may well *induce* more stress rather than reduce it (Lopez 2012; Bazzano 2013a, b, c).

In this context, the notion of *awakening*, pervasive in Buddhist teachings, may come to signify the realization of the unfathomable nature of reality rather than the more customary meaning attributed to it, namely the certification of having achieved a zenith in one's psychological and spiritual development, a place where the mysteries of existence have been duly eviscerated and resolved. The latter reading testifies the current predominance afforded to consciousness, which is in turn an aspect of the *Prometheanism* of our times. Many will be familiar with the character of Prometheus who in Greek mythology was one of the Titans. The Titans were primordial, powerful deities that ruled during the legendary pre-Olympian Golden Age. Prometheus is both praised and condemned for stealing fire from the gods. Our contemporary world worships Prometheus; we admire the boldness with which he transgresses a prohibition; his self-assurance reminds us of the pride of the self-made person. The notion of Prometheanism had his fair share of interpreters. Rowan Williams recently gave it a rather unconvincing, and distinctly labour-intensive, Protestant ethic-style twist, for he sees Prometheanism as a fitting symbol for 'wanting to steal divinity from God' as well as a poor substitute for our task of 'labouring at being human' (Williams 2016, p. 15). A much more convincing and in fact truly captivating interpretation—one that would not be out of place within a Zen perspective—is found in the writings of someone whom Williams himself summons as an inspiration, Thomas Merton. For Merton (2003), our contemporary human predicament is Promethean in the sense that we paradoxically want to steal what is freely available. Why? Because we assume that God is keeping something good from us. But the hidden treasure is there for all to see. In the Zen tradition, similarly we learn that there is no need to buy water by the river. 'One of the real reasons—Merton writes—why Prometheus is condemned to be his own prisoner is because he is incapable of understanding the liberality of God … [T]he fire he thinks he has to steal is after all his own fire … But Prometheus, who does not understand liberality since he has none of it himself, refuses the gift of God' (Merton 2003, p. 24).

What is even more crucial here is that Merton sees the Promethean permeating spirituality and theology. His is a confrontational claim that reinforces the suspicion that Prometheanism, far from being solely the province of dogmatic claims associated with science, is in fact so pervasive as to be even at times the driving force behind religion, an area supposedly steeped in humility and untainted by human hubris. The latter makes of spirituality itself another ego ornament or, as I believe may well be the case with the mindfulness movement, a tool for bolstering the domain of the ego instead of a strategy for its eventual decentring. Merton sees Promethean spirituality as 'obsessed with 'mine' and 'thine'—on the distinction between what is mine and what belongs to God' (2003, pp. 24–25). My contention here is that there is a dominant cultural bias, reflected, as I wrote elsewhere, 'in the ways in which Dharma teachings are currently apprehended' (Bazzano 2013a, b, c, p. 72):

> favouring manifest over latent states of consciousness, and relegating the latter to the purgatorial locus of obstructions (āvaraṇa in Sanskrit), afflictions (kleśa) and imprints (vāsanā or, in the language of western psychology, phylogenetic and trans-generational inheritance). This is in many ways parallel to the predominant reading of the unconscious as *Id* in contemporary psychology culture and the concomitant bypassing of its latent creative and healing possibilities. A worrying tendency, arguably gaining ascendancy at present in the field of mental health, would all too happily relegate the unconscious to the museum of outmoded curiosities in the name of 'progress' (ibid.).

Psychotherapy's Contribution

A couple of examples taken from literature may be useful here in trying to illustrate some of the areas of experience psychotherapy can help uncover and in so doing illumine aspects which are often bypassed in orthodox Buddhist practice and especially in mindfulness.

Whenever Flaubert describes an amorous moment in *Madame Bovary*, he shifts his attention to the description of a painting. At first, I thought this had to do with modesty and even prudishness, as in old-fashioned movies when the focus politely drifts to clothes scattered on the floor, on the curtains, or on a glimpse of outdoor scenery through the window. This is in itself more alluring (and sexier, in my view), than depictions of sweaty, emoting film stars reaching climaxes unknown to common mortals.

I suspect something more important is at play here, found not only in Flaubert's writings, but in great realist literature too. One way of describing this is as a shift from the domain of narrative to that of *affect* (Jameson 2014). When this happens, narrative is interrupted and the writing takes flight. In Flaubert's case, with the storyline lapsing into description, and depiction of paintings taking over, we almost partake of the intensity and ineffability of what is coming to pass. Tenderness, passion, the ecstatic and bewildering feelings experienced by the fated heroin—all carry her into a different dimension where straightforward narrative is simply inadequate.

Yet narrative is clearly useful: it takes us from A to B; it relies on cause and effect; it gives us the frame, the subject matter; it provides us with information; it tells us the context, informing us of the functional reference points we need to have in order to follow what is going on. Narrative is important; even though, overused by politicians and commentators, the term itself has nowadays become a cliché.

In therapy, narrative is also known as *content*. Naturally, it is not a bad idea for a therapist to pay attention to content: at the very least, as a sign of respect towards clients, a way of attending to and taking seriously the presenting issues and concerns they bring. At the same time, do I really have to remember the maiden name of my client's cousin's second wife? A supervisor once asked me, in response to my consternation in being unable to remember such details: What if the client's content is fiction? He had a point. There is more to human experience than the story; there is a lot more to life than a sequence of facts and events. This 'something more' is commonly known in therapy as *process*.

Content refers to the 'what' of therapy. It tells us what the client and the therapist talk about. It addresses the nature of the 'problem'; it includes valuable information. It is undoubtedly an essential aspect of the whole endeavour. Yet most practitioners would agree that to stop at content would be incomplete—something else needs to be taken into account.

Going back to Emma Bovary's romantic interludes with her idealized Rodolphe, Flaubert's lapse from narrative to description signals the upsurge of *affect*, a domain of experience not adequately represented by narrative and plot. It may well be that affect is beyond representation; hence, we can only evoke, suggest or, by a leap in style and expression, register a change in perception, the quickening of our heartbeat, a change in body temperature. Recent research and theory (Gregg and Seigworth 2010; Massumi 1995; Bazzano 2013a, b, c) suggest that affect may denote a level of intensity in not measurable until it gets summarily translated (and diluted) as subjective emotion (Massumi 1995).

It appears that the troubadours of the high Middle Ages knew about this, for their love songs were marked by *tempo rubato*, a music signature literally meaning 'stolen tempo' as well as 'stolen *time*', encouraging expressive and rhythmic freedom, speeding up or slowing down according to how the singer was *affected* and impacted. The tempo (as well as time itself) expands or contracts in such moments; that it to say, the experience of rapture escapes a linear sequence.

Affect is then a realm of experience not readily accessible through discourse, facts and reason but one that may be approached by means of a more diffuse awareness. There is in affect a different logic at play, one that does not rely on cause and effect. For instance, the *relational* element, intrinsic in any encounter, is certainly *part* of affect. Client and therapist co-create the counselling environment, ideally through mutual endeavour and cooperation. But affect also comprises of another element, a more *impersonal* dimension which is then inhabited by the relationship. Marcel (1965) similarly spoke of a given that precedes encounter, the mystery of being which for him is blind knowledge, a sort of blindfold knowledge of being inferred in all particular knowledge.

One could say this has to do with the general atmosphere, with the tonality and texture that permeate the therapeutic encounter. Gaining an insight, or at least an inkling of affect, however tentatively, may give us a sense of the general 'feeling' of our meeting with another. And this in turn may provide us with a deeper understanding of process *beyond* the relational, which in turn can become useful to the therapeutic relationship. Openness to affect (another word for openness would be 'objectivity') assists the therapeutic relationship precisely because when we are attuned to affect we are not *enmeshed* in the relational—hence can perceive the relationship more openly or objectively.

Being attuned to affect and going 'beyond' the relational is not as mystical as it may sound at first. A famous passage from the realist literature may help illustrate this. At one point in his novel *The Belly of Paris*, Emile Zola describes the Parisian market of Les Halles, the narrative exploding in a multiplicity of smells, sounds and textures that are truly disorienting and take the reader into an altogether different dimension. The vast quantity of vegetables described in the long passage, then the meats and blood, the dairy products, the feverish variety of seafood and their strange, even monstrous shapes carry the reader into a space that is also wholly independent of narrative. Rather than being provided with an allegory, or a cluster of symbols placed there just in order to represent and explain something else, the fantastic richness of the description—particularly the bewildering variety of cheeses described, a veritable 'symphony of cheeses' their smells and flavours—makes readers dizzy and presents us with an opening into affect—a space that is different from the narrative dimension of cause and effect. For a moment, we almost feel what it was really like to be there in the food market at Les Halles in nineteenth-century Paris.

The other important component of affect is *multiplicity*: many factors and many characters come together to create this moment. The client walking into the room is a complex assemblage of diverse relations and connections, a relational field that would be missed by too narrow a focus on content.

Perhaps if process can be understood as part and parcel of affect, we may gain greater insights into 'what is going on'. Process refers to the *how* of therapy, but it seems to me that this *how* is not entirely covered by the relationship. It *includes* the relationship between therapist and client; it also describes the *flow* of activities and interactions between the two, the full meaning of which is often beyond the reach of conscious thought. What this requires of us therapists is fine attunement and openness—what Diana Voller fittingly calls 'listening to the music behind the words' (personal communication, 2015).

By paying attention to process, I attend to the impersonal as well as the personal and relational elements at play—I listen to the general 'feeling' of the meeting with another while attending to the client's process and to my own process. A simple and direct way to access process is via the body, as we learned from the great phenomenologist Maurice Merleau-Ponty: direct, uncluttered awareness of our sensations, body posture, feelings, and emotions—a way of being-in-the-world that reminds us of our inescapable limitations (as embodied beings) as well as of our potential for openness (Merleau-Ponty 1969, 1983).

There is another important aspect to this, one that is articulated by Jan Hawkins. Relating her experience of counselling people with learning disabilities, she wonders whether clients or patients who do not conform to the conventional parameters of the talking therapies effectively challenge us to reconsider our boundaries as practitioners and even whether we need to focus almost exclusively on *process* (Pearce and Sommerbeck 2014).

References

Adler, A. (2006). In H. Stein (Ed.), *The general system of individual psychology*. Bellingham. WA: Classical Adlerian Translation Project.

Adorno, T. (2005). *Minima Moralia: Reflections on damaged life*. London: Verso.

Ansbacher, H. L. & Ansbacher, R. R. (1956/1964). *The individual psychology of Alfred Adler: A systematic presentation in selection from his writings*. New York: Harper & Row.

Batchelor, S. (2010). *Confession of a Buddhist Atheist New York*. NY: Spiegel & Grau.

Bazzano, M. (2005). To feel with the heart of another: Notes on Adler and Zen Buddhism. In *Adlerian year book* (pp. 42–54).

Bazzano, M. (2013a). Back to the future: From behaviourism and cognitive psychology to motivation and emotion. *Self & Society an International Journal of Humanistic Psychology, 40*, 42–46.

Bazzano, M. (2013). Mindfulness and the good life. In M. Bazzano, (Ed.), *After mindfulness: New perspectives on psychology and meditation* (pp. 61–78). Basingstoke, HANTS: Palgrave MacMillan.

Bazzano, M. (2013c). In praise of stress induction: Mindfulness revisited. *European Journal of Psychotherapy and Counselling, 15*(2), 174–185.

Bazzano, M. (2016). 'Changelings': The self in Nietzsche's psychology. In M. Bazzano & J. Webb (Eds.), *Therapy and the counter-tradition: The edge of philosophy*. Abingdon, Oxon: Routledge.

Bollas, C. (2001). *The Christopher Bollas Reader*. Abingdon, Oxon: Routledge.

Bollas, C. (2007). *The Freudian Moment*. London: Karnac.

Booth, R. (2015). Mindfulness in the mainstream: An old solution to modern problems. *The Guardian*. http://www.theguardian.com/lifeandstyle/2015/oct/20/mindfulness-in-the-mainstream-an-old-solution-to-modern-problems. 20 October, Retrieved November 27, 2015.

Dawkins, R. (2006). *The God delusion*. Ockham, SRY: Black Swan.

Dōgen. (2002). *The heart of Dōgen's Shōbōgenzō* (Translated and annotated by Waddell, N. & Abe, M). Albany, N.Y.: State University of New York.

Eagleton, T. (2006, October 19). Lunging, flailing, mispunching. In *London review of books* (pp. 32–34).

Forbes, D. (2012). Occupy mindfulness. http://beamsandstruts.com/articles/item/982-occupy-mindfulness. Retrieved September 9, 15.

Goldstein, K. (2000). *The organism*. New York, NY: Zone Books.

Gregg, M., & Seigworth, J. (Eds.). (2010). *The affect theory reader*. Durham, NC: Duke University Press.

Gregoire, C. (2013). The daily habits of these outrageously successful people. In *The Huffington Post*. Retrieved November 26, 2015. http://www.huffingtonpost.com/2013/07/05/business-meditation-executives-meditate_n_3528731.html

Harris, S. (2005). In defense of torture. In *The Huffington post*. Retrieved November 26, 2015. http://www.huffingtonpost.com/sam-harris/in-defense-of-torture_b_8993.html

Hume, D. (2004). *A treatise of human nature*. Chatham, KNT: Dover Publications.

Jameson, F. (2014). *The antinomies of realism*. London: Verso.

Kabat-Zinn, J. (2015). The deeper dimensions of mindfulness. In *The mindfulness summit*. http://themindfulnesssummit.com/sessions/jon-kabat-zinn Retrieved November 15, 2015.

Kierkegaard, S. (1980). *The Sickness Unto Death* Princeton: Princeton University Press.

Kierkegaard, S. (1987). *Either/Or* (Vol. II). Princeton: Princeton University Press.

Lopez, D. S. Jr. (2012). *The scientific Buddha: His short and happy life*. New Haven and London: Yale University Press.

Lorde, A. (2007). *Sister outsider: Essays and speeches*. Berkeley, CA: Crossing Press.

Marcel, G. (1965). *Being and having*. New York: Harper & Row.

Massumi, B. (1995). *The autonomy of affect*. http://www.brianmassumi.com/textes/Autonomy%20of%20Affect.PDF. Retrieved April 2, 2012.

Matte-Blanco, I. (1975). *The unconscious as infinite sets: An essay in bi-logic*. London: Duckworth.

Matte-Blanco, I. (1999). *Thinking, feeling, and being: Clinical reflections on the fundamental antinomy of human beings and the world*. London and New York: Routledge.

Merleau-Ponty, M. (1964). *The primacy of perception and other essays Evanston*. ILL: Northwestern University Press.

Merleau-Ponty, M. (2000). *Humanism and terror* (2nd ed.). London: Transaction Publishers.

Merton, T. (2003). *The New Man*. London: Continuum.

Merleau-Ponty, M. (1969). *The visible and the invisible*. Evanston: Northwestern University Press.

Merleau-Ponty, M. (1983). *The structure of behaviour Pittsburgh*. PA: Duquesne University Press.

Nietzsche, F. (1978). *Beyond good and evil: Prelude to a philosophy of the future*. London: Penguin.

Otto, R. (1950). *The idea of the holy: An inquiry into the non-rational factor in the idea of the divine and its relation to the rational*. London: Oxford University Press.

Pearce, P. & Sommerbeck, L. (Eds). (2014). *Person-centred practice at the difficult edge*. Ross-on-Wye, PCCS-Books.

Purser, R. (2015). Confessions of a mind-wandering MBSR student: Remembering social amnesia. *Self & Society—An International Journal of Humanistic Psychology, 43*(1), 6–14.

Rank, O. (1996). In R. Kramer (Ed.), *A psychology of difference: The American lectures* (Selected, edited and introduced by R. Kramer). Princeton. NJ: Princeton University Press.

Rogers, C. R. (1961). *On becoming a person*. Boston, MA: Houghton Mifflin.

Rosenzweig, F. (1999). *Understanding the sick and the healthy: A view of world, man and god*. Cambridge, MA: Harvard University Press.

Schore, A. (2011). *The science of the art of psychotherapy*. New York, NY: W.W. Norton & Co.

Thanissaro, B. (translator). (2013). Lokavagga: Worlds" (Dhp XIII) *Access to insight (Legacy edition)*. http://www.accesstoinsight.org/tipitaka/kn/dhp/dhp.13.than.html. November 30 Retrieved November 26, 2015.

Voller, D. (2013). Welcome to our world. *Self & Society—International Journal of Humanistic Psychology, 40*(2).

Williams, R. (2016, December 12). A summons to writers. *The Guardian Review*, 14–15.

Constructing the Mindful Subject: Reformulating Experience Through Affective–Discursive Practice in Mindfulness-Based Stress Reduction

Steven Stanley and Charlotte Longden

Introduction

Since the late twentieth century, a 'mindfulness movement' has developed, comprising acceptance- and mindfulness-based therapeutic modalities for cultivating 'well-being' and healing stress, anxiety and depression (Davies 2015; Wilson 2014). Practices of mindfulness meditation, which were once confined to the retreat centres of California or the Buddhist monasteries of Southeast Asia, are now becoming a central element of mainstream societies, cultures and institutions, especially in North America and the United Kingdom. For example, recently, the UK *Mindfulness All-Party Parliamentary Group* campaigned for the UK to become a collection of 'mindful' nations by integrating courses in mindfulness into public policy domains of health care, education and criminal justice.

The mindfulness 'movement' in part involves taking the practices of mindfulness meditation, which were previously part of modernised Buddhist traditions, out of institutional contexts such as medical clinics and psychotherapy consulting rooms and distributing them throughout the society. We now find that mindfulness meditation is being taught and practiced in a seemingly ever-expanding array of unexpected contexts: school classrooms, military

barracks, corporate boardrooms, stock trading floors, prison cells and government offices. Standardised courses of MBSR and *Mindfulness-Based Cognitive Therapy* (MBCT) are increasingly being offered, having garnered attention as a result of evolving scientific evidence of their clinical effectiveness.

In the encounter between American and European cultures and Buddhist traditions, the 'psy-sciences' (Psychology, psychiatry, cognitive science) tend to take the leading role in framing mindfulness meditation, with little evidence of mutually constructive dialogue with the interpretative social sciences or humanities (for an exception, see Williams and Kabat-Zinn 2013). Much mindfulness research is positivist in the sense of comprising hypothetico-deductive studies aiming to predict and control human behaviour through psychometrics, experimentation and randomised clinical trials (RCTs). Researchers often attempt to 'model' the cognitive, affective and behavioural processes explaining 'mechanisms of change' that produce the therapeutic effects of mindfulness meditation. In such research, 'mindfulness' is often conceptualised as an internal state or trait within 'the mind', with behavioural and neural correlates; it is very often taken-for-granted by psychologists and cognitive scientists as an inner and private psychological entity-state or trait-existing within the mind/brain of the individual person (Davidson and Dimidjian 2015).

In this chapter, we take an alternative view of mindfulness, which is informed by interpretative social science and humanities scholarship.

S. Stanley (✉) · C. Longden
School of Social Sciences, Cardiff University, Cardiff, UK
e-mail: StanleyS1@Cardiff.ac.uk

© Springer International Publishing Switzerland 2016
R.E. Purser et al. (eds.), *Handbook of Mindfulness*,
Mindfulness in Behavioral Health, DOI 10.1007/978-3-319-44019-4_20

The Storied World of Mind-Body Medicine

In addition to its therapeutic uses, following Wilson (2014), we would argue mindfulness is also a social/civic/moral movement and therefore should become an object of interpretative research. Such research has the potential to make a vital contribution, theoretically and empirically, to the current professional, public and popular debates about the mindfulness movement: by exposing the social, cultural and historical contingency of practices and insights about 'the mind'. In most professional, scientific and Buddhist literatures, 'the mind' tends to be understood as 'universal': trans-social, trans-cultural and trans-historical. The growth in popularity of MBSR can be partly attributed to the recent historical re-interpretation of Buddhism as a universal 'scientific religion' compatible with the principles of rationalism, evolutionary biology, materialism and psychotherapy (McMahan 2008). Harrington (2008), in her cultural historical study, illustrates how MBSR can be contextualised as a form of 'mind-body' medicine: 'both a mainstream/professional and an alternative/popular body of knowledge and practice' (p. 247). According to a mind-body medicine approach, there is:

> more to physical illness than can be seen just in the body; and more to healing than can be found in pills and shots. Mind matters too: how one thinks, how one feels, what kind of personality or character one has or cultivates (ibid., p. 18).

According to Harrington, mind-body medicine utilises 'a diverse set of cultural resources to make sense of [patient] experiences. Many of these … have strong historical origins in religion' (p. 245). Harrington names MBSR as an 'eastward journeys' narrative tradition which, following Said (1978), relies upon orientalist discourses of recruiting 'eastern wisdom' for modern times. This 'exoticism' narrative contrasts with community building and social support networks; instead, 'it seeks the source of healing not so much in the caring communities we have lost but

in the healing practices of ancient Eastern cultures we have never known' (p. 29).

An often-overlooked historical influence upon MBSR is 'Buddhist modernism' or 'protestant Buddhism', a recent and largely individualised revisioning of Buddhist thought and practice, which emphasises 'inner truths' discovered personally during meditation and resulting in practical 'this-worldly' benefits (Sharf 1995). This modernised Buddhism is the product of complex inter-cultural exchanges between Europe and Southeast Asia, especially in the context of late nineteenth and early twentieth century British colonialism (e.g. Braun 2013). By reasserting the influence of history and culture on mindfulness, we are building upon Harrington's (2008) interest in the meanings of the 'storied world' of mind-body medicine.

Social science researchers of the mindfulness movement have developed this style of approach through empirical research along with critical discussions of the 'psychologising', 'medicalising', 'recontextualising', 'therapising' and 'secularising' of Buddhist meditation, whereby it is frequently presented as a tool with which the individual can better adapt to social circumstances (e.g. Purser and Milillo 2015). Mindfulness features prominently in recent histories of stress research (Jackson 2013). For example, Barker (2014) has recently conducted a critical discourse analysis of popular literature and guided mindfulness meditation audio recordings by Jon Kabat-Zinn. She illustrates how the discourse of mindfulness, while locating the sources of stress in the conditions of modern life, simultaneously asserts the impact of the 'mind' on the 'body', thus making the individual responsible for self-healing. Barker situates MBSR within the context of current trends in 'medicalisation' and argues that mindfulness meditation has been turned into a 'do-it-yourself' practice of 'medicalising every moment': indefinitely extending the realm of health and illness to all areas of life. In his popular writings, Kabat-Zinn (e.g. 1994) diagnoses the citizens of Western democratic societies as suffering from Attention Deficit

Hyperactivity Disorder—especially inattentiveness to the present moment—for which he prescribes mindfulness meditation as a remedy. While Barker (2014), in her textual analysis, suggests mindfulness has become medicalised through MBSR, she acknowledges that she neglects to study the practical conduct of MBSR courses and experiences of participants.

Kabat-Zinn (1994) defines mindfulness as a conscious awareness, naturally and universally occurring within human beings, which can be cultivated through Buddhist-influenced meditation practices. For Kabat-Zinn (1994) mindfulness arises when an individual pays attention in a particular way: on purpose, in the present moment, and non-judgementally to experience, moment-by-moment. This practice of present moment attention is presented as a technique to interrupt the psycho-physiological 'stress response cycle' and thus to heal the individual from within. MBSR programmes generally consist of one 2–2.5 h session per week for eight consecutive weeks. Participants are taught a combination of mindfulness meditation, mindful hatha yoga and psycho-educational content on stress. The first half of the course focuses on training participants to bring attention to their internal experience, while the second half emphasises the application of mindfulness practice to everyday life and the challenges this may bring (Crane 2009).

Barker's critical analysis of the discourse of Kabat-Zinn's popular books and audio recordings, along with recent accounts of changing meanings of mindfulness meditation over time (Sun 2014; Stanley 2013), situate mindfulness as a historically contingent psychological category (Danziger 1997). Indeed, it is possible to frame the turn to mindfulness as a therapeutic culture in the context of the late modern project of the reflexive self (Giddens 1991) and the rise of a 'psy-complex', comprising psychological styles of 'governmentality' (Rose 1998). This would support the notion of 'psychologisation' as including an increasing expansion of therapy, self-care and self-help across modern societies in which state governance and self-governance are combined (Madsen 2014).

It is upon this cultural and historical research that we seek to build our study mindfulness in

action, that is, through analysis of the situated conduct of practices taking place within courses of MBSR. There is a family resemblance between our approach and ethnographic research in anthropology and sociology of religion, for example Cook's (2010) study of meditation in Thai monastic life, Preston's (1988) ethnography of the San Francisco Zen centre, and Pagis' (2009) ethnographic research on *vipassana* (insight) meditation courses which cultivate 'embodied' and 'discursive' self-reflexivity.

The main distinctive quality of the present research is that it concerns interactional analysis of the conduct of MBSR courses. This chapter furthers our previous research, which took a discursive and conversation analytic approach towards the analysis of MBSR and MBCT (Crane et al. 2014; Stanley and Crane 2016). Previously, we investigated the pedagogy of mindfulness teaching in terms of its interactional dimensions. We develop this work here by grounding claims about the 'psychologisation' of modernised Buddhist practices (Cohen 2010) within the actual conduct of MBSR courses.

Affective–Discursive Practices and Technologies of the Self

We present critical psychological research comprising qualitative investigation of mindfulness as it is constructed within social interaction. Mindfulness is conceptualised as existing between people in their encounters: a historically contingent psychological category comprising inter-subjective practices. It is situated and embodied in specific social situations, in the present case, within MBSR course interaction. In contrast to positivist research, which tends to construct artificial situations such as experiments to study mindfulness, instead we recorded, transcribed and analysed the audio of interaction taking place within MBSR classes, according to principles of discourse and conversation analysis.

We adopt an eclectic style of analysis, which draws upon: Foucaldian post-structuralism; conversation analysis; discursive, rhetorical and dialogic psychologies; and recent work on affect.

We find the concept of 'affective–discursive practice' productive for synthesising these contrasting traditions. Wetherell (2013) defines affective–discursive practice as inter-subjective occasions of 'embodied meaning-making' which assemble historical, cultural, somatic and psychological phenomena. It therefore becomes possible to analyse 'embodied positioning' and 'mindful bodies' (Sheets-Johnstone 2015) which are socially and culturally produced.

In analysing affective–discursive practices, we bring attention to pattern, power and context. In this understanding, following the French philosopher–historian Michel Foucault, power relations are productive of knowledge, and discourse is productive of subject positions. For example, a discourse of mindfulness contains objects (meditation cushions, bells) and subjects (the mindfulness teacher, the mindful individual) within it. This perspective is post-humanist in the sense of suspending the principle of a universal self. Instead, we analyse the functions of historically specific power/knowledge relations and practices of 'subjectification' (Hall 2001).

Foucault's approach has been criticised for paying insufficient attention to agency, experience and freedom. Towards the end of his life, he pursued his interest in how 'a human being turns him- or her-self into a subject' by turning to first- and second-century Greco-Roman philosophy and fourth- and fifth-century Christian theology (Martin et al. 1988). The concept of technology of the self, used by Foucault, helps us to understand mindfulness meditation as a method for transforming the self:

> Technologies of the self … permit individuals to effect by their own means, or with the help of others, a certain number of operations on their own bodies and souls, thoughts, conduct, and way of being, so as to transform themselves in order to attain a certain state of happiness, purity, wisdom, perfection, or immortality (ibid., p. 18).

Foucault's technologies of the self can be applied to our current object of study, not least because mindfulness is often articulated as a disciplined, ethicised, 'care of the self' (*askesis*). In 1978, Foucault briefly visited a Zen temple in Kyoto (Japan). Attempting Zen meditation, he

commented that 'if I have been able to feel something through the body's posture in Zen meditation, namely the correct position of the body, then that something has been *new relationships which can exist between the mind and the body and, moreover, new relationships between the body and the external world*' (Carrette 1999, pp. 112–113; emphasis added). While Zen meditation, *zazen*, is not identical with mindfulness meditation, both are body/mind training regimes (Preston 1988). Influenced by this orientation, we can ask 'What are the epistemic regimes of truth involved in learning mindfulness meditation?'

Foucault's genealogical method involves analysis of texts and documents and therefore we follow Wetherell (1998) by integrating post-structuralism with Conversation Analysis (CA). CA is the study of talk-in-interaction. The analyst reveals how interaction is organised through repeated patterns and routines. In the present project, we analyse the sequential organisation of 'inquiry' in MBSR classes, conceptualised as 'institutional' interaction. MBSR is a hybrid discourse. The interactional dynamics within MBSR are not easily characterised as traditional doctor–patient or therapist–client interaction. They better resemble pedagogic teacher–student interaction. However, the term favoured by professional mindfulness teachers and trainers for students/clients/patients of MBSR courses is 'participants', which reflects the principle of 'participatory medicine' in which the client or patient is encouraged to take at least partial responsibility for their healing. MBSR therefore comprises a complex amalgam of therapeutic, clinical and psycho-educational features. The 'therapeutic' aspect of MBSR has been disputed, and it has been claimed these courses involve training in 'universal dharma' rather than (group) psychotherapy (Kabat-Zinn 2011).

From ethnomethodology—the study of people's everyday sense-making practices—we use the concept of 'action orientation' (Heritage 1984). To understand the meaning of a word or utterance, we look at its social functions. This resonates with discursive, rhetorical and dialogic approaches to psychology. We become interested in argumentative patterns of accountability and the negotiation of 'mind/world' relations in talk (Edwards 1997) while allowing for the existence of 'inner' lives that are inter-subjectively produced.

However, people are not free to construct themselves as they please. Ideological dilemmas comprise the contrary themes of common sense (e.g. authority/equality), which enable thought and discussion, while at the same time preserving the societal status quo (Billig et al. 1988).

Conversation Analysis

In the terms of conversation analysis, a course in MBSR can be considered 'institutional interaction'. This does not mean that mindfulness courses necessarily need to take place within social institutions such as schools or hospitals. Rather, Drew and Heritage (1992), following Levin, suggest institutional talk can be characterised as: (i) goal-oriented in institutionally relevant ways; (ii) comprising special and particular constraints on allowable contributions; (iii) comprising specific inferential frameworks and procedures. These features, they argue, are to be found within talk-in-interaction itself as participants' *orientations* to institutional context.

We have chosen to analyse sequences taken from two different courses of MBSR, in which the mindfulness teacher explores with the class difficulties in the practice of mindfulness meditation. In Extract One, the difficulties are brought to class by participants; in Extract Two, the difficulties are initiated by the mindfulness teacher. The audio has been transcribed using a simplified version of Gail Jefferson's transcription notation for CA (see Appendix).

Extract One is taken from *Healing and the Mind* (Grubin 1993), a five-part documentary programme surveying mind-body healing practices in the USA and China, broadcast on Public Broadcasting Service in 1993. Presented by television journalist and political commentator Bill Moyers (BM), the title illustrates the programme's concern with mind-body medicine. The sequence we will analyse is taken from the episode 'Healing From Within' which focuses on

the work of Jon Kabat-Zinn during the early 1990s at the University of Massachusetts Medical School. Harrington (2008) suggests this documentary '[m]ore than anything else … served to focus public attention on this new mix-and-match vision of mind-body healing' (p. 244), with the initial broadcast being viewed by more than twenty-four million Americans. In the companion book of the series, Moyers (1993) writes that 'Two important questions shaped the series: How do thoughts and feelings influence health? How is healing related to the mind?' (p. xiii).

This documentary not only illustrates and promotes mindfulness meditation, but it also works as an instructional tool to those taking courses in mindfulness meditation. It is often shown to participants before or during courses in MBSR to demonstrate aspects of mindfulness, along with its benefits. However, this is an edited documentary created for entertainment and promotional purposes and therefore it does not necessarily illustrate what actually happens in MBSR classes. Nevertheless, the particular sequence we will analyse demonstrates a common interactional patterning, which is found within other MBSR courses we have analysed, and which will concern us in the latter half of this paper.

Extract One, taken from week two of an MBSR course, is a transcribed example of 'inquiry' concerning the 'home practice' guided meditation exercises conducted by the participants in the week since the previous class.

'Inquiry' is a teacher-led interactional sequence, usually immediately following a guided meditation practice, and aiming to fulfil specific institutional objectives (Crane et al. 2014). In terms of its claimed historical precedents, Kabat-Zinn (2011) cites Zen koan-based 'Dharma combat' exchanges between students as an influence upon 'inquiry' in MBSR:

> This form contributed in part to the element of interactive moment-by-moment exchanges in the classroom between teacher and participant in which they explore together in great and sometimes challenging detail direct first-person experience of the practice and its manifestations in everyday life. This salient feature of MBSR and other mindfulness-based interventions has come to be called 'inquiry' or dialogue (p. 289).

Chaskalson and Hadley (2015) suggest that:

> such inquiry is seen as an important way of helping
> MBSR course participants to develop a greater
> curiosity about their inner experience. By being
> invited to describe their experience of any practice
> soon after doing it, a deeper quality of reflection is
> encouraged. Participants may also be invited to
> turn toward what is difficult for them in the
> moment. That can help people to see that what
> they ordinarily avoid may in fact be tolerable – or
> even positively interesting (p. 51).

Kabat-Zinn introduces the edited segment of
the documentary by suggesting he welcomes
discussions of difficulty: 'it's always interesting to
hear what comes up for them'. 'Always' is an
extreme case formulation, which is used to present
the speaker as frequently and consistently
open-minded in encouraging discussion of 'what
comes up' for the participants (Pomerantz 1986).
In the interactional exchange that follows, what
could be seen as 'personal' difficulties with the
practice are made to be not so personal, but rather
are made to be universal or natural. This making of
universality is practically achieved in part through
the teaching of a psychological 'non-expert'
expertise, which comprise the psycho-educational
curriculum of this MBSR course.

Extract 1: *Healing and the Mind* (1993): *Healing From Within*: Week 2 of course in MBSR

```
1    JKZ        after each class they're required to go home and
2    ((voice    practice for forty five minutes every day (1.9) it's
3    over))      always interesting to hear what comes up for them in
4               this first week
```

Figure 1.1: Participant 2

```
5    P1         I can't be a[ware of my breath
6    P2                    [I dunno
7    JKZ        you can't be aware of your breath
8    P1         no
```

Figure 1.2: Participant 1 and Participant 2

```
9    P2         I found it difficult te- sh- keep my thoughts just on
10              (.) my breath
11   JKZ        mmhm (.) the mind was full of
12              thou[ghts coming and going all the time
13   P2             [my mind was alive right (.)
14              [absolutely
15   BM         [I'm with- I'm with him [wherever you were
16   P1                                 [((laughter))
17              [((laughter))
18   BM         [WHEREVER YOU WERE I WAS WITH YOU
19   P1         ((laughter))
```

Figure 1.3: Participant 2 and Bill Moyers

```
20  BM          my mind was a monkey it was [leaping from tree to
21  JKZ                                      [yes
22  BM          =tree I even think it leaped into his tree
23              [while we were meditating
24  P2 &        [((laughter))
25  class
26  JKZ         well lemme- lemme ask you both a question (.) er is
27              this news to you or were you aware of it before
28  P2          I was never aware of how much (.) I think while I'm
29              not (.) talking or [driving or walking or
30  Class                          [((laughter))
31  JKZ         this is an important observation because very often
32              we go through life on a kind of ((turns to address
33              class)) automatic pilot basis and we kind of aren't
34              even awa:re that there're all sorts of thoughts going
35              on they're not subliminal they're just the:re
36              slightly beneath the surface of awareness but they
37              actually drive many of our actions and behaviors
```

Figure 1.4: Jon Kabat-Zinn

Extract 2: Week Five of Course in MBSR

```
1   Teacher      anyone else? Anything you noticed (2.3) particularly
2                around (0.8) if there's anything around difficulty I
3                was inviting you to come towards the difficulty
4                (3.7) might have been a bit of difficulty trying to
5                do that, that can be so
6                (6.2)
7   Participant  I noticed with erm I-I was thinking of a particular
8                problem tht umm, we have with one of the dogs
9                they're puppies
10  Teacher      [oh
11  Participant  [and hhuh it's a training issue, and I immediately
12               felt tension acroHHss heHHHre
13  Teacher      oh that's [rees-
14  Participant            [which I hadn't noticed before so it was
15               nice to act[ually sort of
16  Teacher                 [ahhhh
17  Participant  experiment about (0.5) following that through and
18               breathing into [it which is
19  Teacher                      [so that that sounds like quite a a-
20               f-familiar (0.6) thing the puppies
21  Participant  well-
```

```
22  Teacher      the dogs [but then
23  Participant          [it's it's a tension huh huh huh
24  Teacher      te- you noticed so [you noticed that in the=
25  Participant                     [yeah
26               =I noticed that yeah
27  Teacher      what did- what happened as you noticed that?
28  Participant  uhmm
29  Teacher      what happened next
30               (0.6)
31  Participant  well I-k-I I (0.5) followed through what [we've
32  Teacher                                             [yeah
33  Participant  been doing which is kind of just acknowledge that
34               that's there an it-
35  Teacher      yep
36  Participant  and it- I relaxed [and
37  Teacher                        [yeah
38  Participant  and it wasn't y'know there wasn't any tension (0.6)
39               [and but=
40  Teacher      [so-
41  Participant  =my mind wandered off on to something else after
42               [that so
43  Teacher      [oh did it- okay s-but that first bit was that you
44               noticed that thought um the dog and that you noticed
45               the puppy the te-tension
46  Participant  mm
47  Teacher      and then brought your breathing to that and actually
48               then the tension wasn't there so something shifted
49               for [you
50  Participant      [yes
51  Teacher      that's quite int[eresting isn't it=
52  Participant                  [yeah
53  Teacher      =yeah so for you in that practice as you brought the
54               attention to it actually it shifted, and then your
55               mind went off to something different
56               (0.3)
57  Participant  yes I had to put it back [then
58  Teacher                               [yeah
59  Participant  to the breathing again
60  Teacher      yeah, [yeah
61  Participant        [um
62  Teacher      that's really interesting thank you
```

MBSR as 'Institutional Talk': Inquiry, Third Turns and Formulations

The initiation of question–answer sequences by the teacher and the provision of answers by students are followed by 'third turns' on the part of the teacher. These turns are common to institutional environments and not usually found in everyday interaction. In support of Drew and Heritage (1992), we found talk within sequences of inquiry to be 'predominantly characterized by question-answer sequences in which the professionals largely ask the questions and the lay "clients" respond with answers' (p. 54).

Three-turn sequences are one of the most familiar organisations of pedagogical discourse (Lee 2007). Often referred to as Initiation–Response–Evaluation (IRE) sequences, they take place when the teacher poses a question, listens to a response from a member of the group and

feeds back with what may be viewed, on a superficial level, as a summary or confirmation of what they have heard. However, analysis of the specific qualities of this feedback suggests it is doing much more than confirming understanding; it provides a repositioned version of events whereby the original account is subtly transformed. Formulations have been defined as a method of continuously and actively listening to, evaluating and interpreting talk, while simultaneously creating an opportunity for the speaker to omit certain parts of an account and to emphasise other parts, to advance their own institutional interests (Antaki 2008). Indeed, the fact that formulations are rarely found in ordinary talk suggests they are brought into service within institutional settings for a specific purpose, especially psychotherapy (Peräkylä 2013). Formulations in this third-turn position can therefore be particularly influential in coordinating the flow of interaction.

Towards the end of the extract, Kabat-Zinn occupies a teacher subject position and asks a question (26–27), which is followed by a participant answer (28–29). The interaction then returns to the teacher who provides a formulation, simultaneously reformulating the participants' contribution and summing up what has been learned (31–37). Overwhelmingly it is the teacher who offers both the questions and the third-turn formulations of participant contributions. These practices together evidence the institutional character of the talk; the interactional asymmetry of a mindfulness course, along with its pedagogic quality.

Formulations are the means by which mindfulness teachers transform the accounts given by participants in order that they fit the pedagogic and institutional aims of the course. However, formulations should be distinguished from overt or instructional intervention, as they are integrated into sequential interaction between teacher and participant as a way of managing progress. This means that while they are teacher-led, they are largely collaborative in nature as they depend upon the construction of events provided by participants themselves in the first instance. 'Professional control here manifests itself as a pattern of sequences through which clients may find themselves being led' (Drew and Heritage 1992, p. 45). Indeed, as we shall see, the formulation often contains echoes of the prior turn given by participants themselves (Heritage and Watson 1979).

Turning Difficulties into Discoveries

By analysing in detail the interactional progression of inquiry sequences, we can see how subjectivity is gradually and subtly reformed. The mindfulness practice of 'approaching difficulty' is practically demonstrated (or embodied) by the teacher in the way they respond to declarations of difficulty (Crane 2009). In Extract One, Participant One declares her inability to be aware of her breath, which is echoed word-for-word by Kabat-Zinn, and followed by an affirming response by the participant ('no'). Participant Two then offers a response that is affiliative with

Participant One, thus re-asserting the difficulty of the practice. He concurs with the first speaker, while also displaying his understanding of the purpose of the 'home practice' 45-min guided mindfulness meditation exercise the class were required to do each day during the interim week: 'I found it difficult te- sh- keep *my* thoughts just on (.) *my* breath' (emphasis added; lines 9–10). Through this utterance, the participant suggests he also experienced difficulties with the practice, thus implying shared experience, while at the same time personalising his thoughts and his breath. In this sense affirms Participant One's claimed inability to be aware of her breath while also suggesting that he did not find the practice of breath awareness impossible ('I found it difficult').

After acknowledging this turn, Kabat-Zinn gives another affiliative response which also reformulates Participant Two's turn to '*the* mind was full of thoughts coming and going all the time' (emphasis added). While this is delivered as a repetition or echo of what the participant has said, the utterance functions as a reformulation as it makes several transformations. A container metaphor is used for 'the mind' in which 'the mind' is depicted as a container for thoughts; an implication is that 'the mind' was 'full' in the sense of being crowded out by thoughts entering the mental space. It also reframes the experience using language recognisable of a discourse of mindfulness: 'thoughts coming and going all the time' (i.e. changing). By using the extreme case formulations 'full' and 'all the time', Kabat-Zinn orients to the problematic yet implied to be understandable status of the mind being 'full of thoughts', but at the same time, there is a subtle implication that the mind is not always full ('coming and going'). This reformulation shifts Participant Two's 'thinking about the breath' to the mind being a container, which can be 'full of thoughts', and hence made the subject of awareness. The teacher thereby de-personalises 'my' mind by using the definite article 'the' mind, turning it into a universal container or space.

Following this turn, Participant Two offers an affirmative response, which subtly reformulates Kabat-Zinn's formulation: '*my* mind was alive right (.) absolutely' (emphasis added). The participants' 'my' mind re-personalises his mind and makes it a personal possession which is active. Through this utterance, the participant gives his mind agency and activity. At the same time, the shift from what earlier were 'thoughts' to 'mind' reorients the language closer to the teacher's version.

Following this participant's personalising of his experience, which implies it might have been an idiosyncratic experience unique to himself alone, Bill Moyers offers an affiliative response which aligns him with Participant Two: 'I'm with him wherever you were... WHEREVER YOU WERE I WAS WITH YOU'. Moyers uses metaphorical language to depict a visual scene and uses the metaphor of his mind as a monkey: '*my* mind was a monkey it was leaping from tree to tree' (emphasis added). This utterance continues to personalise '*my* mind' as the participant had done and yet constructs this experience as mutual and shared, at least between two people, rather than being singularly personal. As with Participant Two, Moyer's mind is constructed as having agency and being separate to himself; described as a monkey, his mind is implied to be out of control and undisciplined. The 'monkey mind' is constructed as a problematic barrier or difficulty to the meditation practice. Kabat-Zinn says 'yes' (line 23) thus affirming what is being said and implicitly validating this point as being part of the learning of an MBSR course ('monkey mind' is a common metaphor used within mindfulness and Buddhist discourse about meditation).

At this point, in this televised MBSR course descriptions are given of the activities and qualities of 'my' and 'the' mind. 'My mind' or 'the mind' is made to be an entity which is separate to, or distinct from, the person. The possessive 'my' makes the mind a personal possession, as well as an entity, which can perform actions of its own such as 'leaping' from tree to tree. By contrast, at this moment, Kabat-Zinn also constructs 'the mind' as a container filled with thoughts. One function of these constructions is that 'mind' is made to be the object of discussion and retrospective investigation, as either a container or agent distinct from the self ('my' + 'mind';

'the' + 'mind'). This objectifying of the mind allows it to become the object of careful monitoring and self-governance by the subject themselves, as we shall explore below.

Discovering 'the Wandering Mind': Historical and Interactional Dimensions

Through inter-subjective affective–discursive practices, which comprise 'inquiry' within an MBSR class, the 'problem' of what psychologists and mindfulness teachers call 'mind wandering' (or 'monkey mind') is turned into a 'discovery' of mindfulness meditation practice. That is, the 'problem' is reframed by the teacher and turned into an 'important observation'—a discovery about the nature of 'the mind', as established through mindfulness and mindfulness meditation practice. This illustrates one of the main features of 'inquiry' sequences in MBSR as practical training in a specific psychological 'insight'.

One of the interactional tasks of mindfulness courses appears to be establishing 'the mind' as a universal human psychological/biological property, that is, rather than as a necessarily socially, culturally and historically contingent phenomena. To achieve this goal, personalised constructions of mind ('my mind') are consistently reformulated by the teacher to generate a universality of experience, shared by the group (including the teacher himself; 'we' lines 32–33; 'our', line 37), and in which 'the mind' displays particular generic characteristics (such as 'wandering').

This notion of 'the mind' as a purportedly universal human property is a feature commonly shared by psychological (e.g. humanist, cognitive-behaviourist) and philosophical (e.g. empiricist) traditions (Billig 2008; Dryden and Still 2006). Similarly, the notion that thoughts, which are 'slightly beneath the surface of awareness … drive many of our actions and behaviours', is a basic principle of rationalist and idealist/mentalist philosophy and cognitive-behavioural psychology. It assumes a functionalist theory of mental objects whereby thoughts, understood as mental mechanisms, can function (sub-, pre- or unconsciously) to drive actions and behaviours. This principle was also a key component of the therapeutic culture of the 'mind cure movement' and the 'New Thought' religion in nineteenth century American culture (Rakow 2013).

Kabat-Zinn suggests that while thoughts drive 'our' behaviours, 'we' might not be aware that they are doing this. He thus articulates the learning point for week two of an MBSR course: automatic pilot. Participants are being taught how to be aware of 'the mind' that wanders (Crane et al. 2014). A shift in language use occurs, away from the participant's confession of difficulty in keeping his 'thoughts' on his breath (19–20) (implying that thinking about the breath is the aim of the practice) to what is later revised as an enduring and complete unawareness of his wandering mind ('I wasn't aware …', lines 28–29), at least at this moment in the interaction.

Thus, the activities of 'the' mind are made not only universal rather than personal, but are also made into an object of (at least potentially) visible reflection and investigation and therefore subject to self-surveillance and control. As he says 'automatic pilot', Kabat-Zinn turns to the on-looking (and overhearing) audience of participants, in order to emphasise his core teaching point for this session (Fig. 1.4).

Coming Towards Difficulty

Extract Two is taken from a course in MBSR for the general public, taught by a female mindfulness teacher from the *Centre for Mindfulness Research and Practice* (School of Psychology, Bangor University, Wales), and which took place in the UK. The study was reviewed and approved by the university's research ethics and governance committee. The data are taken from week five of the course in which 23 participants (17 female; six male) took part. In this extract, the teacher explicitly asks participants 'anything you noticed' particularly around 'coming towards difficulty'. The theme at this point in the curriculum is 'acceptance' and 'allowing' or 'letting-be', and it is the first point at which a difficulty is purposefully introduced within a practice to facilitate an

exploration of the effect it may have on body and mind (Crane 2009). Analysing this extract in detail allows us to appreciate some of the careful work involved in constructing and negotiating the subjectivity of the mindful individual.

Formulation as Transformation

The teacher rephrases her question several times to accommodate various levels of reflexivity (lines 1–5). First she asks whether there was anything participants noticed, and after a relatively lengthy pause of 2.3 s, she narrows this line of inquiry by bringing the focus towards difficulty in particular. Another slightly longer pause follows (3.7 s) and the question is once again adapted, this time to address potential difficulty in attempting to approach difficulty itself. This forms the initiation part of an IRE sequence. In repeatedly reformulating her own question, the teacher herself orients to the difficulty in talking about difficulty, and this is exemplified in the aforementioned pauses. Indeed, a full 6.2 s of silence occurs before a response is elicited from the group, suggesting some level of reluctance to be the first to self-select to speak.

In response to the teacher's question, a participant presents a retrospective account of an internal negotiation and management of a particular 'problem' (line 8) using mindfulness-based practices. The teacher feeds back with a series of questions and formulations. A formulation projects an expectation of either agreement or denial; a response from the participant is required at the next turn, influencing the sequential organisation of the interaction and ensuring the participant remains central throughout.

The teacher remains in control by using questions such as 'what happened as you noticed that?' (line 27) to direct the interaction, such that the participant is continuously engaged. Early in the interaction, the teacher's initial formulation of 'familiarity' is rejected. The interruption, beginning at line 21 with a dispreferred response ('well-'), suggests a repair (Schegloff et al. 1977) whereby the participant begins to correct the teacher, repeating that 'it's a tension' for him (line 23), changing the trajectory of the talk. In response, the teacher then leads the participant through a collaborative interaction, where questions function as a form of 'non-authoritarian' authority (explored below), which produces a more institutionalised version of events.

While the teacher does not correct the participant, as may be the case in ordinary teacher–student interaction, she holds the right to use particular turns at talk to steer the direction of the sequence towards more desirable orientations (Gardner 2013). The original account is thereby subtly reorganised into a narrative of sequential events; descriptions of the practice are constructed as a set of step-by-step actions for using mindfulness as a tool to move through difficulty (lines 43–49). Thus, the initial account has undergone some degree of transformation through the interaction; a revised chronology has been attributed to it. At line 50, this chronological formulation is accepted by the participant and acknowledged by the teacher as valuable, 'that's quite interesting isn't it', exemplifying this cooperative process.

Embodying 'Troubles' Receptivity

The participant has been explicitly asked to talk about his experience of 'coming towards difficulty', and it is arguable that he mirrors the teacher's orientation towards the unease of addressing this topic by laughing during his response. According to Jefferson (1984), laughter particles can occur in what she calls 'troubles talk'. While discussing something problematic in some way, the troubles-teller produces an utterance and then laughs, sometimes as they say a particular word, while the troubles-recipient produces a recognisably serious response instead of joining in with this laughter. This phenomenon makes visible the idea that laughter does not always occur in talk because something is funny, but may perform a variety of different tasks. It is interesting that the teacher's initial question encourages an exploration of difficulty (lines 1–6) and this sense of difficulty is indeed demonstrated through laughter particles in the next turn response (lines 11–12). The participant refers to a feeling of 'tension' in relation to the

'problem' he was remembering, and laughter can be heard both directly before this tension is brought up at line 11, and while describing where it was felt in the body at line 12. It is here at line 12 in particular that the placement of the laughter becomes noteworthy: 'I immediately felt tension acroHHss heHHre'. It is fair to assume that at this moment, the participant is simultaneously indicating through gesture where in the body this tension was felt, constituting an embodied orientation. However, without a video record we cannot be sure. Where the laughter occurs could be considered a methodic device in its own right, as the laughter results in the distortion of the words themselves. Arguably, as the participant says 'acroHHss heHHre', the laughter accounts for a level of distortion, making the words difficult to hear yet potentially masking any implication that he may in fact be reluctant to say them (Jefferson 1985).

Following this sequence, the teacher attempts a formulation ('so … that sounds like … quite a … familiar thing … the puppies', line 19), which is rejected by the participant through a repair ('it's it's a tension', line 23), where the participant emphasises this tension for a second time, and laughter follows once again (line 23). This repair takes the form of an interruption, perhaps suggesting this was a particularly sensitive moment in the interaction. In troubles talk, the troubles-teller may not always defer to the onset of the recipient's speech, and this interruption is one example of the participant continuing across the teacher's talk (also see lines 14 & 41). As the troubles-teller, the participant had projected a non-serious trajectory up until this point, and in leaving the recipient to take the next turn, it then falls to them, in this case the teacher, to take the talk in one direction or another (Jefferson 1984). A seriously framed problem may be seen as 'over disclosing' in a group context where there is an observing audience, whereas framing events as laughable may feel more acceptable. Arguably, the participant is doing a particular job here in that he is exhibiting what Jefferson (1984) calls 'troubles-resistance'. By laughing, he implies that while this experience was troublesome for him, it does not warrant concern from the group; he is in a position to take the trouble lightly, having already dealt with it (as he goes on to explain). Indeed, the only point at which the participant utters the word 'tension' and does not laugh is when he is explaining how it was no longer present after the practice (line 38). In declining to join in with the laughter, the teacher demonstrates 'troubles-receptiveness' and is thus shown to be taking the trouble seriously.

While this interaction is taking place between the teacher and an individual participant, it also occurs within a larger group setting, and therefore, the talk can be shown to have a demonstrative application as a teaching device. Re-organising the various thoughts, feelings and actions described in the participant's initial account into a sequential narrative acts to construct the practice of mindfulness as both active and progressive: a step-by-step process resulting in a favourable outcome: 'there wasn't any tension' (line 38).

Teacherly Talk and Mindfulness Student Positioning

A pedagogic style of speaking is particularly prominent in the participant's description of how he approached dealing with the 'problem': 'Well I-k-I I (0.5) followed through what we've been doing which is kind of just acknowledge that that's there' (line 33), which suggests a student orientation, and also in the teacher's affirmative acknowledgement token 'yep' (line 35). This 'yep' has a teacherly quality in the sense that it can be heard as evaluating what is hearable as her students' description of his practice of mindfulness as being correct or good. The participant explains what he means by 'what we've been doing' using the plural pronoun so as to address the group ('we') and to situate himself as a part of the group. By including within his explanation 'kind of just', the sense of risk associated with displaying expertise or authority not in line with the subject position of 'student' is thereby reduced. This mitigation of expertise or authority is similarly demonstrated earlier in the extract when the student comments upon what he has

been doing previously by saying 'so it was nice to actually *sort of* experiment' (lines 14–17; emphasis added). The teacher's later response 'yep' is simple and direct, emphasising her position as the teacher, and thus as the authority figure, through the provision of a distinctly positive confirmation that the student has understood and done the 'right' thing on this occasion.

The potential power differential between teacher and student has implications for the consideration of authority in MBSR, directing attention particularly towards 'epistemic authority' or 'asymmetries of knowledge and rights to knowledge' (Drew and Heritage 1992, p. 49). It is interactionally established that the participant (or student) has epistemic authority with regard their own inner experience, while the teacher neither accepts nor corrects a given account of this experience. In other words, while teachers of mindfulness courses are considered to not have direct access to the inner psychological lives of their students, which are presumed to lie solely within the domain of the participants themselves, at the same time, a mindfulness teacher can reformulate descriptive accounts of students' recent prior experiences. This in turn contributes towards fulfilling certain institutional objectives.

This epistemic asymmetry means that the nature of the interaction in MBSR is not wholly equivalent to a traditional pedagogic teacher/student power dynamic. In offering reformulations, the mindfulness teacher systematically modifies the descriptions given previously by students in specific ways. Professional caution or hesitation is displayed in reformulating participants' inner lives in a way which suggests that while mindfulness teachers have epistemic rights to reformulate the inner worlds of their students, they do not have direct unmediated access to those worlds. Question sequences are initiated by teachers as a means for eliciting discussions of prior experience with participants and on such occasions, the participant occupies the subject position of the 'primary knower' (Lee 2007). This is in direct contradiction to the idea of there being a predetermined, objectively correct answer, which is known in advance by the teacher, as in traditional pedagogic interaction.

At the same time, in an obvious sense, the mindfulness teacher is delivering ideas and practices taken from a pre-established curriculum, which the teacher knows prior to the class being taught, and which the students might or might not be knowledgeable about prior to the class. While there may not be 'correct' or 'right' answers, and mindfulness teachers cannot know what may come up in each session, there are nevertheless demonstrable desirable and undesirable orientations with regards to what is spoken about, in terms of how participants are heard to be *relating* to their experience – at least as it is described or inferred through their accounts. Mindfulness teachers must therefore strike a balance between, on the one hand, adhering to a pre-existing curriculum and, on the other hand, responding in the moment to what actually happens in a mindfulness course, which cannot be predicted in advance. This tension or dilemma has subsequent implications for how authority and expertise are constructed and negotiated within mindfulness courses.

Revealing Subtle Power Dynamics and Ideological Dilemma

While MBSR can be understood as psycho-educational, it does not represent a didactic or traditional classroom interaction based on an ideology of authoritative knowledge. Before Piaget's concept of child-centred learning and the 'little scientist', which began to challenge the idea that education was about imparting wisdom onto children as though they were 'blank slates', the role of the teacher consisted of standing before rows of front-facing desks, facilitating the transmission of ready-made factual information directly from knowledgeable adult to naïve child (Edwards and Mercer 1987). Modern 'progressive' education is much more nuanced than this. Children are positioned as active participants in their own learning, and

education has become less prescriptive and more explorative, changing the nature of what it means to teach.

While MBSR consists of interaction between adults, and so cannot be called 'child-centred', there is great emphasis on learning through experience. In this sense, MBSR courses practically illustrate progressive teaching ideology, with teachers facing many of the same practical considerations as school teachers. They must strike a balance between the transmission of necessary information or knowledge, enabling students to discover on their own, while at the same time ensuring learning of a pre-established curriculum is displayed to the group. A way of doing authority differently to authoritarianism is required if a curriculum must be communicated, but cannot be directly or explicitly taught (Billig et al. 1988). It is arguable then, based on the MBSR courses we have analysed so far, that the educational practice of MBSR itself is dilemmatic and ideological; subtle power dynamics are negotiated by the teacher and students.

Educational ideologies are variants of wider social ideologies, and the position of authority is not straightforward, as it is imbued with social values and political dynamics. The general societal development towards democracy in the Western world calls for 'user' or 'citizen' influence and participation in decision-making. Furthermore, the norms of democracy are fundamentally egalitarian (Ericsson and Lindgren 2011). It makes sense then that ideological dilemmas within education are relevant to broader conflicts between authoritative constraint, equality and the nurturing of individual freedom. The implication here is that while mindfulness teachers can be said to wield power over their students, this power is no longer distinctly authoritarian. Wetherell et al. (1987) have called this 'unequal egalitarianism'.

The mindfulness teachers in the courses we studied rarely explicitly instruct or command participants to act in a certain way or do a certain thing. Instead, their commands are delicately positioned as questions or invitations, presenting participants with a choice and urging them to do 'what is comfortable and feels right for them'. The language is diplomatic, friendly and non-hierarchical. At the same time, the mindfulness teacher remains in control throughout the interaction. Power is still maintained, but authority is performed in more subtle ways and teaching points are led in relation to, or drawn out from, participants' described experiences. The talk is presented as a free and equal exchange, but the teacher remains in an undeniable position of professional authority, which takes the form of a 'non-expert expertise'. This authority is performed through liberal teaching strategies, and thus, the teacher metaphorically 'hunches her shoulders' as an authority figure, as she wields a subtle form of liberal power. Crane et al. (2014) describe this as a 'disciplined improvisation' in which the mindfulness teacher practically negotiates a 'tension between directional leadership and participatory co-construction' (p. 1113).

We also found that in the courses we have studied, democratic semantics, such as the plural pronoun 'we', are consistently used by teachers while discussing students' experience, which along with constructing universality of experience, aims to create a sense of unity in the group. In these moments, the teacher closes the distance between the subject positions of student and teacher.

Producing 'Mindful' Subjectivity: Discourse and Experience

Arguably, the reframing of accounts of prior experience which takes place in formulations given by teachers in MBSR inquiry sessions works to articulate a new way of relating to experience for these participants—a 'mindful' orientation—which is being practically demonstrated and trained through a particular way of talking. In returning to formulations, we can see how this is achieved discursively through the liberal power of the 'hunched-shouldered' authority figure. By using specific ways of speaking about experience, the nature of an account given by the participant is subtly

reformulated by the teacher in order to conform to an orientation towards experience, which includes a certain way of 'seeing the mind'. While both teacher and participant appear to work through the interaction collaboratively, the teacher remains in control and the required orientations are cued by carefully constructed questions and lexical choices. Drew and Heritage (1992) highlight the importance of lexical choice in institutional talk and summarise research identifying specific professional vocabularies where 'the use of such vocabularies can embody definite claims to specialized knowledge and institutional identities' (p. 29). Therefore, the development of a shared vocabulary of 'mindfulness' is part of what makes MBSR recognisably institutional.

The rhetorical aspect of mindfulness teaching is important in this regard, in the sense of the teacher co-creating an argumentative context in which mindfulness is shown to be beneficial. Following van Langenhove and Harré (1999), we can conceptualise formulations in inquiry sequences as 'rhetorical redescription' of retrospective accounts of prior experience. They work to collaboratively produce reconstructed versions of past experience, sequentially organised in the interaction in order to move the session forward by constructing the immediate past in specific, and often beneficial, ways. In these formulations, teachers employ skills of argument to persuade participants of the benefits of a self-reflexive mindful orientation. Returning to the extract from week 5, the teacher's formulation serves to demonstrate how bringing attention to difficulty can help the practitioner to move through it: 'and then brought your breathing to that and actually then the tension wasn't there so something shifted for you' (lines 47–49).

Part of this way of teaching involves a re-appropriation of what the participants are saying through the lens of a specific meta-discourse, or discourse about experience. This new way of speaking may be characterised as a 'language of meta-awareness', as it is explicitly concerned with articulating a position in which the subject is present during the momentary flux of conscious experience and is also then able to retrospectively reconstruct past prior experience in a highly specific way. This substantively involves using a language of 'noticing' experience (Stanley and Crane 2016). It is possible then that part of the claimed transformation amongst participants of MBSR courses manifests as a change in participants' own lexical choices. In the week 5 extract, there is a distinctive repetition of the word 'noticed' by both teacher and participant, occurring repeatedly (lines 1, 7, 14, 24, 26, 27 and 44). The participant's descriptive account of his prior experience consists of embodied orientations and mirrors the teacher's meta-discourse on experience: 'it was nice to actually experiment about following that through and breathing into it' (lines 15–18). This suggests a distinctive style of 'experimental' self-reflexivity has been taken up and articulated by the participant, at least in relation to how he speaks about, or narrates, his experience. It is through developing this form of reflexivity that narratives of transformation begin to be told within the interaction. Through his retrospective description of conducting mindfulness practice, the participant produces a positive evaluation of 'working with difficulty'—'it was nice'—and implies the practice was successful in producing a good outcome for him.

Conclusion: Self-knowledge and Self-care

In summary, we have analysed how a psychologised subjectivity is constructed through 'inquiry' sequences understood as technologies of the self. The power dynamics and institutional practices of the mindfulness courses we analysed are productive of a subject position, which appears to be composed of complex contradictory practices, mixing self-care and self-knowledge. This mindful subject possesses two somewhat contrary orientations to itself:

On the one hand, the mindful subject is a monitoring, regulating and controlling subject: enhancing and developing their self as a choosing agent through practices attention-, emotion- and self-regulation. At first glance, the sequences

of inquiry in the mindfulness courses that we have analysed so far ostensibly produce a 'choosing' subject. This is an individual self who is the agent of change and transformation, especially self-produced inner transformation.

However, the mindful subject, as well as being a choosing subject, is also an 'accepting', 'letting-be' and 'letting-go' subject of self-care; they are taught to kindly accept and acknowledge body and mind in each experienced moment. The MBSR course involves the reshaping of the personal self of the mindfulness practitioner towards increasing warmth and ethically sensitive self-care. The subject must not only 'be' in the moment but 'be well' in the moment (Cederström and Spicér 2015). Thus, as well as being 'obliged to be free', mindful citizens might be being trained through mindfulness to be 'obliged to be well' through the careful and continual self-surveillance and self-governance of their conduct. That is, mindful subjects are being empowered to perpetually monitor and regulate themselves through highly specific and refined disciplinary regimes of the body.

This mindful subject of MBSR is produced in interaction through the adoption on the part of the teacher of a psychological style of 'non-expert' expertise or post-traditional authority in which power relations are presented as being egalitarian but nevertheless can be shown to display a distinctly unequal authority relation —specifically in relation to asymmetries of speaking position and epistemic authority. Inquiry cumulatively reproduces the subject position of the 'universal' mindful individual. The 'mindful gaze' embedded within inquiry and embodied in the authority of the mindfulness teacher is arguably 'turned inwards' as a style of self-surveillance and becomes part of the constitution of the mindful subject.

Acknowledgments Thank you to Rebecca Crane for initiating this project and making available the data. We would like to thank Sara Rees for feedback on this chapter and the mindfulness teachers and participants featured in the extracts. We are also grateful for funding from the *Cardiff University Research Opportunity Placement* scheme and support from *Qualitative Research and Analysis Workshop* participants.

Appendix: Key for Transcription Notation (Based on a Simplified Jefferson-Style Transcription)

Notation	Example	Phenomenon
Rounded brackets enclosing full stop	right (.) so that again is interesting	Pause
Double rounded brackets enclosing text	((lines omitted))	Transcriber comments
Square brackets	[it wanders [it does wander yes	Speaker overlap
Less than followed by greater than signs	>"get back to what we're doing"<	Quicker speech
Double quotation marks	"what am I thinking that for"	Direct reported speech or private thoughts
Hyphen	there's some- is there a bit of judgement	Cut off or repair of word
Comma	it's getting easier,	Continuing intonation
Question mark	does yours?	Questioning intonation
Underlining	mm mm mm mmm	Emphasis
Equals	is that right= =yes	Contiguous words
Colon	so::	Elongated sound

References

Antaki, C. (2008). Formulations in psychotherapy. In A. Peräkylä, et al. (Eds.), *Conversation analysis and psychotherapy*. Cambridge: Cambridge University Press.

Barker, K. K. (2014). Mindfuness meditation: Do-it-yourself medicalization of every moment. *Social Science and Medicine, 106*, 168–176.

Billig, M., et al. (1988). *Ideological dilemmas*. London: Sage.

Billig, M. (2008). *The hidden roots of critical psychology*. London: Sage.

Braun, E. (2013). *The birth of insight*. London: The University of Chicago Press.

Carrette, J. R. (1999). Michel Foucault and Zen: a stay in a Zen temple (1978). In J. R. Carrette (Ed.), *Religion and culture: Michel Foucault* (pp. 110–114). London: Routledge.

Chaskalson, M., & Hadley, S. G. (2015). Mindfulness: historical and contemplative context and recent developments. In J. Reb & P. W. B. Atkins (Eds.), *Mindfulness in organizations* (pp. 42–66). Cambridge: Cambridge University Press.

Cohen, E. (2010). From the Bodhi tree, to the analyst's couch, then into the MRI scanner: The psychologisation of Buddhism. *Annual Review of Critical Psychology, 8*, 97–119.

Cook, J. (2010). *Meditation in modern Buddhism*. Cambridge: Cambridge University Press.

Crane, R. (2009). *Mindfulness-based cognitive therapy*. London: Routledge.

Crane, R., et al. (2014). Disciplined improvisation: Characteristics of inquiry in mindfulness-based teaching. *Mindfulness, 6*(5), 1104–1114.

Cederström, C., & Spicér, A. (2015). *The wellness syndrome*. Cambridge: Polity.

Danziger, K. (1997). *Naming the mind*. London: Sage.

Davidson, R. J., & Dimidjian, S. (2015). Special issue: The emergence of mindfulness in basic and clinical psychological science. *American Psychologist, 70* (7), 581–665.

Davies, W. (2015). *The happiness industry*. London: Verso.

Drew, P., & Heritage, J. (1992). *Talk at work*. Cambridge: Cambridge University Press.

Dryden, W., & Still, A. (2006). Historical aspects of mindfulness and self-acceptance in psychotherapy. *Journal of Rational-Emotive and Cognitive-Behavior Therapy, 24* (1), 3–28.

Edwards, D. (1997). *Discourse & cognition*. London: Sage.

Edwards, D., & Mercer, N. (1987). *Common knowledge*. London: Routledge.

Ericsson, C., & Lindgren, M. (2011). The conditions for establishing an ideological dilemma: antagonistic discourses and over-determined identity in school music teaching. *Discourse: Studies in the Cultural Politics of Education, 32*(5), 713–728.

Gardner, R. (2013). Conversation analysis in the classroom. In J. Sidnell & T. Stivers (Eds.), *The handbook of conversation analysis* (pp. 593–611). London: Wiley-Blackwell.

Giddens, A. (1991). *Modernity and self-identity*. Stanford, CA: Stanford University Press.

Grubin, D. [Producer]. (1993). *Bill Moyers: Healing and the mind* [Video Recording]. New York: Ambrose Video.

Hall, S. (2001). Foucault: Power, knowledge and discourse. In M. Wetherell, et al. (Eds.), *Discourse theory and practice* (pp. 72–81). London: Sage.

Harrington, A. (2008). *The cure within*. London: WW Norton & Co.

Heritage, J. (1984). *Garfinkel and ethnomethodology*. London: Polity.

Heritage, J. C., & Watson, D. R. (1979). Formulations as conversational objects. In G. Psathas (Ed.), *Everyday language* (pp. 123–162). New York: Irvington.

Jackson, M. (2013). *The age of stress*. Oxford: Oxford University Press.

Jefferson, G. (1984). On the organization of laughter in talk about troubles. In J. M. Atkinson & J. Heritage (Eds.), *Structures of social action* (pp. 346–369). Cambridge: Cambridge University Press.

Jefferson, G. (1985). An exercise in the transcription and analysis of laughter. In T. van Dijk (Ed.), *Handbook of discourse analysis* (Vol. 3, pp. 25–34). London: Academic Press.

Kabat-Zinn, J. (1994). *Wherever you go, there you are*. London: Hyperion.

Kabat-Zinn, J. (2011). Some reflections on the origins of MBSR, skilful means, and the trouble with maps. *Contemporary Buddhism, 12*(1), 281–306.

Lee, Y.-A. (2007). Third turn position in teacher talk: contingency and the work of teaching. *Journal of Pragmatics, 39*(6), 1204–1230.

Madsen, O. J. (2014). *The therapeutic turn*. London: Routledge.

Martin, L. H., Gutman, H., & Hutton, P. J. (1988). *Technologies of the self*. Amherst: The University of Massachusetts Press.

McMahan, D. (2008). *The making of Buddhist modernism*. Oxford: Oxford University Press.

Moyers, B. (1993). *Healing and the mind*. London: Doubleday.

Pagis, M. (2009). Embodied self-reflexivity. *Social Psychology Quarterly, 72*(3), 265–283.

Peräkylä, A. (2013). Conversation analysis in psychotherapy. In J. Sidnell & T. Stivers (Eds.), *The handbook of conversation analysis* (pp. 551–574). West Sussex: WileyBlackwell.

Pomerantz, A. (1986). Extreme case formulations: a way of legitimizing claims. *Human Studies, 9*, 219–229.

Preston, D. L. (1988). *The social organization of Zen practice*. Cambridge: Cambridge University Press.

Purser, R. E., & Milillo, J. (2015). Mindfulness revisited: A Buddhist-based conceptualization. *Journal of Management Inquiry, 24*(1), 3–24.

Rakow, K. (2013). Therapeutic culture and religion in America. *Religion Compass, 7*(11), 485–497.

Rose, N. (1998). *Inventing our selves*. Cambridge: Cambridge University Press.

Said, E. W. (1978). *Orientalism*. London: Penguin.

Schegloff, E. A., Jefferson, G., & Sacks, H. (1977). The preference for self-correction in the organization of repair in conversation. *Language, 53*(2), 361–382.

Sharf, R. (1995). Buddhist modernism and the rhetoric of meditative experience. *Numen, 42*(3), 228–283.

Sheets-Johnstone, M. (2015). Phenomenology and the life sciences: Clarifications and complementarities. *Progress in Biophysics and Molecular Biology, 119*, 493–501.

Stanley, S. (2013). 'Things said or done long ago are recalled and remembered': The ethics of mindfulness in early Buddhism, psychotherapy and clinical psychology. *European Journal of Psychotherapy and Counselling, 15*(2), 151–162.

Stanley, S., & Crane, R. (2016). Discourse analysis of naturally occurring data: The relational development of mindfulness. *SAGE Research Methods Datasets,.* doi:10.4135/9781473958845.

Sun, J. (2014). Mindfulness in context: A historical discourse analysis. *Contemporary Buddhism, 15*(2), 394–415.

van Langenhove, L., & Harré, R. (1999). Introducing positioning theory. In R. Harré & L. van Langenhove (Eds.), *Positioning theory* (pp. 14–31). Oxford: Blackwell.

Wetherell, M. (1998). Positioning and interpretative repertoires: Conversation analysis and post-structuralism in dialogue. *Discourse & Society, 9*(3), 387–412.

Wetherell, M. (2013). *Affect and emotion*. London: Sage.

Wetherell, M., Stiven, H., & Potter, J. (1987). Unequal egalitarianism: A preliminary study of discourses concerning gender and employment opportunities. *British Journal of Social Psychology, 26*(1), 59–71.

Williams, J. M. G., & Kabat-Zinn, J. (2013). *Mindfulness: Diverse perspectives on its meaning, origins and applications*. Oxon: Routledge.

Wilson, J. (2014). *Mindful America*. Oxford: Oxford University Press.

Jenny Eklöf

Introduction

In the decade following the official "Decade of the Brain" of the 1990s (Jones and Mendell 1999), the public was fed one single and straightforward message: It is possible to change your brain. In various forms, this message was presented as a personal and moral obligation to *act on* the brain, in different ways and for different reasons. If neuroscience had been able to establish the material basis of the mind—which was often claimed—then the problems of the mind could be dealt with by managing the brain.

Problems of the mind have also taken center stage in the birth of the new "science of mindfulness." Over the course of 10–15 years, medical interventions and therapeutic approaches based on, or informed by, mindfulness meditation emerged and gained traction in clinical psychology, psychotherapy, behavioral medicine, and neuroscience. Although the merging of science with ancient contemplative traditions (particularly Zen and Theravada branches of Buddhism) was not historically new (Lopez 2008, 2012), the growth rate of this emergent field has been unprecedented, and constitutes a kind of academic

takeoff.[1] The science of mindfulness today shows many signs of institutionalization; academic courses and programs have been launched, textbooks have been written for clinicians, mental health professionals, and mindfulness teachers; and international conferences and other professional forums have been established, and a new academic journal—*Mindfulness*—was founded in 2010. In a 2012 special issue of the journal *Social Cognitive and Affective Neuroscience*, the marriage between neuroscience and mindfulness was described as follows: "Mindfulness neuroscience is a new, interdisciplinary field of mindfulness practice and neuroscientific research; it applies neuroimaging techniques, physiological measures and behavioral tests to explore the underlying mechanisms of different types, stages and states of mindfulness practice over the lifespan" (Tang and Posner 2012). Mindfulness neuroscience directs our attention to the brain and to the neural mechanisms involved in meditation and mindfulness training, and is also related to various popular conceptual innovations such as "neurodharma" or "contemplative neuroscience."

Scientific activities in the subfield of mindfulness neuroscience have been explained, promoted, translated, and given a cultural meaning in various public accounts, which in themselves constitute an integrative part of a broader trend,

J. Eklöf (✉)
Department of Historical, Philosophical, and Religious Studies, Umeå University, Umeå, Sweden
e-mail: jenny.eklof@umu.se

[1]A search in WoS shows that the number of publications on mindfulness has grown exponentially since the late 1990s, from around 25 in 2000 to over 900 in 2014. The USA accounts for more than half of these publications.

© Springer International Publishing Switzerland 2016
R.E. Purser et al. (eds.), *Handbook of Mindfulness*,
Mindfulness in Behavioral Health, DOI 10.1007/978-3-319-44019-4_21

what has been variously described as our contemporary "brain culture" (Johnson Thornton 2011), "neuroculture" (Ortega and Vidal 2011), or "therapeutic" and "self-help culture" (Füredi 2004; Illouz 2008). The public appeal of neuroscience's role in personal and societal transformation is manifested in the many best-selling books on the subject (see, e.g., Begley 2007; Doidge 2007; Johnson 2004; Restak 2001; Schwartz and Beyette 1996). The genre, however, has also been problematized (e.g., Choudhury et al. 2009; Satel and Lilienfeld 2013). Indeed, the process by which neuroscientific knowledge claims gain social and cultural meaning is a complex one (Pickersgill et al. 2014). Sociologists Nicolas Rose and Joelle M. Abi-Rached argue that even though it is too early to speak of a radical paradigmatic shift from "psy-" to "neuro-," "neurobiology is undoubtedly reconfiguring some of the ways in which individual and collective problems are made intelligible and amenable to intervention" (Rose and Abi-Rached 2013, p. 227). The borders between academic psychology and neuroscience and their societal applications have always been porous though, as has been the boundary between "real" neuropsychology and "popular" neuropsychology. As with neuroscience more broadly, the scientization of mindfulness has developed in parallel with its popularization.

Many historical case studies testify to this co-production of academic and popular forms of knowledge (Ekström 2004, 2011), and this process might be even more poignant today as the reach and effectiveness of Internet-based (social) media has transformed the public communication of and engagement with science and technology (Allan 2011; Holliman 2010). In terms of the simultaneous processes of scientization and popularization, mindfulness neuroscience is an interesting case, because the cultural and social impact of this field is often taken to be a sign of its prior academic validation. Instead, we can conceive of communication as a continuum, supported by processes of personalization. Mindfulness neuroscience is communicated to the public through channels that are not directly framed by media logics (albeit not completely

separated from them either), and processes of personalization impinge on what is being communicated and how, and also on the role of communicating experts.

Personalization and Science Communications as a Continuum

It has been a repeated claim among science communication scholars that science (in general terms) has developed a more intense relationship with the mass media over the past 30 years or so (Bucchi 2013). The medialization thesis offered by Peter Weingart and colleagues (Rödder et al. 2012) suggests that scientific institutions and their members are increasingly seeking to synchronize their activities with the (mass) media. With expanded and professionalized public relations departments, the messages stemming from traditional knowledge institutions, such as universities, are made to better "fit" the logic of the media and news value criteria. That science and medicine are now more sensitive and responsive to the priorities and needs of extra-scientific institutions is also manifested in relation to industry and policy-making. This relationship is often understood, described, and sometimes criticized as a tendency toward marketization and politicization. Voices have been raised that caution against, for example, increased risk of hype and dishonesty in science communication (Bauer 2008; Caulfield and Condit 2012). Mindfulness neuroscience is no exception in this regard (Heuman 2014).

However, with the growth of online media, scientific organizations as well as individual scientists increasingly bypass traditional mass media channels and address various public audiences on their own terms. New and old media overlap and intermingle in intricate ways, however, and existing social relations are (at least partly) reproduced in digital forums. Nevertheless, scientists can now reach out to different audiences through personal Web sites, blogs, social networks, video hosting sites, Twitter, etc., i.e., communication channels where they can control the timing and content themselves, or at least to a higher degree than in traditional media outlets.[2] As new possibilities for actively addressing different types of audiences have opened up, the medialization of science also entails forging of personalized networked communication and personalized audiences. In such media, scientists are not bound by the dominant constraints of specialist communication or the logics of the mass media. The tone can be more conversational, and there is an emphasis on the process and practice of science, instead of on published findings (Hermida 2010). Therefore, personalization of science communication involves the same two major aspects as personalized political communication (Aelst et al. 2011); it points to the general visibility of individual scientists, facilitated and fueled by both emerging digital media and science's intensified outreach activities, and a focus shift on personal experiences and characteristics.

So, if the sciences are increasingly communication oriented and if the tools available for communication are multiplying, this should make us seriously rethink (again) the inner–outer aspect of science communication. In his much-cited article from 1990, sociologist Stephen Hilgartner argued that this inner–outer idea of science communication is a rhetoric underpinning a professional scientific ideology. He argued that it is, in fact, difficult to empirically distinguish "popular science" from "real science" regardless of which criteria one uses, be it level of simplification, type of audience, etc. Instead, communication can be seen as a continuum and popularization as *a matter of degree* (Hilgartner 1990). That the inner–outer distinction is a rhetorical resource flexibly used by professional scientists makes it important for science communication scholars to investigate, question, and sometimes ignore, in order to be able to identify the flow of meanings across different societal spheres. This stance does not require us to deny that any differences or discontinuities exist between what is commonsensically understood as "real science" and its popularized varieties; rather, it means recognizing that such distinctions

are professionally guarded as a result of the social division of knowledge. Hence, the underlying idea of an "inner" scientific sphere and an "outer" public sphere is often also hierarchized as higher and lower forms of knowledge. Being able to control what is being said about science in public spaces has always been a scientific concern. Whitley's observation (1985) showed that the public communication of science often relies on depictions of "decontextualized" science, i.e., the filtering out of much of the messiness, provisional character, and subtle nuances of the research process, and that this makes scientific claims look like authoritative constellations of "facts." Others have argued that the public communication of science also involves a process of recontextualization, as scientific results are framed to fit different contexts of application.

As incentives for reaching out to wider audiences change, along with the emergence of digital media, the question we can ask is whether this alters both the inner–outer aspects of science communication and its temporal dimension, the sequence in which facts are first produced and then communicated (Bucchi 1998). The conventional view of communication has assumed a two-stage process, and many descriptions of mindfulness neuroscience assume the same; they tell us about how concepts move from the specialized literature into the wider domain or how knowledge leaves the enclosed space of laboratories and clinics to gain traction in the outside world. However, strategies for gaining traction in the outside world might in fact precede or coincide with the formulation of knowledge claims inside science proper. What we know about scientific efforts to gain media attention prior to the process of collegial peer review might suggest that this is often the case, manifested in such phenomena as "science by press conference" (Spyros 1980). Being visible in the media helps scientists to, for example, gain social and political legitimacy, might boost scientific citations (Peters 2013; Peters et al. 2008), and can also function as a way to publicly settle academic disputes. Thus, not only is the inner–outer dimension of science communication overlapping, but also the temporal order in

---

[2]For a discussion on blogs, see Trench (2012).

which scientific knowledge is produced and communicated.

As the communication orientation of science has intensified, and the opportunities for using communication channels circumventing traditional mass media have multiplied, it makes sense to "follow experts around" and pay attention to what they do as experts or in the name of mindfulness neuroscience. Seen in this way, tweeting, posting blog entries, producing social media updates, uploading videos to video hosting sites, etc., as well as writing self-help books for the public at large, is part of what science is about. As a result, the analytical border between "real" science and its public representation, its inner and outer dimension, becomes highly unstable as experts repeatedly transgress it. What is being said in popular accounts about a specific area of study and practice can be treated as representations and extrapolations of meanings that might, or might not, be present in the academic literature, but it is not by its very nature something completely different. Analyzing what experts actually say about a specific field of study, how they make sense of different studies, how the results are made to fit the concerns of different audiences, and how their role as communicators relates to how these messages are being promoted and framed opens up a space for revisiting these sometimes problematic distinctions between inside–outside, higher–lower, before–after, and instead think of science communication as a continuum.

Public representations of mindfulness neuroscience come in many forms. The empirical material used in this chapter has been selected to cover some of that complexity, and focuses on the work of three experts on mindfulness neuroscience, experts who are important figures in the communication of mindfulness neuroscience to the public, that is, neuroscientist Richard Davidson, psychiatrist Daniel Siegel, and clinical psychologist Rick Hanson. Three types of sources have been selected for examination: self-help books, personal and institutional Web sites, and videos displayed on these Web sites and/or through personal and institutional YouTube channels. The selected experts are all authors of

books which include not only accounts of new brain science discoveries, but also some kind of program for action. Thus, these are books that can be categorized as self-help. Self-help is not a clear-cut genre, however, and the term itself seems to have a negative connotation within many academic circles. The term is not used in a derogatory manner here, but simply as a descriptor for a type of literature that provides guidance for personal transformation. According to professor of folklore Dolby (2005), the self-help genre has certain characteristics: It is non-fiction literature aimed toward self-improvement, it is written in an intimate and personal style with an "I" (author) addressing a "You" (reader), it defines problems and proposes solutions to these problems, it is educational and requires (and tries to stimulate) the active involvement of the reader. The books analyzed in this chapter include Daniel Siegel's *Mindsight: Transform Your Brain with the New Science of Kindness* (Siegel 2010); Rick Hanson's *Buddha's Brain: The Practical Neuroscience of Happiness, Love & Wisdom* (Hanson 2009), written with Richard Mendius, M.D.; and Richard Davidson's *The Emotional Life of Your Brain: How to Change the Way You Think, Feel, and Live* (Davidson 2012), written together with science writer Sharon Begley.[3]

The content displayed on these experts' personal Web sites as well as their institutional Web sites (these were heavily interlinked) complements the material in the self-help books.[4] Furthermore, messages put forward through online video clips were also used, many of which were displayed on the aforementioned Web sites as well as distributed through personal and/or institutional YouTube channels. In addition, the Web of

---

[3]I will use the shortened titles *Emotional Life*, *Mindsight*, and *Buddha's Brain* to refer to these books.

[4]The three communicators have personal Web sites at http://www.dansiegel.com, http://www.rickhanson.net, and http://richardjdavidson.com, but are also represented as directors or co-directors of *The Mindsight Institute*, https://www.mindsightinstitute.com/, *The Centre for Investigating Healthy Minds*, http://www.investigatinghealthyminds.org/index.html, and *The Wellspring Institute for Neuroscience and Contemplative Wisdom*, http://www.wisebrain.org/.

Science database was searched for articles about mindfulness in the database category "neuroscience neurology." The top 50 most cited items were selected for analysis, as well as a number of research review articles, and were compared with the content presented in the popular material.

Self-directed Neuroplasticity Through Mindfulness Practice

There are a number of well-cited key studies in the field of mindfulness neuroscience, some of which are featured more prominently in popular contexts than others. Perhaps the most important one to date is an article published in 2003 by Richard Davidson and Jon Kabat-Zinn (with colleagues)—the latter pioneering mindfulness in medicine in the late 1970s. This study tested the effects of the so-called Mindfulness-Based Stress Reduction program on healthy employees. Their brain activity and immune system responses were measured on several occasions, and the results showed significant increases of left-sided activation in the frontal and prefrontal regions of the brain, as well as stronger immune responses (Davidson et al. 2003). Other highly cited studies have concluded that mindfulness practice can lead to increases in regional brain gray matter density (Hölzel et al. 2011), that it improves the ability to regulate attention (Jha et al. 2007), and that it engages distinct neural modes of self-reference (Farb et al. 2007).

In the scientific literature, the existence of knowledge gaps as well as the preliminary character of many research results is often underscored. Evidence "suggests" and "might indicate" this or that, but "additional research is needed." The neural mechanisms underlying reported positive psychological and physiological effects are still seen as largely unknown and poorly understood (Creswell et al. 2007), and the whole neuroscientific study of meditation is "still in its infancy" (Lutz et al. 2008). A systematic review published in 2010 concluded that the effects of mindfulness meditation were hard to establish, due to low-quality research designs and other methodological flaws and limitations

(Chiesa and Serretti 2010), and more methodologically rigorous studies were called for in a *Nature Neuroscience* review in 2015 (Tang et al. 2015). At the same time, mindfulness neuroscience is believed to be full of "cutting-edge discoveries" (Tang and Posner 2012), and we are said to be witnessing something like a paradigmatic revolution in science and medicine. A tension here comes through; the field is both understood as revolutionary, cutting-edge, and having momentum—in this very moment "revolutionizing" science—and at the same time it has not matured, is still dealing with substantial unknowns, and is fraught with definitional and methodological struggles.

This level of uncertainty is not present to the same degree (albeit not absent) in the public representations of mindfulness neuroscience examined here. In self-help books, in online videos, and on public Web sites, we find constant reminders that mindfulness meditation is "science based," and not just based in any science, but in the laboratory-generated, "rigorous" procedures of neuroscience. These reminders might be interpreted as signs of a field that views itself as new, exciting, and pioneering, but which also does not want to be perceived as being "out there." Communicating the science relevant for understanding mindfulness practice is said to be important because, in Richard Davidson's words: "It is rare that the human mind can determine the truths of nature, or even of ourselves, by intuition or casual observation. That's why we have science. Only by methodical, rigorous experiments, and lots of them, can we figure out how the world works—and how we ourselves work" (Davidson 2012, p. xiii). Explaining to the public the truths of nature and of ourselves—as constructed by science—is what Davidson's book *Emotional Life* sets out to do. His own academic contribution—identifying the brain's emotional styles and the six dimensions they are comprised of—emerged out of laboratory studies of brain mechanisms, which is important since "Anything having to do with human behavior, feelings, and ways of thinking arises from the brain, so any valid classification scheme must also be based on the brain" (Davidson 2012, p. xii).

Explaining and representing recent academic achievements in the field is thus an important part of what the public communication of mindfulness neuroscience is about. But more importantly, the public representation of mindfulness neuroscience also transforms the field into a kind of practical or applied neuroscience. We are presented with the tools needed to act on our own brains, to enhance our capacity to live our lives to their fullest and to be able to cope with an increasingly stressful world and stress-related health problems. This integration of mindfulness practices into daily life becomes part of a larger quest to use the neurosciences for the public good. The link between Buddhist mindfulness practices and neuropsychology is often framed in the straightforward message that science has now proven that meditation works and that meditation can change your life, actually your very biological brain, in the long term (Hanson 2009). So while the neuroscientific take on mindfulness directs attention to the biological materiality of the brain, it simultaneously points out that mental techniques and practices can change the brain. In addition to being aligned with the culturally embedded view that we can change our lives by changing our thoughts—as promoted in both cognitive behavioral therapy and the so-called positive thinking literature (Ehrenreich 2009)— this line of reasoning takes a "detour" through the brain. (In the case of mindfulness science, it is not so much about changing your thoughts though, but more about altering your relation to them.) The claim is that by self-regulating our minds we can actively and intentionally change the brain and that these changes in the brain will, in turn, positively change our minds, our whole lives, and the world. Some of the critique leveled at neuroscience more broadly concerns this leap— drastically moving from scientific studies of brain activity and brain function to assumed social and cultural implications. In the words of Choudhury et al., "[…] much of the concern and revolutionary language about the radical changes imposed by neuroscience on society arise from the gap between actual findings in research and the representations of the findings" (Choudhury et al. 2009, p. 63).

The idea that mindfulness neuroscience can have profound societal impact draws on academic work on neuroplasticity and the mutability of the brain over the course of one's life (Davidson and McEwen 2012). The brain is said to have a negativity bias, and it is like "Velcro for negative experiences and Teflon for positive experiences," to use Rick Hanson's expression (Hanson 2009, p. 68). People therefore need techniques for self-regulating neuroplasticity so that the brain can respond better to positive emotions and experiences and "rewire" itself for happiness. In mindfulness neuroscience contexts, the results from these earlier research endeavors on neuroplasticity are often communicated through easy-to-remember catchphrases such as "neurons that fire together, wire together," a phrase attributed to psychologist Hebb (1949). Just as there has been a proliferation of new sciences in this area—the "new science of kindness" and "the science of a meaningful life" being two of them—it is also full of new catchphrases and slogans. On the Centre for Investigating Healthy Minds's Web site, founded by Davidson, one reads "Change Your Mind, Change the World" (Centre for Investigating 2013). Siegel, whose work emphasizes the social aspects of the brain, calls on us to "Inspire to rewire," whereas Hanson's 2013 book is about "Hardwiring happiness" (Hanson 2013b).

Mindfulness meditation is frequently framed as a kind of mental training that will, if adhered to, result in mental or brain "fitness." This is seen in much the same way that lifting weights and doing sit-ups will improve one's physical fitness. Or in Rick Hanson's words, Eastern meditators are "the Olympic athletes of mental training" (Hanson 2009, p. 1). In the self-help literature, the readers are offered sets of tools, practices, exercises, and tests that will, presumably, help them to help themselves. Many of these are no different from what is offered by mindfulness teachers, instructors, therapists, or coaches elsewhere, but here they are presented within the framework of neuroscience as "neural pathways" to achieving the same goals that mindfulness advocates are promoting elsewhere. In *Emotional Life*, Richard Davidson describes and explains how the discovery of emotional styles came about, how emotional styles develop, how they are based in the brain, and, more importantly, how such knowledge can be applied and how ordinary readers can assess their own emotional styles and use that information as a stepping-stone for personal transformation. Working within the hybrid field of so-called affective neuroscience, a field that he himself pioneered, the knowledge about emotional styles presented in Davidson's book is a distillation of his research in the field. In his own words: "Emotional *Style* is a consistent way of responding to the experiences of our lives. It is governed by specific, identifiable brain circuits and can be measured using objective laboratory methods." And he continues "they can be considered the atoms of our emotional lives—their fundamental building blocks" (Davidson 2012, p. xi). Some hands-on practices and tests are therefore more directly influenced by recent academic achievements in the field, such as Davidson's assessment of emotional styles. His Web site provides an online "book teaser" where one can assess oneself in one of the six dimensions comprising an emotional style, on a scale from 1 to 10. This is only the first out of six dimensions, and for a complete assessment, one must read the book. Davidson subjects himself to the same assessment and provides the results in his book (Davidson 2012, p. 64). The idea is that by gaining knowledge about your own emotional style, it can become a stepping-stone for learning how to shift it, for example, through mindfulness meditation practices.

Just as Davidson has coined the term "emotional style," Daniel Siegel advocates his own conceptual innovation called "mindsight." Mindsight has many similarities to mindfulness and refers to our ability to see the mind and to become aware of the content and workings of the mind. In his book *Mindsight*, Siegel explains the science and practice of mindsight through therapeutic case studies. His Web site offers more hands-on practices, such as the "Wheel of awareness" podcast and the "Healthy mind platter" (Dr. Dan Siegel Web site 2013), and he curates a mindsight digital journal in which he collects the latest science in the field. Rick Hanson's book *Buddha's Brain* also shows the reader, according to the description on his Web site, "many effective ways to light up the brain circuits that relieve worry and stress and promote positive relationships and inner peace" (Hanson 2013a). Many of these effective ways are modified versions of mindfulness or lovingkindness (metta) meditation, but they are promoted in other forms such as the Buddha's Brain iPhone app, available through Hanson's Web site. Overall, Davidson, Hanson, and Siegel become bridges between the "inner" and the "outer" spheres of science, between knowledge production and its reception, and between science and the clinical and everyday uses of specific tools, techniques, and practices.

Mindfulness Neuroscience–A Personal Story

A theme that is lacking in the scientific literature on mindfulness neuroscience, but runs like a common theme throughout the popularized material, is how doing research on mindfulness or practicing mindfulness in clinical or everyday settings constitutes an integrative part of the lives and experiences of those conducting or communicating the research. The self-help books, for example, provide many instances where readers can identify with the authors as ordinary people. For example, Richard Davidson's book *Emotional Life* provides, apart from telling a story of academic choices, struggles, dead ends, and breakthroughs, also a personal story of key life events and the "coming out of the closet" as a meditator, a process facilitated by his repeated meetings and friendship with the spiritual head of Tibetan Buddhism, His Holiness the Dalai Lama. Presenting these private aspects of doing research in this field is important, because: "while this book is a story of my personal and scientific transformation, I hope it offers you a guide for your own transformation" (Davidson 2012, xvii). One such point of transformation was when, in

1992, The Dalai Lama challenged him to scientifically study positive qualities of mind, what Davidson calls "virtuous qualities," a challenge which later resulted in the founding of the Centre for Investigating Healthy Minds. Davidson's close relationship with the Dalai Lama is described both in his book and on his Web site. The book *Emotional Life* also presents a story of how Davidson gradually started to be more open and explicit about his interest in meditation (something that was frowned upon in his earlier years), and the thrilling excitement he felt after his first studies showed encouraging results. At one such point, "the field of contemplative neuroscience had just been born" (Davidson 2012, p. 196).

The public communication of mindfulness neuroscience also resonates with what has become the mindfulness movement's main route into people's lives: the portrayal of contemporary life as fraught with stress, overwork, overachieving, multitasking, and exhaustion. Explaining the causes of mental suffering in terms of recognizable, real-life problems makes sense of these particularly conflict-ridden areas of human experience. The need to be able to break out of stress cycles, or at least to be able to handle them, is something that the authors also acknowledge as *their* personal problems. As Davidson writes (2012, p. 184):

> I live what most people would call a stressful, overscheduled life, typically putting in seventy hours of work each week; running a lab with dozens of graduate students, postdoctoral fellows, technicians, and assistants; raising millions of dollars from private and government funders to support everyone; vying for grants; and trying to stay at the top of a competitive scientific field. I believe my ability to juggle all this, with the small amount of equanimity I can muster, is a direct effect of my meditation practice.

What the self-help books achieve in this process is not only to translate, explain, and give meaning and context to specific scientific claims and their supposed applicability in the everyday life of the audience, but also to give these claims a personal touch. It is not only that mindfulness meditation works in the scientific sense of the word, but the authors also bear witness to how it works *for them*. In the words of Hanson:

> Last, if I know one thing for sure, it's that you can do small things inside your mind that will lead to big changes in your brain and your experience of living. I've seen this happen again and again with people I've known as a psychologist or as a meditation teacher, and I've seen it in my own thoughts and feelings as well. You really can nudge your whole being in a better direction every day. When you change your brain, you change your life (Hanson 2009, p. 3).

If the mediation of science in public relies on, in Whitley's (1985) terminology, a process of decontextualization, this form of personalization, in a sense, recontextualizes mindfulness neuroscience, as it brings in a subjective, experiential dimension. In the self-help books, we can find stories about personal failures, or episodes where the experts themselves have been acting in a less-than-ideal way. One such embarrassing moment is described by Siegel, as he shares an episode when his two children were fighting. He describes how a gradual tension was building up within him as he, unsuccessfully, tried to make them stop, and it all ended with him finally "losing his mind." The sharing of such personal stories is motivated, because: "I'm not proud to tell you any of this. But I do feel that since such explosive episodes are quite common, it is essential that we acknowledge their existence and help one another understand how mindsight can diminish their negative impact on our relationships and the world" (Siegel 2010, p. 25).

These are some of the ways that mindfulness neuroscience experts use themselves as case studies, and the audiences can rehearse their own lives by engaging in theirs. The audiences are addressed very directly, an "I" addressing a "You," as described by Dolby (2005), and are encouraged to share their own experiences. As it says in Siegel's welcoming video, on the starting page of his Web site: "So throughout the site you'll find lots of different opportunities to dive deeply into this field, and as you do, I hope you'll find a deep sense of connection and meaning and I look forward to hearing about how these connections are going for you in the future. Welcome" (YouTube channel Dr. Dan Siegel 2013).

Science Communication: A Moral Vocation

A common theme that runs through the material examined here is the explicit statement by each expert of their intention and motivation to help people lead healthier and more fulfilling lives, to find peace, to thrive, or just to be happy. This motivation translates almost seamlessly into world-saving language where it is not just individual lives that are at stake, but the whole of humankind and the future of the planet. This "moralizing" of mindfulness, as Buddhist scholar Jeff Wilson has phrased it, connects to an overall commitment within the mindfulness movement to see "self-healing as the essential first step to larger healing of the body politic" (Wilson 2014, p. 186). A number of quotes can serve as examples. First, we have Davidson who in a video recording of a Google Tech Talk envisions what will have happened were we to be moved to the year 2050. In this future scenario, mental exercise will have become as accepted as physical exercise is today. We will have a "science of virtuous qualities," and teachers and children will be taught better ways to regulate emotions and attention and to cultivate qualities like kindness and compassion. We will "Increase awareness of our interdependence upon others and upon the planet and be more responsible caretakers of our precious environment" (YouTube channel UWHealthyMinds 2013). Another example is taken from Siegel's Web site (Dr. Dan Siegel Web site 2013):

> Welcome to our worldwide conversation about the human mind and the cultivation of well-being. Discover the mindsight approach and the latest science as it emerges in the exciting field of interpersonal neurobiology. Our mission here at the Mindsight Institute is to provide a scientifically grounded, integrated view of human development for mental health practitioners, educators, organizational leaders, parents and others to promote the growth of vibrant lives and healthy minds. Join in

the collaborative journey to bring more kindness, compassion, and resilience into our world!

These experts are conveyors of a message that not only promises to relieve the suffering of stressed-out, multitasking, and overworked people but also promises on a collective level—transgressing all economic, cultural, political, and geographical boundaries—to transform all of humankind. Thus, the communicators in this field assume a kind of dual expertise, being both scientific experts and moral leaders, inviting us to join a movement that holds out to save the world. Hanson is the co-founder of the *Wellspring Institute for Neuroscience and Contemplative Wisdom*, and the institute's mission is to: "offer skillful means for changing the brain to benefit the whole person—and all beings in a world too full of war. It draws on psychology, neurology, and the great contemplative traditions for tools that anyone can use in daily life for greater happiness, love, effectiveness, and wisdom" (Wellspring Institute 2013). Or, as Hanson puts it in *Buddha's Brain*: "As you and other people become increasingly skillful with the mind—and thus the brain—that could help tip our world in a better direction" (Hanson 2009, p. 18).

This connection between mindfulness practices (and their neurological underpinnings) and the creation of "well-being" on a global planet scale seems to be an underlying driving aspiration, but is not presented as evidence-based in the same way as other claims are. No studies are presented to show that the form of brain management proposed in these public outreach activities is changing "the world" for the better. Rather, it is a future vision that provides a meaningful framework for brain management and perhaps protects the field from accusations of self-preoccupation and egocentrism, a common theme in criticisms pertaining to self-help culture, or indeed psychotherapy, overall. The leap taken, from inherently bounded studies in the laboratory or the clinic, to a vision of a world full of virtue, mindfulness and wisdom, is a process enabled by these communicators, but articulated in a universalistic language (humankind, human beings, planet), a type of language that is also a distinguishing feature of the mindfulness movement overall. These claims are not based in unambiguous research results. As is stated in a recently published research review:

> [...] some initial attempts have been undertaken to investigate the brain regions that are structurally altered by the practice of meditation. However, our knowledge of what these changes actually mean will remain trivial until we gain a better understanding of how such structural changes are related to the reported improvements in affective, cognitive and social function. Very few studies have begun to relate findings in the brain to self-reported variables and behavioral measures. (Tang et al. 2015, p. 215)

Nevertheless, in popular discourse, belief in the enormous societal impact of mindfulness neuroscience is coupled with very concrete efforts to actively rearrange and influence actual societal institutions. As it says on the Web site for the Centre for Investigating Healthy Minds, which Davidson directs:

> Your support is needed to advance this work. With a larger body of scientific evidence, we can work toward influencing policy in our schools, health care system and other institutions. With your support, we can help reduce suffering and increase happiness throughout the world. Together, we can, indeed, change the world (Center for Investigating Healthy Minds 2013)

The experts thus almost assume the role of world saviors, supported by institutional slogans, such as Davidson's "Change your Mind Change the World" or Siegel's "Inspire to Rewire." This is not only a process of translating scientific findings to everyday practices that people can engage with, but also to support a transference of excitement for what mindfulness neuroscience can become, its promissory nature. Davidson, Siegel, and Hanson can be seen as the personal bearers of this excitement, and the realization of a better world depends on the very communicational activities that they are themselves immersed in. Hence, science communication becomes a moral vocation. In the words of Richard Davidson:

> One of the aspirations is to enable this work to really live on to produce the kind of change that we so importantly need. What's at stake is nothing

short than the flourishing of humanity and the planet. It is absolutely critical that we get along more cooperatively, and more compassionately. I feel like I am totally dedicated to doing this work for the remainder of my time. I just feel that very deeply in my bones, this is why I'm here. (YouTube channel Healthy Minds 2015)

Concluding Discussion

This article has shown some of the ways in which processes of personalization have influenced how mindfulness neuroscience is communicated to the public, and how they may have played a part in the scientific takeoff and public appeal of the field. Personalization can be identified on several levels. First, communicational channels that by and large bypass traditional mass media outlets are used, which allows for a kind of personal branding of research approaches and results. This is facilitated and fueled by the overall communication orientation of science itself, but also the availability of new Web-based media. Here, experts can directly address audiences through, in this case, interactive personal Web sites and video channels, but also through more "traditional" means, such as self-help books. Second, the communicators become bridges between the inner and the outer spheres of science, between knowledge production and its reception, and between supposedly pure science and the specific tools, techniques, and practices offered for use in medicine and everyday life.[5] This "practical" neuroscience is presented as being evidence-based, even though research in the field is judged to be in its infancy. The personal experiences of the experts therefore play an important role for negotiating whether these practices are "scientific enough." One could say that the experts, and the practices they promote, help settle some of the facts that mindfulness neuroscience is still struggling to

---

[5]However, in the case of mindfulness neuroscience, these communicators are also nodes of knowledge transmission from "the outside" into the scientific sphere, since they mediate how Buddhist contemplative traditions should be understood in the context of neuroscience.

find. Hence, the "first fact—then popularization" model of science communication gets reversed. Third, scientific and medical knowledge is filtered through and made understandable by the personal experiences and stories of the expert communicators themselves. The messages are directed to a "You" to whom the communicators seek to relate by adjusting and translating research results into something that people can use to make sense of their own everyday experiences. Fourth, the audience is asked to invest not only in the scientific credibility of rigorous science, but also in the sincerity and good will of those communicating the science. If you believe their sincerity, you believe in their goals. If it is possible to demonstrate that mindfulness meditation influences the brain, that the brain influences well-being, and that the well-being of the individual can be scaled up to the well-being of the planet, then this secures a legitimate place for experts on mindfulness neuroscience in the public sphere. Communication becomes a moral vocation; it is critical that people engage in (neuropsychologically validated) mindfulness practices, but for that very reason it is equally critical that scientists reach out and talk about them.

This type of message puts the communicators in the role of world saviors in a suspiciously apolitical world, as saviors of the planet as well as of humankind. These universalistic claims come through by framing mindfulness mediation as a kind of mental training, suitable for everyone, by defining problems that are recognizable by "all," and by providing exercises that are—to varying degrees—based on rigorous scientific inquiry into how the mind and brain work. This framework opens up a space for claiming to speak on behalf of humankind, the planet, or the universe.

The analysis presented here also sheds new light on the understanding of science communication as a continuum. It is, at least in the case of Siegel and Davidson, the *same* people who advance scientific claims that also transform these into everyday practices to be used by laypeople. The meaning of mindfulness neuroscience is filtered through these people's personal experiences and is expressed through them. Whitley's observation (1985) that the public communication of science relies on depictions of decontextualized science is here turned on its head. The public face of mindfulness neuroscience involves a kind of recontextualization of science and medicine which allows for a more subjective, personal framing.

In light of this, communication becomes part of mindfulness neuroscience's momentum. It is not outside or temporally separated from it. If experts can convince people of the credibility and validity of scientific findings, and based on that knowledge convince them to act differently, then the findings will have the very real impact and relevance that motivated research in the first place. This will simultaneously lay the groundwork for obtaining more research funding and allow for even stronger claims of (a)political relevance. Mindfulness neuroscience becomes one of many ways in which the neurosciences have made people's problems manageable. This very manageability is supported by mechanisms of personalization and rests on the successful translation of research results into practical tools, on the coupling of expert knowledge and lay experience, and on the belief that individual practice has the power to change the world.

References

Aelst, V. P., Sheafer, T., & Stanyer, J. (2011). The personalization of mediated political communication: A review of concepts, operationalizations and key findings. *Journalism, 13*(2), 203–220.

Allan, S. (2011). Introduction: Science journalism in a digital age. *Journalism, 12*(7), 771–777.

Bauer, M. (2008). Paradigm change for science communication: commercial science needs a critical public. In D. Cheng, M. Claessens, T. Gascoigne, J. Metcalfe, B. Schiele, & S. Shi (Eds.), *Communicating science in social contexts* (pp. 7–25). New York: Springer.

Begley, S. (2007). *Train your mind, change your brain: How a new science reveals our extraordinary potential to transform ourselves*. New York: Ballantine Books.

Dr. Dan Siegel website. (2013a). Resources. http://www.drdansiegel.com/resources/. Accessed on October 23, 2013.

Dr. Dan Siegel website. (2013b). Home. http://www.drdansiegel.com/ Accessed on October 23, 2013.

Bucchi, M. (1998). *Science and the media. Alternative routes in scientific communication*. London & New York: Routledge.

Bucchi, M. (2013). Style in science communication. *Public Understanding of Science, 22*(8), 904–915.

Caulfield, Timothy, & Condit, Celeste. (2012). Science and the sources of hype. *Public Health Genomics, 15* (3–4), 209–217.

Center for Investigating Healthy Minds. (2013). Retrieved from http://www.investigatinghealthyminds.org/cihmCenter.html Accessed on October 23, 2013.

Chiesa, A., & Serretti, A. (2010). A systematic review of neurobiolgoical and clinical features of mindfulness meditations. *Psychological Medicine, 40*(8), 1239–1252.

Choudhury, S., Nagel, S. K., & Slaby, J. (2009). Critical neuroscience: linking neuroscience and society through critical practice. *BioSocieties, 4*(1), 61–77.

Creswell, D. J., Way, B. M., Eisenberger, N. I., & Lieberman, M. D. (2007). Neural correlates of dispositional mindfulness during affect labeling. *Psychosomatic Medicine, 69*(6), 560–565.

Davidson, R. J. (2012). *The emotional life of your brain: How to change the way you think, feel and live*. London: Hodder & Stoughton.

Davidson, R. J., Kabat-Zinn, J., Schumacher, J., Rosenkranz, M., Muller, D., Santorelli, S. F., et al. (2003). Alterations in brain and immune function produced by mindfulness meditation. *Psychosomatic Medicine, 65* (4), 564–570.

Davidson, R. J., & McEwen, B. S. (2012). Social influences on neuroplasticity: Stress and interventions to promote well-being. *Nature Neurosience, 15*(5), 689–695.

Doidge, N. (2007). *The brain that changes itself: Stories of personal triumph from the frontiers of brain science*. New York: Viking.

Dolby, S. K. (2005). *Self-help books: Why Americans keep reading them*. Urbana: University of Illinois Press.

Ehrenreich, B. (2009). *Smile or die: How positive thinking fooled America and the world*. London: Granta.

Ekström, A. (Ed.). (2004). *Den mediala vetenskapen*. Nora: Nya Doxa.

Ekström, A. (Ed.). (2011). *History of participatory media: Politics and publics, 1750-2000*. New York: Routledge.

Farb, N., Segal, Z., Mayberg, H., Bean, J., McKeon, D., Fatima, Z., et al. (2007). Attending to the present: Mindfulness meditation reveals distinct neural modes of self-reference. *Social Cognitive and Affective Neuroscience, 2*(4), 313–322.

Füredi, F. (2004). *Therapy culture: Cultivating vulnerability in an uncertain age*. London: Routledge.

Hanson, R. (2009). *Buddha's brain: The practical neuroscience of happiness, love & wisdom*. Oakland, CA: New Harbinger Publications.

Hanson, R. (2013a). Rick Hanson website retrieved from http://www.rickhanson.net/rick-hanson Accessed on October 23, 2013.

Hanson, R. (2013b). *Hardwiring happiness: The new brain science of contentment, calm, and confidence*. New York: Harmony books.

Hebb, D. (1949). *The organization of behavior*. New York: Wiley.

Hermida, A. (2010). Revitalizing science journalism for a digital age. In D. Kennedy & G. Overholser (Eds.), *Science and the media* (pp 80–87). Cambridge: American Academy of Arts and Sciences. Pdf available at: https://www.amacad.org/multimedia/pdfs/publications/researchpapersmonographs/scienceMedia.pdf

Heuman, L. (2014). Don't believe the hype. *Tricycle blog*, October 1. Accessed on October 9, 2014. http://www.tricycle.com/blog/don%E2%80%99t-believe-hype

Hilgartner, S. (1990). The dominant view of popularization: Conceptual problems, political uses. *Social Studies of Science, 20*(3), 519–539.

Holliman, R. (2010). From analogue to digital scholarship: Implications for science communication researchers. *Journal of Science Communication, 9*(3), 1–6.

Hölzel, B. K., Carmody, J., Vangel, M., Congleton, C., Yerramsett, S. M., Gard, T., et al. (2011). Mindfulness practice leads to increases in regional brain gray matter density. *Psychiatry Research Neuroimaging, 191*(1), 36–43.

Illouz, E. (2008). *Saving the modern soul: Therapy, emotions, and the culture of self-help*. Berkeley: University of California Press.

Jha, A. P., Krompinger, J., & Baime, M. J. (2007). Mindfulness training modifies subsystems of attention. *Cognitive Affective & Behavioral Neuroscience, 7*(2), 109–119.

Johnson, S. (2004). *Mind wide open: Your brain and the neuroscience of everyday life*. New York: Scribner.

Johnson Thornton, D. (2011). *Brain culture: Neuroscience and popular media*. New Brunswick, New Jersey, US, London, UK: Rutgers University Press.

Jones, E. G., & Mendell, L. M. (1999). Assessing the decade of the brain. *Science, 284*(5415), 739.

Lopez, D. S. (2008). *Buddhism and science: A guide for the perplexed*. Chicago: University of Chicago Press.

Lopez, D. S. (2012). *The scientific Buddha: A short and happy life*. Yale: Yale University Press.

Lutz, A., Slagter, H. A., Dunne, J. D., & Davidson, R. J. (2008). Attention regulation and monitoring in meditation. *Trends in Cognitive Sciences, 12*(4), 163–169.

Ortega, F., & Vidal, F. (2011). *Neurocultures: Glimpses into an expanding universe*. Frankfurt am Main: Peter Lang.

Peters, H. P. (2013). Gap between science and media revisited: Scientists as public communicators. *Proceedings of the National Academy of Sciences, 110* (Supplement 3), 14102–14109.

Peters, H. P., et al. (2008). Medialization of science as a prerequisite of its legitimization and political relevance. In D. Cheng, M. Claessens, T. Gascoigne, J. Metcalfe, B. Schiele, & S. Shi (Eds.), *Communicating science in social contexts* (pp. 71–92). New York: Springer.

Pickersgill, M., Martin, P., & Cunningham-Burley, S. (2014). The changing brain: Neuroscience and the enduring import of everyday experience. *Public Understanding of Science*, Online first, April 9: 1–15.

Restak, R. (2001). *Mozart's brain and the fighter pilot: Unleashing your brain's potential*. New York: Harmony Books Three Rivers Press.

Rödder, S., Franzén, M., & Weingart, P. (2012). *The sciences' media connection: Public communication and its repercussions*. Dordrecht: Springer.

Rose, N., & Abi-Rached, J. M. (2013). *Neuro: The new brain sciences and the management of the mind*. Princetone, Woodstock: Princeton University Press.

Satel, S. L., & Lilienfeld, S. O. (2013). *Brainwashed: The seductive appeal of mindless neuroscience*. New York: Basic Books.

Schwartz, J., & Beyette, B. (1996). *Brain lock: Free yourself from obsessive-compulsive behavior: A four-step self-treatment method to change your brain chemistry*. New York: Regan Books.

Siegel, D. J. (2010). *Mindsight: The new science of personal transformation*. New York: Random House.

Spyros, A. (1980). Gene cloning by press conference. *New England Journal of Medicine, 302*, 743–746.

Tang, Y.-Y., Hölzel, B. K., & Posner, M. I. (2015). The neuroscience of mindfulness meditation. *Nature Reviews Neuroscience, 16*, 213–225.

Tang, Y.-Y., & Posner, M. I. (2012). Special issue on mindfulness neuroscience. *Social Cognitive and Affective Neuroscience, 8*(1), 1–3.

Trench, B. (2012). Scientits' blogs: Glimpses behind the scenes. In S. Rödder, M. Franzén, & P. Weingart (Eds.), *The sciences' media connection: Public communication and its repercussions* (pp. 273–289). Dordrecht: Springer.

Wellspring Institute for Neuroscience and Contemplative Wisdom. (2013). http://www.wisebrain.org/wellspring-institute Accessed on October 23, 2013.

Whitley, R. (1985). Knowledge producers and knowledge acquirers. In T. Shinn & R. Whitley (Eds.), *Expository science, forms and functions of popularization* (pp. 3–28). Dordrecht: Reidel.

Wilson, J. (2014). *Mindful America. The mutual transformation of Buddhist meditation and American culture*. New York: Oxford University Press.

YouTube channel Dr. Dan Siegel. (2013). Welcome to Dr. Dan Siegel's website: DrDanSiegel.com. http://youtu.be/0cFoyDayDtU. Accessed on November 5, 2013.

YouTube channel UWHealthyMinds. (2013). Transform your mind, change your brain. https://youtu.be/7tRdDqXgsJ0?list=PLD6BE37B6F07C0DAD. Accessed on November 11, 2013.

YouTube channel Healthy Minds. (2015). Meet Richard Davidson, Founder of the Center for Healthy Minds. https://youtu.be/hyJQ-UdUnkU. Accessed on February 22, 2016.

Websites and YouTube Channels

http://www.rickhanson.net/
http://greatergood.berkeley.edu/
http://www.wisebrain.org/
http://www.youtube.com/user/drrhanson
http://richardjdavidson.com/
http://www.investigatinghealthyminds.org/index.html
http://www.youtube.com/user/UWHealthyMinds
http://drdansiegel.com/
http://www.mindsightinstitute.com/
http://www.youtube.com/mindsightinstitute

The Ultimate Rx: Cutting Through the Delusion of Self-cherishing

Lisa Dale Miller

Introduction

Western and Buddhist psychologies acknowledge the significant role distorted self-narratives play in poor mental health. But these two disciplines hold divergent views on the utility of "cherishing the self." Western psychology claims high self-esteem is a requirement for self-confidence, happiness, and success. Buddhist psychology asserts wisdom and compassion are the forerunners of genuine confidence and sustainable personal and collective well-being. It further states that endemic *self-cherishing*—the habitual reification of distorted hyper-egoic self-narratives—is a primary source of mental and emotional affliction. Yet, Buddhist psychology also affirms the innate capacity of all human beings to awaken from *avidyā*, the mental suffering of self-cherishing.

This chapter explicates Western and Buddhist psychological models of self, Buddhist theories of *not-self*, and *conventional and ultimate self-cherishing*, and outlines a clinical approach that help patients recognize self-cherishing mentation and lessen its deleterious effects. Reducing cognitive–affective fixation on self-narratives of exceptionality or brokenness increases capacity for accurate self-evaluation and self-regulation.

This clinical method focuses on imparting the following skills: cultivating greater meta-awareness and mindful self-reflectivity; engaging in dialogic inquiry to distinguish distorted inner narratives from experiential reality; and sustaining awareness of the actuality of experience through intentional use of *embodied presencing*. The dialectic, somatopsychotherapeutic, and experiential quality of these skills make them easy to learn and self-apply. As the chapter's patient accounts illustrate, cutting through self-cherishing is particularly beneficial for individuals struggling with depressive, anxious, trauma-related symptoms, chronic illnesses, and addictions.

Western Psychology on the Self

William James' seminal chapter "The Consciousness of Self" (1890) launched psychology's phenomenological study of the *self*. That approach was eclipsed in the early twentieth century by psychoanalysis and behaviorism (Leary and Tangney 2003). It took several more decades for innovators from object relations psychology (Horney 1950; Adler 1964), self-actualization psychology (Maslow 1973), and cognitive behavioral psychology (Beck 1979) to reignite investigation into the characteristics of a healthy self, and explore the role distorted self-narratives play in neurosis, anxiety, depression, and psychosis.

By the late twentieth century, three new research methods—systems modeling, examining the self in context, and seeking the neural

L.D. Miller (✉)
Los Gatos, CA, USA
e-mail: ldmil108@gmail.com

© Springer International Publishing Switzerland 2016
R.E. Purser et al. (eds.), *Handbook of Mindfulness*,
Mindfulness in Behavioral Health, DOI 10.1007/978-3-319-44019-4_22

correlates of self—led to the current consensus view of *self as a constructed, process-oriented, and context-adaptive system* (Damasio 2012; Northoff 2013). Though these advances proliferated numerous self-function models and a plethora of terms for describing self-processes and self-identities (see Fig. 22.1), the fact is twenty-first century psychology is no closer to definitively isolating a self.

However, sociologists and psychologists have agreed upon five basic categories of *self*: the whole person or unitary self; a personality (all or parts); an experiencing subject "I" or object "Me"; a collection of perceptions, thoughts, and feelings; and an agentic doer. A further simplification delimits all descriptions of self to three basic mentative processes: attention, cognition, and regulation (Mitchell 2003; Baumeister and Vohs 2012).

Another accepted framework distinguishes *self* from *identity*. *Self* is, "a feeling that something is about 'me'" (Oyserman et al. 2012). *Self-concepts* are the mental supports a person uses to navigate and make sense of the world and their place in it—motivations, goals, expectations, beliefs, and judgments. Self and self-concepts contribute to *identity*—an enduring, yet shifting inner narrative about self (e.g., who I am, was, or will be.) Identity is formed from personal traits, familial/relational characteristics, group memberships, and self-views about personal value and authenticity

over a lifetime (Oyserman and Markus 1998). Most social science theories also accept the stable *and* variable nature of self. For instance, in mid-life, I am the same being as the 10-year-old Lisa, yet I am also an entirely different person.

Buddhist Psychology on the Self

> We all have direct acquaintance with a self, the apparent source of the phenomenal unity of our perceptual and introspective experiences. Yet… it is notoriously difficult to provide an account of just what this thinking, feeling, remembering, planning, experiencing entity called the self *is* (Klein 2012, p. 617).

The self is an intimate enigma. We know it well, yet we know it not. To accomplish direct insight into the phenomenology of mind, the Buddha and successive Buddhist yogi-philosophers trained in various analytical meditation practices, now known as first-person contemplative research tools. The Buddha realized that self is, in actuality, *not-self (anātman)*—a collection of impermanent, interdependently existing cognitive–affective functions (*aggregates*) that together create the appearance of a separate, enduring entity (Bodhi 2000, 2003).

"Buddhist and scientific analyses of consciousness share a certain formal similarity. They both ask how things occur, not what they are—a

Descriptive Terms for Self-processes

Ego	Self-doubt
Ego defense	self-discrepancy
Ego identity	self-esteem
Ego integrity	Self-efficacy
Ego strength	Self-evaluation
Ideal self	Self-identitfication
Identity	Self-worth
Self-acceptance	Self-schema
Self-actualization	Self-perception
Self-appraisal	Self-regulation
Self awareness	Self-deception
Self-assessment	Self-denial
Self-blame	Self-trust
self-concept	Self-conscious
self-control	Self-regard
self-loathing	Self-care

(Adapted from Leary & Tangney 2003)

Descriptive Terms for Self-identities

Psychophysical be-er
Agentive decider and doer
Subjective experiencer
Autonomous subjective entity
Belief generator
Social and familial participant
Meaning-maker
Life determiner
Directional motivator
Intuitive being
Instinctual senser
Rational thinker
Emotional relator
Truth seeker
Spiritual striver
Creator and destroyer

Fig. 22.1 Describing self-process and self-identity

question that is answered by causes and conditions, not by essences or entities" (Waldron 2008, p. 7). Buddhist topology of mind has much in common with theories recently put forth by cognitive neuroscientists and philosophers who envision a self assembled from nested neural-computational layers of increased complexity (Damasio 2012; Siderits et al. 2010).

These models postulate a *minimal self* receiving, processing, and responding to sensory inputs from the body and its environs to accomplish basic life functioning. Much of this activity is unconscious and autonomic. The next level is a *primitive egoic self*: a pre-reflective phenomenal self-awareness capable of determining boundedness (e.g., This is me, that is not me.) This nascent ego is the basis of the fully formed "I" (Vago and Silbersweig 2012).

Buddhist psychology advanced an analogous topology of self, elucidated in great detail in the historical Buddha's teachings on *Dependent Origination* (*paticca-samuppāda*) and in later Yogācāra teachings on the structure of consciousness (Miller 2014). In these models, the minimal self is a *mere subjective receiver* and the primitive egoic self is *manas*, the "I"-"you" delineator and dispositional determiner.

Without contemplative training, the human mind naïvely attributes thingness to the felt-sense of a self (Garfield 2014). And who could blame us for this error? This illusory self appears to really exist inside a body, possess a stable autobiographical identity, have agency and capacity for directed action, and spew forth a continual stream of personal thoughts, feelings, opinions, and desires. And while all this mentation might feel very intimate and true, its self-possession is as illusory as the apparitional entity from which it supposedly emanates.

In the midst of such apparencies, Buddhist psychology questions the usefulness of cherishing the mentative output of what it considers a deluded mind. More importantly, Buddhist psychology considers this fundamental confusion—*avidyā* (*primordial ignorance* or *self-cherishing*)—and its concomitant cognitive–affective afflictions (*kleśas*) and distorted mental proliferations

(*papañca-saññā-sankhā*) the root cause of all human suffering (Tsering 2006; Miller 2014).

That conclusion is grounded in a primary tenet of Buddhist psychology, which holds that all phenomena (internal and external) can be understood both conventionally and ultimately (Garfield 2014). Conventional descriptions of objective reality affirm mundane, broadly agreed-upon properties of entities and objects. Conventional truths are the product of ordinary apperception, or what Buddhist psychology calls *obscured perception*. Ultimate truths represent unobscured, nonordinary ontological insights into the compounded, interdependent nature of objective reality and its myriad manifestations—including the self and its views (Brown and Ryan 2004). "Each of us is constructing our own reality, and understanding how we do this becomes crucial to our ability to experience happiness and meaning in our lives." (Olendzki 2003, p. 17).

Western Psychology on Distortions of Self

Introspection produces self-views—the lens through which we perceive and construe experience (Swann et al. 2003). This skill requires the subjective "I" to reflexively contemplate itself as an object of perception (Northoff 2013). Sincere introspection and accurate self-evaluation are essential for mental health. Conversely, excessive rumination and distorted self-appraisals foster depression, anxiety, egocentrism, and self-and-other harming. Chronically depressed people hold disproportionately aversive narratives about self and world that amplify mental anguish and anxiety, and undermine self-efficacy and life satisfaction (Mendlowicz and Stein 2014; Beck et al. 2011). Individuals with schizophrenia and dissociative disorders suffer tremendously from fragmented self-representations and distorted worldviews (Westin and Heim 2003). Disrupted attachment and childhood trauma can instigate ruptures in the self, causing emotion dysregulation and various personality disorders (van der Kolk 1987; Briere and Scott 2014).

Buddhist Psychology on Distortions of Self

> We must see that the root of all our suffering, all our pain, all our confusion is our own self-clinging, our sense of self–importance (Ponlop 2010, p. 87).

We are fragile beings; subject to the vicissitudes of human existence. It is a stark reality human beings tend to avoid. Though pain and pleasure are an inherent part of life in a human body, we are inclined to meet these experiences reactively rather than responsively. Pain induces aversion and withdrawal. Pleasure stimulates wanting and approach. *Suffering is the mental and emotional distress produced by an agitated mind overwhelmed with reactive resistance and longing.* Restated in Buddhist psychological terms: resistance to pain causes *aversion (dosa)*; longing for pleasure causes *craving (lobha)*; overwhelm arises from not recognizing the pervasive *self-delusions (avidyā)* which give rise to all afflictive mind states (*kleśas*).

No wonder the self's distorted narratives of omniscience, significance, and supremacy seem so enticing. How comforting to believe we are the master controller! "Humans seem to be unique in their preference for such self-delusions. In fact, humans would make better choices… if they did not believe that they personally could control what are, in actuality, chance outcomes" (Baumeister and Vohs 2012). So much human suffering results from this fundamental mis I"dentification, known in Buddhist psychology as *self-cherishing* (Tsoknyi and Swanson 2012).

Conventional Self-cherishing

Do not equate self-cherishing with accurate self-regard or basic human goodness. *Self-cherishing is deeply painful mentalizing; an all-consuming cognitive–affective fixation on distorted, hyper-egoic self-narratives.* As you read these descriptions, bear in mind that though mental health clinicians primarily work with people struggling with severe presentations of self-cherishing, because *avidyā* is intrinsic to the human condition, we all suffer its deleterious effects to a lesser or greater extent.

Conventional self-cherishing is: (1) putting one's importance and interests above that of others; (2) a strong belief in or over-identification with feelings of separateness and/or animosity toward other beings; (3) unawareness of the mechanisms by which distorted self-narratives cause inner and outer harming (Jinpa 2011; Mitchell and Wiseman 2003).

Conventional self-cherishing has both positive and negative manifestations:

Positive self-cherishing is: (1) pervasive self-schemas of arrogance, exceptionality, and entitlement; (2) heedless, reckless, impudent, self-satisfying conduct; (3) obsessive fixation on one's superiority and specialness; and (4) compulsive affirming of inner exceptionality.

Negative self-cherishing is: (1) pervasive self-schemas of brokenness, inferiority, self-loathing, self-blaming, and unworthiness; (2) fearful, over-cautious, acquiescent, self-repudiating conduct; (3) obsessive fixation on one's inferiority and insignificance; and (4) impulsive affirming of inner brokenness.

Is Conventional Self-cherishing Comparable with Self-esteem?

I grapple with this question each time I do an intake with a new patient visibly struggling with negative self-cherishing. Inevitably at some point in our first encounter they will exclaim with great sincerity, "I know all my problems come from low self-esteem. If I had high self-esteem everything would be fine." I understand why they cling to this supposed remedy. For the last twenty-five years Western psychology and the American educational system have touted high self-esteem as a cure for all manner of deficiencies. But is it?

Research has delineated two types of self-esteem: *explicit*—conscious self-evaluation derived from external boosts and prompts, and; *implicit*—unconscious internally derived dispositional self-evaluations.

(1) Though self-esteem has some relationship with psychological well-being, no direct causal link has been found between happiness and high self-esteem (Baumeister et al. 2003). In fact, the pursuit of high

self-esteem turns out to be quite problematic (Neff 2011). Adults who test high in self-esteem claim to be more likable, attractive, and have better relationships than those with low self-esteem. But objective measures disconfirm most of these self-views (Judge and Bono 2001). *That suggests the presence of positive self-cherishing.*

(2) Low explicit self-esteem can predict later depression. But depression has no effect on levels of implicit self-esteem (Orth et al. 2008; Brown 2014). Depressed individuals crave self-esteem boosts, but react with strong aversion when self-esteem prompts are offered. Moreover, excessive wanting of self-esteem is a predictor of poor mental health outcomes (Bushman et al. 2012). *That suggests the presence of negative self-cherishing.*

(3) How about self-esteem in education? Though today's young adults place more value on receiving self-esteem boosts, explicit self-esteem exerts no influence on improving K-12 or later academic performance. Enhanced academic achievement is an outcome of self-discipline and self-regulation (Di Giunta et al. 2013; Valentine et al. 2004).

So what is the actual effect of self-esteem boosting? Research shows a direct causal relationship between high explicit self-esteem and narcissism (Bosson et al. 2008; Campbell et al. 2002). In fact, data on levels of narcissism gathered over a 30-year period from 15,000 American college students showed significantly higher levels of narcissism in those tested in the 2000s (recipients of K-12 self-esteem curricula) than those tested in the 1980s and 1990s (Twenge et al. 2008). Of greatest concern is the finding that high self-esteem in combination with high narcissism produces higher levels of aggression (Bushman et al. 2009).

Based on these findings, it seems reasonable to correlate high and low self-esteem with positive and negative self-cherishing. And if the goal is to help patients free themselves from the suffering of conventional self-cherishing, replacing one self-fixation for another is not the appropriate psychotherapeutic intervention.

A Culture of Self-cherishing

War, crime, pollution, racism, income disparity, exploitation, and hunger thrive in societies where self-evaluation and self-regulation are undervalued (Strauman and Goetz 2012). Americans are now, more than ever, engrossed in the pursuit of personal welfare and the enterprise of indiscriminately generating good feelings about it—no matter the action or result. Our culture's normalized narcissism and assumed entitlement (Konrath et al. 2011) trump empathic concern and wise introspection.

On the one hand, we are imperfect beings; on the other hand, our imperfect perception allows us to experience the awe-inspiring beauty of ordinary existence. Yet by and large we don't. Most modern people live disaffected, disembodied lives, desperately seeking distraction in a miasma of work, relationships, substances, devices, and possessions. It is a sad state of affairs, particularly so, since an ordinary human mind and body are the necessary elements for achieving liberation from suffering. Buddhist psychological methods for transforming delusion into wisdom and self-centeredness into compassion are designed to awaken a *culture of self-cherishing* to its own-and-other suffering (Jinpa 2011).

Cutting Through Conventional Self-cherishing

The Buddha taught that liberation from suffering arises spontaneously in a mind devoid of craving, aversion, and delusion; freed from self-cherishing. If that sounds like a tall order the Dalai Lama reminds us that, "Overcoming these afflictions takes place not in one instance of awakening to not–self… one can gradually overcome… acquired afflictions… that are more naturally and deeply embedded" (Kabat-Zinn and Davidson 2012).

Working from within a frame of sudden *and* gradual awakening expands a clinician's range of beneficial psychotherapeutic interventions. Using tools that alleviate mild-to-moderate discomfort *and* stimulate mindful inquiry, empowers therapist and patient to examine everyday concerns while lessening more endemic and distorted self-processing.

The first step is to help patients recognize the habitual, dysfunctional, self-schemas they tend to orient to and fixate upon. Over time, these self-narratives habituate, reify, and become affectively stickier (Myers and Wells 2015; Hillis et al. 2015). This may explain why most people unquestioningly believe the mind's *self-cherishing content* and blindly follow its bidding; seemingly unaware of how these dysfunctional thoughts and feelings effect their daily life.

Pointing out the difference between distorted thoughts and actual occurrence cuts through experiential fusion and reveals the cognitive reification (Lutz et al. 2015) of believing thoughts are anything more than representational mental events (Condon et al. 2015). While this may sound like decentering and cognitive perspective taking (Butler et al. 2006), those interventions reify the substantiality of "alternative, more accurate" thoughts. Here the therapist helps a patient develop meta-awareness so they can directly perceive the insubstantial nature of all thoughts and land in the actuality of phenomenal experience.

Below are two examples of cutting through conventional self-cherishing in the midst of psychotherapeutic dialogue:

Patient: I was in my exercise class and the teacher mimicked the incorrect way to do the move and then said, "Don't do the exercise like this." That was exactly how I had done it! The thoughts of how I can't get anything right came (*habitual negative self-cherishing*) and I felt that awful melting in my belly and legs (*somatic distress*

response) because I couldn't do what the instructor asked.

Lisa: What was the first thing you remember feeling when you heard him say, "Don't do the exercise like this."

P: I felt fear. (*An automatic response arising from negative self-cherishing*)

L: He scared you? You felt threatened? (*Inviting inquiry into the actuality of experience*)

P: No… I immediately smiled and said, "That's exactly how I did it!" (*A direct authentic response*)

L: That sounds spontaneous, almost childlike… Not fearful. (*Pointing out the discrepancy between distorted narrative and actual experience*)

P: Well that was probably the "good girl" talking (*Reifying the conventional self-cherishing identity*)

L: Even when you said it right now, it sounded and looked spontaneous and genuine. You were smiling and excited. No sign of fearfulness. (*Naming conventional reality as it actually is*)

P: Well, I guess I was just embarrassed… and only listening to my inner story about how I always do everything wrong. (*Cutting through conventional self-cherishing*)

Another example from my textbook on Buddhist psychology (Miller 2014, pp. 79–80):

Patient: I kind of taught myself to meditate from books and I've listened to a couple of CDs. I get so bored with the breath. My thoughts are just so much more fascinating that I end up giving into them instead.

Lisa: (smiling) Yes, the mind can be very attracting! All those fascinating scenarios, ideas, narratives, and images. Who would want to be with ordinary breath? (*Joining in the distorted narrative to increase awareness of it*)

P: Exactly! Breath is so ordinary and boring.

L: Well, the ordinary is pretty extraordinary if we are willing to experience it devoid of our mental constructions about it. (*Naming conventional reality as it is*)

P: What do you mean? I experience breath when I think on breath?

L: Yeah, that's the thing… most people think about breath when they "meditate on breath" instead of experiencing the physical activity of breathing.

P: I don't see the difference. Thinking about something is the same as experiencing it.

L: Pick an object in this room. Any object. (We are not in my office.)

P: Okay… that notebook on the shelf.

L: Which one?

P: The one that says, Codes and Stipulations.

L: Can you describe what you see? (*Inviting them into the actuality of experience*)

P: It is a green notebook and when I see Codes and Stipulations I get really angry because it reminds me of all the years I spent going though pages and pages of regulations for my job. I am sure that notebook is full of pages of grief! (*Positive self-cherishing narrative takes over*)

L: Did you notice how your mind constructed a story about that notebook based upon your past history of dealing with regulations? And how it colored your experience of that object? How convinced you are that you know exactly what it is and that it is filled with pages and pages of grief? (*Directly pointing out the narrative distortions*)

P: Well what else could it be? That's the way they always are. (*Reifying the habitual, distorted self-cherishing*)

L: Shall we look and see what is inside?

P: Why not, I know what is going to be in there. [*Positive self-cherishing*]

L: (I reach for the notebook. When I open it, there is one page with about six sentences.)

P: Wow… I wonder how much I do this with everything else in my life? (*Cutting through conventional self-cherishing*)

Cutting through the distortions of conventional self-cherishing allowed these individuals to loosen internal fixation on false notions of inferiority and grandiosity. Experiential awareness is the great uncoupler and facilitator of a clarity of mind I call *provisional not-self*, which is unobstructed, accurate knowing of conventional reality.

This experience from a patient (2015) with a long history of PTSD illustrates the liberative effect of realizing *provisional not-self*: "For the first time in my life I feel like me as I truly am, without the constant feelings of threat and worry. This week I even had one day when I felt a strong wave of frustration, which before would have frozen me with anxiety. Instead I found myself just getting to know it. It was amazing! That made it possible to simply ask myself why I might be feeling frustrated? It was easy to figure out the reason and then determine what I could and could not do about it. That calmed me down and gave me a direction to follow. *None of the*

negative thoughts about myself came up like they normally would and there was no anxiety."

Embodied Cognition

> Cognition is not an event happening inside the system; it is the relational process of sense-making that takes place between the system and its environment (Thompson and Stapleton 2009, p. 26).

Identity is not just thought-based. Self-schemas are also shaped by repeated nonconscious interactions between the body and its surround (Vago and Silbersweig 2012). Theories of grounded cognition (Barsalou 2008) and enactivism (Colombetti 2010) are beginning to move cognitive neuroscience away from its long-held brain-centric views and toward a contextual approach to thinking, emoting, and sense-making.

Similarly, Western psychology and psychiatry have maintained a mind-brain-centric stance by largely ignoring the vital role bodily systems play in the cause and remediation of mental health disorders. Enactive approaches to mental health like Somatic Experiencing™ therapy (Levine 2010) and integrative psychiatry (Oulis 2013) consider central nervous system dysregulation and gut-brain-microbiota imbalances possible contributing factors in mental health disorders (Porges 2007; Luna and Foster 2015). Complementary approaches and somatosensory awareness techniques empower patients to employ mind-body skills in their daily lives (Payne and Crane-Godreau 2015; Horowitz 2014; Tang 2011; Staples et al. 2011).

I incorporate Somatic Experiencing™ interventions with psychotherapeutic dialogue, and when appropriate, offer targeted qigong practices. Both methods calm autonomic overactivation (Levine 2003; Sawynok and Lynch 2014; Payne et al. 2015) and teach patients to deliberately shift attention from anxious/ruminative mentation to actual objects of awareness (e.g., environmental stimuli and bodily sensations). Mind-body mental training is critical for increased attentional control and self-regulation (Tang and Posner 2012; Schmalzl et al. 2014; Johnson et al. 2014; Clark et al. 2015). Greater capacity for mindful embodiment means more presence and aliveness and that means less conventional self-cherishing.

Embodied Presence

> I've been having strong awareness of being in my body, in my own boundaries. It is a visceral, embodied experience of self-worth rather than a mental or emotional understanding of self-worth. It feels fully contained and deeply self-reflective. — patient comment

The next step in cutting through conventional self-cherishing is facilitating a patient's somatosensory awareness of basic aliveness—what I call *embodied presence*. Somatosensory awareness is a healing balm for the narrative absenting of conventional self-cherishing. Experiential focus (EF) (Farb et al. 2007, 2013) is the primary mechanism of embodied presence. EF is a "deliberate opening to the natural flow of sensory stimuli, body sensations, thoughts, and emotions, during which self-referencing is momentary and contextually integrated within an ever-shifting continuum of experience" (Miller 2014, p. 59).

Interoceptive attending to bodily and mental stimuli, and exteroceptive attending through the five senses to environmental stimuli, shows us what aliveness is like from the body's perspective. The physical system effortlessly navigates each moment *embedded* in, *extended* out into and *enactively* participating in its world (Di Paolo 2009). I call this hypo-egoic aliveness *organismic wisdom*—an innate intercorporeality that confers clarity, openness, confidence, and tranquility; even in the midst of distress.

Here a patient (2012) describes the calming and empowering effects of embodied presencing:

> I woke up about 3:00am with a fast heart rate and thoughts of impending doom. As I became aware of the bodily feeling of anxiety, I realized I was just relaxing into the experience of anxiety. My mind was with the reality of my bed, the warmth and comfort of being in the dark. No threat, no danger; just seeing mind for what it was. Reality was in my body. The thoughts of sudden death, impending doom, and things I'd done wrong, cycled through. But I stayed with them and they felt just like racing thoughts. Then I returned awareness to my body, my refuge.

The greatest benefit of embodied presence is *it empties the mind of self-narratives* (Vago and Silbersweig 2012). A patient (2015) who came to work with me after years of failed talk therapy for childhood sexual trauma and a resulting eating disorder describes it thusly:

> I never understood that what my mind creates isn't a current danger. I knew I was really scared all the time of everything, but didn't understand my mind was using the memory of abuse to flavor everything. Now I can play with or hear how the fear isn't real right now. The mind is the prison, the world is safe, and I am not the craving." That is the blossoming of organismic wisdom.

Another patient (2013) with a history of severe anxiety and depression extolls the benefits of knowing reality through the refuge of embodied presence:

> Though I'd done two years of CBT, when I was emotionally sunk with terror I still couldn't believe my own commonsense thoughts about what I was experiencing. *I was so used to knowing my body through the stories my mind told me about it, not the actual experience of it.* Now I soften into the uncomfortable sensations, calm down and realize it's actually okay. Irrational thoughts and emotions subside and I feel actual relief. I have been using these techniques daily and amazingly the panic episodes and depressive terrors lessened in frequency and intensity and now I don't have them anymore. I no longer automatically believe the stories my mind tells me about myself. *Now I seek out actual experience and trust in that.*

Resistance to impermanence is ego-created and adventitious. The bodily system perceives and willingly participates in the phenomenal interplay of moment-to-moment co-arising, co-existing and co-dissolving. This embodied intersubjectivity (Di Paolo and De Jaegher 2015) enables the system to know its boundaries while perceiving the surround as both other than and a part of itself. This is what I call *organismic compassion* (Fig. 22.2)—an intrinsically hypo-egoic altruistic inclination for equanimity, willingness, and connectedness (Sato et al. 2015; Warneken et al. 2007; List and Samak 2013).

A patient (2014) with chronic depression, anxiety, and a debilitating physical illness describes the interaction of naturally occurring

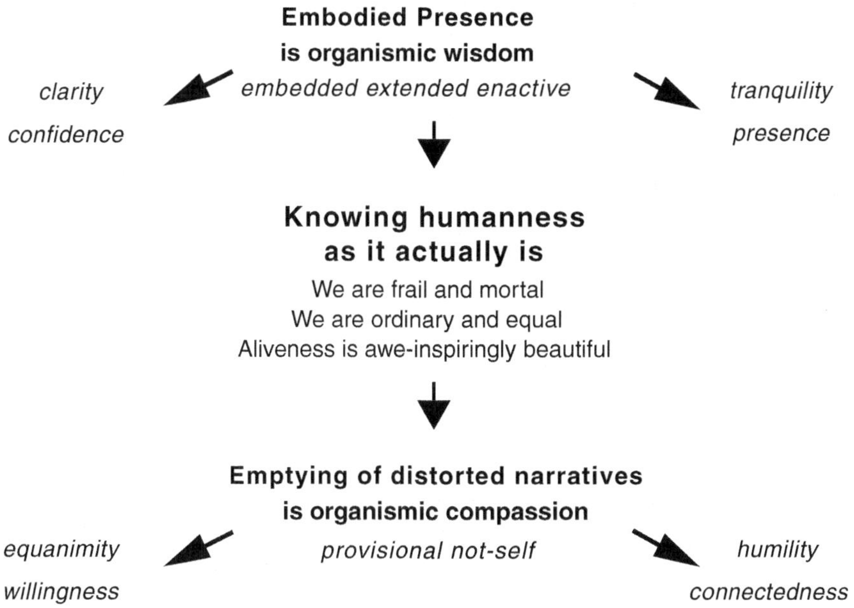

**Embodied Presence
is organismic wisdom**

clarity   *embedded extended enactive*   *tranquility*

confidence   *presence*

**Knowing humanness
as it actually is**

We are frail and mortal
We are ordinary and equal
Aliveness is awe-inspiringly beautiful

**Emptying of distorted narratives
is organismic compassion**

equanimity   *provisional not-self*   *humility*

willingness   *connectedness*

Fig. 22.2 Organismic wisdom and organismic compassion

and intentionally cultivated organismic wisdom and organismic compassion:

> I've had moments today where I've been able to relax into the moment more than in the past. Just now the melancholy tried to swallow me up; I used to always go with it and get drawn in by it. This time I was aware of it and did not get drawn in. I allowed myself to feel the little sadness but did not get hooked by the melancholy. I never realized I had a choice. This is where the new habit really begins. I can feel sad that it's Friday night and my only choice is to hang out alone in the living room with the dog, while my family is out having fun. Or I can feel the safety of this room, look out at the trees, feel the freedom to write, listen to the soft music and enjoy the companionship of a loving and devoted animal. I can feel sympathetic joy as she chews on her bone or nestles in her pillow. I should be able to do this any time, even when I'm not well rested. I just have to deliberately go this way instead of the other older way.

Another patient (2013) with a history of self-medicating severe social anxiety with alcohol had this experience:

> I arrived at the gathering and was not feeling anxious, like I normally do. I poured myself a glass of ice water and started saying hello to people. It was so strange I've known many of them for years and have intense opinions about them—lots of it negative or fearful. But I found myself not judging them. Just being with them as they are. I let them talk and really listened without the stream of anxious negative self talk. I was actually with them! After some time, the hostess asked if I would like a glass of wine. Normally I would have already had at least one or two glasses of wine by that time to lessen the anxiety of talking to people. I said yes, but it didn't really make much of a difference in how I felt about them or me. I realized the story I had been telling myself about my anxiety and what alcohol did for me, why I needed it, was completely false.

Notice the ease of being in a challenging situation emptied of the strain and judgmentality of conventional self-cherishing. A mind with less conventional self-cherishing actively transforms delusion into wisdom and self-centeredness into compassion. But is *provisional not-selfing* the same awakening the Buddha experienced and taught? Conventionally, yes. Ultimately, no.

Ultimate Self-cherishing

The patients who have generously shared their experiences in this chapter are a mix of meditators and nonmeditators. Their accounts suggest that as long as one is cultivating embodied awareness and applying real-time analytical inquiry, conventional self-cherishing is easily identified and cut through in the midst of lived experience.

But to be perfectly clear, the endgame of Buddhist psychology is not less conventional cognitive–affective affliction. Awakening out of *avidyā* or primordial unawareness, is the definitive objective. That can only be accomplished by cutting through the underlying, pervasive delusion of *ultimate self-cherishing*.

Though Western and Buddhist psychologies agree that unconscious drives, impulses, and feelings influence self-schemas and identity formation these two disciplines hold quite different conceptualizations of the unconscious. Fortunately, modern neuroscience is moving Western psychology away from solely equating the unconscious with subliminal urges (Bargh and Morsella 2008) and toward an embrace of a phenomenal, continually constructed unconscious (Eagleman 2011; Damasio 2012).

That notion has long been held by Buddhist yogi/philosophers who, "systematically examined and analyzed how unconscious processes determine the shape of our experiences and delimit the autonomy of our actions… since these observations arose out of traditions that had long before deconstructed any autonomous ego… the loss of an autonomous self proved a gain in understanding of self" (Waldron 2008, p. 1). Such profound self-understanding comes when the mind clearly comprehends unconscious material through direct awareness of its occurrence and results.

At this point, it might help to restate Buddhist psychology envisages an individual, "as a matrix of dependently related events in a state of flux" (Wallace 2009, p. 109) and construes the interdependent co-arising of mind-body with causes and conditions. Furthermore, the felt-sense of a self results from mere conceptual imputation. "That is, on the basis of either some aspect of the body (e.g., "I am tall") or some mental process (e.g., "I am content"), the self is conceptually imputed *upon something it is not*" (Wallace 2009, p. 109). Here Buddhist psychology points

out the basis of self-delusion: the habitual innate reification and mindless imputation of solidity and separateness to a compounded, identityless, entityless self.

Ultimate self-cherishing is: (1) Not recognizing the harmful effects of *innate reification*—the embedded habit of reflexively perceiving self and outer phenomena as separate, permanent, and substantially existent (Garfield 2014); (2) not recognizing all harm perpetrated upon oneself occurs because other beings are similarly suffering the delusion of innate reification; and (3) not recognizing all phenomena, including self, are mere appearances of the basic luminosity of mind (Ponlop 2010).

Buddhist psychology posits the possibility of liberation from ultimate self-cherishing because: (1) Wisdom and compassion are intrinsic characteristics of human consciousness; (2) every human being has the capacity to awaken from primordial unawareness and attain wise understanding of *not-self*; and (3) this is accomplished through steadfast practice of ethical conduct, vigilant contemplation, and direct recognition of the inseparability of awareness and emptiness (deathless, *śūnyatā*, *rigpa*)

Those suppositions might provoke one to question if cutting through ultimate self-cherishing belongs in the context of psychotherapy? The National Institute of Mental Health (2015) describes psychotherapy as, "a way to treat people with a mental disorder by helping them understand their illness. It teaches people strategies and gives them tools to deal with stress and unhealthy thoughts and behaviors." Though that description implies psychotherapists work primarily with conventional self-cherishing, psychotherapy often involves uncovering and healing more deeply entrenched psychological distress.

Recognizing ultimate self-cherishing exposes the unconscious anxiety of self-reification (Dahl et al. 2015)—a mental agitation, so pervasive we rarely question the dualism, egotism, separatism, animosity, and avarice it spawns. That is the very inner unsettledness which gives rise to every form of conventional self-cherishing and this makes the line between conventional and ultimate self-cherishing quite permeable. Cutting through the permeable awareness of self-reification requires a disruptive healing technology: one capable of producing a quiescent, diamond-like mind that clearly comprehends both gross and subtle levels of perception. *Only dedicated contemplative research can accomplish that task.* That means daily practice of formal meditation such as, concentration (*shamatha*), insight (*vipassanā*) and compassion (*karunā*) practices.

Concentration meditation develops attentional focus, mental stability, and serenity (Brefczynski-Lewis et al. 2007; Lutz et al. 2009; Wallace 2011). Insight meditation hones the mind's capacity for inquiry and deconstruction (Tang et al. 2015; Thera 2005; Goldstein 1993). *Shamatha* and *vipassanā* offer direct insight into the compounded, ever-shifting nature of all internal and external objects of perception.

Lovingkindness, *tonglen*, and equanimity meditations (Miller 2014) are critical for decreasing self-fixation (Dambrun and Ricard 2011). These practices lessen mental time spent judging, blaming, and hating others and oneself and create more inner space for the arising of genuine happiness and altruism (Shonin et al. 2014; Hoffman et al. 2011; Desbordes et al. 2012). Knowing the true source of our own suffering catalyzes deeper insight into other suffering (Singer and Klimecki 2014; Halifax 2012). Understanding that other beings may not know the source of their suffering or how to alleviate it, elicits empathy and compassionate motivation to help (Ozawa-de Silva et al. 2012). More than once I have witnessed a trauma survivor or a perpetrator startled by a spontaneous wave of compassion for the other's suffering. These are moving and astonishing experiences, an inspiring example of the power of self-dereification.

Egolessness and Liberative Insight

When we reach the point of having "looked" deeply and extensively into both body and mind and are unable to find the existence of a self, we'll experience a gap. At that point, we can rest our mind in a moment of pure openness, which we call nonconceptual awareness. That's the beginning of our discovery of selflessness (Ponlop 2010, pp. 86–87).

Egolessness in Buddhist psychology does not mean no ego. It means directly perceiving the essence of *not-self*—nonconceptual, pristine awareness. Like a mirror, the clarity and open-heartedness of nonconceptual awareness accurately, unbiasedly reflects whatever comes. Yet awareness remains unperturbed and unaltered, vivid and vibrantly illuminating reality (Mingyur and Swanson 2010). A unified mind, uncoupled from the habit of dulled, distorted perception, readily recognizes, "The entire phenomenal world is nothing other than empty appearance. It is not how we believe it to be—singular, permanent, intrinsic, and solid. That is ignorant mind's designation of things" (Kongtrul and Schmidt 2009).

Patients who regularly practice formal meditation do have direct insight into *not-self*—in and out of psychotherapy sessions. Here is a great example:

I am starting to see directly, or at least glimpse now and then, that the whole constellation of sensation and experience is *not* actually real or external or permanent, but workable.—patient comment

To cut through ultimate self-cherishing, Tibetan Buddhist teacher Thubten (2015) suggests:

Do not seek your problems in the body, your circumstances, or others. In the end you will find all your suffering resides in your own mind. This is the most profound insight one can have. Always be aware of what is occurring in the mind, vigilantly minding the contents, and always keep the benefit of others as the supreme guide for all actions. Many traditions talk about finding the space between thoughts. That is where there is no more storyline about who you were, are or will be; no more fantasies, anxieties, dreadful memories, depressions, hopes or fears. This is the sacred space within; where all struggles cease for a single moment. And for that moment complete belief in the mind is suspended and awe-inspiring reality is apparent. That is the healer for all wounds.

Conclusion

> By *revolution* I mean turning over the system that has made you go into analysis to begin with (Hillman and Ventura, 1992, p. 38).

Buddhist psychology is a revolutionary analytical and experiential therapeutic method. One that seeks to liberate the human mind from its deepest affliction: the delusion of self-cherishing. Facilitating an individual mind to free itself from its own ignorance, greed, and hatred is truly revolutionary when therapist and patient are motivated by selflessness. That psychotherapeutic work has the power to liberate not just the patient, but also all beings with whom that person interacts and by extension the society in which they reside.

That said Buddhist psychology was introduced to the West through the lens of an existing psychotherapeutic tradition more interested in healing individual selves than generating collective healing and societal transformation. This has led to Buddhist psychological interventions that are palatable to a self-focused therapeutic community. Additionally, America's disjointed and paltry mental health system mainly serves those who can afford treatment and are pursuing relief from the distress of modern living. "De-stress, love yourself, be happier" is how the mental health system markets and dispenses clinical mindfulness and clinical compassion interventions and even Buddhist psychology.

I am reminded of a recent much-touted randomized controlled trial of mindfulness-based cognitive therapy (MBCT) that showed MBCT is not superior to maintenance antidepressant treatment for the prevention of depressive relapse (Kuyken et al. 2015). Put that finding in context with the fact that antidepressants have repeatedly been shown to be no more effective than placebo. Moreover, a 2012 study found, "Increased capacity for decentering and curiosity may be fostered during MBCT," but also stated, "It is currently unknown whether the studied mediators and proposed mechanisms such as

mindfulness, rumination, compassion and decentering are unique to MBCT… Other therapies such as CBT [Cognitive Behavioral therapy], IPT [Interpersonal Psychotherapy], and antidepressant medicine (ADM) may also impact these variables" (Bieling et al. 2012).

But let us suppose these mediators and mechanisms do turn out to be unique to mindfulness-based interventions (MBIs) like MBCT. Should we then infer that reductions in reactivity, pain unpleasantness, and brain area activation, or increased awareness, calm, and self-compassion signals a fundamental movement from hyper-egoism to hypo-egoism, or the emergence of profound insights into the causes of self-and-other suffering? In other words, does lessening of mind-body symptoms indicate similar decreases in self-centeredness and self-separateness? I suspect that without delivering an analytical framework for wisdom and compassion, symptom relief becomes just another source of self-cherishing. Definitive answers to these questions might come from comparing MBIs to a standardized intervention that delivered similar meditation practices along with Buddhist psychological and philosophical teachings on impermanence, unsatisfactoriness and not-self.

In conclusion, the purpose of cutting through *conventional self-cherishing* is to eliminate all distorted self-narratives of exceptionality or brokenness and land intentionally and virtuously in intercorporeal engagement with mind and world. Cutting through the delusion of *ultimate self-cherishing* is the work of going beyond all concepts—fearlessly dereifying the self until all that remains is the innate luminosity of nonconceptual awareness. With that realization, one is able to wisely, compassionately and skillfully enact the total workability of each moment. Thus liberated from the suffering of *avidyā*, one's life energy naturally turns toward ending the suffering of all other beings. That revolutionary transformation is the penultimate aim of Buddhist psychology.

References

Adler, A. (1964). *Individual psychology of Alfred Adler* (Vol. 1154). New York: Harper Collins.

Bargh, J. A., & Morsella, E. (2008). The unconscious mind. *Perspectives on psychological science, 3*(1), 73–79.

Barsalou, L. W. (2008). Grounded cognition. *Annual Review of Psychology, 59*, 617–645.

Baumeister, R. F., Campbell, J. D., Krueger, J. I., & Vohs, K. D. (2003). Does high self-esteem cause better performance, interpersonal success, happiness, or healthier lifestyles? *Psychological Science in the Public Interest, 4*(1), 1–44.

Baumeister, R. F., & Vohs, K. D. (2012). Self-regulation and the executive function of the self. In M. R. Leary & J. P. Tangney (Eds.), *Handbook of self and identity* (2nd ed., pp. 180–197). New York: Guilford Press.

Beck, A., Crain, A. L., Solberg, L. I., Unützer, J., Glasgow, R. E., Maciosek, M. V., et al. (2011). Severity of depression and magnitude of productivity loss. *The Annals of Family Medicine, 9*(4), 305–311.

Beck, A. T. (1979). *Cognitive therapy and the emotional disorders*. New York: Penguin.

Bieling, P. J., Hawley, L. L., Bloch, R. T., Corcoran, K. M., Levitan, R. D., Young, T. L., et al. (2012). Treatment-specific changes in decentering following mindfulness-based cognitive therapy versus antidepressant medication or placebo for prevention of depressive relapse. *Journal of Consulting and Clinical Psychology, 80*(3), 365–372.

Bodhi, B. (2000). *The connected discourses of the Buddha: A translation of the Samyutta Nikāya* (p. 325). Boston, MA: Wisdom Publications.

Bodhi, B. (Ed.). (2003). *A comprehensive manual of Abhidhamma: The Abhidhammattha Sangaha*. Onalaska, WA: Pariyatti Editions.

Bosson, J. K., Lakey, C. E., Campbell, W. K., Zeigler-Hill, V., Jordan, C. H., & Kernis, M. H. (2008). Untangling the links between narcissism and self-esteem: A theoretical and empirical review. *Social and Personality Psychology Compass, 2*(3), 1415–1439.

Brefczynski-Lewis, J. A., Lutz, A., Schaefer, H. S., Levinson, D. B., & Davidson, R. J. (2007). Neural correlates of attentional expertise in long-term meditation practitioners. *Proceedings of the National Academy of Sciences, 104*(27), 11483–11488.

Briere, J. N., & Scott, C. (2014). *Principles of trauma therapy: A guide to symptoms, evaluation, and treatment*. Thousand Oaks: Sage Publications.

Brown, J. D. (2014). Self-esteem and self-evaluation: Feeling is believing. *Psychological Perspectives on the Self, 4*, 27–58.

Brown, K. W., & Ryan, R. M. (2004). Perils and promise in defining and measuring mindfulness: Observations from experience. *Clinical Psychology: Science and Practice, 11*(3), 242–248.

Bushman, B. J., Baumeister, R. F., Thomaes, S., Ryu, E., Begeer, S., & West, S. G. (2009). Looking again, and harder, for a link between low self-esteem and aggression. *Journal of Personality, 77*(2), 427–446.

Bushman, B. J., Moeller, S. J., Konrath, S., & Crocker, J. (2012). Investigating the link between liking versus wanting self-esteem and depression in a nationally representative sample of American adults. *Journal of Personality, 80*(5), 1453–1469.

Butler, A. C., Chapman, J. E., Forman, E. M., & Beck, A. T. (2006). The empirical status of cognitive-behavioral therapy: a review of meta-analyses. *Clinical Psychology Review, 26*(1), 17–31.

Campbell, W. K., Rudich, E. A., & Sedikides, C. (2002). Narcissism, self-esteem, and the positivity of self-views: Two portraits of self-love. *Personality and Social Psychology Bulletin, 28*(3), 358–368.

Clark, D., Schumann, F., & Mostofsky, S. H. (2015). Mindful movement and skilled attention. *Frontiers in Human Neuroscience, 9*, 297.

Colombetti, G. (2010). Enaction, sense-making and emotion. *Enaction: Toward a New Paradigm for Cognitive Science*, 145–164.

Condon, P., et.al. (2015). http://epublications.marquette.edu/cgi/viewcontent.cgi?article=1021&context=smv_imf

Dahl, C. J., Lutz, A., & Davidson, R. J. (2015). Reconstructing and deconstructing the self: Cognitive mechanisms in meditation practice. *Trends in Cognitive Sciences, 19*(9), 515–523.

Damasio, A. (2012). *Self comes to mind: Constructing the conscious brain*. New York: Vintage.

Dambrun, M., & Ricard, M. (2011). Self-centeredness and selflessness: A theory of self-based psychological functioning and its consequences for happiness. *Review of General Psychology, 15*(2), 138.

Desbordes, G., Negi, L. T., Pace, T. W., Wallace, B. A., Raison, C. L., & Schwartz, E. L. (2012). Effects of mindful-attention and compassion meditation training on amygdala response to emotional stimuli in an ordinary, non-meditative state. *Frontiers in Human Neuroscience, 6*.

Di Giunta, L., Alessandri, G., Gerbino, M., Kanacri, P. L., Zuffiano, A., & Caprara, G. V. (2013). The determinants of scholastic achievement: The contribution of personality traits, self-esteem, and academic self-efficacy. *Learning and Individual Differences, 27*, 102–108.

Di Paolo, E. (2009). Extended life. *Topoi, 28*(1), 9–21.

Di Paolo, E. A., & De Jaegher, H. (2015). Toward an embodied science of intersubjectivity: Widening the scope of social understanding research. *Frontiers in Psychology, 6*.

Eagleman, D. (2011). *Incognito: The hidden life of the brain*. New York: Random House.

Farb, N. A., Segal, Z. V., & Anderson, A. K. (2013). Attentional modulation of primary interoceptive and exteroceptive cortices. *Cerebral Cortex, 23*(1), 114–126.

Farb, N. A., Segal, Z. V., Mayberg, H., Bean, J., McKeon, D., Fatima, Z., et al. (2007). Attending to the present: Mindfulness meditation reveals distinct neural modes of self-reference. *Social Cognitive and Affective Neuroscience, 2*(4), 313–322.

Garfield, J. L. (2014). *Engaging Buddhism: Why it matters to philosophy*. New York: Oxford University Press.

Goldstein, J. (1993). *Insight meditation*. Boston: Shambhala Publications.

Halifax, J. (2012). A heuristic model of enactive compassion. *Current Opinion in Supportive and Palliative Care, 6*(2), 228–235.

Hillis, K., Paasonen, S., & Petit, M. (2015). *Networked affect*. Boston: MIT Press.

Hillman, J., & Ventura, M. (1992). *We've had a hundred years of psychotherapy–And the world's getting worse*. San Francisco: Harper Collins.

Hofmann, S. G., Grossman, P., & Hinton, D. E. (2011). Loving-kindness and compassion meditation: Potential for psychological interventions. *Clinical Psychology Review, 31*(7), 1126–1132.

Horney, K. (1950). *Neurosis and human growth: The struggle toward self-realization*. New York: WW Norton & Company: New York.

Horowitz, S. (2014). T'ai Chi and Qigong: Validated health benefits. *Alternative and Complementary Therapies, 20*(5), 263–269.

James, W. (1890). *The consciousness of self. The principles of psychology, 8*. New York: Dover.

Jinpa, T. (Ed.). (2011). *Essential mind training*. New York: Simon and Schuster.

Johnson, D. C., Thom, N. J., Stanley, E. A., Haase, L., Simmons, A. N., Pei-an, B. S., … & Paulus, M. P. (2014). Modifying resilience mechanisms in at-risk individuals: A controlled study of mindfulness training in Marines preparing for deployment. *American Journal of Psychiatry*.

Judge, T. A., & Bono, J. E. (2001). Relationship of core self-evaluations traits—self-esteem, generalized self-efficacy, locus of control, and emotional stability—with job satisfaction and job performance: A meta-analysis. *Journal of Applied Psychology, 86*(1), 80.

Kabat-Zinn, J., & Davidson, R. (Eds.). (2012). *The mind's own physician: A scientific dialogue with the Dalai Lama on the healing power of meditation*. Oakland: New Harbinger Publications.

Klein, S. B. (2012). The two selves: The self of conscious experience and its brain. In M. R. Leary & J. P. Tangney (Eds.), *Handbook of self and identity* (2nd ed.). New York: Guilford Press.

Kongtrul, D., & Schmidt, M. B. (2009). *Uncommon Happiness: The path of the compassionate warrior*. Kathmandu: Rangjung Yeshe Publications.

Konrath, S. H., O'Brien, E. H., & Hsing, C. (2011). Changes in dispositional empathy in American college students over time: A meta-analysis. *Personality and Social Psychology Review, 15*(2), 180–198.

Kuyken, W., Hayes, R., Barrett, B., Byng, R., Dalgleish, T., Kessler, D., … & Byford, S. (2015). Effectiveness and cost-effectiveness of mindfulness-based cognitive therapy compared with maintenance antidepressant treatment in the prevention of depressive relapse or recurrence (PREVENT): a randomised controlled trial. *The Lancet*.

Leary, M. R., & Tangney, J. P. (2003). The self as an organizing construct in the behavioral and social sciences. *Handbook of Self and Identity*, 3–14.

Levine, P. A. (2003). Panic, biology and reason: Giving the body its due. *US Association Body Psychotherapy Journal, 2*, 5–21.

Levine, P. A. (2010). *In an unspoken voice: How the body releases trauma and restores goodness*. Berkeley: North Atlantic Books.

List, J. A., & Samak, A. C. (2013). Exploring the origins of charitable acts: Evidence from an artefactual field experiment with young children. *Economics Letters, 118*(3), 431–434.

Luna, R. A., & Foster, J. A. (2015). Gut brain axis: Diet microbiota interactions and implications for modulation of anxiety and depression. *Current Opinion in Biotechnology, 32*, 35–41.

Lutz, A., Jha, A. P., Dunne, J. D., & Saron, C. D. (2015). investigating the phenomenological and neurocognitive matrix of mindfulness-related practices. *American Psychologist, 70*(7), 632–658.

Lutz, A., Slagter, H. A., Rawlings, N. B., Francis, A. D., Greischar, L. L., & Davidson, R. J. (2009). Mental training enhances attentional stability: Neural and behavioral evidence. *The Journal of Neuroscience, 29* (42), 13418–13427.

Maslow, A. H. (1973). *On dominance, self-esteem, and self-actualization*. Maurice Bassett.

Mendlowicz, M. V., & Stein, M. B. (2014). Quality of life in individuals with anxiety disorders. *American Journal of Psychiatry*.

Miller, L. D. (2014). *Effortless Mindfulness: Genuine mental health through awakened presence*. New York: Routledge.

Mingyur, R. Y., & Swanson, E. (2010). *Joyful wisdom*. New York: Three Rivers Press.

Mitchell, D. W., & Wiseman, J. (2003). *Transforming suffering: Reflections on finding peace in troubled times* (pp. 64–81). New York: Doubleday.

Mitchell, R. W. (2003). Subjectivity and self-recognition in animals. In M. R. Leary & J. P. Tangney (Eds.), *Handbook of self and identity* (pp. 567–593). New York: Guilford Press.

Myers, S. G., & Wells, A. (2015). Early trauma, negative affect, and anxious attachment: The role of metacognition. *Anxiety, Stress, & Coping*, (ahead-of-print), 1–16.

National Institute of Mental Health. (2015, September 10). *Psychotherapies*. Retrieved from http://www.nimh.nih.gov/health/topics/psychotherapies/index.shtml

Neff, K. D. (2011). Self-compassion, self-esteem, and well-being. *Social and Personality Psychology Compass, 5*(1), 1–12.

Northoff, G. (2013). Brain and self—A neurophilosophical account. *Child Adolescent Psychiatry and Mental Health, 7*, 28.

Olendzki, A. (2003). Buddhist psychology. In S. R. Segall (Ed.), *Encountering Buddhism: Western psychology and Buddhist teachings*. New York: SUNY Press.

Orth, U., Robins, R. W., & Roberts, B. W. (2008). Low self-esteem prospectively predicts depression in adolescence and young adulthood. *Journal of Personality and Social Psychology, 95*(3), 695.

Oulis, P. (2013). Toward a unified methodological framework for the science and practice of integrative psychiatry. *Philosophy, Psychiatry, & Psychology, 20*(2), 113–126.

Oyserman, D., Elmore, K., & Smith, G. (2012). Self, self-concept, and identity. In M. R. Leary & J. P. Tangney (Eds.), *Handbook of self and identity* (2nd ed., pp. 68–104). New York: Guilford Press.

Oyserman, D., & Markus, H. R. (1998). Self as social representation. In S. U. Flick (Ed.), *The psychology of the social* (pp. 107–125). New York: Cambridge University Press.

Ozawa-de Silva, B. R., Dodson-Lavelle, B., Raison, C. L., Negi, L. T., Silva, B. R. O., & Phil, D. (2012). Compassion and ethics: scientific and practical approaches to the cultivation of compassion as a foundation for ethical subjectivity and well-being. *Journal of Healthcare, Science and the Humanities, 2*, 145–161.

Payne, P., & Crane-Godreau, M. A. (2015). The preparatory set: a novel approach to understanding stress, trauma, and the bodymind therapies. *Frontiers in Human Neuroscience, 9*, 178.

Payne, P., Levine, P. A., & Crane-Godreau, M. A. (2015). Somatic experiencing: using interception and proprioception as core elements of trauma therapy. *Frontiers in Psychology, 6*.

Ponlop, R. D. (2010). *Rebel Buddha: On the road to freedom*. Boston: Shambhala Publications.

Porges, S. W. (2007). The polyvagal perspective. *Biological Psychology, 74*(2), 116–143.

Sato, N., Tan, L., Tate, K., & Okada, M. (2015). Rats demonstrate helping behavior toward a soaked conspecific. *Animal Cognition*, 1–9.

Sawynok, J., & Lynch, M. (2014). Qualitative analysis of a controlled trial of qigong for fibromyalgia: Advancing understanding of an emerging health practice. *The Journal of Alternative and Complementary Medicine, 20*(8), 606–617.

Schmalzl, L., Crane-Godreau, M. A., & Payne, P. (2014). Movement-based embodied contemplative practices: Definitions and paradigms. *Frontiers in Human Neuroscience, 8*.

Shonin, E., Van Gordon, W., & Griffiths, M. D. (2014). Loving-kindness and compassion meditation in psychotherapy. *Thresholds: Quarterly Journal of the Association for Pastoral and Spiritual Care and Counseling, Spring Issue*, 9–12.

Siderits, M., Thompson, E., & Zahavi, D. (Eds.). (2010). *Self, no self?: Perspectives from analytical, phenomenological, and Indian traditions*. New York: Oxford University Press.

Singer, T., & Klimecki, O. M. (2014). Empathy and compassion. *Current Biology, 24*(18), R875–R878.

Staples, J. K., Abdel Atti, J. A., & Gordon, J. S. (2011). Mind-body skills groups for posttraumatic stress disorder and depression symptoms in Palestinian children and adolescents in Gaza. *International Journal of Stress Management, 18*(3), 246.

Strauman, T. J., & Goetz, E. L. (2012). Self-regulation failure and health. In M. R. Leary & J. P. Tangney (Eds.), *Handbook of self and identity* (2nd ed., pp. 247–267). New York: Guilford Press.

Swann, W. B., et al. (2003). Self-verification: The search for coherence. In M. R. Leary & J. P. Tangney (Eds.), *Handbook of self and identity* (pp. 367–383). New York: Guilford Press.

Tang, Y. Y. (2011). Mechanism of integrative body-mind training. *Neuroscience Bulletin, 27*(6), 383–388.

Tang, Y. Y., Hölzel, B. K., & Posner, M. I. (2015). The neuroscience of mindfulness meditation. *Nature Reviews Neuroscience, 16*(4), 213–225.

Tang, Y. Y., & Posner, M. I. (2012). Theory and method in mindfulness neuroscience. *Social Cognitive and Affective Neuroscience*, nss112.

Thera, N. (2005). *The heart of Buddhist meditation: Satipatthāna: A handbook of mental training based on the Buddha's way of mindfulness, with an anthology of relevant texts translated from the Pali and Sanskrit*. Buddhist Publication Society.

Thompson, E., & Stapleton, M. (2009). Making sense of sense-making: Reflections on enactive and extended mind theories. *Topoi, 28*(1), 23–30.

Thubten, A. (2015). *Unpublished lecture*. Richmond: Dharmata Foundation.

Tsering, G. T. (2006). *Buddhist psychology: The foundation of Buddhist thought* (Vol. 3). Somerville: Wisdom Publications.

Tsoknyi, R., & Swanson, E. (2012). *Open Heart, Open Mind: Awakening the power of essence love*. New York: Harmony Books.

Twenge, J. M., Konrath, S., Foster, J. D., Campbell, W. K., & Bushman, B. J. (2008). Further evidence of an increase in narcissism among college students. *Journal of Personality, 76*(4), 919–928.

Vago, D. R., & Silbersweig, D. A. (2012). Self-awareness, self-regulation, and self-transcendence (S-ART): A framework for understanding the neurobiological mechanisms of mindfulness. *Frontiers in Human Neuroscience, 6*.

Valentine, J. C., DuBois, D. L., & Cooper, H. (2004). The relation between self-beliefs and academic achievement: A meta-analytic review. *Educational Psychologist, 39*(2), 111–133.

van der Kolk, B. A. (1987). The separation cry and the trauma response: Developmental issues in the psychobiology of attachment and separation. *Psychological Trauma*, 31–62.

Waldron, W. S. (2008). A Buddhist theory of unconscious mind (ālaya-vijñāna). *Handbook of Indian Psychology*, 105–128.

Wallace, B. A. (2009). *Contemplative science: Where Buddhism and neuroscience converge*. New York: Columbia University Press.

Wallace, B. A. (2011). *Minding closely: The four applications of mindfulness*. New York: Snow Lion Publications.

Warneken, F., Hare, B., Melis, A. P., Hanus, D., & Tomasello, M. (2007). Spontaneous altruism by chimpanzees and young children. *PLoS Biology, 5* (7), e184.

Westin, D., & Heim, A. K. (2003). Disturbances of self and identity in personality disorders. In M. R. Leary & J. P. Tangney (Eds.), *Handbook of self and identity* (pp. 643–664). New York: Guilford Press.

Part IV
Mindfulness as Critical Pedagogy

Critical Integral Contemplative Education

David Forbes

What the Hell Is Water?

Wallace's (2005) commencement address at Kenyon College, known as "This is Water," leads off with the story of two young fish who meet up with an older one. "Morning boys, how's the water?" he asks them. The two swim on for a bit and then eventually one of them looks over at the other and says, "What the hell is water?"

Like the young fish, many of those who practice mindfulness in educational settings do not seem to know what the hell the water is in which they swim. They glide along unwittingly within the powerful undercurrents of biased cultural worldviews, constraining structural inequities, conformist developmental stages, and unchartered harmful emotional forces. Mindfulness educators are often unmindful of the problematic context of mindfulness itself. They need to get outside their own water bubble and critically awaken to, engage with, and tackle the challenges of the swelling seas of which mindfulness by itself cannot be aware.

As a secular program that has severed itself from a morally based tradition, Buddhism, mindfulness in education swims in shallow waters. It flounders with regard to moral princi-

ples and practices of social justice and engagement, inquiry into the development and nature of the self, and reflection on and enactment of everyday cultures and meanings. There is a need to embed mindfulness within critical, integral programs that uncover and resist dominant ideologies and institutions in which we swim and to consciously help us heal and create new relationships that work toward optimal personal development and universal social justice (Ng and Purser 2016). Part of this can be called a critical, civic mindfulness: "Mindfulness in education offers an opportunity to reorient education away from narrowly conceived instrumental ends towards broader ethical and socially-engaged ones" (Ng 2015; see also Healey 2013).

What place and role does mindfulness education have in a shrinking, interdependent world—amidst predatory corporate institutions that generate poverty and inequity, racist and cultural domination and the rollback of civil rights, alarming climate destruction, and global militarism and violence? Mindfulness education programs can be helpful for some individuals: They tend to alleviate stress, promote skills useful for self-success, adjust students and teachers to the pressures and inequities of schooling, and help individuals competitively navigate around high-stakes tests, teacher bashing, and other neoliberal detritus strewn on the surface. Overall many do little to nothing to link agency with social justice and challenge the moral crises of our day that are based on

D. Forbes (✉)
School of Education, Brooklyn College CUNY, Brooklyn, NY, USA
e-mail: dforbes@brooklyn.cuny.edu

© Springer International Publishing Switzerland 2016
R.E. Purser et al. (eds.), *Handbook of Mindfulness*,
Mindfulness in Behavioral Health, DOI 10.1007/978-3-319-44019-4_23

self-attachment, greed, and delusion which fuel the sources of stress in the first place. They tend to unwittingly reinforce rather than challenge the neoliberal individualist practices, culture, and social structures that prime the self for marketability. What the hell indeed.

We need a comprehensive, critical perspective on contemplative education that accounts for the varieties of experiences, worldviews, developmental orders, cultures, and systems, and that stands for optimal development for all. Integral meta-theory (Wilber 2006, 2016; Esbjörn-Hargens 2009) is a good place to start; it is a method of inquiry, a way of seeing things, and a vision of human history that encourages us to consciously evolve toward universal goodness, truth, and beauty. As it turns out, integral meta-theory is arguably not comprehensive or explicit when it comes to social justice (Stein 2015; Corbett, n.d.; Patten and Morelli 2012); for example, see Wilber (2016). I add the call for a universal ethics that brings together the contemplative traditions of the East and the prophetic demand for social justice for all from the Abrahamic traditions of the West (Loy, n.d.; Bodhi 2015; Woods and Healey 2013). Both demand we shed attachment to the self in favor of universal compassion. Both exhort us to realize and enact the inseparability of all aspects of life, including societal institutions. Both together challenge society's self-centeredness: its individualism, commodification, materialism, and the maintaining of the status quo of inequitable power and privilege, for example, around class, race, gender, and sexual orientation, that thwarts optimal development, intrinsic love, mutual relationships, democratic social justice, and universal care. A critical integral approach includes the best of traditional prophetic and contemplative values and practices, modernist scientific methods, knowledge and critical thinking, and postmodern multi-perspectives and inclusivity. It situates mindfulness education within the waters of both inner and outer awareness and enhances it from a more evolved and comprehensive perspective.

A Brief Murky History: MBSR and the Critique of McMindfulness

Mindfulness meditation has its origins in Buddhism in which mindfulness is but one of a number of activities that revolve around ethical and wisdom precepts (the dharma). In Buddhism, mindfulness refers to remembering and reflecting on other previous moments in the mind's life in terms of what is wholesome, and establishing links with what are right thoughts, action, speech, concentration, intention, livelihood, and effort. Mindfulness meditation is an essential part of following the dharma which includes wisdom about the insubstantial nature of the self and the impermanence, interdependence, and non-duality of all things in the universe, the moral demand to promote a compassionate life free of suffering for all beings, and the quest to realize non-duality, enlightenment or awakening.

When mindfulness was secularized, it became severed from its organic connection to its original Buddhist ethical context and purpose to attain awakening. People credit Jon Kabat-Zinn who created the mindfulness-based stress reduction (MBSR) program and whose definition of mindfulness has become the gold standard in secular settings (clinics, hospitals, corporations, schools, the military): mindfulness means "paying attention in a particular way: on purpose, in the present moment, and non judgmentally" (Mindfulnet.org, n.d.). This morally neutral, technical, or instrumental definition of mindfulness gained popularity and became accessible to many people outside of a religious framework. For Buddhists, mindfulness is not about stress reduction or being non-judgmental but is part of the study and practice of the dharma which indeed includes judging and enacting what is right. But for Kabat-Zinn, mindfulness, in his words, is "not about Buddhism, but about paying attention" (Szalavitz 2012). Despite this dismissal, Kabat-Zinn also claims that MBSR is the "universal dharma" (Kabat-Zinn 2011, p. 296). Kabat-Zinn would like it both ways: Calling MBSR "the universal dharma" acknowledges its Buddhist roots and aims; yet by making it about non-judgmental attention and stress reduction, it has little to do with what the dharma teaches, for example, in terms of gaining awareness and understanding about non-duality and the non-existence of self (Kabat-Zinn 2011; Purser 2014).

Instead of grounding mindfulness ethics in a moral tradition, Kabat-Zinn sidesteps the issue and takes a relativist stance at best. He leaves questions of ethics to the quality of the training and background of the individual MBSR instructor (Kabat Zinn 2011, pp. 15–16), some but not all of whom have Buddhist backgrounds. In a dialogue with Angela Davis, Kabat-Zinn frames his statements about ending social injustice in global terms that float above distinctions about race and white privilege. He says that mindfulness is a "transformative practice" that is capable of moving society in a more "human" way and "that we need something that speaks to all humanity"; to which Angela Davis, conscious of white privilege, asks him, who are "we?" (Spirit Rock Meditation Center 2015, April 21). Kabat-Zinn's approach is akin to spiritual by-pass, the appeal to absolute truths as a way to avoid and dismiss painful or difficult everyday needs that require concrete consideration. Elsewhere, as do other mindfulness believers, he even suggests that mindfulness itself can lead to a moral life (Kabat-Zinn 2006, p. 103); Hyland (2016) notes that this evokes that same uneasiness we feel in the face of the Socratic claim that the truly wise person will never act in an evil way.

Kabat-Zinn's brilliantly ambiguous move has allowed secular mindfulness to flourish and become many things to many people. With its eastern, Buddhist caché and relativist, vague, but benign ethos (compassion, non-judgment, happiness, lack of suffering), secular mindfulness generates various interpretations and practices that aim to promote personal well-being in education as well as in medicine, psychotherapy, government, the military, and the corporate workplace.

At the same time, the technical, neutral definition, and relativist lack of a moral foundation has opened up secular mindfulness to a host of dubious uses, now called out by its critics as McMindfulness (Purser and Loy 2013). McMindfulness occurs when mindfulness is used, either with intention or unwittingly, for self-serving and ego-enhancing purposes that run counter to both Buddhist and Abrahamic prophetic teachings to let go of ego-attachment and enact skillful, universal compassion. Instead of letting go of the ego, McMindfulness aims to enhance it and promotes self-aggrandizement; its therapeutic function is to comfort, numb, adjust, and advance the self within a neoliberal, corporatized, individualistic society based on private gain.

In this way, instead of bringing the self into question (Buddhism), or having a moral worldview, or a soteriology—a way out of human suffering, mindfulness becomes a neoliberal technology of the self (Reveley 2015). Rather than a way to attain awakening toward universal love, it becomes a means of self-regulation and personal control over emotions (Ibid). McMindfulness is blind to the present moral, political, and cultural context of neoliberalism. As a result, it does not grasp that it is an individualistic, commodified society that creates distress and that needs to be called out; instead, the best it can then do, ironically, is to offer to sell us back an individualistic, commodified "cure"—mindfulness—to reduce that distress. By refusing to critically discuss actual social context, McMindfulness ignores seeing our inseparability from all others and from social institutions; it thereby abandons the moral demand that follows this insight to enact universal compassion, service, and social justice in all ways and all forms of human endeavor. Calling out McMindfulness is a prophetic critique of greed, ill will, and delusion in concrete, historical terms at both personal and societal levels. McMindfulness critics insist that the personal and the social are inseparable and that mindfulness should contribute to both full development and universal social justice in all areas of life.

Constructive Critique

McMindfulness critics, myself included, have been accused of being too critical. On Facebook pages and in responses to articles I have observed some advocates of mindfulness programs to be defensive and even hostile—they appear unable to mindfully sit with their own discomfort, are quite attached to their own beliefs about mindfulness programs, and project their own intolerance on to the critics. They conflate criticism of how mindfulness is employed with an attack on the value of mindfulness itself. A number declare that critique is just being negative and unhelpful and that being critical serves no purpose. They see social criticism as a waste of time. Some argue that if you have not taken an MBSR course, you have no right to question anything about it, including the social context in which it occurs. In line with the ideology of positive psychology, true believers in mindfulness prefer to just cite programs they think have a positive effect and some argue that everyone should do the same (see Nowogrodzki 2016). They regard mindfulness in individualist and personal terms: It helped these people, it helped me—end of story.

But social criticism not only can and should be defended; it can be turned on its head as a positive force. First, McMindfulness critics (Ng and Purser 2016), following Foucault, point out that critique is not just to say things are not right but to undercut what is considered as self-evident and show that it no longer has to be accepted as such. Social critique is therefore valuable in its own right as a practice of questioning normative ideologies, beliefs, and practices. Critics are not required to come up with proper, predetermined alternatives then and there. Critique can be useful in order to dislodge everyday notions and create a space to consider how things could be different.

Second, the same critics propose a "critical mindfulness" which converts critique into something constructive, a liberating act that employs mindfulness to dismantle attachments to conditioned patterns of dominant beliefs and open possibilities toward more evolved and encompassing perspectives—which themselves need to be critically discussed and enacted. A prophetic, integral stance is critical and discerning and at the same time holds out the possibility for inclusivity, respect, and mutuality.

Rowe (2015) shows that skillful oppositional thinking, even among socially engaged contemplative communities, is crucial to social change; contemplatives need to skillfully deploy oppositional approaches to, for example, toxic fossil fuel companies, while being aware of relative and absolute truths. Yes, we are all interconnected in an absolute sense but in relative terms we also are required to oppose those who seek to harm ourselves, others, and the earth—and that practice can lead to "collective liberation" (ibid.) Let us first critically look at culture, social structure, and development before suggesting an integral contemplative approach.

Culture

We all swim in culturally constructed beliefs, norms, and rules that implicitly frame everyday meanings. For example, some are individualism and consumerism and assumptions about race, class, gender, and sexuality. Within the same shared everyday space, people inhabit different implicit orders of cultural development. Members of traditional cultures tend to operate in terms of eternal truths and believe there is one established way to know what is right. Those who adhere to a more recently developed modernist culture subscribe to scientific evidence, materialism, reason, individualism, and entrepreneurial values. The latest postmodernists dismantle master narratives and seek their own relative truths and interpretations. Without bringing awareness of such cultural frameworks to the foreground mindfulness education programs fall prey to the Myth of the Given (Wilber 2006). This is the belief that events and actions in everyday life are directly perceived as given, objective facts rather than as socially constructed, interpretable, and contested meanings that can be uncovered, discussed, and transformed.

Many educators do not question the problematic, socially constructed nature of schooling and school values in which they offer mindfulness programs. Without this, mindfulness itself becomes an ideology that reinforces the ideology of neoliberal schooling. Students and teachers are directly encouraged to perceive things as given reality; in this way, implicit culturally constructed meanings are glossed over yet still operate in the background. This mystification is compounded by the pseudo-objective language overlaid with a spiritual patina employed in mindfulness practices. Programs encourage students and teachers to "be in the here and now," "see things as they are," and "be with whatever is." The actual social relations that frame the meaning of these terms—how *do* we construe what is happening here and now?—are not critically exposed and discussed but stand around outside conscious awareness like stagnant water.

Let us name a few unexamined cultural myths lurking in the education water. Two predominant beliefs that operate under the surface are that the individual alone is both the source of and solution to unhappiness (neoliberalism) and that therapeutic behavioral change and neuroscience are the means by which the individual attains and proves personal success (scientism). Neoliberal ideology in education posits that stress, lack of attention, and reactivity are problems that lie within the individual, not society, societal institutions, or social relations. The individual by oneself is believed to be responsible to overcome these presumed deficits. One can and should obtain success and happiness by purchasing, owning, and consuming things and by marketing one's self as a personal brand (Giroux 2014; Ravitch 2014). Solutions are achieved through scientific and technocratic approaches: The individual should employ the technology of mindfulness to improve individual wellness, social and emotional skills, academic performance, and self-regulation, and have these confirmed through brain imaging and other "objective" outcome measures such as education audits and test scores (Taubman 2009). Scientism then serves as an ideology through the predominant assumption that in education only measurable, observable phenomena are real and truthful and are the only measures of success, e.g., high-stakes testing, outcomes assessments, self-regulation practices, and data-driven or evidence-based programs.

Other aspects of everyday culture swirl about unaddressed by mindfulness programs. Fromm's (2010) insights into the pathology of normalcy

still resonate, the everyday unconscious acceptance of and adjustment to unhealthy and unethical values, practices, and ways of being. Examples are racist perceptions and attitudes around white norms and privilege that are woven into day-to-day life, as is the acceptance of much white working class *ressentiment* (Sleeper 2014) which is seldom acknowledged and addressed. Trauma and its aftermath, including addictions, is mostly regarded as an individualized phenomenon but is also an unaddressed aspect of many people's everyday culture; Bloom (2013) considers the USA as having much unresolved trauma as a nation of immigrants and formerly subjugated ancestors; many have generational issues of loss and live in a culture that has tried to solve conflicts through violence, militarism, and domination over others that are then papered over with denial. Positive psychology is offered up as a popular therapeutic solution to problems that are seen to lie solely within the individual. Yet an emerging critical literature uncovers its ideological undercurrents and shows how positive psychology, the marketing of spirituality, the therapy industry, and the self-help culture reinforce adjustment to neoliberal values and institutions (Binkley 2014; Carrette and King 2005; Cederstrom and Spicer 2015; Davies 2015; Ehrenreich 2010; Ilouz 2008; Moloney 2013; Rakow 2013). Mindfulness practices along with social emotional learning (SEL) programs in schools share the same approach and play well into reinforcing conformity to the individualist, competitive, and marketing aspects of neoliberal culture. Left behind by mindfulness education programs in the wake of the neoliberal wave is the cultural capital of many schools and communities of color in urban areas. It is rare that mindfulness school programs acknowledge these and work with and within them to discuss and employ shared skills, strengths, and interests.

At a deeper level, we experience a culture of lack—Loy's (2002) term to describe the feelings of emptiness and craving and that one is never enough. These fuel consumerism and addictions, the endless search for external goods or relationships to feel better or complete one self. From an engaged Buddhist perspective, the way through is to realize that because we have no solid self to which to cling we are already complete, and to also work to change institutions such as corporations, the media, and the military that reinforce ego-attachment ("wego") on a cultural scale as well as in personal terms.

Social Structure

Structural and systemic injustices frame the lives of children and teachers and the mindfulness practices in which they engage; they too do not figure into the mindfulness navigation plan. These are shark, capitalism-infested, waters which get naturalized and accepted ("seeing things as they are") as part of everyday life. The system creates painful income, class, and racial inequalities. These contribute to poor neighborhoods of disenfranchised citizens who suffer from poor health and health care, inadequate housing, and chronic unemployment. Impoverished schools without decent resources and programs further inflame students' anger, violence, substance abuse, and despair. The wealthy that benefit from enormous tax breaks send their children to privileged schools that provide enriched learning environments. The children, however, pay their own psychic price, the stress that accompanies the intense pressure to produce and compete for limited elite college slots.

Neoliberalism (Giroux 2014; Harvey 2005; McGuigan 2014) is the dominant ideology and system that impacts education policies and practices. It promotes an individualistic, market-based worldview and structure. It glorifies the private individual who competes for and purchases all of one's needs through the market, which replaces social institutions and the public good. The neoliberal self is self-reliant, a risk-taker, and not dependent on or connected with others; one is motivated by personal gain as a perpetual self-entrepreneur and consumer of choice. Education reformers push neoliberal, market-based ideas, policies, and practices in schools. Neoliberal policy makers in public education are in it to promote world market competition; they are happy to employ mindfulness in the schools as an instrument to better adjust teachers and students to conform to

corporatized high-stakes tests, arbitrary standards, and micromanagement, surveillance, and scripting of classroom lessons—all the while those stressors continue to lurk in the background, unnamed and unchallenged. Policy makers want students and teachers to gain greater self-regulation and adjustment to a neoliberal society, to successfully adapt to stressful and often morally reprehensible situations (Forbes 2015). Not only is there no link made between mindfulness and problematic ethical and social justice values and conditions, there is no consideration that these contribute to serving as the actual sources of stress themselves—which mindfulness, along with social emotional learning (SEL) programs, is then expected to mitigate (Forbes 2012; Hsu 2013, November 4; Zakrzewski 2015).

The impact of neoliberal practices on urban students of color is of particular concern. Because neoliberalism negates the notion of society, it obscures social inequities such as systemic racism and the need to fight them. It dismisses systemic racism as a social, structural, and institutional problem since everything should be a matter of individualized choice and each individual is personally responsible for one's own success and failure (Davis 2013; Robbins 2004). Although racial neoliberalism and unequal structural power relations still exist, they disappear as topics from public discourse and public policy (Enck-Wanzer 2011). Yet mindfulness is employed in a number of impoverished inner-city schools attended by many disaffected, indignant, and at times disruptive students of color. Without a critical understanding of the neoliberal education agenda, mindfulness practices geared toward stress reduction, conflict resolution, emotion regulation, anger management, and focus and concentration serve as functions of social control and reinforce emotional self-regulation that puts the onus back on the individual student. The need to conform to school expectations preempts the issue of why there is so much stress, suspensions, and angry behavior in the first place. It pushes aside structural questions of what needs to better occur in the school and community. Mindfulness of one's anger and frustration in the "here and now" leaves out the social context of the social injustices that many students of color experience. In particular, SEL programs tend to ignore the cultural context and cultural capital of students of color and the impact on them of racism and prejudice (Slaten et al. 2015; Zakrzewski 2016).

Mindfulness programs teach awareness of emotions. But they do not address or analyze how emotional life is inextricably related to the complex, rich, and often problematic social nature of the lives of students, teachers, and community members. Experiences of trauma, addiction, anxiety, and depression of course require healing, dialogue, and support. But mindfulness programs do not see these, along with anger and sadness, as responses embedded within social relations and systems that often require both critical reflection and transformation toward greater caring and justice. Mindfulness instead becomes one more individualist endeavor that excises personal experiences from their social context and adjusts individuals to swim better in the polluted waters. Mindfulness programs and teacher trainings (e.g. Jennings 2015) ignore the structural context of class and racial inequities, competitive individualism, and the neoliberal assault on public education, teachers, and their unions. Social problems that contribute to stress, burnout, and demoralization are obscured and translated into personal concerns in need of psychotherapeutic and/or mindfulness solutions. We need to understand these waters and what lies within them, and work with others to swim toward clearer currents.

Development

Stages or orders of self-development are another crucial medium that, like the unacknowledged water, surround educational mindfulness programs. Developmental models are seldom if ever applied or even acknowledged as a way to inform and help students, teachers, and the schools. Mindfulness practices can create a heightened *state* of awareness; a student or teacher through practice might notice or witness one's feelings, thoughts, and sensations and come to distance one's self from and disidentify with them in a calm manner. However, the developmental *stage* of self and moral development of the student or teacher frames how that state is interpreted and contributes to how the person thinks about how to respond (Wilber 2016). As we develop, we turn our patterns of thinking, experienced from within, into objects of our own awareness from a more inclusive perspective; our subjectivity at each order becomes the object of our awareness at a later stage (Kegan 1994).

A mindful teacher or student may attain an advanced meditative state through mindfulness but one's developmental structure constrains one's worldview and how the experience is interpreted. Educators and students can gain a contemplative experience or state by practicing mindfulness, but many still have at best a conformist and conventional stage mentality. In some mindfulness programs, participants despite their practice still adhere to loyalty to authority, strict rule-following behaviors, and uncritical, conformist thinking. Mindfulness practice by itself does not lead to critical questioning, moral reasoning, or skillful and moral actions. Nor by itself does it lead to later stages of autonomous thinking, the ability to hold ambiguity, and to think on one's feet from a post-conventional cognitive or moral developmental order.

While mindfulness education programs encourage awareness and reflection of emotions and intentions, they steer a middle course through the developmental waters of interiority. Mindfulness is mostly taught and practiced in the service of producing conventionally successful students and teachers who can adjust to the demands of neoliberal society. Unlike depth psychology, they avoid the unchartered realms of unconscious emotional life; unlike Buddhism, they by-pass higher, ego-transcendent states and stages. For full and optimal human development, educators would need to be free to explore the shadow and contemplative aspects of human experience. These require the knowledge and awareness of developmental orders from which one views the world: unconscious (at any level), egocentric, conventional, post-conventional, and ego-transcendent (integral), and a conscious intention and practice to gain higher, more inclusive perspectives.

Mindfulness programs rely on social emotional learning (SEL) curricula to provide the best version of secular ethics with which mindfulness can associate. Yet, SEL programs are unaware of their own self and moral developmental stage and worldview. The competencies or behavioral skills favored by SEL fit in nicely with neoliberal achievement-oriented values designed for conventional levels of success in a competitive, corporatized, market-based society. Employing corporate language, two SEL educators (Brackett and Rivers 2014) approvingly note that leading economists, including a Nobel Prize Laureate, call for these "soft" emotion skills to be taught in schools since they yield the greatest returns on education investments and lead to greater success in life. According to the authors, the Laureate thinks this is a cost-effective way to increase "the quality and productivity of the workforce through fostering workers' motivation, perseverance, and self-control" but are concerned that "[a]s increasing efforts move toward better preparing youth to enter and contribute to a competitive and global workforce, epidemiological evidence suggests that the basic needs of youth still are not being met" (p. 3).

SEL skills are framed in terms that emphasize pseudo-objectivity, self-control, and success. These include the ability to "accurately" recognize one's emotions and thoughts; "accurately" assess one's strengths and limitations; self and stress "management"; attain relationship skills such as cooperating and resisting "inappropriate" social pressure; "responsible decision making"; and with a nod to positive psychology, having "a well-grounded sense of confidence and optimism" (CASEL, n.d.). Teachers and students can be mindful of thoughts and feelings and learn the latest skills that pass as secular ethics, yet continue to swim within conventional and conformist structures that govern and restrict their awareness.

In the absence of a developmental framework, the meaning of a trait such as compassion floats freely without being grounded in any particular social context. A vague but potentially important

term, compassion is a socially desirable skill within SEL programs (Zakrzewski 2015, January 7) and serves as a catch-all buzzword that mindfulness educators favor. Its meaning depends, among other things, on the developmental and moral worldview of the practitioner. There is no developmental framework, let alone any concrete social context, for analyzing and discussing how and why compassion is taught: Is it practiced to please the teacher and because everyone does it? Do students reach a later level of understanding and engage in compassion for the best of intentions and for its own sake?

A significant blind spot for mindfulness educators is their own unacknowledged level of self-development. Rather than stepping outside the individualist and neoliberal educational systems of which they are a part and with which they identify, some mindfulness educators at a fourth order of self-development (Kegan 1994; Murray 2009) may tend to identify with and are attached to their particular school of thought or their own mindfulness programs. As a result, they do not examine and critically challenge these systems from later and more comprehensive perspectives that account for a fuller range of human development and social justice. At the fifth order, people are able to let go of their defensive attachment to their own fourth-order belief systems and reflect upon them with dispassion. They can now disidentify with particular belief systems and experience the self as embodying a variety of evolving beliefs that arise in different contexts.

Toward a Critical Integral Contemplative Education

We can consider paths for new directions that frame mindfulness in education within a critical integral meta-perspective. The challenge is to re-construe the contemplative and the prophetic as part of a broader project in education that revitalizes the wisdom and values of earlier traditions on new ground. It requires that we incorporate the best of contemplative and prophetic traditions, along with modernist knowledge and progressive postmodern awareness of multiple, culturally constructed and developmental perspectives on gender, race, ethnicity, sexual orientation, and other categories. At the same time, we seek to challenge and transcend the current limitations of a society governed by neoliberal, market-based structures, ideologies, and policies.

A critical integral approach helps students and teachers uncover implicit cultural values, interrogate neoliberal educational policies, and act to change them in terms of more encompassing and universal moral stages of self, cultural, and societal development. Students and teachers need to see, study, question, and act on the sources of stress, using mindfulness within an anti-oppressive-inspired context of experiencing and working toward universal educational equality. This is embedded in the everyday culture of those whom are most impacted by stress and oppression. The goal is to contribute to rekindling and enacting values of democratic, quality education within a society that is seeking to create new mutually satisfying relationships and social structures that are healing and fulfilling for all.

Wilber (2016) proposes a model of integral meditation from which we can draw and expand on as a model for what a comprehensive critical education program would entail. Following the integral meta-model, we can look at four areas or quadrants which would provide an overall schema.

Subjective

In the Subjective realm, a program would of course include the individual practice of meditation and mindfulness to gain greater capacity to experience contemplative states and which can lead to the insight of non-duality. Wilber (2016) calls this "Waking up." School community members would meditate not just for stress reduction, self-regulation, or to improve concentration but as part of an inquiry into the nature of the self and to cultivate a relationship with the patterns of their own mind in the context of greater moral and social values and relationships.

In terms of stages of self-development school community members learn about models of development and ask, what developmental stage of mine and ours is interpreting how mindfulness is employed? They aim to promote healthy awareness and practices within one's current stage (translative development) and also to help members when they are ready to develop toward later stages and toward universal compassion and non-duality (transformational development). Mindfulness in part can be valuable as an intentional developmental tool; like insight meditation, developmental growth occurs by witnessing and reflecting on one's subjectivity and converting it into a more encompassing object of awareness (Kegan 1994; Forbes 2004). School community members can move through egocentric, socio-centric, and post-conventional to integral orders of self and moral development. At the later stages, one reaches a stable awareness of unity consciousness or awakening. Wilber (2016) calls this "Growing up."

A third practice within the Subjective quadrant is around psychological awareness and individual relief from emotional suffering and stress. This occurs through mindful individual and group counseling, programs, projects, and workshops that address issues around emotional self-awareness, moral values, trauma, addictions, disorders, and unconscious ("shadow") realms, that is, dissociated factors of one's self. Wilber (2016) terms this "Cleaning up."

Cultural or Intersubjective

In the Culture/Intersubjective realm, there are inside and outside perspectives and practices. From the inside school community members create mindful, healthy, "We-spaces" (Gunnlaugson 2009, June) and relationships. These can be groups, even the school as a whole, that empower and foster support, trust, safety, respect, inclusiveness, caring, compassion, healing, and connectedness among everyone. They engage in mindful practices that explore sensitive issues such as racism and white privilege between community members. They build on local diverse strengths of the community at large and also encourage cultural growth toward more inclusive stages at a later moral stage of development. Wilber (2016) calls working on relationships "Showing up."

From the outside perspective of culture, within groups and as a whole school, members study, uncover, and challenge hidden, implicit cultural biases, assumptions, attachments, practices, and rituals—that is, the cultural constructions of meaning that operate in the school and in education overall. These, for example, would be neoliberal ideology, individualism, materialism, consumerism, ethnocentrism, racism, white privilege, sexism, homophobia, colonialism, even contemplative education itself. The group would ask, what kind of school culture do we have that we are uncovering, what kind of moral culture do we want, and how do we change it? I call this "Wising up."

Behavior or Objective

The school community can make healthy use of findings from neuroscience to enhance the quality of their lives, as opposed to reducing consciousness to a materialist stratum and fetishizing neuroscience itself: A thicker prefrontal cortex is not the goal, becoming a morally evolved, wise person who skillfully acts in the social world, is. Members can study if and how critical integral mindfulness can enhance healthy neural development and vice versa, and how broader cultural and structural realities such as stressful conditions that stem from poverty and other adverse situations may enhance or negatively impact brain development and overall health (Maté 2010).

Teachers and students can employ contemplative practices to deepen and strengthen meaning and connection in learning which is made more whole within a critical integral awareness of educational context. The school community can use data to support but not solely to validate or "drive" wise, skillful, meaningful educational projects. Members can engage in critical mindfulness research that investigates and uncovers hidden norms in everyday culture and local social systems such as consumerism (Stanley et al. 2015) that impede personal and interpersonal development.

With respect to personal action, community members can mindfully enact more evolved compassionate, healthy behaviors such as social emotional learning skills. These, however, are performed critically and in a moral and social context of evolving, caring relationships and values that are relevant to the school community (Slaten et al. 2015; Zakrzewski 2016) rather than reinforcing individualistic, neoliberal attitudes and practices.

Social Structure or Interobjective

School community members can mindfully investigate and uncover barriers in the social structure that impede social justice. They identify, study, and resist together through classroom, groups, and workshops unjust social structures that impact their lives as aspects of a mindful anti-oppressive critical pedagogy (Berila 2016; Hyland 2015; Magee 2015; Orr 2002, 2014; Reveley 2015). These include taking on local school policies, larger neoliberal educational policies (high-stakes testing, Common Core), systemic bullying, and deep-rooted structural barriers such as poverty, income inequity, systemic racism, sexism, homophobia, neocolonialism, and the corporate power structure of school, government, society. This too is "Wising up."

The members of the school community also engage in mindful social action for social justice. They work together and develop alliances across class and race and with like-minded activists. They do so in the school itself, and at local, national, and even global levels in resisting unjust policies, practices, and institutions. Together, they create healthy, more inclusive, socially just, policies, systems, and political arrangements in schools and defend and demand universal quality public education, sustainability, and interdependence. I call this "Acting up."

Back in the Water

A critical integral contemplative approach calls us to touch and see the water in which we swim: to both experience and evolve toward the absolute of contemplative awareness and to engage fully in helping make the relative world into one of universal justice and love. In his commencement speech, Wallace (2005) caught a glimpse of the water and shared it with his college audience. He describes a "spiritual-type" integral, visionary state which reconciles the ego-driven everyday life with the transcendent awareness of non-duality: "It will actually be in your power to experience a crowded, hot, slow, consumer-hell type situation as not only meaningful, but sacred, on fire with the same force that made the stars: love, fellowship, the mystical oneness of all things deep down." Later he adds a description of a similar inspiring state from a highly evolved plane of awareness that infuses sacred compassion into the mundane: the freedom of "being able truly to care about other people and to sacrifice for them over and over in myriad petty, unsexy ways every day." Wallace, sadly, was unable to sustain this state of awareness as a lasting stage of his own development—and could not envision a way out of "consumer-hell" that includes and involves others—but he left us with these inspiring images of the water of non-dual awareness in which we all swim. "This is water," he concluded, "This is water."

References

Berila, B. (2016). *Integrating mindfulness into anti-oppression pedagogy: Social justice in higher education*. New York: Routledge.

Binkley, S. (2014). *Happiness as enterprise: An essay on neoliberal life*. Albany: SUNY Press.

Bloom, S. (2013). *Creating sanctuary: Toward the evolution of sane societies*. New York: Routledge.

Bodhi, B. (2015, May). Modes of applied mindfulness. Unpublished manuscript.

Brackett, M. A., & Rivers, S. E. (2014). Transforming students' lives with social and emotional learning. In R. Pekrun & L. Linnenbrink-Garcia (Eds.), *International handbook of emotions in education* (pp. 368–388). New York: Taylor & Francis. http://ei.yale.edu/wp-content/uploads/2013/09/Transforming-Students%E2%80%99-Lives-with-Social-and-Emotional-Learning.pdf

Carrette, J., & King, R. (2005). *Selling spirituality: The silent takeover of religion*. New York: Routledge.

CASEL. (n.d.). Collaborative for academic, social, and emotional learning. http://www.casel.org/social-and-emotional-learning/core-competencies

Cederstrom, C., & Spicer, A. (2015). *The wellness syndrome*. London: Polity.

Corbett, J. (n.d.). How Ken Wilber and integral theory leave out justice. http://www.decolonizingyoga.com/how-ken-wilber-and-integral-theory-leave-out-justice/

Davies, W. (2015). *The happiness industry: How the government and big business sold us well-being*. New York: Verso.

Davis, A. (2013, May 6). Recognizing racism in the era of neoliberalism. http://www.truth-out.org/opinion/item/16188-recognizing-racism-in-the-era-of-neoliberalism

Ehrenreich, B. (2010). *Bright-sided: How positive thinking is undermining America*. New York: Picador.

Enck-Wanzer, D. (2011). Barack Obama, the tea party, and the threat of race: On racial neoliberalism and born again racism. *Communication, Culture & Critique, 4*, 23–30.

Esbjörn-Hargens, S. (2009, March 12). An overview of integral theory. Integrallife.com. Retrieved from https://integrallife.com/integral-post/overview-integral-theory

Forbes, D. (2004). *Boyz 2 Buddhas: Counseling urban high school male athletes in the zone*. New York: Peter Lang.

Forbes, D. (2012). Occupy mindfulness. http://beamsandstruts.com/articles/item/982-occupy-mindfulness

Forbes, D. (2015, November 8). Mindfulness and neoliberal education. Published as they want kids to be robots: Meet the new education craze designed to distract you from overtesting. *Salon*. http://www.salon.com/2015/11/08/they_want_kids_to_be_robots_meet_the_new_education_craze_designed_to_distract_you_from_overtesting/

Fromm, E. (2010). *The pathology of normalcy*. Brooklyn: Lantern.

Giroux, H. A. (2014, December 30). Barbarians at the gates: Authoritarianism and the assault on public education. *Truthout*. http://www.truth-out.org/news/item/28272-barbarians-at-the-gates-authoritarianism-and-the-assault-on-public-education

Gunnlaugson, O. (2009, June). Establishing second-person forms of contemplative education: An inquiry into four conceptions of intersubjectivity. *Integral Review, 5*(1), 25–50. Retrieved from http://integralreview.org/documents/Gunnlaugson,%20Intersubjectivity%20Vol.%205,%20No.%201.pdf

Harvey, D. (2005). *A brief history of neoliberalism*. New York: Oxford.

Healey, K. (2013, August 5). Searching for integrity: The politics of mindfulness in the digital economy. http://nomosjournal.org/2013/08/searching-for-integrity/

Hsu, F. (2013, November 4). The heart of mindfulness: A response to the *New York Times*. Buddhist Peace Fellowship. http://www.buddhistpeacefellowship.org/the-heart-of-mindfulness-a-response-to-the-new-york-times/

Hyland, T. (2015). The limits of mindfulness: Emerging issues for education. *British Journal of Educational Studies, 63*(3), 1–21. http://philpapers.org/rec/HYLTLO

Hyland, T. (2016). On the contemporary applications of mindfulness: Some implications for education. In O. Ergas & S. Todd (Eds.), *Philosophy east/west: exploring intersections between educational and contemplative practices* (pp. 9–28). Chichester, UK: Wiley Blackwell.

Ilouz, E. (2008). *Saving the modern soul: Therapy, emotions, and the culture of self-help*. Berkeley: University of California Press.

Jennings, P. A. (2015). *Mindfulness for teachers: Simple skills for peace and productivity in the classroom*. New York: Norton.

Kabat-Zinn, J. (2006) *Coming to our senses. Healing ourselves and the world through mindfulness*. New York: Hachette.

Kabat-Zinn, J. (2011). Some reflections on the origins of MBSR, skillful means, and the trouble with maps. *Contemporary Buddhism, 12*(1), 281–206. Retrieved from http://umassmed.edu/uploadedFiles/cfm2/training/JKZ_paper_Contemporary_Buddhism_2011.pdf

Kegan, R. (1994). *In over our heads: The mental demands of modern life*. Cambridge: Harvard.

Loy, D. R. (2002). *A Buddhist history of the west: Studies in lack*. Albany: SUNY Press.

Loy, D. R. (n.d.). What's Buddhist about socially engaged Buddhism. http://www.zen-occidental.net/articles1/loy12-english.html

Magee, R. (2015, May 14). How mindfulness can defeat racial bias. Retrieved from http://greatergood.berkeley.edu/article/item/how_mindfulness_can_defeat_racial_bias

Maté, G. (2010). *In the realm of hungry ghosts: Close encounters with addiction*. Berkeley: North Atlantic books.

McGuigan, J. (2014). The neoliberal self. *Culture Unbound, 6*, 223–240. http://www.cultureunbound.ep.liu.se/v6/a13/cu14v6a13.pdf

Mindfulnet.org. (n.d.). What is mindfulness? Retrieved April 27, 2016 from http://www.mindfulnet.org/page2.htm

Moloney, P. (2013). *The therapy industry: The irresistible rise of the talking cure, and why it doesn't work*. London: Pluto Press.

Murray, T. (2009). What is the integral in integral education? From progressive pedagogy to integral pedagogy. *Integral Review, 5*(1), 96–134.

Ng, E. (2015, June 29). Mindfulness and justice: Planting the seeds of a more compassionate future. http://www.abc.net.au/religion/articles/2015/06/29/4264094.htm

Ng, E., & Purser, R. (2016, April 4). Mindfulness and self-care: Why should I care? *Patheos*. http://www.patheos.com/blogs/americanbuddhist/2016/04/mindfulness-and-self-care-why-should-i-care.html

Nowogrodzki, A. (2016, April 21). Power of positive thinking skews mindfulness studies. *Scientific American*. http://www.scientificamerican.com/article/power-of-positive-thinking-skews-mindfulness-studies/

Orr, D. (2002). The uses of mindfulness in anti-oppressive pedagogies: Philosophy and praxis. *Canadian Journal of Education, 27*(4), 477–490. http://files.eric.ed.gov/fulltext/EJ728316.pdf

Orr, D. (2014). In a mindful moral voice: Mindful compassion, the ethic of care and education. *Paideusis, 21*(2), 42–54.

Patten, T., & Morelli, M. V. (2012, February 20). Occupy integral! http://www.beamsandstruts.com/articles/item/814-occupy-integral

Purser, R. (2014, August 12). The myth of the present moment. *Mindfulness, 6*, 680–686. https://www.academia.edu/8070322/The_Myth_of_the_Present_Moment

Purser, R., & Loy, D. (2013, July 1). Beyond McMindfulness. http://www.huffingtonpost.com/ron-purser/beyond-mcmindfulness_b_3519289.html

Rakow, K. (2013). Therapeutic culture and religion in America. *Religion Compass, 7* (11), 485–497. Retrieved from https://www.academia.edu/6467041/Therapeutic_Culture_and_Religion_in_America

Ravitch, D. (2014). *Reign of error: The hoax of the privatization movement and the danger to America's public schools*. New York: Vintage.

Reveley, J. (2015, January 27). Foucauldian critique of positive education and related self-technologies: Some problems and new directions. *Open Review of Educational Research, 2*(1), 78–93. http://www.tandfonline.com/doi/full/10.1080/23265507.2014.996768

Robbins, C. G. (2004). Racism and the authority of neoliberalism: A review of three new books on the persistence of racial inequality in a color-blind era. *Journal of Critical Education Policy Studies, 2*(2). http://www.jceps.com/wp-content/uploads/PDFs/02-2-09.pdf

Rowe, J. K. (2015, September 29). Learning to love us-versus-them thinking. https://www.opendemocracy.net/transformation/james-k-rowe/learning-to-love-us-versus-them-thinking#

Slaten, C. D., Irby, D. J., Tate, K. & Rivera, R. (2015). Towards a critically conscious approach to social and emotional learning in urban alternative education: School staff members' perspectives. *Journal for Social Action in Counseling and Psychology, 7*(1), 41–62. http://www.psysr.org/jsacp/slaten-v7n1-2015_41-62.pdf

Sleeper, J. (2014, December 5). Our real white male problem: Why Fox News defeats Bruce Springsteen and liberal moralizing every time. *Salon*. http://www.salon.com/2014/12/05/our_real_white_male_problem_why_fox_news_defeats_bruce_springsteen_and_liberal_moralizing_every_time/

Spirit Rock Meditation Center. (2015, April 21). Angela Davis & Jon Kabat-Zinn. https://www.youtube.com/playlist?list=PLGP57y-64pOTYAjGjZda-F0Dr-8tpKTpY

Stanley, S., Barker, M., Edwards, V. & McEwen, E. (2015). Swimming against the stream? Mindfulness as a psychosocial research methodology. *Qualitative Research in Psychology, 12*(1). http://orca.cf.ac.uk/68206/1/Stanley%20Barker%20Edwards%20McEwen%20accepted%20manuscript%20for%20ORCA.pdf

Stein, Z. (2015, June 26). The Integral movement is an anti-capitalist movement: ITC debate preamble. http://

www.zakstein.org/the-integral-movement-is-an-anti-capitalist-movement-itc-debate-preamble/

Szalavitz, M. (2012, January 11). Q&A: Jon Kabat-Zinn talks about bringing mindfulness meditation to medicine. *Time*. Retrieved from http://healthland.time.com/2012/01/11/mind-reading-jon-kabat-zinn-talks-about-bringing-mindfulness-meditation-to-medicine/

Taubman, P. (2009). *Teaching by numbers: Deconstructing the discourse of standards and accountability in education*. New York: Routledge.

Wallace, D. F. (2005). Transcription of the Kenyon commencement address. http://web.ics.purdue.edu/~drkelly/DFWKenyonAddress2005.pdf

Wilber, K. (2006). *Integral spirituality*. Boston: Integral Books.

Wilber, K. (2016). *Integral meditation*. Boston: Shambhala.

Woods, R. H., Jr., & Healey, K, (Eds.). (2013). *Prophetic critique and popular media: Theoretical foundations and practical applications*. New York: Peter Lang.

Zakrzewski, V. (2015, January 7). Social-emotional learning: Why now? http://greatergood.berkeley.edu/article/item/social_emotional_learning_why_now

Zakrzewski, V. (2016, March 31). Why don't students take social-emotional learning home? http://greatergood.berkeley.edu/article/item/why_dont_students_take_social_emotional_learning_home

What Is the Sound of One Invisible Hand Clapping? Neoliberalism, the Invisibility of Asian and Asian American Buddhists, and Secular Mindfulness in Education

24

Funie Hsu

> Children demonstrated faster reaction times while performing tests such as Dr. Diamond's "Flanker Fish" trials. The [sic]correlates to heightened self-regulatory ability.
> —*MindUP™, The Hawn Foundation.*
> All I can see is Buddhist practice—particularly 'mindfulness' and 'loving-kindness' ideals—used to placate resistance from marginalized populations.
> —*Dedunu Sylvia, Turning Wheel.*
> So, we've got to think about, 'What is education for?'.
> —*Grace Lee Boggs, American Revolutionary: The Evolution of Grace Lee Boggs.*

In this chapter, I provide a critical interpretation of secular mindfulness in schools that situates the phenomenon within the broader context of neoliberalism and the interwoven dynamics of race. In doing so, I argue that secular mindfulness requires an ideology of white conquest that makes invisible the enduring efforts of Asian and Asian American Buddhists in maintaining the legacy of mindfulness practices. Thus, neoliberalism's sleight of invisible hand makes invisible the contributions of Asian and Asian American Buddhists in secular mindfulness. My use of the term "secular mindfulness" is an intentional act to denote the manner in which the mainstream, popularized form of mindfulness has strategically distanced itself from the religious foundations of (Asian) Buddhist mindfulness, from which it was derived.

F. Hsu (✉)
San José State University, San José, CA, USA
e-mail: funie.hsu@sjsu.edu

Given this foundation, I argue that when secular mindfulness programs are applied in schools, they often unconsciously advance both the neoliberal marketization of public schools and a curriculum predicated on a system of white superiority. Such curricula then discipline students both through neoliberal self-regulation and through a racial conditioning of white superiority as common, and calm, sense. Consequently, in the face of widespread economic and racial inequality, most secular mindfulness in education programs instructs a sense of individual responsibility and uplift, rather than government accountability and structural change. This chapter is part of a larger exploration of race, power, neoliberalism, and secular mindfulness in schools. It serves as the initial foundation for a deeper investigation of these topics, which I undertake in a separate article.

My intention in this work is to foster a much needed critical discussion on the issues of race, power, and inequality that arise when secular mindfulness is applied without deep reflection on the larger structural forces that shape its environment. I write from the perspective of a scholar

© Springer International Publishing Switzerland 2016
R.E. Purser et al. (eds.), *Handbook of Mindfulness*,
Mindfulness in Behavioral Health, DOI 10.1007/978-3-319-44019-4_24

of color in American studies; a former elementary school teacher in a low-income school; a board member of an Engaged/Applied Buddhism and social justice non-profit organization; and an Asian American Buddhist practitioner raised in a low-income, immigrant Buddhist household. I highlight my background as a political maneuver to explicitly point out the ways in which the perspectives of Asian and Asian American Buddhists have been left out of the conversation on mindfulness in the USA and the West—precisely because of the matters addressed in this chapter. The communities to which I belong have been directly affected by these issues. It is with a deep sense of interbeing with educators and students of the past, present, and future, that I offer this work in the hopes of cultivating urgent conversations.

Neoliberalism and Secular Mindfulness in Education

I begin by discussing how neoliberalism has determined the contours of schooling in the USA and, thus, the milieu of secular mindfulness in education. In this section, I provide a brief overview of neoliberalism as the hegemonic, or dominant, framework of our contemporary period. A detailed review of the topic is beyond the scope and purpose of the chapter. Rather, a short summation of the key points is provided as context for understanding the current educational environment under which secular mindfulness is being integrated into schools. In particular, I associate secular mindfulness education programs and their focus on developing individual well-being with the neoliberal ideal of enhancing human well-being through unfettered economic competition (Harvey 2005, p. 2). I underscore how secular mindfulness education's intention of cultivating student well-being is circumscribed within neoliberalism's economic imperative.

The prevalent paradigm in our current educational era upholds the primacy of market-based reforms. It stems directly from the neoliberal model of economic restructuring popularized in the USA in the late 1970s and catapulted into

dominance by the Reagan administration in the 1980s. Since then, it has continued to dictate policymaking and institutional reforms across a wide-ranging arena of American public life, including education. The basic tenets of neoliberalism include privatization of public resources, free market competition, free trade, and limited government intervention (Harvey 2005, p. 2).

The field of education has served as a crucial site for neoliberal restructuring due to its large market potential and the role of formal schooling in the global marketplace (Kuehn 1999; Ross and Gibson 2006). Through neoliberal policies, education as a former public good has been reduced into components that have been privatized and controlled through the for-profit business model of management. For example, Ross and Gibson (2006) note that educational services that were once provided by the state or federal government are increasingly operated by "for-profit educational management organizations (such as Edison Schools)" (p. 4). Additionally, neoliberal reforms have emphasized a restrictive system of standards based education and related structures of high-stakes accountability, which entail aggressive measurement requirements coupled with punitive consequences.

The most prominent example of the extensive social, economic, and political power of neoliberalism and its authority over the institution of education is the former federal policy, No Child Left Behind (NCLB) (Hursh 2007). The law, signed by George W. Bush in 2002 and in effect until 2015, was premised on the idea of "human capital" and the need to prepare and produce competitive students for the global marketplace. American students, it was argued, were failing to achieve mastery level competency in key content areas of math and language arts, ultimately leaving them underprepared and uncompetitive in the global workforce. The purpose of education in this context, then, is purely economic. By referencing the demands of globalization, NCLB stressed the principle of market competition and institutionalized this value as an unquestioned shared educational objective.

NCLB purported to hold schools accountable for improving student achievement through annual testing to meet designated Adequately Yearly Progress (AYP) targets. These evaluations were "high stakes" in that their outcomes determined whether schools would face disciplinary sanctions, ranging from the withholding of federal funds to school closures (U.S. Department of Education 2010). The penalty of school closure provides a direct example of the neoliberal economic principles of competition applied within the field of education. "Failing" schools were deemed unviable and often subjected to either complete closure or new management by private organizations (Gill et al. 2007), often in the form of publicly funded charter schools. The 2009 federal "Race to the Top" grants further encouraged competition-based reform measures to incentivize school improvement, including the encouragement for states to expand charter school enrollment (U.S. Department of Education 2009). School closures and for-profit educational management transitioned public funds and resources to private organizations, effectively privatizing public education through a process of accumulation by dispossession (Aggarwal et al. 2012; Au and Ferrare 2015; Harvey 2005; Lipman 2011a). Such dispossession disproportionately effected low-income students of color (Aggarwal et al. 2012; Johnson 2013; Lipman 2011a).

NCLB exploded the market paradigm in education. Schools were to reform themselves through competition and choice models of change, demonstrating the neoliberal economic theory of the "invisible hand." This idea asserts that there exists a natural social and economic equilibrium that can best be achieved through individual competition. Indeed, a focal point of both neoliberalism and the invisible hand is the individual as a unit, rather than society as a collective. It argues that we can intentionally create social and economic viability by the way of incentivized rivalry. It further asserts that society benefits from the intangible forces of the free market to a much greater extent than it would through conscious government intervention and protections. These values have become so prominent that they are now a generally unchallenged part of our social consciousness and logic, what the Italian Marxist scholar Gramcsi (1971)

referred to as "common sense." As common sense, invisible hand neoliberalism has increasingly exerted a stronghold on the organization of education in the last four decades, emphasizing a narrative of individual selfhood and individual power over public good and common interests (Apple 2001; Hursh 2005; Lipman 2011b; Ross and Gibson 2006).

Secular Mindfulness Education in the Context of Neoliberalism

The principles of invisible hand, free market economics are intended to serve as a method for achieving maximum human potential and well-being. "Neoliberalism is in the first instance" notes Harvey (2005), "a theory of political economic practices that proposes that human well-being can best be advanced by liberating individual entrepreneurial freedoms and skills within an institutional framework characterized by strong private property rights, free markets and free trade" (p. 2). As Margaret Thatcher once exclaimed, "Economics are the method; the object is to change the heart and soul" (as cited in Butt 1981). The neoliberal invocation of human well-being echoes the objectives of secular mindfulness programs in education which seek to enhance educational and personal prosperity. However, secular mindfulness education is subsumed within a broader neoliberal educational paradigm that views students as human capital; a labor force, in the first instance. Attempts at cultivating secular mindfulness and wellness within an unchallenged framework of neoliberalism, then, often substantiate the primacy of the economic base, serving to produce a well-acclimated workforce for global capital.

NCLB and neoliberal reforms have successfully created a culture of high pressure, competition, fear, and disciplinary action. For schools serving low-income student populations, historical and social factors such as neighborhood segregation, poverty, and inadequate access to social services significantly inhibit the possibility of achieving AYP targets, despite the individual efforts of educators and students. With the threat of high-stakes sanctions, such as school closures and reorganization, teachers and students working within these conditions have become progressively strained and stressed. It is within this educational context that secular mindfulness has developed as a phenomenon that has been increasingly adopted in classrooms across the USA.

In order to understand the recent fervor for implementing secular mindfulness in schools, we must examine the historical and racial process by which it emerged as a practical, data-driven educational technique.

Secular Mindfulness and the Invisibility of Asian and Asian American Buddhists

An unexamined aspect of secular mindfulness in the time of neoliberalism is the commodification of Buddhist mindfulness practice and the ensuing erasure of Asia and Asian American Buddhist heritages. This development parallels in many ways the issues around race, cultural appropriation, and yoga (Antony 2014; Puustien and Rauteniemi 2015; York 2001). As Cheah (2011) demonstrates, Buddhism in the USA is situated within a framework of white racial power and white supremacy. I argue, then, that secular mindfulness, as explicitly developed from Buddhist practice, is similarly entrenched within this structure of white supremacy. This poses a direct implication for secular mindfulness education in schools, since, unexamined, such programs can impart an unconscious and potent racial hierarchy of white superiority, especially given the fact that the majority of secular mindfulness in education programs is lead by white instructors (Brown 2015).

This section provides a brief examination of the oppressive racial dynamics of power in neoliberal secular mindfulness, especially in regard to education. I include the critiques of other Asian American Buddhists as an intentional move to provide space for our voices in the conversation. This is a necessary intervention to the dominant narrative which has excluded our histories and perspectives. I explore this critical issue of race, Asian Americans, and secular mindfulness in education in more detail in a separate article.

During the period of neoliberal economic expansion, the secular mindfulness movement has rapidly grown in popularity in the USA and other Western capitalist societies. Though mindfulness as a general concept can be traced to many spiritual traditions, the particular brand of secular mindfulness that has been popularized in the USA and other Western capitalist countries draws its lineage from Buddhism (Sharf 2015; Sun 2014) through Kabat-Zinn's Mindfulness-Based Stress Reduction (MBSR). Indeed, secular mindfulness is most commonly defined through Jon Kabat-Zinn's (1994) work: "Mindfulness means paying attention in a particular way; on purpose, in the present moment, and nonjudgmentally" (p. 4). This definition serves both to articulate a working meaning for secular mindfulness and to mark Kabat-Zinn's role in parsing mindfulness from its Buddhist form and function and aligning it with Western science, thus establishing secular mindfulness within mainstream American culture. For example, in 2014, *Time* magazine dedicated a special issue on what it termed, "The Mindful Revolution" (Pickert 2014), extolling the virtues of secular mindful meditation to a far-reaching audience.

Race and Science

The success of MBSR is due in large part to its relationship to Western science and the manner in which the Buddhist practice of mindfulness was translated into the positivist language of research and "evidenced based" technique. Developed in 1979 at the University of Massachusetts Medical Center, Kabat-Zinn describes MBSR as "a well-defined and systematic patient-centered educational approach which uses relatively intensive training in mindfulness meditation as the core of a program to teach people how to take better care of themselves and live healthier and more adaptive lives" (as cited in Mindfulness-Based Stress Reduction 2014, n.p.).

Indeed, MBSR draws both from Buddhist practice and teachings and prevailing standards in evidence-based science:

> MBSR spans a confluence of epistemologies and practices from two very distinct and until recently, divergent lineages, both committed to empirical investigation, albeit utilizing very different methodologies: that of science, medicine, and psychology, on the one hand, and that of Buddhist meditative traditions and their teachings and practices, known collectively as the Dharma, on the other (History of MBSR 2014, para. 6).

Though MBSR borrows directly from Buddhist spiritual practice, there is a simultaneous distancing from this tradition through the employment of science as a method of validation. The rhetoric of science has enabled MBSR to garner attention from a wide audience, many of whom would not have been receptive to its offerings otherwise. Science provided a language and mechanism by which to rationalize mindful meditation within the familiar discourse of Western society and, relatedly, neoliberalism, especially with the latter's focus on measures of effectiveness and accountability. This helped to establish what the University of Massachusetts Center for Mindfulness identifies as Buddhist mindfulness's "universal applicability" (History of MBSR 2014, para. 6) and enabled both the spread of mindfulness and its secularization. Through MBSR, both Buddhist mindfulness and the broader teachings of Buddhism are expressed as universal traits; "In the present context, to recognize the universal character of the dharma, we use the term with a small 'd'" (History of MBSR 2014, para. 6).

However, the universalizing of Buddhism in this manner demonstrates Cheah's assertion of the white supremacist context of American Buddhism and illuminates the unconsciousness of white privilege in articulating a universality for mass consumption (participants must pay a fee for MBSR training) without recognizing the imbalance of power—especially the historical processes of white conquest—that were involved in the West's procurement of Buddhism. "The fact is," Edwin Ng (Ng and Purser 2015) points out,

"that mindfulness entered Western modernity by the way of the colonial legacy of 'Buddhism'." The rationality behind this universalizing mirrors the common claim that "the Buddha was not Buddhist"; both stem from a racial logic of white superiority that erases the violence of white conquest and plunder. As Ng explains, it "easily effaces the longstanding relations of domination and exploitation that allow one to receive the gift of the Dharma in the first place, a gift inherited from generations upon generations of non-Western, non-white others who have dutifully maintained the teachings for millennia (Ng and Purser 2015)." The universalizing of Buddhist practice through MBSR and the secularization of mindfulness also erases the sustained efforts of Asian and Asian American Buddhists in maintaining the teachings over time and colonized/oppressive spaces, often in the face of immense discrimination. This erasure and the use of positivism and Western science to "discover" a new validity of non-white cosmologies is situated in a larger system of power and history of European colonialism (Smith 1999); and thus, intimately tied to the accumulation of capital. In particular, the secularization of Asian and Buddhist mindfulness demonstrates a neoliberal tendency to commodify cultural and racial identities for white economic and personal profit.

Neoliberal Marketing and Racial Invisibility

As detailed above, in the process of becoming mainstream, secular mindfulness has strategically been distanced from Buddhism, Asia, and Buddhist Asian America. This racial disavowal enables the neoliberal commodification of mindfulness as a product for mass consumption under the auspices of white hierarchy. "Until recently, I thought of meditation as the exclusive province of bearded swamis, unwashed hippies, and fans of John Tesh music," "Nightline" anchor Dan Harris confided in his 2014 book on mindfulness meditation (p. xiv). That Harris evoked "bearded swamis" illuminates the racial dynamics of power that have long cast Asian spiritual practices as strange and exotic, while also hinting at an important aspect of the cachet of secular mindfulness: its very taint of other worldliness, or racial otherness. However, the mass appeal of secular mindfulness can only be secured insomuch as secular mindfulness can establish itself in opposition to this peculiar otherness. "Meditation suffers from a towering PR problem," Harris continued, "If you can get past the cultural baggage, though, what you'll find is that meditation is simply exercise for your brain. It's a proven technique for preventing the voice in your head from leading you around by the nose" (2014, p. xiv). To many secular mindfulness advocates, such as Harris, the more one can erase the implied Buddhist and Asian "cultural baggage" from secular mindfulness, the more one can individually profit (literally—his book was a New York Times bestseller) from the practice.

In fact, the title of his book provides an apt example of the intersecting function of the measures of Western science and the neoliberal focus on efficiency and the self in erasing the historical process of white racial dominance: *10 % Happier: How I Tamed the Voice in My Head, Reduced Stress Without Losing My Edge, and Found Self-Help That Actually Works-A True Story*. Here, the legacy of Asian and Asian Americans Buddhists is rendered invisible, replaced by Harris' attention to himself and his success in taming the voice in his head. His attainment of happiness is attributed to an individuated, scientific process divorced from the over 2000-year history of mindfulness cultivation in Asia. Moreover, in this uncoupling, Harris' regulation of stress is also disconnected from an awareness of the system of neoliberalism which has significantly structured mass political, economic, and social insecurity since the 1970s, as well as the individuation of people from a collective sense of society and history.

Harris' reference of the cultural baggage of mindfulness echoes an earlier term, "Baggage Buddhism," employed by Nattier (1997, p. 73) to categorize Asian Buddhist immigrants who packed up their religion and brought it with them to the USA, as opposed to the "import" or "Elite Buddhism" of convert White Buddhists (p. 73).

Nattier's use of the term baggage mirrors the same racial innuendo exemplified by Harris; both function to highlight the ways in which Asian and Asian Americans have been depicted as perpetual foreigners (Kim 1999) as part of their racial othering. To be sure, the realm of religion has served as a historic site for marginalizing Asians and Asian Americans from mainstream America. In addition to ongoing attacks at Buddhist temples (Violence and Vandalism, n.d.), the most oppressive historical example of this was the state-sponsored incarceration of Japanese Americans during World War II (Williams 2002), where Buddhist priests were specifically targeted for questioning and arrest by the FBI.

Secular mindfulness, therefore, is rife with cultural and racial appropriation. Sri Lankan American Buddhist activist, Dedunu Sylvia, lists but a few examples while highlighting urgent concerns:

> Countless 'mindfulness' books and workshops and trainings at heavy costs. Glorified retreats for White, able-bodied, thin, cis, straight, and class-privileged peoples. Images and films focused almost exclusively on the attainment of nirvana by the White man. Histories of generational attachment to colonialism, slavery, genocide, and conquest, all unapologetically glossed over through exotified ventures to the "third world." All I can see is Buddhist practice — particularly "mindfulness" and "loving-kindness" ideals — used to placate resistance from marginalized populations. Upheld to weaponize model minority myths of Asian passivity in contrast to Black liberation. Exercised in the service of corporate, capitalist, and militarized agendas (Sylvia 2015).

An example of the type of appropriation Sylvia critiques is demonstrated in Jerry Kolber's, "The Branding of Buddhism," in which he advocates for the elimination of the Buddha as a marketing strategy for Buddhism. "When most folks see Buddha, they see a foreign and unfamiliar face that speaks of mysterious eastern religions—oooooo, Buddhists," he argues. Demonstrating a common misconception of the history of American Buddhism, Kolber asserts, "Buddhism in America is at the long end of the initial boom sparked in the 60s among intellectuals and artists who craved that elite connection with the east." "Now it's time for Buddhism to be cool just because regular contemplative practice is cool—it means you know better who you are and how to be in the world," he concludes, adding; "Image is everything, and unless we figure out a way to make the image of the Buddha hip and cool, we'd be better off figuring out some other way to present the techniques without the awesome smiling face of our Eastern inspiration" (Kolber, n.d., para 4).

Here, Kolber assumes the white privilege of promoting the invisibility of any trace connection to Asian history or lineage. The conditioning of white supremacist hegemony has allowed him to remain unconscious of the contributions of Asian American Buddhists in developing and maintaining Buddhist practice in the USA. Arun Likhati, who's blog "Angry Asian Buddhist" has critiqued the lack of Asian American Buddhist representation in the media since 2009, offers a deft analysis of the problematic nature of Kolber's branding proposal.

> With a single sentence, he dons the hat of a historical revisionist and wipes American Buddhist history clean of its Asian affliction. The author disregards the basic fact that Buddhism in America enjoys an unbroken history that stretches back over 100 years. For all those years, it is Asian Americans who have constituted the outright numerical majority of Buddhist Americans—even today, we are still the majority. Plain and simple, Buddhism in America wouldn't be half of what it is without its Asian American members, and for Jerry Kolber to patently neglect our contributions with utter impunity smacks entirely of excessive hegemonic privilege (Likhati 2009, para. 5).

Indeed, Asian American Buddhist such as Rev. Ryo Imamura, an 18th generation priest, and his family have played pivotal roles in keeping Buddhism and mindfulness practice alive in the USA. For example, his parents, Rev. Kanmo Imamura and Jane Imamura, welcomed the young beat poets, Jack Kerouac, Gary Snyder, and Philip Whalen, into their study group at the Berkeley Buddhist Temple (Tabrah 2006, p. 683), effectively shaping the contours of American Buddhism and profoundly influencing American literature at the same time.

Both Harris and Kolber's rebranding of Buddhist mindfulness and practice evidence a co-constituting function of white supremacy and neoliberalism in reducing an Asian system of knowledge/being into the dominant logics of white rationality through science and self-focused improvement. They exemplify how secular mindfulness benefits from an Asian exoticness that infuses it with commodity value as a new product for mainstream consumption within the neoliberal marketplace, while also violently erasing the histories, traditions, and contributions of Asian and Asian American Buddhist communities in maintaining mindfulness practice. This is a crucial context for understanding secular mindfulness education in schools as they illuminate the urgency for such programs to reflect on the broader systems of neoliberal and racial power that they stem from and can unconsciously propagate.

Racial Curricula and the Neoliberal Marketization of Public Schools

"In striving to present mindfulness in a purely secular way, there is the risk of (further) marginalizing the peoples and cultures that have contributed so much to this gift to humanity through many generations," notes Dr. Dzung Vo, author of *The Mindful Teen* (2016, para. 4). Vo calls attention to the ways in which secular mindfulness programs in education can perpetuate a racist system of domination that renders invisible the influences of Asian and Asian American Buddhist communities. "This needs to be considered," he argues, "in a larger, complex context of historical colonial legacies, imbalances of power, and privilege" (2016, para. 4). His sentiment highlights the fact that the mainstream mode of instructing secular mindfulness in schools evades any recognition of the Asian and Asian American influence, let alone a critical analysis of such histories and their invisibility. Secular mindfulness programs in education, therefore, often silently promote an unquestioned racial curriculum that at its foundation asserts white dominance and conquest through the erasure of Asian and Asian American Buddhists. This racial curriculum enables the further marketization of public schools through the purchase of secular mindfulness programs and the instruction of embedded pedagogies for complying with the stresses of the neoliberal system.

Here, again, science proves fundamental in promoting a white washed, secular version of Buddhist mindfulness. Brown (2015) argues that the language of science has been paramount in permitting secular mindfulness organizations and initiatives to enter schools. Because of legal mandates against the integration of religious practice with publicly funded instruction, secular mindfulness advocates have stressed the scientific validity of mindfulness and aggressively distanced it from Buddhism. In detailing the manner in which proponents of secular mindfulness in schools invoke the power of science to sell secular mindfulness, Brown focuses on Goldie Hawn and her MindUP curriculum as a primary example. "The MindUP 'script' tells the story that Buddhist mindfulness meditation is really 'secular' neuroscience," Brown notes, adding that "The MindUP curriculum consists in large part of simplified (and not always accurate) lessons in brain anatomy ('reflective, thinking prefrontal cortex' = good, 'reflexive, reactive amygdala' = bad)..." (2015, para. 8–9).

Whereas Brown highlights the use of science to mask intentions of promoting Buddhism in schools, I argue that there is a racial dimension to this mask. It is not that Buddhism—an inherently Asian tradition—is being advocated, but rather a white-dominated form of secular Buddhism that is entangled with conquest and notions of racial superiority. This secular, scientific mindfulness then renders Buddhist practice suitable for mass consumption in the neoliberal educational marketplace of public schools. Moreover, the majority of these programs in education are lead by white instructors (Brown 2015), further erasing the fundamental role of Asian and Asian American Buddhists in maintaining mindfulness practice and instilling the idea of white expertise. Secular mindfulness in education then becomes more than a commodity, but also a curricular mechanism to perpetuate white superiority through a pedagogy of calm acceptance. Who becomes elevated to the status of mindfulness instructor, through the neoliberal process of fees-based certification and training, is an important cite for the critical analysis of white privilege and domination. Relatedly, who becomes subject to such curriculum—in many cases, low-income students of color—demonstrates a racial power imbalance where the supposed superiority of white knowledge (acquired from the conquest and commodification of Asian Buddhist ways of knowing/being) is imposed to perpetuate the colonial legacy of benevolent uplift of students of color (Hsu 2015).

The science of secular mindfulness has not only paved a popular pathway for its integration in schools, but a lucrative one at that. Congressman Tim Ryan of Ohio, a vocal advocate of secular mindfulness and author of *A Mindful Nation*, for example, allocated nearly $1 million in public funds to implement mindfulness and Social Emotional Learning (SEL) programs in schools in his district (Ryan 2013, para. 7). Secular mindfulness in education, with its scientific turn, is constantly framed within, and limited by the economic imperative of neoliberalism. The manner in which the corporate sector has taken to valorizing secular mindfulness demonstrates this relationship (Purser and Loy 2013). Forbes notes that education policymakers have turned to secular mindfulness as a way of infusing "corporatized culture and market values" into the public sphere of schooling and to promote the neoliberal management of education (2015).

The science of secular mindfulness has been especially marketable in schools in regard to its perceived effectiveness in managing stress and increasing student achievement. Many students "grow up in low-income, high crime areas," notes Ryan. "They often live in households that have experienced job loss or in which one or both parents are abusive. Over time these stressors have a debilitating effect" (2013, para. 5). In commenting on mindfulness meditation in New York City schools, Chancellor Carmen Fariña stated, "We're putting it in a lot of our schools... because kids are under a lot of stress" (Harris 2015). To manage the stress levels of students, and teachers, mindfulness and other meditation-based programs have been rapidly ushered into the classroom, bolstered by research that demonstrates measureable differences in stress reduction (Barnes et al. 2003; Kerrigan et al. 2011; Mendelson et al. 2010; Van de Weijer-Bergsma et al. 2014).

While secular mindfulness programs that target student and educator stress may provide a service in teaching techniques to regulate the effects of pressure and anxiety, they lack an important contextual analysis of the systems of power that create these very stressors. Indeed, the principles of neoliberalism have shaped schools and schooling in regard to the attendant anxieties of the imposed neoliberal lifestyle. Policies such as NCLB are directly responsible for creating the motivations for the interventions, of which secular mindfulness programs are among the most popular, to reduce stress. Without an understanding of the neoliberal and racial frameworks which have influenced schools and American society, mindfulness programs in education will only assist students and educators in coping with an ever-compounding crisis. Such programs treat the symptoms while enabling the social disease. This mystifies the reality that such conditions are produced as a direct result of the structural forces of neoliberalism and racism. *Inequality—and therefore, the resulting traumas and stresses associated with it—is a requisite element of a system that argues for the elimination of public services in favor of competition and private ownership.* Neoliberalism and secular mindfulness' shared goal of enhancing human well-being, then, is reserved for the economically fittest among us who can acquire (through competition) enough resources to purchase it.

Low-income communities, high crime, joblessness, and the associated stresses that Ryan writes of are not accidental. To the contrary, they have been created to benefit the well-being of predominantly white individuals, at the expense of historically marginalized communities. The aforementioned social issues occur most frequently in racially and economically segregated neighborhoods which were developed through the legacies of "separate but equal" legislation, Jim Crow laws, and discriminatory housing covenants. These neighborhoods were further impoverished in the early neoliberal period through the War on Drugs which sought to criminalize the black male population (Baum 2016, p. 1). As a result, African American males have been incarcerated at disproportionate rates (Gilmore 2000, 2007; Davis 2003). While their imprisonment has lead to increased instances and durations of poverty in their families and neighborhoods, the growth of what's been termed the prison industrial complex, coupled with the private business management of prisons, has ensured that some people are able to enhance their well-being in quite lucrative ways (Davis and Shaylor 2001).

These are the neighborhoods from which many of our low-income students of color are coming. Their stresses, therefore, are not a result of individual failures, so-called cultures of poverty, or pathology; rather, they are direct ramification of white supremacy and neoliberal accumulation and dispossession. Secular mindfulness programs in education will only succeed in creating long-term stress reduction if they take into consideration these structural realities. If not they, they run the risk of being exercises in complacency.

Instructing a Neoliberal Sense of Self

When we are unconscious of the historical roots and neoliberal causes of student and educator stress and trauma, we perpetuate the neoliberal technique of emphasizing individual culpability and responsibility (Jankowski and Provezis 2014; Reveley 2015) while understating the need for state intervention. Secular mindfulness programs in schools, then, can ultimately serve to propagate the idea of the neoliberal self by optimizing students and educators to better regulate themselves and their performance (Forbes 2015). "Capitalism in its latest transformation," explains Goddard (2010), "requires selves which are endlessly adaptable to the levels of change and insecurity, to the personal and social instability generated by a globalised economy" (p. 353.) Secular mindfulness education may serve, therefore, to prepare learners and teachers to better acclimate themselves to the uncertainties of the neoliberal educational landscape. Ng and Purser (2016) employ Michel Foucault's theory of governmentality to highlight the way in which secular mindfulness and its rhetoric of self-care and wellness promote an intimate alignment of individuals with the larger systems of oppressive power. Secular mindfulness in education, then, can operate as an insidious—even if unintentional—tool for training students to submit to the inhumane and inequitable conditions of neoliberalism.

A commonly cited objective and benefit of secular mindfulness educational programs is their utility in extending students' capacity to focus and pay attention in class (Napoli et al. 2005). MindUP's curriculum was found to "increase executive function" in its elementary school participants: "Children demonstrated faster reaction times while performing tests such as Dr. Diamond's 'Flanker Fish' trials. The [sic]correlates to heightened self-regulatory ability" (Research 2016, para. 6). While focus and self-regulatory skills are important in academic success, they are being developed in an educational context where measures for achievement have become decidedly incentivized and consequential (e.g., the former policy of NCLB). Training students to enhance their attention, therefore, translates easily into neoliberal applications. It can, for example, serve to produce more efficient test takers (Forbes 2015; Hsu 2013), thereby conditioning students into young "knowledge workers," to borrow a term from Cortada (1998, p. ix), ready to validate the growth of a testing industry and the needs of the global market. To be clear, I am not advocating for unfocused students. Rather, I am calling attention to the urgent need for deeper awareness and critical, public discourse around the political, economic, and social contexts in which secular mindfulness focus is being cultivated and the ends it serves. Without careful and critical attention to the ways in which secular mindfulness might inadvertently support neoliberal, racist practices and the business of education, such programs may cause students extended

suffering—enabling them to tolerate the demands and consequences of high-stakes testing that much longer—for the gain of a privileged few. Meanwhile, the pressures of the neoliberal schooling framework continue to exert stress upon students and their bodies.

Conclusion

The promotion of secular mindfulness in schools further advances the market paradigm of education established by NCLB, even though these programs are often employed to intervene in the consequences of neoliberal competition. Secular mindfulness interventions, then, are both a response to neoliberalism and an integral part of the neoliberal structure. While secular mindfulness programs in education might enable students to achieve a moment of calm, they do not tend to the systemic issues that will continue to cause them distress. Moreover, the focus on achieving a scientifically based inner tranquility belies a close examination of the external historical, racial, and neoliberal forces that disproportionately impose these conditions upon low-income students of color. The particular brand of well-being promoted by secular mindfulness programs in education instructs students to tranquilly accept and adapt to the stressors imposed by neoliberal competition and privatization of public goods and services.

Additionally, the seemingly well-intended goals of stress reduction and performance enhancement may produce an uncritical, ahistorical consciousness through secular mindfulness' emphasis on non-judgmental, present moment awareness. This runs the risk of students learning to become acquiescent to structural inequalities by further internalizing these conditions through secular mindfulness. There is a real danger, then, in losing the value of resistance as resilience. As Dedunu Sylvia noted, the embedded racial dimensions of secular mindfulness can and has been used to "placate resistance from marginalized communities." Thus, instead of schools serving as sites for imparting critical thinking and analysis, unconscious secular mindfulness programs in education transforms classrooms into spaces for validating the neoliberal focus on individual competition and the economic imperative.

If we are to take seriously the task of developing a critical awareness of structures of power and systemic inequality, we could be faced with glaring concerns in regard to the efficacy of secular mindfulness programs in our schools. "It definitely doesn't address poverty, and it may not work for everybody," notes Patricia Jennings, University of Virginia professor of Education and author of *Mindfulness for Teachers* (Harris 2015). However, being able to fully confront the realities of our neoliberal, racialized educational institution and the ways in which secular mindfulness programs might enhance this oppressive framework is a duty we owe to our students. In awakening from the mystification, or delusion, as Buddhists would put it, of neoliberal promises of well-being through economic imperatives and the common sense of white superiority, we can begin to imagine more just educational alternatives.

One such alternative would be to dismantle the popular educational reform model of white imposed secular mindfulness in schools. As Forbes (2015) has pointed out, there are other educational programs that have proven effective in reducing student stress and increasing academic achievement; "While some students benefit from mindfulness, however, they are just as likely to benefit from any good education or counseling program-still sorely lacking in many schools…" (n.p). This calls attention to the need for critical inquiry around the sudden vehemence in implementing secular mindfulness in schools, and for an examination of those who stand to profit from the secularization of Asian Buddhism and the subsequent marketization of scientific mindfulness in public schools. An obvious alternative to secular mindfulness programs would be to re-appropriate resources to enhancing defunded educational services, such as counseling. The viability of this option provides both an equitable resource to all students and a direct intervention to the neoliberal, white commodification of Asian Buddhism.

In a scene from the documentary film, *American Revolutionary: The Evolution of Grace Lee Boggs* (Lee 2014), the late Asian American, centenarian activist, Grace Lee Boggs chats about education reform with the notable African American actor, Danny Glover. Boggs argues for a paradigm shift in education. They sit in the living room of her Detroit home, in a city that has experienced drastic economic displacement and heightened racial inequality as a result of the dominance of neoliberal policies. Danny Glover reasons that investing in math reforms provides an important avenue for educational change and equity, but Grace Lee Boggs disagrees. "That's basically what we're trying to tell our kids to do: 'You've got to learn mathematics, you've got to learn technology, and so we can compete on the world market," she argues, illuminating the ideals of neoliberalism that have become so infused with our notions of common sense and modern rationality (Lee 2014). "That's what kids are rejecting. And I don't blame them," she continues, "So, we've got to think about 'What is education for?'" (Lee 2014). In contemplating secular mindfulness programs for our classrooms, we would do well to consider this most fundamental question. "What is education for?"

Secular mindfulness in education can only be helpful to the extent that it aids in cultivating a paradigm shift in education; one that enhances the value of education as a public good, recognizes the validity of diverse cosmologies of knowing/being, and enables the collective well-being of all living beings. Ultimately, even more than we need secular mindfulness programs in schools, we are in urgent need of a new conceptualization of education and human progress.

References

Aggarwal, U., Mayorga, E., & Nevel, D. (2012). Slow violence and neoliberal education reform: Reflections on a school closure. *Peace and Conflict: Journal of Peace Psychology, 18*(2), 156.

Antony, M. G. (2014). It's not religious, but it's spiritual: Appropriation and the universal spirituality of yoga. *Journal of Communication & Religion, 37*(4).

Apple, M. (2001). *Educating the 'right' way: Markets, standards, God, and inequality*. New York, NY: Routledge.

Au, W., & Ferrare, J. J. (2015). Neoliberalism, social networks, and the new governance of education. In W. Au & J. J. Ferrare (Eds.), *Mapping corporate education reform: Power and policy networks in the neoliberal state* (pp. 1–22). New York, NY: Routledge.

Barnes, V. A., Bauza, L. B., & Treiber, F. A. (2003). Impact of stress reduction on negative school behavior in adolescents. *Health and Quality of Life Outcomes, 1* (10), 1–7.

Baum, D. (2016, April). Legalize it all. *Harper's Magazine*. Retrieved from http://harpers.org/archive/2016/04/legalize-it-all/

Brown, C. (2015, February 4). Mindfulness meditation in public schools: Side-stepping supreme court religion rulings. *Huffpost Education*. Retrieved from http://www.huffingtonpost.com/candy-gunther-brown-phd/mindfulness-meditation-in_b_6276968.html

Butt, R. (1981, May 3). Interview for sunday times. *Sunday Times*. Retrieved from http://www.margaretthatcher.org/document/104475

Center for Mindfulness in Medicine, Health Care and Society. (2014). *Mindfulness-based stress reduction standards of practice*. Shrewsbury, MA: Saki Santorelli.

Cheah, J. (2011). *Race and religion in American Buddhism: White supremacy and immigrant adaptation*. Oxfor, UK: Oxford University Press.

Cortada, J. (1998). *Rise of the knowledge worker*. Boston, MA: Butterworth-Heinemann.

Davis, A. Y. (2003). Race and criminalization: Black Americans and the punishment industry. *Criminological Perspectives: Essential Readings*, 284.

Davis, A. Y., & Shaylor, C. (2001). Race, gender, and the prison industrial complex: California and beyond. *Meridians, 2*(1), 1–25.

Forbes, D. (2015, November 8). They want kids to be robots: Meet the new education craze designed to distract you from overtesting. *Salon*. Retrieved from http://www.salon.com/2015/11/08/they_want_kids_to_be_robots_meet_the_new_education_craze_designed_to_distract_you_from_overtesting/

Gill, B., Zimmer, R., Christman, J., & Blanc, S. (2007). State takeover, school restructuring, private management, and student achievement in Philadelphia. RAND & Research for Action. Retrieved from http://www.rand.org/content/dam/rand/pubs/monographs/2007/RAND_MG533.pdf

Gilmore, K. (2000). Slavery and prison—understanding the connections. *Social Justice*, 273(81), 195–205.

Gilmore, R. W. (2007). *Golden gulag: Prisons, surplus, crisis, and opposition in globalizing California* (Vol. 21). Berkeley, CA: University of California Press.

Goddard, R. (2010). Critiquing the Educational Present: The (limited) usefulness to educational research of the Foucauldian approach to governmentality. *Educational Philosophy and Theory, 42*(3), 345–360.

Gramsci, A. ([1971]/1992). Prison Notebooks (J. A. Buttigieg & A. Callari, ed. Trans.) (Vol. 1). New York, NY: Columbia University Press.

Harris, D. (2014). *10% happier: How I tamed the voice in my head, reduced stress without losing my edge, and found self-help that actually works—A true story*. New York, NY: HarperCollins.

Harris, E. (2015, October 23). Under stress, students in New York schools find calm in meditation. *New York Times*. Retrieved from http://www.nytimes.com/2015/10/24/nyregion/under-stress-students-in-new-york-schools-find-calm-in-meditation.html?_r=0

Harvey, D. (2005). *A brief history of neoliberalism*. Oxford, UK: Oxford University Press.

History of MBSR. (2014). Center for mindfulness in medicine, health care, and society. Retrieved from http://www.umassmed.edu/cfm/stress-reduction/history-of-mbsr/

Hsu, F. (2013). The heart of mindfulness: A response to the New York Times. *Turning Wheel*. Retrieved from http://www.buddhistpeacefellowship.org/the-heart-of-mindfulness-a-response-to-the-new-york-times/

Hsu, F. (2015). The coloniality of Neoliberal English: The enduring structures of American Colonial English instruction in the Philippines and Puerto Rico. *L2 Journal, 7*(3).

Hursh, D. (2005). Neo-liberalism, markets and accountability: Transforming education and undermining democracy in the United States and England. *Policy Futures in Education, 3*(1), 3–15.

Hursh, D. (2007). Assessing no child left behind and the rise of neoliberal education policies. *American Educational Research Journal, 44*(3), 493–518.

Jankowski, N., & Provezis, S. (2014). Neoliberal ideologies, governmentality and the academy: An examination of accountability through assessment and transparency. *Educational Philosophy and Theory, 46*(5), 475–487.

Johnson, A. W. (2013). "Turnaround" as shock therapy race, neoliberalism, and school reform. *Urban Education, 48*(2), 232–256.

Kabat-Zinn, J. (1994). *Wherever you go, there you are: Mindfulness meditation in everyday life*. New York, NY: Hyperion.

Kerrigan, D., Johnson, K., Stewart, M., Magyari, T., Hutton, N., Ellen, J. M., et al. (2011). Perceptions, experiences, and shifts in perspective occurring among urban youth participating in a mindfulness-based stress reduction program. *Complementary Therapies in clinical practice, 17*(2), 96–101.

Kim, C. J. (1999). The racial triangulation of Asian Americans. *Politics & Society, 27*(1), 105–138.

Kolber, J. (no date). The branding of Buddhism. *Beliefnet*. Retrieved from http://www.beliefnet.com/columnists/onecity/2009/09/the-branding-of-buddhism.html

Kuehn, L. (1999). Responding to globalization of education in the Americas: Strategies to support public education. In IDEA conference in quito, ecuador. Retrieved from http://www.vcn.bc.ca/idea/kuehn.htm

Lee, G. (2014). American revolutionary: The evolution of Grace Lee Boggs (Motion Picture). USA: LeeLee Films.

Likhati, A. (2009). Asian Free Buddhism. *Angry Asian Buddhist*. Retrieved from http://www.angryasianbuddhist.com/2009/09/asian-free-buddhism.html

Lipman, P. (2011a). *The new political economy of urban education: Neoliberalism, race and the right to the city*. New York, NY: Routledge.

Lipman, P. (2011b). Contesting the city: Neoliberal urbanism and the cultural politics of education reform in Chicago. *Discourse: Studies in the cultural politics of education, 32*(2), 217–234.

Mendelson, T., Greenberg, M. T., Dariotis, J. K., Gould, L. F., Rhoades, B. L., & Leaf, P. J. (2010). Feasibility and preliminary outcomes of a school-based mindfulness intervention for urban youth. *Journal of Abnormal Child Psychology, 38*, 985–994.

Napoli, M., Krech, P. R., & Holley, L. C. (2005). Mindfulness training for elementary school students. *Journal of Applied School Psychology, 21*(1), 99–125.

Nattier, J. (1997). Buddhism comes to main street. *The Wilson Quarterly (1976), 21*(2), 72–80.

Ng, E., & Purser, E. (2015, October 2). White privilege and the mindfulness movement. *Turning wheel media*. Retrieved from http://www.buddhistpeacefellowship.org/author/edwin-ng-and-ron-purser/

Ng, E., & Purser, E. (2016, April 4). Mindfulness and self-care: Why should I care? *Patheos*. Retrieved from http://www.patheos.com/blogs/americanbuddhist/2016/04/mindfulness-and-self-care-why-should-i-care.html

Pickert, K. (2014, February 3). The mindful revolution. *TIME magazine, 3*, 34–48.

Purser, R., & Loy, D. (2013). Beyond McMindfulness. *Huffington Post, 1*(7), 13.

Puustinen, L., & Rautaniemi, M. (2015). Wellbeing for sale: Representations of yoga in commercial media. *Temenos, 51*(1), 45–70.

Research. (2016). The Hawn foundation. Retrieved from http://thehawnfoundation.org/research/

Reveley, J. (2015). Foucauldian critique of positive education and related self-technologies: Some problems and new directions. *Open Review of Educational Research, 2*(1), 78–93.

Ross, E. W., & Gibson, R. J. (Eds.). (2006). *Neoliberalism and education reform*. Cresskill, NJ: Hampton press.

Ryan, T. (2013). Bringing mindfulness to our public schools. Retrieved from https://www.eomega.org/article/bringing-mindfulness-to-our-public-schools

Sharf, R. H. (2015). Is mindfulness Buddhist? (and why it matters). *Transcultural psychiatry, 52*(4), 470–484.

Smith, L. T. (1999). *Decolonizing methodologies: Research and indigenous peoples*. Zed books.

Sun, J. (2014). Mindfulness in context: A historical discourse analysis. *Contemporary Buddhism, 15*(2), 394–415.

Sylvia, D. (2015). 5 responses to the awkwardly titled "New Face of Buddhism." *Turning wheel*. Retrieved from http://www.buddhistpeacefellowship.org/5-responses-to-the-new-face-of-buddhism/

Tabrah, R. (2006). Religions of Japanese immigrants and Japanese American communities, In R. S. Keller, R. R. Ruether, & M. Cantlon (Eds.), *Encyclopedia of women and religion in North America: Women and religion: Methods of study and reflection* (Vol. 1, pp. 680–687). Bloomington, IN: Indiana University Press

U.S. Department of Education. (2009). Race to the top executive summary. Retrieved from https://www2.ed.gov/programs/racetothetop/executive-summary.pdf

U.S. Department of Education. (2010). Guidance on school improvement grants under section 1003(g) of the Elementary and Secondary Education Act of 1965, revised June 29, 2010. Washington, DC.

van de Weijer-Bergsma, E., Langenberg, G., Brandsma, R., Oort, F. J., & Bögels, S. M. (2014). The effectiveness of a school-based mindfulness training as a program to prevent stress in elementary school children. *Mindfulness, 5*(3), 238–248.

Violence and Vandalism. (n.d.). The pluralism project. Retrieved from http://pluralism.org/encounter/todays-challenges/violence-and-vandalism/

Vo. D. (2016). Is mindfulness religious? *Mindfulness for teens*. Retrieved from http://mindfulnessforteens.com/blog/

Williams, D. R. (2002). Camp Dharma: Japanese-American Buddhist identity and the internment experience of World War II. *Westward Dharma: Buddhism Beyond Asia*, 191–200.

York, M. (2001). New age commodification and appropriation of spirituality. *Journal of Contemporary Religion, 16*(3), 361–372.

Terry Hyland

Mindfulness, Education and Therapeutic Transformation

In their investigation of the various accounts of the relationship between therapy and education, Smeyers et al. (2007) identify three principal 'climates of thought'. They observe that:

> First, there is the conception of therapy as an obvious good, a practice that helps people lead more fulfilled and less unhappy lives…Second, and partly in reaction to the first, there is increasing scepticism, even hostility, towards therapy and its influence…Therapy is charged with encouraging a debilitating climate of dependence to which it then presents itself as a solution. Third, it may seem to some that the only essential and important questions concerning therapy are whether or not it can be proved to be effective and if so how to do it (p. 1).

The writers go on to justify their rejection of all three approaches in favour of a 'more balanced and nuanced treatment of therapy and its connections with education' and which argues against 'the idea that a sharp conceptual division can be made between education and therapy' (ibid., p. 1).

In explaining and justifying his conception of education as the initiation into worthwhile activities, the philosopher of education, Peters (1966) makes use of an analogy between activi-

ties of 'education' and those of 'reform'. He argues that education is like reform in that it 'picks out no particular activity or process' but, rather, it 'lays down criteria to which activities or processes must conform'. It is suggested that:

> Both concepts have the criterion built into them that something worthwhile should be achieved. 'Education' does not imply, like 'reform', that a man should be brought back from a state of turpitude into which he has lapsed; but it does have normative implications…It implies that something worthwhile is being or has been intentionally transmitted in a morally acceptable manner (p. 25).

I suggest that a similar sort of analogy holds in respect of education and therapy. Neither process picks out any specific method or technique, yet both imply the achievement of a desirable state of mind. In the case of education, this state involves the development of knowledge and understanding, and in the case of therapy, there is the goal of enhancing mental health and well-being by, for instance, removing delusions, breaking harmful habits or developing more wholesome or nourishing thoughts and actions.

Indeed, it would seem that the therapy is even closer to education than the notion of reform since —at least, in the mindfulness-based approaches recommended here—there is a mutual interest in the Socratic method of 'mental midwifery' (Hyland 2003, p. 75). The spirit is captured perfectly in the dialogues in which Socrates functions as a mouthpiece and representative of the philosophical views of Plato. In *The Republic*, Plato rejects the 'conception of education professed by

T. Hyland (✉)
Free University, Dublin, Ireland
e-mail: hylandterry@ymail.com

© Springer International Publishing Switzerland 2016
R.E. Purser et al. (eds.), *Handbook of Mindfulness*,
Mindfulness in Behavioral Health, DOI 10.1007/978-3-319-44019-4_25

those who say they can put into the mind knowledge that was not there before'; the business of the educator is not that of implanting sight but, rather, of 'ensuring that someone who had it was turned in the right direction and looking the right way' (1965 edn., p. 283).

It would be useful to look more closely at Peters' specific criteria of education as a way of further elaborating the extent of the connections between therapy and education. These are as follows:

(1) That education implies the transmission of what is worthwhile to those who become committed to it;
(2) That education must involve knowledge and understanding and some kind of cognitive perspective, which are not inert;
(3) That education at least rules out some procedures of transmission, on the grounds that they lack wittingness and voluntariness on the part of the learner (p. 45).

Taking each of these in turn, we can determine how far therapy—and especially mindfulness-based strategies—satisfies the criteria.

1. Therapeutic activity may be considered worthwhile insofar as it involves the progression from a less to a more desirable state of mind and being. Certainly, direct transmission or teaching is less common in a therapeutic relationship than it is in education (though intrinsically valuable autodidacticism is prized in both spheres), but the significant point is that worthwhile learning is still taking place. A central purpose of mindfulness-based therapy is to free the mind from automatic, ruminative thought and action, and this is similar to the Socratic method of freeing the mind from delusions and error in order to pave the way for genuine learning. Mindfulness cultivates the awareness—especially that which 'emerges through paying attention on purpose, in the present moment, and non-judgmentally to things as they are' (Williams et al. 2007, p. 47)—which is a prerequisite for meaningful and productive teaching and learning. As Schoeberlein and Sheth (2009) explain about their experience of using mindfulness strategies in American schools:

Mindfulness and education are beautifully interwoven. Mindfulness is about being present with and to your inner experience as well as your outer environment, including other people. When teachers are fully present they teach better. When students are fully present, the quality of their learning is better (p. xi).

Both spheres involve the attention to and modification of consciousness and modes of thinking, and both aim at a form of enlightened awareness which pays due attention to values and feelings. Peters was perhaps the most distinguished and foremost advocate of a traditional liberal education grounded in forms of knowledge (Cuypers and Martin 2009), but he took great care to leave room for individual development and personal relationships in teaching and learning. He observes that:

the ability to form and maintain satisfactory personal relationships is almost a necessary condition of doing anything else that is not warped or stunted. If the need to love and be loved is not satisfied the individual will be prone to distortions of belief, ineffectiveness of lack of control in action, and unreliability in his allegiances. His attempt to learn things will also be hampered by his lack of trust and confidence. A firm basis of love and trust, together with a continuing education in personal relationships, is therefore s crucial underpinning to any other more specific educational enterprise (1966, p. 58).

2. Can therapy be said to incorporate the knowledge, understanding and active cognitive perspective required by Peters' second criterion? Certainly, there are clear differences in types of knowledge utilised and exemplified in the fields of education and therapy. These can be illustrated by examining the 'forms of knowledge' which, Hirst claims, cover the whole domain of human endeavour and provide the foundations of a liberal education, which has been traditionally viewed as a 'process which frees the mind from error' (Schofield 1972, p. 154). Originally, seven (or eight, depending on the particular interpretation) disciplines or forms—distinguishable from each other by their conceptual and logical frameworks, methodology and truth criteria—were identified by Hirst (1965):

'mathematics, physical sciences, human sciences, history, religion, literature and the fine arts, and philosophy' (p. 131), in addition to theoretical and practical fields of knowledge which combined elements from the forms (and also incorporated morality). In later versions, these were revised; history was subsumed into the human sciences, mathematics and logic are called symbolics and literature and the fine arts are labelled aesthetics, doubt is cast about whether religion is a genuine form, and a new area 'awareness and understanding of our own and other people's minds' (Hirst and Peters 1970, p. 63) is identified.

In later comments on the forms of knowledge, Hirst (1974) was concerned to stress that the forms do not exhaust the aims or content of educational practice. He observed that:

> much commonsense knowledge and many forms of experience, attitudes and skills may be regarded as lying outside all the disciplines we have…Many forms of education, including liberal education in my sense, will have objectives some of which come from within the disciplines and some of which do not (p. 98).

The knowledge and understanding which guides many therapeutic processes may be characterised as that 'commonsense knowledge', experience, attitudes and skills referred to by Hirst. However, there is also clear evidence within therapeutic practice of the utilisation of aspects of the human and physical sciences and, especially, of the area labelled 'awareness and understanding of our own and other people's minds'. In addition, the aims of education and therapy in terms of freeing the mind from error and delusion to make way for creativity and openness in learning are in close harmony. As will be explained later, the MBIs I am concerned to recommend are necessarily grounded in the *dharma* which lies at the core of Buddhist contemplative traditions. To conclude this section, it is worth considering the view of one such practitioner, Thich Nhat Hanh, particularly his conception of mindfulness and work. He advises us to:

keep your attention focused on the work, be alert and ready to handle ably and intelligently any situation which may arise – this is mindfulness. There is no reason why mindfulness should be different from focusing all one's attention on one's work, to be alert and to be using one's best judgment. During the moment one is consulting, resolving, and dealing with whatever arises, a calm heart and self-control are necessary if one is to obtain good results…If we are not in control of ourselves but instead let our impatience or anger interfere, then our work is no longer of any value. Mindfulness is the miracle by which we master and restore ourselves (Hanh 1991, p. 14).

This message applies to any form of work, including the 'work' of learning, teaching and education.

3. The final criterion which needs to be satisfied for any activity to be educational concerns the use of methods which respect the wittingness and autonomy of learners. Obviously this rules out certain therapeutic practices such as hypnotherapy and behaviour modification but, equally, it rules out many educational practices involving coercion, punishment and indoctrination (Hirst and Peters 1970).

Smeyers et al. (2007) point out that—although those forms of therapy which consist of 'doing things to people' in a manipulative manner—may be ruled out on educational grounds —'many therapists in fact are concerned precisely to distinguish therapy as a relationship between autonomous human beings from therapy as a set of techniques' (pp. 1–2).

Versions of therapy which respect personal autonomy are, I would suggest, well to the fore in mainstream psychotherapeutic mindfulness approaches. Investigating the links between Buddhist practice and psychoanalysis, for example, Rubin (2003) explains the 'similarities between both traditions' and observes that:

> Both are concerned with the nature and alleviation of human suffering and each has both a diagnosis and 'treatment plan' for alleviating human misery. The three other important things they share make a comparison between tem possible and potentially productive. First, they are pursued within the crucible of an emotionally intimate relationship between either an analyst-analysand or a teacher

and student. Second, they emphasise some similar experiential processes – evenly hovering attentions and free association in psychoanalysis and meditation in Buddhism. Third, they recognise that obstacles impede the attempt to facilitate change (pp. 45–46).

This account is strikingly similar to the sort of learning and teaching encounter favoured in open and progressive education which emphasises student-centeredness, autonomy and independent learning with the teacher acting as a facilitator, guide and resource person (Hyland 1979; Lowe 2007). In addition, we can detect the idea of removing obstacles to learning, the freeing the mind from error that is characteristic of liberal education.

Staying within the contemplative tradition, Salzberg and Goldstein (2001) explain how the 'function of meditation is to shine the light of awareness on our thinking'. The educational implications are brought out clearly in their description of how:

The practice of bare attention opens up the claustrophobic world of our conditioning, revealing an array of options. Once we can see clearly what's going on in our minds, we can choose whether and how to act on what we're seeing. The faculty used to make those choices is called discriminating wisdom...the ability to know skillful actions from unskillful actions (p. 48).

MBIs can have a potential impact on both the means and ends of education. Not only do they provide the foundations for productive learning, but also offer a blueprint to guide the direction of that learning. As Thich Nhat Hanh observes:

Mindfulness helps us look deeply into the depths of our consciousness...When we practice this we are liberated from fear, sorrow and the fires burning inside us. When mindfulness embraces our joy, our sadness, and all our mental formations, sooner or later we will see their deep roots...Mindfulness shines its light upon them and helps them to transform (Hanh 1999, p. 75).

As explained below, a number of contemporary MBIs are, arguably, failing to meet both educational and Buddhist foundational criteria, and it is important to examine the ways in which the commodification and marketisation of mindfulness exemplified in the McDonaldization model has operated as a result of the 'mindfulness revolution' (Boyce 2011) which has swept virus-like through academia, public life and popular culture over the last decade or so. Mindfulness is a now a meme, a product, a fashionable spiritual commodity with enormous market potential and—in its populist forms—has been transmuted into an all-pervasive 'McMindfulness' phenomenon. As Purser and Loy (2013) argue:

While a stripped-down, secularized technique – what some critics are now calling "McMindfulness" – may make it more palatable to the corporate world, decontextualizing mindfulness from its original liberative and transformative purpose, as well as its foundation in social ethics, amounts to a Faustian bargain. Rather than applying mindfulness as a means to awaken individuals and organizations from the unwholesome roots of greed, ill will and delusion, it is usually being refashioned into a banal, therapeutic, self-help technique that can actually reinforce those roots (p. 1).

The McDonaldization of MBIs

The process by which McMindfulness has been produced—McDonaldization—was originally conceived and developed by Ritzer (2000) in the construction of a model informed by Weber's writings to describe and explain the increasing technical rationalisation and standardisation of more and more aspects of social, economic, political life and culture. As a form of policy analysis, Ritzer's model has been used extensively to critique developments in education (Hartley 1995; Hyland 1999) and other spheres of public life and culture (Alfino et al. 1998) and its main stages can be usefully employed to map the emergence of McMindfulness. There are four main elements, and they are worth examining in some detail in relation to the evolution of the commodified versions of mindfulness practices.

Efficiency

Defined by Ritzer (2000) as 'choosing the optimum means to a given end' (p. 40), efficiency results in streamlining, standardisation and simplification of both the product and its delivery to customers. In terms of items sold under the mindfulness label, this process is relatively simple. If you want to maximise sales of a colouring book, you just put mindfulness on the front cover (e.g. Farrarons 2015), and the same principle applies to all cultural products such as self-help and health/well-being manuals (arguably, the most lucrative sphere) and leisure activities such as cooking, gardening and sport (see Burkeman 2016). When it comes to mindfulness courses, the standardisation process is greatly helped by having handy bite-sized MBSR/MBCT programmes ready for delivery to potential consumers. Such courses are, of course, the original core vehicles for employing mindfulness practice to deal with depression, addiction, pain and general mind/body afflictions. It is not suggested here that they are typical examples of McMindfulness. However, their 8-week structure—particularly as this is reduced, condensed and transmuted into 'apps' and online programmes (see 'Control' element below)—clearly lends itself to these efficiency conditions and is undoubtedly complicit if not directly responsible for the exponential growth of MBIs and the McMindfulness brand over the last decade or so.

Calculability

This element of the process involves 'calculating, counting, quantifying' such that this 'becomes a surrogate for quality' (Ritzer 2000, p. 62). Ritzer describes how the business of reducing 'production and service to numbers'—examples of higher education, health care and politics are offered in illustration (ibid., pp. 68–77)—results in regression to mediocre and lowest common denominator production and produce. The competence-based education and training (CBET) techniques informed by behaviourist principles provide a graphic illustration of how this obsession with measuring outcomes—at the expense of process and underlying principles—can distort, de-skill and de-professionalise education and training from school to university learning (Hyland 1994, 2014a). In a similar way, the drive to measure the outcomes of

mindfulness has led to similar negative transmutations. Since the exponential development of the mindfulness industry, Grossman (2011) has been forceful in his criticisms of mindfulness measurement scales, particularly those relying upon self-reports by MBI course participants. The key weaknesses are that they de-contextualise mindfulness from its ethical and attitudinal foundations, measure only specific aspects of mindfulness such as the capacity to stay in the present moment, attention span or transitory emotional state and, in general terms, present a false and adulterated perspective on what mindfulness really is. Such developments are of precious little benefit to any of the interested parties whether they are, learners, teachers, mindfulness practitioners or external agencies interested in the potential benefits of MBIs. The position is summed up well by Grossman.

Our apparent rush to measure and reify mindfulness—before attaining a certain depth of understanding—may prevent us from transcending worn and familiar views and concepts that only trivialize and limit what we think mindfulness is. The scientific method, with its iterative process of re-evaluation and improvement, cannot correct such fundamental conceptual misunderstandings but may actually serve to fortify them (2011, p. 1038).

The proliferation of mindfulness scales which has accompanied the exponential growth of programmes has exacerbated this denaturing of the original conception, and it is now no longer clear precisely what is being measured. As Grossman and Van Dam (2011) note, such developments may prove counter-productive and unhelpful to all those working in the field. They argue further that:

> Definitions and operationalizations of mindfulness that do not take into account the gradual nature of training attention, the gradual progression in terms of greater stability of attention and vividness of experience or the enormous challenges inherent in living more mindfully, are very likely to misconstrue and banalize the construct of mindfulness, which is really not a construct as we traditionally understand it in Western psychology, but at depth, a way of being (ibid., p. 234).

Along with the gradualness of mindfulness development, this 'way of being' is not susceptible to summative psychological testing. Instead, Grossman and Van Dam recommend formative assessment techniques employing longitudinal interviews and observations of MBI participants in specific contexts. More significantly, they go on to make the eminently sensible suggestion that 'one viable option for preserving the integrity and richness of the Buddhist understanding of mindfulness might be to call those various qualities now purporting to be mindfulness by names much closer to what they actually represent' (ibid., p. 234). The dangers and pitfalls of summative measurement are returned to in later sections in relation to MBIs in educational contexts.

Predictability

In order to produce uniformity of outcomes in line with customer expectations, systems must be reasonably predictable and, to achieve this, a 'rationalized society emphasizes discipline, order, systematization, formalization, routine, consistency, and methodical operation' (Ritzer 2000, p. 83). The standardisation of MBSR/MBCT programmes fully satisfies these predictability criteria. Kabat-Zinn's original 8-week course has been modified slightly over the years but remains essentially similar to the 1979 MBSR version. This includes—as Williams and Penman (2011) describe—the standard ideas about switching off the autopilot, moving from 'doing' to 'being', and so on, realised through breath meditation, body scan, noting pleasant/unpleasant thoughts and feelings, and the like. Similar 'predictability' elements can be discerned in the strict control of teacher training for all those wishing to deliver such programmes (McCown et al. 2011). Of course such 'routinisation' and standardisation is ultimately justified in pragmatic terms of what has been shown to 'work' in the sense of preventing relapse in depression sufferers, alleviating suffering for patients with chronic pain, and the other positive outcomes claimed for course participants. However, there is too little analysis of why it is just *these* standards and routines which need to be implemented and not potential alternatives. Why, for instance, is a course 8 but not 6 or 12 weeks long, and why so little attention given to the positive benefits of illness and the darker aspects of the human condition (Kashdan and Biswas-Diener 2014)? There are also missed opportunities here for introducing therapeutic practices other than mindfulness, and for warning participants—as recent studies indicate (Foster 2016)—about the potential negative impact of practices. Moreover, from an educational point of view, it may be more conducive to effective learning if flexibility of content and methods was allowed in accordance with the fostering of learner independence. Inflexibility linked to the strict adherence to prescribed routines, for example, has been cited as one of the reasons for the failure by the American Philosophy for Children programme to make any substantial impact on European educational systems (Murris 1994; Hyland 2003).

Control Through Non-human Technology

The chief aim of this control element is to diminish the 'uncertainties created by people' and 'the ultimate is reached when employees are replaced by nonhuman technologies' (Ritzer 2000, p. 121). On the face of it, MBIs seem to be quite some way from this form of control since they aim to foster values and dispositions which enhance human agency. However, the use of mindfulness in the military—particularly in the form of mindfulness-based mind fitness training (Purser 2014a, b)—is, arguably, a clear case of control of human capabilities directed towards particular purposes, in this case the production of efficient national warriors. Allied with the increasing use of non-human drone technology, it is entirely possible that mindfulness can be implicated here in the production of more effective killing machines, obviously in direct contradiction of core ethical precepts (Kabat-Zinn 1990). Similarly, the use of mindfulness techniques by employers to influence employee attitudes and behaviour may be discerned in certain workplace applications (Hyland 2015b; Purser and Ng 2015). Moreover, the increasing use of mindfulness 'apps' such as 'Buddhify', 'Smiling Mind' and

'Headspace' (http://www.independent.co.uk/extras/indybest/the-10-best-meditation-apps-8947570.html)—along with increasing use of online versions of MBSR/MBCT programmes—provides ample evidence of the full satisfaction of Ritzer's fourth McDonaldization criterion.

The emergence of the McMindfulness phenomenon in recent years closely follows and fully satisfies the Ritzer model of the increasing technical rationalisation of all aspects of life. The pseudo-spirituality of McMindfulness approaches has proved an invaluable vehicle—with far wider applications and purposes than its forerunner in the Protestant Ethic—for contemporary capitalist exploitation. Harvey (2014) has described in graphic detail how the voracious appetite of neoliberal capitalism has come to devour all aspects of public and private spheres bringing about the total commodification of everyday life. The scramble by large corporations to jump on the mindfulness bandwagon has direct parallels with the expropriation of the Protestant Ethic to serve capitalist interests during the Industrial Revolution (Weber 1930/2014). On the current model, the capitalisation of mindfulness has produced an ideal consumer product with a handy dual purpose which, on the one hand, promises to alleviate stress in employees—often in organisations whose ruthless and draconian working conditions have caused such stress in the first place (see Purser and Ng 2015)—and, on the other, a commodity with infinite sales potential in a spiritually impoverished culture shot through with attention deficit disorder and late-capitalist angst.

It will be crucial for committed practitioners to combat such developments, especially those who, like Stephen Batchelor, abhor a 'dharma that is little more than a set of self-help techniques that enable us to operate more calmly and effectively as agents or clients, or both, of capitalist consumerism' (2015, Kindle edition, loc. 340).

McMindfulness in Education and Work

The marketisation of mindfulness described above can be readily understood in terms of standard capitalistic motivations. It is not difficult to appreciate, for example, how the use of the mindfulness label would enhance the sales potential of colouring books for adults. Similar explanations would apply to all the various mindfulness 'apps' referred to earlier. The commodification and reductionist process in relation to MBSR/MBCT programmes and to schools and workplaces, however, demands a more nuanced analysis.

Schools and Colleges

As mentioned above in relation to the measurement of the educational outcomes of MBIs, there is a clear sense in which mindfulness in educational contexts has been unduly influenced by an inappropriate conception of academic research and endeavour. Certainly, the process has tended to fall some way short of the worthwhileness criteria outlined by Peters and advocates of liberal education and learner autonomy. As Ergas (2015) has argued, the very language of 'curricular interventions' seems inevitably to target either economic problems of schooling such as teacher burnout and fitness for practice or operational issues such as classroom discipline and academic progress. This form of discourse is, of course, perfectly aligned with 'the interests of many policymakers and principals, particularly in the midst of a climate in which accountability and standardisation hover over decision-making processes' (p. 206).

Ergas goes on to suggest that:

> The scientism surrounding mindfulness and contemplative practice in the curriculum can be seen to be nested within a broader discourse – the contemporary and highly contested debate around the attempt to mobilise education towards becoming a 'hard' science that matches the rigour of the natural sciences…Current educational practice tends to be obsessed with assessing performativity, far more than with the 'selves' (or no-selves) behind the performance or providing students with meaningful reasons for why they should perform at all…(ibid., pp. 206–207).

The empirical research on mindfulness in schools is shot through with these scientist and performative assumptions. A review of Australian research on teaching mindfulness in schools, for example, concluded with the comment that 'mindfulness practices have been shown to help teachers: reduce their stress levels; assist with behaviour management strategies and improve self-esteem' (Albrecht et al. 2012, p. 11). Similarly, UK research linked to the *Mindfulness in Schools* (Misp or .b) project describes the outcomes of mindfulness lessons in secondary schools in terms of reducing 'negative emotion and anxiety' in students and contributing 'directly to the development of cognitive and performance skills and executive function' (Weare 2012, p. 2). The recent meta-analysis of work in this field by Zenner et al. (2014) concluded by noting that 'analysis suggests that mindfulness-based interventions for children and youths are able to increase cognitive capacity of attending and learning by nearly one standard deviation' (p. 18).

Such research does, of course, also include much anecdotal talk about enhancing emotional well-being and general mind/body health for both teachers and students (Schoeberlein and Sheth 2009; Burnett 2011), but the overriding impression is that mindfulness practice has in many instances been co-opted to achieve strategic instrumentalist ends in the pursuit of purely academic outcomes. This obsession with training attention and focus through mindfulness in way which detaches it from foundational ethical principles has been noted by a number of writers concerned with MBIs in education (O'Donnell 2015: Lewin 2015).

Almost all the educational benefits claimed for the introduction of MBIs in educational contexts—enhanced attention span and ability to maintain focus, greater emotional resilience and improved well-being (Langer 2003; Schoeberlein and Sheth 2009; Kuyken et al. 2013)—stem from the efficacy of practices such as breath and movement meditation in maintaining attention to and awareness of the present moment. Although this capacity for mindful attention is clearly beneficial for many learning activities, it will not be sufficient to achieve the wider goals of mindfulness practice concerned with cultivating the moral qualities of compassion and equanimity, and realising what Batchelor (2014) has

called 'the experience of the everyday sublime' (p. 37). Access to this deeper dimension of spirituality requires the constant renewal of connections between techniques for establishing present moment awareness and the use of such awareness in disclosing aspects of the human condition which militate against mind/body flourishing.

Maintaining awareness of the present moment may, as Peacock (2014) argues, be very effective in enhancing focus but *sati* (the original Pali word for mindfulness, *smriti* in Sanskrit) is wider than this and 'functions in a much more dynamic way than the simple non-judgemental observation of experience' (p. 9). In this more expansive perspective, mindfulness can be used to develop 'introspective awareness' which allows us to note the differences between wholesome and unwholesome states, enabling us to 'recognize them and cultivate the skillful and wholesome states, whilst relinquishing the unskillful and unwholesome' (ibid., p. 10). This wider interpretation of mindfulness practice has important implications for learning.

Bodhi (2013) explains that the original *sati* meant memory or recollection as originally interpreted by Rhys Davids the founder of the Pali Text Society in 1910. Another layer of meaning relating to 'lucid awareness' using all the senses was added later, and this forged the connection between the 'two primary canonical meanings: as memory and as lucid awareness of present happenings' (ibid., p. 25). There are two aspects of the secular therapeutic conception of mindfulness—as 'bare attention' and non-conceptual, non-judgmental awareness—which require explanation in terms of their difference from Buddhist traditional notions. Buddhist accounts of the awareness involved in *sati* indicate an awareness which is cognitive, discursive and goes beyond bare attention to include the 'perception of the body's repulsiveness, and mindfulness of death'. Moreover, there is 'little evidence in the Pali canon and its commentaries that mindfulness *by its very nature* is devoid of conceptualization' (Bodhi 2013, p. 28, original italics).

In addition, the work of Dreyfus (2013) on the cognitive dimensions of mindfulness has suggested that the non-judgmental features of the modern mainstream interpretation need to be modified in the light of original Buddhist emphases. Echoing aspects of Bodhi's analysis, Dreyfus contends that the 'understanding of mindfulness/sati as present-centred non-evaluative awareness is problematic for it reflects only some of the ways in which these original terms are deployed' (p. 45). Using Buddhagosa's commentaries, he concludes that:

> Mindfulness is then not the present-centred non-judgmental awareness of an object but the paying close attention to an object, leading to the retention of the data so as to make sense of the information delivered by our cognitive apparatus. Thus, far from being limited to the present and to a mere refraining from passing judgment, mindfulness is a cognitive activity closely connected to memory, particularly working memory...(ibid., p. 47).

Although many modern representations of mindfulness in the context of MBSR and mindfulness-based cognitive therapy (MBCT) programmes (Williams et al. 2007; Williams and Kabat-Zinn 2013) implicitly contain this additional active dimension of awareness, Dreyfus is concerned to foreground and emphasise the important cognitive features of meditation. Through constant practice, such insightful awareness uses evaluation of mental states to 'gain a deeper understanding of the changing nature of one's bodily and mental states so as to free our mind from the habits and tendencies that bind us to suffering' (Dreyfus 2013, p. 51). The crucial importance of developing such deeper insights into the nature of suffering is present in the literature on MBSR/MBCT but, as Teasdale and Chaskalson (2013) argue, they deserve much greater attention.

Gethin (2013) suggests that contemporary secular therapeutic mindfulness approaches could be said to portray a 'minimalist' account of the process and that the:

> traditional Buddhist account of mindfulness plays on aspects of remembering, recalling, reminding and presence of mind that seem underplayed or even lost in the context of MBSR and MBCT (p. 275).

In a similar vein, Purser (2014a, b) has referred to the 'myth of the present moment' in describing the limitations of many contemporary mindfulness strategies in terms of their emphasis on techniques of calming the mind and improving attention. Drawing on the Theravadan insight meditation tradition, Purser suggests that the central spiritual project of relieving suffering in ourselves and others requires us to go beyond the present moment to examine the nature of all aspects of conditioned existence.

From an educational point of view, such criticisms and the advocacy of wider conceptions of the mindfulness project are of the first importance. Enhancing awareness and fostering stillness in the present moment are not ends in themselves but need to be seen as providing the necessary conditions for engaging with the broader enterprise of cultivating the moral and spiritual virtues which can assist us in dealing with the challenges of everyday life (Bazzano 2014; Teasdale and Chaskalson 2013). Such instrumentalism needs to be challenged—not just because it is a distortion of foundational mindfulness principles—but because it fails to realise the full potential of mindfulness in opening up the crucial affective domain of education which is all too often neglected in mainstream education (Hyland 2011).

Workplaces

MBIs in relation to mind/body health—the main sphere of operations for MBSR/MBCT courses —manage, in the main, to stay close to foundational principles since they are concerned in one way or another with the relief and transformation of suffering. In education, on the other hand, this central therapeutic/transformational aim tends to be obscured by the tendency towards the achievement of *operational* objectives connected with mainstream learning goals. Examples of such standardised outcomes were noted earlier but—when MBIs are taken out of mainstream educational contexts and employed in organisations and workplace sites—the instrumentalism tends to become even more dominant as mindfulness is expropriated in the service of task-specific institutional missions.

Given the growing empirical evidence about the effectiveness of MBIs in educational settings, it was just a matter of time before such approaches found their way into vocational education and training (VET) and employee training and development. Chaskalson (2011) has investigated the increasing use of MBIs in workplace settings, and, as noted earlier, a number of corporations are now showing an interest in introducing mindfulness training for their employees (see Aetna.com 2014). Although Chaskalson initially appears to be examining the applications of mindfulness to training and work in general, the analysis is restricted mainly to the links between the efficacy of MBIs in promoting emotional intelligence (EI) at the level of management and leadership. Much is made of Goleman's work in this sphere, particularly its applications in the workplace. In the light of his theories of emotional intelligence (Goleman 1996, 2001) had originally analysed data from competence models used in leading companies such as IBM, British Airways, Credit Suisse, as well as public sector organisations in the attempt to discover those personal capabilities that underpinned optimal performance at all levels. Chaskalson (2011) summarises these findings in observing that:

> The results of the analysis were remarkable. As one might expect, intellect was to some extent a driver of outstanding performance. But the higher the rank of those considered to be star performers, the greater their level of emotional intelligence. When the comparison matched star performers against average performers in senior leadership positions, around 85 % of the difference in their profiles was attributable to emotional-intelligence factors rather than to purely cognitive abilities such as technical expertise (p. 113).

Chaskalson goes on to cite a broad range of studies which indicate the importance of EI in teamwork, creative thinking, leadership and innovation in different work environments before explaining how MBSR programmes may contribute to the development of EI through the cultivation of insight, focus, concentration and empathy.

In a concluding section (ibid., pp. 164–165), there is a summary of the key research findings about the typical impact of the eight-week MBSR course on participants. These include:

- A reduction in participants' levels of stress
- An increase in their levels of emotional intelligence
- Increased interpersonal sensitivity
- Lower rates of health-related absenteeism
- Enhanced communication skills
- Increased concentration and attention span
- Higher levels of well-being and overall work and life satisfaction.

Similarly, Glomb et al. (2011) summarise the research on the effectiveness of mindfulness strategies in the workplace in terms of three main areas.

First, mindfulness is associated with factors expected to influence relationship quality. Second, mindfulness is linked to processes indicative of resiliency. Third, mindfulness is linked with processes expected to improve task performance and decision-making (p. 139).

The researchers go on to claim that these three elements represent 'distinct work-related outcomes' (ibid.) Some of the pitfalls of measurement in this field—particularly when tailor-made psychological tests based on self-reporting are used—will be discussed in more detail below. At this stage, it is worth pointing a number of related problems involved in applying standardised mindfulness courses such as MBSR/MBCT in work settings.

Chaskalson (ibid; pp. 168–170) fully recognises the logistical problems of organising and delivering MBSR courses in the workplace. They are costly in time and effort to both employers and employees and—unless adapted to specific work environments—may seem remote from everyday working practices. More significantly —as may be discerned by the potential benefits listed above—they tend to be used primarily to develop skills and traits linked to productive workplace outcomes whether or not these are representative of foundational mindfulness principles. It is true that the affective aspects of learning have tended to be seriously neglected at all levels, and there are direct connections between educating the emotions and cultivating mindfulness (Siegel 2010; Hyland 2015a). However, if emotional literacy and not general mindfulness becomes the primary goal—as seems to be the case in many current programmes—the viability of full eight-week MBSR courses for either employers or trainees becomes questionable.

For this reason, workplace training in this sphere generally takes the form of short subject- or trait-specific courses (typically over one or two days) aimed at enhancing leadership, management or teamworking qualities (Aetna.com 2014; mindfulnessatwork.com 2014). Moreover, the outcomes sought are almost always designed with any eye to increasing productivity as the overriding question becomes 'Can mindfulness increase profits?' (http://mindfulnet.org/,2014). Glomb et al. (2011) also point to the tendency for MBIs in the workplace to converge on specific traits and outcomes which—though valued by employers—may be quite different from mindfulness qualities linked to individual well-being and employee development. Major corporations will relish staff development and training which encourages employees—naturally through mindful present-moment awareness—to say 'yes' to all aspects of their experience no matter how painful and unpleasant (Amaranatho 2015). Such 'training' will guarantee a docile workforce in which there are few challenges to the status quo and which is claimed to lead to 'improved productivity, improved creativity, less absenteeism, better communication and interpersonal relating' (ibid.). Now we can appreciate fully why Google has invested so much in mindfulness-based activities (Bush 2014).

All this is perhaps both predictable and in the nature of the economics of VET and trainee development, though it is does raise the question of how such programmes are related to the broad concept of mindfulness outlined earlier. There seems to be very little scope here for the longitudinal cultivation of values and traits in keeping with the ethical and attitudinal components of mindfulness. It is worth emphasising that the

standard eight-week MBSR course is itself only intended to be an introduction to and preparation for the lifelong challenge 'to make calmness, inner balance and clear seeing a part of everyday life' (Kabat-Zinn 1990, p. 134). Consequently, when even this short programme is further reduced to a few day or afternoon sessions, the inadequacies of quick-fix McMindfulness programmes are fully revealed. To have any lasting impact on trainees and employees, mindfulness strategies need to be woven into work-based learning (WBL) regimes and general working practices. There is, after all, a wide range of WBL and apprenticeship models to draw upon (Hyland 2015b), and some of the lifelong learning research in this field (Unwin 2012) dovetails neatly with MBI applications in the workplace.

Mindfulness and the Affective Domain of Education

The connections between education, therapy and foundational mindfulness principles explored earlier can be usefully located within the affective domain of education. It has been argued above that MBIs in education and the workplace fall some way short of the Peters' criteria for a liberal education in line with the development of learner autonomy and human flourishing (more extreme commercial or corporate McMindfulness models would fail to meet *any* of the relevant criteria). In a recent critique of the mindfulness 'mania' in the publishing world of self-help books, Burkeman (2016) suggests that most of the material is trivial, shallow and irrelevant to the real lives of people and problems experienced in contemporary society. We live in a society shot through with attention deficit disorder and late-capitalist angst in which even the therapeutic strategies designed to deal with mind/body ill-being arising from all this are themselves commodified and thus serve to exacerbate the problems MBIs are seeking to alleviate (Hyland 2016).

Contemporary schooling is obsessed with cognitive learning resulting in the neglect of a whole area of human experience—the affective domain concerned with personal, moral, social and aesthetic development (Weare 2010; Hyland 2014b)—which would provide an ideal counterbalance to the shallow and meretricious popular culture that bombards us from all sides. In an age of social media where virtual reality stands in for the genuinely authentic in terms of relationships, values, and culture, young people are especially vulnerable and susceptible to harmful influences (Foley 2010). Depression and suicide rates for young people continue to rise year on year (http://www.thementalhealthblog.com/2013/10/teenage-depression-and-suicide-statistics/), and the incidence of drug abuse, smoking and alcohol consumption amongst adolescents continues to increase at a terrifying pace (http://www.natcen.ac.uk/our-research/research/survey-of-smoking,-drinking-and-drug-use-among-young-people-in-england/).

There has never been a more urgent time for educators to foreground the affective domain education and seek to deal with the fundamental causes of the angst and despair which prompts all such irrational and harmful behaviour. Programmes of personal and social education which are firmly rooted in the fundamental ethical precepts which underpin mindfulness practice could provide a solid basis for such an affective/therapeutic re-affirmation of this crucial dimension of the educational endeavour (Siegel 2010). The implementation of mindfulness in education with a view to enhancing standardised academic outcomes and competences criticised earlier will fail to address such problems and, consequently, needs to be challenged as a disastrously flawed experiment. MBIs informed by an ethically grounded affective programmes can provide a renewed and robust form of secular spirituality (Hyland 2013) which is so sadly missing from contemporary education systems.

The outstanding success of MBIs in the health and social care sector—in which a secularised Buddhism detached from its origins has prevailed—has served to distort the essential ethical purpose resulting in a set of purely instrumental and utilitarian practices. Calling for a return to the ethical roots of mindfulness, O'Donnell (2015) observes that:

> an unintended consequence of providing the scientific evidence to demonstrate the effectiveness of mindfulness as an intervention is the impoverishment of the ways in which the practice is communicated and its value explained, in particular when it is instrumentalised as a technique primarily focused on the self rather than as part of an ethical practice and way of life (p. 195).

In education, such instrumentalism has too often reduced mindfulness practice to a device for creating calmer classrooms, enhancing self-esteem or improving academic performance, with little attempt to locate all aspects of learning within a moral community inspired by the loving kindness and compassion which are fundamental to the universal *dharma* within which mindfulness has its origins.

All those concerned with applying mindfulness in education would do well to note Kabat-Zinn's (2015) warning that 'it can never be a quick fix' and that there are grave dangers in ignoring 'the ethical foundations of the meditative practices and traditions from which mindfulness has emerged' (p. 1). There can be few clearer statements of what those foundations are than the Buddha's own words from the *Mahavagga*: 'Come, friends…dwell pervading the entire world with a mind imbued with lovingkindness…compassion…altruistic joy…equanimity…without ill will' (Bodhi 2000, p. 1608). Moral and affective education has little need of any higher aims or sources of inspiration.

References

Aetna.com. (2014). *Mindfulness at work and Viniyoga stress reduction programs*; http://www.aetna.com/index.html. Accessed December 11, 2015.

Albrecht, N. J., Albrecht, P. M., & Cohen, M. (2012). Mindfully teaching in the classroom: A literature review. *Australian Journal of Teacher Education, 37*(12), 1–14.

Alfino, M., Caputo, J. S., & Wynyard, R. (1998). *McDonaldization revisited: Essays on consumer culture*. London: Praeger.

Amaranatho. (2015). *Learn to say 'Yes'—Mindfulness in the workplace*, September 8, 2015 http://www.rhhr.com/learn-to-say-yes-mindfulness-in-the-workplace/. Accessed December 3, 2015.

Batchelor, S. (2014).The everyday sublime. In A. Bazzano (Ed.), op.cit. (pp. 37–48).

Batchelor, S. (2015). *After Buddhism: Rethinking the Dharma for a secular age*. New Haven: Yale University Press (Kindle edition).

Bazzano, A. (Ed.). (2014). *After mindfulness*. London: Palgrave Macmillan.

Bodhi, B. (2000). *The connected discourses of the Buddha*. Boston, MA: Wisdom Publications.

Bodhi, B. (2013). What does mindfulness really mean? A canonical perspective. In J. M. G. Williams & J. Kabat-Zinn (Eds.), op.cit. (pp. 19–39).

Boyce, B. (Ed.). (2011). *The mindfulness revolution*. Boston, MA: Shambhala Publications.

Burkeman, O. (2016). Colour me calm. *Guardian Review, 9*(1/16), 5.

Burnett, R. (2011). Mindfulness in schools: Learning lessons from the adults—Secular and Buddhist. *Buddhist Studies Review, 28*(1), 79–120.

Bush, M. (2014). What's it like to take Google's mindfulness training? *Mindful: Taking Time for What Matters*.

Chaskalson, M. (2011). *The mindful workplace*. Oxford: Wiley-Blackwell.

Cuypers, S. E., & Martin, C. (Eds.). (2009). *Reading R.S. Peters today: Analysis, ethics and the aims of education*. London: Wiley-Blackwell [*Journal of Philosophy of Education, 43*(1)].

Dreyfus, G. (2013). Is mindfulness present-centred and non-judgemental? A discussion of the cognitive dimensions of mindfulness. In J. M. G. Williams & J. Kabat-Zinn (Eds.), op.cit. (pp. 41–54).

Ergas, O. (2015). The deeper teachings of mindfulness-based 'interventions' as a reconstruction of 'education'. *Journal of Philosophy of Education, 49*(2), 203–220.

Farrarons, E. (2015). *The mindfulness colouring book*. London: Boxtree.

Foley, M. (2010). *The age of absurdity*. London: Simon & Schuster.

Foster, D. (2016). Kind of blue. *The Guardian Weekend, 23*(1/16), 46–49.

Gethin, R. (2013). On some definitions of mindfulness. In J. M. G. Williams & J. Kabat-Zinn (Eds.), op.cit. (pp. 263–279).

Glomb, T. M., Duffy, M. K., Bono, J. E., & Yang, T. (2011). Mindfulness at work. *Research in Personnel and Human Resources Management, 30*, 115–157.

Goleman, D. (1996). *Emotional intelligence*. London: Bloomsbury.

Goleman, D. (2001). An EI based theory of performance. In C. Cherniss & D. Goleman (Eds.), *The emotionally intelligent workplace* (pp. 27–44). San Francisco: Jossey-Bass.

Grossman, P. (2011). Defining mindfulness by how poorly I think I pay attention during everyday awareness and other intractable problems for psychology's (re)invention of mindfulness: Comment on Brown et al. (2011). *Psychological Assessment, 23*, 1034–1040.

Grossman, P., & Van Dam, T. (2011). Mindfulness, by any other name...: Trials and tribulations of *Sati* in Western psychology and science. *Contemporary Buddhism, 12*(1), 219–239.

Hanh, T. N. (1991). *The miracle of mindfulness*. London: Rider.

Hanh, T. N. (1999). *The heart of the Buddha's teaching*. New York: Broadway Books.

Hartley, D. (1995). The McDonaldization of higher education: Food for thought?. *Oxford Review of Education, 21*(4), 409–423.

Harvey, D. (2014). *Seventeen contradictions and the end of capitalism*. London: Profile Books.

Hirst, P. H. (1965). Liberal education and the nature of knowledge. In R. D. Archambault (Ed.), *Philosophical analysis and education* (pp. 113–138). London: Routledge & Kegan Paul.

Hirst, P. H. (1974). *Knowledge and the curriculum*. London: Routledge & Kegan Paul.

Hirst, P. H., & Peters, R. S. (1970). *The logic of education*. London: Routledge & Kegan Paul.

Hyland, T. (1979). Open education—A slogan examined. *Educational Studies, 5*(1), 35–41.

Hyland, T. (1994). *Competence, education and NVQs: Dissenting perspectives*. London: Cassell.

Hyland, T. (1999). *Vocational studies, lifelong learning and social values*. Aldershot: Ashgate.

Hyland, T. (2003). Socrate est-il un andragogue ou un pedagogue? In J. Ferrari, et al. (Eds.), *Socrate Pour Tous* (pp. 73–92). Paris: Librairie Philosophique J. Vrin.

Hyland, T. (2011). *Mindfulness and learning: Celebrating the affective dimension of education*. Dordrecht: The Netherlands, Springer Press.

Hyland, T. (2013). Buddhist practice and educational endeavour: In search of a secular spirituality for state-funded education in England. *Ethics and Education, 8*(3), 241–252.

Hyland, T. (2014a). Competence. In D. C. Phillips (Ed.), *Encyclopedia of educational theory and philosophy* (Vol. 1, pp. 166–167). London: Sage.

Hyland, T. (2014b). Mindfulness-based interventions and the affective domain of education. *Educational Studies, 40*(3), 277–291.

Hyland, T. (2015a). On the contemporary applications of mindfulness: Some implications for education. *Journal of Philosophy of Education, 49*(2), 170–186.

Hyland, T. (2015b). McMindfulness in the workplace: Vocational learning and the commodification of the present moment. *Journal of Vocational Education and Training, 67*(2), 219–234.

Hyland, T. (2016). Review of mindful nation UK. *Journal of Vocational Education and Training,* *68*(1), 133–136.

Kabat-Zinn, J. (1990). *Full catastrophe living.* London: Piatkus.

Kabat-Zinn, J. (2015). Mindfulness has huge health potential—But McMindfulness is no panacea. *The-Guardian,* October 20, 2015; http://www.theguardian.com/commentisfree/2015/oct/20/mindfulness-mental-health-potential-benefits-uk. Accessed November 3, 2015.

Kashdan, T., & Biswas-Diener, R. (2014). *The upside of your dark side.* New York: Hudson Street Press.

Kuyken W., Weare, K., Ukoumunne, O. C., Vicary, R., Motton, N., Burnett, R., et al. (2013). Effectiveness of the mindfulness in schools programme: Non-randomised controlled feasibility study. *British Journal of Psychiatry, 203*(2), 126–131.

Langer, E. (2003). A mindful education. *Educational Psychologist, 28*(1), 43–50.

Lewin, D. (2015). Heidegger East and West: Philosophy as educative contemplation. *Journal of Philosophy of Education, 49*(2), 221–239.

Lowe, R. (2007). *The death of progressive education: How teachers lost control of the classroom.* London: Routledge.

McCown, D., Reibel, D., & Micozzi, M. S. (2011). *Teaching mindfulness: A practical guide for clinicians and educators.* Dordrecht: Springer.

MindfulnessAtWork. http://mindfulnessatwork.com. Accessed March 10, 2015.

Mindfulnet.org. (2014). *Mindfulness in the workplace*; http://mindfulnet.org/. Accessed April 2, 2014.

Murris, K. (1994). Not now, socrates.... *Cogito, 8*(1), 80–86.

O'Donnell, A. (2015). Contemplative pedagogy and mindfulness: Developing creative attention in an age of distraction. *Journal of Philosophy of Education, 49*(2), 187–202.

Peacock, J. (2014). *Sati* or mindfulness? Bridging the divide. In M. Bazzano (Ed.), op.cit. (pp. 3–22).

Peters, R. S. (1966). *Ethics and education.* London: Allen & Unwin.

Plato (1965 edn). *The Republic.* Harmondsworth: Penguin.

Purser, R. (2014a). The militarization of mindfulness. *Inquiring Mind,* Spring; www.inquiringmind.com. Accessed December 7, 2015.

Purser, R. (2014b).The myth of the present moment. *Mindfulness,* 5(4). doi:10.1007/s12671-014-0333-z)

Purser, R., & Loy, D. (2013). Beyond McMindfulness. *Huffington Post,* July 14, 2015, http://www.huffingtonpost.com/ron-purser/beyond-mcmindfulness_b_3519289.html. Accessed July 14, 2015.

Purser, R., & Ng, E. (2015). Corporate mindfulness is Bullsh*t: Zen or no Zen, you're working harder and being paid less. *Salon.*

Ritzer, G. (2000). *The McDonaldization of society: New century edition.* Thousand Oaks, CA: Pine Forge Press.

Rubin, J. R. (2003). Close encounters of a new kind: Toward an integration of psychoanalysis and Buddhism. In S. R. Segall (Ed.), *Encountering Buddhism: Western psychology and Buddhist teaching* (pp. 31–60). Albany: State University of New York Press.

Salzberg, S., & Goldstein, J. (2001). *Insight meditation.* Boulder, Colorado: Sounds True.

Schoeberlein, D., & Sheth, S. (2009). *Mindful teaching and teaching mindfulness.* Boston: Wisdom Publications.

Schofield, H. (1972). *The philosophy of education—An introduction.* London: Allen & Unwin.

Siegel, D. J. (2010). *Mindsight.* Oxford: Oneworld Publications.

Smeyers, P., Smith, R., & Standish, P. (2007). *The therapy of education.* London: Palgrave MacMillan.

Teasdale, J. D., & Chaskalson, M. (2013). How does mindfulness transform suffering? In J. M. G. Williams & J. Kabat-Zinn (Eds.), op.cit. (pp. 89-1).

Unwin, L. (2012). A critical approach to work: The contribution of work-based learning to lifelong learning. In D. N. Aspin, J. Chapman, K. Evans & R. Bagnall (Eds.), *Second international handbook on lifelong learning* (Vol. 2, pp. 787–800).

Weare, K. (2010). Mental health and social and emotional learning: Evidence, principles, tensions, balances. *Advances in School Mental Health Promotion, 3,* 5–17.

Weare, K. (2012). *Evidence for the impact of mindfulness on children and young people.* (.b Mindfulness in Schools Project/Exeter University Mood Disorders Centre).

Weber, M. (1930/2014). *The Protestant ethic and the spirit of capitalism.* Bowdon: Stellar Books.

Williams, J. M. G., & Kabat-Zinn, J. (Eds.). (2013). *Mindfulness: Diverse perspectives on its meaning, origins and applications.* Routledge: Abingdon.

Williams, M., & Penman, D. (2011). *Mindfulness: An eight-week plan for finding peace in a frantic world.* New York: Rodale.

Williams, M., Teasdale, J., Segal, Z., & Kabat-Zinn, J. (2007). *The mindful way through depression.* London: The Guilford Press.

Zenner, C., Hermleben-Kurz, S., & Walach, H. (2014). Mindfulness-based interventions in schools—A systematic review and meta-analysis. *Frontiers in Psychology, 5,* 1–20.

Education as the Practice of Freedom: A Social Justice Proposal for Mindfulness Educators

Jennifer Cannon

There is no such thing as a neutral educational process. Education either functions as an instrument that is used to facilitate the integration of the younger generation into the logic of the present system and bring about conformity to it, or it becomes "the practice of freedom," the means by which men and women deal critically and creatively with reality and discover how to participate in the transformation of their world (Richard Shaull in the Introduction to Pedagogy of the Oppressed).

The field of mindfulness is growing rapidly and mindfulness discourse is also evolving. As secular mindfulness fully enters the neoliberal market and becomes further separated from its foundation of Asian Buddhism, there is a growing need to redefine the practice, and the language of the practice (Chiesa 2012; Greenberg and Mitra 2015; Heffernan 2015; Monteiro et al. 2014; Purser 2014; Valerio 2016). What does it mean to practice or teach mindfulness? What is mindfulness when divorced from the context of Buddhism and the attendant emphasis on interdependence, non-harming, and compassion, or right livelihood and renunciation? Are the mindfulness practices being taught to Google and Monsanto executives, to military personnel, the same mindfulness practices being used by social work practitioners and public school educators? As a critical scholar and social justice educator, I am left with more questions than answers.

In this chapter, I offer a constructive critique of the mindfulness education movement through a social justice and antiracist lens. In so doing, I join the growing call for a socially engaged mindfulness. I begin by briefly introducing the work of critical scholars who are forging new categories to understand mindfulness as a social justice practice. Next, I situate the field of mindfulness education within a broader critique of corporate mindfulness, highlighting concerns about the ways mindfulness is being marketed as a technique to increase standardized test scores and manage student behavior in K-12 schools. I then explore the racialized discourse prevalent in mindfulness education and examine the ideology of white dominance. An example of this is provided by critically analyzing a film that extols the virtues of mindfulness education but unwittingly demonstrates the white savior trope. Finally, I present a social justice framework for mindfulness educators to consider—a framework that shifts the deficit discourse of "school failure" and "troubled communities" to one of collective responsibility. By shifting accountability, we remove the focus on behavior management of "problem kids" to critically examine the social conditions that create suffering for our children and youth. With an integration of antiracism and critical pedagogy, mindfulness educators can ensure that mindfulness is utilized as a practice of freedom (Freire 1976; Hooks 1994) rather than a technology of compliance.

J. Cannon (✉)
Department of Teacher Education and Curriculum Studies, University of Massachusetts Amherst, Amherst, MA, USA
e-mail: jenwcannon@gmail.com

© Springer International Publishing Switzerland 2016
R.E. Purser et al. (eds.), *Handbook of Mindfulness*,
Mindfulness in Behavioral Health, DOI 10.1007/978-3-319-44019-4_26

Critical Interventions in the Field of Mindfulness

A critical discourse has recently emerged providing a critique of corporate mindfulness (Purser and Ng 2015, 2016), neoliberal mindfulness (Forbes 2015), and commercial mindfulness (Heffernan 2015). Critical scholars are creating new categories to distinguish different mindfulness orientations, or ideological frameworks, such as radical mindfulness (Hick and Furlotte 2010), integral mindfulness (Forbes 2012, 2015), socially transformative mindfulness (Bodhi 2015), socially responsible mindfulness (Duerr 2015), civic mindfulness (Healey 2013; Ng 2015), and critical mindfulness (Hsu 2013; Ng 2015; Ng and Purser 2016). In addition, there is promising scholarship that highlights socially engaged mindfulness interventions. The field of critical social work has produced *Radical Mindfulness Training* (Hick and Furlotte 2010), *Mindfulness-based Critical Social Work Pedagogy* (Wong 2004), and *neurodecolonization*, a conceptual framework that integrates mindfulness research and practices with traditional indigenous contemplative practices to help alleviate trauma created by colonization (Yellow Bird 2013). *The Mindfulness Allies Project* (Blum 2014) presents a model for teaching mindfulness in low-income communities that includes antiracism dialogue and creates models for grassroots community partnerships. Several publications discuss the possibilities for utilizing mindfulness with antiracism and anti-oppression work (Berila 2016; Forbes 2004; Lueke and Gibson 2014; Magee 2015; Orr 2002; Patel et al. 2013; Vacarr 2001). Rhonda Magee, Professor of Law at the University of San Francisco and also a contributor to this volume, has developed *Mindfulness-Based ColorInsight Practices*, an approach to teaching that combines mindfulness-based practices with curriculum focused on race, bias, privilege, and historical conditions of oppression. These critical interventions, both theoretical and applied, provide concrete examples of an emerging socially engaged mindfulness pedagogy.

There are growing concerns about the field of mindfulness and some of the ways the practice is being co-opted and commercialized, creating a new phenomenon referred to as "McMindfulness" (Purser and Loy 2013). Mindfulness has become mainstream, appearing on the cover of *TIME* magazine (Feb 3, 2014), showing up in the corporate business world, the health and wellness industry, the medical profession, professional sports, the US military, prisons and juvenile detention centers, the fields of psychology, social work, addiction treatment, organizational development, and K-16 education (Boyce 2011; Forbes 2012; Purser and Loy 2013). There has been a rapid expansion of Mindfulness-Based Interventions (MBIs) that draw inspiration from Mindfulness-Based Stress Reduction (MBSR) created by Jon Kabat-Zinn. Offshoots of MBSR include Mindfulness-Based Stress Reduction for Teens (MBSR-T), Mindfulness-Based Cognitive Therapy (MBCT), Mindfulness-Based Relapse Prevention (MBRP), Mindfulness-Based Childbirth and Parenting (MBCP), Mindfulness-Based Elder Care (MBEC), Mindfulness-Based Art Therapy (MBAT), Mindfulness-Based Eating Awareness Training (MB-EAT), and Mindfulness-Based Mind Fitness Training (MMFT) for use in the military or police academy (Boyce 2011; Cullen 2011). MBIs are popping up worldwide including a graduate program in Mindfulness-Based Cognitive Therapy at Oxford University and the prominent Centre for Mindfulness Research and Practice at Bangor University in the UK. One can earn a Master of Arts in Mindfulness Studies at Lesley University or a Master of Education with a concentration in Mindfulness for Educators at Antioch University New England. While Mindfulness-Based Interventions boast numerous tangible benefits (less stress, decreased reactivity, improved health, concentration, attention, greater empathy), there is also a growing critique about how mindfulness is being used as an instrument of neoliberalism to help people adjust to oppressive social conditions (Forbes 2012, 2015; Purser and Loy 2013; Purser and Ng 2015). Many large corporations (Google, General Mills, Nike, Target, AETNA[1]) are providing mindfulness classes for their employees in order to increase focus and attention in the workplace, and decrease on-the-job stress so workers can function at their optimum level of productivity.

As mindfulness becomes trendy and readily accessible to corporations and the US military, many are left asking questions about how the secular practice is being decontextualized from its historical and ethical foundation, namely Buddhism. Without a foundation of social ethics, mindfulness is often reduced to a "banal, therapeutic, self-help technique" (Purser and Loy 2013). Marketed as a health and wellness tool that can alleviate everything from ADHD, high blood pressure, chronic pain, stress, anxiety, depression, eating disorders, insomnia, and PTSD (Baer 2003; Smalley and Winston 2010), mindfulness is seen as a route to ameliorate personal suffering. While the alleviation of personal suffering is a vital goal, how does the modern mindfulness movement become accountable to collective forms of suffering caused by systemic oppression?

Individualizing the practice can be understood as a form of Western colonization, and runs counter to Buddhist concepts of interconnection, interdependence, and liberation for all sentient beings (Bodhi 2015; Dalai 1999; Hanh 1991; Miller 2014). Most Buddhist traditions offer meditation training within a framework of *sila*, or ethics. Meditation students in this context are asked to take vows of ethical conduct that include an agreement of non-harming. *Sila* is not only a practice of non-violence or non-harming, but also a cultivation of wholesome mind states such as loving-kindness, compassion, sympathetic joy, and equanimity, otherwise known as the *brahma viharas* (Cullen 2011). As of yet, there is no

governing ethics board for secular mindfulness teachers in the United States, nor is there a national mindfulness association to provide best practices, teacher training, and continuing education (see the Yoga Alliance for an example of this in the field of yoga). If such a board were to manifest, the field of mindfulness could be grounded in specific language around ethics and non-harming, including foundational principles such as interdependence, serving the common good, and fostering collective forms of liberation. A mindfulness ethics board could also provide frameworks for teaching mindfulness in diverse communities, including in-depth teacher training regarding cultural competency and antiracism.

Mindfulness, Yoga, and Colonization

Drawing parallels with the secular mindfulness movement, the introduction of yoga in the west produced a similar colonizing effect as yoga became popularized, secularized, and commodified for the Western consumer market (Das 2013). While there are certainly many yoga teachers who are committed to teaching yoga as a spiritual practice, it is fair to say that millions of Westerners have no awareness of yoga's historical, ethical, and liberating foundation based in Hindu philosophy. There is no question that countless Westerners have benefitted from yoga as a physical exercise and means of stress reduction, but at what cost to the core principles of the practice? In 2013, a handful of Southern California parents protested an Encinitas district-wide yoga program, claiming the practice was anti-Christian and exposed innocent children to Hindu indoctrination (Whitlock 2013). A lawsuit followed, Sedlock versus Baird, and on July 1, 2013, it was determined that yoga could continue to be taught in public schools as long as it remains divorced from any religious or spiritual language, including the use of any Sanskrit words. "The successful defense team rested their argument on the assertion that yoga is now more American than it is Indian, a practice as secular as aerobics" (Singh 2013). While proponents of yoga in public schools celebrated this victory, some South Asian activists

---

[1]See "Mindfulness: Getting Its Share of Attention" (NY Times, Nov. 1, 2013) and "The Mind Business" (Financial Times, Aug. 24, 2012).

mourned the further colonization of a deeply held spiritual practice. South Asian activist and yoga educator, Roopa Singh, described her mixed emotions when the verdict arrived. While very glad that American children will have access to the benefits of yoga in K-12 schools, Singh advocates for a more nuanced discussion that "might focus on the reality of American, or Christian American, discomfort with the power of yoga, the power of South Asian cultural production, and South Asians in general" (Singh 2013). She continues by expressing her concern that in the limited amount of time yoga has existed in the United States, it could be "severed from the culture that created and sustained it for at least 4000 years" (Singh 2013).

Purser and Loy (2013) argue that individualistic, self-help forms of mindfulness serve a corporate agenda by subduing employee unrest, promoting a tacit acceptance of the status quo, and keeping attention focused on institutional goals.

> There is a dissociation between one's own personal transformation and the kind of social and organizational transformation that takes into account the causes and conditions of suffering in the broader environment. Such a colonization of mindfulness also has an instrumentalizing effect, reorienting the practice to the needs of the market, rather than to a critical reflection on the causes of our collective suffering (Purser and Loy 2013).

Mindfulness is not only being used to serve the needs of the capitalist market but also to fortify colonialism and imperialism. Using mindfulness to train US soldiers (Associated Press 2013) to become calmer, more focused and grounded during combat, should highlight grave concerns about how mindfulness is being separated from its ethical foundation (Forbes 2012).

Mindfulness in K-12 Schools

The increasing popularity of mindfulness in K-12 public schools is raising similar ethical concerns among critics. Mindfulness in education has grown rapidly in recent years; evidence of this can be seen in dozens of organizations dedicated to the topic, newly released books, peer-reviewed journal articles, YouTube videos, and several films about mindfulness and youth. Federal grant money is being used to fund mindfulness in schools initiatives, and there are several mindfulness in education annual conferences. Mindfulness has enjoyed a recent endorsement from Congressman Tim Ryan who advocates for mindfulness programs in all K-12 schools (Ryan 2012). Educators are excited about the potential of mindfulness to increase focus and attention, help children self-regulate their behavior, increase empathy, and reduce bullying and aggression. In the 2011–12 school year, *Mindful Schools* (arguably the leading mindfulness education organization in the United States) partnered with the University of California, Davis, to conduct the largest randomized-controlled study to date on mindfulness and children, involving 937 children and 47 teachers in 3 Oakland public elementary schools. The *Mindful Schools* curriculum produced statistically significant improvements in behavior versus the control group with just 4 h of mindfulness instruction for the students (Mindful Schools, n.d.). Current research studies suggest the utility of mindfulness in teacher education and the benefits for stress reduction, emotion regulation, and teacher retention (Jennings 2015, 2016; Miller and Nozawa 2002; Roeser et al. 2012).

Despite this growing trend in mindfulness education, some advise caution. Forbes (2012) warns that mindfulness can be used as a technocratic tool in K-12 education, specifically as a means to increase standardized test scores or manage student behavior. He argues that mindfulness should not be used in the school system to promote acceptance of a neoliberal reform agenda that is causing harm to our children and teachers (Forbes 2012, 2015). The Department of Education has provided substantial funding for K-12 mindfulness initiatives, due in large part, to preliminary research that mindfulness practices increase attention, focus, and test scores (Jennings 2012; O'Brien 2012; Parker 2009). Do we want our children learning to meditate so they can perform better on high-stakes tests? Hsu (2013) asks us to critically examine our intentions for teaching mindfulness in K-12 public schools.

As educators seeking to incorporate mindfulness into formal education, we must be acutely aware of how we apply our mindfulness so that it does not serve to delude us from the persistence of suffering around us. Integrating mindfulness in schools is a commitment to engaging with the systems of power and domination that contribute to suffering in our communities. Let's think about how we can do this in schools, not just to make calm test takers, but to enliven our students' hearts so that they are stirred to creating the world that they deserve (Hsu 2013).

Is it irresponsible to teach mindfulness in marginalized communities without also exploring the causes and conditions of suffering (institutional forms of oppression)? Mindfulness, taught as self-improvement, self-regulation, or a technology of the self (Reveley 2015), may prove to be harmful if not coupled with critical inquiry about systemic causes of oppression. Mindfulness alone will not help youth of color experiencing the traumas of our criminal (in) justice system, police surveillance and repression, poverty, lack of access to jobs, gentrification, and housing dislocation—to name but a few examples of historical and institutional racism. As mindfulness educators, we run the risk of fostering a belief that students are the problem (i.e., unable to self-regulate, control behavior in the classroom, or focus during high-pressure tests) rather than viewing their challenges within a sociopolitical context. In this framework, mindfulness is presented as a tool to help students overcome individual shortcomings, rather than a transformative pedagogy to explore collective forms of suffering and liberation.

Ideology of White Dominance

There is a troubling racialized discourse that can be found in many mindfulness education initiatives. As a long-time social justice educator and student of the dharma, I was initially excited about the nascent mindfulness education movement. My love of meditation, and my firm belief in its liberating potential, led me to attend many contemplative education symposiums and mindfulness conferences. I also volunteered to work on several mindfulness retreats for teens. The

level of unconscious racism and white privilege prevalent in these spaces left me feeling deeply troubled. The organizers of these mindfulness conferences were white, the keynote speakers were primarily white, conference participants were majority white, and perhaps most importantly and insidiously, the content and framework was steeped in a white dominant ideology (Ng and Purser 2015). When people of color were included, it was rarely from a position of power or leadership. Polite liberal ideology prevailed at these conferences, where diversity and inclusion were celebrated without addressing structural racism or systemic oppression. There was also a hierarchy of knowledge, privileging academic and scientific knowledge about mindfulness above the lived experience of mindfulness practitioners, including community activists and educators.

While attending a contemplative education conference at the Garrison Institute in November 2011 (Advancing the Science and Practice of Contemplative Teaching and Learning), I could not help but notice all of the primary speakers were white academics with PhDs, while the African American and Latino speakers from the Holistic Life Foundation (a yoga and mindfulness program for Baltimore youth of color) were relegated to lead the morning contemplative practice sessions. These young men of color were praised and applauded for their work with "inner-city" Baltimore youth, yet their theoretical insights, developed out of many years of field work, were not deemed scholarly enough for one of the many keynote talks delivered over three days by "experts" in the field of contemplative education.

At a March 2011 Mindfulness in Education conference in Washington DC, sponsored by the Mindfulness in Education Network, a panel of white academics described the transformative impact mindfulness programs are having on "inner-city" schools. Conference panelists provided ample statistics that highlighted the deficits of urban communities (poverty, percentage of students receiving free lunch, violence in schools, dropout rates, drugs, crime). Community strengths or assets were not discussed. While race was not explicitly mentioned, the racialized discourse was clear. "Inner-city" kids meant poor and brown. Urban neighborhoods were synonymous with drugs, violence, crime, and broken homes. Mindfulness classes were portrayed as a necessary intervention in these needful communities. The "service" being offered was mindfulness, delivered by an outsider to the community, a white mindfulness expert. One of the conference panelists (a white PhD academic) told a story that invited audience members to imagine that we each held the ability to spread hope, via mindfulness, in "neglected" communities, as demonstrated by her power point slides that showed children of color peacefully meditating. Looking around the mostly white audience, the call to spread hope in economically disenfranchised communities of color landed like a missionary intervention.

I was immediately reminded of community service-learning (CSL) initiatives, and the ensuing critique of well-intentioned college students entering poor and working-class communities of color with an attitude of benevolence, often referred to as charity service-learning (Morton 1995). In this paradigm of CSL, the community receiving the volunteer services is described in terms of their deficits and needs. Statistics are offered to highlight the poverty and social ills of the community. The cultural capital (Yosso 2005) and assets of the community are not explored or valued. Primarily white middle-class college students commute into low-income communities of color to provide community service and often do not forge long-lasting relationships or sustainable partnerships (Morton 1995). They are outsiders, dropping into a "foreign" community to provide a service, and then they leave. Poor and working-class communities are depicted as needful, dependent on aid, voiceless, lacking agency, and power (Illich 1968; McKnight 1989). The dynamic of community service elevates the helper, the one offering the services. Volunteers are seen as having agency, voice, cultural capital, intelligence, and perhaps most importantly, good intentions.

Antiracist scholars and activists have thoroughly critiqued this model as patronizing and reinforcing white privilege and white superiority. From this collective critique, a new discipline was born called critical service-learning (Mitchell 2008) that interrogates dynamics of race and racism, power and privilege, seeks to highlight community assets over deficits, and identifies community residents as having something invaluable to teach college student volunteers. With critical service-learning, the power dynamic becomes horizontal, and all participants benefit from the partnership. There is also an effort to analyze systems of oppression and institutional structures that create conditions of inequality and marginalization. In this vein, critical service-learning seeks to create allies for social justice, not benefactors of charity that provide temporary band-aid relief. Critical service-learning programs encourage students to see themselves as agents of social change (Mitchell 2008).

Perhaps it is time for a critical mindfulness paradigm (Hsu 2013; Ng 2015; Ng and Purser 2016). As social justice educators, are we doing all we can to eliminate deficit discourses in the mindfulness education movement? Let us think carefully about which communities are being solicited for mindfulness education and why. Who is determining the need for mindfulness education in communities of color? Who has agency in the mindfulness field and who is seen as the expert? Much has been written about white teachers in communities of color and the lack of cultural competency and antiracism training in teacher education (Delpit 1995; Gay 2010; Howard 2006; Ladson-Billings 2001; Picower 2009). Similar attention must be given to the training of mindfulness educators, beyond a superficial nod toward diversity and inclusion. If we are to embrace mindfulness as a liberating practice, then we must integrate mindfulness education with antiracism and social justice pedagogy. An antiracism framework should include a commitment to eliminate the use of mindfulness practices as a behavior management tool. While educators across the nation struggle with behavior issues from students (regardless of race and class privilege), we must be vigilant about not using mindfulness practices to quell feelings of anger, discontent, or even rage, without also creating pathways to critically analyze the causes and conditions that give rise to such feelings in our students. Youth today are overpathologized and overmedicated, and are all too often flagged as trouble-makers, especially youth of color. Without an integrated critical pedagogy, mindfulness alone can be seen as the latest tool to pacify unruly students. In the following section, I provide an example of this by analyzing a film about a mindfulness educator working for the organization, *Mindful Schools*.

Room to Breathe

The trailer for *Room to Breathe* (a film that aired on PBS in 2013) depicts out of control, aggressive, and disrespectful youth of color who are constantly getting in trouble at school, until a white mindfulness teacher intervenes to teach the youth how to meditate. Despite good intentions, *Room to Breathe* reinscribes a racialized discourse about "troubled" youth of color and introduces a white mindfulness instructor as the teacher-hero.

> The film presents a hopeful story of transformation, following a young mindfulness teacher, Megan Cowan, who spends several months attempting to teach the technique to troubled kids in a San Francisco public middle school that tops the district in disciplinary suspensions. Cowan is confronted at first by defiance and contempt. But under her guidance, the students begin to learn the mindfulness technique and eventually use it to take greater control over their lives, decrease stress, and better focus in class and at home (Room to Breathe, n.d.).

This film is reminiscent of the well-known subgenre of urban public school films (Bulman 2005) with a familiar plot line of "defiant" inner-city youth of color who are saved from their fate of poverty and school failure by a caring, kind, and devoted white teacher. Films in this subgenre include *Freedom Writers* and *Dangerous Minds* and can also be classified in the "white savior film" genre (Giroux 2002; Hughey 2014).

The heroine in *Room to Breathe* is a benevolent white woman who teaches the youth that hard work, effort, and self-discipline (through mindfulness) will produce the results they need to be successful in school and in life. The mindfulness curriculum does not include a critical inquiry about the student's sociopolitical context; there is no critical analysis of poverty, institutional racism, the school-to-prison pipeline, or the multitude of reasons teenagers might be bored or disruptive in school. There is also no exploration of contemplative practices the youth may already utilize in their home life or communities. While students may not have previously been exposed to formal meditation, they may already be practicing various forms of mindfulness or ways to pay attention in the present moment. The heroine of *Room to Breathe* demonstrates a banking model of education (Freire 1970) as opposed to a collaborative inquiry about mindfulness. She also elevates herself as the authority figure and disciplinarian in the class. Despite the disapproval of the African American Assistant Principal, she removes several disruptive students from her mindfulness class for the benefit of the other students. The Assistant Principal informs her in a meeting that it is not the practice of the school to exclude students from participation; however, Cowan makes the final decision to remove the students and they are never reintegrated during the film.

Critical scholar and urban educator, Duncan-Andrade (2009), describes his classroom as a micro-ecosystem where every child contributes to a delicate balance of interdependency where "pain and healing are transferable from person to person inside the classroom" (p. 9). He argues that when teachers exclude "disobedient" children from the classroom, it introduces social stressors into the micro-environment. "From this perspective, the decision to remove a child from the classroom, rather than to heal her, is not only bad for the child but is also destructive to the social ecosystem of the classroom" (p. 9). He goes on to describe how trust and hope are preconditions for positive educational outcomes and when a child is removed from class, it can erode that sense of trust for all the students who witness the removal (p. 10).

Room to Breathe reinscribes white dominance by valorizing the knowledge and wisdom of the mindfulness teacher, by elevating her status as disciplinarian and authority figure, and by highlighting the deficits of the youth rather than their strengths and talents. By the end of the film, our teacher-hero has helped transform the rowdy and disrespectful youth of color into calm, centered, meditating students. The transformation is remarkable, and as Bulman (2005) describes is typical of the subgenre, the "audience is left feeling triumphant and optimistic about the potential for improvement in urban public schools" (p. 74). The American urban public school film strongly suggests that "the answer to the students' problems is revealed to primarily be an individual one—to reform the individual student, not the educational system or wider society" (p. 69). Unfortunately, *Room to Breathe* falls right in line with this reformist framework—using mindfulness as a tool to create calm and compliant students without a structural critique or critical exploration of the social conditions in their lives.

My point here is not to launch an attack on *Mindful Schools*, a leading mindfulness education organization in the United States with an excellent reputation. Rather, I am attempting to open a dialogue about the way mindfulness education is marketed in *Room to Breathe*—as an educational tool to improve behavior management and focus among kids with a disciplinary track record. By using a deficit framework to describe the youth, highlighting the ways in which they are "problem kids," and erasing the structural racism and classism endemic to the US public school system, the film fails to grapple with the real obstacles faced by students and teachers alike in that system. The film also misses the opportunity to represent the potential of mindfulness to be a transformative or critical pedagogy that might support the youth in making meaning of their lives and creating sites of resistance to oppression. I have shown the trailer of this film numerous times for undergraduate students, pre-service teachers, and Education faculty. Each time the trailer has elicited a strong response, especially from people of color who are often disturbed by the film. *Mindful Schools* has the honor and distinction of being a leader in mindfulness education, and therefore, they also carry a greater ethical responsibility to set standards in the field.

Mindfulness Education for Social Justice

What might mindfulness education look like if developed from a foundation of antiracism and social justice? As a starting point, mindfulness educators can make a commitment to represent the racial and ethnic diversity of the student populations they are trying to serve. Guided by this basic principle, mindfulness teachers of color would take the lead within communities and schools where children of color are the majority population. White mindfulness teachers would co-facilitate with a mindfulness teacher of color or partner with a parent, administrator, or teacher of color in the school district. While some may claim this is too difficult to accomplish, with a dedicated commitment, it certainly can be done. Interracial co-facilitation is common practice in the field of social justice education (Brigham and Williams 2013; Zúñiga et al. 2007) and models power sharing in the classroom, effectively interrupting the hierarchy of whiteness. Urban school districts and communities of color are already overrun with white teachers, administrators, and social workers. Mindfulness educators, if teaching from a foundation of antiracism and social justice, will begin to pay careful attention to power dynamics associated with race and class, including the fraught dynamic of the white teacher-hero or white savior.

Once an appropriate mindfulness teaching team is established, the co-facilitators can build on the cultural funds of knowledge (Gonzalez et al. 2005) already present in the classroom, highlighting the strengths, assets, and talents of the students, rather than starting with a deficit-based approach to teaching mindfulness. Let us take one final look at the film.

Room To Breathe is a surprising story of transformation as struggling kids in a San Francisco public middle school are introduced to the practice of mindfulness meditation. Topping the district in disciplinary suspensions, and with overcrowded classrooms creating a nearly impossible learning environment, overwhelmed administrators are left with stark choices: repeating the cycle of trying to force tuned-out children to listen, or to experiment with timeless inner practices that may provide them with the social, emotional, and attentional skills that they need to succeed. The first question is whether it's already too late. Confronted by defiance, contempt for authority figures, poor discipline, and more interest in "social" than learning, can a young mindfulness teacher from Berkeley succeed in opening their minds and hearts? (Room to Breathe: A Documentary Film About Transformation through Mindfulness Meditation in Public Schools, n.d.)

This is a perfect example of a deficit discourse; all the descriptors used to label the youth are negative. In her article, *Pushing Past the Achievement Gap: An Essay on the Language of Deficit*, teacher educator Ladson-Billings (2007) challenges the deficit-laden framework of an educational system that "constructs students as defective and lacking" (p. 321). She asks educators to move toward a discourse that holds us all accountable for school failure and proposes a shift in language from "achievement gap" to "education debt" (Ladson-Billings 2006). This strategic change in language effectively removes the blame of underachievement from students and families and places responsibility on our collective shoulders.

> When we speak of an education debt we move to a discourse that holds us all accountable. It reminds us that we have accumulated this problem as a result of centuries of neglect and denial of education to entire groups of students. It reminds us that we have consistently under-funded schools in poor communities where education is needed most. It reminds us that we have, for large periods of our history, excluded groups of people from the political process where they might have a say in democratically determining what education should look like in their communities (Ladson-Billings 2007, p. 321).

Let us take up the challenge presented by Ladson-Billings and abandon the deficit discourse so prevalent in mindfulness education. This would include challenging funders of mindfulness programs, and mindfulness research, who insist on labeling communities and school districts based on lack, need, and failure.

As a final comment about the *Room to Breathe* film, the only person described as having agency is the white mindfulness teacher from Berkeley who will hopefully "succeed" in opening the students' hearts and minds. In a social justice mindfulness paradigm, the youth would also have agency. The students would be encouraged to engage the social conditions of their lives through critical inquiry, youth-led dialogue, peer mindfulness facilitation, popular education, or youth participatory action research. Rather than being told that they need social, emotional, and attentional skills in order to succeed, youth would have the opportunity to define for themselves what skills they need to develop in order to thrive in their own communities.

In a social justice framework, the mindfulness organization would build grassroots community alliances by partnering not only with schools, but also with neighborhood cultural centers, after-school programs, and parent organizations. If mindfulness is to be seen as a useful and transferable practice (i.e., not just a tool to use in school but also beneficial at home and in one's neighborhood), then mindfulness must be understood within the cultural context of the student's lives. Parallel critiques, highlighting a lack of culturally relevant pedagogy, have recently been made about Social Emotional Learning (SEL) curriculum, especially in urban public schools (Forbes 2012; Slaten et al. 2015; Zakrzewski 2016). As mindfulness educators and critical scholars, we need to interrupt the notion of importing mindfulness into communities, especially marginalized communities, as if mindfulness practices do not already exist in their own cultural manifestations. *Mindfulness Without Borders*, a mindfulness education nonprofit, encourages interested applicants to "get certified to bring mindfulness to youth" (Mindfulness Without Borders, n.d.) once again conferring agency on the mindfulness instructor. Our choices around language reveal our ideological and political commitments. Do we want to encourage a new cadre of mindfulness educators to "bring" mindfulness to youth, as a tool to be imported, or do we want mindfulness teachers to help youth explore and discover mindfulness practices and capacities they already have within themselves and their own cultures? Let us explicitly say what we mean, and be clear about the mindfulness paradigm from which we are teaching.

In an effort to maintain integrity and rigorous training standards for mindfulness educators, we should be wary of new markets opening in the field of mindfulness education. There are dozens of certificates available online, and many do not require prior exposure to meditation, experience working with youth, or any form of cultural competency or antiracism training. *Mindfulness Without Borders* offers a Mindful Educators Certification Training that consists of eleven 90-min online training sessions (levels one and two combined); however, only nine sessions are required for completion of the certificate (Mindfulness Without Borders, n.d.). While online courses open up the possibility for people all over the world to participate, we might collectively ask ourselves what is lost when mindfulness training does not require in-person mindfulness instruction. Clearly, equitable access to mindfulness training is important and online courses increase access to certain kinds of course material. However, a close look at the limitations of online education suggests that the training of mindfulness instructors necessitates a meaningful in-person component and cannot be effective solely through online programs. It is also sobering to consider the ways in which mindfulness education has been taken up as a for-profit endeavor. We should be wary of entrepreneurs who view mindfulness education as a business opportunity, seeking to profit from the suffering of children and youth, whether that is suffering induced from the test-taking industry or suffering caused by institutional racism and oppression. As social justice educators, it is our collective responsibility to develop ethical guidelines for working with youth.

Perhaps most importantly, in our efforts to create a social justice mindfulness paradigm, we must utilize mindfulness as an anti-oppressive pedagogy (Berila 2016; Kumashiro 2000; Orr 2002) and as a humanizing pedagogy (Freire 1970; Patel et al. 2013). This means using mindfulness practices as a pathway for deepening connection through vulnerability and authenticity—allowing us to witness each other in the fullness of who we are, with our pain, rage, confusion, fear, hopes, and collective longings. This also means utilizing mindfulness practices to uncover issues of power and privilege, including racism and other forms of oppression. To this end, mindfulness education can be framed as an exploration of embodied knowledge, an inner language that arises out of contemplative practice. This is a pedagogical orientation that reminds students they are already carriers of knowledge, that they have the capacity to develop insight and awareness by listening deeply to their own wisdom. True awareness requires clear seeing of the causes and conditions that give rise to the present moment, including structural oppression, societal pain, and institutional injustice. This is a stark contrast from introducing mindfulness as an instrumental tool, something foreign that is imported into communities and school districts.

Divorced from a social justice foundation, mindfulness education can be manipulated as a tool of the neoliberal market to improve test scores, increase attention, or control student behavior. At best, mindfulness education can help provide our students with a foundation of inner peace, calm and centered awareness, and offer a reprieve from self-defeating thoughts. If we integrate principles of compassion, interconnection, and solidarity along with concrete pathways to enact these principles in service to community empowerment and social justice, then we are birthing a new paradigm in mindfulness education. We have mentor-teachers to guide us, veteran educators who have created liberatory pedagogies that integrate critical hope (Duncan-Andrade 2009; Freire 1992) with a deep commitment to an ethics of solidarity (Freire 1970). Freire describes solidarity as standing with oppressed communities in their own liberation struggle, rather than extending "false generosity," acts of charity or service that perpetuate domination. Let us work together to ensure that mindfulness education is not being used as a tool to pacify our students or force them into compliance with an unjust schooling system. Let us walk in the footsteps of Freire, hooks, and many other critical educators who have taught us that education can be the practice of freedom.

> The classroom with all its limitations remains a location of possibility. In that field of possibility we have the opportunity to labour for freedom, to demand of ourselves and our comrades, an openness of mind and heart that allows us to face reality even as we collectively imagine ways to move beyond boundaries, to transgress. This is education as the practice of freedom (Hooks 1994, p. 207).

References

Associated Press. (2013, January 23). U.S. marine corps members learn mindfulness meditation and yoga in pilot program to reduce stress. *New York Daily News.* Retrieved from http://www.nydailynews.com/lifestyle/health/u-s-marines-learn-meditate-stress-reduction-program-article-1.1245698

Baer, R. A. (2003). Mindfulness training as a clinical intervention: A conceptual and empirical review. *Clinical Psychology: Science and Practice, 10,* 125–143.

Berila, B. (2016). *Integrating mindfulness into anti-oppression pedagogy.* New York, NY: Routledge.

Blum, H. (2014). Mindfulness equity and Western Buddhism: Reaching people of low socioeconomic status and people of color. *International Journal of Dharma Studies,* 2(10).

Bodhi, B. (2015, May). Modes of applied mindfulness. Unpublished manuscript.

Boyce, B. (2011). *The mindfulness revolution: Leading psychologists, scientists, artists, and meditation teachers on the power of mindfulness in daily life.* Boston: Shambhala.

Brigham, E., & Williams, T. (2013). Developing and sustaining effective cofacilitation across identities. In L. Landreman (Ed.), *The art of effective facilitation: Reflections from social justice educators.* Sterling, VA: Stylus Publishing.

Bulman, R. (2005). *Hollywood goes to high school: Cinema, schools, and American culture.* New York: Worth Publishers.

Chiesa, A. (2012). The difficulty of defining mindfulness: Current thought and critical issues. *Mindfulness, 4*(3), 255–268.

Cullen, M. (2011). Mindfulness-based interventions: An emerging phenomenon. *Mindfulness, 2,* 186–193.

Dalai, L. (1999). *Ethics for the new millennium.* New York: Riverhead Books.

Das, K. (2013, July 30). A new initiative seeks to restore yoga's South Asian heritage [Web log message]. Retrieved from http://theaerogram.com/a-new-initiative-seeks-to-restore-yogas-south-asian-heritage/

Delpit, L. (1995). *Other people's children: Cultural conflict in the classroom.* New York: New Press.

Duerr, M. (2015, July 15). Toward a socially responsible mindfulness. Retrieved from http://www.wildmind.org/blogs/on-practice/toward-a-socially-responsible-mindfulness

Duncan-Andrade, J. (2009). Note to educators: Hope required when growing roses in concrete. *Harvard Educational Review,* 79(2).

Forbes, D. (2012, June 30). Occupy mindfulness. Retrieved from http://beamsandstruts.com/articles/item/982-occupy-mindfulness

Forbes, D. (2015, November 8). Mindfulness and neoliberal education. Retrieved from http://www.salon.com/2015/11/08/they_want_kids_to_be_robots_meet_the_new_education_craze_designed_to_distract_you_from_overtesting/

Forbes, D. (2004). *Boyz 2 Buddhas: Counseling urban high school male athletes in the zone.* New York: Peter Lang.

Freire, P. (1992). *Pedagogy of hope.* New York: Continuum International Publishing Group.

Freire, P. (1970). *Pedagogy of the oppressed.* New York: Continuum International Publishing Group.

Freire, P. (1976). *Education, the practice of freedom.* London: Writers and Readers Publishing Cooperative.

Gay, G. (2010). *Culturally responsive teaching: Theory, research, and practice.* New York: Teachers College Press.

Giroux, H. (2002). The politics of pedagogy, gender and whiteness in 'dangerous minds'. In *Breaking into the movies: Film and the culture of politics.* Hoboken, NJ: Wiley-Blackwell.

Gonzalez, N., Moll, L., & Amanti, C. (2005). *Funds of knowledge: Theorizing practices in households, communities and classrooms.* New York: Routledge.

Greenberg, M. T., & Mitra, J. L. (2015). From mindfulness to right mindfulness: The intersection of awareness and ethics. *Mindfulness, 6*(1), 74–78.

Hanh, T. N. (1991). *Peace is every step: The path of mindfulness in everyday life.* New York: Bantam Books.

Healey, K. (2013, August 5). Searching for integrity: The politics of mindfulness in the digital economy. Retrieved from http://nomosjournal.org/2013/08/searching-for-integrity/

Heffernan, V. (2015). The muddied meaning of 'mindfulness.' *New York Times.* http://www.nytimes.com/2015/04/19/magazine/themuddied-meaning-of-mindfulness.html?_r=0

Hick, S. F., & Furlotte, C. (2010). An exploratory study of radical mindfulness training with severely economically disadvantaged people: Findings of a Canadian study. *Australian Social Work, 63*(3), 281–298. doi:10.1080/0312407x.2010.496865.

Hooks, B. (1994). *Teaching to transgress: Education as the practice of freedom.* New York: Routledge.

Howard, G. (2006). *We can't teach what we don't know: White teachers, multiracial classrooms.* New York: Teachers College Press.

Hsu, F. (2013, November 4). *The heart of mindfulness: A response to the New York Times*. Retrieved from http://www.buddhistpeacefellowship.org/the-heart-of-mindfulness-a-response-to-the-new-york-times/

Hughey, M. (2014). *The white savior film: Content, critics and consumption*. Philadelphia: Temple University Press.

Illich, I. (1968). To hell with good intentions. In J. Kendall (Ed.), *Combining service and learning* (pp. 314–320). Raleigh: National Society for Internships and Experiential Education.

Jennings, P. (2012). *Building an evidence base for mindfulness in educational settings*. Retrieved from https://www.garrisoninstitute.org/building-an-evidence-base-for-mindfulness-in-educational-settings

Jennings, P. (2015). *Mindfulness for teachers: Simple skills for peace and productivity in the classroom*. New York: Norton.

Jennings, P. (2016). CARE for teachers: A mindfulness-based approach to promoting teachers' social and emotional competence and well-being. *Mindfulness in Behavioral Health Handbook of Mindfulness in Education*, 133–148. doi:10.1007/978-1-4939-3506-2_9

Kumashiro, K. (2000). Toward a theory of anti-oppressive education. *Review of Educational Research, 70*(1), 25–53.

Ladson-Billings, G. (2001). *Crossing over to Canaan: The journey of new teachers in diverse classrooms*. San Francisco: Jossey-Bass.

Ladson-Billings, G. (2006). From the achievement gap to the education debt: Understanding achievement in U. S. schools. *Educational Researcher, 35*(7), 3–12. doi:10.3102/0013189x035007003.

Ladson-Billings, G. (2007). Pushing past the achievement gap: An essay on the language of deficit. *The Journal of Negro Education, 76*(3).

Lueke, A., & Gibson, B. (2014). Mindfulness meditation reduces implicit age and race bias: The role of reduced automaticity of responding. *Social Psychological and Personality Science, 6*(3), 284–291. doi:10.1177/1948550614559651.

Magee, R. (2015, May 14). How mindfulness can defeat racial bias. Retrieved from http://greatergood.berkeley.edu/article/item/how_mindfulness_can_defeat_racial_bias

McKnight, J. (1989). Why servanthood is bad. *The Other Side, 25*(1), 38–42.

Miller, J. P. (2014). *The contemplative practitioner: Meditation in education and the workplace*. Toronto: University of Toronto Press.

Miller, J. P., & Nozawa, A. (2002). Meditating teachers: A qualitative study. *Journal of In-Service Education, 28*, 179–192.

Mindful Schools: Online mindfulness training for educators. (n.d.). Retrieved April 15, 2016, from http://www.mindfulschools.org/

Mindfulness Without Borders. (n.d.). Retrieved April 15, 2016, from http://mindfulnesswithoutborders.org/

Mitchell, T. (2008). Traditional vs. critical service-learning: Engaging the literature to differentiate two models. *Michigan Journal of Community Service Learning, 14*(2), 50–65.

Monteiro, L. M., Musten, R., & Compson, J. (2014). Traditional and contemporary mindfulness: Finding the middle path in the tangle of concerns. *Mindfulness, 6*(1), 1–13. doi:10.1007/s12671-014-0301-7.

Morton, K. (1995). The irony of service: Charity, project, and social change in service-learning. *Michigan Journal of Community Service Learning, 2*(1), 19–32.

Ng, E. (2015, June 29). Mindfulness and justice: Planting the seeds of a more compassionate future. Retrieved from http://www.abc.net.au/religion/articles/2015/06/29/4264094.htm

Ng, E., & Purser, R. (2015, October 2). White privilege and the mindfulness movement. *Buddhist Peace Fellowship*. http://www.buddhistpeacefellowship.org/white-privilege-the-mindfulness-movement/

Ng, E., & Purser, R. (2016, April 5). Mindfulness and self-care: Why should I care? Retrieved from http://www.huffingtonpost.com/edwin-ng/mindfulness-and-selfcare-_1_b_9613036.html

O'Brien, D. (2012, April 13). Ryan takes 'mindfulness' to inner-city schools. *The Business Journal*. Retrieved from http://businessjournaldaily.com/education/ryan-takes-'mindfulness'-inner-city-schools-2012-4-13

Orr, D. (2002). The uses of mindfulness in anti-oppressive pedagogies: Philosophy and praxis. *Canadian Journal of Education/Revue Canadienne De L'éducation, 27*(4), 477. doi:10.2307/1602246.

Parker, A. (2009). Mindfulness-based academic achievement program for middle school. Retrieved from http://ies.ed.gov/funding/grantsearch/details.asp?ID=733

Patel, L., Atkins-Patterson, K., Healy, D., Haralson, J. G., Rosario, L., & Shi, J. (2013). Mindfulness as method: Teaching for connection in a dehumanizing context. *Poverty & Race, 22*(3), 5–13.

Picower, B. (2009). The unexamined whiteness of teaching: How white teachers maintain and enact dominant racial ideologies. *Race, Ethnicity and Education, 12*(2), 197–215.

Purser, R. (2014). Clearing the muddled path of traditional and contemporary mindfulness: A response to Monteiro, Musten, and Compson. *Mindfulness, 6*(1), 23–45. doi:10.1007/s12671-014-0373-4.

Purser, R., & Loy, D. (2013, July 1). Beyond McMindfulness. Retrieved from http://www.huffingtonpost.com/ron-purser/beyond-mcmindfulness_b_3519289.html

Purser, R., & Ng, E. (2015, September 27). Corporate mindfulness is bullsh*t: Zen or no Zen, you're working harder and being paid less. Retrieved from http://www.salon.com/2015/09/27/corporate_mindfulness_is_bullsht_zen_or_no_zen_youre_working_harder_and_being_paid_less/

Purser, R., & Ng, E. (2016, March 22). Cutting through the corporate mindfulness hype. *Huffington Post*.

Retrieved from http://www.huffingtonpost.com/ron-purser/cutting-through-the-corporate-mindfulness-hype_b_9512998.html

Reveley, J. (2015). Foucauldian critique of positive education and related self-technologies: Some problems and new directions. *Open Review of Educational Research, 2*(1), 78–93.

Roeser, R. W., Skinner, E., Beers, J., & Jennings, P. A. (2012). Mindfulness training and teachers' professional development: An emerging area of research and practice. *Child Development Perspectives, 6,* 167–173.

Room to Breathe. (n.d.). Retrieved April 15, 2016, from http://www.videoproject.com/roomtobreathe.html

Room to breathe: A documentary film about transformation through mindfulness meditation in public schools. (n.d.). Retrieved April 15, 2016, from http://roomtobreathefilm.com/

Ryan, T. (2012). *A mindful nation: How a simple practice can help us reduce stress, improve performance, and recapture the American spirit.* Hay House Inc.

Singh, R. (2013, July 9). Yes we won and what we lost: Sedlock vs. Baird decision allows yoga in public schools [Web log message]. Retrieved from http://saapya.wordpress.com/2013/07/09/yes-we-won-and-what-we-lost-sedlock-vs-baird-decision-allows-yoga-in-public-schools/

Slaten, C. D., Irby, D. J., Tate, K., & Rivera, R. (2015). Towards a critically conscious approach to social and emotional learning in urban alternative education: School staff members' perspectives. *Journal for Social Action in Counseling and Psychology, 7*(1), 41–62.

Smalley, S., & Winston, D. (2010). *Fully present: The science, art, and practice of mindfulness.* Philadelphia, PA: Da Capo Press.

Vacarr, B. (2001). Voices inside schools: Moving beyond polite correctness: Practicing mindfulness in the diverse classroom. *Harvard Educational Review, 71*(2), 285–296. doi:10.17763/haer.71.2.n8p0620381847715.

Valerio, A. (2016). Owning mindfulness: A bibliometric analysis of mindfulness literature trends within and outside of Buddhist contexts. *Contemporary Buddhism,* 1–27. doi:10.1080/14639947.2016.1162425

Whitlock, J. (2013, November 14). Man protesting yoga program outside of EUSD schools. *Coast News Group.* Retrieved from http://yogaencinitasstudents.org/

Wong, Y. R. (2004). Knowing through discomfort: A mindfulness-based critical social work pedagogy. *Critical Social Work, 5*(1).

Yellow Bird, M. (2013). Neurodecolonization: Applying mindfulness research to social work. In M. Gray, J. Coates, M. Yellow Bird & T. Hetherington (Eds.), *Decolonizing social work* (pp. 293–310). Burlington, VT: Ashgate Publishing Company.

Yosso, T. (2005). Whose culture has capital? A critical race theory discussion of community cultural wealth. *Race, Ethnicity and Education, 8*(1), 69–91.

Zakrzewski, V. (2016, March 31). Why don't students take social-emotional learning home? *Huffington Post.* Retrieved from http://www.huffingtonpost.com/vicki-zakrzewski-phd/why-dont-students-take-so_b_9586008.html

Zúñiga, X., Nagda, B. A., Chesler, M., & Cytron-Walker, A. (2007). *Intergroup dialogue in higher education: Meaningful learning about social justice* (ASHE-ERIC Higher Education Report Vol. 32, Num. 4). San Francisco, CA: Jossey-Bass.

The Curriculum of Right Mindfulness: The Relational Self and the Capacity for Compassion

Joy L. Mitra and Mark T. Greenberg

There has been substantial discussion in the past decade about the secular decontextualization of mindfulness as it has been applied in medicine, education, business, and other settings. In this chapter, we explore the Buddhist understanding of right mindfulness and how new secular frameworks are emerging that are focused on the nurturing of compassion. We focus on how these emerging models of mindfulness and compassion can be nurtured in applied settings with youth, teachers, parents, and others to support understanding of the nature of mind, to sharpen attention and awareness, and to promote compassion. In doing so, we draw on insights of Bodhi (2011) on mindfulness, Varela et al. (1991) on mind and compassion, and Vago and Silbersweig (2012) on neuroscientific models of the mechanisms of mindfulness.

The ongoing debate regarding the role of ethics in contemporary models of mindfulness interventions has focused on the issue of decontextualization of mindfulness in secular society and on the relationship between mind-

fulness and ethical action (Bodhi 2011; Grossman and Van Dam 2011; Monteiro et al. 2015). We agree that this is a central and compelling question in the cultural translation of mindfulness and of meditative practice when utilized in Western secular settings. Here, we address some foundational issues relating to the practices of mindfulness in both personal and interpersonal contexts, and of the association of mindfulness with secular ethics, wisdom, and compassion. In doing so, we consider mindfulness to be not only a specific set of meditative practices, but also an integral component of the Eightfold Path in which right speech, right action, right livelihood, etc., are intrinsically linked.

In this chapter, we do not address the vigorously debated issues of spiritual materialism, nor concerns regarding the utilization of meditative techniques in the service of self-advancement, combat, or other more controversial contemporary contexts. However, we share the concerns of the more traditional Buddhist approach that a delimited approach to mindfulness that does not include a clear and explicit ethical foundation is misappropriated and questionable. We also propose that it may be helpful to begin to articulate principles and criteria: (i) To draw a clear line to distinguish mindfulness from right mindfulness in secular interventions and (ii) to serve as scaffolding for the development of an ethical framework as the basis of education for nonviolent and compassionate forms of action at the universal level.

J.L. Mitra (✉)
Edna Bennett Pierce Prevention Research Center,
Penn State University, University Park,
16802 PA, USA
e-mail: jlm891@psu.edu

M.T. Greenberg
College of Health and Human Development, Penn State University, University Park, 16802 PA, USA
e-mail: mxg47@psu.edu

© Springer International Publishing Switzerland 2016
R.E. Purser et al. (eds.), *Handbook of Mindfulness*,
Mindfulness in Behavioral Health, DOI 10.1007/978-3-319-44019-4_27

Historical Foundations of Ethical Practices and Current Secular Extensions

Historically, ethical practices have often been directly bound to one's faith or religious perspective. From the standpoint of our secular Western culture, it is instructive to review the elements common to most of the world's spiritual practices (Judeo-Christian, Buddhist, Islamic, Religious Society of Friends, etc.). In each of these traditions, there are at least three elements: practices, ethical beliefs/teachings, and community. The first element, practices, can include prayer, pilgrimage, chanting or singing, ritual, meditative techniques, witnessing, and many others (see the Tree of Contemplative Practices http://www.contemplativemind.org/practices/tree as one model). Second, there is typically a set of ethical beliefs, principles, or precepts that are delivered in the form of oral sermons, homilies, dharma talks, as well as through prayers, songs, and actions. In many different spiritual traditions, followers look toward their leaders (priests, rabbis, guru, master) for the everyday embodiment of these principles. Third, these perspectives and principles are traditionally "held" by a like-minded group that supports the group's norms and reinforces their beliefs. This "formula" of practices, beliefs/ethical teachings, and a community that "holds" these practices and teachings has been central to spiritual traditions throughout history, traditions that fostered motivating belief systems and that encouraged and enforced ethical actions.

Our ultimate focus is on the extension of mindfulness practice via ethical action to the enactment of compassion at the interpersonal level. It is driven by a sense of urgency directed toward new modes of secular education and the development of nonviolent and sustainable communities. Further, we believe it will be useful to consider the essential importance of each of these three elements—practices, beliefs/principles, and community—in the extension of mindfulness interventions to include cultivation of wisdom and compassion.

Joining the Wings of Wisdom and Compassion

Our view involves a broader translation of the term "mindfulness," in accord with that of Bodhi (2011), in which mindfulness necessarily encompasses ethical speech and compassionate action as part of a complex set of interrelated processes, including discernment, discrimination, remembrance, and imagination. This expanded definition moves us from mindfulness toward "right mindfulness" by incorporation of an ethical foundation that may be explicitly taught and/or implicitly communicated through the embodiment of such values in a teacher.

The argument for an extension of mindfulness-based practice to broader interpersonal applications necessarily rests on the cultivation of insight and wisdom through sustained practice. It is increasingly supported by evidence-based studies based on the neurobiological processes associated with human cognition and the experience of suffering. (Demasio 2000; Decety and Chaminade 2003; Legrand and Ruby 2009; Vago and Silbersweig 2012; Ataria et al. 2014). In the S-ART framework, Vago and Silbersweig (2012) have provided a conceptual framework and neurobiological model for understanding the potential mechanisms by which mindfulness reduces the propensities related to self-processing that can lead to "a distorted or biased sense of oneself." (p. 2). They categorize the model of nonjudgmental awareness in the present moment as being a contemporary definition primarily associated with stress reduction. Our concern with this more limited contemporary definition, especially within educational contexts, is that it tends to isolate awareness from related practices traditionally associated with mindfulness which serve as "the guarantor of correct practice of all the other path factors" (Bodhi 2011, p. 26). The call for an integrative approach that incorporates both contemporary and traditional models necessitates reliance on a broader framework with emphasis on ethical conduct and compassion, as well as the cultivation of bare awareness.

The argument for expanding the model of mindfulness practices and measures in educational settings is not intended to minimize the significant contributions that have resulted from the secularization of concepts and practices introduced in MBSR programs. Rather, we intend to build on this strong foundation, both to expand its scope (enlisting the engagement of other essential and universal human faculties, such as memory, discernment, reason, and imagination) and to extend its reach to practices associated with the development of enactive compassion.

In the Vipassana tradition, the metaphor of the two wings of the Brahmavihara yokes wisdom and compassion in service of the well-being of the broader community. Emphasis on practices such as meditation or cultivation of awareness and attention is a necessary basis for achieving wisdom, since dharma includes both formal teachings and the evidence of lived experience. Interrelationship with others in a community is both the ground and the motivation for extension of insight and wisdom to the field of compassionate action.

This proposal for the broader scope of secular mindfulness reflects not only the relational nature of mental processes, including mindfulness, but also the growing body of evidence regarding the essential relational nature both of consciousness and of "the self." For Varela mindfulness was not an abstraction: "There has to be something to be mindful of, aware of, to realize the intrinsic goodness of and to be compassionate for." (1991, p. 234) Building on these seminal insights, Siegel's systems-based approach (2012) applies at the individual level with regard to mental processes, as well as at the level of interrelationships between individuals in communities and collectives.

Such integrative approaches are supported by parallel developments in philosophy, cognitive science, and social neuroscience that have led to the recognition that suffering and the relief of suffering can be understood only by going beyond the illusion of the separate self and the associated realization of interdependence. To facilitate the goal of acting with compassion in the world, we invite consideration of an extended path for contemplative practice that proceeds from mindfulness and self-compassion to the practice of cultivation of broader, generalized compassion with and for others. In this respect, Varela et al. (1991) described generalized compassion as a disposition that evolves after significant and sustained mindfulness practice, as a stage at which "the street fighter mentality of watchful self-interest can be let go somewhat to be replaced by interest in others." He concluded that "the tradition of mindfulness/awareness offers a path by which this may actually be brought about." (p. 247). Bodhi (2011) argued that "the practitioner of mindfulness must at times evaluate the mental qualities and intended deeds, make judgments about them and engage in purposeful actions" (p. 26). While there is considerable controversy regarding a deeper interpretation of the "secular definition," Grossman and Van Dam (2011) have argued that "attention and awareness are at most aspects that serve as preconditions, rather than equivalents of mindfulness" (p. 223). H. H., the Dalai Lama (2011), has underlined the importance of the yoking of insight and wisdom in the development of compassion: "The idea is that when compassion is complemented and reinforced by the faculty of wisdom, the individual has the ability not only to empathize, but also to understand the causes and conditions that led to that suffering, and to envision the possibility of freedom from that state" (p. 255). During the past two decades, the dominant narrative of mindfulness in service of personal stress reduction and enhancement of individual well-being has at times subordinated the potential for the cultivation of engaged compassionate action resulting from the sustained practice of mindfulness. Varela's earlier urging toward enactive compassion was not inconsistent with practices of inwardly directed mindfulness. In fact, he positioned traditional meditation practice as merely the first phase in a longer development ultimately leading to what he refers to as "non-naïve" compassion. This later stage goes beyond instinctive or spontaneous reactions of kindness and is expressed in ethical action resulting from the eventual perceived dissolution of apparently impenetrable boundaries between self and others.

Given Western culture's emphasis on the centrality of the individual and its heightened emphasis on the value of personal independence and self-sufficiency, the recent emphasis on self-compassion and on personal healing may be requisite foundations for supporting broader interpersonal applications of right mindfulness. Currently, multiple sources of evidence in various disciplines point to the wisdom and necessity of bridging cultivation of awareness and attention at the individual level with enactment of compassion at the interpersonal level.

Next-Generation Mindfulness: Cultivation of Enactive Compassion

During the past thirty years, the mindfulness movement has grown exponentially. Unfortunately, during that same time period, youth self-report of compassion (defined as "self-reported concern for the welfare of others and the relief of their suffering") has significantly declined. In a study of 13,000 college students assessed in 1979 and 2009, Konrath found that self-reported concern for the welfare of others has been steadily dropping and is at the lowest rate in the past 30 years (cited in deSteno 2015, p. 2). The dilemma presented by the parallel phenomena of the growth of the mindfulness movement and the decreasing rates of compassion in youth provides a compelling argument for placing greater emphasis on developing explicit criteria for the cultivation of wisdom and compassion within the curricula of mindfulness interventions and for measuring its impact.

It is clear that without the wisdom and inspirational guidance of Jon Kabat-Zinn's groundbreaking work, which has provided direction for much of the contemporary practice of mindfulness, the current debate would be situated within a much narrower audience. Yet it is also clear that important work remains to be done by all who are ultimately concerned with directing these practices toward building a peaceful and sustainable future.

What should be the emphasis in the next decades? What extensions of current practices should be made to address the apparent devolution of concern for the welfare of others? What steps can help to broaden the scope of current secular practices in order to place stronger emphasis on loving-kindness practices (*metta*) and compassion practices (*karuna*) that represent other-directed forms of awareness and intentionality?

In responding to the typical association in Western psychology between positive and negative emotions, those that result in pleasure and pain, H. H., the Dalai Lama (2011), asserted that a primary distinction in Buddhist psychology is between those states that are beneficial and those that are harmful. In other words, the focus shifts from one of interpersonal pleasure (i.e., hedonic) and pain to the impact of one's actions in the world—as being beneficial or harmful. We agree with Flanagan (2003) that it is important to go beyond a utilitarian focus on pleasure for the many (especially if secured at the expense of increased suffering for the few), as exemplified in the dramatically increasing inequality gap in most Western countries, and to contextualize wholesome actions not only as personal, but as interpersonal and social in nature (Goleman 2015). This imperative reaches beyond the ethical principle "of do no harm" to a focus on the "motivation to remove sources of suffering" whenever possible. Bodhi (2015) recently observed that without the broader model of the Eightfold Path, the real danger in "bare [i.e., decontextualized] mindfulness" is the risk of its being "turned into a mere adornment to a comfortable lifestyle" and reduced to "an assortment of therapeutic techniques." It is clear that progress in understanding the nature of "the self," as well as recognition of the universal ground of interdependence, provides the foundation for developing contemporary practices associated with enactive compassion.

Expanding the Context of Mindfulness: The Wisdom of Interdependence

Increasing scientific evidence of "the self" as neither solid nor separate, but rather as a locus of perspective grounded in a field of interdependence, supports an extension in the depth and scope of models for education in mindfulness. Classical teachings regarding dependent origination and emptiness are now supported by evidence-based research demonstrating the porous nature of the boundaries between self and other (Decety and Chaminade 2003; Dor-Ziderman et al. 2013; Legrand and Ruby 2009; Vago and Silbersweig 2012). Traditional dharma teachings are now substantiated by a new generation of philosophers, cognitive scientists, and neuroscientists who have examined the absence of a centralized locus for the relational capacity conventionally referred to as the self. Although the concept of a solid and separate self is a conceptual frame that is continuously reinforced in Western education and literature, it is difficult to locate "the self" experientially and through neuroimaging evidence other than in the perspective that one has toward others, i.e., the first person perspective (Ruby and Legrand 2007).

Arguably the first negation of what classical Buddhism refers to as primordial mind, or substrate consciousness, occurs with the development of language and, with it, the encapsulation of sensory experience into patterns and constructs frequently identified with the Narrative Self (NS) (Gallagher 2000; Dennett 1992). This first negation is accompanied by the construct of the sense of a solid separate self, as the Narrative Self (NS) strives *to protect itself and to confirm its own solidity.*

Across disciplines, there now exists convergent evidence of the distinction drawn between the present-moment self as experienced in meditation (analogous to Gallagher's Minimal Self,

Damasio's core self, and EPS, the Experiential Phenomenological Self in the S-ART framework) and the Narrative Self that typically directs our experience of ordinary, everyday life (Dennett 1992; Dor-Ziderman et al. 2013; Gallagher 2000; Vago and Silbersweig 2012). The Narrative Self (NS), which is caught up in projecting, wishing, ruminating, etc., is the locus in the thinking mind of the hindrances or obstacles which mindfulness meditation has the capacity to dissolve.

Vago and Silbersweig (2012) build on the conceptualizations of self earlier introduced by Gallagher, Demasio, and Legrand to further develop the notion of Narrative Self, characterized as "the evaluative self-as-object, reflecting the autobiographical narrative reconstructed from the past or projected into the future." (p. 6) Dennett (1992) explained the Narrative Self as one that exists in the dimensions of time and history and that accompanies our experience of personal agency, arguing that the Narrative Self is more akin to a fictional character: "We ... do not consciously and deliberately figure out what narratives to tell and how to tell them; like spiderwebs, our tales are spun but for the most part we don't spin them; they spin us" (p. 418).

These distinctions take on increasing significance in the S-ART model in which Vago and Silbersweig (2012) asserted that *"biased self-processing can be understood as a contemporary model for suffering."* The S-ART model describes the limiting role of this Narrative Self (NS):

> The narrative one creates about oneself in terms of self-reflection or future projection becomes increasingly more rigid as it is conditioned over time through a causal chain of repetition. Each trajectory of self-development represents a repeatedly reconstructed, reinforced, and reified NS with reliable patterns of subject-object relations that are relatively stable and accessed during self-specifying processes. (p. 20)

Since the Narrative Self is reliant on "language, episodic/autobiographical memory and imagination," it is closely associated with mind-wandering and the activity of the default network (Dor-Ziderman et al. 2013, p. 1). In their study of long-term mindfulness meditators, Ataria et al. (2014) examined the neural correlates of MS and NS in contemplative practitioners. These practitioners developed the capability of (i) shifting voluntarily between the modes of the Minimal Self, characterized by present-moment awareness, and the self-narrating, episodic mode of Narrative Self, and (ii) dissolving the sense of boundaries between self and other (correspondent with the prosocial or self-transcendent phase of the S-ART framework in which the practitioner realizes the nature of self as being "co-dependent with the relations to objects it experiences") (Vago and Silbersweig 2012, p. 23).

Phenomenological Studies of the Experience of Self

In mindfulness training, the practice of resting in the present moment functions to negate the mode of experiencing characteristic of the Narrative Self. The practitioner rests in the present moment, experiencing each moment independently, outside of linear passage of time and often with an open and fluid sense of space. Damasio (2000) described this present-moment self as "a transient entity, ceaselessly recreated with each and every object with which the brain interacts," and explained that "in this way it implements a self/non-self distinction" (p. 17). That is, the Minimal Self is characterized by self-specifying processes in which the first person perspective is operative and anchored in the sense of agency and ownership.

The experience of the Minimal Self fits well within the MBSR definition of mindfulness as present-moment, nonjudgmental awareness. In this mode, one can direct awareness and attention to the present moment while necessarily still being heavily invested in concepts of agency, ownership, and the sense of separation between subject and object. This enables the beginning meditator to rest (albeit briefly) in the open space of the present moment and *to begin* to pierce the illusion of the solid separate self. In this phase of

practice, exercises in loving-kindness may beneficially impact relationships between subject and object without yet approaching Varela's desired end state of enactive compassion, a broader dimension characterized by the dissolution of boundaries between subject and object through sensed interdependence.

Consequently, as in MBSR, this cultivation of the present-moment mode does alter one's sense of living within the body and of the relationship to personal pain. Working with illness and stress in this way can open the subject to a sense of more flexible boundaries, with a diminished sense of the solidity of the distinction between self and other and a decrease in rumination and experiential avoidance. The experience of present-moment mindfulness, however, does not *necessarily* facilitate transcendence beyond these illusory boundaries that separate "self" from others.

In the scientific dialogue held at the 13th Mind and Life Institute regarding the possible healing or liberating effects of meditation, Kabat-Zinn et al. (2011) questioned whether the experience of pure awareness is in itself physically or mentally healing. In response, His Holiness, the Dalai Lama (2011), explained that meditation leading to pure awareness may divert attention from physical pain and result in a sense of healing or freedom, but "it might also act as a tranquilizer." H. H. continued: "One of the characteristics of compassion is that it immediately opens your heart outward to a much more expansive field… If we find some way to get out of this prison of self-centeredness and reach out to the wider common humanity, truly this will have an impact." (p. 56)

Moving from Present-Moment Awareness to Softening the Boundaries of the Solid Self

Because the sustained practice of mindfulness results in decreasing reliance on constructs associated with fixed boundaries between self and others, it often results in an awareness that is difficult to express in language other than by the negation of constructs that we take for granted. It is therefore not surprising that many Buddhist concepts are expressed in negative terms, e.g., when "a" or "an" is used as a prefix in Sanskrit, it negates the term or construct that follows. So, for example, *ahimsa* is "not harming," and *annata* (*or anatman*) refer to "not self." Although much of the language of contemporary mindfulness research describes the nature of the observing self and of nonjudgmental awareness, less attention has been paid to the developing evidence that supports the Buddhist notion of *Annata*—or "not self"—as the foundation for awareness of interdependence. Watson (2000) situates this understanding of the illusory concept of a solid self within the traditional Buddhist context: "The Self then, which in Buddhism, is to be negated, is an illusion; it is the imposition of a container self with attributes of independence and permanence upon the foundation of the conventional or transactional self of ever-changing mind states" (p. 33). This nondual sense of 'self' as being at once both real and illusory is echoed in imaging studies in which there are wide areas of overlap in neural processes, such that it is difficult to differentiate between representations of self and other (Ruby and Legrand 2007).

After sustained and repeated mindfulness practice, the possible experience of selflessness or boundarylessness corresponds to what Vago and Silbersweig described as self-transcendence and to the meditative experience that one research subject compared to the "air pocket"—his experience of flipping between the awareness of the present moment and the breath to the sense of having no body or breath, "as if I disappeared" (Berkovich-Ohana 2015, p. 7).

Does the sustained practice of mindfulness then enable one to go beyond the experience of the self as a separate entity and lay the groundwork for understanding interdependence? In a state of profound awareness of interdependence does the softening of boundaries between self and others cancel or overtake the force of the Narrative Self directed at protecting and defending a separately bounded unitary self? By developing this wise awareness over time, can one pass beyond the concerns of one's own body, beyond the medical or clinical bases for suffering, beyond stress reduction, beyond the perceived need to preserve and protect the Narrative Self, to the cultivation of "long-distance compassion"—the ability to care for those who are remote? (Singer et al. 2011, p. 191). Can an extension in the scope of practice, forged through a combination of insight and wisdom, result in the generation of compassion and in an imperative for ethical social action?

Generation of Prosocial Intentionality Through Mindful Awareness

The progression beyond the thoughts of the Narrative Self to development of a present-moment, nonjudgmental awareness is now increasingly evidenced in studies of mindfulness. The phenomenological inquiry regarding our capacity to voluntarily shift between the mode of the Narrative Self and immersion in the present moment articulates the lived experience of many practitioners who draw upon insights gained "on the cushion," as they return to the world of *chronos*, of time, history, and purposeful action.

In order to negate the reification of the Narrative Self, it is necessary to move not only to reflective awareness but also to invoke intentionality. Kabat-Zinn (1990) said that "attention and intention work beautifully together to further the possibility to waking up to the actuality of one's experience, which you could call the ground of being—or the groundlessness of being." (p. 62) Echoing Merleau-Ponty's proposition that intentionality gives our lives purpose, Bodhi (2011) argued that if mindfulness is to qualify as right mindfulness, it must be connected to a "web of factors that give it direction and purpose" (p. 26). When mindfulness is situated as a part of this broader scaffolding, it serves the transitive function not only of relating subject to object but also eventually to the dynamic web of interconnectedness of all subjects; in this way, mindfulness enables a natural progression from the release of suffering at an individual level to broader levels of wisdom and compassionate action. In this model, the practice of right mindfulness is premised on a clear intention to generate well-being both at the individual and at the universal, species-wide level, consistent with H. H. the Dalai Lama's (1995) definition of compassion as being "a *sensitivity* to the suffering of self and others, with a *deep commitment* to try to relieve it" (p. 16).

In the systems-based S-ART framework, the repeated practice of mindfulness is a path that leads one to transcend self-focused needs and to increase one's prosocial disposition. For Tirch (2010), "this can be construed as an abundant compassion emerging from the intellectual and experiential knowledge of the intimate interconnection of all phenomena. Compassion and mindfulness may be viewed as *co-creating* one another" (p. 121, italics added).

In contrast with exercises premised on the use of extrinsic forms of motivation, the curriculum of right mindfulness is oriented toward methods and content that soften the perceived solid boundaries between self and other. Its emphasis is not on personal independence and differentiation of the abilities between self and other, so much as on learning to let go of the need to control both self and others.

Arguably the notion of suffering cannot arise without the concomitant notion of interdependence or connectedness. If individuals were in fact self-sufficient, independent entities, there would be no experiential basis for suffering. All forms of suffering stem from our dependence upon others for existence—at the level of physical nourishment, emotional well-being, and intentional actions. We know that human infants require social interactions for survival and for the development of healthy, secure attachments to others. The Buddhist concept of dependent origination articulates this fundamental dependence of one's own being upon that of all other beings. While Westerners often find it difficult to grasp *sunyata* in the abstract, most find it difficult, when challenged, to cite an example of suffering that is not a function either of an individual's connections with others or, alternatively, the isolation that results from perceived lack of connection with others.

In summary, how would grasping and attachment arise if one were in fact self-sufficient? How would kindness, violence, aggression, and stress occur if each individual were an inviolable, solid, and sustainably permanent unit?

Perspective-Taking as a Core Capability for the Generation of Compassion

In the loving-kindness practices as taught by Salzberg (2002), Halifax (2012), and others, a goal is to extend a sense of compassion not only to oneself but to all other sentient beings. With regard to this essential activity, Decety and Chaminade (2003) speculated that the ability to take perspective or to imagine oneself in the place of another may be a core capacity that distinguishes humans from other primates. This ability enables us to understand the intentions and mental lives of others and to project the thoughts and feelings of others, to feel empathy, sympathy, and compassion for others.

A curriculum directed toward right mindfulness and the end of suffering is inextricably bound to the realization of interdependence, which in a dualistic frame appears contradictory, since inter-being is both a source of *dukkha* and also the key to the release from suffering via generation of compassion. Thus, we see the more narrowly defined concept of secular mindfulness, which has been shown to reduce personal stress and to improve health as a first step, rather than an end state. It is one step on the path toward the generation of compassion based on the dharma of inter-being.

Considerations for Constructing a Curriculum for Right Mindfulness

Contemporary mindfulness approaches that do not directly address the limitations of resting within the illusory boundaries of the separate self are therefore necessarily incomplete. That is, a sole focus on the cultivation of concentration and attention may be insufficient if the goal is the construction of interpersonal or organizational interventions aimed at reducing suffering or eliminating inequity.

At the level of cognition, this view reflects movement from attention and awareness through the related mental factors of discernment, intention, imagination, and reason toward the ends of developing wise understanding and engendering beneficial or wholesome outcomes (Varela et al. 1991). At the level of intentionality and action, this view extends the initial emphasis on stress reduction to the cultivation of self-compassion and the generation of a broader transactional form of compassion as the ground for enactment of mindfulness in our interpersonal interactions.

The majority of our time is spent not on the cushion, but as actors in the arena of time, history, and society. In the movement between the cushion and the domain of everyday action, we can join awareness and deeper insight into the nature of interdependence with the wisdom of Right Action in support of the well-being of the community.

Foundational Principles of Secular Ethics

Here, we are searching for a basis for articulation of a secular ethics that can serve as an implicit or explicit guide for the practice of right mindfulness. The foundations for this as a species-wide capacity have been increasingly documented in the works of social neuroscience. One purpose of the practice of nonjudgmental, present-moment awareness is to enable us to disengage from the conceptual attachments, discursive scripts, and automatic reactions that are root causes of suffering. Not only does this grounding in present-moment awareness yield clarity of vision, but it can also serve as the "nutriment" for emergent intentionality and discernment, linking insight with the skillful practices of right speech and actions (Wallace and Bodhi 2006).

Attitudinal Foundations for Compassion

To facilitate the development of right mindfulness, it is also instructive to consider possible *attitudinal foundations for compassion*. Critical inquiry involved in the original teachings on mindfulness is relevant to applications of mindfulness to interpersonal settings. The practices of loving-kindness (*metta*) and cultivation of compassion (*karuna*) extend beyond concern for individual uncertainty and stress; they are fundamental to nurturing children and to the creation and development of nonviolent and sustainable communities. To position this perspective more clearly, we are seeking grounds for applying mindfulness within educational and other learning contexts to propel learners beyond the fixed boundaries of the illusory separate self to nurture an awareness of greater connectedness at the interpersonal level.

Some foundational principles that can support such a proposed ethical frame are (i) nonharming (ahimsa), as in the Hippocratic Oath; (ii) understanding of interdependence, as increasingly manifested across all fields of natural and social sciences, (e.g., quantum mechanics, ecological environmental and economic theories, technology and proliferation of social media, development of planetary tools for reducing egotism (Scharmer and Kaufer 2013)); and (iii) the Golden Rule, one way of expressing the crux of various formulations of virtuous action.

In right mindfulness, the embodiment of a broad and inclusive ethical framework is suggested as an alternative approach to methods that stop short of considering the implications of interdependence. Just as mindfulness is intended to secularize the practice of meditation, the conceptual elements of the Eightfold Path can serve as a platform for the development of a secular ethical frame that can be used across social and cultural contexts. With right mindfulness, the definition of community includes all living beings—not just the members of one classroom, nor of a single neighborhood, ethnicity, corporation, or nation-state—and the practice of mindfulness aligns with the wisdom of the Third Noble Truth.

Contemporary practices that focus only on present-moment awareness and the relief of personal distress and which limit the role of remembrance, discernment, and intentionality may unwittingly lead to unquestioning acceptance of whatever oppressive forms are dominant in the present moment (Wallace and Bodhi 2006). Without a clear commitment to alleviating individual suffering in a manner that does not increase suffering for others, might we be reinforcing passivity and maintaining oppression for teachers, students, and others?

The push toward universal and species-friendly ethical principles necessitates inclusion of a normative component in the practice of mindfulness in the arenas of speech and action. This approach suggests a possible bright-line test for differentiating contemporary notions of mindfulness (*sati*) from the practice of right mindfulness (*samma sati*); that is, if an intentional action results in the reduction of suffering for an individual but increased suffering for others, then such practice may not be considered an exercise of right mindfulness. It would therefore be interesting to test these foundational principles through programs incorporating implicit and explicit secular ethical frames which can be evaluated via both qualitative and quantitative outcomes.

The linkage of inner and outer values that results in intentional actions directed toward the highest good is a common factor that unites many faith traditions. In this argument for the development of criteria for secular ethics, the central normative affirmation must be that actions should result in well-being for both the self and for others—neither being sufficient on its own. As such, we believe that it is possible to outline a framework of broad criteria that are neither moralistic nor sectarian.

The Relational Self and the Eightfold Path

If mindfulness practice is used merely as a tool for alleviating personal discomfort, we can forget that the original purpose of this practice involves arriving at broader grounds for the enactment of compassion. Clearly, Kabat-Zinn (1990) never perceived of mindfulness practice in such a limited manner; at the outset, he argued "One way to think of this process of transformation is to think of mindfulness as a lens, taking the scattered and reactive energies of your mind and focusing them into a coherent source of energy for living, for problem solving and for healing" (p. 11).

The path of right mindfulness involves a dialectical progression and opens a pathway for the cultivation of engaged compassion. One moves away from the Narrative Self, enters the present-moment awareness of the Minimal Self, then returns to the domain of time and history as an observing self with enhanced capacity for perspective-taking.

This relational view of self is supported in many traditions in which mindfulness is considered to be a preparatory activity, first for meditation and concentration, and successively from that point to the causal unfolding of the steps on the Eightfold Path. As Bodhi (2015) recently observed, "Spurning social engagement, contemplative practice may turn into an intellectual plaything of the upper middle class or a cushion to soften the impact of the real world."

The case for right mindfulness is intended as an extension of the current definitions and measures of mindfulness in a manner consistent with Varela et al. (1991), as a frame for awakening "the expression of the human capacity to cultivate that which is not self-centered and ego-based." This generative sense of well-being is also at the heart of the Buddhist notion of "sukkah" (often translated as happiness), which Ricard et al. (2011) described as being more akin to "the depths of the ocean … a depth of being that can remain stable … and that includes a sense of wisdom in understanding the qualities of that state and distinguishing it from pleasure" (p. 195).

The curriculum of right mindfulness outlined here assumes the development of ethical awareness and compassion as a natural, but not inevitable, extension of the contemporary practice of mindfulness. As the student develops the ability to pay attention within the present moment, a transformation in the subject's relationship to the Narrative Self and to others is also initiated. Concurrently, the faculty of judgment is increasingly deactivated, as the capacity for discernment and perspective-taking is enhanced.

It is our conviction that empathy and compassion are core dimensions of human nature and can be nurtured. Further, when nurtured, these capacities enhance one's personal growth and health, as well as the health and well-being of others and of the natural, physical environment. In taking a mindful approach to both noticing and detaching from destructive emotions, such as anger, greed and hatred, we reconize these as forms of ego-cilnging and as distortions of the actual experience at hand (Ricard 2003). The accelerating experience of the global interdependence of the human condition and the downsides of the complex modern narratives associated with global warming, environmental degradation, and endless warfare convince us of the need to develop programs that foster the essential human capacities of wisdom and compassion. With this goal in mind, it is important to remind ourselves that induction into the Buddhist cannon is not necessary for understanding the Four Noble Truths (including the nature of suffering and how to relieve it). Aside from teachings that are specifically Buddhist, the unfolding of the Eightfold Path occurs in a natural, causal progression, reflecting the intricate and essential connections between human cognition, ethics, and behavior.

Deriving Right Action and Right Speech from Awareness of Interdependence

Einstein's famous dictum that "You cannot simultaneously prevent and prepare for war" might lead to the less eloquent corollary that you cannot practice right mindfulness and at the same time engage in violent speech and actions premised on the view that 'self' is essentially separate and fundamentally different from 'other.' This mistaken view is the first step toward *miccha sati* and provides the underlying rationale for all forms of aggression, combat, and for human and planetary destruction. While the sustained exercise of right mindfulness does not erase the distinction between self and other, it supports the softening of boundaries that otherwise reinforce and perpetuate the grounds for suffering.

The link between mindful awareness and adoption of prosocial values, such as increased democratization, human equality, and interdependence, is present in the original Dharmic association of mindfulness with wisdom and ethics. Interventions directed at liberation or transformation at the interpersonal level therefore necessarily involve encounters within the dimensions of *chronos* and intentionality, including exploration of conflict, incorporation of systematic learning, and engagement in many-sided debates.

In working on mindfulness with families and in classrooms, subjects encounter many such "fixed" boundaries that initially appear as impermeable. Parents sometimes begin by asserting hard rules intended for the protection of their children, only to recognize that granting increased levels of autonomy and independence to adolescents may be the most effective means for maintaining the integrity of the parent-child relationship, as well as for creating a dynamic basis for safety.

Working in middle school and high schools, we observe that these environments are fraught with many partitions enclosing students within the confines of in-groups and cliques. In a recent study within a semi-urban high school setting, one student remarked, "I have to constantly be proving how cool I am. Well, I am a cool dude. But the pressure of being in front of other people is a great stressor" (personal communication, June, 2015). Although the power of peer pressure can operate like the invisible air pressure that prevents a house from collapsing, ultimately the defensive energies that it generates support the reification of a cardboard cut-out persona, while limiting access to the full dimensionality of one's whole being.

Imagining the Future

If one were to attempt to titrate the original teachings of classical Buddhism to extract the quintessential components of a curriculum in right mindfulness, the resulting imperatives would include not only stress reduction, but also perspective-taking and the exercise of building peaceful and egalitarian communities. Kabat-Zinn (1990) argued "when we are able to mobilize our inner resources to face our problems artfully, we find we are usually able to orient ourselves in such a way that we can use the pressure of the problem itself to propel us through it, just as a sailor can position a sail to make the best us of the pressure of the wind to propel the boat" (p. 3). Beginning with adolescence, the nurturing of this combination of awareness, attention and mobilization of inner resources can be a potent prescription for personal growth and ethical action.

If mindfulness practice for teens, even for a few brief moments, allows them to experience extension beyond the constraints and boundaries of the "cardboard stick figure" of the solid self, might it also enable an enhanced sense of agency as they face the pressure-reinforced environment of their peers? Can it lead to recognition that these experiences are not solely individual and personal but part of a shared space, one that others experience with equal discomfort and with equal potential for good? In this way, education on the nature of interdependence has the potential to dissolve judgmental reactivity and to build new forms of collaboration with our neighbors and with our planet.

A central challenge in education is nurturing appreciation for the interdependence of all beings. Facilitating such awareness and action in youth can also support environmental awareness that deepens commitment and caring for the natural environment, for preservation of forests and wildlife and for reversing the trend of global warming.

Promoting secular, universal interventions with families, teachers, and youth in classrooms and community centers may provide new types of learning opportunities. Elsewhere we have elaborated on how nurturing the interpersonal aspects of mindfulness with parents (Coatsworth et al. 2014, 2015) and teachers (Jennings et al. 2013) can facilitate aspects of right mindfulness, including caring and compassion. Nurturing intra- and interpersonal skills, through practices such as loving-kindness (Fredrickson et al. 2008), Tong-len (the Tibetan practice of giving and receiving), and "just like me" (Broderick 2013) provides the opportunity to reach beyond our individual finite selves. Utilizing secular ethical principles can be effective in directing the chain of cause and effect toward beneficial outcomes across diverse social and cultural contexts.

Here, we offer some preliminary thoughts about secular principles and practices that may contribute to developing a common framework for such interventions (Greenberg and Mitra 2015):

(i) discernment exercises that extend the "what is" to "what is beneficial" when developing human capacity;

(ii) creation of a learning or developmental space to encourage inquiry into and critical evaluation of the causes of suffering;

(iii) skillful examination of the grounds of human motivations and intentions;

(iv) cultivation of respect for serious study and learning essential to development on the path of wisdom;

(v) removing the hindrances and obstacles to safe, healthy, and democratic modes for organizing human communities;

(vi) including exercises to foster taking the perspective of the other;

(vii) leading with the premise that people are more alike than different, yet mindfully exploring, recognizing, and honoring differences;

(viii) cultivating nonaggression and peaceful modes of action,

(ix) developing mutual tolerance and respect for all faiths; and

(x) watering the seeds of compassion.

Conclusion

As educators, we suggest and evoke the future in our words and in our actions. Undoubtedly, both teachers and parents play a critical role in unlocking capacities of youth—capacities that can result not only in reducing suffering, but also in releasing the fruits of human potential in ways that go beyond what we, as individuals, can now imagine. The current debate is an opportunity to connect our shared sense of urgency for unlocking human capacity with the temperate and systematic articulation of ethical principles and the application of the lived experience of *dharma*. If successful, this fusion of *sati, sila,* and *prajna* can bear fruit in a curriculum of right mindfulness—in a manner that is tested, evidence-based, and scientific—but that points inevitably toward the end of suffering.

References

Ataria, Y., Dor-Ziderman, Y., & Bergovich-Ohana, A. (2014). *Consciousness and Cognition, 37,* 133–147. Retrieved from www.elsevier.com/locate/concog/, http://dx.doi.org/10/1016/concog.2015.09.002

Bergovich-Ohana, A. (2015). A case study of a meditation-induced altered state: Increased overall gamma synchronization. *Phenomenology and the Cognitive Sciences,* 1–16.

Bodhi, B. (2011). What does mindfulness really mean? *Contemporary Buddhism, 12*(1), 19–39.

Bodhi, B. (2015). Facing the great divide. Retrieved from http://secularbuddhism.org.nz/resources/documents/facing-the-great-divide/

Broderick, P. (2013). *Learning to breathe: A mindfulness curriculum for adolescents to cultivate emotion regulation, attention, and performance.* CA: New Harbinger

Coatsworth, D., et al. (2014). The Mindfulness-enhanced strengthening families program: Integrating brief mindfulness activities and parent training within an evidence-based prevention program. *New Directions for Youth Development, 142,* 45–58.

Coatsworth, D., et al. (2015). Can mindful parenting be observed: Relations between observational ratings of mother-youth interactions and mothers' self-report of mindful parenting. *Journal of Family Psychology, 29*(2), 276–282.

Dalai Lama, H. H. (1995). *The Power of Compassion.* New Delhi, India: Harper Collins.

Dalai Lama, H. H. (2011). Session 1. In J. Kabat-Zinn & R. Davidson (Eds.), *The mind's own physician: A scientific dialogue with the Dalai Lama on the healing*

power of meditation (pp. 21–66). CA: New Harbinger.

Damasio, A. (2000). *The feeling of what happens: Body and emotion in the making of consciousness.* NY: Mariner Books.

Decety, J., & Chaminade, T. (2003). When the self represents the other: A new cognitive neuroscience view on psychological identification. *Consciousness and Cognition, 12,* 77–596.

Dennett, D. (1992). The self as a center of narrative gravity. In F. Kessel, P. Cole, & D. Johnson (Eds.), *Self and consciousness: Multiple perspectives.* Hillsdale, NJ: Lawrence Erlbaum.

deSteno, D. (2015). The kindness cure. Retrieved from http://www.theatlantic.com/health/archive/2015/07/mindfulness-meditation-empathy-compassion/398867/

Dor-Ziderman, Y., Berkovich-Ohana, A., Glicksohn, J., & Goldstein, A. (2013). Mindfulness-induced selflessness: A MEG neurophenomenological study. *Frontiers in Human Neuroscience. 7*(582), 1–17. doi:10.3389/fnhum.2013.00582

Flanagan, O. (2003). The Western perspective. In D. Goleman (Ed.), *Destructive emotions: How can we overcome them? A scientific dialogue with the Dalai Lama.* New York: Bantam.

Fredrickson, B., Cohn, M., Coffey, K., Pek, J., & Finkel, S. (2008). Open hearts build lives: Positive emotions, induced through loving-kindness meditation, build consequential personal resources. *Journal of Personality and Social Psychology, 95*(5), 1045–1062. doi:10.1037/a0013262

Gallagher, S. (2000). Philosophical conceptions of the self. *Trends in Cognitive Science, 4*(1), 14–21.

Goleman, D. (2015). *A force for good: The Dalai Lama's vision for our world.* NY: Bantam.

Greenberg, M. T., & Mitra, J. L. (2015). From mindfulness to right mindfulness: The intersection of awareness and ethics. *Mindfulness, 6,* 74. doi: 10.1007/s12671-014-0384-1

Grossman, P., & Van Dam, N. (2011). Mindfulness, by any other name: Trials and tribulations of sati in Western psychology and science. *Contemporary Buddhism, 12*(1), 219–239.

Halifax, J. (2012). A heuristic model of enactive compassion. Retrieved from http://www.supportiveandpalliativecare.com

Kabat-Zinn, J. (1990). *Full catastrophe living: Using the wisdom of your body and mind to faces stress, pain, and illness.* NY: Dell.

Kabat-Zinn, J., et al. (2011). Session 1. In J. Kabat-Zinn & R. Davidson (Eds.), *The mind's own physician: A scientific dialogue with the Dalai Lama on the healing power of meditation* (pp. 21–66). CA: New Harbinger.

Jennings, P. A., Snowber, K. E., Frank, J. L., Coccia, M. A., & Greenberg, M. T. (2013). Improving classroom learning environments by cultivating awareness and resilience in education (CARE): Results of a randomized controlled trial. *School Psychology Quarterly, 28,* 374–390.

Legrand, D., & Ruby, P. (2009). What is self-specific? Theoretical investigation and critical review of neuroimaging results. *Psychological Review, 116*(1), 252–282.

Monteiro, L., Musten, R., & Compson, J. (2015). Traditional and contemporary mindfulness: Finding the middle path in the tangle of concerns. *Mindfulness, 6*, 1–13. doi:10.1007/s12671-014-03101-7

Ricard, M. (2003). A Buddhist psychology. In D. Goleman (Ed.), *Destructive emotions: How can we overcome them? A scientific dialogue with the Dalai Lama*. New York: Bantam.

Ricard, M. in Singer, W., et al. (2011). Session 5. In J. Kabat-Zinn & R. Davidson (Eds.), *The mind's own physician: A scientific dialogue with the Dalai Lama on the healing power of meditation* (pp. 173–205). CA: New Harbinger.

Ruby, P., & Legrand, D. (2007). Neuroimaging the self? In Y. Rossetti, Y. Haggard, & M. Kawato (Eds.), *Sensorimotor foundations of higher cognition (22nd attention & performance meeting)* (pp. 293–318). Oxford: Oxford University Press.

Salzberg, S. (2002). *Lovingkindness: The revolutionary art of happiness*. Boston: Shambala.

Scharmer, O., & Kaufer, K. (2013). *Leading from the emerging future: From ego-system to eco-system economies*. San Francisco: Berrett-Koehler Publisher.

Siegel, D. (2012). *The developing mind: How relationships and the brain interact to shape who we are* (2d ed.). NY: Guilford Press.

Singer, W., et al. (2011). Session 5. In J. Kabat-Zinn & R. Davidson (Eds.), *The mind's own physician: A scientific dialogue with the Dalai Lama on the healing power of meditation* (pp. 173–205). CA: New Harbinger.

Tirch, D. (2010). Mindfulness as a context for the cultivation of compassion. *International Journal of Cognitive Therapy, 3*(2), 113–123.

Vago, D., & Silbersweig, D. (2012). Self-awareness, self-regulation, and self-transcendence (S-ART): A framework for understanding the neurobiological mechanisms of mindfulness. *Frontiers in Human Neuroscience, 25*, 1–58. Retrieved from http://dx.doi.org/10.3389/fnhum.2012.00296

Varela, F., Thompson, E., & Rosch, E. (1991). *The embodied mind: Cognitive science and human experience*. MA: MIT Press.

Wallace, B., & Bodhi, B. (2006). The nature of mindfulness and its role in Buddhist meditation. A correspondence between B. Alan Wallace and Bhikkhu Bodhi. Retrieved from http://shamatha.org/sites/default/files/BhikkuBodhiCorrespondence.pdf

Watson, G. (2000). I, mine and views of the self. In G. Watson, S. Batchelor, & G. Claxton (Eds.), *The psychology of awakening* (pp. 30–39). ME: Samuel Weiser Inc.

Community-Engaged Mindfulness and Social Justice: An Inquiry and Call to Action

Rhonda V. Magee

Introduction

In this chapter, I argue that mindfulness, which I define here as a state of awareness with compassion that may be cultivated by human beings (and the variety of practices that engender this state), must include practices and teachings that make explicit the links between mindfulness and social justice. Drawing on my experience within the fields of mindfulness teaching, law teaching, and contemplative pedagogy, in the first part of this chapter, I discuss how the practices we call mindfulness tend to cultivate a felt sense not only of interconnectedness and compassion but also of *solidarity*—relative agreement in feeling or action (especially among individuals with a common purpose)—among practitioners, that assist us in working together for a more just world.

Notwithstanding strong indications of the power of mindfulness to promote positive social engagement, mindfulness teachers tend to focus on the personal practices of mindfulness, and perhaps the importance of group practice for deepening awareness, but eschew a focus on the potential for mindfulness practice to support social justice projects in the world. In the next part, I show how failing to explore the systemic justice issues that arise both in mindfulness practice

communities *and* in the broader social systems within which we practice renders the practices of limited value across a range of contexts and social groups, including perhaps especially traditionally marginalized people and communities. Since individuals and communities tend to suffer greatly not only due to attachment, aversion and ignorance but also as a result of systemic oppression and structural violence, we are often drawn to practices that inspire compassionate action to alleviate systemic and structural suffering as well. Thus, I suggest that mindfulness teachers and practitioners should explore, embrace, offer, and remain open to receiving the gifts of what might be called "*community-engaged*" mindfulness practices across all of the settings in which mindfulness is currently taught. Toward that end, I describe and present an exploratory Case Study in the elaboration of such a set of practices, reflect on challenges and opportunities presented there, and suggest future directions for research and teaching among students and practitioners of Contemplative Studies and Science.

Mindfulness and Social Justice: An Inquiry

For some, mindfulness practice inherently raises awareness of our inherent interconnectedness. For others, such awareness must be specifically cultivated. Whether inherently so or not, in my personal experience, and as I have observed

R.V. Magee (✉)
School of Law, University of San Francisco, San Francisco, CA, USA
e-mail: rvmagee@usfca.edu

© Springer International Publishing Switzerland 2016
R.E. Purser et al. (eds.), *Handbook of Mindfulness*,
Mindfulness in Behavioral Health, DOI 10.1007/978-3-319-44019-4_28

among my own students, mindfulness supports increasing awareness of my interconnectedness with so-called Others in both the human and beyond-human worlds.

That mindfulness may increase our lived experience of interconnectedness is important. Many social justice theorists believe that the sense of interconnectedness is the central insight that supports compassionate action in the world. They share this insight with many long-term practitioners of contemplative practice. Thus, it is worth exploring whether and to what extent contemplative practices aid in the development of consciousness that best supports the sustained will to work with others on behalf of self-with-and-for-others.

In the United States, mindfulness practice is most often presented as an individual, *personal* practice, with an emphasis on its capacity to increase well-being and enhance psychological flexibility and executive functioning. Indeed, in most places where secularized mindfulness is taught, there is little if any emphasis on the relational dimensions of mindfulness, or interpersonal mindfulness. Even where mindfulness is offered in more traditional settings, to the degree that the sangha is an important dimension of practice, little is explored in the way of collective action for social justice.

The failure to explore the relational and systemic dimensions of mindfulness in most settings may be attributed to the common reliance on Buddhist modernist adaptations of teachings of the fifth century B.C. teacher known as the Buddha. As presented in the West, those trainings emphasize personal practice. While even the early teachings of the Buddha offered counsel on such issues as the proper distribution of wealth, maintaining social harmony, and interpersonal practice, modernist adaptations have for the most part ignored the social and ethical dimensions that appear to have been as important as meditation and mindfulness training to the early teachers in the Buddhist tradition.

And yet, the ethical commitments that have long supported development along the path—the precepts that call for refraining from harming living beings, stealing, sexual misconduct, lying,

and becoming intoxicated—are not merely personal but interrelational and interpersonal commitments. They call upon practitioners to bring awareness to the ways that our relations with other beings and people affect the quality of our own experience: our own experience of suffering and that of others in the world.

While the precise formulation of the Four Noble or Ennobling Truths may vary, the received teachings focus on raising awareness of and comprehension of suffering; of the arising of suffering; of the cessation of suffering and the freedom it brings; and of our capacity to cultivate the path to liberation (Batchelor 2015). This focus on suffering, its causes, and our capacity to end suffering by ceasing reactivity applies not only to our personal and interpersonal experience, but also to our work within systems that create and maintain systemic suffering and structural violence. As Stephen Batchelor put it:

> To ground mindfulness in the fourfold task means to keep these ideas in mind and apply them to illuminate whatever is taking place in our experience at a given time and place.... When *The Grounding of Mindfulness* describes mindfulness as the "direct path to nirvana," it affirms that paying attention to life leads to a falling away of habitual patterns....Nirvana is reached by paying close, uncompromising attention to our fluctuating, anguished bodies and minds and the physical, social, and cultural environments in which we are embedded.[1]

Moreover, the sangha—and the community of practitioners of mindfulness whose efforts to live well together serves as a means of awakening—is clearly central to the Buddha's teachings. For example, in the well-known story of the disciple Ananda's conversation with the Buddha regarding the role of friendship, in which Ananda posited as "half of the holy life," the Buddha reportedly corrects Ananda, saying that "friendship is all of the holy life."

What if mindfulness practice were universally presented as a means of deepening both personal and interpersonal well-being and liberation? For example, what if the "all of the holy life" teaching story was seen and presented as an

---

[1]Batchelor (2015), p. 240.

invitation to inquire deeply into the nature of the "friendship" referenced therein? Surely, there may be dimensions of the notion of friendship meant to be conveyed by the term that resist the ready definitions for the term that emerge in the Western mind. This seems especially so given the great degree of emphasis and importance that the Buddha seemed to be placing on the notion of friendship here. If this is plausible, two questions seem naturally to arise:

(1) Might the term refer to something more akin to solidarity than to that which we think of as "friendship?"
(2) And if even possibly so, what might that entail for mindfulness-inspired social justice projects in the world?

As suggested above, according to the Oxford English Dictionary, solidarity means "unity of agreement in feeling or action, especially among individuals with a common interest or members of a group." Given the nature of the common engagement with the dharma that arises naturally in a sangha, at least some degree of "unity of agreement in feeling *or* action" seems highly likely to arise in sanghas as a matter of course. And this is likely true not only of traditional sanghas, but of the less traditional "sanghas" by which group mindfulness is often defined in the West.

Whatever the logic of such a proposition, in contemplative inquiry, it fails the test of what might be called truth to the degree that it does not comport with our lived experience. Fortunately for me as I reflect on this set of questions, I can draw on personal experience with many sanghas—formal and informal—in which such a feeling emerged.

Indeed, sangha practice tends to generate among participants a feeling of support and well-being unlike most anything else. This is so whether the "sangha" has been working together for many years or is instead a temporary community of practice and learning, such as that arises among participants in relatively short-lived mindfulness-based workshops and retreats. In setting after setting, participants report that group practice tends to generate a sense of the strength and re-awakening to the power of community. At appropriate times, I highlight and amplify the community dimension as it arises in the contemplative or mindfulness teaching and learning communities I am fortunate to guide and support in secular settings. For example, in the context of a course on mindfulness for law students taught at the University of San Francisco, a student shared the following with me in an email after a class discussion:

> In response to the opinion that mindfulness is useless against, i.e. the KKK, you said that those who are mindful, who are capable of personal development, can have a community of our own, help who we can (or at least that's what I remember).
>
> I'm still kind of floored by how positive a revelation this is to me. I think my original stance, as expressed in my reflection, was mostly based out of a sense of loss of community and a deep fear of what I expect will be an escalation in violence as the white community further unravels. I'll spare you the personal details, but, the white community hardly felt like a community in the first place, and this apparently "official" split leaves me "between communities" in the sense that one is "between jobs" after mass layoffs.
>
> So with all that… I wanted to thank you for responding to my fear and ignorance by reminding me that "community" is not limited to one's bio family and childhood friends.

Thus, we should expect mindfulness teachers at least on occasion and perhaps as a matter of course to bring mindfulness to that dimension of individual and group experience—to suffering and to solidarity—and to actively explore ways of bringing mindfulness to bear on real problems in real life.

Unfortunately, such explorations are not common. A survey of mindfulness teachers, a sampling of mindfulness training venues, and a visit to any center of practice would reveal that social justice is often not associated as core to the practices of mindfulness. Instead, social justice is often seen as something separate from, and optional to, mindfulness practice and mindfulness in the world.

This fact is exacerbated or perhaps pre-figured by the contexts within which Western mindfulness emerged: predominantly among white, male, and upper class students of Buddhism with a dream of taking the practices into the world. Given the relatively privileged backgrounds of many of the original teachers and practitioners of mindfulness in the West, it is easy to see why the practices have become largely if not primarily associated with personal well-being and productivity, and not social justice.

For this reason, mindfulness practices are often perceived as more or less unavailable to or unhelpful for members of traditionally marginalized communities. In my experience teaching and facilitating groups discussing the use of contemplative pedagogy or mindfulness, I am often approached by one of the few attendees of color who poses a question like, "How might I take these into the community where I teach/live/work?" followed by a signifier that implicitly or explicitly suggests a host of cultural and social challenges. Most recently, for example, the question was put to me by a Latina professor who was exploring bringing mindfulness into her sociology classes to deepen students' learning and development together. When she wondered about "how to take this to my students," she included a reference to the place where she lived: "San Jose." Similar questions have been raised to me by people seeking to take these practices into low-income communities of color; Native American/American Indian communities; and, indeed, to groups of lawyers or teenagers.

That these questions arise again and again appears to confirm that culture and context matter to the teaching of mindfulness. Context and culture matter deeply to the development and delivery of practices to be brought to bear to assist suffering individuals, families, and other social groups within communities of any kind—suffering whether it be existential or social/structural in nature. It is especially troubling, however, that these questions of culture and context may render practically inaccessible the practices of traditional mindfulness to those communities who might need them the most.

Community-Engaged Mindfulness as a Response

In contrast, some teachers in the West have discussed an approach to Buddhism that focuses on engagement with the problems of the social world. For example, Vietnamese master Thich Nhat Hahn emphasized the ways that Buddhism supported social action aimed at alleviating suffering in the context of the Vietnam war and found common ground with Martin Luther King in the fight against the triple evils of militarism, capitalism, and racism. Sometimes using the term "Engaged Buddhism," teachers have been known to advocate taking the practices of mindfulness into direct engagement with issues of social justice and inequality in the world, and/or, to communities where the practices are not typically available. Elaborated by Donald Rothberg (as "Engaged Spirituality"), Fleet Maull, and others, engaged approaches explore means of interpreting the dharma and spirituality more generally not merely as a paths to personal awakening, or inner work, but to awareness that extends to engagement in the world aimed at redressing social suffering and injustice through structural oppression (Rothberg 2006).

Relatedly, Jon Powell and others have examined some of the ways that social justice itself is spiritual practice and as such may inform traditions of practice not typically seen as being "about" social justice—including, perhaps, Western mindfulness (Powell 2003). At the 2014 International Symposium for Contemplative Studies, Powell encouraged students and researchers in the Contemplative Studies community to explore ways of examining justice and injustice in the contexts in which we find ourselves, exploring our own lived experience of othering and belonging, and opening to the dimensions of spirituality and ethical engagement that exist in diverse communities.

In the work of advocates such as these, the "engagement" dimension may be said to operate simultaneously on three levels: personal, interpersonal, and systemic. Personal practices increase awareness and ethical approaches to life that lead to empathy, compassion, sympathetic joy, and equanimity and extend to awareness of the experiences of those around us (the interpersonal dimension). Ultimately, the commitments to non-harming and right relationship that are at the core of Buddhist ethics tend to heighten the sense of interconnectedness upon which compassionate action often arises, and appear to support work against structural inequality and oppression in our midst (Batchelor 2015). One way to distinguish the more socially conscious approach to teaching and practicing mindfulness that may result, and to link it to efforts to engage specific communities in need is by use of the term "community-engaged mindfulness."

I define community-engaged mindfulness as the discipline and practice of bringing mindfulness—awareness with compassion—into engagement *in community*, using and adapting mindfulness and compassion practices as aids in community-engaged, social justice work. Community engagement—working with real people in various geographic or otherwise loosely bound collectives to support awareness of community resources and to deepen capacity to work and thrive together—provides the opportunity simultaneously to the following:

- develop the personal dimension of our own capacity to work with and learn from our own experiences, including experiences of social suffering and to learn about the structural nature of the suffering of others (the *personal* dimension);
- offer and receive supportive practices and collaborate across lines of real and perceived cultural, racial, and other differences (the *interpersonal* dimension); and
- work with others to relieve suffering at all levels, including the material and structural–institutional (the *systemic* dimension).

As long-term practitioners may readily sense, what I'm describing here may not be a new approach to mindfulness at all. Instead, it may be seen as an elaboration of an orientation that might naturally evolve from the practice of mindfulness in ways that gradually expand our circles of compassionate concern for others. It accords with Stephen Batchelor's vision for an ethical, secular Buddhism grounded in, among others, a "commit[ment] to an ethics of care, founded on empathy, compassion, and love for all creatures" in which "[p]ractitioners seek to understand and diminish the structural violence of societies and institutions as well as the roots of

violence that are present in themselves."[2] And yet, because we so seldom encounter discussions of these dimensions of the mindfulness perspective in mainstream settings, it may be helpful to set forth community-engaged mindfulness as a dimension of the practice and invite consideration of what such a dimension might look like. In the following pages, I describe one preliminary effort at offering and exploring the efficacy of a form of community-engaged mindfulness in response to a traumatic incident involving race and policing in a major American city.

Community-Engaged Mindfulness: An Exploratory Case Study

In March 2015, the San Francisco press revealed that a number of San Francisco police officers had engaged in an exchange of racist text messages.[3] Shortly thereafter, the San Francisco District Attorney's office called a Community Meeting to discuss the D.A. office's concerns and to hear the concerns of the community. As a resident of San Francisco's Black community with a history of local community service, I was invited to attend this meeting.

The majority of the invitees and attendees were African American. A smaller number were Latino. At that meeting, members of the community asked for a community-wide series of

---

[2]Batchelor (2015), p. 322.

[3]For the official document containing the reported offensive messages, see "Government's Opposition to Defendant Furminger's Motion for Bail Pending Appeal," U.S. Dist. Ct., N.D. Ca., CR 14-0102 (March 13, 2015), available here: https://drive.google.com/a/usfca.edu/file/d/0B4pdvMvLhJfdQXNKTUt0R04tUUU/view. *See also*, "The Horrible, Bigoted Text Messages Traded Among San Francisco Police Officers," Gawker, March 18, 2015 (reporting, for example, the following messages obtained from the official record: "We got two blacks at my boys [sic] school and they are brother and sister! There cause dad works for the school district and I am watching them like hawks;" and, in response to a text saying, "Niggers should be spayed," [Former San Francisco Police Officer] Furminger wrote "I saw one an hour ago with 4 kids," and, "in response to a text saying," All niggers must fucking hang, "Furminger wrote" Ask my 6 year old what he thinks about "Obama.").

meetings to raise the community's awareness of the depth of the problems of bias in law enforcement and to develop means of addressing these problems and turning the tide. I was ultimately asked to assist the District Attorney's (DA's) office in facilitating these discussions.

The San Francisco DA was concerned about the possibility/probability that racism in policing leads, in a structured way, to bias in their prosecutions. This calls into question the level of bias infecting the whole system. For this reason, the DA elected to investigate the connections between bias and law enforcement. The facilitated discussions would be part of that investigation. The DA and his office hoped to obtain greater understanding of the issues as seen from the standpoint of the Black community, ultimately to assist them in confirming that they are not simply taking the biased reporting and policing of the police and building cases on them that lead to further subordination of brown and black people in SF. (A number of DAs spoke directly about that concern during the first facilitated discussion and healing circle, described below.) While members of the police department were invited to the sessions, the sessions were structured to give a sense of safety to the members of the community and not feel a sense of obligation to present a "balanced" hearing: these sessions were meant to be community centered.

I co-led the session with the District Attorney's Director of Community Relations, Assistant District Attorney Marisa Rodriguez, a USF Law graduate, and former student in my Torts class and in my Race, Law and Policy class. Marisa (and others in the DA's office who attended) had some experience in mindfulness and law, having participated with several of her colleagues in the DA's office in a series of mindfulness classes offered by a colleague in the mindfulness and law world, Judi Cohen. Other members of the community were not polled on their background in mindfulness. This was intentional, as it has been my experience that when offering these practices as a means of supporting work in community and challenging issues, it helps to look for ways of introducing them that seem consistent with community needs, not pre-planned or "canned." In pre-meetings with Marisa and various constituents within the community (religious leaders, DA's office, etc.), I determined that many in these intersecting communities were seeking a way of dealing with these issues that might support a sense of healing in the community and the building of capacity to deal with difficult issues.

Two sessions were subsequently conducted: an opening, 5-h "Healing Circle," and, several weeks later, a follow-up two-h "Working-Together Session." Each of these were voluntarily attended by a broad cross-section of 40+ members of the San Francisco community with an interest in working together to address issues at the intersection of race and law enforcement. The practices introduced were more or less specific practices that I have developed to assist in Mindful Facilitation of Group Dialogue. These include the creation of a norm of pausing and creating space together; the setting of intentional guidelines for our discussion, such as mutual respect; listening with the intention of inviting the truth; fully transitioning into our space; and inviting silence as a support. In the Healing Circle Workshop, these included very lightly guided invitations to:

- "Mindfully Transition into and Enter, and Be Present in the Space,"
- "Mindfully Co-Create a Space for Respectful, Mutual Healing Community," and
- Engage in "Mindful Speaking and Listening."

In the "Working Together Workshop," we included the practices above, as well as a set of instructions for:

- "Mindful Small Group Work,"
- a Simple "Moment of Gratitude," and
- "Closing Awareness" Practice.

Summary of Outcomes

As the following few pages will reveal, when measured by the desire to infuse mindfulness practices unobtrusively but in ways that would support dialogue and community building, these interventions provide promising indications of potential successful interventions based on community-engaged mindfulness. When brought to bear with respect for the assets of the community and respect for their needs, these and similar practices may be introduced to assist in meeting communities in distress mindfully where they are and meeting them in ways that mindfully support them in strengthening and moving forward (Blum 2014).

Workshop Design and Participant Response

In collaboration with members of the DA's office, I developed two workshops. Each infused mindfulness practices in what might best be referred to as "stealth" or "low-threat" ways. The following is a brief overview of the design of each of these two workshops.

> "Opening Healing Circle": Mindfully Creating a Space for Individual and Community Healing.

The first of the two workshops was defined as a "Healing Circle" and was scheduled for 5 h (10 am–3 pm) on a Sunday. The intention with this session was to provide an opportunity for participants to come together for the healing engagement necessary to form an effective learning-and-working community in times of distress. For the purpose of this session, I defined healing as an opportunity to: (1) come to terms, compassionately, with things as they are (Kabat Zinn 2010) and (2) re-experience the self in a way that might promote an opening of the heart and a turning toward new possibility (Magee 2016).

Co-facilitator Assistant DA Marisa Rodriguez and I agreed on the desirability of paying attention to a variety of what might be called "Mindful Space" details. In terms of the basics, we agreed to arrive early to greet each participant with genuine welcome. And we arranged for light food (fresh fruit and small pastries, coffee, and water) on arrival and a buffet lunch (fresh salad and sandwiches) to be available by noon. In addition, we planned to facilitate mindfully throughout the day—pausing and creating space for all voices, amplifying inner wisdom as it arose, and "reading the room" to ensure that identity-safety was a reality.

I see the core of the work of creating Mindful Healing Spaces as being about the marriage of awareness and compassion, compassionate engagement with things as they are, including who we are, what we are feeling, and the ways we are feeling cutoff from our essential beauty, wonder, and nobility. Thus, I included a "Centering and Honoring" practice. Using candles, for their symbolic and energetic impact, I lit a candle and invited each participant to think of the candle as a reminder of their own inner light, of the sacredness of their own voices, and of the sacredness of our own space.

Following this, I briefly described a set of what we called "Ground Rules." (I have at other times referred to these as "Contemplative Commitments," or "Contemplative Community Agreements.") The purpose here was to invite a conversation about how we communicate with one another—with respect, giving equal time, etc.—and to invite "buy-in" with this idea through the process of discussing and naming what we refer to as "agreements" by which these commitments are made manifest.

Poetry is a contemplative practice often used in the teaching of mindfulness. To honor the creative instincts of this community, and to include the energetic voice of the young, we included a published, spoken word artist, who offered an emotionally powerful piece of poetry (see Appendix).

I then invited members of the circle to silently reflect on what brought them to the event. I invited a quiet centering on the breath, on the sense of the body in this space, and on the support in the room for each of them. From this place of support, I invited them to hold the following questions, aimed at increasing self- and interpersonal awareness, and a focus on the assets in the group and to notice what arose in response:
1. "What brought you here?"
2. "What resources could you share in working together to address those concerns which lie behind your motivations?"

Following this period of reflection, I invited participants to speak into the circle. They would be timed, so that no one would speak longer than

2 min, and we would use a mindfulness bell to keep the voices moving around the circle. Although I indicated that one might pass, nearly everyone took the opportunity to speak into the circle, and we ended up spending the entire Workshop in the Circle format engaged in structured and timed Mindful Speaking and Listening.

For reasons of confidentiality and concern for creating a sense of complete safety, I chose not to record the proceedings electronically, and avoided taking copious notes. However, the notes I did take revealed common concerns about race-based policing, such as the following:

> While crime exists in all communities, [it's] mostly Black/Brown people being prosecuted.
>
> [Noting experiences of] random stops because of stereotypes
>
> As a member of the DA's office, it's challenging doing the work knowing that the system has bias, knowing that there is racial injustice and still doing the work of a prosecutor.
>
> This is a painful period. Feeling the need of community.
>
> [As a] "formerly incarcerated," I have an abolitionist perspective. We need less policing, less incarceration. Because the prison system perpetuates racism. [At San Quentin] as soon as you arrive, you have to state your race. The whole time you're there, you are categorized by race. You are housed with other people according to your race. It's the same as Jim Crow: everything is segregated. I tried to push back against this while in prison.
>
> [As a white woman] I am committed to working on exposing whiteness, and white privilege. Struck by a recent report of a 5-year-old Black child in kindergarten who had had the police called on him 3 times. Something has to be done. This is why I'm here.
>
> [Appreciating the sense that] There are a lot of gifts here. Excited to be here.
>
> I'm a defense attorney. I bring lawsuits for people who've been injured by the police. When the issue finally reaches the court, or in trial, a lot of this stuff can't be spoken about. Need to climb upstream to begin the work.
>
> [A prosecutor] I think we are on the brink of a civil rights/human rights movement.
>
> [Community member] Colorblind is a lie.
>
> [Community member] Cops believe Black Lives Don't Matter.
>
> I became an assistant D.A. because of the issue of underrepresentation of Blacks in this part of the system. But at the same time, I am part of this system, and it needs to be fair. There's extreme bias in this system.

> I came to be here because my wife asked me to come. But now I realize my voice needs to be in this conversation. [Revealed racist texts] brought out sadness. We have to take these issues more seriously. I'm also a former victim of a crime. I was shot right here in this City over a cell phone by another Black man. Surgery [required efforts to address these issues] must take place all across the board.
>
> There was a White male police officer who was able to get to know all of the people in his [predominantly Black] beat. Want to know and share how that can be done.

At lunchtime, we took a short, half-hour lunch break and transitioned back into the circle with the benefit of another poem. After lunch, we returned to the Healing Circle format and continued creating space for sharing until all who wanted to speak had been heard and closed with a reflection on what felt gratitude engendering and healing about the experience and an inspiration (based on a Maya Angelou quote) to go forward and offer healing to someone else.

> "Working Together Workshop": Building Capacity for Engagement. Through Mindfulness-Based Personal and Interpersonal Practices.

This workshop was structured as a follow-up to the first session, and only those who particulated in the first session were invited to attend the second. It was organized and promoted as a "Working Dinner" and ran from 6:00 pm to 8:00 pm on a Tuesday evening. We provided a very light dinner of pizza and salad. The first half hour was intentionally designed to permit welcoming, support in settling into the space, and gentle instruction in mindful eating. From 6:30 pm to 6:45 pm, the District Attorney, Assistant DA, and the Dean, on behalf of the law school, framed the conversation and the effort. From 6:45 pm to 7:15 pm, participants were to reflect together according to small group self-facilitation instructions (choose a note-taker, choose someone to report out at the end, choose someone to ensure everyone gets a chance to speak) and guided by specific questions that were both spoken and available at each seat in a written document. The audience was divided into groups of 5 (using a "counting off" process to enable efficient but "random" group formation). Each group was given large format Post-It paper, and blank sheets of 8 × 11 in. paper, on which to take notes. The questions were the following:

- What problems do you see at the intersection of racial bias and law enforcement (profiling? Cultural issues? Disrespect? Others?) that need to be addressed?
- What steps would you suggest be taken next to organize and identify community views and needs?
- What specific policies or practices would you suggest be called for from City leadership?

From 7:30 pm to 7:50 pm, we facilitated sharing out from each group, and at 7:50 pm we turned toward closing comments and a brief gratitude reflection, aimed at having each person identify some aspect of the evening for which they were grateful, bring it to consciousness, and allow that to serve as inspiration for the continued work ahead.

Participant Responses

We built time into Session One to obtain written feedback during our time together. Based on feedback from participants, the first session accomplished the goal of helping people come to terms with things as they are and turn toward the work to be done with a sense of possibility. We were not able to get the scheduled "Working Lunch"—due to the decision to permit the "Healing Circle" to continue until everyone who wanted to speak could speak. Mindfulness-based practices, while often somewhat stealth in their delivery, seemed critical to the success of this event. The practices seemed especially effective in assisting attendees to feel supported and to feel that their voices mattered (as indicated by the feedback comments, below).

To underscore the importance of the participants' feedback on the day, we set aside 15 min toward the end of the session to enable participants to complete evaluation forms. In consideration of the short format of this Chapter, I have included only a representative sample of the responses to two of the most relevant questions, below.

Question [2]: "Thinking about the Sessions (Opening Circle, Working Lunch, and Closing Share-Out) what was the most significant take-away for you?"

The following are the first 5 and several of the remaining most poignant of the responses:

(1) Opening circle
(2) Continued dialogue and brainstorming
(3) That we still have a lot of work to do
(4) Opening the circle—strong, positive, and constructive; you set boundaries and guidelines to facilitate meaningful dialogue
(5) The respect we share in the room was significant.

And:

(1) That each person wants to make a difference and be the change we want to see.
(2) Sometimes, it is important to just have a space to speak from heart and not try to move a structured agenda.
(3) I know my vision to young black…they are from a beautiful people… They should be more about do not be afraid of being a man "boy." Thank you.
(4) Meeting comrades-in-arms for justice.
(5) "You are not alone." The open circle was healing to hear everyone express how they feel.
(6) That SFDA (San Francisco District Attorney's office) and USF (the University of San Francisco) are collaborating to create a medium for listening to community members' concerns.
(7) The most significant take-away is how the open circle supported and encouraged personal and sometimes painful experiences.
(8) There is an abundance of hurt/pain that has to be tucked away and lost within the criminal justice system but we have a chance to change it.
(9) This is the start of a new process. Wondering what will become of it.
(10) Obviously—circle.
(11) I am not alone! Please, if you can, keep doing this.

Question [4]: "What did you like best/enjoy the most about the Forum?"

The following are the first 8 of 31 similar comments taken verbatim from participant evaluation forms:

(1) Having this event in the first place.

(2) The facilitator and methods to facilitate this dialogue; collaborations with the D.A. office and USF; location and venue.

(3) I enjoyed listening to everyone's story and the reasons they came to the forum.

(4) Everything.

(5) The feelings in the room were palpable. I appreciated the candor and vulnerability in the room; the need for continued discussion regarding the intersectionality of race, poverty, and law enforcement.

(6) Appreciated the circle format; flexibility with energy level and pacing were very thoughtful.

(7) Range of folks present; opportunity to sit in large group and witness/hear each other a good beginning.

(8) That people were allowed to be honest and truthful and to speak from their personal experiences....

Feedback forms indicate that of the more than 40 people in attendance, only one or two people were disappointed that we took the time to go around the Circle and allow everyone to be heard. Most of the group found the time devoted to mindful listening and speaking to be essential, valuable, and even healing. But indeed, doing so took a significant amount of time, and the will to do so emerged from deep concern for the community. Mindfulness-based interventions of this sort require both time and genuine compassion, indeed love, for the participants.

As one final indication of how well the first workshop went, a great majority of those who attended the first workshop responded to our next call and attended the second. At the second, too, subtle, deep mindfulness seemed most effective in helping frame the work of the evening so as to build on community resources with respect. We ended up scheduling the second session as a Working Dinner to take up that part of the scheduled work that we were not able to accomplish in the first 5-h session. The group dialogues went extremely well, yielding both thoughtful notes on commentary regarding a difficult topic and a sense of hope, indicated by the will to continue meeting again.

During the second (shorter) session, we circulated feedback forms afterward (rather than during the session), which unfortunately were not sent out until a week after the event. As might be predicted, we received relatively few returned forms, but among those, the responses were quite good.

The following were taken verbatim from one of the six small group's (Group Two's) submitted notes:

Group Two Notes

A. Problems

1. [Community members feel] disrespected, and fear
2. Police feels unstoppable—[community members feel] owned
3. [Police] Feel above and feel that people are beneath
4. Assumptions [Bias]
5. Chance [risk] mindset
6. [Outsiders are] In denial because it does not affect them

[Proposed Solutions]

1. Find root problem
2. [Address] racial hatred
3. [Illegible] evaluation/questionnaire
4. Race is a Factor
5. Make officers do community service
6. Divers[ity] race[s] among/between police partners
7. [Subject police to] psychiatric evaluation
8. Screenings
9. Closely watch new officers when hired
10. Examination of [Police Commission] charter
11. Evaluate officers → see whether there are any patterns
12. [Biased questioning] "Are you on parole or probation?"
13. Policies—beliefs
14. [Policies] revised 10/11 years ago—needs [sic] to be revised again
15. Subpoena police officers

As with the first session, we concluded with an exercise intended to bring people back together, to center, and to reflect with gratitude on the experience.

Among the three feedback forms returned via email were the following two representative sets of responses:

Participant Feedback One:

Law Enforcement, Race, and Justice Forum: Part Two Feedback Form

Date: July 28, 2015

Take a moment to provide a short set of responses to assist us in improving our offerings.

1. What did you like best/enjoy the most about the Forum?

   I liked the structure and use of our limited time that was used to generate excellent input from the community members present.

2. What is the most significant take-away for you?

   I was encouraged by the commitment I saw from many in the room to help lead change in policing. I was very grateful for the DA's opening remarks that created strong support and emphasis on community leadership.

3. What topics raised today, or related topics, Would you like to see the SFDA and/or USF focus on in future events?

   I would like to hear how we can change the culture of the Police. I would ideally like explore a reliable feedback system for the Community that would provide specific feedback on the quality of each police contact with a community member. This process would be intended to provide feedback on service (Procedural Justice) and not intended to elicit complaints against the police. We would begin to learn which officers create positive contacts with community members and those who consistently miss doing that. It would provide critical information that would be both qualitative and quantitative. We now have the technology to do this at a reasonable cost. This effort would begin to place quality service as a top priority. Focus would first be to implement it in diverse communities.

4. What, if anything else, would you suggest as areas for improvement?

   Just more time and participation of more community members.

Participant Feedback Two:

1. What did you like best/enjoy the most about the Forum?

I liked that the Forum was held in a collegiate setting with a cross-section of people from different walks of life (attorneys, students, religious and community leaders, professors, and other concerned citizens) and ethnicities sharing their experiences and points of view regarding this issue. I also liked that the fact that we, as a group, recognized that the "text messages" that initially brought this Forum together is a symptom of larger discriminatory behavior fueled by bias.

2. What is the most significant take-away for you?

The most significant take-away for me is that people care enough about this issue and are willing to find meaningful and substantial ways to address it.

3. What topics raised today, or related topics, would you like to see the SFDA and/or USF focus on in future events?

Include implicit bias training for officers (if it is not being done already): recognizing that we are all guilty of this in some respect, yet being aware of it so judgment in various situations is not clouded or blinded by it. Also, someone in my work group brought up a good suggestion: In addition to law enforcement officials sitting on an oral board for PD [police department] applicants, have a citizen or community member included as well (if this is not being done already). This person would be able to pose questions to applicants.

4. What, if anything else, would you suggest as areas for improvement?

None at the moment. This is a good foundation to build on.

In short, each of these first two sessions indicates the value of mindfulness- and compassion-based interventions to support the mindful healing and working together of members of communities in distress. The first Session's intersection of mindfulness with "spoken word" or "rap" performance seemed to meet the need for a space in which to cultivate skills of mindful communication and the healing experience of speaking from the heart, in ways that felt authentic, and being listened to in a large group setting. The second Session seemed to have benefited from the grounding in mindfulness accomplished in the first and resulted in suggestions that were turned over to the DA's office and provided a basis for further investigation into the nature of the problems, from the perspective of community members, and some assistance in sorting through possible effective responses. The feedback from Session One indicates that the approach led to a felt sense of compassion, empathy ("The respect we share in the room was significant," said one participant) and solidarity ("I am not alone!" said one participant). The feedback for Session Two indicated that the approach supported the renewed commitment to working together to diminish the structural violence threatening this community. Future research, including pre- and post-intervention surveys of participants with specific reference to the value of the stealth mindfulness practices deployed, will be important next steps in confirming the value of these interventions. The possibility for continued engagement in this project continues. Moreover, as discussed below, additional exploratory community-engaged mindfulness interventions will permit the exploration of means of deepening the delivering of appropriately tailored mindfulness or other contemplative practices suitable to culture and context.

Reflections and Suggestions for Future Research

The mindfulness-based interventions infused in the workshops explored in this partnership between the San Francisco DA's office and the University of San Francisco School of Law demonstrate the potential efficacy of mindfulness-based interventions to support healing and strengthening of communities. If made available to a wider population, these practices could provide the basis for deep healing and the new beginnings at the intersection of race and justice that so many so deeply need.

The above intervention demonstrates one model by which community-engaged mindfulness might be elaborated. Another approach might include the development of a mindfulness-inspired program for a specific community. As this Chapter goes to print, community-engaged mindfulness projects are in the planning stages for communities as diverse as San Francisco, California, where I will offer an introduction to mindfulness to community advocates at the African American Art and Cultural Center later this year, and, at the University of Virginia, where I am presently collaborating with the Contemplative Sciences Center to develop and offer a program targeted to address the needs of African American students. In each of these, I intend to experiment with additional ways of supporting mindfulness-based practices that deepen the sense of meaningful engagement while drawing on the contemplative practice experience and commitments that already exist within and strengthen diverse communities.

Conclusion

Community-engaged mindfulness provides one framework for exploring the social justice implications of mindfulness practices, including those derived from the Buddhist tradition. The exploratory study described here provides ample encouragement for the work of expanding our circle of mindful community and our engagement with projects aimed at ameliorating structural and systemic suffering. May it also provide a window into some of the ways that community-engaged mindfulness—and the concomitant infusion of alternative, diverse practices, commitments, and voices into more mainstream approaches to mindfulness—can deepen all of our mindfulness teaching and practices and create pathways to alleviating suffering in the world.

Appendix

ANTHONY BRANDON SMITH, **by Javier Reyes,** delivered by the author at Session One:

STREET MAN

This is how killers get made!

ENSEMBLE

Made!

STREET MAN

This is how killers get played!!

ENSEMBLE

Played!!

BOTH

Fast life, fast life, one man dies, one man cries

Don't think I didn't warn you 'bout that genocide

Kill a brotha, kill a sista, pull a trigga

Now your community is laying dead like an action figure.

MIDDLE WORLD

I once knew this boy named Anthony Brandon Smith, born May **1**, 1997, the oldest of **2** children. Abandoned at the age of **3**, staring **4** stories out of his project window,

Saw **5** bullets bully their way into his uncle's chest,

And crescendo with his death.

STREET MAN

6 months passed before his auntie collapsed,

Pronounced dead of a heart attack

After **7** days of shuffling,

He's struggling,

At an orphanage with his god-sister and little brother,

Who after **8** brutal months of abuse, he finds bloody and without color.

9 years since

And as a consequence,

Anthony won't be the same

He passes the time playing video games

The doctor gives him Prozac to slow down his brain.

STREET WOMAN

10 years go by with no glimpse of love,

He misses the kisses and hugs, cries himself to sleep and is slowly growing numb.

He learns addition,

Counting the **11** bruises his foster-father gives him.

Understands the language of anguish,

Listens as knuckles rip him,

Blood glistens, on battered skin,

Can't understand why he's in the hell he's in.

STREET MAN

12 midnight comes and he runs away with a vengeance,

But a violent instance has him serve a **13**-month sentence.

Juvie hall proves to be more friendly than home,

Anthony Brian Smith **14** years old and full grown.

STREET WOMAN

Now tearless, can't recall the last time he shed his pain,

Living in a foster home where they barely know his name.

15 minutes in his shoes,

And you'll have more than the blues.

16 years ago he was abandoned,

Drug addicted mother couldn't stand him,

So her sister raised him, as if she made him,

Uncle loved him, fed him and bathed him,

Why did those bullets have to take him?

STREET MAN

Now **17** years old,

Ice cold.

He made a plan to take control.

Being overlooked was getting old.

Temper tantrums, had his lungs expanding.

Screaming cursing no one understands him.

So he loaded...

Aimed. But they he heard a voice in his head say/This is how I killer gets saved, this is how a killer gives praise, this is how a killer will change his ways, Amen

References

Batchelor, S. (2015). *After Buddhism: Rethinking the Dharma for a secular age.*

Blum, H. (2014). Mindfulness equity and Western Buddhism, reaching people of low socioeconomic status and people of color. *International Journal of Mindfulness Studies, 2,* 10.

Kabat Zinn, J. (2010). Jon Kabat Zinn and the healing power of mindfulness. *Tricycle,* October 10, 2010.

Magee, R. (forthcoming 2016). The way of colorinsight: Understanding race and law more effectively through mindulness-based colorinsight practices. *Georgetown Law Journal of Modern Critical Race Perspectives,* 8, 2.

Powell, J. (2003). Lessons from suffering: How social justice informs spirituality. *U. St. Thomas L.J.,* 1, 102–127.

Rothberg, D. (2006). *Engaged spirituality: A Buddhist approach to transforming ourselves and our world.*

Natalie Flores

Defining Mindfulness in Early Childhood Education Settings

This chapter is dedicated to the critical exploration of three questions that I, as an educator and researcher, have developed over the years: (1) What is mindfulness? (2) What purpose does mindfulness serve in early childhood education settings, and what is its connection to school readiness and schoolification? and (3) What are the implications of standardizing and measuring approaches to mindfulness?

To begin, mindfulness as often described to young children, educators, and families, is the practice of paying attention in a very special way. To some degree, it teaches children to build an awareness of oneself, others, and one's surroundings. More specifically, the mindfulness that has began to trend and spread across school settings is generally rooted in mindfulness-based stress reduction (or MBSR): a popular, secular, scientifically researched approach to Buddhist practices constructed by Jon Kabat-Zinn in the late 1970s. As a professor of medicine, Kabat-Zinn developed MBSR training to help his patients cope with psychological and psychosomatic problems. (Naturally, I begin to wonder why mindfulness practices that were once used to assist adults with psychological and psychosomatic issues is now being utilized in early childhood education settings. However, I will return to this speculation later in the chapter.)

A variety of programs that utilize mindfulness are in existence, such as mindfulness based stress reduction (MBSR), mindfulness based cognitive therapy (MBCT), dialectic behavior therapy (DBT), as well as acceptance and commitment therapy (ACT) (Burke 2009). Each of these approaches offers unique ways of introducing the most common understanding of "westernized mindfulness," which is defined as monitoring, in real time, experiences and doing so in a non-judgmental way (Kabat-Zinn 1994). Examples of this include practicing meditation, utilizing breathing exercises, as well as applying focused attention (Jennings 2015). MBSR, the largest and non-secular form of mindfulness practice, not only has quickly transitioned from being used in the treatment of psychological and psychosomatic problems, but also has now become a foundation for the application of mindfulness in schools (Purser and Milillo 2014).

What Purpose Does Mindfulness Serve in Early Childhood Education Settings?

Mindfulness, a method that has gained recent attention, aims to provide just what is needed in order for children, even in the early childhood years, to develop a greater sense of emotional

N. Flores (✉)
Department of Curriculum and Teaching, Teacher's College, Columbia University, New York, NY, USA
e-mail: nyf2104@tc.columbia.edu

© Springer International Publishing Switzerland 2016
R.E. Purser et al. (eds.), *Handbook of Mindfulness*,
Mindfulness in Behavioral Health, DOI 10.1007/978-3-319-44019-4_29

understanding and self-regulation. As stress dysregulation is likely to significantly interfere with learning (Farran 2011), especially in demanding environments such as early childhood classrooms, mindfulness emphasizes incorporating approaches to pro-social actions, which have been shown to aid in the decrease of challenging behaviors (Holtz et al. 2009). However, because mindfulness is a relatively new phenomenon, limited research has been conducted regarding its presence in early childhood settings, as well as the implications it holds for very young children, educators, and families.

Exploring Mindfulness Programs in Early Childhood Education Settings

In order to further understand the mindfulness movement's presence in early education settings, I examine three mindfulness "intervention" programs: Mindful Awareness Practices (MAP), Mindful Schools (MS) program, and the MindUP curriculum for K-2.

Mindful Awareness Practices (MAP)
Mindful Awareness Practices are structured group programs based on MBSR/MBCT models, which include sitting, movement, and body scan meditations, taught by experienced instructors (Burke 2009). Findings assert that based on parent and teacher reports, 44 4–5-year-old children in early childhood education settings improved in executive functioning skills, as well as social skills and temperament by utilizing the MAP interventions. More notably were the results which provided preliminary indications that young children are capable of participating in group mindfulness meditation practices.

Mindfulness Schools Program (MSP)
Conceptualized as a naturalistic field evaluation, the Mindful Schools (MS) program was implemented in a public elementary school, which consisted of 409 student participants and 17 teachers in 17 different classrooms. The MS curriculum was delivered to all classrooms for 15 min, 3 times a week, for a total of 5 weeks.

Moreover, the MS+ curriculum included the same curriculum and an additional once-weekly class, for a total of 7-week intervention (Burke 2009). Mindful behaviors evaluated consisted of paying attention, self-control, participation in activities, and caring and respect for others. This large intervention trial yielded results which included improvements in behavior, paying attention, calmness and self-control, an increase in activity participation, as well as caring and respect for others (Black and Fernando 2013).

MindUP Curriculum for K-2
In conjunction with Columbia Center for New Media Teaching and Learning, the MindUp curriculum, offered through Scholastic, has quickly been adopted in 29 U.S. states seeking to enhance mindfulness in participating classrooms (The Hawn Foundation 2015). Although the study by Oberle et al. (2011) focused on implementing the MindUp curriculum in 4th and 5th grade classrooms, the results have been adopted to fit a developmentally appropriate curriculum for children in kindergarten through 2nd grade. MindUp aims to foster children's self-regulation, optimism, and empathy, which may lead to improvements in executive functioning skills, school absenteeism, grades, and social-emotional competence. Moreover, Oberle et al. contend that this "simple-to-administer mindfulness-based education program consists of 12 lessons taught approximately once a week, with each lesson lasting approximately 40–50 min" (p. 6). The core mindfulness practices are administered for 3 min, 3 times a day, and consist of focusing on breathing and attentive listening to a single resonant sound. Each component of the program included sensory and cognitive experiences, ending with a reflection based on gratitude and kindness (Oberle et al. 2011).

School Readiness and Schoolification in the Early Years

Although the results from these trials seem favorable, very little criticism has been given to the actual purpose of positioning mindfulness in early childhood settings. In taking a critical approach, it seems that mindfulness programs aim to provide educators with the tools in order to implement a phenomenon known as "school readiness," and in turn, "schoolify" (or "academicize") the education of young children. The gain of tangible results seems to coincidentally align quite well with the underlying intentions of schoolification and school readiness.

School readiness is a notion that is grounded in a child's ability to comply with the behavioral demands of school as an institution; this includes a child's ability to follow directions and behave in accordance to the classroom rules. In order to illustrate this, imagine yourself as a 4 year old in a group of about 12 other children that are of similar age range. It is about noon, and you have had a long day of playing outside on a hot day, singing songs with the teacher, practicing your block-building skills, and are beginning to feel rather tired. Now imagine that instead of doing what your body and mind might want you to do, which is to find a cozy space, lay down, and take a nap, your teacher announces that it is "clean up time". As you look around the room, and the area that you have been playing in, you realize that cleanup is going to take quite a while. You begin to feel overwhelmed and emotional. What begins as feeling sleepy suddenly turns into feeling cranky and you begin to externalize your negative emotions, and in no time, you have manifested a full blown temper tantrum because you are simply not ready, capable, or willing to clean up. And then, as your teacher asks you to "please clean up the blocks," you throw a triangle shaped wooden block in protest. Your teacher reprimands you by restating the classroom rule of "no throwing blocks" and you are asked to spend a moment in "time out". These are the types of behavioral demands associated with school readiness, which often consist of displaying appropriate emotional responses, practicing cooperation, and paying attention to teacher directed activities to name a few (Farran 2011). However, as we have seen, becoming "school ready" can be a very complex and emotional task.

Additionally, because the externalization of some challenging behaviors are often perceived as disruptive or obstreperous to the school readiness agenda, many teachers and school administrators are relying on behavioral interventions. Such interventions range in diversity and purpose: from giving a young child a warm hug and encouraging words, to incorporating a "shadow"—where an adult, perhaps family member or educator, is given the task of personally and *physically* monitoring the child in order to prevent any further harm to themselves or others during moments of emotional stress. Although behavioral interventions are generally supportive in assisting the young child to find emotional stability in the face of impassioned environmental responses, many classrooms are now also adopting mindfulness interventions such as those discussed above in order to further implement school readiness practices.

Interestingly, a large component of the school readiness agenda asserts that young children adhere to the institutional norms seen in older grades, such as compliance and prolonged focused attention (Farran 2011). This becomes an even more complex issue when considering the trend of *schoolification* seen in early childhood education settings.

Schoolification can best be described as the phenomenon of early childhood education settings moving toward more academic curriculums, requiring distinctive goals and standards in order to measure student's achievements. More technically, Doherty describes it as "an emphasis on the acquisition of specific pre-academic skills and knowledge transfer by the adult rather than a focus on broad development[al] goals such as social-emotional well-being and the gaining of understanding and knowledge by the child through direct experience and experimentation" (Doherty 2007, p. 7). Thus, the "formal" learning mostly seen in K-12 settings is now being pushed down into what was once play-based pedagogies found in preschools. Moreover, due to the schoolifcation trend, less attention is being given to the caring and emotional nurturing dimension of young children. Schoolification supports the idea of young children's learning to be measured and enhanced (Kamerman 2005), which is problematic considering that young children, specifically those in the preschool years are episodic learners, making the measurement of their knowledge and skills very difficult.

Connecting Mindfulness to School Readiness and Schoolification: A Social Justice Issue

So how do the school readiness and schoolification agendas relate to the presence of mindfulness in early childhood classrooms? And why is not mindfulness being utilized as a tool to resist the top-down academic approaches now visible in early childhood settings?

Well, many studies report that practicing mindfulness may increase executive functioning and self-regulation skills, which are the mental processes that allow us (and young children) to focus, plan, remember, and multitask. However, it is important to note that the early developmental years for many young children aged 0–8 are comprised of several emotions. From celebrating essential accomplishments such as potty training, tying shoes independently, and establishing meaningful relationships with peers to temper tantrums, hyperactivity, and noncompliance (Holtz et al. 2009). It is no doubt that the spectrum of these externalized emotions can be extensive and for the most part deemed healthy and necessary for social-emotional growth.

Additionally, Black and Fernando (2013) posit that one of the "benefits" of incorporating mindfulness into classrooms is that it is a trainable skill, which improves self-regulation and attentional control which are associated with school readiness, pro-social behavior, and academic achievement. In response to the diversity of "disruptive" behaviors, some early childhood settings have begun to introduce mindfulness strategies which report significant improvements in some executive functioning abilities, temperament, and social skills (Burke 2009).

Thus, after much research and reflection, it has become simple for me as an ECE educator, advocate, and researcher to speculate the calculated and underlying motive of practicing mindfulness in schools: If the goal of schoolification

in conjunction with school readiness is to increase "normalized" behavior, fewer emotional outbursts, and teach children to accept the frustration and hardships they may be enduring than mindfulness has become the tool to achieve it. Although mindfulness curriculums and advocates praise its benefits related to cooperation and compliance, it seems rather clear that perhaps one alternative intention of using mindfulness in early childhood classrooms would be the emphasis it has on sublimating strong emotions such as anger, for example.

Simply put, certain mindfulness practices could send unintended messages about not speaking up in the face of wrongdoing and injustice, which can have serious ramifications for children's later participation in social activism. Teaching mindfulness that lacks a critical lens may perpetuate and promote what Ronald Purser and Joseph Milillo call "institutional blindness," which helps to maintain the status quo rather than encouraging transformative change of power structures. Such learning environments may encourage children to become peaceful and passive in their acceptance of hardships, rather than questioning, or holding an oppositional stance to inequities of social class, race, or gender. Teaching children to accept the frustration and hardships they endure, instead of taking a critical approach to understanding them, has been and still is a problematic issue for K-12. However, it now seems that it is becoming a trend in early childhood settings.

What Are the Implications of Standardizing and Measuring Approaches to Mindfulness?

Measured Mindfulness

There are several implications in standardizing and measuring mindfulness. To begin, there are at least nine different questionnaires that claim to define and measure mindfulness (Stanley 2013). Furthermore, according to Grossman and Van Dam (2011), no standard of reference exists which can be used to evaluate questionnaires that purport measuring mindfulness. Similarly,

Kabat-Zinn (2003), the founder of MBSR, admitted that "mindfulness cannot be accurately measured using survey based instruments" (Purser and Milillo 2014, p. 13). Simply put, several schools are standardizing this approach to well-being, and several mindfulness curriculums are indeed measuring and evaluating young children's ability to remain "zen".

Alternatively and critically, the consequence of standardizing and measuring mindfulness establishes itself in the problematic notion of separating mindfulness from its Buddhist roots, resulting in a myopic overemphasis of technique (Purser and Milillo 2014). Hence, mindfulness as "technical spirituality" (Driscoll and Wiebe and Wiebe 2007) becomes a mechanism for improving productivity, efficiency, and gaining tangible results—aspects that are not only far removed from the Buddhist concept of mindfulness, but also work in opposition to it.

It is unfortunate that mindfulness, in many cases, is being separated from its holistic foundation in order to ensure that children behave "normally" as the academic year progresses, and decrease any stress caused by the increasing schoolification expectations and high-stakes assessments. Considering what can be deemed the problematic nature of standardizing and measuring mindfulness, I align myself with Farran (2011) in agreeing that that just because something can be measured, does not necessarily mean that it should be measured.

Conclusion

It is important to note that what I have argued for in this chapter is not the complete removal of mindfulness in classrooms. It is quite the opposite. I, as an educator and researcher, firmly advocate for the use of *critical approaches* to mindfulness in classrooms of all ages, specifically those which cater to younger children.

Ethically and honestly, the promotion of mindfulness in early childhood settings offers several positive outcomes for children, specifically for those who may benefit from "trainable" self-regulation skills. For example, imagine again that you are your 4-year-old self, experiencing a temper tantrum during clean up time. Some aspects of mindfulness could have been positively utilized. To begin, your teacher could have used empathy and understanding as a technique toward mindfulness. Instead of escorting you to time out, he or she could have had the opportunity to practice breath work with you in order to bring oxygen flow to the brain, and relax your body's physical response to frustration and anger. More importantly, once you were calm, your teacher could have had a supportive conversation in understanding what was really upsetting you. And even better, a critical approach could have been taken to facilitating a dialogue around why you felt injustice as you were asked to clean up the blocks. Your 4-year-old self would have walked away feeling understood and cared for. Additionally, your 4-year-old self could now have the opportunity to practice this new technique of relaxing your mind and body in order to begin a critical dialogue with your teacher, explaining why you felt so upset. Therefore, mindfulness can in fact be utilized as a positive approach to further practicing and developing states of well-being and critical consciousness.

However, when mindfulness, as a tool to develop social–emotional awareness, becomes a non-transparent device toward marketing meritocracy, the Buddhist foundation and purpose of practicing awareness becomes non-existent. Moreover, the "awareness" that the westernized mindfulness movement has colonized may easily become a distraction from the recognition that children, even as young as preschool, have the conscious capabilities of adopting a critical lens toward social injustices. Practicing mindful awareness as a means toward academic improvement and encouraging normalized behavior should not come at the cost of neglecting an awareness of the current state of oppression that many marginalized groups endure.

Recommendations

In an effort to encourage educators and families, I recommend that mindfulness continue to be utilized as a tool toward deepening the understanding oneself, others, and the environment in which we exist. More specifically, I advocate for the use of mindfulness to be a supportive tool in recognizing and championing beliefs of social justice, activism, and equity. Some may wonder why young children should be encouraged to become agents of change. What happened to playdough and puzzles? Recognition of injustices and social activism is indeed developmentally appropriate, and as advocates for the well-being and education of young children, we need to equip these children with both insight and problem solving, equanimity as well as social responsibility, and the commitment and courage to advocate for a better society.

As an early childhood educator with my own mindfulness practices, I am inclined to explore the positive aspects of bringing these practices into the classroom. However, I also encourage and recommend that those who do so to consider *how* and *why* they may be using such mindfulness strategies. As we embark on this journey toward cultivating more unprejudiced classroom environments, it is my final recommendation that you ask yourself a critical question: Does your (or your child's) classroom mindfulness strategy promote well-being as a way to inadvertently sedate activism and social justice? Or does it empower young children to think critically as agents of change?

References

Black, D. S., & Fernando, R. (2013). Mindfulness training and classroom behavior among lower income and ethnic minority elementary school children. *Journal of Child and Family Studies, 23*, 1242–1246.

Burke, C. A. (2009). Mindfulness based approaches with children and adolescents: A preliminary review of current research in an emergent field. *Journal of Child and Family Studies, 23*(2), 133–144.

Doherty, G. (2007). *Conception to age: The foundation of school-readiness.* Paper presented at The Learning Partnership's Champions of Public Education across Canada: Partners in Action—Early Years Conference, Toronto, January 25–26.

Driscoll, C., & Wiebe, E. (2007). Technical spirituality at work. *Journal of Management Inquiry, 16*, 333–348.

Farran, D. (2011). Rethinking school readiness. *Exceptionality education international, 21*(2), 5–15.

Grossman, P., & Van Dam, N. T. (2011). Mindfulness by any other name: Trials and tribulations of sati in western psychology and science. *Contemporary Buddhism, 12*, 219–239.

Holtz, C. A., Carrasco, J. M., Mattek, R. J., & Fox, R. A. (2009). Behavior problems in toddlers with and without developmental delays: Comparison of treatment outcomes. *Child and Family Behavior Therapy, 31*(4), 292–311.

Jennings, P. A. (2015). *Mindfulness for teachers: Simple skills for peace and productivity in the classroom.* New York, London: W.W. Norton & Company.

Kabat-Zinn, J. (1994). *Wherever you go, there you are: Mindfulness meditation in everyday life.* New York: Hyperion.

Kabat-Zinn, J. (2003). Mindfulness-based interventions in context: Past, present, and future. *Clinical Psychology: Science and Practice, 10*, 144–156.

Kamerman, S. B. (2005). Early childhood education and care in advanced industrialised countries: Current policy and program trends. *Phi Delta Kappa, 87*(3), 193–195. doi:10.1177/003172170508700307.

Oberle, E., Lawlor, M., Thomson, K., Abbot, D., Oberlander, T.F., Diamond, A., et al. (2011). *Enhancing cognitive and social-emotional development through a simple-to-administer school program.* Vancouver, BC, Canada: Department of Educational and Counseling Psychology, and Special Education, University of British Columbia.

Purser, R. E., & Milillo, J. (2014). Mindfulness revisited: A Buddhist-based conceptualization. *Journal of Management Inquiry, 24*(1), 3–24.

Stanley, S. (2013). "Things said or done long ago are recalled and remembered": The ethics of mindfulness in early Buddhism, psychotherapy and clinical psychology. *European Journal of Psychotherapy & Counseling, 15*, 151–162.

The Hawn Foundation. (2015). Retrieved from http://thehawnfoundation.org/mindup/

A "Mechanism of Hope": Mindfulness, Education, and the Developing Brain

30

Joshua Moses and Suparna Choudhury

Mainstreaming Mindfulness: Shadows and Light

The rise of the mindfulness "movement" and its diffusion into Western society roughly during the last 20 years (Harrington and Dunne 2015) raises a number of questions. For example, what social conditions account for the fervent reception of mindfulness-based activities in institutions ranging from schools to clinics to workplaces? How exactly have state-of-the-art neuroscience and mindfulness meditation come together in this movement, and with what effects? In line with the framework of critical neuroscience (Choudhury and Slaby 2012), an approach that investigates the historical, scientific, and cultural contexts make possible zeitgeist trends revolving around brain science; we explore the emergence of brain-based mindfulness in educational curricula and propose reasons for the enormous excitement and investment surrounding it. We address these questions below, but first of all, more broadly, we will attend to the question of how we should understand the recent acceptance of Buddhist-oriented practices and theory in British and North American institutional settings. Buddhism has only recently become mainstream in America. Although 1960s counterculture helped introduce eastern practices to the West, it was not until the last 10 or 15 years or so that we have seen a proliferation of "secularized" versions of Buddhist practices represented in the mainstream press—i.e., in popular publications like John Kabat Zinn's *Full Catastrophic Living* (1990); Zinn's well-attended mindfulness-based stress reduction (MBSR) courses; and the Mind/Life society's widely publicized neurological studies of Tibetan monks (Lopez 2009).

There are a number of likely reasons for the popularity of these practices. Historian Harrington (2005) and others suggest several. With the embrace of eastern philosophy by 1960s counterculture and Herbert Benson's scientific studies of meditation and the "relaxation response" in the 1970s, meditation was introduced to a much broader audience in the West than ever before, primarily in the form of Asian practices that had been secularized through the jettisoning of "cultural trappings" (Lopez 2009). More recently, Bill Moyers' *Healing and the Mind* has publicized practices like *chi kung* and meditation. Joan Halifax, famous for her work on death and dying and friend to Hollywood celebrities and billionaires, has helped promote the idea that Buddhism has something unique to teach the West about dying. A glance at various New Age and Buddhist publications will reveal a vast menu of trainings dealing with all manners of pain,

J. Moses (✉)
Haverford College, Haverford, PA, USA
e-mail: jmoses@haverford.edu

S. Choudhury
Division of Social and Transcultural Psychiatry,
McGill University, Montreal, QC, Canada
e-mail: suparna.choudhury@mcgill.ca

© Springer International Publishing Switzerland 2016
R.E. Purser et al. (eds.), *Handbook of Mindfulness*,
Mindfulness in Behavioral Health, DOI 10.1007/978-3-319-44019-4_30

suffering, stress, and anxiety. That mindfulness has been accepted by the US military as a practice designed to promote resilience (Meyers 2015) provides some indication of the widespread appeal of these practices.

In the "master narrative of anti-modernism," (Rosenberg 2007) religion and spirituality play a key role in critiquing what have been seen as the alienating forces of modernity. This master narrative runs through American history going back to the Romantic era, when Thoreau and Emerson were already exploring Asian spirituality (Lears 1994). Since then, Buddhism has become the spiritual practice of choice for many in the USA. Given this context, we can understand the phenomenon discussed in this paper as but a single instance in the historical ebb and flow of tensions between holism and mechanization.

In the USA, Buddhism is typically discussed within a broader discourse on holism, and the concept of "mindfulness" is similarly framed. What Anne Harrington (2008) calls "Eastward Journeys"—the late twentieth century melding of eastern philosophy and Western medicine—comprises a potent cocktail of spirituality, scientific research, and oriental mystique that resonates with what she and many others consider the "existential deficiencies" (ibid.) of contemporary Western medicine—a domain in which mindfulness practices found an institutional fit earlier than in educational settings. The current success and popularity of Buddhism and mindfulness in the USA rests largely on its ability to leverage the currency of holism in general (Rosenberg 2007). But another factor contributes to their success as well.

While the colonial gaze has long played a role in rationalization of Asian traditions (Lopez 2009), recent years have seen an intensification of these processes, as well as peculiar twists that have allowed versions of once counter-cultural mysticisms to become institutionalized. These unprecedented developments are evidenced by a widespread acceptance of mindfulness practices across a variety of institutions (McMahan 2008), as suggested above.

Several skeptics have interpreted the mindfulness boom to be a kind of "antipolitics machine" (Ferguson 1994). Slavoj Zizek has been a prominent critic, representing the voices of skeptics who see the popularization of Western Buddhism, with its focus on cultivating individual mental well-being, as "the most efficient way for us to fully participate in capitalist dynamics while retaining the appearance of mental sanity" (2001).

This article examines whether or not these critiques are justified in the case of mindfulness-based education in the USA, which has significantly proliferated across the nation over the past several years. We conclude by suggesting that mindfulness-based education does not in fact lend itself easily to critiques such as Zizek's, partly due to the way in which its execution is inextricably linked to US educational politics. Perhaps unsurprisingly, the reality of implementing programs, and the intellectual sophistication and self-reflexive tendencies of many advocates of mindfulness practices in schools, belies attempts to paint them as dupes for neoliberal reforms.

This paper was catalyzed by conversations increasingly fervent uptake of mindfulness in various arenas of behavior management. Working at the intersection of anthropology and cognitive neuroscience, our initial exploration of this area raised a number of concerns. While the application of mindfulness practices in the various educational settings that we looked at appeared to have beneficial effects, the rapidly expanding use of mindfulness as a brain-based intervention still made us initially skeptical.

To explore our intuitions, we began preliminary qualitative research within an emerging project on religion, education, and neuroscience: (i) surveying the growing literature as the field develops; (ii) talking to key advocates of mindfulness-based education; and (iii) following online listservs on mindfulness and education. Following this research, our initial reflexes of skepticism shifted toward a greater openness and curiosity about spaces that may helpfully cultivate self-awareness, and in doing so, potentially laid the groundwork for the development of socially engaged citizenship. This is not to say, however, that our skepticism is no longer present at all. Below, we lay out preliminary findings from our explorations, highlighting in particular features of mindfulness practices that manifest tension and promise.

The Specter of the Brain and Mindfulness as a "Neurointervention"

The field of mindfulness-based education has grown considerably in the last few years, progressing from being viewed as a "new age" marginal phenomenon on the fringes of educational practices to being increasingly accepted by the mainstream, with funding agencies recently investing significantly in randomized controlled trials to analyze its long-term benefits on developing children and adolescents. Mindfulness-based education has become an institution, generating texts written by scientists, workshops on neuroscience, and mindfulness for educators, new educational curricula, and networks in the USA, Canada, and the UK, and the establishment of non-profit organizations in the USA that contend with poverty and behavioral problems among youth by using mindfulness and yoga.

In the last few years, new curricula, new courses, and even new schools have been explicitly established with the goal of equipping children and teenagers with the ability to practice mindfulness, and to develop this ability as a tool for reducing stress, improving focus, regulating emotions, and building resiliency.

For example, the Blue School, a "progressive independent school" in lower Manhattan for children aged 2–9 years, has aroused much interest in view of its "revolutionary approach" to education, which places as much emphasis on "nurturing compassion, the human spirit and human relationships" as it does on learning to read. At the Blue School, learning is understood as a "social act" rather than an individual one, and social and emotional skills are privileged above all. The founders describe the establishment of the school as a response to the "unsustainable and disharmonious world" in which

young children will "graduate into," and for which they will need to be equipped with the skills of creativity, innovation, and emotional resilience.[1] Similarly, the .b program (which stands for Stop, Breathe and Be!), a 9-week course crafted by teachers in the UK in 2007 and introduced in schools across the UK as well as other European countries, helps children and adolescents cope with the stresses of school life such as those induced by tests and bullying; improve their social interaction with teachers and peers; enhance focus and attention; and achieve greater happiness, calm, and fulfillment.

The "7/11" and "beditation" techniques are becoming familiar exercises for about 3000 students in Britain who have been introduced to mindfulness teachings. Similarly, "the breathing song" and "elevator breath" are phrases pupils in California are starting to use in schools. Mindfulness in Education teachers report using bells, reading stories, offering relaxation practices, yoga, centering, and breath focus in their classes. Teachers describe weaving these experiences into their curricula, interrupting "regular" classes and activities to engage in mindfulness practices. As Sydney, one of these teachers said:

> Our theory is that you give underlying tools for people to do non-cognitive work and that will increase their ability to do math and reading in the long run and will give them the ability to deal in our society and may undermine the education that just allows you to function in a market.

Adolescent Neuroplasticity: Teen Brains "Under Construction"

Many of the programs described above emphasize the benefits of mindfulness-based activities for the still-developing adolescent brain. For example, the developers of "Learning2Breathe," a mindfulness curriculum for adolescents,

---

[1]http://blueschool.org/. See video presented on website that discusses the goals behind this school project. Note that recent tabloid press articles in the UK and USA have reported widespread scrutiny about the school, citing complaints that children are not being taught to read properly.

describe the program as research-based, citing evidence that "adolescent brains are still under construction" and that teaching mindfulness can help cultivate skills for "emotion management while neural pathways are being shaped." To date, most of the scientific literature on the subject cites associations between areas of the brain that are still developing in adolescence and those that are activated through mindfulness.

Researchers at Harvard and Cambridge are currently conducting RCTs to attempt to evaluate the causal effects of mindfulness on the brain. In a recently published White Paper on Integrating Mindfulness into K-12 Education in the USA, which reviews the links between research on the developing brain, stress physiology, and the effects of mindfulness training, the connection between mindfulness practices and the developing brain are also addressed, with the paper arriving at the conclusion that "the brain regions that are impacted by mindfulness training are implicated in executive functioning (EF) and the regulation of emotions and behavior" (Meiklejohn et al. 2012).

Such insight has become almost self-evident as dialogue between Buddhism and neuroscience continues to develop, particularly in consciousness studies in cognitive neuroscience. Several research groups in laboratories around the world have focused their investigations on the "meditating brain": Using neuroimaging techniques, these groups are primarily concerned with empathy, attention networks, the brain's "default mode," the self, and mind-wandering. Furthermore, the growing trend of following embodied and extended mind approaches in neuroscience converges with, and is even inspired by, theories from Buddhism (Stanley 2012). We intend to pursue the confluence of these lines of inquiry further as we develop our project in the future, but for now it is important to point simply to the fact that existing research on mindfulness training and brain activity/structure suggests a link between mindfulness on the one hand, and, on the other, increased neural activity and gray matter volume in the prefrontal, frontoinsular, and limbic regions of the brain (Hölzel et al. 2010).

Developmental cognitive neuroscientists have used magnetic resonance imaging (MRI) and functional MRI to point to structural and functional developments in these areas of the brain during adolescent development. Indeed, it was largely the successful use of magnetic resonance imaging (MRI) technology in the first set of cross-sectional and longitudinal studies of children, adolescents, and adults in the 1990s that motivated the expansion of adolescent brain development as a field of inquiry in neuroscience: MRI provided a non-invasive and relatively easy method for collecting large amounts of data, which scientists were able to use to expand on a small set of postmortem findings from the 1970s that demonstrated cellular changes in adolescence (Huttenlocher 1979). MRI results from a number of studies have demonstrated that gray and white matter changes over developmental time. In particular, white matter density increases with age while gray matter decreases; this phenomenon peaks around puberty (Gogtay et al. 2004).

In light of the postmortem findings from the 1970s, these more recent neuroimaging data have been interpreted to reflect the processes of axonal myelination and synaptic pruning, which occur most in the areas of the brain that are associated with executive functions (i.e., behavioral control) and social cognition, such as the prefrontal cortex. The correlation between structural and functional developments in these brain regions on the one hand, and the performance of cognitive tasks that tap into capacities such as impulse control, empathy, planning, and multi-tasking on the other hand, has led many neuroscientists and public commentators to frame the stereotypical behaviors that are commonly associated with teenagers in terms of the brain.

For example, in light of recent findings in functional MRI papers, risk-taking, lack of emotional regulation, and impulsivity have been attributed to developmental processes in the brain that occur during adolescence (Steinberg 2008). Such characterizations of adolescence are not new. At the turn of the twentieth century, Hall described teenagers as "ships without sails," unable to adequately direct their own behavior

(Hall 1904). Cognitive neuroscience has reframed this description of adolescence as a period of "vulnerability and opportunity" for the plastic brain, characterizing teenagers as "engines without skilled drivers" (Dahl and Spear 2004). New neuroimaging (MRI and fMRI) data have demonstrated that middle childhood to late adolescence represents a "sensitive period" of malleability—or neuroplasticity—during which environmental input can have particularly profound effects on brain development (Blakemore and Frith 2005).

Neuroplasticity has been a central theme in a wealth of literature, particularly in the fields of social policy, self-help, and psychology. It has provided a biological backbone for programs as varied as juvenile crime, work productivity, and parenting, in addition to being discussed in detail elsewhere (Ortega and Vidal 2011; Rose 2007). For our current purpose, we would like to point out that brain plasticity not only explains anecdotally familiar adolescent behaviors. It also provides a substrate for measuring cognitive development and identifying at-risk candidates for early interventions. In other words, the notion that the brain is going through structural change accounts for problem behaviors such as impulsivity or risk taking; it also constitutes the basis for intervening socially or clinically during development in measurable ways, because the brain is capable of responding to particular inputs during this period of reconstruction. As Dr. Jean Clinton, a behavioral neuroscientist at McMaster University, puts it, "the teen brain is still undergoing a period of active construction… during this time, teenagers are more reactive. Mindfulness allows them to pay attention to their feelings rather than being their feelings."

Modulating the Brain Through Neurotalk

The literature and findings outlined above generally posited adolescent brains as "brains-in-construction." As previously stated, many programs with mindfulness-centered curricula cite this

impressionability as a rationale for their teaching strategies. It has in fact become common for educators in such programs to be literate in neuroscience terms, and for the programs themselves to explicitly define their pedagogical approaches in terms of neurological science. Children at the Blue School, for example, are taught impulse control, empathy, and emotional regulation in a curriculum that draws heavily on research from cognitive neuroscience (the school's board members include David Rock, CEO of the Neuroleadership Group, and Dan Siegel, Clinical Professor of Psychiatry at UCLA). Other programs go a step further: Not only are their pedagogical strategies informed by brain development research, but the students are also taught neurological vocabulary as part of the educational process.

> On 5 April 2013 04:55, Paloma (*sic.*) <paloma@gmail.com> wrote:
>
> I am hoping that someone can clarify some information about the brain for me.
>
> I presently teach Mindfulness in public elementary schools in Tacoma, WA. I use the picture of the brain from the MindUp Curriculum. As I was talking about the amygdala, a fifth grader asked me if there was one or two amygdala in the brain. I have always believed that there is only one amygdala, but the picture from MindUp clearly shows two? Is this because the illustration is confusing or is the amygdala bilateral?

This excerpt from an email that was distributed over the "AME" listserv—the mandate of which is to connect educators of mindfulness—demonstrates how teachers trained to teach curricula such as MindUP are required to educate children about brain regions and their functions. MindUP prescribes the employment of a hand model (i.e., using the fingers and thumb) to visually demonstrate how frontal regulation of limbic areas, or what it designates the "upstairs brain" and the "downstairs brain," can contribute to the management of social, emotional, and moral behaviors. A facility with the language of the brain is central to intervention goals.

Alvin, one of our respondents who runs an educational training program that teaches mindfulness to public school teachers, insists on the importance of teaching children how to conceptualize their emotions in neurological terms—a skill that also facilitates the ease with which they can perform the type of meditation that they are taught with mindfulness-focused curricula. The ability to attribute names to different areas of the brain facilitates a child's activation of "spaciousness"—i.e., the gap between a child's emotional state and her understanding of it. As Alvin puts it:

> Young kids now are talking about how their amygdala gets over active. I think that's good. In a way that level of critical distance and spaciousness between one part of mind and other can be very powerful. I think kids can and do get it. It's an interesting question about how it interacts with developmental process and their sense of self. I do think it can be narrative useful to kids.

Mark, another one of our respondents and a psychiatrist and advisor to a number of education programs, believes that learning about the brain is much more than a useful metaphor. The following is an excerpt from a conversation between Mark and Suparna.

> Mark: We used to think we don't know much about the brain, which we don't know but the little we do know you can actually teach and, you know, for example you can teach children the difference between left and right, you can teach them when a brainstem is being reactive, in a fight-flight-freeze response versus being more receptive and reflective, you can teach the role of the tenth cranial nerve in mediating a whole set of responses, you can teach kids about when their prefrontal cortex you know, isn't integrating the cortex, the limbic area, the brainstem, the body and even the social world, when it's flipped your lid, when you've flipped your lid, kids can feel what that feels like, they can see when they themselves or when other kids are flipping their lids. We've had kids as young as five years old come to their teacher with this model and say "I need to take a break, 'cause I'm going to do things I don't wanna do. I need a time out." They take a break, they get more integrated—literally—this prefrontal region integrates in their brain and they can sense when these things happen.

> SC: So is the brain here a metaphor? Is it a way for children to communicate about themselves or identify an emotion or are they really connecting with a neural process? What I'm asking is what does the brain… at that level of description, what's

added for the child and how they're able to modulate their own behaviours?

> Mark: Well, attention, we know attention can direct which areas of the brain are activated. We know that. When you offer them a model or a drawing, it's a metaphor. But whether it's a metaphor or mechanism… Well, this is based on a mechanism. So you're asking kids to directly drive energy flow through different circuits of their brain than they would otherwise, because they're being taught the metaphor yes, but it's a symbol based on the science of the brain. The way you drive attention differentially activates the brain. Otherwise there's no reason to know about the brain unless you could do something about it. Kids can learn how to strengthen their attention and specify how to use their attention.

Here, Mark describes how neuroscience serves as more than just an evidence base for mindfulness-based interventions. He posits that the language of the brain is also instrumental in teaching children how to directly modulate their brain function toward more regulated states. Interestingly, our interviewees are not consistent about whether the brain is being used as metaphor or mechanism, with Alvin seemingly settling on an understanding of the brain as a "symbol based on the science of the brain." In this formulation, it is the symbolic power of the brain that not only legitimizes the integration of mindfulness practices into the education system, but that also provides a tool—a language—for children to use in their regulatory process. Mark, however, suggests that the vocabulary of the brain is not merely symbolic: Its very use "drives attention" in ways that can directly modulate brain mechanisms.

Regulating the Effects of Childhood Adversity: Mindfulness Programs and Urban Youth

Urban US schools are constantly depicted as cite of crisis. Sources of seemingly endless research on achievement, gaps, violence, barriers to success are focal points for larger ethical debates, frequently seen as symbols for much of what has gone wrong in the USA. Entering this fraught arena, recently mindfulness-based initiatives have emerged with a specific focus on "urban" or "inner city" youth. In 2001, for instance, Andres Gonzales and brothers Ali and Atman Smith founded the non-profit Holistic Life Foundation in Baltimore after returning to the city from college and realizing that "something had to be done about [the sense] that things were wrong with the world." The organization works with young people from some of the most impoverished neighborhoods in the city through programs "that are anchored in values of unity and interconnectedness," including ones that teach yoga and mindfulness, and ones that provide social mentoring.

The premise on which the Holistic Life Foundation was founded—i.e., that impoverished youth could particularly benefit from mindfulness-based pedagogy—is substantiated by recent research. "Adolescence Matters," a quarterly brief published by John Hopkins University, includes a summary of a recent study on "poverty-related stress." In addition to the observation that "poverty stresses out youth," the study presented evidence demonstrating that "childhood adversity triggers neurobiological events that can alter brain development, potentially impairing the stress-response systems... being exposed to multiple poverty-related risks increases the chance that children will have more difficulty controlling their emotions." The study included 97 early adolescents from the Baltimore area and found that a 12-week mindfulness intervention that included yoga-based activity, breathing techniques, and guided mindfulness sessions generated improvements in involuntary stress responses, relationships with peers, and depressive symptoms. This improvement was determined by tracking the changes that participants made in self-reported questionnaire scores.

An initiative in New York City called the Lineage Project similarly endeavors to use mindfulness meditation and yoga "to break the cycle of poverty, violence, and incarceration." Established by Andrew Getz and Soren Gordhamer, two meditation teachers who had studied the work of Jon Kabat-Zinn at the University of Massachusetts, the project offers classes to young people in juvenile detention centers and alternative-to-incarceration community programs; children in foster care and in alternative public schools; and children with academic and behavioral difficulties. Teachers at the Lineage have reported significant improvements among youth after they have practiced yoga or meditation, particularly in terms of their ability to deal with rage, reduce detachment from others, develop empathy, and consolidate peace of mind through self-awareness and self-regulation practices.

Broken Futures, Plastic Brains, and the Soteriology of Mindfulness

How can we understand the mounting popularity of mindfulness as a tool for intervening in youth poverty? Why has mindfulness achieved widespread cultural resonance and why has it done so now, at this particular juncture in the US history? These questions have been vexing for both of us. Thus far, we have not found a satisfying answer in the literature. The best cultural explanation we have come up with ventures that the potential for change offered by the combination of neuroplasticity *and* mindfulness meditation resonates with a particular American desire for optimism and hopeful horizons (Peale 1952; Seligman 2002; Ehrenreich 2007). In the USA, despair and anxiety are currently widespread affects, felt in relation to politics, the economy, and the state of the environment. One might imagine that such a state of affairs would have motivated a galvanizing political project or collective civic engagement (Lasch 1984; Lear 2006). However, there is in fact a current lament over the *lack* of such a project. Mindfulness attends to this cultural need for a belief in possibilities, for a future where people can grow and change. In the case of youth and poverty, this is particularly important. What could be more heartbreaking than the oft-cited statistic that nearly one quarter of American children live in poverty (IRP)? And what could be more hopeful—and have greater emotional and moral resonance—than the notion that children are not permanently impacted by their conditions and can intervene in their own brains in ways that would help them develop into compassionate members of society, and moreover beat the odds of the structural inequalities that have come to characterize the USA? It is this sense of possibility, this expansion of the horizons of the future which gives force to endeavors to market mindfulness-based interventions for young people.

Indeed, the perceived transformative potential that had lent Asian meditative practices their widespread appeal in the USA in the 1960s continues to drive many contemporary advocates of mindfulness programs for at-risk-youth. To evaluate the social and political stakes of privileging such programs over more conventional strategies for addressing the issues of at-risk youth—i.e., by contending with economic inequalities, activating reforms in school financing, and evaluating exam outcomes—it would be useful to consider the insights of those who have firsthand experience in the field of mindfulness-based education. The professionals that we interviewed hold diverse and nuanced perspectives about mindfulness and its effects, but they are generally of the opinion that the educational framework has the capacity to generate significant social change.

For instance, Alvin shares Zizek's apprehension that mindfulness is vulnerable to being co-opted by what many consider to be a broken capitalist system. That being said, his experience with the inefficacies of more conventional educational frameworks has motivated him to explore the potential of mindfulness practices: "kids are coming in at risk and we are not doing well. [...] Our current system, it's making us stressed out and unhappy and not achieving the instrumental goals we say we want to achieve. Maybe it's better to do yoga and meditation and put art class back in."

Alvin acknowledges that the mindfulness movement (and his use of the word "movement" is significant) does not engage the conventional issues that are addressed in educational policy, but he posits that this does not mean that the movement is lacking in fundamental critiques of the "system." While Zizek might see Alvin as having been duped by an ideology that works to reproduce complacency in addition to furthering a neoliberal agenda, our investigation of how the movement has manifested in education practices elucidates the possibility of instrumentalizing its resources toward radical ends. As stated earlier, mindfulness can help its practitioners develop the mental tools that can facilitate engagement in social action and political criticism. Moreover, the movement is in fact largely rooted in a radical vision that advocated the upheaval of existing social mores. Perhaps the way in which this vision has manifested as a movement to cultivate the individual, rather than as one that advocates for more tangibly structural changes, is born from a despair of old school confrontational politics. Nevertheless, there is a strongly articulated sense of hope, one that has come a long way from Marxist-infused leftist agendas or the revolutionary utopias of Haight-Ashbury, and to suggest that mindfulness advocates are merely pawns of capitalism is to disregard the experiences of our respondents.

All that being said, Alvin still cautions against taking a wholly uncritical stance on the movement. As we demonstrated earlier, the increasing popularity and legitimization of mindfulness can be largely attributed to the way in which its alleged benefits have recently been reinforced by scientific research. However, Alvin reminds us that science is itself not invulnerable to bias and co-option:

> We went from positivist world-view and optimism in early 20th that we could solve everything and then we deteriorated into a complex postmodern depressing place and this [mindfulness] offers a return of positivism links back to ideas in education, to something that can be seen and measured by science. For a long time education has been struggling with that kind of question. Education is so tied into politics. It's a $500 billion in dollars a year industry–but how do you move it. You have to make research substantiated claims so you shift to paradigm of medical clinical gold standards of randomized control and the more biological it sounds there's a tremendous appeal. There's a lot of value to that on some level but to me that paradigm when you talk about health has problems too like these biomedical models that don't serve population but serve drug companies. We risk that in education too, trying to do these RCTs. It can shift conversation in good ways but problematic also in way to talk about culture.

Conclusion

> Mindsight is a learnable skill. It is the basic skill that underlies what we mean when we speak of having emotional and social intelligence. When we develop the skill of mindsight, we actually change the physical structure of the brain. (www.drdansiegel.com)
>
> And the data came back on it now, and it's absolutely blown our socks off. We have about 60 % optimism rise in our class, therefore we call it the optimistic classroom.
>
> (Goldie Hawn, transcript from TEDMed talk on MindUp, 2009).

Our preliminary observations suggest that the coupling of neuroscience and mindfulness succeeds as an intervention strategy for at-risk youth in multiple ways. One of the reasons that neuroscience and mindfulness complement one another is that they both recognize the significance of affect, social circumstances, and cultivating emotional intelligence (EI) in education, the workplace, and therapeutic contexts (McKinney et al. unpublished). Goleman's very popular book *Emotional Intelligence* (1995) draws on a body of data from cognitive neuroscience experiments and attempts to overcome what he calls Cartesian distinctions between reason and emotions. Specifically, neuroscientist Antonio Damasio has argued that emotions are core to the human ability to make decisions or to "reason," and central to our very sense of self, challenging the prevailing belief in Western thought that privileges rational thinking over emotions. Aware of the demonstrable importance of emotions in this recent body of neuroscience, Goleman believes that it is emotional intelligence that "makes us more fully human" (Goleman 1995, p. 45). He proposes that emotional literacy can help us to achieve the right balance between the rational and emotional mind to avoid "emotional hijackings" and subsequent inappropriate actions. This focus on EI resonates with the conceptions of human nature ("Human Nature 2.0") from social neuroscience that render human beings "hardwired to be social," and naturally empathic or benevolent. Not only is developing the capacity for EI integral to the development of

healthy personal relationships and good mental health; it has also been linked to future economic success, and constitutes an important resource for dealing with an increasingly insecure and uncertain global market (Fricke and Choudhury 2011). It is the areas of the brain associated with EI and social cognition that are undergoing the most pronounced change at adolescence. Thus, the right time to develop EI is during the most neuroplastic period of life—adolescence.

We suggest that the convergence of mindfulness and neuroscience explored in this paper is an extension of a trend across literature promoting self-optimization and risk-avoidance—namely the citation of data on neuroplasticity. The mounting popularity of neuroplasticity in general suggests a move away from biological reductionism, emphasizing as it does the way in which the brain is socially and culturally situated and therefore inflected (although see Choudhury et al. 2012). The prominence of neuroplasticity in the mindfulness-in-education project in particular has functioned to promote "neuronal selfhood" as part of a purported "therapeutic turn" in education. Indeed, the focus on EI, self-esteem, and compassion has been vehemently criticized by some education researchers for the way in which it promotes passivity and implicitly espouses the notion that the individual student is capable of contending with circumstances that are in fact beyond their control (Ecclestone 2004). For such educators, these teachings ultimately "anaesthetiz [e] people in times of risk and uncertainty" (Hyland 2009, p. 122) and detract from the active citizenship and tools needed to engage with structural inequalities—the symptoms of which many mindfulness education programs in fact seek to remedy. However, as many researchers in education and Buddhism as well as some of our interviewees have suggested, mindfulness in education can in fact help children to cultivate the skills that will enable them to become more socially engaged and to critically examine their circumstances (Thompson 2007). As Alvin said, "People [in the mindfulness movement] *are* thinking about individuals versus communal." The good ones are saying things along the lines: "you cannot do contemplative learning without touching on interconnectedness." Most of the serious people in the field are non-instrumentalists. I don't know how much it shines through." The incorporation of mindfulness into education is also a move away from the market model of education, which prioritizes teaching science- and market-based vocations, turning instead toward a project of re-instilling humanistic values into students from the earliest stages of their education. As Nel, an education activist said, "The mindfulness project has a radical goal in shifting the curriculum away from the current narrowing tendencies driven by the market."

Adolescence as a life stage, education as an institutional space, and neuroplasticity as a biological process—together, they constitute a charged moral arena in which societal projections play out about the characteristics we want to cultivate in the adults of the future. The plastic brain is presented as a substrate in the making, a process of becoming, and a space of potentiality through which to effect change and cultivate certain appropriate social, emotional, and moral behaviors, the end goals of which remain to be debated (adjustment or active engagement?). On top of that, as Alvin said, "People like the hard cold numbers. If you walk into the principal's office, someone who has a few seconds, and say, here are hard cold numbers and you give him a picture of the brain, he might say, 'Oh you're changing the brain.' That's good." The neurobiological dimension adds a sense of durability to these goals, while also maintaining objectivity, and stripping the project of religious connotations. According to one neuroscientist, the scientific framing is not only necessary for legitimizing the widespread introduction of mindfulness in schools, but even sufficient for maintaining a set of values: "moral mindsets [which] arise out of biological propensities… are shaped by experience during sensitive periods" (Narvaez 2009). Despite the scientific guise of mindfulness approaches, there are implicit moral assumptions about what sort of person one ought to be. In other words, even the scientized versions of mindfulness provide a secularized spirituality and leave the realm of fact for the realm of values (Weber 1958) by prescribing moral mindsets.

We began this paper by discussing our own ambivalence about the role of mindfulness practices in US educational settings. While we still remain ambivalent, we hope to have illuminated both the origins of these practices and the reasons for their current cultural resonance. The potent combination of brain sciences and the secularized spiritual–moral discourse of mindfulness intertwines with deeply held American beliefs about hope and optimism (Crapanzano 2003; Rozario 2007; Seligman 2002). Given the current political climate, the retreat of federal funding, and the fracturing of political horizons, mindfulness practices, particularly in relationship to children who might otherwise be considered broken and unredeemable, fills a critical niche—one that allows its advocates to imagine a world where people can change, become more compassionate, resilient, reflective, and aware. A world with a viable future. Far from the poverty traps depicted in recent research, where overwhelming structural forces threaten to cause despondency, mindfulness practices promise at least a partial way out.

The recent boom in literature on resilience (Ungar 2004) also fits well with this secular narrative of hope. In contexts of the collective disasters that characterize our "New Age of Anxiety" (Moses 2009, 2010), resiliency provides a version of scientized hope, an ecumenical discourse that resonates with some religious views of redemption, as well as American-style bootstrap grit, which also has its origins in religious sensibilities (Rozario 2007). As Francine, a mindfulness teacher, says:

> It's hopeful and a way of studying who and what we are—it's aligned with a scientific cosmology. Brain science is an intersection of our cosmology and meaning-making. It's the best cutting-edge science but the problem with science is it lacks a narrative about how we fit in and what we are about and all of those things in culture. It provides a hopeful narrative.

Ironically, the convergence of mindfulness with neuroscience, arguably a science laden with meaning, values, and prescriptive goals further expands the hopeful scientific narrative about human nature—which posits human beings as social, benevolent, and evolving toward better futures.

References

Blakemore, S.-J., & Frith, U. (2005). *The learning brain: Lessons for education.* Blackwell, UK: Wiley.

Choudhury, S., McKinney, K. A., & Merten, M. (2012). Rebelling against the brain: public engagement with the 'neurological adolescent'. *Social Science and Medicine, 74*(4), 565–573.

Choudhury, S., & Slaby, J. (Eds.). (2012). Critical neuroscience: A handbook of the social and cultural contexts of neuroscience. Chichester: Wiley-Blackwell.

Crapanzano, Vincent. (2003). Reflections on hope as a social and psychological category. *Cultural Anthropology, 18*(1), 3–32.

Dahl, R. E., & Spear, L. P. (2004). Adolescent brain development: Vulnerabilities and opportunities. *Annuals New York Academy of Science, 1021*, 1–22.

Ecclestone, K. (2004). Learning or therapy? The demoralisation of education. *British Journal of Educational Studies, 52*(2), 112–137.

Ehrenreich, B. (2007). Notebook. *Harper's Magazine, 314*(1881), 9–11.

Ferguson, J. (1994). *The anti-politics machine development, depoliticization, and bureaucratic power in Lesotho.* Minneapolis: University of Minnesota Press.

Fricke, L., & Choudhury, S. (2011). Neuropolitik und plastische Gehirne – Eine Fallstudie des adoleszenten Gehirns. *Deutsche Zeitschrift für Philosophie, 59*(3), 391–402.

Gogtay, N., Giedd, J. N., Lusk, L., Hayashi, K. M., Greenstein, D., Vaituzis, A. C., et al. (2004). Dynamic mapping of human cortical development during childhood through early adulthood. *Proceedings of the National Academy of Sciences of the United States of America, 101*, 8174–8179.

Goleman, D. (1995). *Emotional intelligence.* New York: Bantam.

Hall, G. S. (1904). *Adolescence: Its psychology and its relations to physiology, anthropology, sociology, sex, crime, religion and education.* New York: D. Appleton and company.

Harrington, A. (2005). Uneasy alliance: The faith factor in medicine, the health factor in religion. In Robert D. Proctor (Ed.), *Science, religion and the human experience* (pp. 287–307). Oxford: Oxford University Press.

Harrington, A. (2008). *The cure within: A history of mind-body medicine.* New York: W.W. Norton.

Harrington, A., & Dunne, J. (2015). When mindfulness is therapy: Ethical qualms, historical perspectives. *American Psychologist, 70*(7), 621–631.

Hölzel, B. K., Carmody, J., Evans, K. C., Hoge, E. A., Dusek, J. A., Morgan, L., et al. (2010). Stress reduction correlates with structural changes in the amygdala. *Social Cognitive and Affective Neuroscience, 5*(1), 11–17.

Huttenlocher, P. R. (1979). Synaptic density in human frontal cortex—developmental changes and effects of aging. *Brain Research, 163*, 195–205.

Hyland, T. (2009). Mindfulness and the therapeutic function of education. *Journal of Philosophy of Education, 43*(1), 119–131.

Lasch, C. (1984). *The minimal self: Psychic survival in troubled times.* New York: W.W. Norton.

Lear, J. (2006). *Radical hope: Ethics in the face of cultural devastation.* Cambridge, Mass.: Harvard University Press.

Lears, J. (1994). *No place of grace: Antimodernism and the transformation of American culture, 1880–1920.* Chicago: University of Chicago Press.

Lopez, D. (2009). *Buddhism and science: A guide for the perplexed.* Chicago: University of Chicago Press.

McMahan, D. L. (2008). *The making of Buddhist modernism.* Oxford: Oxford University Press.

Meiklejohn, J., Phillips, C., Freedman, M.L., Griffin, M.L., Biegel, G., Roach, A., et al. (2012). Integrating mindfulness training into K-12 education: Fostering the resilience of teachers and students, *Mindfulness, 3*(4), 291–307.

Meyers, M. (2015). Improving military resilience through mindfulness training. United States Army Medical Research and Material Command. http://www.army.mil/article/149615/. Accessed March 10, 2016.

Moses, J. (2009). Religion, mental health and disaster response in a new age of anxiety (unpublished dissertation).

Moses, J. (2010). An anthropologist among disaster caregivers. In G. Brenner, D. Bush & J. Moses (Eds.), *Creating spiritual & psychological resilience: Integrating care in disaster relief work* (pp. 19–25). New York: Routledge Press.

Narvaez, D. (2009). Triune ethics theory and moral personality. In D. Narvaez & D. K. Lapsley (Eds.), *Moral personality, identity and character: An interdisciplinary future* (pp. 136–158). New York: Cambridge University Press.

Ortega, F., & Vidal, F. (2011). *Neurocultures: Glimpses into an expanding universe.* NY: Peter Lang.

Peale, N.V. (1996) [1952]. The power of positive thinking. New York: Ballantine Press.

Rozario, K. (2007). *The culture of calamity: Disaster and the making of modern America.* Chicago: University of Chicago Press.

Rose, N. (2007). *The politics of life itself: Biomedicine, power, and subjectivity in the twenty-first century.* Princeton: Princeton University Press.

Rosenberg, C. E. (2007). *Our present complaint: American medicine, then and now*. Baltimore: The Johns Hopkins University Press.

Seligman, M. (2002). *Authentic happiness: Using the new positive psychology to realize your potential for lasting fulfillment*. New York: Free Press.

Stanley, S. (2012). Intimate distances: William James' introspection, Buddhist mindfulness, and experiential inquiry. *New Ideas in Psychology, 30*, 201–211.

Steinberg, L. (2008). *A Social Neuroscience Perspective on Adolescent Risk-Taking Development Review, 28*(1), 78–106.

Thompson, J. (2007). Changing ideas and beliefs in lifelong learning? In D. Aspin (Ed.), *Philosophical perspectives on lifelong learning*. Dordrecht, The Netherlands: Springer.

Ungar, M. (2004). *Nurturing hidden resilience in youth*. Toronto: University of Toronto Press.

Weber, M. (1958). *From Max Weber: Essays in sociology*. Oxford: Oxford University Press.

Zizek, S. (2001). From western Marxism to western Buddhism. *Cabinet.* http://www.cabinetmagazine.org/issues/2/western.php. Accessed May 23, 2013.

Using a Mindfulness-Oriented Academic Success Course to Reduce Self-limiting Social Stereotypes in a Higher Education Context

Adam Burke

Disparity in Higher Education

Education is a key to changing disparities in income, health, and social opportunity. There is, for example, an approximate 100 % lifetime earning disparity between individuals with bachelor's degrees and those without postsecondary education (Carnevale 2015). Unfortunately, educational underachievement and disengagement is a growing contemporary issue, with significant social costs. Lower educational attainment is associated with poverty, limited employment opportunities, poor health status, lowered life expectancy, and reduced economic productivity and competitiveness (Backlund et al. 1999; Kubzansky et al. 1998; Ogbu 1994). Lower educational attainment also affects the economic productivity and competitiveness of the nation. A recent report by a California State University Education Research Institute noted that without significant improvements in graduation from California colleges, especially for the growing Latino population, state worker education levels would decrease and per capita income

would fall below national averages (Shulock and Moore 2007). At the national level, a review of educational trends found that only 44 % of Americans (aged 25–34) had a college degree, with 11 other countries having the same or higher numbers, such as South Korea with 66 % (OECD 2014).

Retention and Graduation Rates

Although enrollments in institutions of higher education have been increasing over the past two decades, many students are inadequately prepared to meet the intellectual and social demands of college and consequently do not achieve their educational goals. High school graduation rates for Hispanic/Latino and African-American students are lower than those for Whites and Asians (75, 71, 87, and 89 %, respectively). Rates for students from low-income families are also approximately 15 percentage points below the national average (DePaoli et al. 2015). Of students who do advance to college, degree completion shows similar disparities. Minority students continue to be underrepresented both in college degree completion and in access to top-tier universities (Bowen et al. 2005). According to ACT, the retention rate for first-time, full-time students at two-year colleges is 60 % (Kena et al. 2015). The dropout rate is proportionally highest in these first two years for ethnic and racial minority groups, especially

A. Burke (✉)
Institute for Holistic Health Studies, San Francisco State University, San Francisco, CA, USA
e-mail: aburke@sfsu.edu

© Springer International Publishing Switzerland 2016
R.E. Purser et al. (eds.), *Handbook of Mindfulness*,
Mindfulness in Behavioral Health, DOI 10.1007/978-3-319-44019-4_31

Latino and African-American students, and students from lower socioeconomic status families (Clinedinst and Hawkins 2011). Finally, for all students in a typical four-year college, only 61 % have completed their degree within six years (Franke et al. 2011). Although graduation rates for minority students at four-year institutions have improved in the past decade, it is still 14 % lower than that of Whites (Engle and Lynch 2009). Reasons for student dropout include lack of family support for education (motivational or financial support), negative learning experiences, and disenchantment with school (Field et al. 2007).

Career Inequity

For those students who do enroll and graduate from college, there is additional inequity related to career opportunity. Specifically, universities continue to show significant underrepresentation of women, minorities, and students with disabilities in science and engineering (STEM) majors. The National Institutes of Health 'recognizes a unique and compelling need to promote diversity in the biomedical, behavioral, clinical, and social sciences research workforce' (NIH 2013). Inequity in the sciences is problematic for many reason, one of which relates to changing national demographics. Hispanics, for example, are the largest minority in the United States (54 million, 17 %), yet represent only 5 % of physicians (Llopis 2013). In California, Hispanics represent 36 % of the population with significantly more Hispanics living in medically underserved areas, yet only 5 % of physicians in the state are Hispanic (UCLA 2012). For females, although women earned 50.4 % of bachelor's degrees in the sciences in 2012, their representation in specific areas was significantly lower, such as computer science (18.2 %) and engineering (19.2 %). Even more problematic is the fact that only 11.2 % of minority women earned a science and engineering BA degree in 2012 (NSF 2015).

Institutional/Community Factors Contributing to Inequity

There are many potential factors contributing to the complex social issue of inequity in education and educational attainment. One that has been shown to be crucial is institutional support, from elementary education through college. A national review noted the problem of inadequate college preparation of high school students, with only 51 % of high school seniors having college-ready reading abilities. College faculty contributing to the report indicated inadequate student preparation for college-level math and writing as well (Ferguson 2004). The Institute for Higher Education Leadership and Policy at Sacramento State University cited the need for additional resources and support services to help students meet their goals. They noted that despite the need for services there was limited campus spending on student support due to state regulations and campus spending priorities (Shulock and Moore 2007).

Financial challenges are another contributor. Student financial aid has been shown to have a positive effect on retention. A recent study found that a majority of students who dropped out reported insufficient financial assistance from family or from student financial aid programs (scholarships or loans). Financial aid reduces the need to work full time and thereby reduces time away from school (stop out or dropout). Having to work is a major predictor of lower retention rates (Johnson and Rochkind 2009; Lotkowski et al. 2004; Orozco and Cauthen 2009). By reducing the need to work, financial support also allows for increased contact with other students and faculty and better integration into the social and academic milieu of the university. Work by Tinto (1987) noted that the more students were integrated into college academic and social life, the more likely they were to graduate. Pascarella and Terenzini (1979) also found that students who had more contact with faculty members inside and outside the classroom were more likely to graduate than those who did not.

Social Psychological Factors Contributing to Inequity

In addition to institutional factors, individual social psychological issues can also contribute to retention and graduation rates. One relevant challenge is stereotype threat and corresponding self-limiting beliefs. Stereotype threat has been defined as a social psychological phenomenon in which an individual feels threatened by a potentially confirming negative stereotype related to his/her social group identity. Some characteristic of an event amplifies the salience of a negative social stereotype and puts the individual in the performance spotlight. The result is a hypothesized performance interference (Steele 1997). For women, minorities, and lower-income students, those negative social group stereotypes may include beliefs about academic abilities. Relevant stereotypes for these social groups include lower interest or aptitude for science and engineering, lack of interest in educational attainment, lower intelligence, lower motivation, lower commitment to quality academic work, or gender–career mismatch (Eccles 2007).

If those stereotypes were implicitly or explicitly conveyed in a university context, then those social group-identified students would be more susceptible to the performance impairment evoked by such stereotypical beliefs. Pioneering research by Steele and Aronson (1995) showed how stereotype threat can impede intellectual performance among minority college students. Results from a wide range of studies have provided evidence of stereotype threat related to gender, racial, and socioeconomic stereotype activation (Aronson et al. 1999; Croizet and Claire 1998; Inzlicht and Ben-Zeev 2000; Spencer et al. 1999; Steele et al. 2002; Steele and Aronson 1995; Wheeler and Petty 2001). A meta-analysis of over 100 studies on stereotype threat examined moderators proposed in the theory. The observed interplay of stereotype relevance, domain identification, and test difficulty highlight both the social significance and complexity of the construct (Nguyen and Ryan 2008). It has been proposed that the underlying mechanisms of stereotype threat may

include increased disruptive mental workload (negative rumination), excess monitoring, stress arousal, emotional dysregulation, and negative expectancy (Beilock et al. 2006; Cadinu et al. 2003; Croizet et al. 2004; Keller and Dauenheimer 2003; O'Brien and Crandall 2003).

What Is Needed

Solutions to educational inequity will require institutional reform, application of best practices in classrooms and communities, and appropriate allocation of resources (OECD 2014). The ultimate goal is to create and promote policies, practices, and conditions that enable achievement. Indeed, institutional reform is essential, from removing the structural inequities that limit opportunity to the provision of needed resources, services, and relevant instruction. An ACT report, *What Works in Student Retention*, found the most helpful practices for four-year public colleges included academic advising for select student populations, first-year programs such as a freshman seminar, and learning support resources (Habley and McClanahan 2004). These are indeed important, but may be difficult to attain at times due to institutional priorities and limited resources (Shulock and Moore 2007).

Application of best practices in the classroom is another strategy that can help contribute to positive change. Existing courses can be used to instill beneficial attitudes, beliefs, and skills to help students become more able to navigate their educational paths and find the resources they need to succeed. Although a focus on the individual, via classroom instruction, may not necessarily change institutional policies and practices, it does at least have the potential to provide students with tools that can be used productively in many life circumstances and environments. It is also an approach that can be implemented relatively easily, as it only requires the intention of one faculty member to put it into action. As the sage Shantideva wrote, it is easier for an individual to wear sandals than it is to cover the whole earth with leather.

Coping and Self-efficacy

On the individual level, teaching students how to deal more effectively with the challenges of college would be a useful skill set that students could carry with them throughout their academic career. One place to start would be coursework to improve student coping and academic self-efficacy. Academic self-efficacy has been defined as the personal judgment of one's ability to attain specific educational outcomes (Bandura 1997; Schunk 1989). Academic self-efficacy has been found to be associated with learning and achievement (Campbell and Hackett 1986; Wood and Locke 1987). It has been shown to affect motivation to learn and persistence with learning tasks and to be related to increased performance (Gist et al. 1989; Schunk and Hanson 1985; Schunk 1981). A meta-analysis of over 100 college outcome studies looked at the impact of academic self-efficacy, general self-concept, academic-related skills, achievement motivation, academic goals, perceived social support, social involvement, institutional commitment, and contextual influences. GPA and retention were evaluated in light of these predictors. Retention was moderately associated with academic self-efficacy, academic goals, and academic-related skills. GPA was predicted by academic self-efficacy and achievement motivation (Robbins et al. 2004).

In a study of 107 nontraditional largely immigrant and minority college freshmen at a large urban commuter institution, self-efficacy was found to be a robust predictor of GPA and first-year retention (Zajacova et al. 2005). Several studies have shown that first-generation college students or minority students may hold beliefs or attitudes that can affect their expectations for college success. McWhirter et al. (2007), for example, found that Mexican American students anticipated more barriers to higher education in terms of preparation, ability, support, and motivation, compared to White students. In a study of 170 Latino college students, self-efficacy was directly associated with persistence intentions.

Intervening in the Classroom

Many things can take place in the classroom to empower students toward greater academic success (Lorsbach and Jinks 1999). This could include opportunities for meaningful engagement with the instructor, such as being a teaching assistant; general encouragement and positive interpersonal student–faculty interactions; teaching about social inequity and stereotyping; and providing gender–race–SES relevant role models through media or guest speakers, or use of peer mentoring. Another strategy is to offer courses specifically designed to teach academic self-efficacy skills, providing the students with sandals rather than trying to cover the earth with leather. This is a solution within the hands of almost any teacher. Even if structural barriers persist within a specific institution, individual faculty members have the power to teach skills that can affect their students' self-perception and worldview. Ultimately, considering the negative impact of factors such as stereotype threat on academic self-efficacy and educational attainment, the goal of finding a variety of effective educational interventions from the individual to the institutional level is imperative, particularly for underrepresented students.

An Academic Success Course

To this end, a lower-division general-education 16-week academic/life skills course was designed to increase student academic self-awareness and positive self-appraisal, coping skills, and resilience/problem solving. The course, *Holistic Approaches to Academic Success*, is built on the principles of self-regulated learning, a social cognitive learning theory-based approach to academic success. Self-regulated learning has been defined as a strategy to help students become more motivationally, metacognitively, and behaviorally engaged in their own learning (Zimmerman 1986). Self-regulated learning emphasizes three important elements: use of strategies that support academic goal persistence, self-regulation, and academic self-efficacy (Zimmerman 1989).

The *Academic Success* course teaches concepts and strategies that are hypothesized to impact the underlying mechanisms of stereotype threat on academic achievement, such as performance monitoring, stress arousal, and emotional dysregulation. The course is built on three core practices: (1) goal-oriented mental imagery; (2) kaizen continual improvement principles; and (3) mindful self-awareness. It is hypothesized that the course content and skill training will affect stereotype threat, mitigating performance interference and its resulting impact on course/career aspirations. This is accomplished through increasing recognition of threat activation, increasing positive reappraisal and use of self-regulatory skills (cognitive, affective, and behavioral), and increasing capacity for effective problem solving and persistence toward goal completion.

Course Innovation

The course is unique for a variety of reasons. As with more traditional student achievement courses it covers information on university culture, academic skills, major and career selection, and health. More importantly, however, it integrates a variety of unique elements, rooted in evidence-based social cognitive learning theory principles, cognitive neuroscience, and cross-cultural wisdom traditions, to enhance academic self-efficacy and positive expectancy for university and personal success. Courses that teach students how to underline a textbook and take notes in class are fine, but profoundly inadequate for addressing long-standing, largely unconscious self-limiting beliefs that impede success and happiness. For that simple reason, the *Academic Success* course and textbook are built on the core concept that all humans are learners—if you have not succeeded yet, you can in time. The essential strategy is simple—reappraisal of threat through reappraisal of self. The course mantra is, 'I don't know how to do that yet, but I will learn' (Burke 2016). Finally, because this is a lower-division general-education course on academic skill development, it has the potential to reach a diverse body of students, including females, minorities, disabled individuals, and lower SES students, who have not yet committed to a major and who potentially perceive threats to their social group identity within the university context.

Course Concepts and Materials: Learning Life—the Textbook

The course uses a single textbook, *Learning Life* (Burke 2016). The text has 18 chapters organized into four sections. The first section—*The Foundation*—provides instruction on the three core perspectives/techniques employed throughout the course: (1) goal-oriented mental imagery; (2) kaizen continual improvement/problem solving; and (3) mindful awareness. The reason for using academic goal-oriented mental imagery is to cultivate the habit of goal clarification and goal setting, reinforce positive expectancies and self-efficacy, provide opportunities for mental rehearsal and vicarious success experiences, encourage positive self-talk, and provide a constructive coping resource for managing stress. The purpose of the kaizen/continual improvement core is to cultivate a problem-solving orientation. The kaizen philosophy helps students understand and embrace quality as a guiding principle to enhance lifelong learning and personal growth. Students are encouraged to recognize the importance of efficiency, effectiveness, incremental change (small steps—enactive mastery), and commitment. The kaizen perspective also provides the basis for ongoing strategic problem solving to address perceived threats and enhance resilience and goal persistence. The intention of mindful awareness training is to enhance presence, self-awareness and reflection, emotional self-regulation, self-acceptance, threat perception and reappraisal, task focus, memory, and well-being, as well as supporting nonjudgmental performance monitoring.

The second section—*Learning Strategies*—is comprised of three chapters focused on core academic learning strategies starting with knowledge/skill acquisition through integration and display. Section three—*Skillful Means*—provides five chapters related to academic/life self-management skills, including critical thinking and problem solving, time management, stress and equanimity, emotional literacy, decision making, and habit change. The final section —*Applications*—brings all the ideas together with the goal of increasing capacity in the universally important areas of health, love and relationship, career and income, and spirituality–community service (the goal of contributing to a greater social good is thematic to this course). These final chapters emphasize the importance of human interdependence, social contribution, and planetary responsibility. Woven throughout the entire book is important information on the neurological and psychological mechanisms underlying human learning and behavior. This information provides the basis for insights into habits and the recognition that what was learned can be relearned and changed.

The Core Practices

1. *Goals, Goal-Oriented Mental Imagery, and Priming*

Working with goals and goal-oriented mental imagery is one of the three core practices in the course and text. Mental imagery involves the internal production of an experience that resembles the actual event. It can include rich sensory detail, emotions, and movement (Finke 1989; Lang 1979). Imagery is believed to play a role in a variety of human cognitive functions, such as encoding memory, navigating through the physical environment, and in social interactions (Pearson et al. 2008). There is substantial evidence supporting the value of mental imagery for health and healing, sports, performance arts, education, goal achievement, and other domains (Barnes et al. 2004; Driskell et al. 1994; Galyean 1983; Martin and Hall 1995). Imagery can be used to help shape personal attitudes and beliefs, to improve performance, to increase motivation and persistence, and to sharpen goal focus. In this course and its text, mental imagery is used as a

tool to help enhance clarity of vision, commitment to success, and ultimate goal attainment. One specific technique used throughout the courses is called Priming (Burke 2004). Students are encouraged to start the day with a Priming exercise to envision their desired outcomes. If used as instructed, this method supports a reflection on daily goals in relation to longer-term vision and priorities.

2. *Continual Improvement/Kaizen—Quality*

In addition to goal-oriented mental imagery, the second core practice we work with in the course is continual improvement and kaizen. Continual improvement is a philosophy and practice that emphasizes ongoing problem identification and solution. It is a cornerstone to a life characterized by growth and success. The specific continual improvement philosophy we focus on is called kaizen. The word *kai* means change and *zen* means good, so *kaizen* refers to change for the good, or improvement. Brilliantly, kaizen is not about working harder, but rather, working smarter. A primary goal is to reduce all forms of waste, including wasting one's own time and energy. One of the simple ways students are encouraged to cultivate this habit of working smarter is to bring more quality into the things they do. A focus on quality will naturally change the relationship to work tasks as well as impact the final product. A continual improvement orientation and a focus on quality will naturally lead to greater mindfulness during activities of importance to the individual. Kaizen also places an emphasis on small steps, rather than radical change. Social learning theory research shows that self-efficacy tends to grow with small successes over time, enactive mastery (Bandura 1997). The sections on continual improvement, quality, and kaizen were influenced by the works of Masaaki (1986), Deming (2000), and others.

3. *Mindful Awareness*

Mindful awareness is the third core practice employed in the course and book. The history of the Buddhist mindfulness tradition is presented to acknowledge the source of ideas for this element. The actual methods, however, are modified for use in a secular context, one that does not espouse Buddhist philosophy or practices specifically. For example, as a way to teach active ongoing mindfulness, the Three-Spheres Model (Burke 2016) was developed to represent the skill of momentary awareness, but to do so framed in a more contemporary social psychological perspective, and one that is clearly related to stereotype threat. Other key traditional Buddhist concepts fit naturally into the course content, such as impermanence (time/change management), metta (self-acceptance, gratitude, equanimity), suffering (dopamine reward mechanisms, aversion, habit/addiction), and no self (transitory mind–body states, urge surfing).

The purpose of using mindful awareness in the course is to help students become more present with their momentary experiences, both internal and external, in order to increase their ability to engage with their academic and personal goals, notice facilitative/obstructive thoughts and feelings, recognize environmental triggers, and work toward greater quality in relevant tasks. The basic rationale provided for practicing mindful awareness is that it is much harder to make changes in life, to learn, or to build capacity, if one is not aware of what they are doing, thinking, feeling, or saying. By becoming aware of responses to life, in the moment, the individual can begin to recognize habitual patterns, conditioned body–mind at work, including limiting stereotype beliefs.

Mindfulness is framed in the course as both a meditation practice (a specific mental training), and as a self-monitoring practice, a self-awareness technique that can be used at any point throughout the day. Weekly class sessions include opportunities for mind–body practices, including mindfulness sitting meditation. Weekly homework assignments regularly include mindfulness activities. For example, a homework assignment on study skills integrates a mindfulness technique called the Return Method as the means to monitor on-task/off-task study time. In this way, mindful awareness is used as a metacognitive practice for reflection on personal study strategies. Another homework assignment on time management uses mindfulness to observe patterns of attraction and aversion related to procrastination (using the mindful foundation of *vedana*, the mindfulness of pleasant, unpleasant, and neutral).

Both the sitting and active mindfulness methods focus on being aware of personal experience in the moment, paying attention to what is happening here and now, such as paying attention to whether off task or on task with a class lecture or writing a term paper. Mindfulness also directly supports the two other core practices of goal imagery and continual improvement. It can be used for observing one's relationship with goals, how they are enacted or avoided. It can be the basis for finding opportunities for change, mindfulness of moments for growth, mindfulness with the intention to improve.

The Learning Life Three-Sphere Model

Key to the *Learning Life* approach to mindful awareness is the three-sphere model (Burke 2016). The model is influenced by the work of scholars from diverse fields, including Lewin (1951) in psychology, Wiener (1948) in cybernetics, von Bertalanffy in biology (1968), and especially the concepts of social cognitive learning theory and reciprocal determinism in psychology (Bandura 1986). The model provides a system view of our life and a holistic view.

At the heart of the model are three interacting, interdependent spheres—Person, Behavior, and Environment. These three interacting spheres are contained within a larger single sphere—our momentary awareness/experience—all of which is moving through the matrix of time and space. The *Person* represents everything that is happening inside of us, our thoughts, emotions, and physiological processes. The *Environment* is the world we live in. Our *Behavior* is what the world sees and how we are known.

This model provides a useful and intuitive way to describe processes that facilitate or impede learning. For example, the stereotype threat concept can be represented quite comprehensively and simply: (1) *Future*—educational aspirations and choices; (2) *Past*—roots of stereotype threat; (3) *Present*—learning opportunity; (4) *Environment* (situational stereotype cues, models, supportive resources, institutional policies, faculty and peers); (5) *Person* (cognitive processes of perception, interpretation, group identification/schema, appraisal of threat/resources, physiological arousal); and (6) *Behavior* (impaired performance, avoidance, or coping and persistence/success). The model is used to cultivate awareness of this ongoing life dynamic and the skills needed to work constructively toward goal attainment, including personal goal commitment, self-regulatory skills, problem solving, and the vicarious/imagined experience of ideal goal outcomes. Students are encouraged to be mindful of these three interactive and interdependent elements throughout the semester in order to work with them effectively for insight and growth.

The AIR Strategy

Another important mindful awareness resource used in the course is the AIR Strategy (Burke 2016). The purpose of this technique is to teach students a simple way to think about mindfulness and quality improvement. The acronym AIR stands for *Awareness*, *Inquiry,* and *Response*. *Awareness* is momentary mindfulness, being conscious of mind, body, environment, and behavior. It is a consideration or reflection on what one is thinking, feeling, and doing right now. Awareness in its own right is powerful, but it may be very slow or insufficient to promote change if that is all that happens. To move toward improvement and problem solving, the next step following awareness is *Inquiry*. This is a reflection on reasons for the observed thoughts, feelings, or

behaviors. The final step, if appropriate, is a *Response*. The response is ideally an appropriate and effective way to deal with what was observed and considered. Used in this way, the AIR Strategy is an in-the-moment opportunity to consider and potentially develop alternative ways of thinking, feeling, or behaving that could be more constructive over time, in other words, learning.

Course Content

The course is taught over 16 weeks, one day per week, for three hours. The lessons cover the following material:

1. Introduction—university achievement
   The introductory lesson provides an overview of the program and a description of the benefits and applications. The lesson also provides an overview of university life and what is needed to succeed in the university environment.
2. Setting Goals—academic and personal goal vision
   This lesson focuses on goal setting as a critical facet of academic and personal success. Exercises focus on personal values and life goals, including earning a bachelor's degree as a realistic and attainable goal.
3. Imaging Success
   Evidence-based information on the use of mental imagery for performance enhancement is covered. The Priming method is taught.
4. Mindful Learning—the power of self-awareness
   This lesson introduces the practice of mindfulness as applied toward educational goals.
5. Excellence and Continual Improvement
   This lesson introduces the concepts of continual improvement and kaizen, and describes how excellence in education can be the key to a successful college experience.
6. Scholastic Skills and Strategies
   Strategies are provided for efficient and effective study, reading, note taking, test

taking, memory and recall, and research. On-task study is used as a focus of mindfulness practice.

7. Managing Life and Time
Strategies are offered for effective time management, including realistic time allocation for study and sufficient time for rest and recreation. Procrastination is used as a focus for academic mindfulness practice.

8. Life Purpose and Career Clarity
This lesson focuses on the process of career selection and the relationship between courses, majors, and life work. Mindful awareness is used to notice interests and aptitudes.

9. Optimism and Emotional Literacy
Emotional literacy, self-acceptance/metta, gratitude, optimism, and other concepts are taught. The AIR Strategy is used to support awareness of thoughts, feelings, and behaviors.

10. Behavior Change
This lesson teaches a suite of tools and strategies for shaping new patterns of thinking and acting. Mindfulness of the three interacting spheres of life is a foundation practice.

11. A Healthy Lifestyle
Information on diet, exercise, rest, drugs and alcohol, and other issues relevant to having a successful and healthy college experience is provided.

12. Stress Reduction and Successful Coping
Essential information on stress and coping, including coping styles, and effective long-term coping techniques is covered. Meditation is a key quieting practice.

13. Decision Making
Analytical and intuitive methods of decision making are taught. Students are instructed to be mindful of choice.

14. Social Support—friends, family, and the world
The lesson provides information on the value of positive social support, and ways to foster it. Respect of self and others, kindness and acceptance, are encouraged.

15. Finances and Financial Freedom
This lesson looks at money management, college financial aid resources, and the relationship between income and education.

16. Global Citizenship
This lesson covers the role each individual plays in creating a civil society and a sustainable socioeconomic system.

Mind–Body Practices

At the end of each class, there is a weekly experiential mind–body practice. These are often mindfulness-oriented exercises related to the homework or readings for the week.

Preliminary Findings

The HH200 course is currently being evaluated using a multi-year matched cohort comparison, looking at a number of outcomes, including impact on stereotype threat and differences in retention and graduation rates. One set of data has been analyzed. The preliminary findings suggest positive effects on a number of key student variables (Burke 2012). The study sample consisted of students enrolled in the course, primarily first- and second-year students from San Francisco State University's racially, ethnically, and socioeconomically diverse student body. SFSU is an ideal campus to explore and develop strategies that can impact limiting factors such as stereotype threat. It is a dynamic urban university serving 24,000 undergraduate and 6,000 graduate students (70 % students of color, 61 % women, and 50 % receiving financial assistance), with a significant number of students being the first in their families to attend college. Pre–post survey data was collected in several sections of the class. The survey instrument was a composite of three measures developed by Schwarzer and colleagues (2016). These included the *General Perceived Self-Efficacy Scale*, the *Self-regulation Scale*, and the *Procrastination Scale*. There were statistically significant positive changes on all items.

Student Responses

Student qualitative responses to the course and text were also collected. A sample is provided here.

> As cliché as it sounds, this book changed my life. I learned how to be more mindful of everyday activities, work efficiently, and take care of my overall well-being. This book has helped me make the rocky transition into college smoother.
>
> ~ RK, English/Freshman

> *Learning Life* helped me to set up my goals in small easy-to-manage steps. The book has definitely made me a better, more successful student. I am much more organized than I was before and I have become more effective at planning ahead. I am able to accomplish tasks with greater ease and efficiency. The semester that I applied these techniques was the same semester that I got straight A's, which was a feat that I haven't accomplished in years.
>
> ~ MC, Computer Science/Senior

> I always knew I could change for the better, but never actually took the steps to do it. The strategies in this book didn't force me to change, they made me want to change, and have helped me personally and academically. Simple things like thinking of 'time management' as 'change management' have made me look at things differently.
>
> ~ JT, Pre-Nursing/Sophomore

> This book is an excellent guide to finding the life that each human is granted. It is not to be read once, but to be used as a reference for direction throughout life.
>
> ~ TB, Kinesiology/Junior

> *Learning Life* taught me to STOP (stop, take some breaths, orient, and press on). It has helped me to become more self-aware. The book is complete with methods to help reinforce change and logs to keep track of the progress you are making.
>
> ~ JR, Psychology/Sophomore

> Dr. Burke's holistic approach to academic success has provided me with a variety of tools that I can use to improve my study habits. I was able to successfully earn my first 4.0 in college and my GPA has improved greatly compared to previous semesters. I have learned new ways to approach and view problems (not only academic ones) from different perspectives.
>
> ~ JD, Microbiology/Junior

> Although the book appeared to be geared to help new students adapt healthy working habits to aid them throughout college, it was actually written in a way that could foster growth no matter what phase of a person's educational experience. For me it worked perfectly.
>
> ~ SS, Psychology/Senior

> *Learning Life* is a tool. I have used it many times in my own life with much success. I have used priming and visualization to land my dream job. The stress management techniques Dr. Burke offers in this text have helped me through the toughest semesters. If you are given the chance to read this text, take advantage of it! Your work, school, and personal life will be transformed.
>
> ~ TT, Recreation Administration/Junior

Conclusion

Given the personal and social costs of educational inequity, there is a growing need to change educational practices. As the pace of change from within institutions may be slow, another strategy is a grassroots approach, from the classroom up. Here, the individual instructor has the power in his/her hands to change the story. It may be a small nudge, but drops of water create the ocean of life. Integrating course content that can help students grow in their academic self-efficacy is essential. Helping students recognize their own self-limiting ruminations and behaviors is a key step in that process. Mindfulness in its various manifestations is a logical element in any student academic self-efficacy intervention. Simple is good—notice, accept, plan, and improve. Put on the sandals. Take a small step. One day at a time, a life is changed.

References

Aronson, J., Lustina, M. J., Good, C., Keough, K., Steele, C. M., & Brown, J. (1999). When white men can't do math: Necessary and sufficient factors in stereotype threat. *Journal of Experimental Social Psychology, 35* (1), 29–46.

Backlund, E., Sorlie, P. D., & Johnson, N. J. (1999). A comparison of the relationships of education and income with mortality: The national longitudinal mortality study. *Social Science and Medicine, 49* (10), 1373–1384.

Bandura, A. (1986). *Social foundations of thought and action: A social cognitive theory.* Prentice-Hall, Inc.

Bandura, A. (1997). *Self-efficacy: The exercise of control.* Macmillan.

Barnes, P. M., Powell-Griner, E., McFann, K., & Nahin, R. L. (2004). Complementary and alternative medicine use among adults: United States, 2002. *Advance Data, 27*(343), 1–19.

Beilock, S. L., Jellison, W. A., Rydell, R. J., McConnell, A. R., & Carr, T. H. (2006). On the causal mechanisms of stereotype threat: Can skills that don't rely heavily on working memory still be threatened? *Personality and Social Psychology Bulletin, 32*(8), 1059–1071.

Bowen, W., Kurzweil, M., & Tobin, E. (2005). From bastion of privilege to engines of opportunity. *Chronicle of Higher Education, 51*(25), B18–B18.

Burke, A. (2004). *Self hypnosis demystified: New tools for deep and lasting transformation.* Ten Speed Press/Random House.

Burke, A. (2012). Preliminary evaluation of a novel academic achievement program. *Hawaii International Conference on Education,* Honolulu, HI, January 5–8.

Burke, A. (2016). *Learning life: The path to academic success and personal happiness.* Rainor Media.

Burke, A., Shanahan, C., & Herlambang, E. (2014). An exploratory study comparing goal-oriented mental imagery with daily to-do lists: Supporting college student success. *Current Psychology, 33*(1), 20–34.

Cadinu, M., Maass, A., Frigerio, S., Impagliazzo, L., & Latinotti, S. (2003). Stereotype threat: The effect of expectancy on performance. *European Journal of Social Psychology, 33*(2), 267–285.

Campbell, N. K., & Hackett, G. (1986). The effects of mathematics task performance on math self-efficacy and task interest. *Journal of Vocational Behavior, 28* (2), 149–162.

Carnevale, A. P. (2015). *The economic value of college majors executive summary 2015* (pp. 1–44). McCourt School of Public Policy: Georgetown University Center on Education and the Workforce.

Clinedinst, M. E., & Hawkins, D. A. (2011). *State of college admission.* Washington, DC: National Association for College Admission Counseling.

Croizet, J. C., & Claire, T. (1998). Extending the concept of stereotype threat to social class: The intellectual underperformance of students from low socioeconomic backgrounds. *Personality and Social Psychology Bulletin, 24*(6), 588–594.

Croizet, J. C., Després, G., Gauzins, M. E., Huguet, P., Leyens, J. P., & Méot, A. (2004). Stereotype threat undermines intellectual performance by triggering a disruptive mental load. *Personality and Social Psychology Bulletin, 30*(6), 721–731.

Deming, W. E. (2000). *The new economics: For industry, government, education.* MIT press.

DePaoli, J. L., Fox, J. H., Ingram, E. S., Maushard, M., Bridgeland, J. M., & Balfanz, R. (2015). Building a grad nation: Progress and challenge in ending the high school dropout epidemic. Annual update 2015. *Civic Enterprises.*

Driskell, J. E., Copper, C., & Moran, A. (1994). Does mental practice enhance performance? *Journal of Applied Psychology, 79*(4), 481–492.

Eccles, J. S. (2007). *Where are all the women? Gender differences in participation in physical science and engineering.* American Psychological Association.

Engle, J., & Lynch, M. (2009). *Charting a necessary path: The baseline report of public higher education systems in the access to success initiative.* Education Trust.

Ferguson, R. (2004). *Crisis at the core: Preparing all students for College and Work.* Prepared for ACT, information for life's transitions. Retrieved January 18, 2005.

Field, S., Kuczera, M., & Pont, B. (2007). *No more failures. Ten steps to equity in education. Summary and policy recommendations.*

Finke, R. A. (1989). *Principles of mental imagery.* Cambridge: The MIT Press.

Franke, R., Hurtado, S., Pryor, J. H., & Tran, S. (2011). *Completing college: Assessing graduation rates at four-year institutions.* Los Angeles: Higher Education Research Institute, Graduation School of Education and Information Studies, University of California.

Galyean, B. (1982–1983). The use of guided imagery in elementary and secondary schools. *Imagination, Cognition and Personality, 2*(2), 145–151.

Gist, M. E., Schwoerer, C., & Rosen, B. (1989). Effects of alternative training methods on self-efficacy and performance in computer software training. *Journal of Applied Psychology, 74*(6), 884.

Habley, W. R., & McClanahan, R. (2004). *What works in student retention? Four-year public colleges.* ACT, Inc.

Inzlicht, M., & Ben-Zeev, T. (2000). A threatening intellectual environment: Why females are susceptible to experiencing problem-solving deficits in the presence of males. *Psychological Science, 11,* 365–371.

Johnson, J., & Rochkind, J. (2009). With their whole lives ahead of them: Myths and realities about why so many students fail to finish college. *Public Agenda.*

Keller, J., & Dauenheimer, D. (2003). Stereotype threat in the classroom: Dejection mediates the disrupting threat effect on women's math performance. *Personality and Social Psychology Bulletin, 29*(3), 371–381.

Kena, G., Musu-Gillette, L., Robinson, J., Wang, X., Rathbun, A., Zhang, J., … & Velez, E. D. V. (2015). *The condition of education 2015. NCES 2015-144.* National Center for Education Statistics.

Kubzansky, L. D., Berkman, L. F., Glass, T. A., & Seeman, T. E. (1998). Is educational attainment associated with shared determinants of health in the elderly? Findings from the MacArthur studies of successful aging. *Psychosomatic Medicine, 60*(5), 578–585.

Lang, P. J. (1979). A bio-informational theory of emotional imagery. *Psychophysiology, 16*(6), 495–512.

Lewin, K. (1951). *Field theory in social science.* New York, NY, US: McGraw-Hill.

Llopis, G. (2013). Healthcare industry must mirror the growing hispanic population to authentically educate and serve the community. *Huffington Post: Latino Voices*. August 30.

Lorsbach, A., & Jinks, J. (1999). Self-efficacy theory and learning environment research. *Learning Environments Research, 2*(2), 157–167.

Lotkowski, V. A., Robbins, S. B., & Noeth, R. J. (2004). *The role of academic and non-academic factors in improving college retention. ACT policy report.* American College Testing ACT Inc.

Martin, K. A., & Hall, C. R. (1995). Using mental imagery to enhance intrinsic motivation. *Journal of Sport and Exercise Psychology, 17*, 54.

Masaaki, I. (1986). *Kaizen: The key to Japan's competitive success.* New York: McGraw-Hill.

McWhirter, E. H., Torres, D. M., Salgado, S., & Valdez, M. (2007). Perceived barriers and postsecondary plans in Mexican American and white adolescents. *Journal of Career Assessment, 15*(1), 119–138.

National Science Foundation, National Center for Science and Engineering Statistics. (2015). *Women, minorities, and persons with disabilities in science and engineering: 2015.* Special report NSF 15-311. Arlington, VA.

Nguyen, H. H. D., & Ryan, A. M. (2008). Does stereotype threat affect test performance of minorities and women? A meta-analysis of experimental evidence. *Journal of Applied Psychology, 93*(6), 1314.

O'Brien, L. T., & Crandall, C. S. (2003). Stereotype threat and arousal: Effects on women's math performance. *Personality and Social Psychology Bulletin, 29*(6), 782–789.

OECD. (2014). *Education at a glance. OECD Indicators.*

Ogbu, J. (1994). Racial stratification and education in the United States: Why inequality persists. *The Teachers College Record, 96*(2), 264–298.

Orozco, V., & Cauthen, N. K. (2009). Work less, study more and succeed: How financial supports can improve postsecondary success. *Demos, 12*(5), 10.

Pascarella, E. T., & Terenzini, P. T. (1979). Student-faculty informal contact and college persistence: A further investigation. *The Journal of Educational Research, 72*(4), 214–218.

Pearson, J., Clifford, C. W. G., & Tong, F. (2008). The functional impact of mental imagery on conscious perception. *Current Biology, 18*(13), 982–986.

Robbins, S. B., Lauver, K., Le, H., Davis, D., Langley, R., & Carlstrom, A. (2004). Do psychosocial and study skill factors predict college outcomes? A meta-analysis. *Psychological Bulletin, 130*(2), 261.

Schunk, D. H. (1981). Modeling and attributional effects on children's achievement: A self-efficacy analysis. *Journal of Educational Psychology, 73*(1), 93.

Schunk, D. H. (1989). Social cognitive theory and self-regulated learning. In *Self-regulated learning and academic achievement* (pp. 83–110). New York: Springer.

Schunk, D. H., & Hanson, A. R. (1985). Peer models: Influence on children's self-efficacy and achievement. *Journal of Educational Psychology, 77*(3), 313.

Schwarzer, R. (2016). *Psychometric scales.* http://userpage.fu-berlin.de/health

Shulock, N., & Moore, C. (2007). *Rules of the game: How state policy creates barriers to degree completion and impedes student success in the California community colleges.* California State University Sacramento, Institute for Higher Education Leadership and Policy.

Spencer, S. J., Steele, C. M., & Quinn, D. M. (1999). Stereotype threat and women's math performance. *Journal of Experimental Social Psychology, 35*(1), 4–28.

Steele, C. M. (1997). A threat in the air: How stereotypes shape intellectual identity and performance. *American Psychologist, 52*, 613–629.

Steele, C. M., & Aronson, J. (1995). Stereotype threat and the intellectual test performance of African Americans. *Journal of Personality and Social Psychology, 69*(5), 797–811.

Steele, C. M., Spencer, S. J., & Aronson, J. (2002). Contending with group image: The psychology of stereotype and social identity threat. In M. P. Zanna (Ed.), *Advances in experimental social psychology* (pp. 379–440). San Diego, CA: Academic Press.

Tinto, V. (1987). *Leaving college.* Chicago, IL: University of Chicago Press.

UCLA. (2012). *UCLA international medical graduate program. Program rationale.* http://fm.mednet.ucla.edu/IMG/about/about.asp

von Bertalanffy, L. (1968). *General system theory: Foundations, development, applications.* New York: George Braziller.

Wheeler, S. C., & Petty, R. E. (2001). The effects of stereotype activation on behavior: A review of possible mechanisms. *Psychological Bulletin, 127*(6), 797.

Wiener, N. (1948). *Cybernetics, or communication and control in the animal and the machine.* Cambridge: MIT Press.

Wood, R. E., & Locke, E. A. (1987). The relation of self-efficacy and grade goals to academic performance. *Educational and Psychological Measurement, 47*(4), 1013–1024.

Zajacova, A., Lynch, S. M., & Espenshade, T. J. (2005). Self-efficacy, stress, and academic success in college. *Research in Higher Education, 46*(6), 677–706.

Zimmerman, B. J. (1986). Development of self-regulated learning: Which are the key subprocesses. *Contemporary Educational Psychology, 16*(3), 307–313.

Zimmerman, B. J. (1989). A social cognitive view of self-regulated academic learning. *Journal of Educational Psychology, 81*(3), 329.

Part V
Commentary

Meditation Matters: Replies to the Anti-McMindfulness Bandwagon!

Rick Repetti

To Be Mindful or Not to Be Mindful, that Is the Question

As this volume makes clear, many individuals have leveled objections against mindfulness and/or 'McMindfulness.'[1] I will address some of them separately, below, but I will address some in terms of each other. First, however, I'll briefly describe what I think mindfulness is and what McMindfulness is supposed to be. Second, I'll share a sampling of some objections to mindfulness or McMindfulness, followed by some observations. Then, I'll reply to some of these representative objections.

Mindfulness is only contingently related to practices designed to cultivate mindfulness that happen to go by the same name, whether those practices are or are not considered orthodox within Buddhism. Buddhism, it should be noted, is considered unorthodox within the larger context of Indian philosophy, from which it emerged.

By analogy, strength is only contingently related to various bodily disciplines designed to cultivate strength, such as resistance training and calisthenics.

Mindfulness is easily understood by contrast with its opposite, mindlessness, the meaning of which is intuitively comprehensible to most, namely the state of mind or quality of consciousness characterized by not paying attention to what one is doing, thinking, perceiving, experiencing, etc. Mindfulness, then, is the opposite, namely the state of mind or quality of consciousness characterized by paying attention to what one is doing, thinking, perceiving, experiencing, etc. Mindfulness and mindlessness, therefore, are states of mind or qualities of consciousness.

Mindlessness is any state of mind that is marked by an absence of metacognitive awareness of itself, as exemplified in cases where an individual's attention is scattered and dominated by its objects, with little or no sense of its own characteristics. Examples include ineffective multitasking (though there may be relatively effective forms of multitasking or attention-shifting that are not mindless), scattered or dispersed attentional focus, and certain repetitive activities one engages in without attention, such as eating popcorn while watching an action movie.

Mindfulness is a state of mind or quality of consciousness, independent of meditation practices that are specifically designed to cultivate

---

[1] Apart from the many examples in this volume, see also Žižek (2012), Bodhi (2011), Purser and Loy (2013), Forbes (2012), Sharf (2012), Thompson (2014), Flanagan (2012), Moore (2014), Manthorpe (2015), and Edwards (2015).

R. Repetti (✉)
Department of History, Philosophy, and Political Science, Kingsborough Community College, Brooklyn, USA
e-mail: rickrepetti@aol.com; Rick.Repetti@kbcc.cuny.edu

© Springer International Publishing Switzerland 2016
R.E. Purser et al. (eds.), *Handbook of Mindfulness*,
Mindfulness in Behavioral Health, DOI 10.1007/978-3-319-44019-4_32

such a state of mind or qualities of conscious-ness, which practices are also called 'mindful-ness,' but which practices are distinct from the qualities or states they aim to cultivate.[2] Again, by analogy, 'strength' is distinct from strength training exercises. (The analogy might seem imperfect because the word 'strength' is not used homologously to describe strength-cultivating practices, except perhaps by a few exercise enthusiasts who might describe what they are working on at the gym on a particular occasion as 'strength' as opposed to, say, 'cardio' or 'speed,' but this is an irrelevant accident of lan-guage.) By contrast, one may engage in simple activities, intentionally or spontaneously, in such a way that one experiences, and attends in full awareness to, their phenomenological character-istics, being conscious that one is aware of what one is experiencing: Such states of mind are appropriately described as characterized by mindfulness.

There are degrees of mindfulness, and the distinction between mindfulness as a state and mindfulness as a practice might be indeterminate or opaque in certain cases. For example, one may practice yoga mindfully, fully consciously engaged in and noticing the kinesthetic, somatic, and other proprioceptive elements of the experi-ence (bodily sensations generated from moving in and out of poses, while remaining in them, correlated with muscular efforts, etc.), energetic elements of the experience (sensations or feeling tones associated with breathing, mentally direct-ing what is believed to be one's life-force energy, etc.), intentional elements of the experience (e.g., effortful elements and teleological elements, aimed at attaining the alleged benefit of the pose), and other features of the experience (var-ious soothing qualities, massage-like bodily sensations, the extent to which awareness moves fluidly with each such element as it arises or fades out periodically), consciously attending to the fact that one is doing so; alternately, one may

practice yoga without any such self-reflexive or metacognitive qualities present, absorbed instead in the competitive ego dynamics that sometimes emerge within a particular social setting marked by an excessive emphasis on vanity, body-sculpting, and other features of the aesthetics of the practice and thus forcing one's body to mimic what other practitioners are exhibiting, so as to keep up with them, ignoring one's pains, one's mental strain, and so on. (One such yogini, female yoga practitioner, that I know personally, claimed she was ranked as the number three yogini in the Western hemisphere in the Pattabhi Jois Ashtanga Yoga community; she had seri-ously injured her lower back when trying to do a backward bend, for reasons I attribute to the sort of competitive ambition that typically pushes against and blots out proprioceptive awareness, increasing the likelihood of injury.) As a practi-tioner of yoga for 43 years, and an instructor for 17 years, I can affirm that whether, and the extent to which, yogis are in the poses mindfully varies from moment to moment.

This is true not only of yoga, but of any activity. Some activities seem to lend themselves to mindfulness, and others seem to make it more difficult to remain mindful. For example, cook-ing, washing dishes, gardening, drawing, jog-ging, swimming, and a host of activities like them may be conducive to mindful engagement in them, particularly for those who find these activities intrinsically enjoyable, relaxing, or otherwise wholesome. Conversely, multitasking, texting, playing violent video games, watching action movies, rushing to catch a train or plane, skim reading headlines, surfing the Web, and a variety of similar activities tend to reduce the extent to which many of us are inclined to be mindful. Of course, this all depends on how one conceives such activities, the speed at which one performs them, the extent to which one finds them relaxing, and the sort of general modus operandi of the particular person. What one person finds meditative, another finds stressful.

Because yoga is, at least traditionally, sup-posed to be a form of bodily meditation, akin to walking meditation, tai chi, or chi gung, it fol-lows (from my description of the spectrum of

---

[2]Edwards (2015) and Langer (1991); see Latham (2015) for an impressive analytic account of mindfulness oper-ationalized and differentiated from practices designed to cultivate it.

mindfulness to mindlessness that is possible during yoga practice) that the extent to which one is mindful even when practicing mindfulness meditation varies similarly. Thus, one may practice the mindfulness-cultivating practice of mindfulness mindlessly or mindfully. For example, a beginner might do what she can to follow the instructions, but what is actually going on behind her eyelids might involve no metacognitive awareness of the phenomenological features of her own experience during the session, but rather she might be so engaged in the flow of her thoughts—about failing to be detached and about how she cannot bring herself to not be so judging and how nonetheless the fact that she is practicing mindfulness makes her better than her coworkers and so on—that she is better described as mindless of the major contents and processes occupying her awareness during the session; alternately, she might be so caught up in the struggle not to cough and reveal to her peers her lack of bodily control that she has absolutely no mental freedom whatsoever during her feigned meditation.

Conversely, someone may experience a mindfulness meditation session with exactly the sort of phenomenological self-mirroring that exhibits the quality of being mindful of the contents of one's own mental state, even if the contents of her mental state are significantly parallel to those just described. For seasoned meditation practitioners, the mere fact that the mind in a particular meditation resembles a raging river of powerful mental contents and currents, so to speak, need not guarantee that the practitioner is not in a state of mindfulness meditation. The difference is analogous to the difference experienced between two individuals who have ingested LSD, only one of which remembers that her experiences are mental fabrications and the other of which is likely to 'freak out.' The former is able to frame or bracket her experiences in such a way that they do not uproot her at all; the latter is unable to do so and thus fully identifies with the experiences. This functional difference between the two LSD users is similar to the functional difference between the two meditators, only one of which is able to detach from the raging stream of consciousness and the other of which is pulled, pushed, and tossed about by an otherwise identical stream of mental contents.

Although this is anecdotal, I have a friend who is a personal trainer, who admittedly described his one-on-one yoga session teaching style with his upper echelon New York City clientele as 'vanity yoga.' By this, he meant the sort of yoga that will help clients to body sculpt, for purely aesthetic reasons (perhaps, I speculate, while at the same time rationalizing to themselves that they are yogis). Of course, many of the criticisms of McMindfulness are analogous to criticisms of 'McYoga.'[3] There is no doubt a major soteriological, evaluative difference between people who take LSD, practice yoga, or meditate in order to attain spiritually advanced states of consciousness and those who engage in these same activities for what are on analysis various forms of ego enhancement. But—and here is my basic response to all such criticisms of these practices—anyone who engages in them for any reason whatsoever (except, perhaps, the LSD…) is likely to experience some mental freedom thereby and that is inherently good. It is likely a truism about human nature and spirituality that anyone who engages in spiritual practices was originally motivated by self-interest. Buddhism is, after all, a philosophy prompted by the desire to escape from suffering and attain happiness (the supreme happiness, either for oneself or for the sake of all sentient beings). Everyone wants to be happy, as Aristotle noted. In that regard, Jon Kabat-Zinn claims that the Dalai Lama asked him to try to develop a secular version of meditation so that the multitude of suffering Westerners could increase their potential for the sort of happiness that Buddhist meditation promises to bring about. I doubt the

---

[3]For an insider's informal expose of what has analogously been dubbed 'McYoga,' see Seligson (2015); see also Guthrie (2002), for an early use of the term 'McYoga,' indicating that the 'Mc' prefix attached to spiritual practices gone viral had been used several years before Purser and Loy (2013) coined the term 'McMindfulness.' See also Marchildon (2012), for a recent defense of McYoga.

Buddha would object, whether the Dalai Lama asked for this or not.

As an exercise, 'mindfulness'-cultivating practices involve various forms of effortful focusing of attention on phenomenological details of the objects of consciousness, such as the breath and the flow of thoughts, with varying focal scopes (broad or narrow, on one object of consciousness or a series of objects, etc.), specifically exercising attentional focus in various ways specifically for the purpose of cultivating the quality or state of mindfulness. There are many variations on the theme regarding techniques of mindfulness-cultivating practices, from various Buddhist traditions. Their differences are irrelevant here, as I will argue below.

The term 'McMindfulness' has the same sort of perjorative anti-neoliberal capitalist connotations as the term 'McDonaldization,' used to critique global capitalist consumerist culture, despite the fact that the evils of globalized McDonaldization are of immense proportions relative to the alleged evils of McMindfulness. The difference is so great that it seems *prima facie* obvious that the term 'McMindfulness' alone betrays hyperbole approaching sophistry, in my view. The term 'McMindfulness' is intended to highlight the widespread application and alleged cultural appropriation of a supposedly watered down, inaccurate extraction of elements of traditional Buddhist mindfulness-cultivating exercise techniques, in contexts that are far removed from traditional Buddhist traditions, lineages, monasteries, teachers, etc., which contexts are viewed as morally questionable for various reasons. For example, the military use of mindfulness is suspected of making more effective soldiers, something that might go against the non-violence norm in Buddhism, and the corporate use of mindfulness is suspected of anaesthetizing corporate executives, management, and employees from any moral sense about their presumably ethically questionable actions by way of the detachment and non-judgment encouraged by some forms of the practice.

Mindfulness is thus a state of mind or quality of a mental state, analogous to the way in which

strength is a bodily state or a quality of the body, both of which may be cultivated through practices that are thought to cause those states or to increase those qualities. Thus, to object to the cultivation of the state of mind that is mindful outside the cultural context in which such cultivation was first explicitly practiced is akin to objecting to the cultivation of the state of body that is characterized as strength outside the cultural contexts in which such cultivation was first explicitly practiced. Suppose strength cultivation was one of the key features of yoga; it is certainly cultivated through rigorous yoga practice. Suppose yoga was the oldest form of strength training and that no other forms of strength cultivation emerged until McYoga spread yoga's benefits far and wide. Critics of McYogic strength would object that it was missing the mark of true yoga, was a form of cultural appropriation, and would be used to evil ends. They would argue that strength-cultivating yoga poses are only a small subset of yoga poses and that while such subset does in fact produce some kind of strength, strength per se is not the goal and privileging it misrepresents the spirit of yoga and so on.

Even if these were all true, it seems rather intuitive that nobody has a cultural copyright on any techniques that improve human attributes that we already possess naturally in varying degrees and value for their own sakes, like strength or mindfulness. It seems equally intuitive that it does not matter if McYogic strength differs from yogic steadiness, say, or if the sort of mindfulness cultivated by McMindfulness differs in some ways from that aspired to in traditional Buddhism. So what? The reply is usually that then it should not be called 'mindfulness,' but that is like saying the McYogic emphasis on poses should not be called 'yoga' because true yoga is so much more than mere poses. It is still yoga, however, and it is still mindfulness meditation.

We will turn shortly to the four main objections to mindfulness or McMindfulness noted above: that mindfulness (a) fails to attempt to try to change the world, (b) is guilty by association with other ends to which its use is applied,

(c) ought not to be separated from its Buddhist ethical framework, and, among others, (d) is not as important to Buddhism as Westerners think it is, as evidenced by the relatively few Buddhist practitioners. But there are others. Some other objections (some mentioned *en passant* above, others not) include (e) McMindfulness involves some form of cultural appropriation, (f) mindfulness is not the same technique as what is being taught in the McMindfulness craze, (g) mindfulness makes people apolitical, apathetic, detached, and thus easily controlled by more aggressive, unjust institutional structures, forces, and individuals, (h) some individuals with certain forms of antecedent mental instability might suffer more from being encouraged to dwell on their thoughts, given that their particular pathologies already involve obsessive absorption in repetitive thought sequences, (i) as a spiritual practice, mindfulness threatens to bypass the sorts of therapeutic catharsis that can only come from other, more aggressively intervening forms of engagement with one's issues, (j) meditators are universally advised within spiritual traditions not to take up the practice without the guidance of a skilled teacher, (k) in its emphasis on non-judgmental witnessing, McMindfulness threatens to morally desensitize practitioners, cutting them off from their otherwise healthy senses of indignation and injustice, and, among others, (l) McMindfulness turns all matters of judgment inward, so that instead of objecting to and advocating against external structures of inequality and injustice, practitioners are encouraged to weaken their own ego responses.

Now, let me make some observations about a sampling of some of these objections; I take it that the observations I make about the objections in the sample apply to a certain extent to all of them, but I cannot go through each individually. There is something *prima facie* odd about the collection of objections against McMindfulness, as some of them seem obviously at odds with others. For example, one objection from the traditional Buddhist community is that the technique is not even what Buddhists do, but a distortion. Another objection, from the same community, is that it is a form of cultural appropriation. Yet another objection is that it fails to maintain its essential embeddedness within the broader ethical and soteriological web of beliefs, practices, and traditions in which it is to be properly understood and experienced as such. But these objections have contradictory implications: If it is a different technique, then it cannot be cultural appropriation, on the one hand, but if it was the same technique, then it would be cultural appropriation. Likewise, if it is not properly embedded within its larger context of origin, then it is not cultural appropriation, on the one hand, but if those elements were also replicated in the new applications, then it would be cultural appropriation. The bulk of the objections against McMindfulness have this character: If they are valid in one regard, then they are invalid in another and vice versa. Thus, one cannot remedy one of these objections without guaranteeing to violate the other. What this suggests is that the collection of criticisms is not only internally inconsistent, but probably incoherent.

Admittedly, some of the individuals who make these objections make valid points, my rebuttals notwithstanding, and their concerns are all nobly intended to preserve the meaning and value of mindfulness, to protect the venerable contemplative traditions that have delivered it to us, and to ward off potential misunderstandings, distortions, abuses, and applications. I do not take issue with these fine elements of their concerns. I, too, however, wish to protect a certain understanding of the value of meditation from possibly misleading objections.

My approach is *entirely informal*, and mostly analogical, as may already be evident, for I take it that the arguments at issue may be best understood and evaluated by comparison with analogous cases in which the same sorts of objections are more transparently weak, false, or absurd. Analogies are not proofs, but rather they function as guides to intuition insofar as, when successful, they facilitate noticing something hitherto not visible. They may be particularly useful when attention to particular details distracts attention from the big picture, so to speak. My 'big picture' perspective on this subject is rooted in over 43 years of serious meditation and yoga practice, with instruction from many meditation teachers from different lineages and traditions, with as many years intentionally avoiding teachers and doctrines, and with as many years of personal and/or professional philosophical inquiry into and teaching of meditation and yoga. Many of those years of practice involved the techniques that some of the objectors would identify as those taught by McMindfulness disseminators and which some object are misguided, distorted, culturally appropriated, watered down, dangerous, best practiced only under the guidance of a Buddhist teacher within a formal lineage, etc. I reject all such claims.

Just as it was Hindus who first came here peddling watered down versions of secularized yoga with the vested interest of seeking devotees, etc., so too it is Buddhists who originally peddled these techniques, who watered them down and marketed them to have secular appeal, and who had vested interests in wanting Americans and other Westerners to adopt them as their gurus (Purser 2013). I'm only one person and this is purely anecdotal, but I've found the practices of yoga and mindfulness and other forms of meditation transformative and more so in the absence of teachers with vested interests. So do many others. But, just as the numbers do not validate whether, say, Bernie Sanders truly represents the interests of Main Street, but rather only whether he will get to actually represent them as the President, so too the number of individuals who share my view is irrelevant to its validity.

We have sampled a fairly representative variety of objections to mindfulness or McMindfulness and noted how they are analogous to objections to yoga or McYoga, among other interesting parallels, and I have offered a fairly representative sampling of my responses to them. Now, let us take a closer look at the four main objections I've chosen to focus on at some length, beginning with what seems to be a rather popular one.

The 'Meditation Fails to Change the World' Objection

Among increasingly many articles like it, an article appeared in the *Guardian* very clearly exhibiting this objection. 'Mindfulness Is All about Self-help. It Does Nothing to Change an Unjust World,' by Suzanne Moore, is essentially a polemic about the many non-unjust-world-changing shortcomings of mindfulness (Moore 2014). As the title reveals, however, this polemic is analogous to the following complaint: 'Journalism Is All about Selling News. It Does Nothing to Change an Unjust World.' Or this one: 'Brushing Teeth Is All about Hygiene. It Does Nothing to Change an Unjust World.'

Somebody, surely, has to report the news, as objectively as possible, which is a distinct task from that of trying to alter what will become the news. Is it really a shortcoming of journalism that it merely reports on the injustices in the world but makes no concerted efforts to change them? It is no more a shortcoming of journalism that it fails to try to change an unjust world into a just one than it is a shortcoming of sports that they do not try to change an unjust world into a just one, any more than it is a shortcoming of psychotherapy that it fails to try to change an unjust world into a just one, to mention just some of the indefinitely many more analogies that may be used to make the point. The list of analogous things about which it is not a shortcoming about those things that they do not try to change an unjust world into a just world includes, but certainly is not exhausted by, the following: interest in or the collection of stamps, interest in or the collection of recipes, interest in or the collection of aquarium fish or related pets, interest in or devoting one's life to understanding and teaching mathematics or linguistics or philosophy of language or aesthetics or medieval literature or Egyptology or semiotics or the Feldenkrais method or cryptology or hermeneutics or aerodynamics or Wittgenstein or Fritz Perls or gardening, karate, chess, poetry, art, hydraulics, classical music, Zumba, dentistry, roofing, rolfing, carpentry, wine-making, embroidery,

Pilates, hotel management, automotive mechanics, vinyl repair, bicycling, marathons, accounting, skydiving, fishing, deep sea diving, theater, prestidigitation, photography, calligraphy, food trucks, method acting, writing criticisms, or defenses about mindfulness—the list goes on indefinitely, but hopefully the absurdity of claims of the form *it is a shortcoming of x that x does nothing to attempt to change an unjust world into a just world* is by now abundantly clear.

The exceptions to this line of rebuttal would include anything that it is inherently in its nature or explicitly in its mission to attempt to change an unjust world into a just world, e.g., certain political organizations, activist groups, and non-governmental organizations. Of course, anything can be criticized for failing to attempt to make the world a better place, as can anyone who is not doing so at any particular time, whether they are part of an organization dedicated to changing the world or not. For example, partisan objections are frequently made against presidents from the opposing political party to the effect that the president went golfing or vacationing just after some tragedy or crisis, and similar objections are often made by partisan individuals against members of any opposing group for enjoying or maintaining life and thus for failing to do something after some tragedy—objections that have nothing to do with meditation. To a certain extent, everyone is collectively responsible for failing to do enough to make the world a better place, including those who specialize in criticizing McMindfulness, but these sorts of criticisms do not carry any particularly relevant or inherently problematic implications for meditation.

It is a distortion to think the reason most individuals meditate or promote meditation is in order to avoid attempting to make the world a better place, or to prevent others from doing so, although some individuals may use meditation as a way of shielding themselves from the stresses generated by their daily encounters with the harsh edges of reality. For them, perhaps, attempting to make the world a better place is above their metaphorical pay grade or outside their metaphorical bandwidth, and it is enough of

a struggle just to cope with the stresses generated by their encounters with the rough edges of reality. Rather, for them, meditation might be part of a package of survival mechanisms, coping strategies, or ways of attempting to make themselves better individuals, indirectly making their lived part of the world a better place. Rather than criticizing anyone who meditates for this sort of reason, compare meditation with the billions who take alcohol, antidepressants, and/or other mind-altering chemicals that, unlike meditation, desensitize them to the problematic dimensions of reality. To the contrary, one very accurate way of describing what happens with mindfulness meditation is that it functions as a form of existential digestion: It facilitates the digesting of experience. Meditation typically helps individuals process the stresses that accompany encounters with the rough edges of reality, but it does not typically do so by chemically blocking them from cognitive processing.

Just as anything in the world can become the object of philosophical inquiry and examination under the right circumstances, even a ham sandwich (why is it, really, that pork is kosher?), so too any item on the list of non-world-changing things above (about which it is no shortcoming that they are not world-changers) could, in principle, play a role in changing the world in some significant way, under certain hypothetical scenarios and, even if not, should be treated as if engagement in it and devotion of one's life to it or a significant amount of one's time and efforts to it matter ethically and thus are subject to inquiry about whether such efforts might be better placed in the Facebook protesting of fracking, of tax laws that benefit the 1 %, or of the Syrian refugee crisis—as if Facebook activism even purports to amount to a world-changing activity. (Indeed, the illusion that Facebook activism, so to speak, counts as activism might prevent real activism: I posted against x! Have you?)

But surely those are all individual choices—for example, whether one ought to repost a political meme (and then have to deal with the obligation to debate with ideological fundamentalists) or watch another episode of *Breaking Bad* (and release some of the stress of life, entertaining a fantasy)—best made by individuals situated in the embedded spaces in which their skills, wherewithal, economic lives, functional matrices, and overall horizons of reasons for action are enmeshed. Each of us must decide for ourselves how much of our lives (and when and how) is appropriate for us to dedicate to the furtherance of the common good, to the best of our abilities, given our limitations, and how much of our lives are appropriately devoted to our obligations to our own minds and bodies, loved ones, and other circles of commitment expanding outward from the center of our being toward all other sentient beings and the rest of the cosmos.

As I noted above, most of my arguments are *arguments by analogy* of the form: That does not make sense, because that is like this other thing that obviously does not make sense. The problem with analogies is that for something to count as one there must be both (i) some elements in common between the two things being compared (or else there would be no basis for the comparison) and (ii) some elements not in common (or else the two things to be compared would actually be one and the same thing). This requirement of a difference between items constituting an analogy opens the door to charges that the items being compared are not properly analogous because some of their features are not shared, not shared sufficiently, or not shared in the right way. To assess whether such a charge of faulty analogy is apt, it is not enough that there be *some* difference between the items being compared, for, again, that is a constitutive requirement of any analogy. Thus, for an analogy to be faulty, an item in the pair of compared items must lack the feature that is *the point* of the analogy. I have made many analogies here between the complaint that it is a shortcoming of mindfulness that it fails to attempt to change an unjust world into a just world and similar objections of the general form *it is a shortcoming of x that x fails to attempt to change an unjust world into a just world*, where my list of instances or values of x is quite long (weightlifting, stamp collecting, Pilates, chess, and so on). Do any of these items constitute faulty analogies?

Before we answer that, note that it is enough if one of them is not faulty. I could examine each analogy to assess whether it is faulty, but logic and considerations of space suggest that I only assess one analogy. Take weight lifting, then, which Ron Purser has suggested might be a faulty analogy.[4] Here is why I think it is not faulty. Many people practice weightlifting for any or all of the following expected benefits: Working out relaxes them, it relieves stress, if done with others or at a gym it supports a form of social bonding, it improves immune functioning, flexibility, bone density, strength, resilience, weight regulation, metabolism, self-esteem, etc. Meditation is practiced by many because it is touted as yielding many of these benefits. Thus, meditation and weight lifting share many motivational factors, making them somewhat analogous, on the one hand. On the other hand, one might object that it is not the case that weight lifting is *all the rage*, so to speak, like mindfulness is, and thus that there is a disanalogy. But the unshared elements here—one item being all the rage and the other not being all the rage—are irrelevant to *the point* of the analogy, and that is, it is a shortcoming of x that x fails to attempt to change an unjust world into a just world.

Technically, even the shared elements mentioned here, which may be summarized as expected (personal and interpersonal) improvements associated with motivations for engaging in both cases of x (meditation and weight lifting), are technically irrelevant to the *validity* of the analogy. For that, all that is needed is that both items are instances of the formula *it is a shortcoming of x that x fails to attempt to change an unjust world into a just world*. However, the more the elements in common in the pair of items that constitute an analogy, the more likely it is to achieve its goal, which is not so much to prove a point as to facilitate understanding by showing how some feature that is not fully transparent in the original case may be seen more readily in the analogous case, which then ought to make that

---

[4]Personal (Facebook) communication (August 8, 2014).

feature more visible in the original case. Analogies are not technical proofs, but guides to understanding. Thus, even an analogy that proves technically faulty in some instances on careful analysis might nonetheless guide understanding, if even only in one instance.

And we can always patch up an analogy that seems faulty on some such technical grounds, by imagining hypothetical conditions to make the two compared cases more analogous. For example, weight lifting does not come to us from a millennia-old spiritual tradition the way mindfulness has come to us from Buddhism, so that is a seemingly significant disanalogous element that someone might appeal to in order to deny the point of—and thus dismiss—the analogy. But we could imagine a hypothetically different past, a possible history in which Shaolin Buddhist monks first invented and institutionalized the practice of weight lifting as a form of yogic bodily control (connected with martial arts), which then spread to all of Buddhism, and for various other hypothetically imaginative reasons never took hold in Western society until, like McMindfulness, it was culturally appropriated from Buddhism, studied by neuroscience, applied in the military, Google, grammar schools, and so forth. Then, the analogy would be more perfect: Traditionalist Buddhist weight lifters could complain about the dangers of becoming muscular and strong outside the Buddhist ethical framework, non-Buddhists could complain that weight lifting (within or outside Buddhism) fails to try to change the world, and Buddhists and other lovers of weight lifting could complain that empirical studies about weight lifting threaten to miss its true meaning, as if strength was really measurably located in the muscles and so on.[5] (This also runs together the non-world-changing objection with the non-Buddhist-ethical-application objection, to be

evaluated shortly, if not also an objection against non-Buddhist cultural appropriation).

In all the analogous cases at hand (karate, stamp collecting, and so on), none of this really matters anyway, for they seem valid insofar as one may understand that a claim about each of them to the effect that it is a shortcoming of that item that it fails to attempt to change an unjust world into a just world would be either simply false or, if true, only trivially true. Thus, even if it is irrelevant whether a certain mundane or otherwise necessary activity like brushing one's teeth does not change the world, it might be arguable that it is a shortcoming of any behavior that it fails to attempt to change the world, but then it would be a shortcoming of almost everything we do, other than efforts specifically designed to try to change the world, that they do not involve efforts to change the world. But the sense in which this would be true, if this is true at all, would be trivial, for then not only brushing one's teeth, but sleeping, falling in love, earning a living, spending time with one's elderly grandparents, tying one's shoes, and so forth, would all be subject to such alleged shortcomings. But clearly they are not seriously subject to such objections. *Reductio ad absurdum*! To the abstract extent any of them are subject to such objections, all of them are relatively equal, including meditation. But then the objections become thoroughly vacuous.

Ironically, from the perspective of Buddhism, it actually is a shortcoming of almost everything most of us ever do that it is done *mindlessly* and not as part of an intentional, mindful effort either to become enlightened (directly or indirectly for the benefit of all sentient beings) or to directly or indirectly reduce suffering in any or all sentient beings. And attaining enlightenment or reducing the suffering of sentient beings is not only one way of changing the world for the better, but from the Buddhist perspective, the most important way. Be that as it may, however, it is not treated as a criticism to be leveled against all unenlightened or non-enlightenment-seeking beings, but rather as a fact to guide the enlightenment-seeking being. From the perspective of Buddhism, the best thing anyone can do is to make

---

[5]Thompson has criticized contemplative neuroscience research for its reductionism in investigating meditation through brain studies, which he sees as analogous to thinking a bird's flight is somehow located in its wings (Heuman 2014). His objection is framed within a rich understanding of the subject, however, as evident in Thompson (2014).

oneself a better being, as nobody can make any-one else a better being—it is hard enough to make oneself a better being. Nothing about this per-spective precludes or disinclines Buddhists from compassionate efforts to reduce suffering and to try to make the world a better place. To the contrary, that is the whole point.

The 'McMindfulness Is Divorced from Buddhist Ethics' Objection

Most serious (long-term) mindfulness practi-tioners believe (reasonably, I think) that *any* emphasis on mindfulness in one's life—no mat-ter what kind of life that is, ethical or otherwise —will naturally foster the blossoming of these noble, altruistic imperatives, as evidenced by the parable of the Buddhist thief. A certain thief, inspired by the Buddhist teaching, asked the Master to accept him as a disciple, but begged to be allowed to continue his life as a thief, insisting he was incorrigible as such. The Master allowed this, advising the thief only to maintain mindful awareness of all his actions while stealing. Three weeks later, the man returned, complaining that he could not both steal and maintain full aware-ness of what he was doing, for while fully aware of all the implications of his actions his com-passion for his victims and what his deeds would cause for them prevented him from stealing.[6]

Those who worry about the use of mindful-ness by corporate capitalism and the military, for example, who suspect mindfulness is being deviously employed to increase production or make more obedient, effective warriors, rather than to address income inequality or other forms of injustice being perpetrated by such institutions (the failure to change the world), ought to be able to see in this parable that things could actually go the other way. Intuitively, more mindful soldiers are less likely to over-react, less likely to be trigger-happy, less likely to misidentify civilians

as enemies, and thus less likely to commit the sorts of mindless, emotionally triggered actions that will not only cause greater suffering for those with whom they come in contact, but which will likely haunt them in the form of postwar PTSD. In fact, recent studies show that mindfulness training may actually help soldiers deal with stress in ways that seem consistent with these intuitions.[7]

Arguably, this parable with the thief illustrates that—contrary to the objection that mindfulness fails to attempt to change an unjust world into a just world—mindfulness stands a far greater chance of changing individuals and thus of changing the world (after all, the world is just the abstract collection of individuals) than its oppo-site does. Completely independently of Bud-dhism or of the Buddhist *practice* of *cultivating* mindfulness, the opposite of mindfulness is *mindlessness*—failing to attend consciously to whatever one is doing or experiencing. Intu-itively, but also empirically, a great majority of all errors in judgment, decision, and action are related to mindlessness, that is, to not paying full attention to what one is doing (Langer 1991). Mindlessness takes many forms, but it is typi-cally associated with either trying to rush through one thing in order to get to another, or trying to do more than one thing at a time, or both. The mindless person ignores the feedback loops from her somatosensory system that metaphorically whisper information to the person about her (literal and metaphorical) sense of equilibrium, pacing needs, etc. When the metaphorically loud noises of mindless multitasking drown out the relatively whispering voices of embodied wis-dom (e.g., the subtle discomfort one feels when being deceived by a fraudulent salesman, or when one senses that one has passed the point of satiation but continues to gouge on treats, alco-hol, pointless chatter, and mind-numbing digital entertainment), the odds of error, accident, and injury increase. An adage of mindful yoga practice that reflects this understanding is: If you listen to your body when it whispers, you will not have to listen to it when it screams.

---

[6]I cannot find an authoritative cite for this tale, but here is one reference: 'Stories from around the World: The Thief (Buddhism),' http://mythologystories.wordpress.com/2013/01/09/thief/.

[7]University of California—San Diego (2014)

The ubiquitous acceptance of the *mindlessness* that is ceaselessly forced upon us by all the competing digital demands, the explosion of biased sources of information, market-research-designed consumerist programming, and related media and other competitive forces in society clearly help to enable the maintenance of the evil status quo. If that is correct, and it seems to me that it makes a lot more sense than the objections against mindfulness do, then we are under a political obligation to put an end to digital multitasking and to take up practices like mindfulness as weapons in our attempt to reclaim ourselves and rescue ourselves from mindless manipulation. The mantra of this mindful political imperative might read: *Occupy your own consciousness!*

It may just be an article of Buddhist faith that meditative skill promotes sensitivity, empathy, compassion, and the like. The conceptual connection between heightened awareness and compassion is not obvious to non-Buddhists, but for Buddhists, it is clear that enlightenment entails the dissolution or at least the loosening or desolidifying of the substantive interpretation of the self/other distinction, which intuitively reduces or eliminates unwholesome self-centered motivations and thus inherently promotes a disposition toward altruism, but a form of altruism that is not self-effacing, as the enlightened being is also a being that matters. But it arguably begs the question in favor of Buddhism to say that enlightenment entails the dissolution of the self/non-self distinction and thus altruism, for non-Buddhists cannot be expected to take it on faith that the Buddhist's eightfold meditative path actually leads to any ego/other-distinction-dissolving form of enlightenment. Of course, Buddhists making the non-Buddhist-ethical-application objection need not concern themselves with this problem, as they tend to accept the transformative power of mindfulness, their preferences that it remained coupled to the broader teachings notwithstanding. After all, the Buddha's modification of the earlier Indian soteriological technique of one-pointedness meditation by the addition of mindfulness is what led to his enlightenment in the course of one evening, and he claimed, in his Discourse on the Foundations of Mindfulness, the *Satipaṭṭhāna Sutta* (MN 10), that the correct practice of mindfulness alone could bring the practitioner to enlightenment in one week's time.

However, even if enlightenment is better understood as a limit case than as an actuality, it does not seem to beg the question to think it is reasonable that long, serious, disciplined, mindfulness practice generates experiential insights into the nature of the self and all mental states as characterized by momentariness or impermanence, impersonality or the interplay of impersonal conditions and factors, and ontological insubstantiality or existential emptiness, which key insights or 'marks of existence' together discourage belief in the value of acting on ego-volitional impulses, craving, and clinging, and thus which foster the falling away of selfishness, and thus which at least indirectly promote spontaneous altruism. One need not believe in the literal attainment of *nirvana* (enlightenment) in order to believe in these things. In this regard, Flanagan (2012) seems right in holding that it is enough, philosophically, for purposes of validating the overall Buddhist world view and teaching that some meditation virtuosos actually instantiate these practices and qualities. Yes, that may suffice, but surely it is no argument against mindfulness being all the rage, so to speak. (I can imagine a similarly invalid objection if 'McPhilosophy' became all the rage in philosophy cafes: Within the history of philosophy, the numbers of actual philosophers are few, and there only need to be a few Aristotles for philosophy to have some sort of instantiated-in-reality validity. Likewise for McPsychology, McScience, McArt, and so on, there have only been, and we only need, a few Freuds, Einsteins, Picassos, and so forth.)

Let us return to the masses of hopefully mindful consumers, soldiers, Google employees, and so forth that the phrase 'all the rage' denotes. If all that a stressed out, underpaid, overworked, overinformed, existentially confused urbanite significantly addicted to social media, to constant checking of her cell phone, and to marathon television series like *Breaking Bad*, *24*, and *Orange Is the New Black* can do to try to take care of herself relatively inexpensively—and without serotonin-reuptake inhibitors, other drugs, or alcohol—is to cultivate a mindfulness practice, who are we to begrudge her because she has not joined the Peace Corps or because she cannot afford the professional friendship of her own Gestalt psychotherapist? And just because someone is, say, from such a socioeconomically and thus educationally challenged background that they saw it as an improvement to join the military, or is bright enough to work at Google, or can afford to wear Prada, or can live off their invested stocks, does that disqualify them from entitlement to enlightenment-fostering practices? Are the otherwise privileged members of the 1 % not also entitled to enjoy the fruits of, say, weight lifting, long-distance running, brushing their teeth, or wholesome psychotherapy sessions? Do we really think the movement toward the improvement of society, away from being unjust and toward being more just, is better served by a majority of citizens who are less mindful, who never exercise, or who never brush their teeth?

Similarly, unfair objections could be leveled against what may be caricatured as sustainable-shopping-bag-toting social media activists, whose chanting of their polemicist mantras and performances of the politically correct equivalences of yogic postures in all of their public actions amount to no more than supporting the appearance (to themselves, if nobody else) of insulating them from the guilt of their limousine liberal lifestyles but which, even when collectively agglomerated, do not put a small dent in global injustice, but function only to deflect attention away from them and onto the targets of their diatribes. Talk about the pot calling the kettle mindless. Are there are better versions of the non-world-changing objection?

Variations on the Theme: Ginsberg and Žižek

Allen Ginsberg thought that meditation is a kind of distraction that prevents the practitioner from changing the world, but he made the objection long before meditation became so popular as to be considered a form of the McDonaldization of the world. And Slavoj Žižek's objection, along similar lines, is that it is worse than that. Ginsberg's objection is the first, historically, of several objections that may be grouped together as different forms of the objection to the effect that meditation has negative social, political, and/or other societal consequences, a kind of objection made most forcefully by Žižek, but also by Sharf, Purser and Loy, and Forbes (and Moore, who we already addressed). Let us first address the objections of Ginsberg and Žižek, as their objections more closely resemble each other than do those of Sharf, Purser and Loy, or Forbes.

I met Allen Ginsberg through Ram Dass, one of my first and most beloved meditation teachers. Ram Dass spoke on many occasions about how Ginsberg often gave him a hard time, objecting that meditation draws people away from political activism, how he encouraged Ram Dass to get off the meditation cushion, stop retreating from the unjust world, and join in with him for political activism, and so forth.[8] Ram Dass used Allen's objections as a springboard for a number of his remarks in some of his talks, about how enlightened protest was not a contradiction, how it involved protesting unjust actions as opposed to those who perform them, how meditation reduces the tendency to polarize protestors into *us* and *them*, and so forth. Years later, when I contacted Allen (in 1989) and asked him to give a talk on Buddhism at the Brooklyn College Philosophical Society (I was then its president), he agreed, but only after giving me a hard time, this time about wasting my time majoring in analytic philosophy, with its excessive concerns for abstract logical points, scientific and mathematical reasoning, and language analysis, on the same grounds that it failed to engage critically with activist attempts to change the structure of society. In a way, Ginsberg inadvertently proved the validity of my analogy: Anything that fails to change the world may be thereby criticized, but that undermines the force of the objection against

---

[8]Ram Dass's lectures are available online at www.ramdass.org.

any particular thing, since it applies to almost everything.

My purely anecdotal account of Ginsberg's objections serves as a priming to the psychoanalyst, philosopher, and social critic Slavoj Žižek's stronger version of the same objection. I also mention Ginsberg's objections because they are not academic, but pressingly personal, first, and because they are an earlier example of the same sort of objection made by Žižek, who has argued, more forcefully than Ginsberg, that meditation is an opiate for the now secular consumerist masses. Like Ginsberg, Žižek's argument is motivated by a view according to which almost everything is guilty of failing to change the world, but he makes a forceful attempt to block the idea that this thereby undermines the objection against meditation, by focusing specifically on the way meditation may be used to fail to improve the world and to maintain the status quo. Nonetheless, as I mentioned earlier, and as I predict we shall see here, an accentuated focus on details often obscures the big picture.

Žižek goes to fairly dramatic lengths to show how Buddhism-inspired but now fundamentally secular Western attitudes of non-judgmentalism, acceptance, passivity, pacifism, tranquility, equilibrium, and the like—the sort of dispositions that are aspired toward as the hopefully attainable consequences of transformative Buddhist meditative practices—fit very conveniently into, and help sustain an unquestioning attitude toward, the status quo of global capitalism, with its vast disparities of wealth and its promulgation of mass consumerism. He suggests that this is the reason mindfulness is all the rage that it now is in so many dimensions of public life and popular culture,[9] almost as if the behind-the-scenes power brokers and strategic planners of global capitalist neoliberalism invented mindfulness precisely for this purpose—to create the metaphorical equivalent in workers and consumers of better milk-producing, obedient cows. His extremely passionate manner of convinced, indignant expression alone seems to count as an additional argument in favor of his rather conspiratorial

interpretation, but, technically, needs to be set aside as an entertaining rhetorical embellishment. In any case, the objection amounts to what I think may be described as the claim that Buddhist meditation is the new opiate of the now secular masses (in their postmodern withdrawal from the balm of religion), functioning essentially to help maintain the unequal capitalist power structures in place globally. It is a chilling view, one that certainly merits serious reflection.

I will say more about it below, but as I suggested earlier, the objections of Sharf, Purser and Loy, and Forbes are roughly in a league with those of Žižek, though not exactly so or as dramatic. Thus, before *further* rebutting Žižek's objection (which, I take it, was already rebutted indirectly in my rebuttal of Moore's non-world-changing objection), we will entertain their versions of the objection. I will nonetheless first sketch here my answer to Ginsberg and Žižek, and that is a combination of Ram Dass's answer and my own. Ram Dass argued that mindfulness and spiritual growth ought to make one a more enlightened and effective protester, when it is appropriate for one to do so (as opposed to when it is appropriate to spend time in reflection), and I argue that even if what Ginsberg and Žižek worry about is well founded, it is not necessarily the case that meditation must function as a means of pacifying potential activists or as an opiate of the masses. In fact, Purser and Loy, and Forbes, argue that it ought not to function that way, and together with Sharf, they all may be said to appeal to the broader Buddhist ethical framework as the remedy to the ill-effects of divorcing meditation from that very ethical framework. Odd as it may sound before hearing my reasons, and despite that their appeals to ethics refute Ginsberg and Žižek, I will ultimately reject the idea that the solution lies in the reconciliation of mindfulness with the ethical dimension of Buddhism.

The 'Unethical-Application' Objection

Robert Sharf, Ron Purser and David Loy, and David Forbes have objected to the extraction of

---

[9]On the claim that it all the rage, see Forbes (2012).

mindfulness from its Buddhist ethical framework. Sharf has objected to the idea, popular among Westerners increasingly drawn toward mindfulness and other forms of Buddhist meditation, to the effect that they can extract these practices from the ethical and broader philosophical and cultural traditions within which they derive their meaning and value.[10] He makes a strong case for the idea that the secularized extraction of mindfulness that is becoming increasingly privileged among Westerners who are drawn to it constitutes a serious distortion of the rather narrow social role and institutional function of the practice within Buddhist communities. He also thinks Westerners have literally distorted the techniques, the emphases on various teaching points connected with them, and their overall place and importance in Buddhist life. His objections all have a range of validity, but I reject the idea that the Western embrace of mindfulness is diminished in any way by these putative facts.

Sharf shares Flanagan's and Thompson's view, to the effect that meditation is not as central to Buddhism as Westerners have come to think it is.[11] Before we turn to that view, however, Ron Purser and David Loy coined the term 'McMindfulness' to capture and dramatize some of the same worries expressed by Sharf and to highlight the many dimensions of the mindfulness explosion in popular culture, and David Forbes has supported the same concerns, but from what may be described as more of an 'integral' philosophical perspective, that is, a perspective that sees the contemplative dimension as only one part of us that needs to be integrated into the many other dimensions of our intrapersonal and interpersonal being—a perspective that seems intuitively correct. As I noted in my opening remarks, all of these thinkers make some valid points, and their concerns seem well intended to preserve the meaning and value of

mindfulness, to protect contemplative traditions that brought it to us, and to ward off distortions, abuses, and counterproductive applications. Again, I do not take issue with these features of their concerns.

What I do take issue with may be responded to in roughly the same way I responded to Ginsberg's and Žižek's objections, their differences notwithstanding. While I think that there might be many great improvements that could result from the teaching of an ethical perspective or framework in connection with the teaching of mindfulness, I do not think it is necessary or ideal, for a few reasons, despite my generally positive attitude toward the teaching and promotion of ethical values. I teach ethics classes, in fact, as part of the philosophy curriculum, but I doubt it would be bad if 'McEthics' became extracted from the philosophy curriculum, popularized, and watered down to the point where everyone found it fashionable to try to justify their actions by casuistry, reasoning from universal principles (like the principle that *like cases ought to be treated alike*), examining the foundational presuppositions of their values, and so forth. If that happened, and a bandwagon of critics emerged objecting to the non-world-changing application of McEthics at Google, in the military, etc., I expect that I'd argue against them, on similar grounds. Although both have likely contributed in some ways to gradual progress in the world, neither the meditation experts within traditional Buddhism nor the professional ethicists within philosophy have seriously changed the world (contrary to the sometimes hyperbolic enthusiasm of their fans within the history of ideas), but it is an empirical question to what extent anything changes anything else, or to what extent any particular element of a thing (e.g., its being embedded in a particular tradition) did change anything else. Maybe McEthics and McMindfulness, due to their mass penetration into the cultural mind-set, might have greater impact.

These are empirical issues, not settled by armchair arguments for or against such matters, which is why empirical research on various forms of meditation is a good thing, contrary to

---

[10]Sharf (2012).

[11]Sharf expressed this aspect of the view in a talk he gave at an NEH Summer Institute that I attended, 'Investigating Consciousness: Buddhist and Contemporary Philosophical Perspectives,' at the College of Charleston (Summer 2012).

the objections from some of the same traditionalists against the reductionist treatment of techniques extracted from their traditional cultural, doctrinal, ethical, and other bearings. This implies that empirical testing ought to be conducted on meditation techniques both within their traditional contexts or packages and in isolation from them, and in various possible combinations in other contexts. It is only by such experimentation, both formally in controlled settings and informally in people's life experiments with such techniques and doctrines, that science can hope to cast aside confirmation bias, rise above the chatter of conflicting emotions and intuitions both for and against various forms of meditation, and begin to settle such matters rationally on the basis of careful interpretation of the evidence.

Contrary to traditionalist Buddhist worries to the effect that mindfulness divorced from Buddhist ethics and soteriology is some power than can wield great harm if not properly harnessed and regulated by ethics, mindfulness is intuitively distinct from ethics per se, in its own nature, on the one hand, and that is as it should remain, on the other. As I mentioned earlier, mindfulness exists outside the Buddhist tradition simply as the opposite of mindlessness. Paying attention to what one is doing, thinking, feeling, perceiving, or experiencing—these are natural abilities of human beings. The *discipline* of training oneself to cultivate these abilities— greater degrees of mindfulness—owes much to Buddhism for developing it, but mindfulness is not the province of Buddhism any more than propositional logic or the scientific method is the province of Western philosophy, though anyone who wishes to cultivate their propositional logical skills or their skills in the use of the scientific method owes a historical debt to Western philosophy and science and may be well served to turn toward them for guidance. Indeed, orthodox Indian philosophers from the Buddha's day could equally object that he divorced the one-pointedness meditation technique from Brahmanism and its deistic metaphysics and arguably transmogrified it by adding mindfulness to it. This Hindu attitude toward non-deistic

meditation survives: One of my meditation teachers in the Hindu tradition, Ma Jaya, claimed that the Indian saint, Neem Karoli Baba (Ram Dass's guru), said never to teach meditation *without God*. However, Buddhists and others who embrace one-pointedness and mindfulness will celebrate this heretical violation, rightly, as far as I can tell.

As Ginsberg made clear when he criticized me for studying philosophy, he had the same view of analytic philosophy that some of our other critics here have of mindfulness, but these objections are analogous to objecting to science because of its unethical uses or applications. It is not science, analytic philosophy, or mindfulness that are problems. Rather, it is their being put to undesirable ends that is the problem in all such cases, and those things have nothing to do with philosophy, science, or mindfulness, per se. Ironically, mindfulness and related meditation practices, if anything, actually tend to increase ethical sensitivity (Davis and Thompson 2015).

Forbes mentions that some proponents of mindfulness, of which I am one, see mindfulness in this regard as subversively revolutionary, ethical, politically engaging, and all the things these critics worry that it might not be, insofar as it tends to produce dispositions to reflect on everything, toward altruistic sensitivity, and so forth, but he worries that in the hands of such manipulators as the military and Google, those tendencies will be downplayed. He may be right, to a certain extent, under certain circumstances. But it is the *uses* of cell phones, hammers, vitamins, exercises, therapies, and just about anything else, as opposed to *those items alone,* that may be negatively evaluated.

My rebuttal to the unethical-application objection, then, is that it is a *misplaced* criticism. It is those *uses* of mindfulness under those circumstances (by the military, Google, and so forth) that would be the problem, if they would be a problem at all, not mindfulness itself. If part of your Samurai, Ninja, or Special Forces training is in the use of meditation in order to make you a better warrior, assassin, or soldier, and your *bushido* (martial code) or your 'training' is so powerfully psychologically, socially, or otherwise compelling that your mindfulness practice will not likely render that code itself the object of your serious ethical reflection, then there seems intuitively to be a problem with the *bushido*, or the 'training' associated with it, not with practices of disciplining attention, per se.

By analogy with mindfulness, the closely related skill of one-pointedness is intuitively an excellent skill for an archer to develop and also for a sniper. Archery for its own sake is a generally harmless sport; sniping is not, but whether or not a particular sniper event is ethical depends on the context: A SWAT team sniper who prevents a terrorist from killing innocent hostages is a world apart from a sniper who is a terrorist or a serial killer. The one-pointedness is technically irrelevant to the evaluation of the sniper event (except that, in support of one-pointedness, a mindless sniper is much more likely to kill non-targeted individuals), just as having a photographic memory, being extremely quick-witted, or highly persuasive are skills that can be put to good or bad ends. Similar rebuttals may easily be surmised for all the other versions of this objection mentioned above, *mutatis mutandis*. There is nothing inherently negative about heightened awareness, or the absence of mindlessness. To the contrary, as Socrates noted (and toward which insight he devoted his life, unto and upon penalty of death), awareness of one's own ignorance contains the very seed of wisdom. And wisdom is probably the greatest virtue, as it inclines one toward all other virtues.

The quite traditional Buddhist monk Bodhi (2011) addresses the question 'whether mindfulness can legitimately be extracted from its traditional context and employed for secular purposes.' He holds such applications of mindfulness not only acceptable, but admirable, because they reduce suffering, so he agrees with the tenor of this paper. He does suggest, however, that the reductionist examination of mindfulness associated with what may be described as contemplative neuroscience threatens to distort our understanding of mindfulness and that for investigators to fully understand it they need to see it as embedded within the religious tradition in which it is rooted. This is a view he shares

with Thompson (2014), but I do not think Thompson is against the idea of *also* researching the phenomenon independently of the religious tradition, in line with the suggestions I made above about how science alone may resolve our questions about the utility of such techniques both within and external to their traditions of origin. Before we turn to Flanagan's and Thompson's claims, however, there is one more analogy I would like to make about the objection regarding extracting mindfulness from its larger Buddhist context.

Let me paint a picture of the background for this analogy, which is about the popular divorce of McYoga from Hinduism in the West, based on my own personal experience. I have been practicing what I have always thought and continue to think was traditional, Patanjali-based, eight-limbed yoga—what Swami Vivekananda dubbed 'Rajah Yoga' (kingly or royal), his enthusiastic description for the eight-limbed path of yoga outlined by the ancient sage Patanjali in his classic treatise on the subject, the *Yoga Sutras*—for as long as I have been practicing meditation, over 43 years now. Indeed, it was my experience of something fantastic, something powerfully mystical and mind-blowing, at the end of my very first yoga session—an unexpected, full-blown out-of-body experience—that led me to take on the practice of meditation as part of my investigation into the spiritual and philosophical dimensions of yoga, for all the texts I could get my hands on recommended meditation as the most powerful key to spiritual growth. I spent the next several years of my life as a very serious *yogi* (practitioner of yoga and meditation) and a member of an extended set of three overlapping *sanghas* (spiritual communities) initially defined by three teachers, Ram Dass, Hilda Charlton, and Ma Jaya (then 'Joya'), with a universalist blend of mostly Hindu, Buddhist, and other spiritual teachings they shared. I was taught the value of traditionalism, and my three teachers never charged us for their services, which included training in yoga, meditation, philosophy, and other forms of esoteric wisdom. I became quite adept at all eight stages of yoga (union, oneness), especially the *asanas* (postures), *pranayama* (breathing exercises), *dhyāna* (attainment of one-pointedness), and *samādhi* (meditative trance absorption) and soon began to lead a private weekly group of *yogis* in these practices.

Once just the *asanas*—but one of the eight limbs of traditional yoga—began to be treated in our society as if they were identical to yoga itself, to appear in gyms alongside aerobics classes and the like, and to become commercialized (and rather successful, at that), I must confess I experienced mixed feelings about the difference between yoga and McYoga. On the one hand, I was happy to see some element of yoga finally catching on in the larger culture, even if it was one of the least spiritually transformative elements of the eight-limbed path of yoga that was catching on (the postures), but on the other hand, I was disheartened to see only that element of yoga being treated as the aspect of yoga that would be absorbed into the larger culture, what may be described somewhat diminutively as merely the body-sculpting, stress-relieving, bodily-flexibility-promoting aspect of McYoga postures.

Note that it is no objection against the widespread popularity of *asanas*-as-yoga that such a practice is not designed to eradicate social injustices, nor against even the totally traditional, full-blown eight-limbed yoga taught by Patanjali. On that note, it may rightly be objected that yoga, meditation, and related spiritual practices, and all the traditions that give rise to them, are essentially *soteriological* or *salvific* in orientation, intention, and philosophy; they are not designed to change the world, but to bring the practitioner into a deeper relationship with it on some level. But then, this objection *levels Buddhism* along with Buddhism's objection against any attempt to extract such practices from Buddhism or its ethics.

My ultimate digestion of this discomfort with the extraction of the *asanas* from the larger eight-limbed tradition of yoga may be reflected in the Buddhist adage: *May a thousand flowers bloom*. In other words: So what? Who cares? Why not? At least the annoying multitude of yoga-mat-toting (and sustainable-shopping-bag-toting) conformists, so to speak, is following *that*, instead of, say, following alcoholism, pugilism, fascism, or nihilism. And some of them are being drawn to traditional, full-fledged, eight-limbed yoga, thanks in no small part to the social acceptability of the popularized, albeit watered down, forms of yoga. There is probably some truth in the historical speculation to the effect that the McYoga-mat swarm helped pave the way for the McMindfulness swarm. Likewise, some of the folks constituting the McMindfulness swarm are being drawn to traditional, full-fledged, Buddhist-eightfold-path-situated mindfulness, rather than being drawn to Ponzi schemes, religious fundamentalism, marijuana, or reality TV.

I may be guilty of confirmation bias, but one thing I feel fairly confident about is that many of the thousands of people I have taught meditation-without-God to, over the many years of my tenure as a non-Buddhist (but heavily Buddhism- and Hinduism-inspired) meditation teacher, yoga instructor, and philosophy professor, have learned something empowering and transformative: they have learned, in varying degrees, that they are capable of investigating their own minds and bodies—their intentions, emotions, bodily sensations, thoughts, beliefs, values, relationships, world views, life hopes, senses of self, hopes, fears, dreams, shadows, hatreds, judgments, and everything and anything else going on within them and without them—directly, personally, clearly, objectively. They have also learned that they can nurture themselves with the same compassion that they can lend to others, creating the conditions for their own healing, self-understanding, self-acceptance, and self-transformation. And they have learned that every moment is one in which mindfulness—conscious presence within one's embodied experience—is the way to fully experience it, absorb it, digest it, and assimilate it, and mindlessness is the way to miss out on it, to fail to absorb, to digest, or to assimilate it. For most of the rest of them, I may be being naively optimistic, but I believe their meditation experiences function as slow-to-sprout seeds, if not effective inoculations against a variety of existential ills.

When asked once what he thought was the greatest thing about meditation in his life after all these years of promoting it, Ram Dass said that its value was best seen in its absence: During those times in his life that his practice lapsed, he found himself walking around with a lot of undigested experiences. If mindfulness helps us digest experiences in this insanely unjust, information-overloaded world, then the complaint that it is a shortcoming of mindfulness that it fails to attempt to change an unjust world into a just one is analogous to the complaint that it is a shortcoming of antacids and probiotics that they fail to attempt to change an unjust world into a just one. Antacids and probiotics help us digest food; mindfulness helps us digest life.

I think we have sufficiently dispensed with the complaint to the effect that it is a shortcoming of mindfulness that it fails to attempt to change an unjust world into a just one, as well as the complaints that mindfulness divorced from Buddhist ethics or in various applications outside the Buddhist context somehow comes up problematically short. Let us turn, then, to the complaint to the effect that meditation plays less of a role in Buddhism than is commonly thought among Westerners and thus is not as important to experts on it as we neophyte Orientalizing enthusiasts would like to think it is.

The 'Meditation Doesn't Matter Much to Buddhists' Objection

Sharf (2012), Flanagan (2012), and Evan Thompson each have claimed that meditation does not really matter to Buddhism the way Westerners have imagined it does and that relatively few Buddhists meditate on a serious basis, as if the latter claim is evidence of the former. As Thompson put it, 'Buddhism isn't reducible to meditation—most Buddhists throughout history haven't practiced sitting meditation' (see Heuman 2014). I do not want to give any false impressions about these otherwise brilliant philosophers, so it should be made clear that they are not really rejecting meditation so much as they are responding to its becoming a fetish among both

Western practitioners of mindfulness and neuroscientists who might be thought to be testing Buddhism in the laboratory, as if meditating brains hold the keys to understanding Buddhism or its magical mindfulness technique. But as valid as these intentions are, and I share them for the most part, this sort of claim itself—that meditation is not widely practiced among Buddhists, and the associated implication that it does not matter that much to Buddhism—still captures my interest the most, perhaps because I have made my own meditation practice the centerpiece of my philosophical life for over 43 years. I may therefore be biased in my motivations, but that ought not to affect the validity of my argument.

Although Flanagan makes the same claim as Thompson about the small number of meditation practitioners in Buddhism, he also tries to hedge against its implication to the effect that meditation does not matter that much in Buddhism by arguing that it only matters that some meditation practitioners become virtuosos and thus instantiate, evidence, and thereby validate those Buddhist claims that revolve around meditation. Fair enough, per se. In light of these and their other sympathetic concerns with meditation, perhaps it is not entirely accurate to saddle either Flanagan or Thompson with the claim that I want to rebut, to the effect that because few Buddhists practice meditation, it does not really matter that much to Buddhism, so I will simply assess that claim on its own, independently of whether this or that individual (or anyone, for that matter) wants to defend it, although some individuals clearly do hold the claim.[12]

I think one analogy alone ought to put an end to the sort of objection to the effect that meditation doesn't really matter even to the traditions that have brought it into great focus, as evidenced by its relatively small number of serious practitioners. Thus, consider this analogous argument:

---

[12]At the NEH 2012 Summer Institute, 'Investigating Consciousness: Buddhist and Contemporary Philosophical Perspectives,' several philosophers and scholars expressed support for some version or another of this claim, most emphatically Robert Sharf.

Relatively few Westerners participate in the practice of scholarly philosophy or scientific research on a regular basis—far less than one percent of the Western population throughout the history of Western society. Thus, philosophy and science do not really matter much, even within the traditions that brought them into great focus.

Hopefully, the fallacious nature of this analogous claim ought to be self-evident, but in case it is not, let us examine it some, in order to render its fallacious nature as perspicuous as possible.

Thus, the mere fact that less than, say, one percent of Westerners now and throughout Western history are or were actually practicing philosophers or scientists surely does not entail that philosophy and science are unimportant in current or earlier periods of Western society. To the contrary, though philosophers and scientists constitute a handful of people relative to the larger population, it is undeniable even that a very small, fractional subset of those philosophers and scientists that do or have practiced their crafts have had a tremendous impact on Western society. Take Plato, Aristotle, Galileo, Copernicus, Newton, Bacon, Descartes, Hume, Darwin, Freud, Marx, and Einstein, to name just a handful of scientists and/or philosophers who have had monumental impact. Does the small percentage of practicing philosophers and scientists undermine the validity or importance of philosophy or science? The idea that it does is absurd. But even if Nāgārjuna was the only philosopher in world history, that would not undermine the validity of philosophy.

If the scientific method and philosophical analysis became all the rage, so to speak, in Buddhist societies the way mindfulness has become all the rage in Western society, the parallel objection to the effect that the scientific method and analytic philosophy are not even all that important or widely practiced in the traditions that birthed and gave prominence to the scientific method and analytic philosophy would be as obviously absurd as the objection to Westerners embracing mindfulness to the effect that mindfulness does not really matter much or wind up actually being practiced much among Buddhists. But we need not wait for that reversal

to happen to see the absurdity of this objection to mindfulness.

Yes, some folks are overzealous, and others are overemphasizing and perhaps hastily embracing overly confident interpretations of the powers and applications of mindfulness. And yes, that bothers some Buddhists the way that the popularity of espresso and latte in Starbuck's might bother some Italians. But so what? Buddhists—of all people—ought to be able to deal with being bothered by such things. And, just as the objection about mindful Ninjas is rightly about Ninjas and not about mindfulness, so too the objection here is rightly about exaggerators and not about mindfulness.

Conclusion: No Buddha Left Behind!

In closing, I should make clear that I respect the otherwise insightful analyses, and I appreciate all the concerns, that the thinkers examined here have expressed about issues connected with mindfulness meditation, apart from those specific aspects of their claims that I have rejected. I reiterate that my objections are restricted to these rather narrow features of their claims. Indeed, many of these critics are respected colleagues, personal friends, or both. My target has been simply to defend mindfulness meditation primarily against the four basic objections to the effect that it does not change the world, cannot rightly stand apart from Buddhist tradition, matters negatively, or does not even matter to those to whom it ought and secondarily against the host of related objections addressed in the course of doing so.

In conclusion, mindfulness meditation practice can increase the disposition to altruistic, civically minded social and political engagement, rather than a retreat from unpleasant aspects of reality, contra Moore, Ginsberg, and Žižek. Power brokers did not invent McMindfulness, even if some of them embraced it because they thought it might benefit them in devious ways, well into its already massively spreading popularity. Nor is it a form of false consciousness unwittingly fostered by consumerism, which is attracted to mindfulness as an adjunct to its addictive myopia. The independence of mindfulness from Buddhism's ethical framework is natural and desirable for various reasons, contra Purser and Loy, Bodhi, and Forbes. The problems of *meditation and x*, so to speak, are problems *of x*, not *of meditation*, per se. And *meditation matters*, despite how few serious practitioners there are or have been (contra Sharf, Flanagan, Thompson, and others) within or outside Buddhism, or how different the techniques of McMindfulness are relative to various Buddhist traditional versions of those techniques. Whether the techniques work or not matters more than where they came from or how they have been modified. Science has jurisdiction over whether such techniques work, Buddhist discomfort with that fact (ironically) notwithstanding.

The opposite of mindlessness is as natural as is mindlessness, but unfortunately not as ubiquitous, and therefore, its introduction into almost every conceivable venue in which mindlessness is the norm is at least as important as is the ubiquity of mindlessness—and therefore actually promises to be world-changing at the grass roots level. To this insight, I add the slogan: *No Buddha left behind!* The awakened consciousness cultivated by mindfulness training is intuitively desirable in almost every conceivable case in which mindlessness is not desirable, just as the opposites of stupidity or of uncritical thinking are intuitively desirable in almost every conceivable case in which stupidity and uncritical thinking are widespread but not desirable. Of course, there are exceptions, but those are, by definition, exceptional.

Mindfulness matters, I conclude, most especially and significantly to those who do actually practice it seriously, who experience its impact directly, and thus who know its power, validity, and import both immediately within and over the course of their lives. And it matters to all of those who are close to them, whose lives are touched by them, whether they realize it or not. Though I've argued that it need not be the purpose of mindfulness to fail to attempt to change the world, every act of kindness, for example, even a friendly facial gesture toward a stranger,

prompted by mindfulness, contributes to the common good, to making the world a better place. To take the case closest to my own experience, the amount of selfish competitiveness, insensitivity, judgmentalism, arrogance, aggression, and violence that would be emanating from me and that would have emanated from me over the course of my life had I not been practicing mindfulness meditation for over four decades certainly counts in the equation to assess the effectiveness of the practice and the extent to which it contributes to the common good or tends toward changing the world for the better, and I am but one person. That many millions are embracing this practice is an overwhelmingly good thing that undoubtedly counterbalances any negative consequences likely not intrinsically connected with the practice anyway, but more likely causally attributable to extraneous factors.

The critics of McMindfulness need to take that in and sharpen the target of their criticisms, from mindfulness because it is associated with x, to x. For many who lack knowledge or experience, but who might otherwise benefit from this practice, might prematurely be detoured from it as a direct result of erroneously targeted criticisms, analogous to the way seriously presented criticisms about an employer's only building a gym for employees in order to maximize efficiency threatens to prevent employees from taking advantage of the opportunity to exercise, which hurts those who need to exercise, whether they realize it or not.

This article was not written so much on behalf of those who know all of this from their years on and off the meditation cushion as it was written on behalf of those who do not know any of this from firsthand experience, that is, for those who are, in a sense, mindless regarding the negative powers of mindlessness and the positive powers and importance of mindfulness. They are the ones analogous to the employees who need the exercise. In light of all these points, it is incredible that otherwise really smart folks should be not only lamenting the explosion of interest in mindfulness, but using the persuasive power of their words in ways that possibly threaten to cause more harm than the alleged harms they purport to be opposing. I am happy to imagine a world in which nearly everyone spends significant time cultivating mindfulness!

References

Bodhi, B. (2011). What does mindfulness really mean? A canonical perspective. *Contemporary Buddhism: An Interdisciplinary Journal, 12*(1), 19–39.

Davis, J. H., & Thompson, E. (2015). Developing attention and decreasing affective bias: toward a cross-cultural cognitive science of mindfulness. In K. W. Brown, J. D. Creswell, & R. M. Ryan (Eds.), *Handbook of mindfulness* (pp. 42–62). New York: Guilford Press.

Edwards, K. (2015). A radical new view of mindfulness. *Huffington Post*, December 1, 2015. url: http://www.huffingtonpost.com/kellie-edwards/a-radical-new-view-of-min_b_8685260.html. Accessed December 19, 2015.

Flanagan, O. (2012). *The Bodhisattva's brain: Buddhism naturalized*. Cambridge, MA: MIT Press.

Forbes, D. (2012). Occupy mindfulness. *Beams and Struts*, June 30, 2012. url: http://beamsandstruts.com/articles/item/982-occupy-mindfulness. Accessed December 15, 2015.

Guthrie, J. (2002). McYoga for the masses: Popular Yogi to Franchise ancient Indian discipline. *SFGate*, November 10, 2002. url: http://www.sfgate.com/news/article/McYoga-for-the-masses-Popular-yogi-to-franchise-2755645.php. Accessed December 19, 2015.

Heuman, L. (2014). The embodied mind: An interview with Evan Thompson. *Tricycle*. url: http://www.tricycle.com/interview/embodied-mind. Accessed December 15, 2015.

Langer, E. J. (1991). *Mindfulness: Choice and control in everyday life*. Collins-Harvill.

Latham, N. (2015). Meditation and Self-control. *Philosophical Studies*, 1–20.

Manthorpe, R. (2015). Mind-wandering: The rise of a new anti-mindfulness movement. *The Long + Short*, December 10, 2015. url: http://thelongandshort.org/society/is-mind-wandering-an-anti-mindfulness-movement. Accessed December 18, 2015.

Marchildon, M. B. (2012). Is McYoga killing yoga? *Origin: The Conscious-Culture Magazine*, October 21, 2012. url: http://www.originmagazine.com/2012/10/21/mcyoga-is-it-killing-real-yoga-origin-columnist-michelle-berman-marchildon/. Accessed December 19, 2015.

Moore, S. (2014). Mindfulness is all about self-help. It does nothing to change an unjust world. *The Guardian*, August 6, 2014. url: http://www.theguardian.com/commentisfree/2014/aug/06/mindfulness-is-self-help-nothing-to-change-unjust-world. Accessed December 15, 2015.

Purser, R. (2013). *Mindfulness and modernity: Cultural accommodation or social transformation? The CUNY mindfulness lecture series*, October 10, 2013.

Purser, R., & Loy, D. (2013). Beyond McMindfulness. *Huffington Post*, July 1, 2013. url: http://www.huffingtonpost.com/ron-purser/beyond-mcmindfulness_b_3519289.html. Accessed December 15, 2015.

Seligson, H. (2015). Yoga teachers behaving badly. *New York Times*, December 16, 2015. url: http://www.nytimes.com/2015/12/17/fashion/yoga-teachers-behaving-badly.html?_r=0. Accessed December 18, 2015.

Sharf, R. H. (2012). *Varieties of mindfulness. Mindfulness or mindlessness: traditional and modern buddhist critiques of 'bare awareness.'* Lecture, Advanced Study Institute, McGill University, June 3–5, 2012. url: https://www.youtube.com/watch?v=c6Avs5iwACs. Accessed December 15, 2015.

Thompson, E. (2014). *Waking, dreaming, being: Self and consciousness in neuroscience, meditation, and philosophy* (p. 2014). New York: Columbia University Press.

University of California—San Diego. (2014, May 16). War and peace (of mind): Mindfulness training for military could help them deal with stress. *Science-Daily*. Retrieved December 15, 2015 from www.sciencedaily.com/releases/2014/05/140516092519.htm.

Žižek, S. (2012). The Buddhist ethic and the spirit of global capitalism. *Lecture, European Graduate School*, August 10, 2012. url: https://www.youtube.com/watch?v=qkTUQYxEUjs. Accessed December 15, 2015.

Criticism Matters: A Response to Rick Repetti

Glenn Wallis

Introduction

Rick Repetti has written a lengthy, somewhat sprawling, rebuttal to four criticisms leveled against contemporary "mindfulness." I offer here my reaction to his text in the form of reader response criticism. I'm not using "reader response" in its technical sense. I just mean to convey that I will not be commenting on each of his complicated meanderings or analyzing his copious analogies or dissecting his various examples. That would be too much. I will instead read through his text, pause at those points that strike me as salient, and then offer my more or less spontaneous response to them.

The [W]hole

To begin, I have some comments about the piece as a whole. As I read the synopsis, I found myself questioning the viability of Repetti's overall argument. That is, I had to wonder whether he was making the right refutations. By "right", I mean refutations that other defenders of contemporary mindfulness would find necessary and significant. To be more specific, would other refuters of the so-called McMindfulness critique

concur that the four objections that Repetti singles out for treatment are indeed the decisive issues to be addressed? If not, what would be the point of responding to his defense of these objections? Mindfulness proponents would simply dismiss my response as an irrelevant straw man argument, even if the straw man was fashioned by one of their own. On reflection, two things occurred to me. First, I have in fact come across these four objections elsewhere, in both formal and informal settings. So, I do think that Repetti is addressing criticisms that mindfulness proponents deem worthy of refutation. Second, it occurred to me that my response will all but certainly be accused of being a flimsy straw man attack *anyway*. Whether they are aware of it or not, mindfulness proponents are fast gaining the reputation of being people who are less than fully open to the full force of the criticism leveled against them. They employ various rhetorical strategies for evading the *brunt* of some critical point. It would be a useful project for someone to chart and analyze these strategies. I was considering whether I should take that approach here; namely, present a kind of rhetorical criticism of mindfulness. Then, it occurred to me: Repetti's piece is valuable not because it defends mindfulness against certain objections, but because it exudes the very spirit of the mindfulness community's engagement with criticism *tout court*. Along the way, Repetti's piece exhibits two stock mindfulness rhetorical responses to criticism. I call these two responses, respectively,

G. Wallis (✉)
Applied Meditation Studies Program, Won Institute of Graduate Studies, Philadelphia PA, USA
e-mail: gw@glennwallis.com

© Springer International Publishing Switzerland 2016
R.E. Purser et al. (eds.), *Handbook of Mindfulness*,
Mindfulness in Behavioral Health, DOI 10.1007/978-3-319-44019-4_33

conceptual shape-shifting and covert idealism. I'll say more about each of these strategies below. The point I am making here is that Repetti's piece is instructive because it *performs* the rabbit hole that is "mindfulness."

The Definition

Repetti helpfully begins by defining the term "mindfulness." As this section's title, "To Be Mindful or Not To Be Mindful, That Is the Question," indicates, Repetti believes that we have a stark choice. We can either cultivate "the state of mind or quality of consciousness" that is mindfulness or can fail to do so and engage in mindlessness. The difference lies in whether or not one is "paying attention to what one is doing, thinking, perceiving, experiencing," via "metacognitive awareness." My first thought on reading Repetti's definition was that it conforms well to Jon Kabat-Zinn's wheel-turning utterance: "Mindfulness is awareness that arises through paying attention, on purpose, in the present moment, non-judgmentally. It's about knowing what is on your mind." (Kabat-Zinn 2016a). Repetti, finally, uses the analogy of bodily strength to convey two crucial facts about mindfulness: (i) it is a natural capacity and (ii) it exists independently of techniques for its development.

My additional thoughts on reading this definition were as follows. Repetti, like Jon Kabat-Zinn himself, is engaging in an equivocation of terms. The critique of mindfulness is not a critique of certain claims regarding cognition. Who would deny that the capacity of "paying attention" and so on is an important human trait? In the most substantive critiques of mindfulness, the term "mindfulness" itself refers to the ideological edifice that has been erected around Kabat-Zinn's founding statement. In offering the definition of mindfulness that he does, Repetti thus obscures and evades the real issue: the identity of mindfulness as a system of thought and practice, one that is, moreover: (i) implicated in a very specific social–economic–political context and (ii) productive of a

very particular subject and world. (From here on, lower-case "mindfulness" refers to the purported cognitive capacity and practice that trains that capacity while upper-case "Mindfulness" refers to the ideological system.) I will have an opportunity to say more about these points in the next section. The main point here is that Repetti's definition establishes at the outset a premise that infects his entire argument. That premise cannot, I believe, avoid the charge of either disingenuousness or obliviousness. Does Repetti really think that a critique is being leveled against the claim for an almost simple-minded quality of human awareness? Repetti is in any case in good company here, for Jon Kabat-Zinn and every other Mindfulness proponent whom I have read or listened to are engaged in the same obscurantism. Perhaps what we are seeing in this (willful?) confusion is a kind of genetic trait of Mindfulness. The trait is an inability to distinguish between mindfulness as a curative fantasy cloned from the existing social formation and Mindfulness as an ideological strategy for *engaging* the existing social formation. If so, that would go a long way in explaining Mindfulness followers' refusal to take criticism seriously. As an aside, Repetti's use of the term "Bandwagon" in his subtitle is symptomatic of this trait. The term suggests that critics are mindlessly mimicking one another's insubstantial talking points. It suggests that criticism is merely a passing fashionable trend, and thus need not be robustly engaged.

Repetti—inadvertently, I suppose—draws attention to a central feature of Mindfulness ideology that normally remains unacknowledged. I am referring to the fact that Mindfulness entails a covert idealism disguised as a materialist phenomenology. Repetti, for sure, is not explicit about this facet of Mindfulness. My impression is that Mindfulness believers, beginning with Jon Kabat-Zinn, are blissfully unaware of, or perhaps indifferent to, this aspect. In any case, we find Repetti ambling perilously close to this timeless Siren song of the spiritualist big Other, to, that is, the pure witnessing consciousness untouched by the contingencies of time, space, and matter. When Repetti speaks of "metacognitive states," "phenomenological self-mirroring," and mindfulness as "a natural capacity," he is marshaling the allies not of an immanental phenomenology, as he seems to believe, but of a transcendental idealism.

The final response that I'll mention here is that Repetti's definition—again, like Kabat-Zinn's—is trivial. In proclaiming that each of us possesses a mental capacity for "being conscious that one is aware of what one is experiencing," we have learned nothing new. That is, the statement, whatever it might mean and whatever profound import it is supposed to have, is tautological. Mindfulness proponents will likely take that comment as evidence of my lack of adeptness in mindfulness. In doing so, however, they are, like Repetti and Kabat-Zinn, confusing Mindfulness with mindfulness.

The First Objection

Repetti next takes on what he terms the "'Meditation Fails to Change the World' Objection" of the so-called McMindfulness critique. His argument seems to boil down to this: It is no more a "shortcoming" of Mindfulness that it fails to change an unjust world into a just world than it is a shortcoming of countless other human activities—"gardening, karate, chess, poetry, art, hydraulics, Classical music, Zumba, dentistry," etc.—that they fail to do so as well. Certainly, his argument continues, no one should blame a mindfulness meditator for seeking a strategy for coping in a stressful world. Indeed, mindfulness meditation is best understood "as a form of existential digestion: it facilitates the digesting of experience. Meditation typically helps individuals process the stresses that accompany encounters with the rough edges of reality, but it does not typically do so by chemically blocking them from cognitive processing."

My first reaction as I read this section was that Mindfulness *does* claim for itself world-changing prowess. This prowess, moreover, comes precisely from the cultivation of mindfulness. The most recent evidence for these two claims comes from Jon Kabat-Zinn's

conversation with Angela Davis in Oakland. The question driving the conversation was whether mindfulness practices can serve the advancement of social justice. Davis, for instance, pointedly asks: "In a racially unjust world, what good is mindfulness?" Kabat-Zinn clearly wants to claim that mindfulness and meditation possess world-changing potential. They are, he says, "transformative practices that are capable of moving the bell curve of the entire society toward a new way of understanding of what it means to be human."[1] I also reflected on the large number of current books with "mindfulness" in their title, such as *Mindful Parenting, Mindful Teaching, Mindful Politics, Mindful Therapy*, and *Mindful Leadership*. I don't think it is a stretch to suggest that the implicit claim of this conglomeration of books is that Mindfulness, and even mere mindfulness, has world-changing implications.

This disagreement of mine, however, was fleeting. What struck me most about Repetti's argument in this section was its valorization of the neoliberal subject, and hence, his argument's reactionary stance. I should add that this stance does not surprise me. Again, Repetti is proving himself to be a faithful Mindfulness subject here. In brief, Repetti seems to assume a subject that has no choice but to accept the "unjust world," adapt to the "rough edges of reality," and engage in practices that foster resilience. Repetti could not paint a clearer portrait of the diminished neoliberal subject. It is a subject that is perpetually vulnerable in the face of global, financial, environmental, political, ad infinitum insecurities. It is a subject that is racked by a

---

[1] "How Can we Bring Mindfulness to Social Justice Movements?" YouTube. http://tinyurl.com/hyrajjw. Retrieved April 18, 2016. Kabat-Zinn's overall answer in this conversation once again reveals the transcendental idealism of Mindfulness. He constantly makes overly simplistic affirmations about the world-altering power of, for instance, attending; being present; heightened awareness; uprooting greed, hatred and delusion, and so on. Davis responds with the anti-idealist argument that social injustices are not a matter of mere personal attitude, much less the lack of attention: Their roots dig deeply into the material structures of our social system.

degree of stress and tension that debilitates the real possibility of robust agency. These characteristics—vulnerability together with the necessity of acceptance, resilience, and adaptation—are classic neoliberal assertions about the human subject. This stance, of course, raises the possibility that Mindfulness is simply an unrepentant ally of neoliberalism. God knows the secret is out on "the long marriage of mindfulness and money."[2] In that case, Repetti's refutation of the "'Meditation Fails to Change the World' Objection" is justified. By all measures, it does indeed appear that Mindfulness is quite content being, as Slavoj Žižek puts it, "the perfect ideology supplement" of a rabid global corporate capitalism (Žižek 2001). So why in the world should we expect it to want to change anything?

My response to this sad conclusion is the following. When Repetti argues for the equivalence of the statements "Mindfulness Is All about Self-Help, It Does Nothing to Change an Unjust World" and "Brushing Teeth Is All about Hygiene, It Does Nothing to Change an Unjust World," he reminds me of Donald Trump talking about abortion. As a recent *Huffington Post* headline put it, "Donald Trump Accidentally Articulates GOP Abortion Stance A Little Too Loudly." In suggesting "that women should face legal sanctions for having abortions," Trump inadvertently sailed "straight into 'here be dragons' territory." Repetti, too, in linking Mindfulness with his "list of analogous things about which it is not a *shortcoming about those things* that they do not try to change an unjust world into a just world"—things like "embroidery, Pilates, hotel management, automotive mechanics, vinyl repair, bicycling, marathons, accounting, skydiving, fishing, deep sea diving"—is taking us into a Mindfulness "unauthorized personnel not allowed" zone. We can paraphrase the Trump article to fit our case: "Repetti was just saying bluntly what the actual implications of longstanding Mindfulness views on the diminished subject are."[3]

---

[2]See, Goldberg (2015).
[3]See, Linkins (2016).

The Second Objection

Repetti next tackles the "'McMindfulness Is Divorced from Buddhist Ethics' Objection." Repetti's point in this section seems to be: *so what? mindfulness is great!*

The section consists of two parts. The first part is simply a paean to mindfulness. It largely repeats and expands on the definitional section that opens the article. To repeat—and there is an awful lot of repetition here—what Repetti means by mindfulness is the capacity "to attend consciously to whatever one is doing or experiencing." As in the earlier section, the supposedly obvious value of mindfulness is highlighted by contrasting it to mindlessness—"failing to attend consciously," etc.—the source of "a great majority of all errors in judgment, decision, and action." So, again, Repetti is not addressing the covertly ideological, subject-forming production that is Mindfulness. He is simply asserting the power and beauty of the *mana*-like quality of consciousness that is mindfulness.[4] What do ethics matter, Buddhist or otherwise, given this power? Then again, come to think of it—Repetti seems suddenly to realize—mindfulness *does* entail an ethics. But, in classic idealist fashion, it is an ethics that ensues naturally from the wondrous wellspring of metacognitive witnessing known as mindfulness. No prescribed ethics can trump the wholesomeness of that indigenous, natural effusion.

The Buddhist parable that Repetti cites to open this section says it all. By merely maintaining "mindful awareness" of the fact that he is stealing, this "incorrigible" thief finds himself incapable of further thievery. His mindfulness produces such an overwhelming gush of "compassion" for his victims that he just can't go on with the thug life. So, critics of Mindfulness should stop already with their unfounded concern that mindfulness in the boardrooms of corporate capitalism and in the war councils of the military-industrial complex will turn out badly for the rest of us. For, as the mindful ex-thief makes irrefutably clear, we can be pretty sure that serious long-term mindfulness

---

[4]See Per Drougge's contribution in the present volume.

practice, like the descent of the Holy Spirit, "will naturally foster the blossoming of…noble, altruistic imperatives."

In the second part, Repetti brings up critiques by Allen Ginsberg and Slavoj Žižek. He doesn't do anything substantial with these critiques. Again, it all seems to boil down to Repetti's earlier argument, repeated here as a rebuttal to Ginsberg's implied objection that "anything that fails to change the world may be thereby criticized." But, Repetti repeats, that objection "undermines the force of the objection against any particular thing, since it applies to almost everything" (think: "theater, prestidigitation [magic tricks], photography, calligraphy, food trucks, method acting," etc.). He lays out some interesting criticisms made by Žižek and seems to consider them important. For instance, Žižek's contention, according to Repetti, that mindful meditation is the new opiate of the masses "is a chilling view, one that certainly merits serious reflection;" but Repetti doesn't grant this, or any other of Žižek's points, any reflection whatsoever. In fact, he simply employs the all-too-common Mindfulness tactic of facile dismissal of criticism. Žižek's objection doesn't really require refutation, Repetti concludes, because, after all, it "was already rebutted indirectly in my rebuttal of Moore's non-world-changing objection." No, it definitely was not.

My most general response to this section is that Repetti is right about his main contention. Critics' objection that Mindfulness errs in not coupling its practices to Buddhist ethics is uninteresting. The objection also misses the point. Mindfulness certainly is a descendant of Buddhism. But, it is a descendant many, many times removed. It is, I would argue, even more closely related to, and bears closer resemblance to, American pop psychology, the 1960s human potential movement, Perennial Philosophy, positive-thinking spirituality, and apocalyptic New Age thought, just to name a few obvious blood relatives.[5] You don't have to look too

closely to see that Mindfulness's most recent progenitors are, of course, Ronald Reagan and Margaret Thatcher. As I mentioned earlier, Mindfulness has the same DNA and was raised on the same values that undergirds today's neoliberal, consumer capitalist social structure (acceptance, resilience, self-help, etc.). So, of course Jon Kabat-Zinn cozies up to corporate CEOs and American military generals.[6] And of course Rick Repetti wants to cancel the warrant on the claims of Buddhist ethics. When Žižek says that "meditation is the ideological form that best fits today's global capitalism," (Žižek and Milbank 2009) he is echoing the point that Repetti is making:

> [Mindfulness meditation] in its abstraction from institutionalized religion, appears today as the zero-level undistorted core of religion: the complex institutional and dogmatic edifice which sustains every particular religion [can be] dismissed as a contingent secondary coating of this core.[7]

My point is that asking Mindfulness to conform to Buddhist ethics (whatever that is—*which* Buddhist ethics, for instance?) is like asking Prosperity Theology to reconcile its ethics with the Babylonian Talmud. So, I agree with Repetti when he rejects the notion that some sort of big fix "lies in the reconciliation of mindfulness with the ethical dimension of Buddhism." The bigger problem for Repetti and his fellow Mindfulness believers, however, is that in invoking "ethics," even as a negation, they inadvertently release the question-genie of just what kind of ethics Mindfulness itself *is* operating by. And, as far as I can make out, *there be dragons.*

The Third Objection

"The 'Unethical-Applications' Objection" is next up for refutation. Strangely, Repetti does not actually present that objection. He certainly does

---

[5]By "apocalyptic New Age thought" I mean beliefs about the end times of the Old World and the coming of a New World, augured not by collective social action but by some sort of "shift in consciousness" from, in Repetti's terms, pervasive mindlessness to widespread mindfulness.

[6]See "The Thousand-Year View: An Interview with Jon Kabat-Zinn," *Inquiring Mind*, *30*(2) Spring, 14–16.

[7]Ibid. "Spiritual meditation" stands in the original.

not present it in any of the relevant and strong forms with which I am familiar, such as William Davies's exemplary *The Happiness Industry* (Davies 2015). Instead, Repetti opines that the objection is a *"misplaced* criticism." The reason he thinks it is misplaced is that "It is those *uses* of mindfulness under those circumstances (by the military, Google, and so forth) that would be the problem, if they would be a problem at all, not mindfulness itself" (emphases in original). That's the core of Repetti's response to the "indictment of the ideology of happiness and its accompanying horrors of mindfulness and well-being"[8] that we find circulating in intelligent, philosophically astute critical circles today. Repetti's position on this issue boils down to a less consequential form of the NRA's dogma that "guns don't kill people—people kill people."

What Repetti does not seem to recognize is that in taking such a position, he is pulling back the curtain on the covertly ideological nature of Mindfulness. An ideology, recall, "represents the imaginary relationship of individuals to their real conditions of existence" (Althusser 2001). So, mindfulness, though presented as a basic natural capacity that, like strength, can be further developed (it is metacognitive awareness, the opposite of mindlessness, paying attention, etc.), in reality functions as a technical term in a discourse of self-actualization (as, in the NRA analogy, "guns" functions as a signifier in a discourse of liberty). "Being mindful" signifies one thing in an old "Leave it to Beaver" episode ("Be mindful of your brother, Wally!"), another in a Jon Kabat-Zinn text ("[being mindful] wakes us up to the fact that our lives unfold only in moments") (Kabat-Zinn 1994), and something altogether different in a Mindfulness-meets-scientism discussion ("being mindful leads to changes in the structural connectivity within the nervous system that would indicate an increase in interoceptive ability") (Siegel 2016). Repetti offers the analogy of "strength" in order to indicate the utter naturalness of "mindfulness." But this analogy is as sheer as the emperor's clothes since the term "strength" itself always occupies a specific

position within a particular chain of signifiers, a position which alone determines its linguistic meaning and ideological function.[9]

That we are dealing here with a central article of Mindfulness faith—with, that is, an ideologically encoded signifier that serves to inscribe the believer into the Mindfulness imaginary—is on full display in Repetti's contention that an "ethical perspective or framework in connection with the teaching of mindfulness" is unnecessary. In any other context, such a claim would sound bizarre at best and irresponsible at worst. In what sort of alien universe might some human action be devoid of ethical implications? Unlike Repetti, apparently, my imagination fails me here. To be mistaken about such a bizarre possibility is irresponsible because it enables the disavowal of the ethical framework that *is* in place when teaching and practicing Mindfulness.

This act of disavowal is bound up with what Žižek calls the "new age spiritual fetish."[10] Repetti's Mindfulness subject embodies the denial of the ethical implications of his or her beliefs and actions *as* interpellated Mindfulness subject. Briefly, a fetish, a used here, is "that which enables you to (pretend to) accept reality 'the way it is.'"[11] Making Mindfulness's ethics explicit is unnecessary because mindfulness is simply that which enables one to view metacognitively the "contents of one's own mental state," and, by implication, the contents of one's world.[12] It is this belief that permits the Mindfulness fetish to function. In

---

[8]Simon Critchley's blurb for Davies's book.

[9]For example, think of the different ways that "strength" functions in the following discourses: Rosy the Riveter; the American cult of masculinity; Alcoholics Anonymous recovery speak; the Catholic Mass (where weakness is strength); national security rhetoric, and so on. The point is not the obvious one about a term's requiring context for its sense. It is, rather, the idea that that sense functions within productive subject and social formations.

[10]See Butler (2014).

[11]Ibid.

[12]Repetti's claim concerning metacognition and "phenomenological self-mirroring" is admittedly a weak, psychologized version of the old spiritualist quest for "things as they are." Jon Kabat-Zinn, however, offers us the strong version: "Coming to terms with things as they are is my definition of healing." See, Kabat-Zinn (2016b).

"practicing" metacognitive awareness (Repetti) or non-judgmental present-moment awareness (Kabat-Zinn), in, that is, simply seeing "things as they are,"[13] the practitioner is able to effectively keep at bay certain truths which, if acknowledged, would reveal Mindfulness to be but an ideological spectacle. Among these truths is, for instance, the fact that the believer's thoughts, actions, and emotions are always inextricably implicated in the human symbolic collective known as society. When the Mindfulness believer is taught to view subjective iterations of the social collective with a putative "awareness" that is open, spacious, non-judgmental, non-reactive, metacognitive, phenomenologically reflective, and so on, he is in fact being taught to see the world through the ideological prism that is Mindfulness. Evidence of this fact is that despite Jon Kabat-Zinn's grandiose proclamation that being thus mindfully present "actually does change the nature of our reality,"[14] the world remains unchanged. The only change that occurs is in the practitioner's perception. The same can be said, of course, for an hallucination. This fact sheds light on another terrible truth that the Mindfulness fetish enables the believer to ignore; namely, the truth that the believer is complicit in sustaining the very social collective that compels him or her to employ the mindfulness remedy in the first place. The Mindfulness fetish fosters the illusion that the believer can live in an ethically neutral, or indeed superior, relation to social reality.

The Fourth Objection

Repetti ends his piece by refuting "The 'Meditation Doesn't Matter Much to Buddhists' Objection." The objection is not really pertinent. Repetti,

indeed, makes it clear that his motivation here is purely personal. He says, the objection "captures my interest…perhaps because I have made my own meditation practice the centerpiece of my philosophical life for over 42 years." He apparently feels a need to resolve the issue for himself.

My response is that Repetti's argument is more interesting for unintentionally highlighting yet another fetishistic aspect of Mindfulness. In this case, the fetish functions to ward off the numerous implications of Mindfulness's relationship to Buddhism. Mindfulness ideology is woefully lacking anything resembling the theoretical apparatus that Buddhists have labored to construct through the ages. As faith-driven, and thus unsatisfactory, as they typically are, Buddhist theories at least attempt to work through the implications of its doctrines in the face of broader epistemological, ontological, and ethical concerns. Much of this theoretical work by Buddhists was prompted by the objections of their critics. As Repetti has revealed to us in each of his previous refutations, Mindfulness thinkers feel no such compulsion to *respond to* as much as to evade or dismiss criticism. Again, Repetti is proving himself to be the good Mindfulness subject in this regard.

A very telling exchange between Jon Kabat-Zinn and Danny Fisher brilliantly highlights the fetish at work, if somewhat frantically so.[15] I will cite the exchange at length, and let my comment on it serve as my conclusion to this response to Rick Repetti.

> Kabat-Zinn: You understand that I myself am not a Buddhist, right? I don't see what I do as Buddhism so much as I see it as Dharma expressing itself in the world in its Universal-Dharma-way.
> Fischer: Well, that might be a good place to start. I know you have a history with…is it the Cambridge Zen Center?
> K-Z: I've studied with a lot of different Buddhist teachers; still do. For a time I actually did consider myself to be a Buddhist. But I realized at a certain point that it was really most important for me to be a human—the fewer affiliations I had, the better.

---

[13]They are, namely, like water cascading off a cliff, inherently empty or content-free, hence not requiring judgment or reaction, etc.

[14]*Inquiring Mind*, "Coming to our Senses: A Conversation with Jon Kabat-Zinn, http://tinyurl.com/jszqm2a. Retrieved April 21, 2016.

[15]"Mindfulness and the cessation of suffering: An exclusive new interview with mindfulness pioneer Jon Kabat-Zinn," *Lion's Roar*. http://tinyurl.com/hb38sag. Retrieved April 22, 2016.

For me personally, that is. Also, I don't think I would have been able to do what I did in quite the same way if I was actually identifying myself as a Buddhist; it inevitably would have been seen as Jon Kabat-Zinn trying to put his Buddhist trip over on other people. I wanted to offer instead a kind of translation of a universal understanding or approach that was never really about Buddhism. The Buddha himself wasn't a Buddhist, and the term Buddhism is an invention of Europeans. And, of course, Buddhists could [sic] really care less because it's all about non-duality: as soon as you start classifying Buddhists and non-Buddhists you're not really a Buddhist anymore. I get quoted on these points a lot. All of this is just a way of giving you a background flavor of what it is that I do.

F: Speaking of quotes of yours, one that appears in the press release about your kickoff of National Breast Cancer Awareness Month at UCLA is about your work as an offering of "the wisdom and the heart of Buddhist meditation without the Buddhism…"

K-Z: Yeah, I have said that. I say those kinds of things a lot. It's not meant to be disrespectful, in that anyone who knows anything about Buddhism will understand that it's not about that—it's about the Dharma. It's about non-attachment to name and form, so to speak.

F: So, in a sense, it's about getting past "Buddhism?"

K-Z: I don't know that it's about getting past it so much as it is about going back to the beginning.

Jon Kabat-Zinn embodies here the fetishistic disavowal that Mindfulness is entangled in a relationship with Buddhism. This disavowal is not a well-reasoned repudiation. It is not a cool-headed rejection of the claim. It functions as a fetish because it enables Kabat-Zinn simultaneously to *distance* himself from the apparently unacceptable truth that Mindfulness is consequentially bound up in Buddhism and to *sustain* the implicit benefits derived from that relationship. Reasons for doing so should be obvious. Repetti's desire to refute the "Meditation Doesn't Matter Much to Buddhists" objection is indicative of these reasons. In short, the reasons constitute a coping mechanism: how to avoid entrapment in the overly determinate network of postulation that is Buddhism while simultaneously asserting mastery over Buddhism's central truth (e.g., Mindfulness captures "the wisdom and the heart of Buddhist meditation without the Buddhism").

If Mindfulness is subsumed under Buddhism, it cannot avoid appearing as an indefensibly simplistic version of its progenitor. For, given Buddhism's complex network of postulates concerning, for instance, the grounding of meditation practice in a robust ethics, Mindfulness is unable to answer adequately for its deviations from traditional norms. If Mindfulness is wholly withdrawn from Buddhism, on the other hand, it then becomes barely indistinguishable from any other unsophisticated, under-theorized self-help cure. Maintenance of this fetishistic disavowal thus enables Kabat-Zinn to view Mindfulness as more essentially Buddhist than is even traditional Buddhism. It enables the delusion that, in his words, Mindfulness is not so much Buddhism as it is "Dharma expressing itself in the world in its Universal-Dharma-way."

We could explore this central facet of Mindfulness ideology through several potent critical theories. We might apply, for instance, Harold Bloom's concept of the anxiety of influence, and find that Mindfulness emerges out of an intentional, if largely unaware, misreading of the Buddhist text. To paraphrase Bloom for our purposes, when two "strong" systematizers meet, the newer one must commit "an act of creative correction that is actually and necessarily a misinterpretation" of the older one. Otherwise, the innovator will be subsumed under the older master, becoming but a mere manager of the traditional status quo. That Kabat-Zinn accomplished this feat with Mindfulness was due in part to his "self-saving caricature of distortion, of perverse, willful revisionism" of Buddhism (Bloom 1997). The fetish aids Kabat-Zinn and his Mindfulness followers in staving off this unacceptable conclusion.

We might also employ the critical tool of Sigmund Freud's "narcissism of minor differences" (Freud 1961). This is a condition that breeds contempt toward that which you perceive as lying too close to you for comfort. Freud defines this as "a convenient and relatively harmless satisfaction of the inclination to aggression, by means of which cohesion between the members of the community is made easier." He gives as an example the way northern

Germans ridicule southern Germans. Any astute observer of the Buddhism/Mindfulness interface is witness to the mutual contempt each has for the other.[16] To see how this condition is narcissistic just consider the manner in which, in the above exchange, Kabat-Zinn gazes into the mirror of Buddhism and sees reflected back to him the "Universal-Dharma-way," the "the wisdom and the heart of Buddhism," that is precisely the face of his very own Mindfulness. It is the fetish that allows this distorted reflection.

We might, finally, apply Jacques Lacan's critical theory of the hysteric's discourse. Doing so, we would see in what ways Mindfulness rhetoric presents an exemplary instance of the hysteric's approach. As in Bloom's theory, the hysteric always stands in relation to a master. For example, Kabat-Zinn reveals in the exchange— as indeed Repetti does throughout his piece— that he values and indeed desires the outcomes of the master discourse, Buddhism. Yet at the same time, he reveals that he is impelled to challenge and resist being a subject of that discourse. As hysteric, he must therefore cast the master project in his own terms. It bears repeating that as chysteric—as, that is, an *alienated* subject of the master—Kabat-Zinn is nonetheless captured by the *desire* for the wisdom and liberation that the master discourse so deftly arouses in him. The

fetish enables him the illusion that he is outflanking today's traditional Buddhists themselves toward that end by "going back to the beginning" and thus becoming more essentially Buddhist than the Buddhists themselves.[17]

We might apply these and numerous other critical tools to better understand Mindfulness as the ideological system that it is. To call Mindfulness an ideological system is not to dismiss it out of hand as a product of false consciousness or devious manipulation. It is, rather, to acknowledge Mindfulness as being uniquely productive. It is productive of a quite particular subject, one that imagines his or her relation to the world in quite particular ways. The value of criticism is that it enables us to make explicit the operations of an ideological system that the system itself keeps implicit—its unstated assumptions; its unspoken values; its relationship to existing social, economic, and political formations; and, perhaps most importantly, its tacit formation of individual actors in the world. Marjorie Gracieuse sums up this task of criticism perfectly: It "consists in wresting vital potentialities of humans from the artificial forms and static norms that subjugate them" (Gracieuse 2012). It might be true that, as Repetti believes, "meditation matters" for humanity. But, it is equally true that criticism matters for humanity. Without it, we can't distinguish a vital human potentiality from a self-serving prescription of a covertly ideological program. Rick Repetti's refutation of four objections to Mindfulness is valuable in stimulating further critical work toward this end.

---

[16]I say "astute" observer because this contempt is often veiled in the passive aggressive niceties of "right speech." "Right speech" has emerged as a decisive discursive strategy in contemporary Western Buddhism. That is, it has become an apparently indispensable mechanism for normalizing certain voices, ideas, and behaviors, while excluding others. It has, in other words, become a censorious tool of subjugation. As such, it deserves further study. It certainly has burgeoned with the coming of the internet, no doubt due to the potentially unruly democratic nature of online discussions. I know of no Western Buddhist (*pace* Kabat-Zinn and Repetti, that moniker includes Mindfulness) blog, podcast, Web site, or Twitter account that does not invoke "right speech" either implicitly or explicitly, quite often the latter. The result is that Western Buddhist discourse is bland and predictable. Western Buddhist webmasters have apparently figured out that "The smart way to keep people passive and obedient is to strictly limit the spectrum of acceptable opinion, but allow very lively debate within that spectrum." (Chomsky 1998). So-called right speech is a crucial element toward this end.

References

Althusser, L. (2001 [1971]). *Lenin and philosophy and other essays* (p. 109) (B. Brewster, Trans.). New York: Monthly Review Press.

Bloom, H. (1997 [1973]). *The anxiety of influence: A theory of poetry* (p. 30). London: Oxford University Press.

Butler, R. (2014). *The Žižek dictionary* (p. 207). London: Routledge.

---

[17]For a fuller account of the Lacan's four discourses, see Wallis (2016).

Chomsky, N. (1998). *The common good* (p. 43). Berkeley: Odonian Press.

Davies, W. (2015). *The happiness industry: How the government and big business sold us well-being*. London: Verso.

Freud, S. (1961). *Civilization and its discontents* (p. 61) (J. Strachey, Trans. and Ed.). New York: W. W. Norton.

Goldberg, M. (2015, April 18). The long marriage of mindfulness and money. *The New Yorker Blog*. http://www.newyorker.com. Retrieved April 18, 2016.

Gracieuse, M. (2012). Laruelle facing Deleuze: Immanence, resistance and desire. In J. Mullarkey & A. Paul Smith (Eds.), *Laruelle and non-philosophy* (p. 42). Edinburgh: Edinburgh University Press.

Kabat-Zinn, J. (1994). *Wherever you go, the you are: Mindfulness meditation in everyday life* (p. 4). New York: Hyperion.

Kabat-Zinn, J. (2016a). *Peel back the onion*. http://www.mindful.org/jon-kabat-zinn-peel-back-the-onion. Retrieved April 18, 2016.

Kabat-Zinn, J. (2016b). *The healing power of mindfulness*. http://tinyurl.com/god4wu6. Retrieved April 21, 2016.

Linkins, J. (2016, April 1). Donald Trump accidentally articulates GOP abortion stance a little too loudly. *Huffington Post*. http://tinyurl.com/zs5sfgf. Retrieved April 18, 2016.

Siegel, D. (2016). *The healing power of mindfulness*. http://tinyurl.com/god4wu6. Retrieved April 20, 2016.

Wallis, G. (2016). "Spectral discourse," at the blog *Speculative Non-Buddhism*. http://tinyurl.com/hhsbjs5. Retrieved April 23, 2016.

Žižek, S. (2001). From Western Marxism to Western Buddhism. *Cabinet Magazine*. http://www.cabinetmagazine.org/issues/2/western.php. Retrieved April 18, 2016.

Žižek, S., & Milbank, J. (2009). In C. Davis (Ed.), *The monstrosity of Christ: Paradox or dialectic?* (p. 27). Cambridge: The MIT Press.

Index

© Springer International Publishing Switzerland 2016
R.E. Purser et al. (eds.), *Handbook of Mindfulness*,
Mindfulness in Behavioral Health, DOI 10.1007/978-3-319-44019-4

Printed by Printforce, the Netherlands